KU-196-141

Constructed Wetlands for Wastewater Treatment

Municipal, Industrial, and Agricultural

Donald A. Hammer

A CRC Press Company
Boca Raton London New York Washington, D.C.

Library of Congress Cataloging-in-Publication Data

Constructed wetlands for wastewater treatment : municipal, industrial, and agricultural / [edited by] Donald A. Hammer.
p. cm.
"Proceedings from the First International Conference on Constructed Wetlands for Wastewater Treatment held in Chattanooga, Tennessee on June 13-17, 1988"-Pref.
Includes bibliographical references.
ISBN 0-87371-184-X
1. Constructed wetlands-Congresses. 2. Sewage -Purification-Biological treatment. I. Hammer, Donald A. II. International Conference on Constructed Wetlands for Wastewater Treatment (1st : 1988 : Chattanooga, Tenn.)
TD756.5.C66 1989
628.3'5—dc20 89-12897

Direct all inquiries to CRC Press LLC, 2000 N.W. Corporate Blvd., Boca Raton, Florida 33431.

Lewis Publishers is an imprint of CRC Press LLC

International Standard Book Number 0-87371-184-X
Library of Congress Card Number 89-12897
Printed in the United States of America 12 13 14 15 16 17 18 19 20
Printed on acid-free paper

Preface

This volume contains the proceedings from the First International Conference on Constructed Wetlands for Wastewater Treatment held in Chattanooga, Tennessee on June 13–17, 1988. Although wetlands treatment of municipal wastewater has been the topic of previous conferences, only limited coverage has been given to other applications of this technology, including treatment of acid mine drainage, urban runoff, agricultural wastes, and industrial effluents.

Though some attendees and readers have only single-use interests, many others can potentially apply the technology to different wastewater types. All can benefit from an understanding of basic principles of wetland ecology, hydrology, chemistry, and microbiology. Furthermore, the relatively large data base from municipal systems has considerable application to the recent use of wetlands for acid mine drainage, and metals removal in municipal, urban runoff, and industrial systems has much to gain from information on acid drainage systems. Finally, basic types and configurations, construction and planting techniques, plant species, and many operating procedures are common to constructed wetlands for treating many types of wastewaters.

This volume presents general principles of wetlands ecology, hydrology, soil chemistry, vegetation, microbiology, and wildlife in Chapters 1–7 followed by case histories of specific types of constructed wetlands and applications to municipal wastewater, home sites, coal and noncoal mining, coal-fired electric power plants, chemical and pulp industry, agriculture, landfill leachate, and urban stormwaters in Chapters 8–21. Chapters 22–36 provide construction and management guidelines beginning with policies and regulations through siting and construction and ending with operations and monitoring of constructed wetlands treatment systems. Recent theoretical and empirical results from operating systems and research facilities, including new applications, e.g., nutrient removal from eutrophic lakes and urban stormwater treatment within highway rights-of-way, are included in Chapters 37–42. The reader with a specific application or interest should read the general principles, applicable case histories, construction and operating guidelines, and the appropriate chapter on recent results.

These proceedings are from the first comprehensive conference on constructed wetlands for water quality improvement. While it represents the state of the art in 1988, as in all rapidly developing fields, new developments will likely modify or revise this information in the future.

Widespread use of constructed wetlands may provide a relatively simple and inexpensive solution for controlling many water pollution problems facing small communities, industries, and agricultural operations. Yet adoption of this technology has been inhibited by a lack of guidelines and instructions supported by adequate information on important system components and basic wetlands ecology. The goal of this volume is to provide information to improve acceptance and increase application of constructed wetlands for water quality improvements.

In addition to the individuals and organizations identified in the following pages, the list of other contributors is so large as to be impractical to present. Suffice it to say that a successful conference and these proceedings would not have happened without substantial contributions of many dedicated people. We are very thankful for their gracious efforts.

A few, however, require special recognition. I would like to thank the staff and volunteers of Bicentennial Volunteers, Inc., for expeditious handling of the myriad of financial and logistical tasks that facilitated planning, preparation, and conduct of a large conference. Special appreciation goes to Sandra Hughey, who played a major role in every aspect of planning and conduct of the conference and preparations of manuscripts for these proceedings. Finally, I am grateful for Janet Tarolli's conscientious and efficient technical editing of every word in this volume.

Donald A. Hammer

Donald A. Hammer is Projects Manager in the Waste Technology Program, Valley Resource Center of the Tennessee Valley Authority. Dr. Hammer received his BSc from the University of North Dakota; his MSc from South Dakota State University; and his PhD in Ecology from Utah State University. He has worked for state agencies and the U.S. Fish & Wildlife Service and taught at the University of Maine before starting with the Tennessee Valley Authority in 1972. His work experience and publications include all aspects of wetlands ecology. He has managed large natural wetlands for multipurpose uses, developed methods to restore or create new wetlands areas, and investigated important components of wetlands ecosystems. He received Tennessee's Wildlife Conservationist Award for his innovative techniques to restore raptor and wading bird populations and The National Institute for Urban Wildlife's Outstanding Conservationist Award for his work in developing constructed wetlands wastewater treatment technology.

Dr. Hammer has held a variety of offices in professional organizations and is currently North American Correspondent for the IAWPRC Specialist Group on Macrophytes for Water Pollution Control and a member of the Specialist Technical Group on Appropriate Waste Management Technologies. He began working on wastewater treatment with constructed wetlands in 1978, and since 1983 most of his activities and publications have been on developing the technology for treatment of municipal, acid drainage, agricultural, and industrial wastewaters.

Acknowledgments

CONFERENCE STEERING COMMITTEE

Billy J. Bond
Tennessee Valley Authority

Greer C. Tidwell
U.S. Environmental Protection Agency

CONFERENCE PLANNING COMMITTEE

Robert K. Bastian
U.S. Environmental Protection Agency

Cynthia R. Britt
Tennessee Valley Authority

Gregory A. Brodie
Tennessee Valley Authority

John G. Doty
Tennessee Valley Authority

Robert J. Freeman
U.S. Environmental Protection Agency

Richard E. Green
Tennessee Valley Authority

Donald A. Hammer
Tennessee Valley Authority

Myron L. Iwanski
Tennessee Valley Authority

Robert J. Pryor
Tennessee Valley Authority

Gerald R. Steiner
Tennessee Valley Authority

David A. Tomljanovich
Tennessee Valley Authority

James T. Watson
Tennessee Valley Authority

Dale V. Wilhelm
Tennessee Valley Authority

We gratefully acknowledge the assistance provided by these sponsors, affiliates, and cooperators.

CONFERENCE SPONSORS

Tennessee Valley Authority
U.S. Environmental Protection Agency
Waste Management, Incorporated
U.S. Office of Surface Mining
Appalachian Regional Commission
Electric Power Research Institute
U.S. Soil Conservation Service
National Aeronautics and Space Administration (NSTL)
U.S. Army Corps of Engineers

CONFERENCE AFFILIATES

Lombardo and Associates, Incorporated
U.S. Bureau of Mines
USDA Forest Service
Peabody Coal Company
Mississippi Bureau of Pollution Control

CONFERENCE COOPERATORS

Society of Wetland Scientists
Wildlife Management Institute
Florida Dept. of Environmental Regulation
National Association of Home Builders
Alabama Dept. of Environmental Management
Kentucky Division of Water
Water Pollution Control Federation
Pennsylvania Electric Company
National Wildlife Federation
Tennessee Dept. of Health and Environment

We are thankful that the following gave graciously of their time and expertise to review manuscripts for this volume.

Hollis H. Allen
Waterways Experiment Station
U.S. Army Corps of Engineers

Robert K. Bastian
Office of Municipal Pollution Control
U.S. Environmental Protection Agency

Hans Brix
Botanical Institute
University of Århus, Denmark

Gregory A. Brodie
Power Operations
Tennessee Valley Authority

Joseph C. Cooney
Aquatic Biology & Fisheries
Tennessee Valley Authority

Paul F. Cooper
Water Research Centre
Stevenage Laboratory, United Kingdom

C. J. Costello
Lough Gara Farms, Ltd., Ireland

James N. Dornbush
Civil Engineering Department
South Dakota State University

Paul Eger
Minnesota Department of Natural Resources

J. Scott Feierabend
Fisheries and Wildlife
National Wildlife Federation

Robert A. Gearheart
Environmental Resources Engineering
Humboldt State University

Michelle A. Girts
CH2M HILL

Bill Good
Louisiana Geological Survey

Glenn Guntenspergen
Department of Botany
Louisiana State University

Raimund Haberl
Institut für Wasserwirtschaft
Vienna, Austria

Donald A. Hammer
Waste Technology Program
Tennessee Valley Authority

Robert S. Hedin
Environmental Technology
U.S. Bureau of Mines

John A. Hobson
Water Research Centre
Stevenage Laboratory, United Kingdom

Robert G. Hoffman
Ducks Unlimited, Inc.

Charles Horn
Water Division
Alabama Dept. of Environmental Management

Beverly B. James
James Engineering, Inc.

Robert H. Kadlec
Department of Chemical Engineering
University of Michigan

Robert L. P. Kleinman
Environmental Technology
U.S. Bureau of Mines

Robert L. Knight
CH2M HILL

Donald K. Litchfield
Environmental Quality and Safety
Amoco Corporation

Pio S. Lombardo
Lombardo Group
Dames & Moore

Gordon E. Miller
JE Hanna Associates

William H. Patrick, Jr.
Center for Wetland Resources
Louisiana State University

Ralph Portier
Institute for Environmental Studies
Louisiana State University

Burline Pullin
Waste Technology Program
Tennessee Valley Authority

Sherwood C. Reed
Cold Regions Research and
Engineering Lab
U.S. Army Corps of Engineers

K. R. Reedy
Soil Science
University of Florida/IFAS

J. Henry Sather
School of Graduate Studies
Western Illinois University

Phillip Scheuerman
Department of Environmental Health
East Tennessee State University

Larry Schwartz
Florida Department of Environmental
Regulation

Gary S. Silverman
Environmental Health Program
Bowling Green State University

Robert L. Slayden, Jr.
Tennessee Dept. of Health
and Environment

Maxwell Small
Maxec, Inc.

Ward W. Staubitz
Water Resources Division
U.S. Geological Survey

Forest Stearns
Department of Biological Sciences
University of Wisconsin

Gerald R. Steiner
Water Quality Program
Tennessee Valley Authority

Rudolph N. Thut
Weyerhaeuser Technology Center

David A. Tomljanovich
Aquatic Biology & Fisheries
Tennessee Valley Authority

James T. Watson
Water Quality Program
Tennessee Valley Authority

W. Alan Wentz
Kansas Dept. of Wildlife & Parks

R. Kelman Wieder
Biology Department
Villanova University

Jack A. Wilson
Kentucky Division of Water

We also thank the session moderators for their service in managing smooth and on-time technical sessions.

Roosevelt T. Allen
Waste Technology Program
Tennessee Valley Authority

Robert K. Bastian
Office of Municipal Pollution Control
U.S. Environmental Protection Agency

Billy J. Bond
River Basin Operations
Tennessee Valley Authority

J. Carroll Duggan
Waste Technology Program
Tennessee Valley Authority

J. Scott Feierabend
Fisheries and Wildlife
National Wildlife Federation

Robert J. Freeman
Water Management Division
U.S. Environmental Protection Agency

Catherine G. Garra
Region V/Water Management Division
U.S. Environmental Protection Agency

Steven Jenkins
Alabama Department of Environmental Management

Conrad J. Kirby
Waterways Experiment Station
U.S. Army Corps of Engineers

Pio S. Lombardo
Lombardo Group
Dames & Moore

Robert L. Slayden, Jr.
Tennessee Dept. of Health and Environment

Robert Smedburg
Waste Management, Inc.

George Tchobanoglous
Department of Civil Engineering
University of California

Charles Terrell
Ecological Sciences Division
U.S. Soil Conservation Service

Gary Wade
Forestry Research Laboratory
U.S. Forest Service

Alfred E. Whitehouse
Program and Technical Support
Office of Surface Mining Reclamation and Enforcement

Dale V. Wilhelm
Environmental Quality Staff
Tennessee Valley Authority

Jack Wilson
Kentucky Division of Water

Donald J. Wilzbacher
Environmental Affairs
Peabody Coal Company

Bill C. Wolverton
National Space Technology Industry
National Aeronautics and Space Administration

Measurement Units—Abbreviations and Conversion Factors

MULTIPLY	BY	TO OBTAIN
acre, ac	4.05×10^3	square meter, m^2
acre, ac	0.405	hectare, ha (10^4 m^2)
acre, ac	4.05×10^{-3}	square kilometer, km^2 (10^6 m^2)
acre-foot, ac-ft	11,616	cubic meter, m^3
British thermal unit	1.06×10^3	joule, J
calorie	4.19	joule, J
cubic yard	0.765	cubic meter, m^3
cubic foot, ft^3	0.0283	cubic meter, m^3
cubic foot, ft^3	28.3	liter, L (10^{-3} m^3)
cubic foot, ft^3	7.48	gallon, gal
cubic inch, in^3	1.64×10^{-5}	cubic meter, m^3
degree (angle)°	1.75×10^{-2}	radian, rad
dyne	10^5	newton, N
foot, ft	0.305	meter, m
gallon, gal	3.79	liter, L (10^{-3} m^3)
gallon per acre, gal/ac	9.35	liter per hectare, L/ha
gram per cubic centimeter	1.00	megagram per cubic meter, Mg/m^3
liter/min, L/min	0.264	gallon per minute, gal/min
m^3/sec	264	gallon per second, gal/sec
m^3	1.31	$yard^3$
metric tonne/hectare, MT/ha	2.24	ton/acre
micron	1.00	micrometer, μm (10^{-6} m)
mile	1.61	kilometer, km (10^3 m)
mile per hour, mph	0.447	meter per second, m/s
milligram per square decimeter hour (apparent photosynthesis)	0.0278	milligram per square meter second, $mg/m^2/s$ (10^{-3} $g/m^2/s$)
milligram per square centimeter second (transpiration)	10,000	milligram per square meter second, $mg/m^2/s$ (10^{-3} $g/m^2/s$)
million gallon/day, mgd	3785	cubic meter per day, m^3/day
ounce	28.4	gram, g (10^{-3} kg)
ounce (fluid)	2.96×10^{-2}	liter, L (10^{-3} m^3)
pint (liquid)	0.473	liter, L (10^{-3} m^3)
pound, lb	454	gram, g (10^{-3} kg)
pound per acre, lb/ac	1.12	kilogram per hectare, kg/ha
pound per acre, lb/ac	1.12×10^{-3}	megagram per hectare, Mg/ha

pound per cubic foot, lb/ft^3	16.02	kilogram per cubic meter, kg/m^3
pound per cubic inch, lb/in^3	2.77×10^4	kilogram per cubic meter, kg/m^3
pound per square foot, lb/ft^2	47.9	pascal, Pa
pound per square inch, lb/in^2	6.90×10^3	pascal, Pa = g/cm^2
quart (liquid), qt	0.946	liter, L (10^{-3} m^3)
quintal (metric)	10^2	kilogram, kg
square centimeter per gram, cm^2/g	0.1	square meter per kilogram, m^2/kg
square foot, ft^2	9.29×10^{-2}	square meter, m^2
square inch, in^2	645	square millimeter, mm^2 (10^{-6} m^2)
temperature (°F - 32)	0.556	temperature, °C
temperature (°C + 273)	1	temperature, K
tonne (metric), MT	10^3	kilogram, kg
ton (2000 lb)	907	kilogram, kg
ton (2000 lb)	2.24	megagram per hectare, Mg/ha

ADDITIONAL COMPARISONS

Person equivalent, pe—in Europe is approximately 100–180 liters/person/day
—in U.S. is approximately 380 liters/person/day

Hydraulic application rates:
1 cm/day = 10 L/m^2/day = 100 m^3/ha/day = 10,700 gal/ac/day = 93.5 ac/mgd
20 acres/mgd = 4.68 cm/day

Contents

Preface iii

Acknowledgments vii

Measurement Units—Abbreviations and Conversion Factors xiii

SECTION I
GENERAL PRINCIPLES

1. Wastewaters: A Perspective, *by A. J. Smith* 3

2. Wetlands Ecosystems: Natural Water Purifiers? *by D. A. Hammer and R. K. Bastian* 5

3. Hydrologic Factors in Wetland Water Treatment, *by R. H. Kadlec* 21

4. Physical and Chemical Characteristics of Freshwater Wetland Soils, *by S. P. Faulkner and C. J. Richardson* 41

5. Wetland Vegetation, *by G. R. Guntenspergen, F. Stearns, and J. A. Kadlec* 73

6. Wetlands Microbiology: Form, Function, Processes, *by R. J. Portier and S. J. Palmer* 89

7. Wetlands: The Lifeblood of Wildlife, *by J. S. Feierabend* 107

SECTION II
CASE HISTORIES

8. Constructed Free Surface Wetlands to Treat and Receive Wastewater: Pilot Project to Full Scale, *by R. A. Gearheart, F. Klopp, and G. Allen* 121

9. The Iselin Marsh Pond Meadow, *by T. E. Conway and J. M. Murtha* 139

10. Integrated Wastewater Treatment Using Artificial Wetlands: A Gravel Marsh Case Study, *by R. M. Gersberg, S. R. Lyon, R. Brenner, and B. V. Elkins* 145

11. Sewage Treatment by Reed Bed Systems: The Present Situation in the United Kingdom, *by P. F. Cooper and J. A. Hobson* 153

12. Aquatic Plant/Microbial Filters for Treating Septic Tank Effluent, *by B. C. Wolverton* 173

13a. Creation and Management of Wetlands Using Municipal Wastewater in Northern Arizona: A Status Report, *by M. Wilhelm, S. R. Lawry, and D. D. Hardy* 179

13b. Land Treatment of Municipal Wastewater on Mississippi Sandhill Crane National Wildlife Refuge for Wetlands/Crane Habitat Enhancement: A Status Report, *by J. W. Hardy* 186

14. Waste Treatment for Confined Swine with an Integrated Artificial Wetland and Aquaculture System, *by J. J. Maddox and J. B. Kingsley* 191

15. Treatment of Acid Drainage with a Constructed Wetland at the Tennessee Valley Authority 950 Coal Mine, *by G. A. Brodie, D. A. Hammer, and D. A. Tomljanovich* 201

16. Constructed Wetlands for Treatment of Ash Pond Seepage, *by G. A. Brodie, D. A. Hammer, and D. A. Tomljanovich* 211

17. Use of Wetlands for Treatment of Environmental Problems in Mining: Non-Coal-Mining Applications, *by T. R. Wildeman and L. S. Laudon* 221

18. Constructed Wetlands for Wastewater Treatment at Amoco Oil Company's Mandan, North Dakota Refinery, *by D. K. Litchfield and D. D. Schatz* 233

19. Utilization of Artificial Marshes for Treatment of Pulp Mill Effluents, *by R. N. Thut* 239

20. Use of Artificial Wetlands for Treatment of Municipal Solid Waste Landfill Leachate, *by N. M. Trautmann, J. H. Martin, Jr., K. S. Porter, and K. C. Hawk, Jr.* 245

21. Use of Wetlands for Urban Stormwater Management, *by E. H. Livingston* 253

SECTION III
DESIGN, CONSTRUCTION AND OPERATION

22. Use of Wetlands for Municipal Wastewater Treatment and Disposal—Regulatory Issues and EPA Policies, *by R. K. Bastian, P. E. Shanaghan, and B. P. Thompson* 265

23. States' Activities, Attitudes and Policies Concerning Constructed Wetlands for Wastewater Treatment, *by R. L. Slayden, Jr., and L. N. Schwartz* 279

24. Human Perception of Utilization of Wetlands for Waste Assimilation, or How Do You Make a Silk Purse Out of a Sow's Ear? *by R. C. Smardon* 287

25. Preliminary Considerations Regarding Constructed Wetlands for Wastewater Treatment, *by R. K. Wieder, G. Tchobanoglous, and R. W. Tuttle* 297

26. Selection and Evaluation of Sites for Constructed Wastewater Treatment Wetlands, *by G. A. Brodie* 307

27. Performance Expectations and Loading Rates for Constructed Wetlands, *by J. T. Watson, S. C. Reed, R. H. Kadlec, R. L. Knight, and A. E. Whitehouse* 319

28a. Ancillary Benefits of Wetlands Constructed Primarily for Wastewater Treatment, *by J. H. Sather* 353

28b. Overview from Ducks Unlimited, Inc., *by R. D. Hoffman*.... 359

29. Configuration and Substrate Design Considerations for Constructed Wetlands Wastewater Treatment, *by G. R. Steiner and R. J. Freeman, Jr.* 363

30. Hydraulic Design Considerations and Control Structures for Constructed Wetlands for Wastewater Treatment, *by J. T. Watson and J. A. Hobson*.......................... 379

31. California's Experience with Mosquitoes in Aquatic Wastewater Treatment Systems, *by C. V. Martin and B. F. Eldridge* 393

32. Constructing the Wastewater Treatment Wetland—Some Factors to Consider, *by D. A. Tomljanovich and O. Perez* ... 399

33. Considerations and Techniques for Vegetation Establishment in Constructed Wetlands, *by H. H. Allen, G. J. Pierce, and R. Van Wormer* 405

34. Operations Optimization, *by M. A. Girts and R. L. Knight*... 417

35. Pathogen Removal in Constructed Wetlands, *by R. M. Gersberg, R. A. Gearheart, and M. Ives* 431

36. Monitoring of Constructed Wetlands for Wastewater, *by D. B. Hicks and Q. J. Stober*............................ 447

SECTION IV
RECENT RESULTS FROM THE FIELD AND LABORATORY

37. Dynamics of Inorganic and Organic Materials in Wetlands Ecosystems

37a. Decomposition in Wastewater Wetlands, *by R. H. Kadlec* 459

37b. Thermoosmotic Air Transport in Aquatic Plants Affecting Growth Activities and Oxygen Diffusion to Wetland Soils, *by W. Grosse*.. 469

37c. Nitrification and Denitrification at the Iselin Marsh/Pond/Meadow Facility, *by R. L. Davido and T. E. Conway* 477

37d. Denitrification in Artificial Wetlands, *by E. Stengel and R. Schultz-Hock* 484

37e. Nitrogen Removal from Freshwater Wetlands: Nitrification-Denitrification Coupling Potential, *by Y.-P. Hsieh and C. L. Coultas* 493

38. Efficiencies of Substrates, Vegetation, Water Levels and Microbial Populations

38a. Relative Radial Oxygen Loss in Five Wetland Plants, *by S. C. Michaud and C. J. Richardson* 501

38b. Potential Importance of Sulfate Reduction Processes in Wetlands Constructed to Treat Mine Drainage, *by R. S. Hedin, R. Hammack, and D. Hyman*................ 508

38c. Evaluation of Specific Microbiological Assays for Constructed Wetlands Wastewater Treatment Management, *by R. J. Portier* 515

38d. Secondary Treatment of Domestic Wastewater Using Floating and Emergent Macrophytes, *by T. A. DeBusk, P. S. Burgoon, and K. R. Reddy* 525

38e. Amplification of Total Dry Matter, Nitrogen and Phosphorus Removal from Stands of *Phragmites australis* by Harvesting and Reharvesting Regenerated Shoots, *by T. Suzuki, W. G. A. Nissanka, and Y. Kurihara* 530

38f. Domestic Wastewater Treatment Using Emergent Plants Cultured in Gravel and Plastic Substrates, *by P. S. Burgoon, K. R. Reddy, and T. A. DeBusk* 536

38g. Aquatic Plant Culture for Waste Treatment and Resource Recovery, *by J. B. Kingsley, J. J. Maddox, and P. M. Giordano* 542

38h. Bacteriological Tests from the Constructed Wetland of the Big Five Tunnel, Idaho Springs, Colorado, *by W. Batal, L. S. Laudon, T. R. Wildeman, and N. Mohdnoordin* 550

38i. Use of Periphyton for Nutrient Removal from Waters, *by J. Vymazal* 558

39. Management of Domestic and Municipal Wastewaters

39a. Danish Experience with Sewage Treatment in Constructed Wetlands, *by H. Brix and H.-H. Schierup* 565

39b. Man-Made Wetlands for Wastewater Treatment: Two Case Studies, *by J. Jackson* 574

39c. Research to Develop Engineering Guidelines for Implementation of Constructed Wetlands for Wastewater Treatment in Southern Africa, *by A. Wood and L. C. Hensman* 581

39d. Constructed Wetlands: Design, Construction and Costs, *by K. J. Whalen, P. S. Lombardo, D. B. Wile, and T. H. Neel* 590

39e. Wastewater Treatment/Disposal in a Combined Marsh and Forest System Provides for Wildlife Habitat and Recreational Use, *by B. B. James and R. Bogaert* 597

39f. Root-Zone System: Mannersdorf—New Results, *by R. Haberl and R. Perfler* 606

39g. Constructed Wetlands for Secondary Treatment, *by T. J. Mingee and R. W. Crites* 622

39h. Hydraulic Considerations and the Design of Reed Bed Treatment Systems, *by J. A. Hobson* 628

39i. Use of Artificial Cattail Marshes to Treat Sewage in Northern Ontario, Canada, *by G. Miller* 636

39j. Some Ancillary Benefits of a Natural Land Treatment System, *by A. L. Schwartz and R. L. Knight* 643

39k. Performance of Solid-Matrix Wetland Systems, Viewed as Fixed-Film Bioreactors, *by H. J. Bavor, D. J. Roser, P. J. Fisher, and I. C. Smalls* 646

39l. Fate of Microbial Indicators and Viruses in a Forested Wetland, *by P. R. Scheuerman, G. Bitton, and S. R. Farrah* .. 657

39m. Wastewater Wetlands: User Friendly Mosquito Habitats, *by C. H. Dill* 664

40. Treatment of Nonpoint Source Pollutants—Urban Runoff and Agricultural Wastes

40a. Development of an Urban Runoff Treatment Wetlands in Fremont, California, *by G. S. Silverman* 669

40b. Urban Runoff Treatment in a Fresh/Brackish Water Marsh in Fremont, California, *by E. C. Meiorin* 677

40c. Design of Wet Detention Basins and Constructed Wetlands for Treatment of Stormwater Runoff from a Regional Shopping Mall in Massachusetts, *by P. Daukas, D. Lowry, and W. W. Walker, Jr.* 686

40d. Creation of Wetlands for the Improvement of Water Quality: A Proposal for the Joint Use of Highway Right-of-Way, *by L. C. Linker* 695

40e. Wetlands Treatment of Dairy Animal Wastes in Irish Drumlin Landscape, *by C. J. Costello* 702

40f. Potential Role of Marsh Creation in Restoration of Hypertrophic Lakes, *by E. F. Lowe, D. L. Stites, and L. E. Battoe* 710

41. Applications to Industrial and Landfill Wastewaters

41a. Utilization and Treatment of Thermal Discharge by the Establishment of a Wetlands Plant Nursery, *by M. S. Ailstock* 719

41b. Experiments in Wastewater Polishing in Constructed Tidal Marshes: Does It Work? Are the Results Predictable? *by V. G. Guida and I. J. Kugelman* 727

41c. Potential Use of Constructed Wetlands to Treat Landfill Leachate, *by W. W. Staubitz, J. M. Surface, T. S. Steenhuis, J. H. Peverly, M. J. Lavine, N. C. Weeks, W. E. Sanford, and R. J. Kopka* 735

41d. Natural Renovation of Leachate-Degraded Groundwater in Excavated Ponds at a Refuse Landfill, *by J. N. Dornbush*.... 743

42. Control of Acid Mine Drainage Including Coal Pile and Ash Pond Seepage

42a. Biology and Chemistry of Generation, Prevention and Abatement of Acid Mine Drainage, *by M. Silver* 753

42b. Design and Construction of a Research Site for Passive Mine Drainage Treatment in Idaho Springs, Colorado, *by E. A. Howard, J. C. Emerick, and T. R. Wildeman* 761

42c. Manganese and Iron Encrustation on Green Algae Living in Acid Mine Drainage, *by S. E. Stevens, Jr., K. Dionis, and L. R. Stark* 765

42d. Determining Feasibility of Using Forest Products or On-Site Materials in the Treatment of Acid Mine Drainage in Colorado, *by E. A. Howard, M. C. Hestmark, and T. D. Margulies* 774

42e. Use of Wetlands to Remove Nickel and Copper from Mine Drainage, *by P. Eger and K. Lapakko* 780

42f. Windsor Coal Company Wetland: An Overview, *by R. L. Kolbash and T. L. Romanoski* 788

42g. Wetland Treatment of Coal Mine Drainage: Controlled Studies of Iron Retention in Model Wetland Systems, *by J. Henrot, R. K. Wieder, K. P. Heston, and M. P. Nardi* 793

42h. Tolerance of Three Wetland Plant Species to Acid Mine Drainage: A Greenhouse Study, *by W. R. Wenerick, S. E. Stevens, Jr., H. J. Webster, L. R. Stark, and E. DeVeau* 801

42i. Control of the Armyworm, *Simyra henrici* (Lepidoptera: Noctuidae), on Cattail Plantings in Acid Drainage Treatment Wetlands at Widows Creek Steam-Electric Plant, *by E. L. Snoddy, G. A. Brodie, D. A. Hammer, and D. A. Tomljanovich* 808

List of Authors 813

Poster and Other Presentations 818

Index 819

Constructed Wetlands for Wastewater Treatment

Municipal, Industrial and Agricultural

SECTION I

General Principles

CHAPTER 1

Wastewaters: A Perspective

Al J. Smith

The water quality protection field is undergoing major changes. The Federal Water Pollution Control Act (FWPCA) of 1948 was one of the first national legislative efforts to deal with water quality problems. Under the FWPCA, the states were primarily responsible for controlling water pollution. The Federal Water Pollution Control Act Amendments passed in 1972 (PL-500) are the basis for the current Clean Water Act (CWA), which made a commitment to a federally focused and funded water program and shifted primary responsibility away from the states and local entities. The CWA mandated water pollution control for both municipal and industrial point source dischargers, initiated a federal program for nonpoint source water pollution control, and required control of toxic pollutants.

To date, under the CWA the federal government has provided over $50 billion in grant funds and states and localities have added about $17 billion for the planning, design, and construction of municipal wastewater treatment works. These funds went into construction of primary, secondary, and advanced municipal treatment plants, interceptor sewers, and collection systems in unsewered areas. Industrial treatment systems, systems for federal facilities, and most nonpoint source treatment costs are not included and would substantially increase the total.

The 1987 Water Quality Act Amendments to the CWA have established a new direction. Under this legislation, the focus of water pollution control is turned back to the states. The federal grants program for municipal wastewater treatment works will be phased out in 1990 and a State Revolving Loan Program funded in its place through 1994. As a result, we face a future of declining federal funds for municipal pollution control, but with huge construction needs to protect water quality. The magnitude of the remaining problem is illustrated in the following:

- Treatment works construction needs for small communities (under 10,000 population)—with the least ability to pay—amount to between $10 and $15 billion nationwide and exceed $1.3 billion in the eight states of Region IV alone.

- Agricultural activities generate large sources of nonpoint source water pollution and are causing serious water quality problems in at least 24 states.
- Approximately one-third of existing lakes and reservoirs have water quality impairment from nonpoint source pollution, principally from agricultural activities.
- Large treatment needs exist in the industrial sector. For example, acid mine drainage affects over 11,800 miles of streams in Appalachia alone.
- Leachate from landfill disposal of both industrial and municipal solid waste is affecting water quality.
- Industrial toxic chemicals and hazardous waste treatment and disposal have become major water quality concerns, with high potential costs and uncertain future treatment requirements and alternatives.

Both the reduction in available federal dollars and increasing focus on water quality underscore the need for a continual effort to identify and encourage technologies that provide effective, low-cost treatment. This book focuses on one type of treatment system that may effectively balance the need for reliable wastewater treatment with the need for minimal cost—wetland systems.

Wetlands have received wastewater discharges in numerous situations in the past, but only recently have they been recognized as a potentially cost-efficient treatment system. Studies over the last few years have shown that both natural and constructed wetland systems can provide high-quality wastewater treatment at relatively low cost. The growing interest in wetland systems is in part due to recognition that natural treatment systems offer advantages over conventional concrete-and-steel, equipment-intensive mechanical treatment plants. When the same biochemical and physical processes occur in a more natural environment instead of reactor tanks and basins, often the resulting system consumes less energy, is more reliable, requires less operation and maintenance—and, as a result, costs less.

In addition, creation of a constructed wetland for treatment, or preservation and enhancement of a natural wetland treatment system, is usually preferred from the standpoint of locating the facility next door. A wetland system can be very beneficial as a wildlife habitat or can even serve as a nature center.

This book evaluates these systems by pulling together state-of-the-art information so that we all can better determine when and where wetland treatment systems would be appropriate and how to develop them. The International Conference on Constructed Wetlands for Wastewater Treatment, a joint effort of the U.S. Environmental Protection Agency and the Tennessee Valley Authority, is a major first step in what I hope will be an expanding effort of our interagency cooperation to become much more active in providing technically competent, financially sound advice and consultation to municipalities, industries, agricultural representatives, and others facing new and more restrictive environmental requirements.

CHAPTER 2

Wetlands Ecosystems: Natural Water Purifiers?

Donald A. Hammer and Robert K. Bastian

Everyone has a vague idea of what constitutes a wetland, but not everyone has the same idea. Just as the predominant types of wetlands vary from one region to another, names commonly used to describe them also vary regionally. A *slough* is a freshwater marsh in the Dakotas, a brackish marsh along the West Coast, and a freshwater swamp on an old river channel in the Gulf Coastal plain, while the vernacular *bog* covers anything wet and difficult to cross.

Obviously, wetlands are not continuously dry lands. On the other hand, they need not be continuously wet. Many types of wetlands are wet only after heavy rains or during one season of the year. At other times they may be very dry, as many people are surprised to learn when they visit Everglades National Park during the dry season.

Wetlands are an *ecotone*—an "edge" habitat, a transition zone between dry land and deep water, an environment that is neither clearly terrestrial nor clearly aquatic. Because land and water can merge in many ways, it can be frustrating to attempt to define wetlands or determine where wetlands begin or end strictly on the basis of wetness or dryness.

There is no single "correct" definition of wetlands for all purposes. Several definition and classification systems have been devised for differing needs and purposes. Most tend to skirt the how-wet-is-wet question by identifying wetlands in terms of soil characteristics and the types of plants these transitional habitats typically support, since shallow standing water or saturated soil soon cause severe problems for all plants except hydrophytes, which are specifically adapted for these conditions.

In 1979, the U.S. Fish and Wildlife Service developed a definition and classification system capable of encompassing and systematically organizing all types of wetland habitats for scientific purposes.[1] It broadly recognizes wetlands as a transition between terrestrial and aquatic systems, where water is the dominant factor determining development of soils and associated biological communities and where, at least periodically, the water table is at or near the surface, or the land is covered by shallow water. Specifically, it requires that wetlands meet one or more of three conditions:

- areas supporting predominantly hydrophytes (at least periodically)
- areas with predominantly undrained hydric soil (wet enough for long enough to produce anaerobic conditions that limit the types of plants that can grow there)
- areas with nonsoil substrate (such as rock or gravel) that are saturated or covered by shallow water at some time during the growing season

The classification system is much like the hierarchical system scientists use for classifying plants and animals. It starts with five large systems: these are progressively divided into 10 subsystems, 55 classes, and 121 subclasses, which are then characterized by examples of dominant types of plants or animals.

This system provides a consistent standard of terminology for use among scientists and specialists throughout the country. It is now being used by the U.S. Fish and Wildlife Service as the basis for the National Wetlands Inventory, a comprehensive identification and mapping of wetlands that will greatly assist wetland assessment and management.[2]

However, various legislation and agency regulations define wetlands in more general terms. With minor variations, most describe wetlands as areas flooded or saturated by surface water or groundwater often and long enough to support those types of vegetation and aquatic life that require or are specially adapted for saturated soil conditions. Such descriptions can be stretched, if necessary, to accommodate many of the scientific classifications. At the same time, they more comfortably conform to popular conceptions of what constitutes wetlands—salt and freshwater swamps, marshes, and bogs and perhaps a few subclassifications of these basic types.

In popular usage, shallow-water or saturated areas dominated by water-tolerant woody plants and trees are generally considered *swamps;* those dominated by soft-stemmed plants are considered *marshes;* and those with mosses are *bogs.*

Our principal saltwater swamps are mangrove wetlands along the southern coast of Florida. Mangroves are among the very few woody plants adapted to tolerate saltwater environments.

Freshwater swamps contain a variety of woody plants and water-tolerant trees. Southern swamps typically contain bald cypress (*Taxodium*); tupelo gum (*Nyssa*); water, willow, and swamp white oak (*Quercus*); and river birch (*Populus*) (Figure 1). Northern swamps are more likely to include alder (*Alnus*), black ash (*Fraxinus*), black gum (*Nyssa*), northern white cedar (*Thuja*), tamarack (*Larix*), red maple (*Acer*), and willow (*Salix*).

Coastal salt marshes are dominated by salt-tolerant herbaceous plants, notably cordgrass (*Spartina*), and blackrush (*Juncus*) or other rushes along extensive areas of the eastern and southern coasts or cordgrass and glasswort (*Salicornia*) along the west coast. Less familiar are the inland salt marshes of the

Figure 1. A southern cypress swamp, a wetland dominated by trees and other woody plants.

intermountain west where high evaporation rates from shallow lakes and playas concentrate the salt content.

Freshwater marshes are dominated by herbaceous plants. Submerged and floating plants may occur, often in abundance, but emergent plants usually distinguish a marsh from other aquatic environments. Familiar emergents include cattails (*Typha*), bulrush (*Scirpus*), reed (*Phragmites*), grasses, and sedges (*Carex*). A wet meadow may be only intermittently saturated or flooded with very shallow water but it also supports marsh species, especially sedges and wet grasses. A common type of freshwater marsh, the prairie potholes of the northern Great Plains, are shallow depressions formed by glaciers (Figure

2). Those that hold water year-round, seasonally, or following heavy rains often support luxúriant marsh vegetation. Although most are small, there are many of them (810,000 ha in North Dakota alone), and collectively they constitute an important wetland resource, especially for waterfowl nesting.

Bogs form primarily in deeper glaciated depressions, mainly in "kettle holes" in the northeastern and north-central regions (Figure 3). Bogs are dependent upon stable water levels and are characterized by acidic, low-nutrient water and acid-tolerant mosses. Other bog plants such as cranberry (*Vaccinium*), tamarack, black spruce (*Picea*), leatherleaf (*Chamaedaphne*), and pitcher plant (*Sarracenia*) may be rooted in deep, spongy accumulations of dead *Sphagnum* moss and other plant materials only partially decomposed under bog conditions. In the same region, fens have water nearer to neutral and are dominated by sedges.

There is not a reliable quantification of the original wetland resource of this country or of the portion that has been lost; even estimates vary widely. However, one frequently cited estimate made in the mid-1970s was that only about 40 million ha, or about 46%, remained from a wetland heritage of some 87 million ha.[2] Continuing wetland losses were estimated to average 185,000 ha annually, 87% of which was attributed to conversions to agricultural land.

WETLANDS DYNAMICS

In a geological sense, natural wetlands are an ephemeral component of the landscape, highly dependent upon disturbance whether as long-term, large-scale tectonic forces or localized phenomena such as annual or daily flooding and drying, fire, or storm events. Without tectonic or hydrologic disturbance, wetlands gradually progress through a succession of stages to relatively dry upland-type ecosystems.

On a large scale, extensive wetlands resulted from recent glaciation, mountain-building, and changes in sea level that interrupted or destroyed drainage patterns. Once erosional forces have reestablished streams and rivers, most wetlands are soon drained. Extensive North American wetlands were created by continental glaciation (e.g., midwestern pothole marshes), valley glaciers in the Rocky Mountains, and changing sea levels along the coastal plain in the East and Southeast. Extensive bottomland hardwood swamps associated with the Mississippi and other large river systems in the coastal plain were created by a combination of changing sea levels, silt deposition from upland erosion, and natural channel alterations. Rising lake and sea levels may turn river swamps into bays, whereas falling levels cause greater relief and fast-flowing rivers in narrow V-shaped valleys with few wetlands.

Locally, without disturbance, a New England bog or a prairie marsh will gradually fill in, becoming a forest or prairie. Without seasonal flooding, the great river swamps are soon lost to terrestrial forest types. All are dependent upon disturbance or cyclic fluctuations in local hydrology.

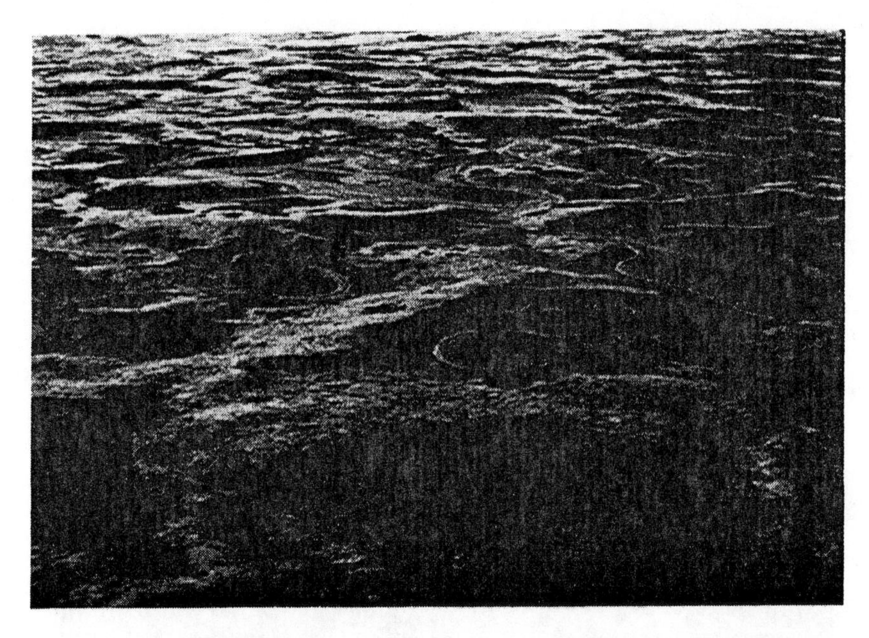

Figure 2. Prairie potholes, highly productive freshwater marshes dominated by herbaceous cattail and bulrush plants.

Internally, natural wetlands are also dynamic, undergoing species compositional changes, seasonally and annually. Early waterfowl managers soon learned that stable water levels throughout a growing season and/or year after year reduced marsh productivity almost as much as the lack of water during the drought years of the 1930s. Kadlec[3] later showed that periodic drying recirculated critical nutrients that were unavailable under constantly flooded and anaerobic conditions. Flooding and drying events will favor some plants and inhibit others, with consequent changes in microbial, invertebrate, and larger animal populations. In addition, major shifts in species composition over time are inevitable. Only 10–20 species were planted in many large National Wildlife Refuges that now have hundreds of plant types. Two years after four species were planted in a mine drainage treatment wetlands, over 40 species were present.[4] Concurrently, substrates, water flow, and microbial and animal population changes have occurred.

The dependency of wetland communities on hydrologic patterns is most obvious in changes in species composition resulting from alterations in water depths and flow patterns. Deliberate modifications of depths and flow can maintain a desirable mix of plant species or inhibit the establishment of others. Some information on the requirements of various species is available in the literature,[5,6] but much practical management knowledge is located in the internal reports and experience of regional wetlands managers of the State Fish and Wildlife Agencies or the National Wildlife Refuges.

Figure 3. A northeastern bog. Note the large expanse of moss surrounded by shrubs grading into trees.

FUNCTIONS OF NATURAL WETLANDS

The productivity of many wetlands far exceeds that of the most fertile farm fields (which in many cases are former wetlands). Wetlands receive, hold, and recycle nutrients continually washed from upland regions (Figure 4). These nutrients support an abundance of macro- and microscopic vegetation, which converts inorganic chemicals into the organic materials required—directly or indirectly—as food for animals, including man.

In addition to their vegetative productivity, wetlands team with zooplankton, worms, insects, crustaceans, reptiles, amphibians, fish, birds, and mammals, all feeding on plant materials or one another. Other animals are drawn from nearby aquatic or terrestrial environments to feed on plants and animals at the highly productive "edge" environment of wetlands, and they in turn become prey for others from a greater distance, thus extending the productive influence of wetlands far beyond their borders.

Sport and commercial hunters and fishermen have called public attention to the economic value of wetlands fish and wildlife. They were first to note the direct relationship between wetland destruction and declining populations of valuable species of fish, shellfish, birds, reptiles, and fur-bearing animals that

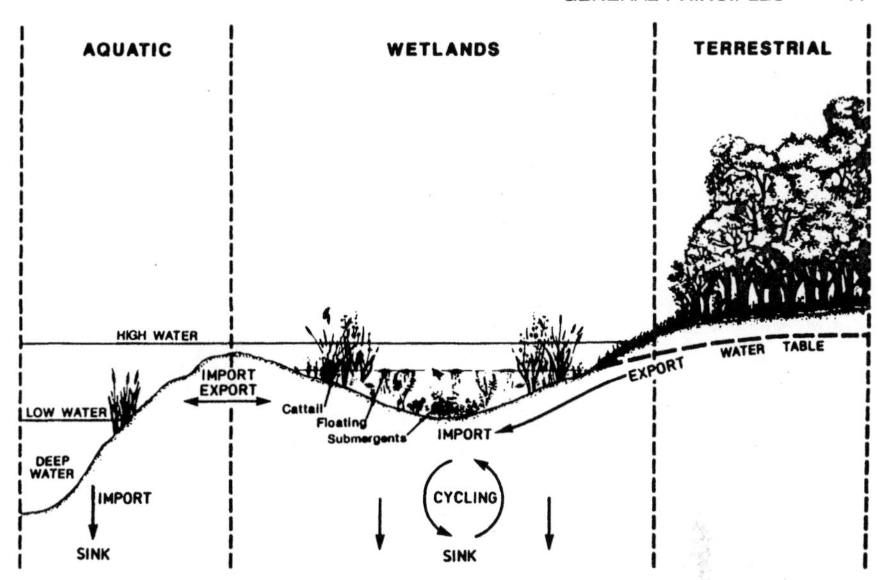

Figure 4. Wetlands are transition zones between terrestrial and aquatic environments and benefit from nutrient, energy, plant, and animal inputs from neighboring systems.

are dependent on certain types of wetland habitats during part or all of their lives. Many studies have now linked the destruction of summer breeding wetlands and winter feeding wetlands to shifts or declines in populations of migratory waterfowl and other birds. Wetland destruction can be especially significant in regions where such habitat is least common and alternative sites may be unavailable.

Because of the diversity of habitats possible in these transition environments, the nation's wetlands are estimated to contain 190 species of amphibians, 270 species of birds, and over 5000 species of plants.[7] Many wetlands are identified as critical habitats under provisions of the Endangered Species Act, with 26% of the plants and 45% of the animals listed as threatened or endangered either directly or indirectly dependent on wetlands for survival.

Wetlands along coasts, lakeshores, and riverbanks have recently begun receiving increased attention because of their valuable role in stabilizing shorelands and protecting them from the erosive battering of tides, waves, storms, and wind. One of the greatest benefits of inland wetlands is the natural flood control or buffering certain wetlands provide for downstream areas by slowing the flow of floodwater, desynchronizing the peak contributions of tributary streams, and reducing peak flows on main rivers.

Some wetlands may function as groundwater recharge areas, allowing water to seep slowly into underlying aquifers. At other times wetlands serve as discharge areas for surfacing groundwaters, allowing stored groundwater to sustain base flow in streams during dry seasons.

Perhaps the most important but least understood function of wetlands is water quality improvement. Wetlands provide effective, free treatment for many types of water pollution. Wetlands can effectively remove or convert large quantities of pollutants from point sources (municipal and certain industrial wastewater effluents) and nonpoint sources (mine, agricultural, and urban runoff) including organic matter, suspended solids, metals, and excess nutrients. Natural filtration, sedimentation, and other processes help clear the water of many pollutants. Some are physically or chemically immobilized and remain there permanently unless disturbed. Chemical reactions and biological decomposition break down complex compounds into simpler substances. Through absorption and assimilation, wetland plants remove nutrients for biomass production. One abundant by-product of the plant growth process is oxygen, which increases the dissolved oxygen content of the water and also of the soil in the immediate vicinity of plant roots. This increases the capacity of the system for aerobic bacterial decomposition of pollutants as well as its capacity for supporting a wide range of oxygen-using aquatic organisms, some of which directly or indirectly utilize additional pollutants (Figure 5).

Many nutrients are held in the wetland system and recycled through successive seasons of plant growth, death, and decay. If water leaves the system through seepage to groundwater, filtration through soils, peat, or other substrates removes excess nutrients and other pollutants. If water leaves over the surface, nutrients trapped in substrate and plant tissues during the growing season do not contribute to noxious algae blooms and excessive aquatic weed growths in downstream rivers and lakes. Excess nutrients from decaying plant tissues released during the nongrowing season have less effect on downstream waters.

It is no secret that natural wetlands can remove iron, manganese, and other metals from acid drainage—they have been doing it for geological ages. In fact, accumulations of limonite, or *bog iron*, were mined as the source of ore for this country's first ironworks and for paint pigment. Limonite deposits are most common in the bog regions of Connecticut, Massachusetts, Pennsylvania, New York, and elsewhere along the Appalachians. Wetlands were abundant in parts of the Tennessee Valley during past ages, and significant bog iron deposits are found in Virginia, Tennessee, Georgia, and Alabama. Although now of limited economic importance in the United States, bog iron is still a significant source of iron ore in northern Europe.

Similarly, mixed oxides of manganese, called *wad* or *bog manganese*, are the product of less acidic wetland removal processes. Often these wad deposits also contain mixed oxides of iron, copper, and other metals.

CONSTRUCTED WETLANDS FOR WASTEWATER TREATMENT

We define *constructed wetlands* as a designed and man-made complex of saturated substrates, emergent and submergent vegetation, animal life, and

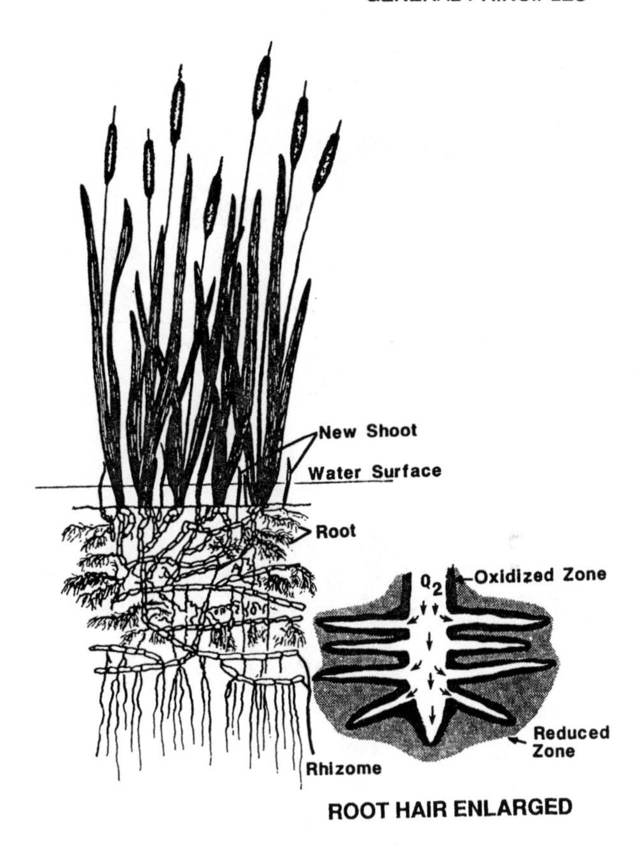

Figure 5. Wetlands plants have the unique ability to transport oxygen to support their roots growing in anaerobic substrates.

water that simulates natural wetlands for human use and benefits. Synonymous terms include *man-made*, *engineered*, and *artificial* wetlands. Although bogs and swamps have been constructed or used for wastewater treatment and consequently are included in the above definition, most constructed wetlands for wastewater treatment emulate marshes. Marshes with herbaceous emergent and, perhaps, submergent plants have the most promise for wastewater treatment (Figure 6). Water-tolerant woody plants in swamps may require 5–20 years for development and full operational performance. Bogs dominated by mosses are difficult to establish, have limited retention capacity[8] and limited adaptability to fluctuating water levels, and are likely to become marshes if water and nutrient inputs are increased.[9] Conversely, marshes with cattail, bulrush, rush, or giant reed are adapted to fluctuating water and nutrient levels and are more tolerant of high pollutant concentrations.[10]

For our purposes, most natural or constructed wetlands have five principal components:

Figure 6. A dense stand of cattail in the Benton, Kentucky municipal wastewater treatment wetlands.

1. substrates with various rates of hydraulic conductivity
2. plants adapted to water-saturated anaerobic substrates
3. a water column (water flowing in or above the surface of the substrate)
4. invertebrates and vertebrates
5. an aerobic and anaerobic microbial population

Microbes—bacteria, fungi, algae, and protozoa—alter contaminant substances to obtain nutrients or energy to carry out their life cycles.[11] The effectiveness of wetlands managed for wastewater treatment is dependent on developing and maintaining optimal environments for desirable microbial populations. Fortunately, these microbes are ubiquitous, naturally occurring in most waters and likely to have large populations in wetlands and contaminated waters with nutrient or energy sources. Only rarely, with very unusual pollutants, will inoculation of a specific group or strain of microbes be required. In addition, many naturally occurring microbial groups are predatory and will forage on pathogenic organisms.

Wetland plants appear to have two important but indirect functions. (1) Within the water column, stems and leaves significantly increase surface area for attachment of microbial populations. (2) Wetland plants have the ability to transport atmospheric gases including oxygen down into the roots to enable their roots to survive in an anaerobic environment. Some incidental leakage occurs, producing a thin-film, aerobic region called the *rhizosphere* surrounding each roothair. Doubtless, some chemical oxidation occurs in this microscopic region, but more important, the rhizosphere supports large microbial populations that conduct desirable modifications of nutrients, metallic ions, and other compounds. The juxtaposition on a microscopic scale of an aerobic region surrounded by an anaerobic region multiplied by the almost astronomical area of rhizosphere boundary is crucial to nitrification-denitrification and numerous other desirable pollutant transformations. Nutrient or other substance uptake by plants is generally insignificant except for a few systems that incorporate periodic plant harvesting to physically remove plant biomass. However, these systems tend to have high operating costs.

Fortunately many wetland plants are widely distributed in suitable environments throughout the northern hemisphere. Some are cosmopolitan—giant reed occurs on every continent except Antarctica, as do various species of cattail. Naturally occurring plant species are adapted to local climate and soil conditions and are much more likely to succeed and provide treatment in a constructed wetlands. Exotic plant species may simply fail to survive or may perform poorly, and could become serious pests in natural waterways.

From the standpoint of wastewater treatment, certain plant/substrate combinations appear to be more efficient in constructed wetlands treatment systems,[12] and others may be more tolerant of high pollutant concentrations.[13] Consequently, many projects have included a single plant/substrate combination. But maintaining a monoculture may require unnecessary operational expenses or even be undesirable since an insect pest outbreak could seriously damage a monoculture system,[14] whereas a mixed-species system may be more resistant to pest attack and fluctuating loading rates and may remove a broader variety of pollutants.

Additionally, fostering growth of a single species may require operational practices with substantial changes in water levels that would essentially remove part or all of the system from operation. Conversely, three plants—cattail, bulrush, and giant reed, commonly used in wetlands treatment systems—tend to create and/or maintain single-species stands by inhibiting or out-competing other plants.

Substrates—various soils, sand, or gravel—provide physical support for plants; considerable reactive surface area for complexing ions, anions, and other compounds; and attachment surfaces for microbial populations.

Surface and subsurface water transports substances and gases to microbial populations, carries off by-products, and provides the environment and water for biochemical processes of plants and microbes.

Constructed wetlands appear to have very broad applicability as wastewater treatment systems for an array of water pollution problems. This is more likely to be true for naturally occurring organic and inorganic substances but may be extended to some, if not most, anthropogenic compounds. A number of factors are involved. Wetland complexes naturally occur at topographic lows receiving runoff waters from various sources and have done so throughout geological time. Wetlands have adapted to substances carried by runoff, using them to help support some of the highest known productivity rates. Many wetlands have higher rates of carbon fixation, biomass production, or other measures of productivity than the most intensively managed agricultural fields.[15] In essence, the high productivity of wetland systems results from inputs of waterborne nutrients and energy sources.

In addition, wetlands microbial populations that conduct critical processing of pollutants have short generation times, high reproductive rates, and considerable genetic plasticity, all of which permit these organisms to rapidly adapt to and exploit new nutrient or energy sources.

However, many mechanisms that modify and/or immobilize pollutants, especially toxic substances, are poorly understood. Some wetland systems appear to remove or modify even complex toxics,[16,17] but plant harvesting and subsequent incineration or other ultimate disposal methods may be necessary. Long-term accumulation of heavy metals or unaltered toxic compounds in wetland vegetation or sediments may reduce widespread distribution in the environment, but the concentrated deposits may contribute to detrimental effects of bioaccumulation and/or biotransport, may require periodic recovery/recycling procedures, or may restrict other uses of these areas.

SUMMARY AND CONCLUSIONS

Our information indicates that constructed wetlands offer an economical, largely self-maintaining, and therefore preferred alternative to conventional treatment of a variety of types of contaminated water. There is every reason to expect that these systems can continue to function reliably for long periods of time just as natural wetlands have, but only long-term operating data will confirm this expectation.

If small communities are to meet wastewater treatment requirements of the future, they must have treatment systems that are not only effective and reliable but also simple and inexpensive to build and operate. Constructed wetlands, which appear to meet all of these criteria, offer a promising alternative to conventional treatment plants. Constructed wetland systems (1) are relatively inexpensive to construct and operate; (2) are easy to maintain; (3) provide effective and reliable wastewater treatment; (4) are relatively tolerant of fluctuating hydrologic and contaminant loading rates; and (5) may provide indirect benefits such as green space, wildlife habitats, and recreational and educational areas.

Disadvantages of constructed wetlands for wastewater treatment relative to conventional systems include: (1) relatively large land area requirements for advanced treatment; (2) current imprecise design and operating criteria; (3) biological and hydrological complexity and our lack of understanding of important process dynamics; and (4) possible problems with pests. Steep topography, shallow soils, a high water table, or susceptibility to severe flooding may also limit their use. Mosquitoes or other pests could be a problem with wetlands systems that are improperly designed or managed.

Furthermore, operation of constructed wetland treatment systems may require two or three growing seasons before optimal efficiencies are achieved. Completion of earth moving, concrete and pipe installation, and vegetation planting does not translate into full operational status. Since treatment efficiencies generally improve as above- and below-surface plant densities increase, full operational status will likely require several years after construction and planting are completed.

Probably the greatest single problem is the lack of detailed information from long-term experience with these systems. Although research and demonstration projects have shown that wetlands can provide effective treatment, this treatment option remains generally unknown outside the scientific community. Available information has been applied on a practical scale in only a few cases, and there have been few attempts to effectively document and communicate the details of project design, construction, operation, maintenance, and performance needed by regulatory officials, engineering consultants, developers, and community leaders. Much available information is sketchy, at times contradictory, and generally inadequate for optimizing process variables, treating other types of wastewaters at other sites, or achieving various effluent limits under specific conditions at minimal costs. Conversely, the existing data base is adequate for *conservative* design and operation of constructed wetlands to meet most discharge limits for household, municipal, and acid drainage wastewaters.

As a final note, we urge caution in attempting to reduce wetland treatment systems to minimal components and treatment areas or simply implementing the most efficient combination of substrate/vegetation/loading rates. The contaminant processing mechanisms in constructed wetlands are likely similar, if not identical to, microbial transformations present in package treatment plants, lagoons, or other conventional wastewater treatment systems. The latter require large inputs of energy, operating procedures, and subsequent costs to maintain optimal environmental conditions for suitable microbial populations within a relatively small treatment area. The low capital and operating costs, efficiency, and self-maintaining attributes in wetland treatment systems result from the complex of plants, water, and microbial populations in a large enough land area to be self-sustaining without significant energy or other maintenance inputs. It may be less costly to construct a minimum size, least-component wetlands treatment system, but operational costs to maintain that system could easily negate initial cost savings. For small communities, farms, mines, and some industries, a conservatively designed and biologically complex system may provide more efficient treatment, greater longevity, and reduced operational requirements and costs.

Although some *natural* wetlands have been effectively used for water quality improvement, we do not wish to encourage additional use. We have recently become aware that natural wetlands are valuable resources that must not be wasted. Much remains to be learned about their many values and functions and the long-term consequences of wetlands destruction. However, enough is known to conclude that it is not worth risking the unnecessary loss of any remaining natural wetlands without a better understanding of their important roles in biological productivity, fish and wildlife habitat, flood protection, groundwater recharge and discharge, base flow stabilization of rivers, and water quality improvement. On the other hand, *constructed* wetlands may provide a relatively simple and inexpensive solution for controlling many water pollution problems without detrimentally affecting our natural wetlands resources. Although all of the processes are not well understood, constructed wetlands are capable of moderating, removing, or transforming a variety of water pollutants while also providing wildlife and recreational benefits commonly associated with natural wetlands systems.

REFERENCES

1. Cowardin, L. M., V. Carter, F. C. Golet, and E. T. LaRoe. "Classification of Wetlands and Deepwater Habitats of the United States," U.S. Dept. of Interior, FWS/OBS-79/31 (1979).
2. Tiner, R. W., Jr. "Wetlands of the United States: Current Status and Recent Trends," U.S. Dept. of Interior (1984).
3. Kadlec, J. A. "Effect of a Drawdown on a Waterfowl Impoundment," *Ecol.* 43:267–281 (1962).
4. Brodie, G. A., D. A. Hammer, and D. A. Tomljanovich. "Constructed Wetlands for Acid Drainage Control in the Tennessee Valley," in *Proceedings Wetlands—Increasing Our Wetlands Resources,* J. Zelazny and J. S. Feierabend, Eds. (Washington: National Wildlife Federation, 1987), pp. 173–180.
5. Schulthorpe, C. D. *The Biology of Vascular Plants* (London: Edward Arnold Ltd., 1967).
6. Stephenson, M., G. Turner, P. Pope, J. Colt, A. Knight, and G. Tchobanoglous. "The Environmental Requirements of Aquatic Plants," in *The Use and Potential of Aquatic Species for Wastewater Treatment,* Publication No. 65, California State Water Resources Control Board, Sacramento (1980), Appendix A.
7. Mitsch, W. J., and J. G. Gosselink. *Wetlands* (New York: Van Nostrand Reinhold Company, 1986).
8. Wieder, R. K. "Determining the Capacity for Metal Retention in Man-Made Wetlands Constructed for Treatment of Coal Mine Drainage," Bureau of Mines Information Circular IC 9183 Pittsburgh, PA (1988).
9. Kadlec, R. H. "Northern Natural Wetland Water Treatment Systems," in *Aquatic Plants for Water Treatment and Resource Recovery,* K. R. Reddy and W. H. Smith, Eds. (Orlando, FL: Magnolia Publishing Inc., 1988).
10. Small, M. M. "Data Report—Marsh/Pond System," USERDA Report BNL 50600 (1976).
11. Alexander, M. *Introduction to Soil Microbiology* (New York: John Wiley & Sons, Inc., 1967).
12. Gersberg, R. M., B. V. Elkins, S. R. Lyon, and C. R. Goldman. "Role of Aquatic

Plants in Wastewater Treatment by Artificial Wetlands," *Water Res.* 20:363–368 (1986).
13. Brix, H., and H. Schierup. "Danish Experience with Sewage Treatment in Constructed Wetlands," Chapter 39a, this volume.
14. Snoddy, E. L., G. A. Brodie, D. A. Hammer, and D. A. Tomljanovich. "Control of the Armyworm, *Simyra henrici* (Lepidoptera: Noctuidae), on Cattail Plantings in Acid Drainage Treatment Wetlands at Widows Creek Steam-Electric Plant," Chapter 42i, this volume.
15. Odum, E. P. *Fundamentals of Ecology,* 3rd ed. (Philadelphia: W. B. Saunders Company, 1971).
16. Dornbush, J. N. "Natural Renovation of Leachate-Degraded Groundwater in Excavated Ponds at a Refuse Landfill," Chapter 41d, this volume.
17. Portier, R. J., and S. J. Palmer. "Wetlands Microbiology: Form, Function, Processes," Chapter 6, this volume.

CHAPTER 3

Hydrologic Factors in Wetland Water Treatment

Robert H. Kadlec

INTRODUCTION

A wide variety of wastewaters have been treated with wetland systems. To at least some degree, most have been successful. Further advance of this technology must be based on a coordinated set of design principles. First among these is understanding wetland hydrodynamics and its influences on data acquisition and design. Three features of water movement in these systems distinguish them from conventional concrete-and-steel wastewater treatment facilities. First, because of large surface area together with outdoor location, wetland systems interact strongly with the atmosphere via rainfall and evapotranspiration. Second, because of high vegetation density, plants influence water movement, both overland flow and evapotranspiration. Finally, because contact times are long with respect to the time scales of rainfall and microbial dynamics but short with respect to seasonal processes, the operation of these systems may not be viewed as a steady-state process.

The purpose of this chapter is to set forth current information on key hydrologic processes, and to examine implications on process performance. Because wetland systems are usually isolated from surface runoff, that process requires little attention. The most important effects are increased catchment area created by dikes or berms forming the wetland, and perhaps some lateral wicking to berms during dry, unfrozen conditions. Water consumption for biomass production is a tiny fraction of total flow, and will not be mentioned further.

Snow and ice phenomena are basically unstudied. Existing results suggest that ice layers retard but do not stop water movement under the ice. Most snowfall is stored until spring thaw, when a significant portion of the melt water exits as over-ice flow. Treatment efficiency is reduced; hence, some degree of winter water storage is necessary in northern climates. Water irrigation can be conducted in winter without freeze-up, provided modest precautions are taken. The primary focus of this work is on the unfrozen seasons, or the entire year for southern locations.

The short-term dynamics of wetland treatment systems are poorly understood. Clearly, waves of concentration and water pass through these systems in response to unsteady pumping activities and atmospheric events. Descriptive tools are available for such behavior,[1,2] but the details are too lengthy to be included here.

FLOW RESISTANCE

Wetland system flow may be separated into overland and underground components, although the dividing line is arbitrary since ground level is difficult to define in wetlands. Hydraulic conductivity changes sharply between the vegetation and litter and the soils below. More important are changes in the fundamental rules that govern flow. All such rules relate flow rate to depth, travel distance, and slope of the water surface. Refer to Figure 1 for terminology.

For flow in fully saturated, fine grained soils, sands, and gravels, Darcy's law is expected to hold:

$$v = -k \frac{\partial P}{\partial x} \tag{1}$$

where v is linear velocity, x is travel distance, and P is pressure expressed in liquid head. The constant, k, is the hydraulic conductivity. This equation holds approximately for peat, but uniformity and directional preferences are questionable. This is the preferred calculation for infiltration.

For lateral free surface flow through coarse gravels and rocks, Ergun's equation is appropriate:

$$\rho g S = 150 \frac{\mu v (1 - \epsilon)^2}{D_p^2 \epsilon^2} + 1.75 \frac{\rho v^2 (1 - \epsilon)}{D_p \epsilon} \tag{2}$$

where
S = the slope of the free surface
ρ = density
g = gravitational acceleration
μ = viscosity
ϵ = porosity
D_p = particle diameter
v = velocity = superficial velocity/ϵ

This equation adds a term to Darcy's law for turbulent flow, which can occur for the larger particle sizes. Further, it is "turned around" and is explicit in the slope of the water surface, not in the velocity. This equation calculates the extent of water mounding that may occur in wetland systems with underground lateral flow (the gravel bed systems). Figure 2 shows the expected

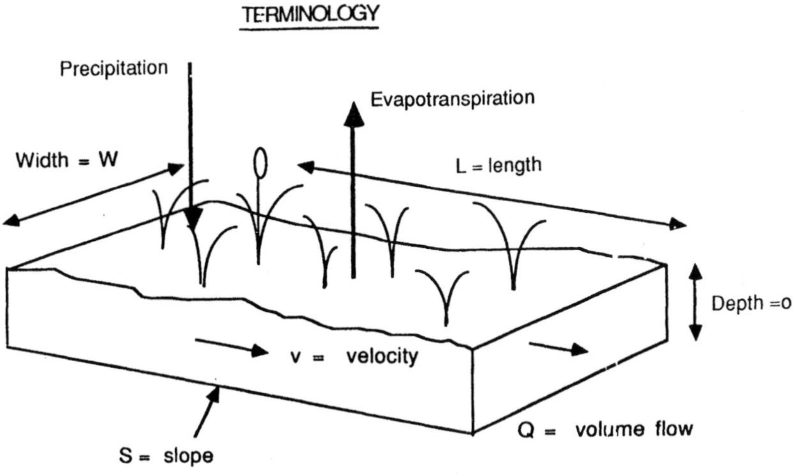

Figure 1. Terminology for wetland treatment systems.

mound height for an outlet depth of 50 cm and a porosity of 30%. As rock diameter and flow rate increase, turbulence may be present; the second term in the Ergun equation will contribute. Turbulence occurs when the particle Reynolds number, $D_p\rho v/\mu$, exceeds a value of 10. Significant water mounds can develop with certain choices for diameter and flow rate, and mound tilt can be

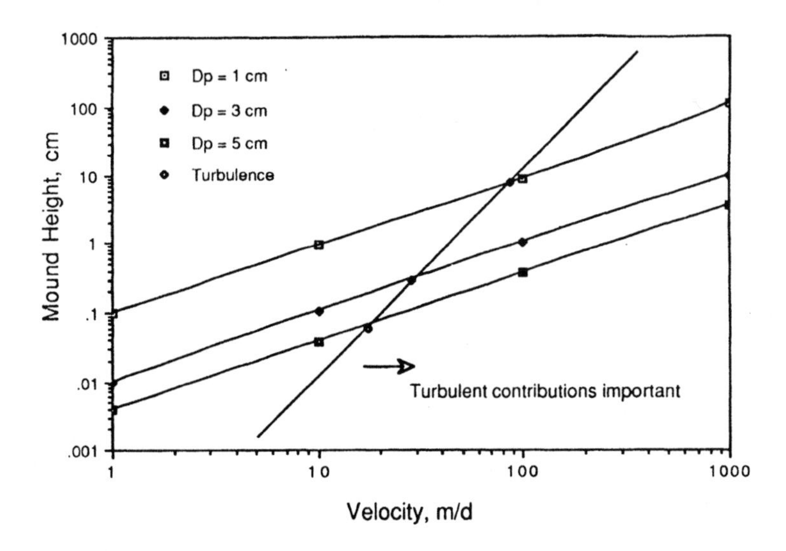

Figure 2. Water mounding potential for rock bed wetland treatment systems for different rock sizes and velocities. Based on exit depth of 50 cm and length of 100 m.

great enough to cause overland flow in the upstream sections of a rock bed wetland.

Contact times for subsurface flow wetlands depend on depth and flow rate. In the absence of mounding, these parameters may be set independently, since velocity does not depend upon depth.

For lateral overland flow through wetland vegetation, an empirical flow law applies, similar in appearance to the Manning equation:

$$Q/W = a\, d^b\, S^c \tag{3}$$

where Q/W = volumetric flow per unit width
d = depth
S = slope of water sheet
a,b,c = constants

This equation is derived from the Ergun equation together with a depth-variable stem density and a depth-variable soil elevation distribution. A turbulence term is included because overland flow in a wetland is transitional between streamline and turbulent flow. However, it is then convenient to recorrelate the model to the empirical form shown above.[1]

This form of the equation recognizes that such flows are partly turbulent and partly laminar.[3] The exponent $b = 3.0$, and $c = 0.338$ for grassy slopes;[4] for the Houghton Lake wetland, $b = 2.5$–3.0 and $c = 0.7$–1.0.[1,5] The coefficient a is site-specific, but equal to 4×10^6 in meters and days for Houghton Lake for $b = 3$ and $c = 1$. All wetland data sets show approximately the same depth dependence, but slope dependence is not as well defined. In any case, calculations are not sensitive to the slope term power.

Contact times (τ) are calculated by dividing the water volume in the wetland by the volumetric flow rate Q:

$$\tau = \frac{LWd\epsilon}{Q} \tag{4}$$

The wetland overland flow law may be used to eliminate the depth:

$$\tau = \frac{L\epsilon}{a^{1/b}\left(\frac{Q}{W}\right)^{1-1/b} S^{c/b}} \tag{5}$$

The power on flow is approximately ⅔, and on slope approximately ⅓. The recommended overland flow equations for terrestrial ecosystems[6] are similar in form but were calibrated for higher slopes and are seriously in error for wetland environments.

Attempts to set wetland operating depths may fail if vegetative resistance creates a significant slope to the wetland water sheet. Water will mound near the inlet to provide the necessary head to drive water through the vegetation.

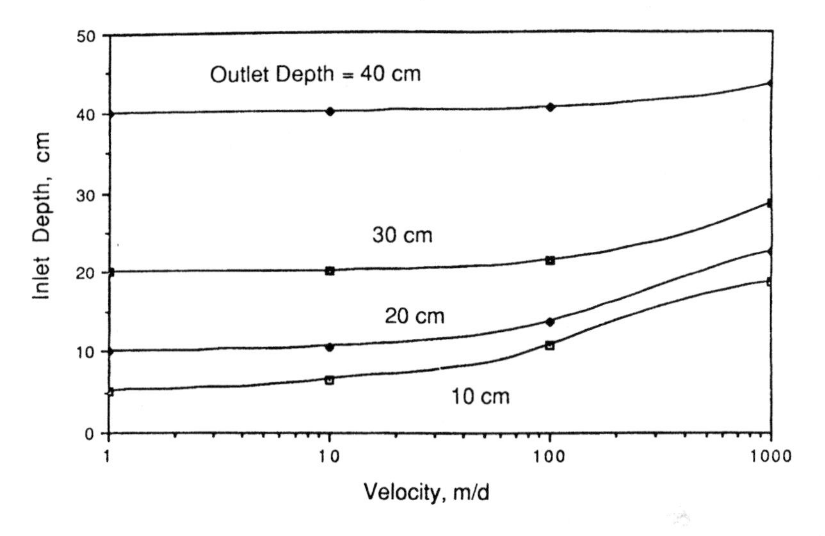

Figure 3. Size of the water mound for aboveground water flow through a vegetated wetland. Based on data from the Houghton Lake system and a length of 100 m.

Using results from the Houghton Lake, Figure 3 illustrates the size of this mound under different conditions of flow and outlet water depth.

EVAPORATION AND TRANSPIRATION

Atmospheric water losses from a wetland occur from the water and soil (evaporation), and from emergent portions of plants (transpiration). The combination is termed *evapotranspiration.* Many reports of wetland vaporization losses include the reviews of Linacre[7] and Ingram.[8] All have attempted to correlate data with ecosystem and meteorological variables and, in most cases, have compared data to open water evaporation. The results form a large and confusing literature, but a coherent set of principles may be drawn that is here restricted to "reedswamp" wetlands, i.e., those containing nonwoody emergent macrophytes, such as *Typha, Scirpus,* and *Phragmites.*

Vaporization energy comes principally from the sun, so the energy balance for the wetland is the logical framework to use for data interpretation. With reference to Figure 4:

$$Q_{rn} + Q_{ai} = Q_e + Q_h + Q_s + Q_{ao} \quad (6)$$

Net radiation (Q_{rn}) is the incoming solar energy less the back radiation and reflection. In turn, incoming radiation depends on cloud cover, time of the year, and latitude. Reflection depends on wetland albedo. Literature values are 0.1–0.4, with low values from low sparse vegetation and high values from

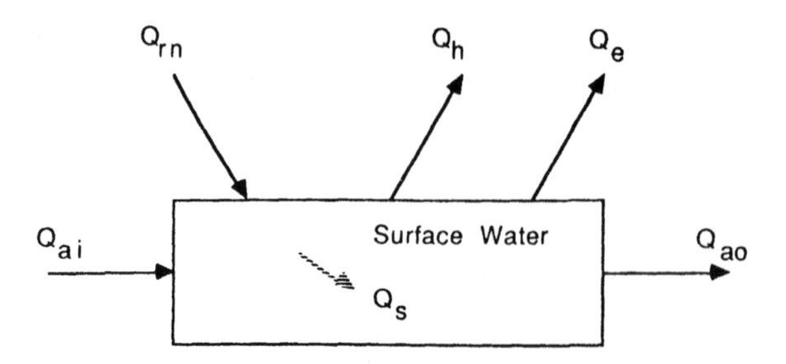

Figure 4. Energy flows which drive evapotranspiration. Net radiation and advective inputs result in vaporization, convection to the air, advective output and storage (rn = net radiation; ai = advective input; ao = advective output; s = storage; h = heat loss; e = evaporative loss).

dense, high vegetation. Transport of heat and water vapor from the surface is governed by local air swirls and is in a fixed proportion called the Bowen ratio. Wind speed influences both, as does vapor pressure and water temperature. Advection terms incorporate air current influences from adjacent environments.

Given sufficient meteorological data, the energy balance can calculate evaporation or evapotranspiration.[9] The text of Eagleson[10] is one of several sources of extensive explanation. About a dozen pieces of information are needed to use this method, some of which are not available *a priori,* such as albedo. Although the calculation is intended only for evaporation, it is frequently used to estimate evapotranspiration.

Data requirements of the energy balance estimation procedure led to many simpler, more empirical procedures.

The following propositions are supported by available data.

1. *The presence of vegetation retards evaporation.* Vegetation increases shade and humidity and reduces wind near the surface. A litter layer can create a mulching effect. A sampling of percentages of open water evaporation is as follows: Bernatowicz et al.,[11] 47%; Koerselman and Beltman,[12] 41–48%; Kadlec et al.,[13] 30–86%. However, the wetland does not necessarily conserve water, because transpiration can equal or exceed the difference.
2. *Wetland evapotranspiration, over the growing season, is represented by 0.8 times Class A pan evaporation from an adjacent open site.* The Class A pan integrates effects of many meteorological variables, with the exception of advective effects. This result has been reported in several studies, including: western Nevada,[13] northern Utah,[14] and southern Manitoba.[15] Stipulation of a time period in excess of the growing season is important, because short-term effects of vegetation can invalidate this simple rule of thumb. The effect of climate is apparently small, since annual data for a wastewater treatment wetland at Clermont, Florida were 0.78 times the Class A pan data from the nearby Lisbon station.[16] This multiplier is the same as for the potential evapotranspiration from terrestrial systems.[17]

   Class A pan data are tabulated monthly and annually in *Climatological*

Data, published by the U.S. National Oceanic and Atmospheric Administration, Asheville, North Carolina.

3. *Wetland evapotranspiration and lake evaporation are roughly equal.* This is a corollary to *2,* since Class A pan evaporation is 1.4 times lake evaporation. Roulet and Woo[18] report this equality for a low arctic site, and Linacre's[7] review concludes: "In short, rough equality with lakes is probably the most reasonable inference for bog evaporation." Vegetated potholes lost water 12% faster than open water potholes,[19] but Virta[20] (see also Koerselman and Beltman[12]) found 13% less water loss in peatlands. There is a seasonal effect which can invalidate this in the short term.
4. *About half the net incoming solar radiation is converted to water loss on an annual basis.* Reported values include 0.51,[13] 0.47,[14] 0.64,[18] and 0.49.[21] If the radiation data from the Clermont wetland is used to test the concept, the value is 0.49, based on Zoltek et al.[16]

   Incoming radiation at the top of the atmosphere is tabulated in several texts, including Thibodeaux,[22] who also gives ground level data for selected cities for 1971. Ground level radiation shows effects of cloud cover according to:

$$Q_{nr} = Q_{ar}\left[0.803 - 0.340n - 0.458n^2\right] \tag{7}$$

   where n = fractional cloud cover (available in *Climatological Data).*
5. *Seasonal variation in evapotranspiration shows effects of both radiation patterns and vegetation patterns.* The seasonal pattern of evapotranspiration resembles the seasonal pattern of incoming radiation (Figure 5). However, attempts to apply *4* (above) on a monthly basis do not produce a constant multiplier. During the course of the year, wetland reflectance changes, the ability to transpire is gained and lost, and a litter layer fluctuates in a mulching function. Christiansen and Low,[14] computed a crop coefficient to account for vegetative effects. In addition to effects due to radiation, wind, relative humidity, and temperature, this is the ratio of wetland evaporation to lake evaporation. The result is a growing season enhancement followed by winter reductions (Figure 6). A similar coefficient derived for the Clermont wetland shows different features. Seasonality is shifted, presumably due to climatic differences; enhanced growth due to nutrients may lead to greater stem densities and biomass, amplifying the vegetative effect and increasing the mulching effect.
6. *Very small wetlands will react strongly to the surrounding microclimate.* Linacre[7] calls this the "clothesline effect," citing several studies with enhanced evapotranspiration for what amounts to potted plants. Since treatment wetlands tend to be small, it is reasonable to enquire at what size this effect becomes important, but little information is available. A wetland of "less than one hectare" displayed minor differences from similar studies on larger wetlands.[12] At Listowel, Ontario,[23] lake evaporation reasonably estimated evapotranspiration for 0.1 and 0.4 hectare wetlands. However, as size decreases, the advective terms in the energy balance become important; Penman methods are no longer adequate; and ratios to pan and lake evaporation, and to radiation, would not hold.
7. *Type of vegetation is not a strong factor in water loss determination.* Bernato-

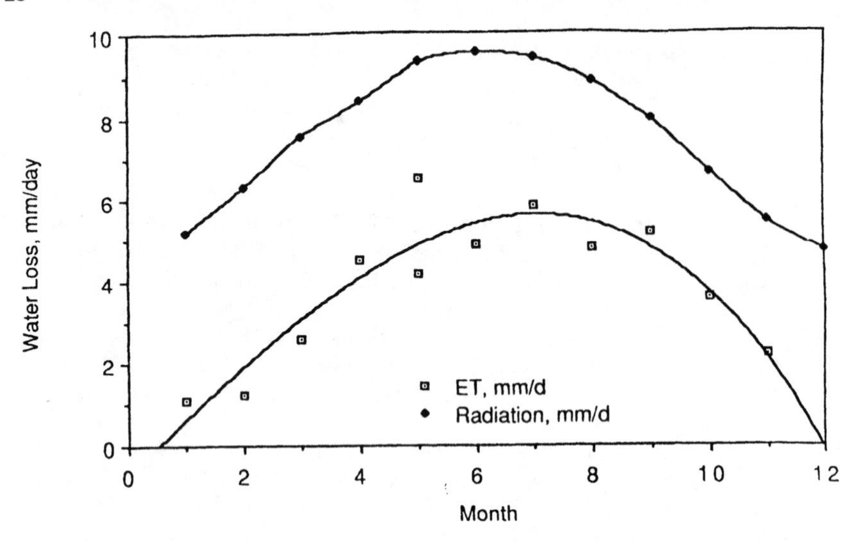

Figure 5. Radiation at ground level and measured evapotranspiration for the Clermont, Florida site. Net radiation is expressed in units of potential evaporation.

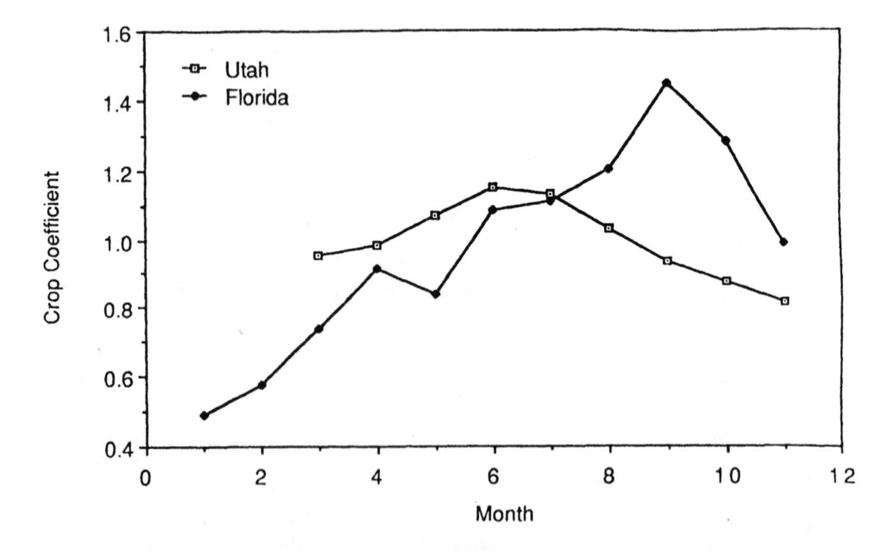

Figure 6. Effect of vegetation as a modifier of potential evaporation. This crop coefficient alters the distribution of water loss throughout the year but not the annual total.

wicz et al.[11] found small differences among several species, including *Typha*. Little difference was found among two *Carex* species and *Typha*,[12] and Linacre[7] concludes: "[I]t appears that differences between plant types are relatively unimportant."

8. *Energy associated with incoming wastewater is not likely to be a strong addition to solar energy in the summer, nor the source of large water losses in the winter.* Simply calculate the energy associated with the temperature reduction of incoming water and compare it to net incoming solar radiation. The rate of water loss due to dissipation of ΔT degrees of incoming water temperature is given by:

$$\text{loss} = (Q/LW)(c\Delta T/\lambda) \tag{8}$$

where loss = cm/day
(Q/LW) = loading rate, cm/day
$(c\Delta T/\lambda)$ = (sensible heat/latent heat) ratio
= $\Delta T/585$

Summer temperature drops were 10°C for the Listowel system (Figure 7), contributing only a 2% loss of applied water. This is, however, a 5–10% augmentation of the evapotranspiration rate of 5 mm/day for Listowel.

ATMOSPHERIC AUGMENTATION

Flow through a wetland system is augmented by precipitation and (negatively) evapotranspiration. Precipitation records are available for nearby sites, and the above outlines methods for estimating evapotranspiration. This section addresses what effect these gains or losses have on wetland hydrodynamics. It is important to note the range of additions or losses expected for typical operations. Figure 8 shows annual data for two northern wetland treatment systems, Bellaire and Houghton Lake Michigan. Even on a long-term basis, rain and vaporization have significant effects. Over one to two months, these wetlands have operated with total evaporation of added wastewater and, at other times, with ratios of rain to wastewater of greater than 1. An expected range of fractional augmentation is ±1.00.

Evapotranspiration slows water flow and increases contact times, whereas rainfall has the opposite effect. For a wetland operated at constant depth, the actual contact time is given by:

$$\tau_a = \tau \left[\frac{1}{\alpha}\ln\left(\frac{1}{1-\alpha}\right)\right] \tag{9}$$

where α is fractional augmentation and τ nominal contact time, based on nominal depth and wastewater addition rate. Evaporation has a strong influence on contact time (Figure 9). Rain, characterized by negative values of augmentation, has a lesser effect. Because normal climatic time sequences combine gains and losses over the span of typical contact times, it is informa-

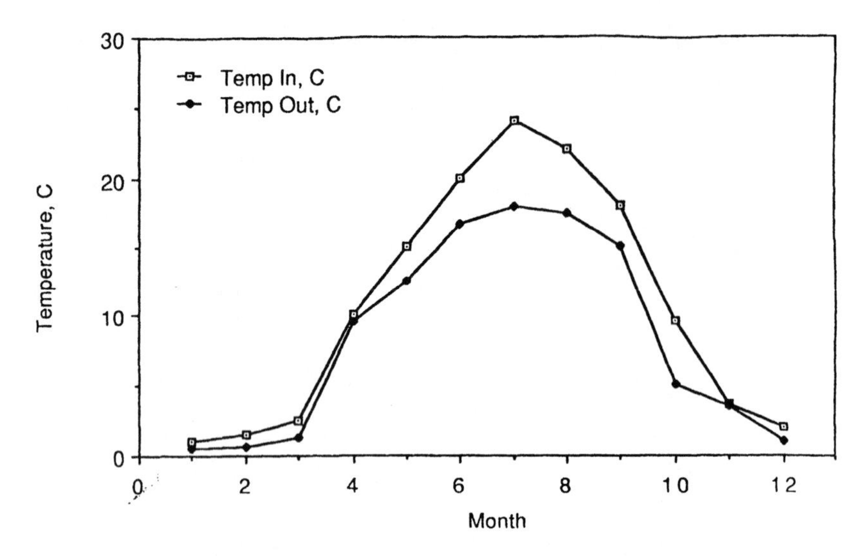

Figure 7. Temperatures of influent and effluent from the Listowel system.

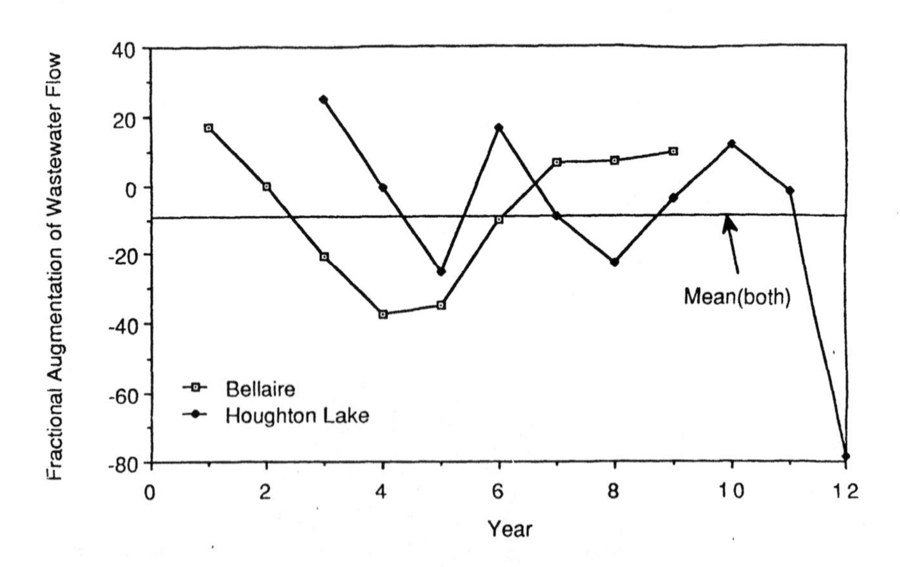

Figure 8. Additions and subtractions from annual wastewater flows to the Bellaire and Houghton Lake systems from rain and evapotranspiration.

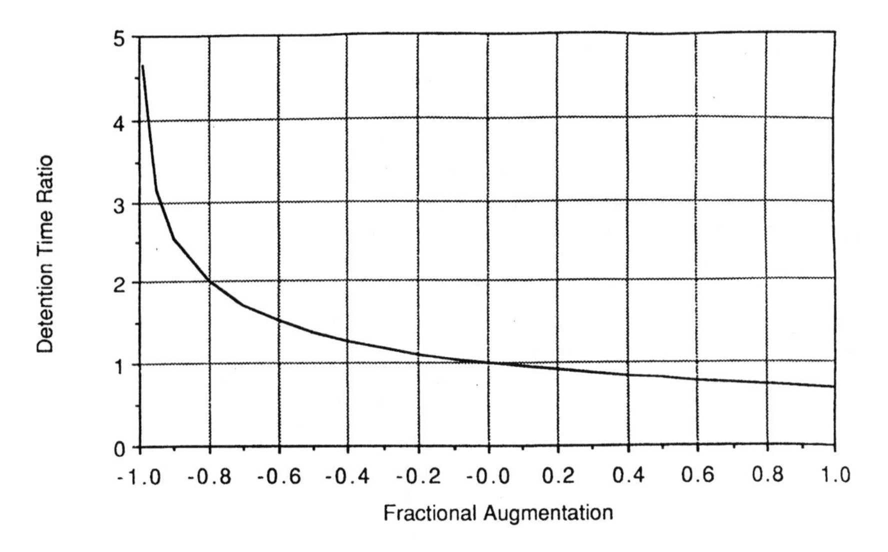

Figure 9. Effect of atmospheric augmentation on detention time. The nominal detention time is computed from the inflow.

tive to look at ratios of measured contact time to nominal contact time. Data for Houghton Lake, determined from water balance procedures, are shown in Figure 10; Figure 11 gives data from dye tracer experiments for Listowel. The range is 40–250% in both cases. The mean for Listowel, spanning all seasons over four years, is 126%. The Houghton Lake mean is the nominal 100%, since precipitation equalled evapotranspiration for the summer period.

The effect on depths and flow rates is more complicated because of wetland storage capacity and different operating modes. The general transient, depth-variable situation has been described and modelled by Hammer and Kadlec.[24] Results from three differently operated systems are given in Figures 12, 13, and 14. Figure 12 shows depth and average flow behavior for one summer season at Houghton Lake. This system is operated with intermittent irrigation, several days on followed by several days off. Flow is overland and controlled by topography of the natural wetland and the vegetation resistance. As a result, depth and flow vary strongly in response to addition patterns as well as atmospheric augmentation. Flow dependence on depth to the power of 2.0 produces greater flow fluctuations than depth fluctuations. In contrast, the Clermont, Florida system[16] applied wastewater one day in seven, and outflow was by infiltration to underlying peat and sand. Depth variations were much greater than outflow variations, as illustrated in Figure 13. The third mode of operation, for Listowel, Ontario, was constant depth (although there may have been some vegetation control for shallow depth experiments) and constant inflow. All atmospheric augmentation, therefore, showed up immediately at the wetland outflow (Figure 14). Thus, the time sequence in Figure 14

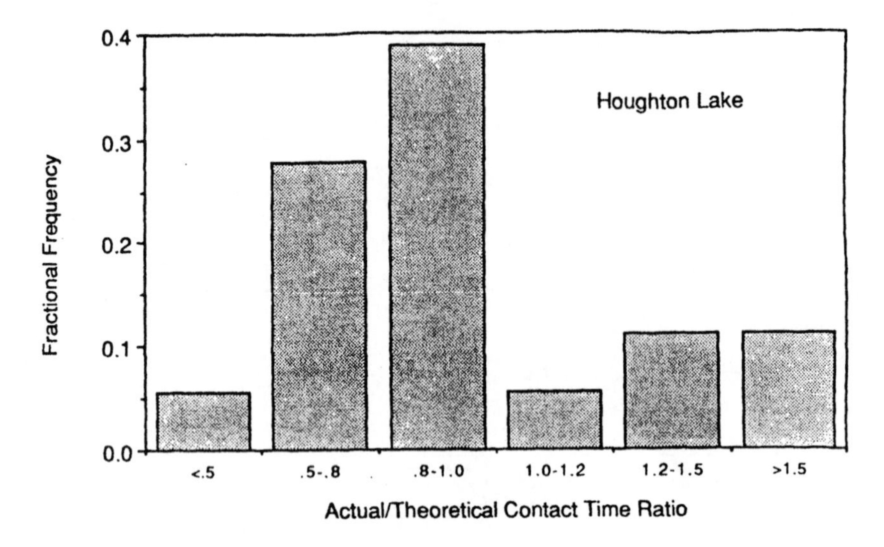

Figure 10. Distribution of contact times for the Houghton Lake system.

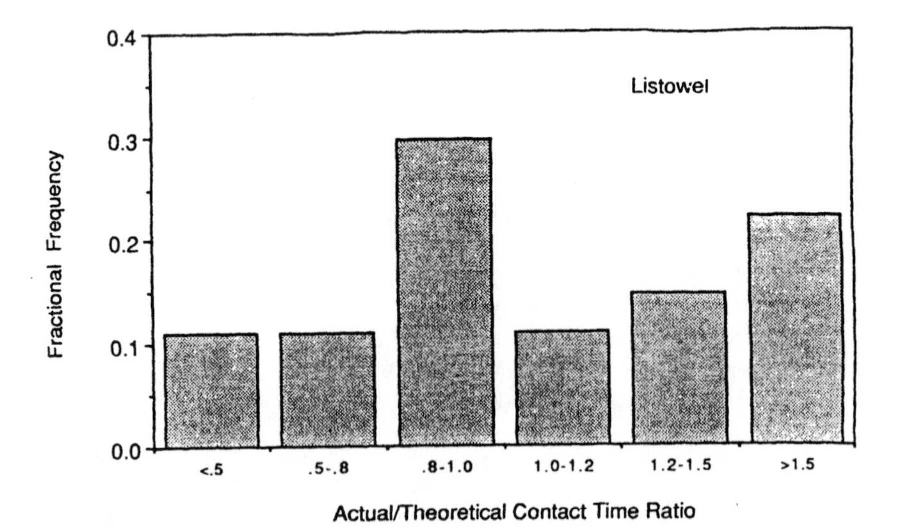

Figure 11. The distribution of contact times for the Listowel system.

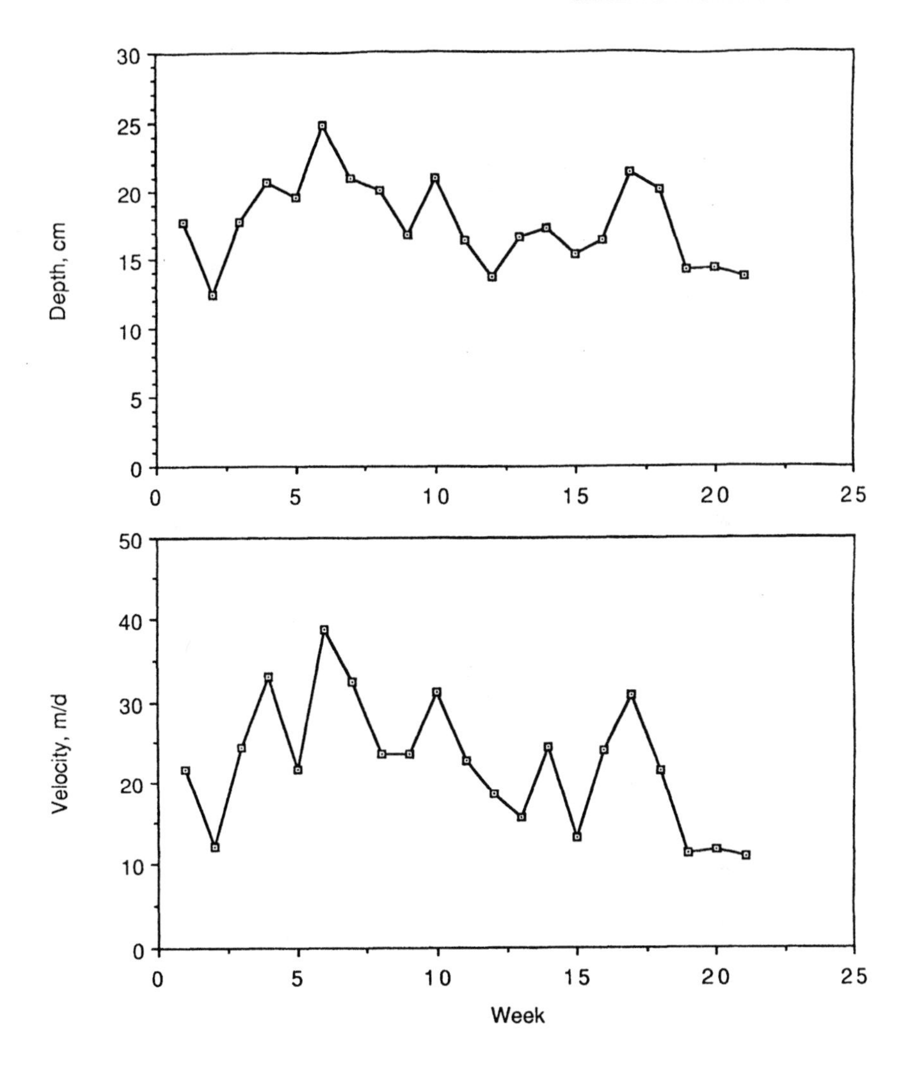

Figure 12. Weekly progression of depths and velocities for the Houghton Lake system, with overland flow through vegetation.

could be sampled to give a contact time distribution comparable to the composite of Figure 11. Clearly, the dye tracer studies missed some brief periods after rains, when average contact times were less than one-third the nominal contact time.

Given the complexity of hydrodynamics, it is very tempting to adopt averaging procedures for data analysis and for design. The next section estimates errors associated with averaging, in terms of the associated water quality parameters.

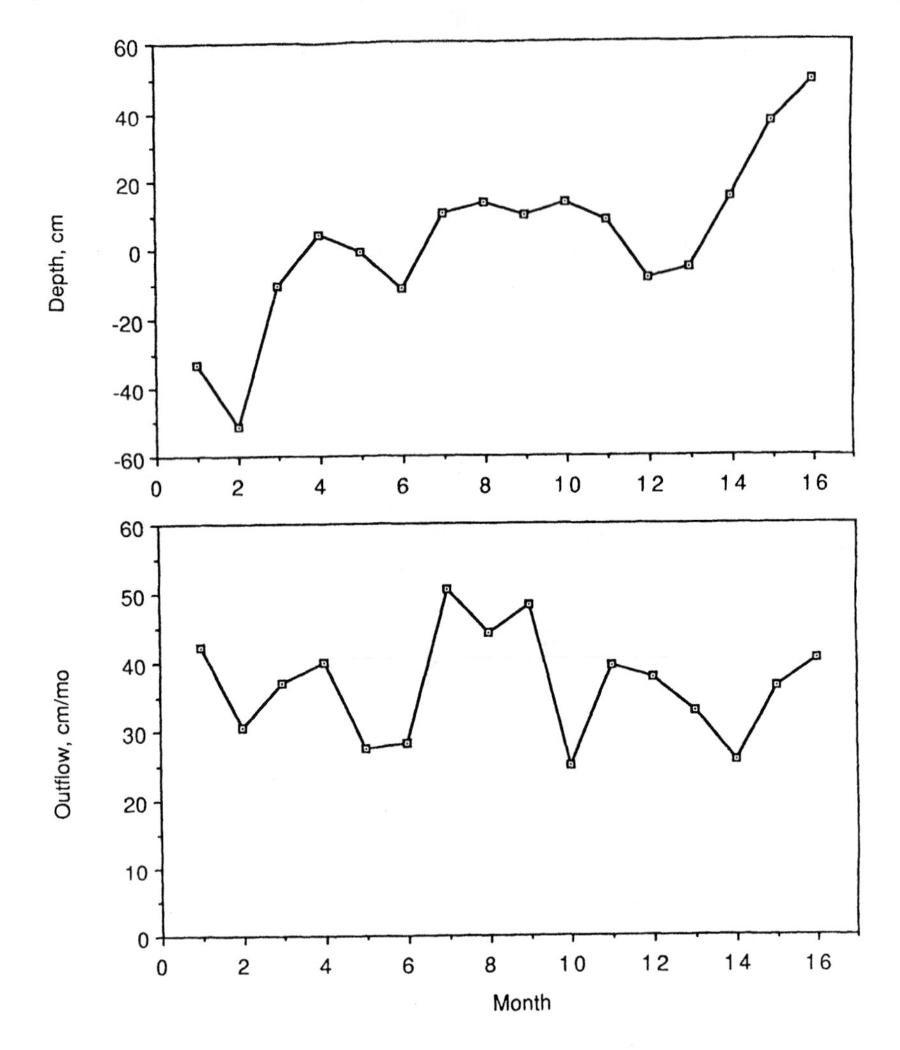

Figure 13. Monthly progression of depths and flow rates for the Clermont system, with vertical infiltration.

WATER QUALITY CONSEQUENCES

Water quality improvement in wetlands may follow first order kinetics, and operating conditions may be predicted from the rate constant and contact time within the system:

$$\frac{C}{C_0} = \exp(-k\tau) \tag{10}$$

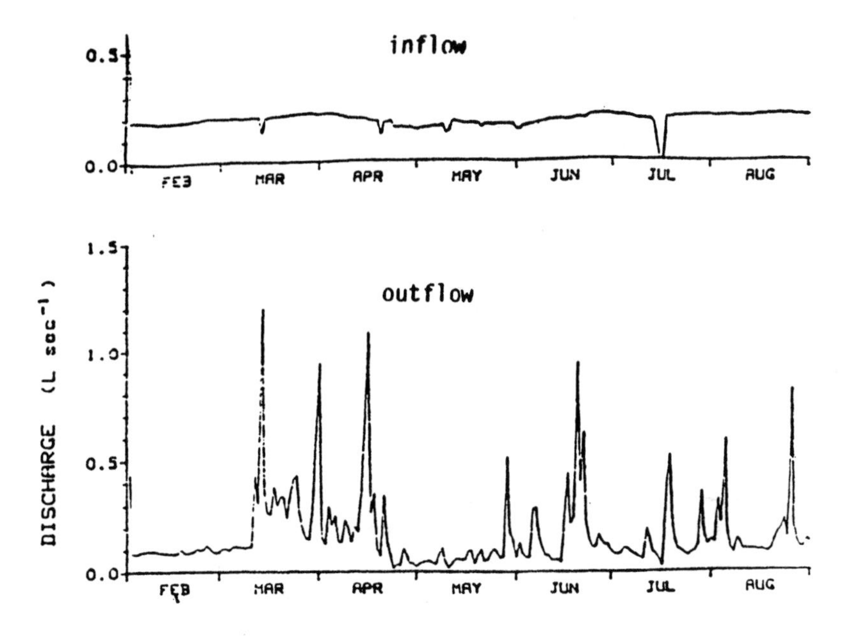

Figure 14. Progression of inflows and outflows for the constant depth Listowel system.

where C = concentration
k = rate constant
τ = contact time

Because of sizeable changes in contact time due to atmospheric conditions, errors may occur in rate constants or in designs, based on zero augmentation.

If data from the wetland system are processed to extract the rate constant, it would be logical to measure concentrations along transects down the wetland length to obtain the contact time effect. In the absence of rain or evapotranspiration, this is accurate. However, rain will cause dilution and speed flow, whereas evapotranspiration will cause concentration and slow the flow. Under either condition:

$$\frac{C}{C_0} = \left[1 - \alpha \frac{x}{L}\right]^{(k\tau/\alpha - 1)} \tag{11}$$

where x/L is fractional distance to outlet, α is augmentation ratio, and τ is nominal contact time. If the resultant data are plotted on semilog coordinates in an erroneous attempt to fit the process with the simple first-order formula, the results in Figure 15 are obtained. The curve fit is excellent for the simple model, but the slopes are not the correct rate constants. Evaporation gives too low a value, while rain gives too high a value. Table 1 gives values for resultant

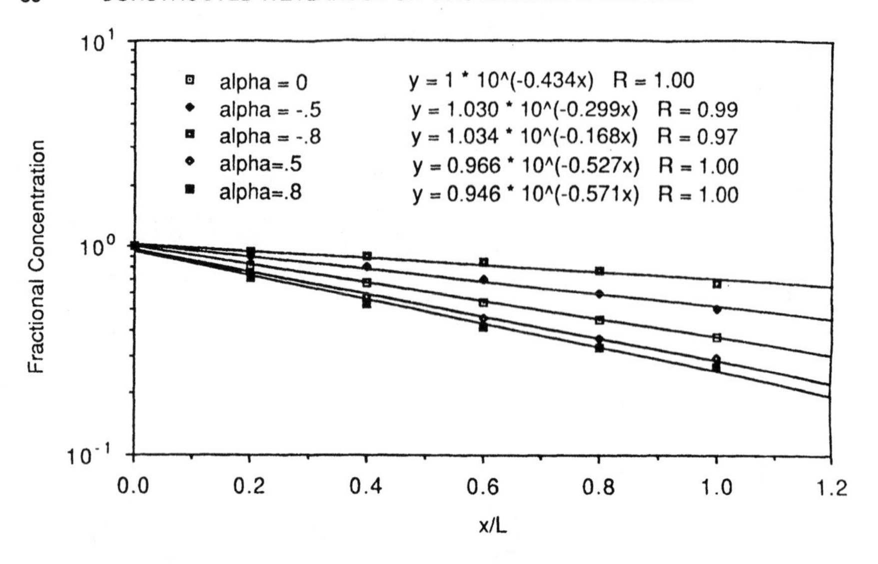

Figure 15. Regression of augmented flow concentration data for first-order kinetics.

errors in the rate constant for the case of nominal $k\tau = 1$ for various choices for the residence time (nominal, exit flow, and average flow contact times).

If the wetland infiltrates, similar types of errors can be made in determination of the rate constant. Water loss occurs as in evaporation, but the concentrating effect is not present. Figure 16 shows concentration transects may be fit by the first-order model, although the correct profiles are given by:

$$\frac{C}{C_0} = \left[1 - \alpha\frac{x}{L}\right]^{(k\tau/\alpha)} \tag{12}$$

In this equation, α represents the fraction of incoming flow that infiltrates. Again, any of three contact times are candidates for data interpretation. Table 2 shows that significant errors in rate constant determination may result under conditions of infiltration for the case of the nominal $k\tau = 1$.

Errors are also possible in design. If accurate values of rate constants are used and no augmentation is assumed, calculated area may be less than required under some operating conditions, while under other conditions the

Table 1. Errors (%) in the Rate Constant for a First-Order Reaction Occurring Under (Ignored) Conditions of Atmospheric Augmentation

Fractional Augmentation		Nominal τ	Exit τ	Average τ
(evap)	−0.8	+32	+137	+69
	−0.5	+21	+82	+46
	0.0	0	0	0
	+0.5	−31	−65	−54
(rain)	+0.8	−61	−92	−87

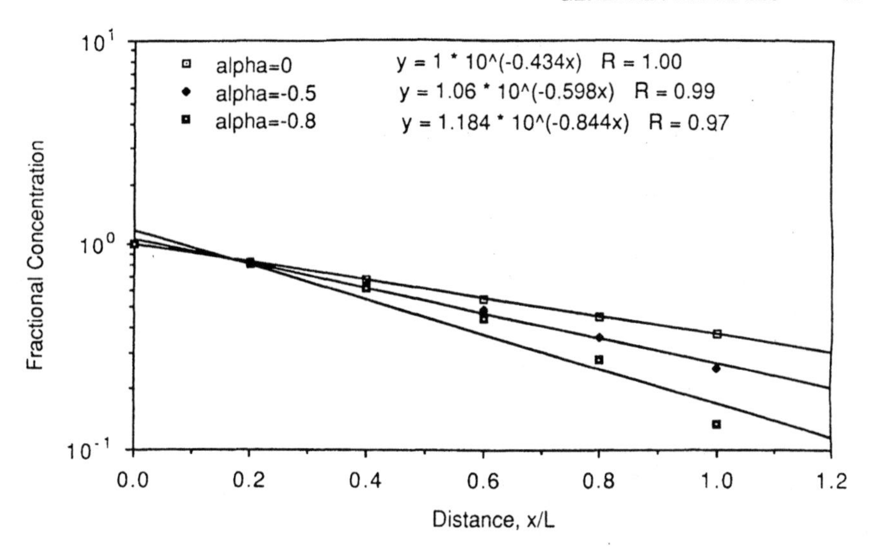

Figure 16. Regression of partially infiltrating flow data for first-order kinetics.

system may have excess capacity. Evaporation can increase concentrations, which defeats the reactive consumption, or dilution can reduce concentrations and slow reaction rates. Table 3 presents these effects for various nominal design efficiencies. This is in some sense a "no-win" situation, because either too much land was used, or the system fails to perform up to design when subjected to atmospheric phenomena. These examples illustrate the possibility of errors due to the neglect of atmospheric phenomena.

Table 2. Errors (%) in the Rate Constant for a First-Order Reaction Occurring Under (Ignored) Conditions of Infiltration

Infiltration	Nominal τ	Exit τ	Average τ
–0.8	+94	–61	–53
–0.5	+38	–24	–8
0.0	0	0	0

Table 3. Errors (%) in the Area Required for a First Order Reaction Occurring Under (Ignored) Conditions of Atmospheric Augmentation

		Design Percent Concentration Reduction		
Fractional Augmentation		63	87	95
(evap)	–0.8	77	112	131
↑	–0.5	82	103	113
	0.0	100	100	100
↓	+0.5	137	102	94
(rain)	+0.8	178	104	91

Note: Entries are the percent of the required area which is available.

CONCLUSIONS

Hydrological complexity makes design and data interpretation difficult for wetland treatment systems. It is not safe to ignore water exchanges with the atmosphere, because they can significantly contribute to total water flow. Rainfall causes two opposing effects: (1) dilution of wastewater, reducing concentrations; and (2) increased velocities, reducing retention times within the wetland. The result will be reduced exit concentrations, which can be interpreted as erroneously high rate constants for the process. The impact of a rain event on velocity is larger for a depth-controlled system than for a vegetation flow-controlled wetland, because the former lacks a surge damping mechanism.

Evapotranspiration can be approximated Weather Service Class A pans, multiplied by 0.8. Similarly, half the net solar radiation received also estimates the long-term average. On a short-term basis, the vegetation effects cannot be ignored, due to growing season enhancement and off-season mulching effects of litter. The impacts of evaporative processes on a wetland treatment system are not trivial. Even in northern climates, *all* applied wastewater can be evaporated during a dry summer season, as occurred twice in 10 years at Houghton Lake. Vapor losses affect water quality by increasing concentrations by evaporation and slowing the water and allowing more time for reaction. Apparent rate constants derived from evaporating systems can be significantly lower than true values. A wetland system designed without regard for atmospheric augmentation may be under- or overdesigned, depending on degree of treatment and local climatic conditions in terms of water gains or losses. An overland flow wetland leaking to groundwater (infiltrating) will apparently have better treatment when rated on the basis of the overland flow effluent, due to increased contact time. Whether the overall performance is better or worse depends on the treatment provided by underlying substrates.

Frictional effects associated with water flow through a gravel or rock substrate differ from flow through stems and litter aboveground. In the former case, system overloads can cause emergent overland flow near the entrance. In the latter, high vegetation densities can increase depths above planned weir settings. Both effects are forms of water mounding. The wetland will display both depth and flow variations in response to input dynamics in any case.

It seems likely that the above features of wetland systems, coupled with the fast dynamics not considered here, will manifest themselves as "site-specific performance" until the hydrological features are acknowledged in data interpretation.

REFERENCES

1. Kadlec, R. H. "Vegetation Resistance in Wetland Water Flow," in preparation (1989).
2. Gilliam, J. W., G. M. Chescheir, R. W. Skaggs, and R. G. Broadhead. "Effects of

Pumped Agricultural Drainage Water on Wetland Water Quality," paper presented at National Symposium on Wetland Hydrology, Chicago, IL (1987).

3. Horton, R. E. "The Interpretation and Application of Runoff Plat Experiments with Reference to Soil Erosion Problems," *Soil Sci. Soc. Am. Proc.* 3:340–349 (1938).
4. Chen, C. "Flow Resistance in Broad Shallow Grassed Channels," *J. Hydraul., Div. Am. Soc. Civ. Eng.* HY3:307–322. (1976).
5. Kadlec, R. H., D. E. Hammer, I.-S. Nam, and J. O. Wilkes. "The Hydrology of Overland Flow in Wetlands," *Chem. Eng. Comm.* 9:331–344 (1981).
6. "Process Design Manual, Land Treatment of Municipal Wastewater" USEPA-CERI, Cincinnati, OH (1981).
7. Linacre, E. T. "Swamps," in *Vegetation and Atmosphere, Vol. 2: Case Studies,* J. L. Monteith, Ed. (London: Academic Press, 1976), pp. 329–347.
8. Ingram, H. A. P. "Hydrology," in *Mires: Swamp, Bog, Fen and Moor. Ecosystems of the World, Vol. 4A,* A. J. P. Gore, Ed. (Amsterdam: Elsevier Science Publishing Co., Inc., 1983), pp. 67–158.
9. Penman, H. L. "Natural Evapotranspiration from Open-Water, Bare Soil and Grass," *Proc. Roy. Soc. Acad.* 193:120–145 (1948).
10. Eagleson, P. S. *Dynamic Hydrology* (New York: McGraw-Hill Book Company, 1970).
11. Bernatowicz, S., S. Leszczynski, and S. Tyczynska. "The Influence of Transpiration by Emergent Plants on the Water Balance in Lakes," *Aquat. Bot.* 2:275–288 (1976).
12. Koerselman, W., and B. Beltman. "Evapotranspiration from Fens," unpublished manuscript, Department of Plant Ecology, University of Utrecht (1987).
13. Kadlec, R. H., R. B. Williams, and R. D. Scheffe. "Wetland Evapotranspiration in Temperate and Arid Climates," in *Ecology and Management of Wetlands,* D. D. Hook, Ed. (Beckenham: Croom Helm, 1987), pp. 146–160.
14. Christiansen, J. E., and J. B. Low. "Water Requirements of Waterfowl Marshlands in Northern Utah," Publication No. 69–12, Utah Division of Fish and Game (1970).
15. Kadlec, J. A. "Input-Output Nutrient Budgets for Small Diked Marshes," *Can. J. Fish Aquat. Sci.* 43(10):2009–2016 (1986).
16. Zoltek, J., S. E. Bagley, A. J. Hermann, L. R. Tortora, and T. J. Dolan. "Removal of Nutrients from Treated Municipal Wastewater by Freshwater Marshes," report to City of Clermont, FL, Center for Wetlands, University of Florida (1979).
17. Penman, H. L. *Vegetation and Hydrology* (Farnham Royal: Commonwealth Agricultural Bureau, 1963).
18. Roulet, N. T., and M. K. Woo. "Wetland and Lake Evaporation in the Low Arctic" *Arct. Alp. Res.* 18:195–200 (1986).
19. Eisenlohr, W. S., Jr. "Water Loss from a Natural Pond Through Transpiration by Hydrophytes," *Water Resour. Res.* 2:443–453 (1966).
20. Virta, J. "Measurement of Evapotranspiration and Computation of Water Budget in Treeless Peatlands in the Natural State," *Phys.-Math. Soc. Sci. Fenn.* 32(11):1–70 (1966).
21. Bray, J. R. "Estimates of Energy Budgets for a *Typha* (cattail) Marsh," *Science* 136:1119–1120 (1962).
22. Thibodeaux, L. J. *Chemodynamics* (New York: John Wiley & Sons, Inc., 1979).

23. Hershkowitz, J. "Listowel Artificial Marsh Project Report," Ontario Ministry of the Environment, Water Resources Branch, Toronto (1986).
24. Hammer, D. E., and R. H. Kadlec. "A Model for Wetland Surface Water Dynamics," *Water Resour. Res.* 22(13):1951–1958 (1986).

CHAPTER 4

Physical and Chemical Characteristics of Freshwater Wetland Soils

Stephen P. Faulkner and Curtis J. Richardson

OVERVIEW

Soils are complex assemblages of inorganic and organic material at the earth's surface that reflect long-term environmental changes. Specifically, any particular soil is a function of parent material acted on by organisms and climate and conditioned by relief over time.[1] The modern soil classification system used in the United States recognizes 10 soil orders distinguished by the presence or absence of diagnostic horizons and features that reflect differences in the soil-forming processes mentioned above. These soil orders range from recently formed Entisols with few diagnostic horizons to highly weathered Oxisols to organic Histosols.[2] Chemical and physical attributes among soils in different orders (and even within a given order) vary widely, and these differences must be considered in the construction and operation of wetlands for wastewater treatment.

Wetland soils are dominated by anaerobic conditions induced by soil saturation and flooding. Freshwater wetland soils can generally be distinguished from upland, nonwetland soils by two interrelated characteristics: (1) an abundance of water and (2) accumulation of organic matter. Excess water causes many physical and chemical changes in soils, and wetland hydrologic regimes can range from nearly continuous saturation (swamps) to infrequent, short-duration flooding (riparian systems). The most significant result of flooding is the isolation of the soil system from atmospheric oxygen, which activates several biological and chemical processes that change the system from aerobic and oxidizing to anaerobic and reducing.

This chapter reviews the chemical and physical parameters of soils, particularly freshwater wetland soils, that influence their ability to effectively treat wastewater. We will not cover all aspects of soil chemistry but will confine our discussion to the more important chemical processes and soil attributes, their role in the retention and transformation of specific wastewater constituents,

Table 1. Comparison of Physical and Chemical Attributes of Organic and Mineral Soils

Parameter	Mineral Soil	Organic Soil
Organic content (%)	Less than 12–20	Greater than 12–20
pH	6.0–7.0	Less than 6.0
Bulk density	High	Low
Porosity	Low (45–55%)	High (80%)
Hydraulic conductivity	High (except clays)	High (fibric) Low (sapric)
Water holding capacity	Low	High
Nutrient availability	Generally high	Often low
Cation exchange	Low, dominated by major cations	High, dominated by hydrogen ion

Source: Adapted from Mitsch and Gosselink.[125]

unique characteristics of wetland soils relative to these processes, and application of these concepts in the design and construction of wetland systems.

GENERAL SOIL PHYSICAL AND CHEMICAL PROPERTIES

The presence of organic as opposed to mineral soil constituents has an important impact on soil chemical characteristics. A soil is classified as a Histosol (organic soil) if it has more than 12–20% organic matter (actual percentage depends on clay content and saturation). In general, organic soils have a lower pH, bulk density, and nutrient availability than mineral soils (Table 1). The chemical and physical differences between mineral and organic soils play a large role in determining the suitability of a particular soil for a specific wastewater treatment system.

Physical

Soils are a matrix of mineral and organic solids, water, and open pore spaces. Spatial arrangement of these phases determines soil structure and pore size distribution. Both attributes significantly affect hydraulic conductivity (K), the major determinant of water movement through the soil. In wastewater treatment, the effectiveness and capacity of a soil to remove/retain contaminants is a function of soil-wastewater contact. Sandy or gravelly soils have high K values, and water moves rapidly through the soil (Table 2) providing little opportunity for soil-water contact. In contrast, fine-textured silty or loamy soils permit more soil-water contact. Mixing sand or gravel with impermeable clay soils can improve soil water movement.

Clayey soils (particularly 2:1 interlayers with high shrink-swell potential) with low K values may have large, interconnected pores, permitting rapid water movement with little soil-water contact, a process known as *short-circuiting*.[3,4] Saturated flow in well-decomposed organic soils (sapric peat) and most clays is slow (Table 2). Due to hydraulic conductivity differences between sapric and fibric peat, a peat soil may exhibit within-profile variability that results in lateral water flows through the upper fibric zone as opposed to

Table 2. Range of Physical Characteristics for Peat and Mineral Soils

Soil Type	Total Porosity (%)	Hydraulic Conductivity (m/d)	Bulk Density (g/cm)
Peat			
Fibric	>90	>1.3	<0.09
Hemic	84–90	0.01–1.3	0.09–0.20
Sapric	<84	<0.01	>0.20
Mineral			
Gravel	20	100–1000	~2.1[a]
Sand	35–50	1–100	1.2–1.8
Clay	40–60	<0.01	1.0–1.6

Source: Peat data from Boelter and Verry;[5] mineral data from Brady[126] and U.S. EPA.[127]
[a]Calculated based on porosity.

vertical penetration into the lower sapric zone.[5,6] Lateral flow patterns and short-circuiting may result in nutrient removal capacities that are less than those calculated from laboratory isotherms where soil-water contact is maximized.[7]

Another important physical parameter is soil bulk density. Bulk density measures dry weight of soil per unit volume and is primarily a function of the kinds and arrangement of soil solids. Organic soils have much more pore space (usually filled with water in wetland soils) in a given volume than mineral soils and, therefore, much lower bulk densities (Table 2).

Chemical

Surface area and surface charge of soil particles account for most of the reactivity of a soil. Charge development is intimately associated with clay-sized particles and organic matter.[8] Almost all temperate zone soils have a net negative charge on the soil solids, providing electrostatic bonding sites for positively charged cations. These ionically bonded cations on the solid surface can exchange with other cations in the soil solution, hence the term *cation exchange*. Cation exchange capacity (CEC) measures the soil's capacity to hold cations on exchange sites and varies widely among different soils.

The total charge capacity of a soil can be divided into permanent and pH-dependent, or variable, charge. Isomorphous substitution of Mg^{2+} for Al^{3+} or Al^{3+} for Si^{4+} within the clay lattice structure leaves unbalanced oxygen and hydroxyl groups, yielding a net negative charge. These are permanently charged sites unaffected by pH changes and are the source of most of the charge in temperate mineral soils. There are three major sources of variable charge in soils. One is the dissociation of hydroxyl (–OH), carboxyl (–COOH) and phenolic (C_6H_4OH) functional groups. As pH rises, these groups deprotonate yielding more negatively charged sites. In phyllosilicate clays, this occurs at the crystal-lattice edge and is a minor source of charge in most soils. However, it can be 50% of the total charge in kaolinitic clays because (1) functional groups on kaolinite dissociate at relatively low pH's (low pK_a) and (2) the edge

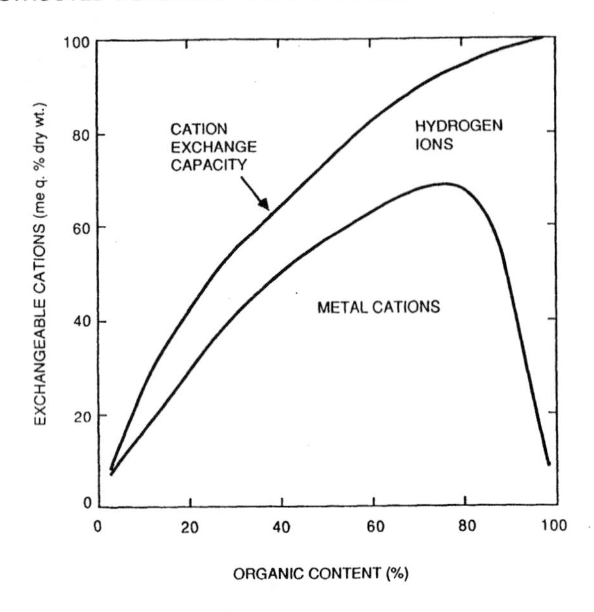

Figure 1. Cation exchange capacity as a function of organic matter content. (Adapted from Gorham.[129])

surface of the mineral is large relative to the planar surface.[8] Aluminum and iron oxides and hydrous oxides are a second source of variable charge, but are relatively unimportant except in highly weathered soils.

Organic matter is the third major source of variable charge in soils, accounting for 20–100% of the total CEC. Cation exchange capacity in high organic matter soils can change from 10–20 cmol(+)/kg at pH 3.7 (ambient soil) to over 100 cmol(+)/kg at pH 7.[9] Not only do organic soils have higher CECs than mineral soils, but the major cations present on the exchange sites are also different. Metal cations (Ca^{2+}, Mg^{2+}, Na^{+}) dominate in mineral soils, while H^{+} dominates at high organic contents (Figure 1). These factors may play major roles in retention of specific wastewater components and in pH changes during treatment periods.

Soils may remove wastewater constituents by (1) ion exchange/nonspecific adsorption; (2) specific adsorption/precipitation; and (3) complexation.[10] Ion exchange was just discussed, i.e., electrostatic bonding between cations and soil solids, which may be easily reversible. This mechanism also functions for anions, most notably NO_3^-, SO_4^{2-}, PO_4^{3-}. Specific adsorption reactions are dominated by ligand exchange where a ligand (mostly PO_4^{3-}) actually occupies a position within the coordination sphere of the cation. Complexation refers primarily to metals binding with soil organic matter. These mechanisms are discussed in later sections.

WETLAND SOIL CHEMICAL PROCESSES

There are four general chemical reactions in natural systems:[11]

1. No proton or electron transfer — $Fe_2O_3 + H_2O \rightarrow 2FeO{\cdot}OH$
2. Protons only — $H_2CO_3 \rightarrow H^+ + HCO_3^-$
3. Electrons only — $Fe^{2+} \rightarrow Fe^{3+} + e^-$
4. Proton and electron transfer — $FeSO_4 + 2H_2O \rightarrow SO_4^{2-} + FeO{\cdot}OH + 3H^+ + e^-$

Reactions 2, 3, and 4 affect either the pH or the oxidation-reduction (redox) system; however, we are primarily interested in *4* (above) because most reactions in wetland systems involve both proton and electron transfer. Redox potential (Eh) quantitatively measures the tendency of soils to oxidize or reduce susceptible substances.[12] *Oxidation* is the loss of electrons, and *reduction* is the addition of electrons. A generalized reaction is given in Equation 1,

$$Ox + ne^- + mH^+ = Red + H^+ \tag{1}$$

and Eh can be calculated by Equation 2,

$$Eh = E^\circ - 0.059(m/n)pH + (0.059/n)\log[(oxidant)/(reductant)] \tag{2}$$

where E° is the standard potential (ability of a given redox couple to exchange electrons under standard conditions); m is the number of protons in the reaction; n is the number of electrons; and activities of the oxidants and reductants are inserted. Note that according to Equation 2, pH changes are determined by the ratio of protons consumed to electrons consumed and not only the number of protons.

Effects of Flooding on Redox

Once a soil is flooded, the oxygen present is quickly consumed by microbial respiration and chemical oxidation.[13] Subsequently, anaerobic microorganisms use a variety of substances to replace oxygen as the terminal electron acceptor during respiration.[12] This electron transfer causes significant changes in the valence state of the chemical species used and the overall soil reduction. Reduction of a saturated soil is a sequential process governed by the laws of thermodynamics.[12,14] This sequence is shown in Figure 2 along with the range of redox potentials for both wetland and upland soils.

Nitrate is the first soil component reduced after oxygen, though this process can proceed before oxygen is completely consumed. Manganic manganese (Mn^{4+}) closely follows NO_3^- in the reduction sequence, even before NO_3^- has completely disappeared.[14–16] While the preceding reactions can and do overlap, the subsequent sequential reactions of ferric iron (Fe^{3+}) to ferrous iron (Fe^{2+}), sulfate (SO_4^{2-}) to sulfide (H_2S), and carbon dioxide (CO_2) to methane (CH_4) will not occur unless the preceding component has been completely reduced.

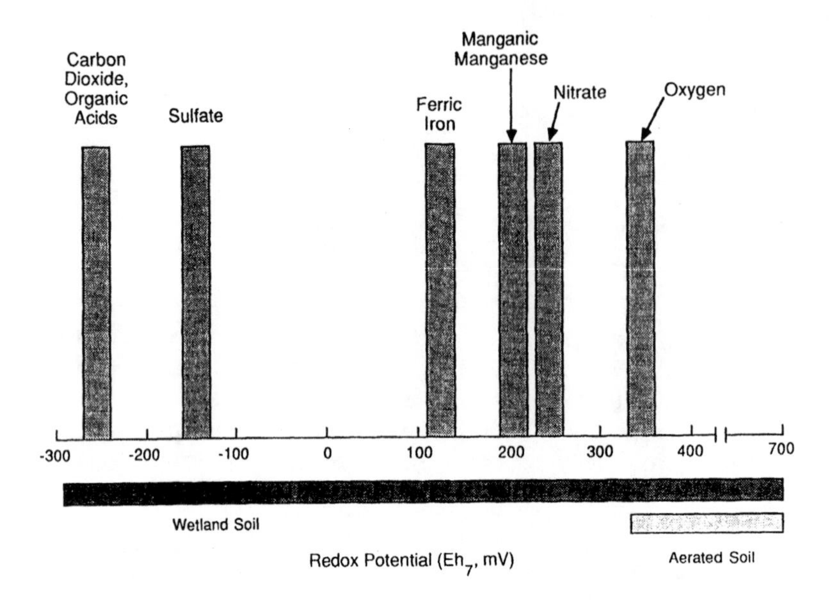

Figure 2. Redox stability for the major soil redox systems. Note the wide Eh range for wetland soils compared to typical upland, aerated soils. (Modified from Gambrell and Patrick.[15])

The greater range of redox potentials for flooded soils versus aerobic soils (Figure 2) is important. Natural wetland systems maintain a wider range of redox reactions than upland soils, and their most important function may be as chemical transformers. Wetlands are often the major reducing ecosystem on the landscape and, as such, have great potential for processing nutrients and other materials.[17]

Figure 3 graphically represents changes in a soil following flooding. After one day of flooding, oxygen content rapidly declined, and redox potential steadily decreased. Nitrate was completely gone after 2.5 days with subsequent significant increases in exchangeable Mn.[14] Ferrous iron (Fe^{2+}) increased significantly by day 5.

Eh/pH Relation

The relation between Eh and pH manifests itself in chemical speciation. One useful tool to study these changes is the Eh/pH stability diagram. These diagrams are geochemical equilibrium models derived from the stability constants and equilibrium concentrations of given reactants and products. They are based on thermodynamic energy relationships, but specific reactions and reaction kinetics are not inferred—only what is thermodynamically stable at the given limits of the system is shown. An Eh/pH diagram for Fe and Mn is given in Figure 4. Solid boundary lines between Fe and Mn forms are derived from

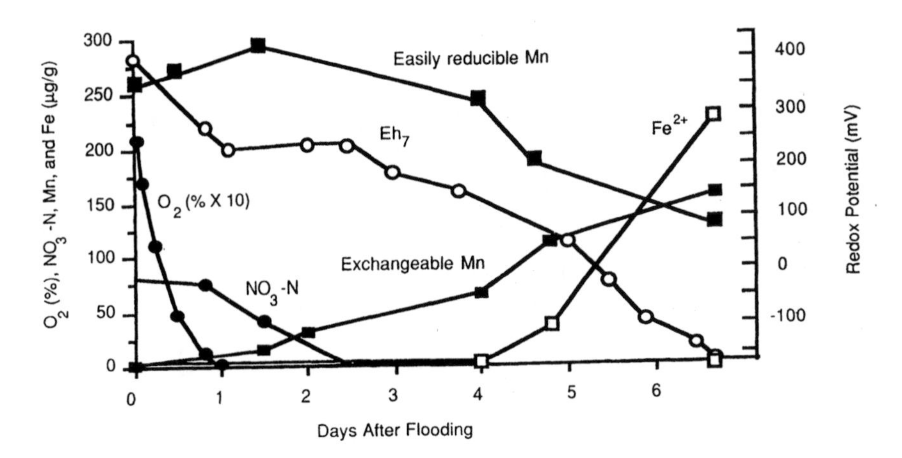

Figure 3. Soil redox changes following flooding. (Modified from Turner and Patrick.[14])

the appropriate equations, and dashed lines represent experimental precipitation reactions.[18] Dashed equilibrium boundary lines between soluble and solid species (Figure 4) confirm that at low Eh (reducing conditions) the predicted pH level necessary to precipitate Fe or Mn is much higher than at higher Eh levels. Acidic and neutral solutions can maintain high concentrations of Fe^{2+} and Mn^{2+}, and Mn^{2+} precipitates only at a higher Eh or pH than Fe^{2+}.[18] This follows the sequence in Figure 2 and has important implications in acid mine drainage systems where both Fe^{2+} and Mn^{2+} are often present and need to be removed, i.e., Mn removal is more difficult than Fe removal and requires higher pH or Eh (Figure 4). Usefulness of stability diagrams to predict various Fe oxides is limited by nonequilibrium conditions resulting from kinetic preferences of formation; metastable phases do exist for pedogenic time spans.[19,20]

Another consequence of flooding and reduction of an aerobic soil is the change in pH (Figure 5). The overall effect of flooding is to decrease the pH of alkaline soils and increase the pH of acid soils until they converge around 6.7 to 7.2.[12] The pH decrease of submerged alkaline soils results from the buildup of CO_2; however, the pH increase in acid soils is mostly from the reduction of Fe^{3+} oxyhydroxides (because of their abundance in most soils):

$$3Fe(OH)_3 + H^+ + e^- \leftrightarrow Fe_3(OH)_8 + H_2O \qquad (3)$$

The importance of Fe in pH changes is shown in Figure 5, where soil 3 had sufficient Fe to increase pH from 3.4 to 6.0, while soil 4 (with 42% less Fe) only rose from 3.5 to 5.0.[12] Depending on the contaminant-removal mechanisms and soil type, pH changes of this magnitude could have significant impacts on pollutant removal/retention performance. Conversely, most acid organic soils are low in Fe, and submergence would not cause large increases in

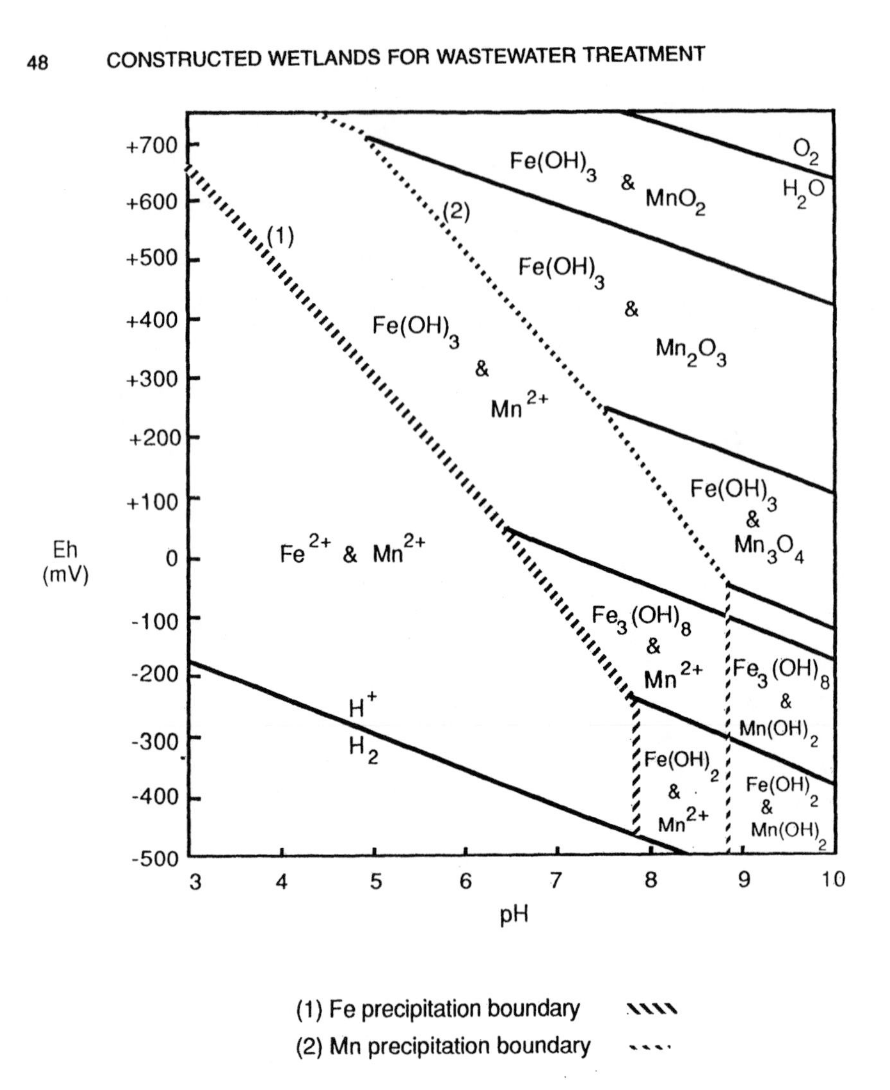

Figure 4. Composite Eh-pH iron and manganese stability diagram. (Adapted from Collins and Buol.[18])

pH. Thus, flooding would have little impact on increasing cation exchange capacity in peat soils.

This general discussion of soil chemistry, and wetland soils in particular, reveals the dynamic nature of chemical processes and wide variation of particular attributes among different soils. Substrate selection for constructed wetlands can play a major role in determining the success or failure of the system. Selection of a soil type or combination of types will depend on the removal/retention mechanisms for a particular contaminant.

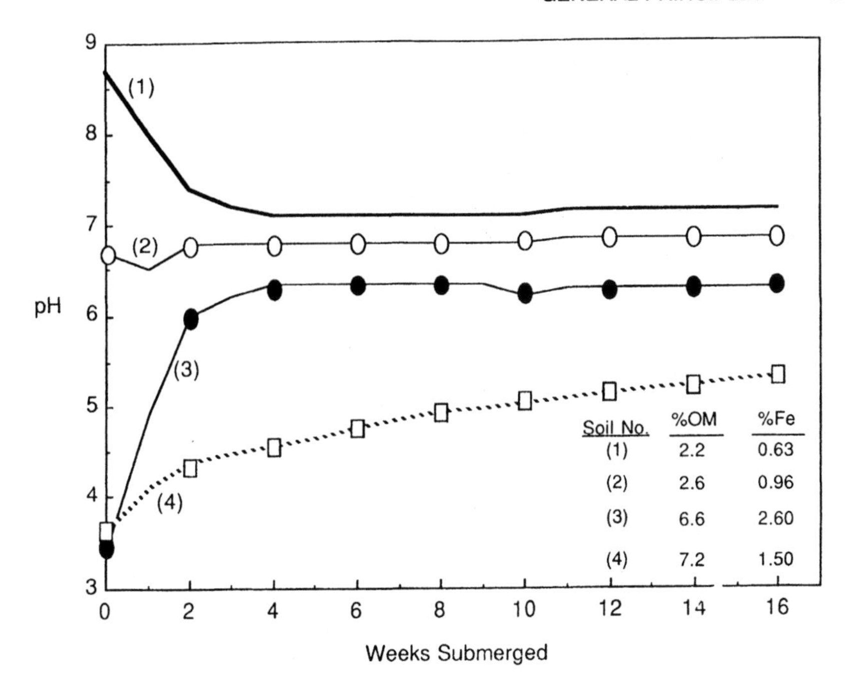

Figure 5. Changes in soil pH following submergence as affected by iron and organic matter content. (Modified from Ponnamperuma.[12])

WETLAND ECOSYSTEM PROCESSES AND REMOVAL/RETENTION MECHANISMS

To maximize the utility of constructed wetlands to treat wastewater, they must be designed and built to emulate natural wetland ecosystem processes. Wetland processes can often enhance removal of a given wastewater constituent, and specific goals should be identified and design criteria developed to meet those goals.[21,22] We must also understand what removal/retention mechanisms are involved and the conditions necessary to optimize those mechanisms.

Chemical Differences Among Wetland Soils

Wetland soils vary considerably in both form and structure, resulting in functional and/or kinetic differences in nutrient retention or chemical transformations. Wetland soils can range widely in many chemical parameters, particularly soil organic matter (17–77%) and pH (3.9–6.5) (Table 3). Nitrogen and P content vary as much as 100–200% among wetland types. Exchangeable Ca and oxalate-extractable Al differ by an order of magnitude, which has important consequences for P retention.[23]

Soil bulk densities listed in Table 3 range from 0.07 to 0.55 g/cm^3. The effect

Table 3. Comparison of Various Soil Chemical Parameters (0–20 cm) Among Wetland Types

					Exchangeable Cations		Oxalate Extractable				
Site	%OM	pH	%N	%P	Ca (μg/g)	Mg (μg/g)	Fe (μg/g)	Al (μg/g)	Fe (μg/cm^3)	Al (μg/cm^3)	Bulk Density (g/cm^3)
Fen (MI)	55	6.0	2.20	0.09	8,120	906	4,924	2,295	1,083	505	0.22
Pocosin (NC)	77	3.9	1.40	0.03	1,033	—	2,370	814	166	57	0.07
Bog (MD)	68	4.5	1.69	0.10	710	477	5,710	6,400	628	704	0.11
Forested swamp (MD)	59	4.5	1.41	0.14	706	517	5,410	7,600	1,352	1,900	0.25
Marsh (WI)	41	6.5	1.77	—	12,730	2,300	—	—	—	—	—
Swamp (NC)	17	4.1	0.57	0.08	4,630	499	1,301	2,280	716	1,254	0.55

Source: Richardson.[44] Marsh data from Klopatek.[128]

of this variation is particularly striking for the pocosin and bog sites, which had lower oxalate-extractable Fe and Al amounts on a volume basis. This is not surprising because these two sites had the highest percentage of organic matter and lowest bulk densities. On a weight basis, the swamp has less Fe and Al than the bog site, but when calculated per cm^3, the swamp actually has more iron and almost twice the Al. Exchange capacity and retention of ions such as PO_4^{3-} is directly related to the amount of Al and Fe per unit volume of soil; therefore, retention is more accurately predicted by including bulk density differences. Also, for high organic matter soils, volume measures provide a much more realistic number for comparative purposes.

Nutrient Pools, Transformations, and Cycles

The following sections provide an overview of the major pools, intrasystem transformations, and fluxes for N, P, and S in freshwater wetlands along with specific applications involving these and other important contaminants. Representative examples of wetland soils and chemical processes were chosen because it is beyond our scope to exhaustively review all wetland systems and possible processing mechanisms for every contaminant.

Nitrogen

Nitrogen has a complex biogeochemical cycle with multiple biotic/abiotic transformations involving seven valence states (+5 to −3). The major N pools in natural freshwater wetlands are total sediment N (mostly organic N), total plant N, and available inorganic N in sediments.[24] The total sediment N pool is the largest, ranging from 100 to 1000 g N/m^2. The total plant N pool is roughly an order of magnitude less than total sediment N, while inorganic sediment N is another order of magnitude less than the plant pool. Major N inputs for a bog in Massachusetts were from atmospheric deposition (NO_3^- and NH_4^+) and N_2 fixation.[25] Eighty percent of the annual input was retained within the bog system through peat deposition or recycling via mineralization and uptake.

Much of the following discussion on N transformations is synthesized from Gambrell and Patrick[15] and Reddy and Patrick[16] who, along with many coworkers, pioneered this field in flooded soils and sediments. Nitrogen transformations in wetland soils are a complex assortment of microbially mediated processes strongly influenced by the redox status of the soil. These transformations are superimposed on a generalized diagram of oxidized and reduced zones in a flooded soil in Figure 6. Most flooded soils have a thin, oxidized layer at the surface caused by proximity to the atmosphere or the higher dissolved oxygen concentrations in overlying floodwater. Reduction processes dominate below this oxidized layer.[13,26-29]

In both reduced and oxidized layers, organic N is mineralized to NH_4^+. In the reduced layer, NH_4^+ is stable and may be adsorbed to sediment exchange sites or used by both plants and microbes. The thin, oxidized layer in flooded

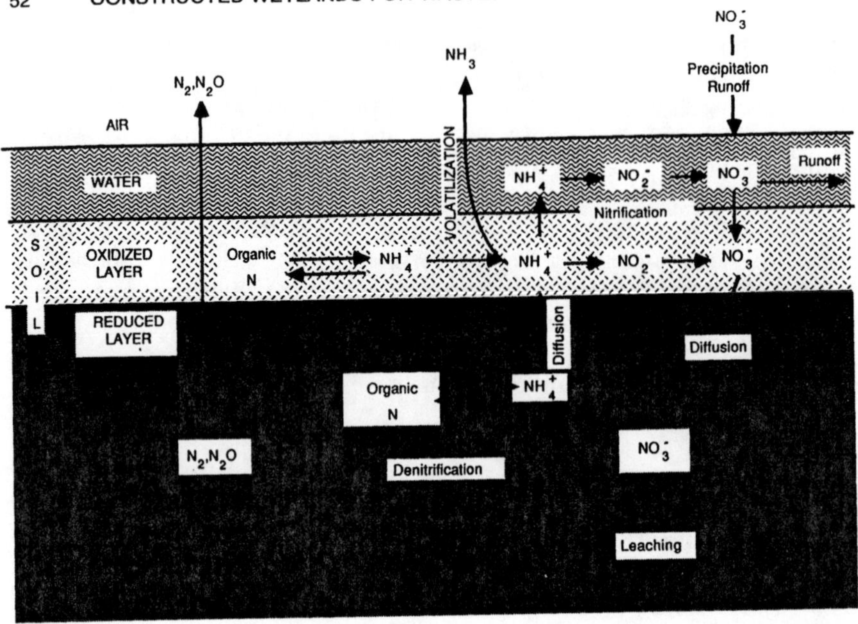

Figure 6. Nitrogen transformations in wetland soils. (Adapted from Reddy and Patrick.[16])

soils is important in N transformations because NH_4^+ is oxidized to NO_3^- by chemoautotrophic bacteria (nitrification) in this layer. Depletion of NH_4^+ in the upper, oxidized layer causes NH_4^+ to diffuse upward in response to the concentration gradient. This diffusion process may be effective from 4 to 12 cm deep.[30,31]

Nitrate is unstable in reduced zones and is quickly depleted via assimilative reduction, denitrification, or leaching. This again sets up a concentration gradient for diffusive flow of NO_3^- from oxidized to reduced zones (Figure 6). Redox potential, pH, moisture content, labile C source, and temperature control the rate of NO_3^- reduction.[16,32] For example, at pH < 6 the reduction of N_2O to N_2 is strongly inhibited.[33] The sequential processes of mineralization, nitrification, and denitrification dominate wetland N cycling and potentially process 20 to 80 g N/m²/yr.[24] Not only is this sequence important at the water-sediment interface, but there is growing evidence of its importance at the root-sediment interface. Oxidation of the root rhizosphere by wetland plants provides the same favorable conditions present at the soil surface, i.e., an oxidized zone immediately adjacent to a reduced zone (Figure 7). Radial oxygen loss from wetland plant roots can range from 100 to 400 mg O_2/m²/h.[34] While few data are available to quantify N loss via this mechanism, a recent study reported 25 to 30 kg/ha of N lost as N_2 and N_2O with an appreciable amount diffusing through aerenchymal structures into the atmosphere.[35]

The major N flux from natural wetlands is gaseous loss of N_2 through denitrification (Figure 8), although it has been suggested that most data repre-

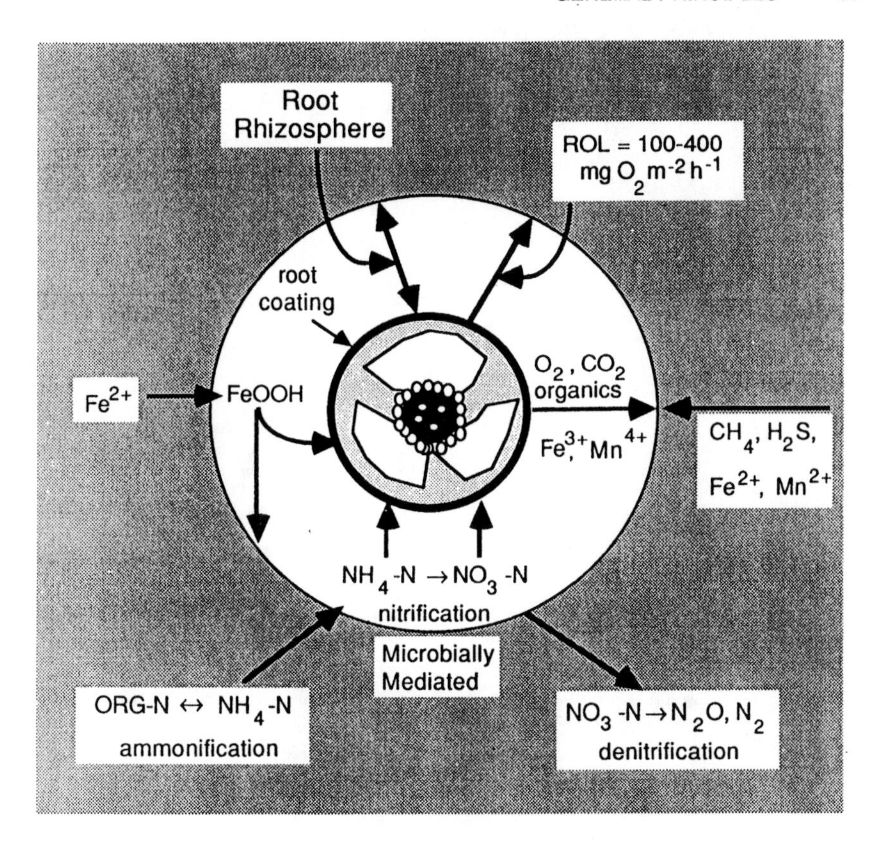

Figure 7. Dynamics of the root-sediment interface. (Adapted from Good and Patrick.[130])

sent potential rather than actual denitrification rates.[24,36] Minor exports include ammonia (NH_3) volatilization and hydrologic losses. Overall, natural freshwater wetlands tend to cycle most of their nitrogen within the system via uptake and mineralization with losses dominated by denitrification (Figure 8).

Anthropogenic inputs of N in sewage or other applications to wetland ecosystems result in 70% removal efficiencies at loading rates up to 20 to 30 g N/m²/yr, but efficiency decreases at higher loading rates (Figure 9). Kelly and Harwell[37] concluded that 43% of total N inputs were released as output. However, their data were for total N and did not include gaseous losses of N_2 to the atmosphere. Note that wetlands receiving cumulative N applications for many years at lower input levels continue to maintain removal efficiencies over time (Figure 9). This suggests that as long as NO_3^- is not highly loaded (~15–25 g N/m²/yr), denitrification should continue to remove N at the same rate unless the supply of labile C (typically available in a wetland) becomes limited.

Constructed wetlands can be managed to enhance removal processes. Deni-

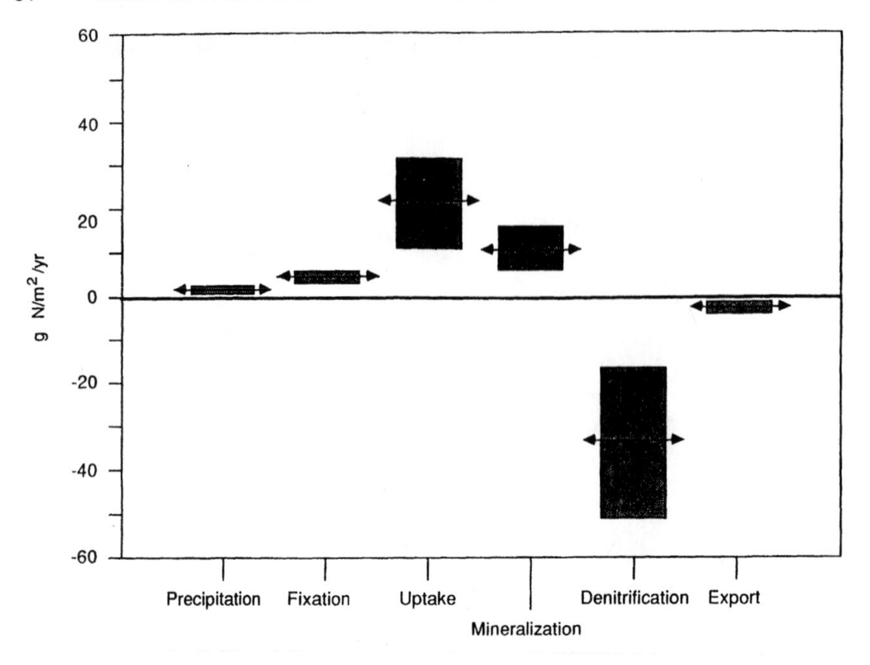

Figure 8. Internal and external wetland nitrogen fluxes. Mean value (arrow) ± one standard error. (Adapted from Bowden.[24])

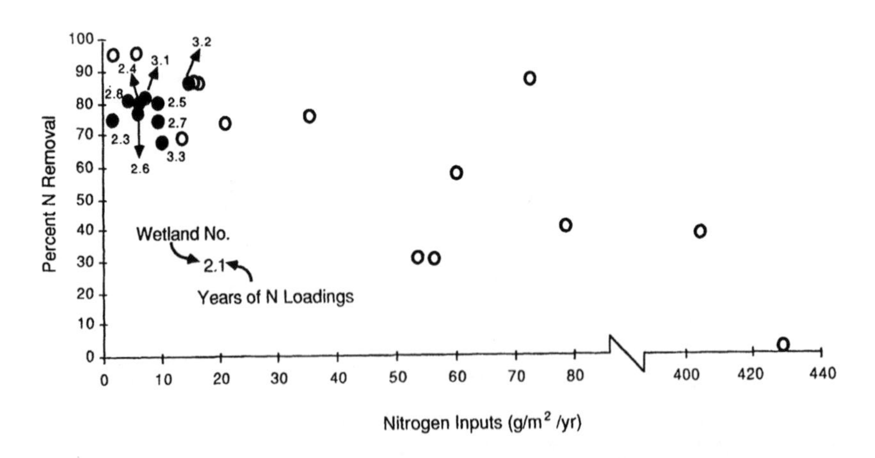

Figure 9. Nitrogen removal efficiency by wetlands as a function of loading rate. Multiple-year data points for the same site are filled. Data are for wetlands receiving wastewater. (Adapted from Richardson and Nichols[21] and Knight et al.[131])

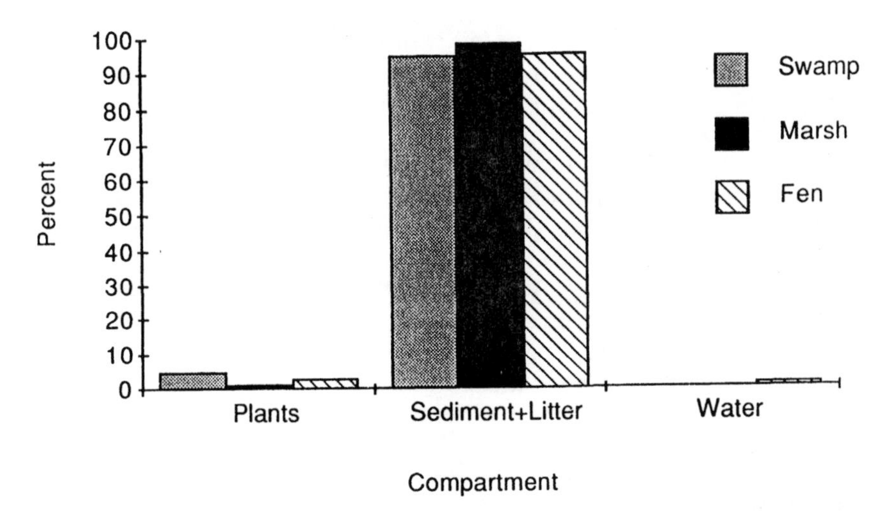

Figure 10. Phosphorus storage by compartments in wetlands. (Data from Verhoeven.[43])

trification in mineral soils is usually limited by available C;[38] this is not a problem in natural wetlands. However, selection of nonwetland, mineral soils for a constructed wetland system may limit the system's capability to process N through denitrification because of low C availability. An available C source may be added to newly constructed wetlands to overcome this problem until litter sources are available. In natural wetland systems, NO_3^- availability appears to limit denitrification rates.[31,39] Nitrate availability can be improved by alternating oxidizing and reducing conditions to promote the sequence of mineralization, nitrification, and denitrification.

Phosphorus

The soil P cycle is fundamentally different from the N cycle. There are no valency changes during biotic assimilation of inorganic P or during decomposition of organic P by microorganisms. Phosphorus has no gaseous phase, and it has a major geochemical cycle.[40] Soil P primarily occurs in the +5 (oxidized) valency state, because all lower oxidation states are thermodynamically unstable and readily oxidize to PO_4 even in highly reduced wetland soils.[41]

The sediment-litter compartment is the major P pool (>95%) in natural wetlands, with a much lower plant pool and little in the overlying water (Figure 10). Most soil P (>95%) in peatland systems is in the organic form[7,42,43] with cycling between pools controlled by biological forces (i.e., microbes and plants). The percentage of organic P is generally lower in wetlands with mineral substrates.[44]

Inorganic phosphorus transformations, subsequent complexes, and P retention in wetland soils are controlled by the interaction of redox potential, pH, Fe, Al, and Ca minerals, and the amount of native soil P. In acid soils, inorganic P is adsorbed on hydrous oxides of Fe and Al and may precipitate as insoluble Fe-phosphates (Fe-P) and Al-phosphates (Al-P). Precipitation as insoluble Ca-phosphates (Ca-P) is the dominant transformation at pH's greater than 7.0.[40,41]

Even though the oxidation state of P is unaffected by redox reactions, redox potential is important because of Fe reduction. The form of Fe-P known as reductant-soluble phosphorus (RS-P) is especially significant in wetland systems. Reductant-soluble P is a poorly-crystalline Fe compound that is stable under oxidized conditions, but releases adsorbed and occluded P when Fe^{3+} is reduced to Fe^{2+} following submergence.[15] This sediment pool may release large amounts of native P into solution under flooded conditions.[45,46] The practical implications are obvious: using a soil high in RS-P for polishing or tertiary treatment may result in increased effluent P levels as the wastewater passes through the wetland.

Interaction of pH and Eh also affects P transformations in wetland soils. Although Eh controls Fe^{3+} reduction, pH controls dissolution and subsequent reprecipitation of reduced compounds.[46,47] Holford and Patrick[46] found that soil reduction caused native soil P releases over a range of pH levels; however, vivianite $[Fe_3(PO_4)_2 \cdot 8H_2O]$, ferrous hydroxide $[Fe(OH)_2]$, or Ca phosphates were the precipitation products at pH 5.0, 6.5, and 8.0, respectively. Sah and Mikkelson[48] concluded that flooding caused a shift from Al-P to Fe-P forms and this was a function of mineral stability; variscite $[Al(OH)_2H_2PO_4]$ was unstable under reducing conditions and was transformed to vivianite, which is theoretically the most stable P mineral in flooded soils.

Inorganic P is retained by Fe and Al oxides and hydroxides, calcite ($CaCO_3$), organometallic complexes, and clay minerals.[49] The most important retention mechanisms are ligand exchange reactions, where phosphate displaces water or hydroxyls from the surface of Fe and Al hydrous oxides to form monodentate and binuclear complexes within the coordination sphere of the hydrous oxide.[23,45,50-53] Richardson[23] determined that P sorption capacity of wetland soils was best predicted by oxalate-extractable (amorphous) Al and Fe. The P sorption capacity of an oxidized soil may increase following flooding and reduction due to amorphous ferrous hydroxides, which have a greater surface area and more sorption sites than the more crystalline, oxidized, ferric forms.[45,49,52]

The literature is unclear whether specific adsorption via ligand exchange or precipitation reactions are the major P removal mechanism[8,51,54-56] Most results show rapid P removal from solution indicating adsorption,[52,55] but this fast reaction is followed by a continuous, slow removal of P into a less exchangeable form.[54,57,58] This slow reaction may be precipitation of insoluble phosphates or conversion from monodentate to binuclear complexes or both. Precipitation reactions have particular implications for wastewater treatment because reported P removals by soils receiving long-term wastewater applications were much greater than predicted by adsorption maxima.[55,59-61] This was attributed to rejuvenation of exchange sites by precipitation during alternating

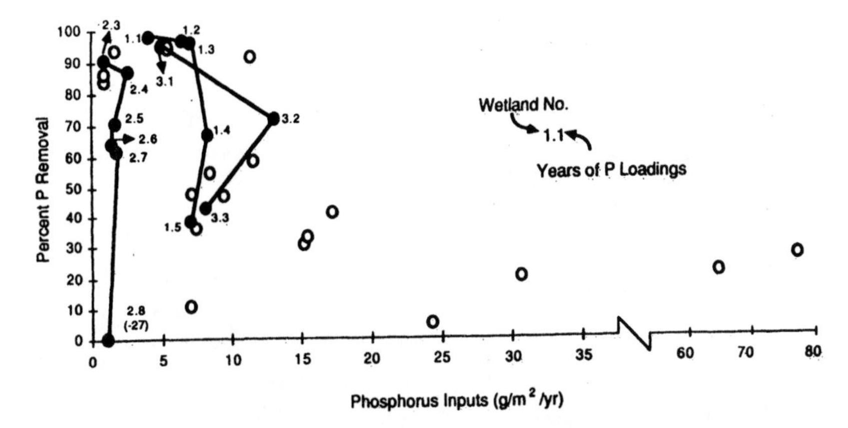

Figure 11. Phosphorus removal efficiency by wetlands as a function of loading rate. Multiple-year data points for the same site are filled and connected. Data are for wetlands receiving wastewater. (Adapted from Richardson and Nichols[21] and Knight et al.[131])

oxidizing soil conditions. However, the interaction of pH, redox, Fe, Al, adsorption, and precipitation often confounds interpretation of results from P removal studies. Much more work in this area is still needed to isolate cause and effect among these interactions.

Sediment processes control the long-term P removal capability of wetland ecosystems.[7,22,61,62] There is little direct uptake of phosphate from the water column by emergent wetland vegetation because the soil is the major source of nutrients.[63] Growing vegetation is a temporary nutrient-storage compartment resulting in seasonal exports following plant death.[42,64–69] The long-term role of emergent vegetation is to transform inorganic P to organic forms. Microorganisms play a definite role in P cycling in wetlands; however, the microbial pool is small and quickly saturated by wastewater P additions.[7]

Finally, processing efficiency of added P among wetland systems varies by an order of magnitude in percentage of retention, and wetlands are less effective at P removal than terrestrial systems.[23,37] Available data indicate P removal efficiency is strongly dependent on loading rate, with 65–95% removal at loading rates of less than 5.0 $g/m^2/yr$ (Figure 11). However, removal efficiency decreases to 30–40% or less when P loadings are greater than 10 to 15 $g/m^2/yr$. In addition, cumulative loadings of P into wetlands have resulted in large exports of P and much lower removal efficiency within a few years.[7,21,23] Initial P removal rates are often in excess of 90% but decline sharply after four to five years of cumulative P additions. This cumulative impact is well illustrated by wetland number 2 in Figure 11, where even low-level P inputs (<2.7 $g/m^2/yr$) resulted in a significant decline in percentage of retention over time and ultimately the wetland became a P source (–27% removal). A similar trend is noted for wetland numbers 1 and 3, but at higher loading rates. In contrast to N (Figure 9), there is a much wider range of P

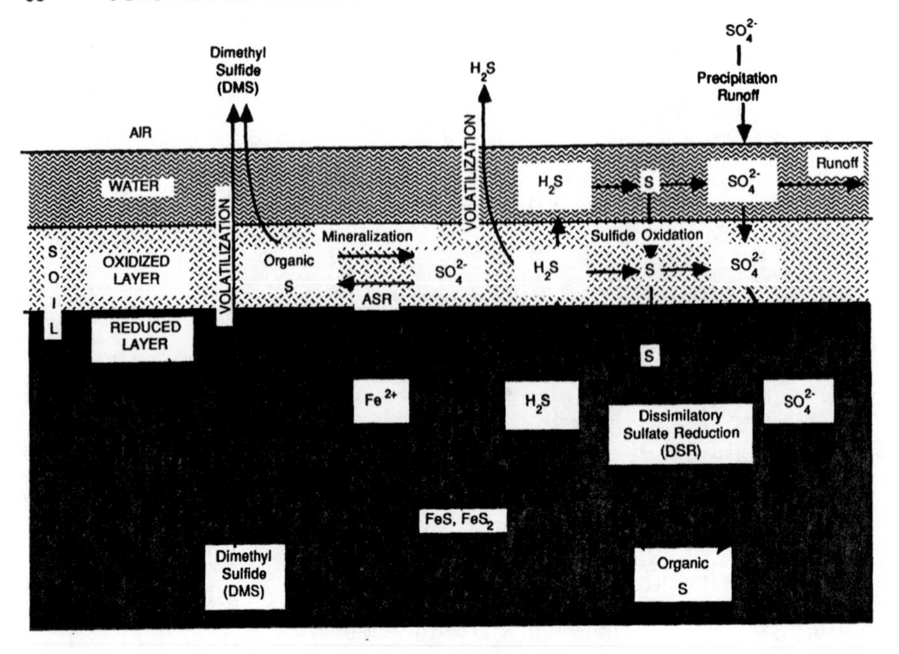

Figure 12. Sulfur transformations in wetland soils.

removal efficiency among wetlands at lower input levels (<10 g/m²/yr), indicating a greater dependence on soil mechanisms and the importance of substrate type for P removal from wastewater.

Sulfur

In contrast to N and P, there are few studies on S cycling and retention in freshwater systems, but available data indicate that most soil S is organic. Wieder and Lang[70] determined that over 90% of the total S in a *Sphagnum* bog was in the organic form. The remaining inorganic fraction was distributed as FeS_2 (4.5%); FeS (2.7%); elemental S (1.2%); and SO_4^{2-} (0.4%). In a comparison of several peat bogs, average organic S content ranged from 78 to 90% of the total.[71] Most studies of S in natural wetland systems have found that carbon-bonded (as opposed to ester-sulfate) organic S is the dominant fraction of the total S soil pool.[71-76]

Sulfur transformations are biologically mediated and, like P and N transformations, are affected by redox and pH interactions. Major transformations in oxidized environments are assimilatory sulfate reduction (ASR), inorganic sulfide and elemental sulfur oxidation, and mineralization of organic S to inorganic SO_4^{2-}. Under reducing conditions, dissimilatory sulfate reduction (DSR) transforms SO_4^{2-} to H_2S during respiration by obligate anaerobic bacteria (Figure 12).[12,15] The H_2S formed by DSR can be released to the atmosphere or react with organic matter providing another pathway for converting inor-

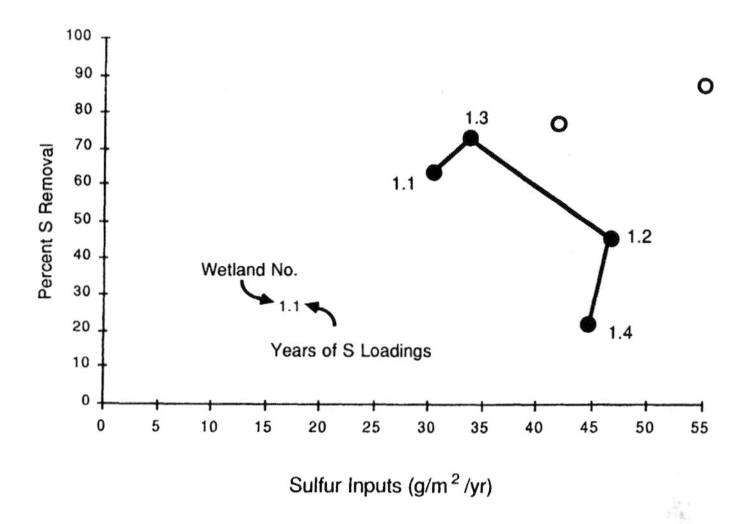

Figure 13. Sulfur removal efficiency by wetlands as a function of loading rate. Multiple-year data points for the same site are filled and connected. (Data from Bayley et al.,[76] Hemond,[83] and Winter and Kickuth.[85])

ganic S to organic S.[74,77,78] With SO_4^{2-} reduction and sufficient Fe, iron sulfides (FeS, FeS_2) can form; pyrite (FeS_2) formation requires alternating (either temporally or spatially) anaerobiosis with limited aeration.[79] These reactions between Fe and S may be important in acid mine drainage systems.

Despite the small size of the inorganic pool, this fraction is the most important for S cycling, retention, and mobility. Fluxes through the inorganic pool dominate S cycling in wetlands with high SO_4^{2-} inputs. Wieder and Lang[70] calculated that 3.5 to 4 times as much inorganic S was processed (through alternating SO_4^{2-} reduction and sulfide/sulfur oxidation) as compared to the organic S pool. This has important implications for wetland S cycles because S inputs are primarily SO_4^{2-} from atmospheric deposition and either natural or amended hydrologic sources. Sulfate retention by aerobic, mineral soils is dominated by the same adsorption mechanisms involved in PO_4^{3-} retention. However, adsorbed SO_4^{2-} is displaced by PO_4^{3-} on the exchange sites, but PO_4^{3-} is not displaced by SO_4^{2-}.[49,80] Sulfate may be readily desorbed by water[80] and most terrestrial systems are SO_4^{2-}-sources not sinks.[81,82]

Many wetlands function as effective S sinks (Figure 13). A bog in Massachusetts retained 77% of the atmospheric SO_4^{2-} input and a northern Minnesota peatland sequestered 56% of the annual SO_4^{2-} input.[84] Utilizing the root-zone process, Winter and Kickuth[85] reported 87% removal of total S from wastewater effluent. Annual sulfate retention by a fen ranged from 22–77% over a four-year period (Figure 13, closed circles).[76] Changes in the hydrologic regime decreased S retention efficiency with the wetland becoming a source rather than a sink. Bayley et al.[76] concluded that aerobic soil conditions during dry summer periods oxidized reduced S to SO_4^{2-} which was flushed from the system during autumn rainfall events (years 2 and 4). These seasonal fluctuations were partly responsible for variable removal efficiencies in wetland number 1 (Figure 13), although higher removal rates coincided with lower inputs. Similar oxidation effects on SO_4^{2-} exports were reported by Weider.[86]

The preceding discussion on S retention did not address the issue of gaseous losses, which were not determined in any of the cited studies. Significant fluxes of S or N to the atmosphere indicate that the wetland is functioning as a transformer as opposed to a true sink. However, quantifying gaseous emissions is a difficult endeavor. Castro and Dierberg[87] calculated a mean H_2S release rate of 80 mg/m²/yr for some freshwater marshes in Florida, which was comparable to the 60 mg/m²/yr average reported by Aneja et al. (as cited by Castro and Dierberg[87]) for North Carolina freshwater wetlands. Hemond[83] did not measure H_2S gaseous losses but concluded that "much sulfur was returned to the atmosphere in gaseous form."

There is increasing evidence that H_2S may not be the primary biogenic sulfide gas; the organic S gases methyl sulfide (MS) and dimethyl sulfide (DMS) can be of equal or greater importance.[88] Organic sulfide concentrations were two orders of magnitude higher than H_2S in gaseous sulfur emissions from an anaerobic sewage lagoon.[89] The average DMS emission rate from some Canadian wetlands was 81 mg/m²/yr with a range of 25–184 mg/m²/yr.[90]

Long-term S retention/removal depends on anaerobic, strongly reduced conditions to enhance SO_4^{2-} reduction, storage in organic forms, and volatilization as H_2S or other organic S gases. Wetlands constructed to remove/store S should be designed to promote these processes and avoid hydrologic release mechanisms associated with oxidized substrates.

Iron and Manganese in Acid Mine Drainage

Drainage from coal deposits and mining activities is a severe environmental problem that seriously degrades water quality when discharged into natural water bodies. Referred to as acid mine drainage (AMD), it is characterized by low pH and high Fe and Mn concentrations. Several review articles and symposia proceedings are available describing the overall environmental problems and geochemistry of AMD.[91–95]

Pyrite (FeS_2) associated with coal deposits oxidizes via an initiation reaction and a propagation or catalytic reaction.[96] These reactions can release large quantities of Fe and SO_4^{2-}, and the associated H^+ lowers pH, which solubilizes associated metals such as Mn. Measured AMD concentrations range from 70 to 5089 μmol Fe/L, 158 to 983 μmol Mn/L, 2812 to 27,270 μmol SO_4^{2-}/L, and 2.7 to 6.3 standard pH units.[97,98]

Ferric iron and O_2 are the two oxidants involved in pyrite oxidation. Oxidation by Fe^{3+} is much faster kinetically, so the oxidation of Fe^{2+} to Fe^{3+} is the rate-limiting step.[96] Further investigations have revealed that microorganisms, including *Sulfolobus* sp., *Metallogenium* sp., *Leptothrix* sp., *Gallionella* sp., *Siderocapsa* sp. and, particularly, the chemoautotrophic bacteria *Thiobacillus ferrooxidans* catalyze this reaction to the extent that the oxidation rate is 6 orders of magnitude greater than abiotic oxidation.[96,99,100]

There are several possible mechanisms for metal retention by soils: adsorption on cation exchange sites, precipitation, and complexation with soil organic matter.[10,101,102] While several studies[97,103-105] have successfully used constructed wetlands in ameliorating acid mine drainage, questions remain concerning the mechanisms involved and long-term functional efficiency. Few studies have investigated removal mechanisms or differentiated between physical/chemical and biological components. Gerber et al.[104] concluded that both microbial oxidation and adsorption mechanisms were involved in Fe and Mn removal by *Sphagnum* moss. However, they did not test the microbial hypothesis directly, but drew their conclusions from indirect observations. Most of the published results involve *Sphagnum* peat; there are few data on Fe and Mn sorption in chemically different mineral soils.

In reduced soils and sediments, metal sulfide formation is the most effective retention mechanism because of insolubility.[15,106,107] However, the few data available do not indicate significant Fe-sulfide accumulation.[70,75] No data are available on Mn-sulfides in wetlands receiving AMD. Sulfide production in freshwater systems is generally limited by low SO_4^{2-} levels,[108,109] but this should not be a limiting factor in AMD wetlands since several studies have found high rates of SO_4^{2-} reduction.[110-112]

Uptake by vegetation represents a minor and unimportant source of Fe and Mn removal in wetland systems.[102,113,114] Sencindiver and Bhumbla[115] determined that *Typha* removed less than 1% of the Fe input into a constructed wetland. Using maximum retention capacities (greatest measured biomass accumulation and plant Fe concentration) for a hypothetical constructed *Sphagnum* wetland, less than 0.1% (0.06%) of the Fe loadings would be taken up by plants.[102] Even if significant amounts of Fe and Mn were taken up by vegetation, these pools represent transformations to organic forms released back to the wetland surface following dieback at the end of the growing season. Other problems include decreased biomass accumulation and death, presumably due to toxic metal levels in the wetland.[115,116]

Few data exist concerning Fe and Mn cycling and processing by freshwater wetland ecosystems receiving AMD; most quantitative studies are input/output, treating the wetland as a black box.[117,118] Wieder and Lang[71] compared Fe and Mn accumulation in four *Sphagnum* peatlands representing a qualitative gradient of increasing metal loadings. They determined that organic Fe was the largest pool, representing 47–90% of the total. The oxide fraction was the next largest (21–56%), and iron sulfides (FeS and FeS_2) represented the smallest fraction (1–8%).

Analysis of the Wieder and Lang[71] data set may provide some insight into Fe transformations and retention mechanisms at these sites. Although Fe inputs to all four peatlands were not measured, the qualitative gradient was based on

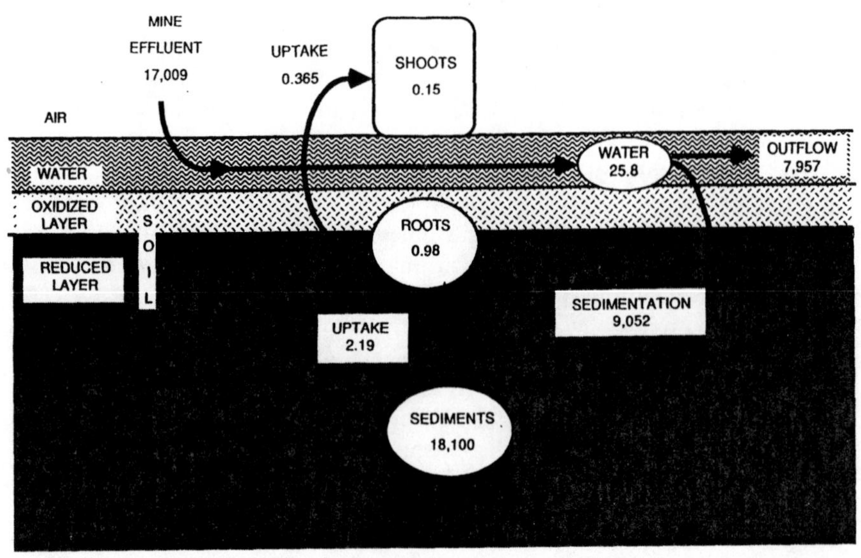

Figure 14. Iron fluxes ($g/m^2/yr$) and pools (g/m^2) in a constructed wetland receiving acid mine drainage. (Adapted from Fennessy and Mitsch.[114])

ombrotrophic (rain-fed) versus minerotrophic (groundwater) wetlands, plus one site (Tub Run Bog) that had been receiving AMD. Distribution of organic Fe and oxide Fe at Tub Run Bog was equivalent (484 and 485 μmol/g), indicating saturation of the peat adsorption capacity and greater oxide formation. Iron distribution at the three sites not receiving AMD was primarily organic Fe ($\bar{x}$ = 63%) over oxide forms ($\bar{x}$ = 33%).

The nature of this oxide material is still unknown; it is possibly ferrihydrite ($5Fe_2O_3 \cdot 9H_2O$), a highly disordered Fe(III)-oxide. There are several lines of evidence supporting this hypothesis. The poor crystallinity of ferrihydrite results from rapid formation and/or inhibited crystallization.[20] Organic matter has a strong inhibitory effect on Fe oxide crystallization, favoring the formation of poorly ordered, amorphous Fe oxides.[119,120] This inhibitory effect is a likely mechanism in highly organic wetlands. Campbell and Schwertmann[121] determined that Fe crystallization was hindered by organic matter and concluded that when the ratio of oxalate-extractable Fe (Fe_o) to dithionite-extractable Fe (Fe_d) was greater than 0.5, ferrihydrite was the dominant oxide. Calculated Fe_o/Fe_d ratios for the Wieder and Lang[71] data set are all greater than 0.5. The Fe_o/Fe_d ratios for the strongly minerotrophic Big Run Bog and the AMD-affected Tub Run Bog were 0.94 and 0.81, respectively.

A preliminary Fe mass balance for a constructed wetland receiving AMD emphasizes the small role of vegetation and the dominant role of sediments in removing and storing Fe from drainage water influent (Figure 14).[114] It provides little insight into specific mechanisms or processes responsible for retaining Fe in the sediment compartment. Current information suggests there are two somewhat contrasting approaches for Fe and Mn removal in constructed wetlands: biologically mediated oxidation and reduction processes. Because of different conditions required to support these two processes, wetlands are the only ecosystems where oxidation and reduction reactions interact. Constructed wetlands could be designed to enhance these two *complementary* processes. Most operating systems stress the importance of the vegetative/microbial complex for oxidizing Fe^{2+} and Mn^{2+} with some combination of subsequent insoluble oxide precipitation, organic complexation, or adsorption.[97,103,104,122,123] Reduction processes as removal mechanisms have not been fully explored in constructed wetlands for treating AMD, but successful empirical evidence is available.[110-112,124] However, questions remain concerning the exact mechanisms, controlling factors, and long-term functional capabilities of these processes.[70,102]

Knowledge of metal removal processes is growing, but we lack a sound, quantitative data base for optimizing removal mechanisms. Unfortunately, the nature of AMD is site specific (i.e., concentrations of H^+, Fe, Mn, and SO_4; flow rates; varying weather events; and seasonal climate differences all interact to produce unique site-specific problems). The next stage is to go beyond simple input/output studies and combine controlled, manipulative field and laboratory experiments to identify the major mechanisms controlling Fe and Mn retention, relative capabilities and capacities of various substrates, and pathways of biogeochemical iron and manganese cycling. From these data, design parameters necessary to enhance successful removal mechanisms can be implemented at each site.

SUMMARY

The physical and chemical nature of wetland soils and sediments varies among soil types and even within a given soil. These differences will have a major impact on the success or failure of a particular system depending on pollutant type, pollutant removal goals, and removal time.

Soil chemical reactions in natural wetlands are dominated by anaerobic conditions, and wetlands are the major reducing system on the landscape. However, few freshwater wetland ecosystems are permanently flooded, so wetlands maintain the widest range of oxidation-reduction reactions of any ecosystem on the landscape. This allows them to function as effective transformers of nutrients and metals. Often these reactions transform inorganic inputs to organic outputs or cause retention or complexation within the wetland.

We need to apply our understanding of biogeochemical cycling in natural wetlands to enhance transformation and storage. Often this will require different processes for different materials:

- For N removal, enhancement of denitrification by alternating oxidizing and reducing conditions will maximize nitrification during aerobic periods, which supplies NO_3^- for denitrification during anaerobic and reducing conditions.
- Phosphorus is removed by soil sorption processes with a finite P capacity, an entirely different mechanism than N removal. Maintaining contact with soils high in Ca or oxalate-extractable (amorphous) Fe and Al is preeminent for wastewater P removal. Maximum soil capacity varies widely, and selecting the correct substrate is critical for P removal. Alternating oxidizing and reducing conditions can recharge sorption sites, allowing greater P removal than under permanently reducing conditions. In this case, operational procedures for N and P are compatible. Increased peat accretion can also result in higher P storage.
- Sulfate removal from most effluents requires permanent reducing conditions for SO_4^{2-} reduction and incorporation into sediment storage. Alternating with an oxidation cycle would oxidize reduced S compounds to SO_4^{2-}, which is mobile and easily flushed from the system. Therefore, optimizing S removal is not compatible with maximum N or P removal.
- The database for Fe and Mn removal from acid mine drainage is insufficient to develop hard and fast guidelines for maximizing removal/retention in constructed wetlands. Current methods of enhancing geochemical and biological oxidation mimic the success of traditional chemical techniques; however, the exact mechanisms and long-term viability of these systems are unclear despite their reported effectiveness. Reduction processes with "permanent" sediment storage may be the preferred long-term solution, but successful implementation of these mechanisms has not yet been achieved.

The challenge facing us in applied ecology is to construct wetlands with optimal retention capacities for removing and storing contaminants. Environmental problems, such as wastewater treatment, are difficult to solve due to the tremendous natural variation inherent in disturbed ecosystems. Our job is to define common variables among these individual conditions, identify mechanisms responsible for observed improvements in water quality, and incorporate these into logical and scientifically sound criteria for the design and operation of constructed wetlands. Soil and sediment biogeochemical processes play a critical role in the successful application of the emerging technology of constructed wetlands for wastewater treatment.

REFERENCES

1. Jenny, H. "Derivation of the State Factor Equations of Soil and Ecosystems," *Soil Sci. Soc. Am. Proc.* 25:385–388 (1961).
2. "Soil Taxonomy: A Basic System of Soil Classification for Making and Interpreting Soil Surveys," USDA-SCS Agricultural Handbook No. 436 (1975).
3. Bouma, J., and L. W. Dekker. "A Case Study on Infiltration into Dry Clay Soil. 1. Morphological Observations," *Geoderma* 20:27–40 (1978).
4. Bouma, J., L. W. Dekker, and J. C. F. M. Haans. "Measurement of Depth to Water Table in a Heavy Clay Soil," *Soil Sci.* 130:264–270 (1980).

5. Boelter, D. H., and E. S. Verry. "Peatland and Water in the Northern Lake States," USDA For. Serv. Gen. Tech. Rep. NC-31, NC For. Exp. Stn. (1977).
6. Daniel, G. C. "Hydrology, Geology, and Soils of Pocosins: A Comparison of Natural and Altered Systems," in *Pocosin Wetlands*, C. J. Richardson, Ed. (Stroudsberg, PA: Hutchinson Ross Publishing Company, 1981), pp. 69–108.
7. Richardson, C. J., and P. E. Marshall. "Processes Controlling Movement, Storage, and Export of Phosphorus in a Fen Peatland," *Ecol. Monog.* 56:279–302 (1986).
8. Bohn, H. L., B. L. McNeal, and G. A. O'Connor. *Soil Chemistry* (New York: John Wiley & Sons, Inc., 1985).
9. Dolman, J. D., and S. W. Buol. "A Study of Organic Soils (Histosols) in the Tidewater Region of North Carolina," N.C. Agric. Exp. Stn. Tech. Bull. 181 (1967).
10. Keeney, D. R., and R. E. Wildung. "Chemical Properties of Soils," in *Soils for Management of Organic Wastes and Waste Waters*, L. F. Elliot and F. J. Stevenson, Eds. (Madison, WI: Soil Science Society of America, 1977), pp. 75–97.
11. Baas Becking, L. G. M., I. R. Kaplan, and D. Moore. "Limits of the Natural Environment in Terms of pH and Oxidation-Reduction Potentials," *J. Geol.* 68:243–284 (1960).
12. Ponnamperuma, F. N. "The Chemistry of Submerged Soils," *Adv. Agron.* 24:29–96 (1972).
13. Howeler, R. H., and D. R. Bouldin. "The Diffusion and Consumption of Oxygen in Submerged Soils," *Soil Sci. Soc. Amer. Proc.* 35:202–208 (1971).
14. Turner, F. T., and W. H. Patrick, Jr. "Chemical Changes in Waterlogged Soils as a Result of Oxygen Depletion," *Trans. 9th Intern. Cong. of Soil Sci.* 4:53–65 (1968).
15. Gambrell, R. P., and W. H. Patrick, Jr. "Chemical and Microbiological Properties of Anaerobic Soils and Sediments," in *Plant Life in Anaerobic Environments*, D. D. Hook and R. M. M. Crawford, Eds. (Ann Arbor, MI: Ann Arbor Science Publishers, Inc., 1978), pp. 375–423.
16. Reddy, K. R., and W. H. Patrick, Jr. "Nitrogen Transformations and Loss in Flooded Soils and Sediments," *CRC Crit. Rev. Environ. Control* 13:273–309 (1984).
17. Richardson, C. J. "Wetlands as Transformers, Filters and Sinks for Nutrients," in *Freshwater Wetlands: Perspectives on Natural, Managed and Degraded Ecosystems*, Univ. of Georgia, Savannah River Ecology Lab., Ninth Symposium, Charleston, SC (in press).
18. Collins, J. F., and S. W. Buol. "Effects of Fluctuations in the Eh-pH Environment of Iron and/or Manganese Equilibria," *Soil Sci.* 110:111–118 (1970).
19. Williams, R. J. P. "An Introduction to Biominerals and the Role of Organic Molecules in Their Formation," *Phil. Trans. Royal Soc. Lon.* 304:411–424 (1984).
20. Schwertmann, U. "Occurrence and Formation of Iron Oxides in Various Pedoenvironments," in *Iron in Soils and Clay Minerals*, J. W. Stucki, B. A. Goodman, and U. Schwertmann, Eds. (Dordrecht, Holland: D. Reidel Publishing Company, 1988), pp. 267–308.
21. Richardson, C. J., and D. S. Nichols. "Ecological Analysis of Wastewater Management Criteria in Wetland Ecosystems," in *Ecological Considerations in Wetlands Treatment of Municipal Wastewaters*, P. J. Godfrey, E. R. Kaynor, S.

Pelczarski, and J. Benforado, Eds. (New York: Van Nostrand Reinhold Company, 1985), pp. 351–391.

22. Richardson, C. J., and J. A. Davis. "Natural and Artificial Wetland Ecosystems: Ecological Opportunities and Limitations," in *Aquatic Plants for Water Treatment and Resource Recovery,* K. R. Reddy and W. H. Smith, Eds. (Orlando, FL: Magnolia Publishing Inc., 1987), pp. 819–854.
23. Richardson, C. J. "Mechanisms Controlling Phosphorus Retention Capacity in Freshwater Wetlands," *Science* 228:1424–1427 (1985).
24. Bowden, W. B. "The Biogeochemistry of Nitrogen in Freshwater Wetlands," *Biogeochemistry* 4:313–348 (1987).
25. Hemond, H. F. "The Nitrogen Budget of Thoreau's Bog," *Ecology* 64:99–109 (1983).
26. Mortimer, C. H. "The Exchange of Dissolved Substances Between Mud and Water in Lakes," *J. Ecol.* 29:280–329 (1941).
27. Mortimer, C. H. "The Exchange of Dissolved Substances Between Mud and Water in Lakes," *J. Ecol.* 30:147–201 (1942).
28. Alberda, T. "Growth and Root Development of Lowland Rice and Its Relation to Oxygen Supply," *Plant and Soil* 5:1–28 (1953).
29. Patrick, W. H., Jr., and R. D. DeLaune. "Characterization of the Oxidized and Reduced Zones in Flooded Soil," *Soil Sci. Soc. Am. Proc.* 36:573–576 (1972).
30. Chen, R. L., D. R. Keeney, D. A. Graetz, and A. J. Holding. "Denitrification and Nitrate Reduction in Wisconsin Lake Sediments," *J. Environ. Qual.* 1:158–161 (1972).
31. Reddy, K. R., W. H. Patrick, Jr., and R. E. Philips. "Ammonium Diffusion as a Factor in Nitrogen Loss from Flooded Soils," *Soil Sci. Soc. Am. J.* 40:528–533 (1976).
32. Avnimelech, Y. "Nitrate Transformation in Peat," *Soil Sci.* 111:113–118 (1971).
33. Wijler, J., and C. C. Delwiche. "Investigations on the Denitrifying Process in Soil," *Plant and Soil* 5:155–169 (1954).
34. Moorehead, K. K., and K. R. Reddy. "Oxygen Transport Through Selected Aquatic Macrophytes," *J. Environ. Qual.* 17:138–142 (1988).
35. W. H. Patrick, Jr. Personal communication (1988).
36. Howard-Williams, C. "Cycling and Retention of Nitrogen and Phosphorus in Wetlands: A Theoretical and Applied Perspective," *Freshwater Biol.* 15:391–431 (1985).
37. Kelly, J. R., and M. A. Harwell. "Comparisons of the Processing of Elements by Ecosystems. I: Nutrients," in *Ecological Considerations in Wetlands Treatment of Muncipal Wastewaters,* P. J. Godfrey, E. R. Kaynor, S. Pelczarski, and J. Benforado, Eds. (New York: Van Nostrand Reinhold Company, 1985), pp. 137–157.
38. Broadbent, F. E., and R. E. Clark. "Denitrification," in *Soil Nitrogen,* Am. Agron. Soc. Mon. No. 10. (Madison, WI: American Agronomy Society, 1965), pp. 347–359.
39. Firestone, M. K., M. S. Smith, R. B. Firestone, and J. M. Tiedje. "The Influence of Nitrate, Nitrite, and Oxygen on the Composition of the Gaseous Products of Denitrification in Soil," *Soil Sci. Soc. Am. J.* 43:1140–1144 (1979).
40. Stevenson, F. W. *Cycles of Soil* (New York: John Wiley & Sons, Inc., 1986).
41. Lindsay, A. L. *Chemical Equilibria in Soils* (New York: John Wiley & Sons, Inc., 1979).
42. Richardson, C. J., D. L. Tilton, J. A. Kadlec, J. P. M. Chamie, and W. A.

Wentz. "Nutrient Dynamics of Northern Wetland Ecosystems," in *Freshwater Wetlands: Ecological Processes and Management Potential,* R. E. Good, D. F. Whigham, and R. L. Simpson, Eds. (New York: Academic Press, Inc., 1978), pp. 217–241.

43. Verhoeven, J. T. A. "Nutrient Dynamics in Minerotrophic Peat Mires," *Aquat. Bot.* 25:117–137 (1986).
44. Richardson, C. J. Unpublished data.
45. Patrick, W. H., Jr., and R. A. Khalid. "Phosphate Release and Sorption by Soils and Sediments: Effect of Aerobic and Anaerobic Conditions," *Science* 186:53–55 (1974).
46. Holford, I. C. R., and W. H. Patrick, Jr. "Effects of Reduction and pH Changes on Phosphate Sorption and Mobility in an Acid Soil," *Soil Sci. Soc. Am. J.* 43:292–297 (1979).
47. Patrick, W. H., Jr., S. Gotoh, and B. G. Williams. "Strengite Dissolution in Flooded Soils and Sediments," *Science* 179:565–565 (1973).
48. Sah, R. N., and D. S. Mikkelsen. "Transformations of Inorganic Phosphorus During the Flooding and Draining Cycles of Soil," *Soil Sci. Soc. Am. J.* 50:62–67 (1986).
49. Parfitt, R. L. "Anion Adsorption by Soils and Soil Materials," *Adv. Agron.* 30:1–50 (1978).
50. Fox, R. L., and E. J. Kamprath. "Adsorption and Leaching of P in Acid Organic Soils and High Organic Matter Sand," *Soil Sci. Soc. Am. Proc.* 35:154–156 (1971).
51. Syers, J. K., R. F. Harris, and D. E. Armstrong. "Phosphate Chemistry in Lake Sediments," *J. Environ. Qual.* 2:1–14 (1973).
52. Khalid, R. A., W. H. Patrick, Jr., and R. D. DeLaune. "Phosphorus Sorption Characteristics of Flooded Soils," *Soil Sci. Soc. Am. J.* 41:305–310 (1977).
53. Cuttle, S. P. "Chemical Properties of Upland Peats Influencing the Retention of Phosphate and Potassium Ions," *J. Soil Sci.* 34:75–82 (1983).
54. Ryden, J. C., J. K. Syers, and R. F. Harris. "Phosphorus in Runoff and Streams," *Adv. Agron.* 25:1–45 (1973).
55. de Haan, F. A. M., and P. J. Zwerman. "Pollution of Soil," in *Soil Chemistry A. Basic Elements*, G. H. Bolt and M. G. M. Bruggenwert, Eds. (New York: Elsevier Science Publishing Co., Inc., 1976), pp. 192–271.
56. Bloom, P. R. "Phosphorus Adsorption by an Aluminum-Peat Complex," *Soil Sci. Soc. Am. J.* 45:267–272 (1981).
57. Barrow, N. J., and T. C. Shaw. "The Slow Reactions Between Soils and Anions: 2. Effect of Time and Temperature on the Decrease in Phosphate Concentration in the Soil Solution," *Soil Sci.* 119:167–177 (1975).
58. Van Riemsdijk, W. H., T. A. Westrate, and J. Beek. "Phosphates in Soils Treated with Sewage Water: III. Kinetic Studies on the Reaction of Phosphate with Aluminum Compounds," *J. Environ. Qual.* 6:26–29 (1977).
59. Ellis, B. G. "The Soil as a Chemical Filter," in *Recycling Treated Municipal Wastewater and Sludge Through Forest and Cropland*, W. E. Sopper and L. T. Kardos, Eds. (University Park: Pennsylvania State University Press, 1973), pp. 46–70.
60. Sawhney, B. L., and D. E. Hill. "Phosphate Sorption Characteristics of Soils Treated with Domestic Waste Water," *J. Environ Qual.* 4:342–346 (1975).

61. Nichols, D. S. "Capacity of Natural Wetlands to Remove Nutrients from Wastewater," *J. Water Poll. Control Fed.* 55:495–505 (1983).
62. Bayley, S. E. "The Effect of Natural Hydroperiod Fluctuations on Freshwater Wetlands Receiving Added Nutrients," in *Ecological Considerations in Wetlands Treatment of Municipal Wastewater*, P. J. Godfrey, E. R. Kaynor, S. Pelczarski, and J. Benforado, Eds. (New York: Van Nostrand Reinhold Company, 1985), pp. 180–189.
63. Sculthorpe, C. D. *Biology of Aquatic Vascular Plants* (New York: St. Martin's Press, 1967).
64. Boyd, C. E. "Losses of Nutrients During Decomposition of *Typha Latifolia*," *Arch. Hydrobiol.* 66:511–517 (1970).
65. Klopatek, J. M. "The Role of Emergent Macrophytes in Mineral Cycling in a Freshwater Marsh," in *Mineral Cycling in Southeastern Ecosystems,* F. G. Howell, J. B. Gentry, and M. H. Smith, Eds. (Aiken, SC: ERDA Symposium Series CONF-740513, 1975), pp. 367–393.
66. Klopatek, J. M. "Nutrient Dynamics of Freshwater Riverine Marshes and the Role of Emergent Macrophytes," in *Freshwater Wetlands: Ecological Processes and Management Potential,* R. E. Good, D. F. Whigham, and R. L. Simpson, Eds. (New York: Academic Press, Inc., 1978), pp. 195–216.
67. Richardson, C. J., J. A. Kadlec, W. A. Wentz, J. P. M. Chamie, and R. H. Kadlec. "Background Ecology and the Effects of Nutrient Additions on a Central Michigan Wetland," in *Proceedings: Third Wetlands Conference,* M. W. Lefor, W. C. Kennard, and T. B. Helfgott, Eds. (Storrs, CT: Institute of Water Resources No. 26, 1976), pp. 34–74.
68. Spangler, F. L., C. W. Fetter, Jr., and W. E. Sloey. "Phosphorus Accumulation Discharge Cycles in Marshes," *Water Res. Bull.* 13:1191–1201. (1977).
69. Davis, C. B., and A. G. van der Valk. "Litter Decomposition in Prairie Glacial Marshes," in *Freshwater Wetlands: Ecological Processes and Management Potential,* R. E. Good, D. F. Whigham, and R. L. Simpson, Eds. (New York: Academic Press, Inc., 1978), pp. 99–113.
70. Wieder, R. K., and G. E. Lang. "Cycling of Inorganic and Organic Sulfur in Peat from Big Run Bog, West Virginia," *Biogeochemistry* 5:221–242 (1988).
71. Wieder, R. K., and G. E. Lang. "Fe, Al, Mn, and S Chemistry of *Sphagnum* Peat in Four Peatlands with Different Metal and Sulfur Input," *Water Air Soil Poll.* 29:309–320 (1986).
72. Casagrande, D. J., K. Siefert, C. Berschinski, and N. Sutton. "Sulfur in Peat-Forming Systems of the Okefenokee Swamp and Florida Everglades: Origins of Sulfur in Coal," *Geochim. Cosmochim. Acta* 41:411–420 (1977).
73. Altschuler, Z. S., M. M. Schnepfe, C. C. Sibler, and F. O. Simon. "Sulfur Diagenesis in Everglades Peat and Origin of Pyrite in Coal," *Science* 221:221–227 (1983).
74. Brown, K. "Sulphur Distribution and Metabolism in Waterlogged Peat," *Soil Biol. Biochem.* 17:39–45 (1985).
75. Wieder, R. K., G. E. Lang, and V. A. Granus. "An Evaluation of Wet Chemical Methods for Quantifying Sulfur Fractions in Freshwater Wetland Peat," *Limnol. Oceanog.* 30:1109–1115 (1985).
76. Bayley, S. E., R. S. Behr, and C. A. Kelly. "Retention and Release of S from a Freshwater Wetland," *Water Air Soil Poll.* 31:101–114 (1986).
77. Casagrande, D. J., G. Indowu, A. Friedman, P. Rickert, and D. Schlenz. "H_2S

Incorporation in Coal Precursors: Origins of Sulfur in Coal," *Nature* 282:599–600 (1979).

78. Brown, K. "Formation of Organic Sulphur in Anaerobic Peat," *Soil Biol. Biochem.* 18:131-140 (1986).
79. Van Breeman, N. "Redox Processes of Iron and Sulfur Involved in the Formation of Acid Sulfate Soils," in *Iron in Soils and Clay Minerals*, J. W. Stucki, B. A. Goodman, and U. Schwertmann, Eds. (Dordrecht, Holland: D. Reidel Publishing Company, 1988) pp. 825–841.
80. Johnson, D. W., and D. W. Cole. "Anion Mobility in Soils: Relevance to Nutrient Transport from Forest Ecosystems," *Environ. Int.* 3:79–90 (1980).
81. Likens, G. E., F. H. Borman, R. S. Pierce, J. S. Eaton, and N. M. Johnson. *Biogeochemistry of a Forested Ecosystem* (New York: Springer-Verlag New York, Inc., 1977).
82. Bayley, S. E., and D. W. Schindler. "Sources of Alkalinity in Precambrian Shield Watersheds Under Natural Conditions and After Fire or Acidification," in *Effects of Acid Deposition on Forests, Wetlands, and Agricultural Ecosystems, Proceedings of the Workshop*, T. Hutchinson and K. Meena, Eds. (in press).
83. Hemond, H. F. "Biogeochemistry of Thoreau's Bog, Concord, Massachusetts," *Ecol. Monog.* 50:507–526 (1980).
84. Urban, N. R., S. J. Eisenreich, and E. Gorham. "Proton Cycling in Bogs: Geographic Variations in Northeastern North America," in *Effects of Acid Deposition on Forests, Wetlands, and Agricultural Ecosystems, Proceedings of the Workshop*, T. Hutchinson and K. Meena, Eds. (in press).
85. Winter, M., and R. Kickuth. "Elimination of Nutrients (Sulphur, Phosphorus, Nitrogen) by the Root Zone Process and Simultaneous Degradation of Organic Matter," Utrecht Plant Ecology News Report (December 1985).
86. Wieder, R. K. "Peat and Water Chemistry at Big Bog Run, a Peatland in the Appalachian Mountains of West Virginia, USA," *Biogeochemistry* 1:227–302 (1985).
87. Castro, M. S., and F. E. Dierberg. "Biogenic Hydrogen Sulfide Emissions from Selected Florida Wetlands," *Water Air Soil Poll.* 33:1–13 (1987).
88. Krouse, H. R., and R. G. L. McCready. "Reductive Reactions in the Sulfur Cycle," in *Biogeochemical Cycling of Mineral-Forming Elements*, P. A. Trudinger and D. J. Swain, Eds. (Amsterdam: Elsevier Science Publishing Co., 1979), pp. 315–368.
89. Rasmussen, R. A. "Emission of Biogenic Hydrogen Sulfide," *Tellus* 26:254–260 (1974).
90. Nriagu, J. O., D. A. Holdway, and R. D. Coker. "Biogenic Sulfur and the Acidity of Rainfall in Remote Areas of Canada," *Science* 237:1189–1192 (1987).
91. *Proceedings of the Second Symposium of Coal Mine Drainage Research, May 1968* (Pittsburgh, PA: Mellon Institute, 1968).
92. Burris, J. E., Ed. *Treatment of Mine Drainage by Wetlands* (University Park: Pennsylvania State University Press, 1984).
93. *Symposium on Surface Mining, Hydrology, Sedimentology, and Reclamation* (Lexington: University of Kentucky, 1984).
94. Brooks, R. P., D. E. Samuel, and J. B. Hill, Eds. *Wetlands and Water Management on Mined Lands* (University Park: Pennsylvania State University Press, 1985).

95. "Mine Drainage and Surface Mine Reclamation," Bureau of Mines Information Circular 9183, U.S. Government Printing Office (1988).
96. Singer, P. C., and W. Stumm. "Acidic Mine Drainage: The Rate-Determining Step," *Science* 167:1121–1123 (1970).
97. Kleinmann, R. L. P., and M. A. Girts. "Acid Mine Water Treatment in Wetlands: An Overview of an Emergent Technology," in *Aquatic Plants for Water Treatment and Resource Recovery,* K. R. Reddy and W. H. Smith, Eds. (Orlando, FL: Magnolia Publishing Inc., 1987), pp. 255–261.
98. Hiel, M. T., and F. J. Kerins, Jr. "The Tracy Wetlands: A Case Study of Two Passive Mine Drainage Treatment Systems in Montana," Bureau of Mines Information Circular 9183, U.S. Government Printing Office (1988), pp. 352–358.
99. Lundgren, D. G., and W. Dean. "Biogeochemistry of Iron," in *Biogeochemical Cycling of Mineral-Forming Elements*, P. A. Trudinger and D. J. Swain, Eds. (Amsterdam: Elsevier Science Publishing Co., Inc., 1979), pp. 211–252.
100. Nordstrom, D. K. "Aqueous Pyrite Oxidation and the Consequent Formation of Secondary Iron Minerals," in *Acid Sulfate Weathering,* L. R. Hossner, Ed. (Madison, WI: Soil Science Society of America, 1982), pp. 37–56.
101. Hodgson, J. F. "Chemistry of the Micronutrient Elements," *Adv. Agron.* 15:119–159 (1963).
102. Wieder, R. K. "Determining the Capacity for Metal Retention in Man-Made Wetlands Constructed for Treatment of Coal Mine Drainage," Bureau of Mines Information Circular 9183, U.S. Government Printing Office (1988), pp. 375–381.
103. Brodie, G. A., D. A. Hammer, and D. A. Tomljanovich. "An Evaluation of Substrate Types in Constructed Wetlands Acid Drainage Treatment Systems," Bureau of Mines Information Circular 9183, U.S. Government Printing Office (1988), pp. 389–398.
104. Gerber, D. W., J. E. Burris, and R. W. Stone. "Removal of Dissolved Iron and Manganese Ions by a Sphagnum Moss System," in *Wetlands and Water Management on Mined Lands*, R. P. Brooks, D. E. Samuel, and J. B. Hill, Eds. (University Park: Pennsylvania State University Press, 1985) pp. 365–372.
105. Wieder, R. K., G. E. Lang, and A. E. Whitehouse. "Metal Removal in *Sphagnum*-Dominated Wetlands: Experience with a Man-Made Wetland System," in *Wetlands and Water Management on Mined Lands*, R. P. Brooks, D. E. Samuel, and J. B. Hill, Eds. (University Park: Pennsylvania State University Press, 1985b), pp. 353–364.
106. Engler, R. M., and W. H. Patrick, Jr. "Stability of Sulfides of Manganese, Iron, Zinc, Copper, and Mercury in Flooded and Non-Flooded Soil," *Soil Sci.* 119:217–221 (1975).
107. Stumm, W., and J. J. Morgan. *Aquatic Chemistry: An Introduction Emphasizing Chemical Equilibria in Natural Waters,* 2nd ed. (New York: Wiley Interscience, 1981).
108. Cappenberg, T. E. "Relationships Between Sulfate-Reducing and Methane Producing Bacteria," *Plant and Soil* 43:125–128 (1975).
109. Lovely, D. R., and M. J. Klug. "Model for the Distribution of Sulfate Reduction and Methanogenesis in Freshwater Sediments," *Geochem. Cosmochim. Acta* 50:11–18 (1986).
110. Tuttle, J. H., P. R. Dugan, C. B. MacMillan, and C. I. Randles. "Microbial

Dissimilatory Sulfur Cycle in Acid Mine Drainage," *J. Bacteriol.* 97:594-602 (1969).
111. Herlihy, A. T., and A. L. Mills. "Sulfate Reduction in Freshwater Sediments Receiving Acid Mine Drainage," *Appl. Environ. Microbiol.* 49:179-186 (1985).
112. Spratt, H. G., Jr., M. D. Morgan, and R. E. Good. "Sulfate Reduction in Peat from a New Jersey Pinelands Cedar Swamp," *Appl. Environ. Microbiol.* 53:1406-1411 (1987).
113. Giblin, A. E. "Comparison of the Processing of Elements by Ecosystems II: Metals," in *Ecological Considerations in Wetlands Treatment of Muncipal Wastewaters*, P. J. Godfrey, E. R. Kaynor, S. Pelczarski, and J. Benforado, Eds. (New York: Van Nostrand Reinhold Company, 1985), pp. 158-179.
114. Fennessy, S. and W. J. Mitsch. "Design and Use of Wetlands for Renovation of Drainage from Coal Mines," in *Ecological Engineering: An Introduction to Ecotechnology*, S. E. Jorgensen and W. J. Mitsch, Eds. (in press).
115. Sencindiver, J. C., and D. K. Bhumbla. "Effects of Cattails (Typha) on Metal Removal from Mine Drainage," Bureau of Mines Information Circular 9183, U.S. Government Printing Office (1988), pp. 359-366.
116. Spratt, A. K., and R. K. Wieder. "Growth Responses and Iron Uptake in *Sphagnum* Plants and Their Relation to Acid Mine Drainage Treatment," Bureau of Mines Information Circular 9183, U.S. Government Printing Office (1988), pp. 279-285.
117. Wieder, R. K., and G. E. Lang. "Modification of Acid Mine Drainage in a Freshwater Wetland," in *Symposium on Wetlands of the Unglaciated Appalachian Region* (Morgantown, WV: West Virginia University, 1982), pp. 43-53.
118. Wieder, R. K., and G. E. Lang. "Influence of Wetlands and Coal Mining on Stream Water Chemistry," *Water Air Soil Poll.* 23:381-396 (1984).
119. Schwertmann, U. "Inhibitory Effect of Soil Organic Matter on the Crystallization of Amorphous Ferric Hydroxide," *Nature* 212:645-646 (1966).
120. Kodama, H., and M. Schnitzer. "Effect of Fulvic Acid on the Crystallization of Fe(III)-Oxides," *Geoderma* 19:279-291 (1977).
121. Campbell, A. S., and U. Schwertmann. "Iron Oxide Mineralogy of Placic Horizons," *J. Soil Sci.* 35:569-582 (1984).
122. Brodie, G. A., D. A. Hammer, and D. A. Tomljanovich. "Treatment of Acid Drainage from Coal Facilities with Man-Made Wetlands," in *Aquatic Plants for Water Treatment and Resource Recovery,* K. R. Reddy and W. H. Smith, Eds. (Orlando, FL: Magnolia Publishing Inc., 1987), pp. 819-854.
123. Demko, T. M., and B. G. Pesavento. "A Staged Wetland Treatment System for Mine Water with Low pH and High Metal Concentrations," U.S. Bureau of Mines Information Circular 9183 (1988), p. 401. (Abstract).
124. Hedin, R. S., D. M. Hyman, and R. W. Hammack. "Implications of Sulfate-Reduction and Pyrite Formation Processes for Water Quality in a Constructed Wetland: Preliminary Observations," U.S. Bureau of Mines Information Circular 9183 (1988), pp. 382-388.
125. Mitsch, W. J., and J. G. Gosselink. *Wetlands* (New York: Van Nostrand Reinhold Company, 1986).
126. Brady, N. C. *The Nature and Properties of Soil,* 8th ed. (New York: Macmillan Publishing Co., Inc., 1974).
127. "Protection of Public Water Supplies from Ground-Water Contamination," Cen-

ter for Environmental Research, U.S. EPA Seminar Publication-625/4-85/016 (1985).

128. Klopatek, J. M. "Production of Emergent Macrophytes and Their Role in Mineral Cycling Within a Freshwater Marsh," M.S. thesis, University of Wisconsin, Milwaukee, WI (1974).
129. Gorham, E. "Some Chemical Aspects of Wetland Ecology," Tech. Mem. No. 90, 12th Ann. Muskeg. Res. Conf., Calgary, Alberta, Canada (1967), pp. 20–38.
130. Good, B. J., and W. H. Patrick, Jr. "Root-Water-Sediment Interface Processes," in *Aquatic Plants for Water Treatment and Resource Recovery,* K. R. Reddy and W. H. Smith, Eds. (Orlando, FL: Magnolia Publishing Inc., 1987), pp. 359–371.
131. Knight, R. L., B. R. Winchester, and J. C. Higman. "Carolina Bays—Feasibility for Effluent Advanced Treatment and Disposal," *Wetlands* 4:177–204 (1984).

CHAPTER 5

Wetland Vegetation

G. R. Guntenspergen, F. Stearns, and J. A. Kadlec

INTRODUCTION

Natural wetlands have been used to treat wastewater with varying efficiency.[1] Concern over environmental impacts on natural systems[2,3] and the need for an empirically based wetlands wastewater treatment technology led to emphasis on constructed wetlands. These systems treat a variety of wastewater types; however, any one artificial wetland type may be unable to treat all contaminants.

Wetlands have individual and group characteristics related to plant species present and their adaptations to specific hydrologic, nutrient, and substrate conditions. Because of this, a variety of plant species are used in constructed systems (Table 1). However, few sources describe life history characteristics of wetland species that would facilitate selecting appropriate species for use in this technology. We are only beginning to understand wetland plant adaptations to their environment and more important, their effects on the environment.

In this chapter, we (1) discuss major categories of wetland vegetation and morphological and physiological adaptations to environmental gradients and (2) examine the abilities of plants to affect their environment and transform wastewaters.

CLASSIFICATION

Natural wetlands are populated by different plant types adapted for growth in water or saturated soil. Organization into clear-cut groups is difficult because of ambiguous definitions and the unwieldliness and complexity of the classification schemes. Consequently, we have many terms referring to plants growing along the gradient from terrestrial to aquatic habitats: *hydrophyte, aquatic macrophyte, vascular hydrophyte, aquatic plant,* and *vascular aquatic plant.*

One difficulty derives from previous distinctions between aquatic and wet-

Table 1. Plant Species Tested for Use in Constructed Wetlands for Wastewater Treatment

EMERGENT		FLOATING
Scirpus robustus	*Glyceria maxima*	*Lagorosiphon major*
Scirpus lacustris	*Eleocharis dulcis*	*Salvinia rotundifolia*
Schoenoplectus lacustris	*Eleocharis sphacelata*	*Spirodela polyrhiza*
Phragmites australis	*Typha orientalis*	*Pistia stratiotes*
Phalaris arundinacea	*Zantedeschia aethiopica*	*Lemna minor*
Typha domingensis	*Colocasia esculenta*	*Eichhornia crassipes*
Typha latifolia	**SUBMERGED**	*Wolffia arrhiza*
Canna flaccida	*Egeria densa*	*Azolla caroliniana*
Iris pseudacorus	*Ceratophyllum demersum*	*Hydrocotyle umbellata*
Scirpus validus	*Elodea nuttallii*	*Lemna gibba*
Scirpus pungens	*Myriophyllum aquaticum*	*Lemna* spp.

land plants. Penfound[4] recognized two plant groups adapted to water saturated habitats: *wetland* species found in saturated soils and *aquatic* species present where soils are covered with water. This distinction reflects the diverse evolutionary origin of the different groups and their adaptations to these transitional habitats. In reality we commonly find both aquatic species and wetland species growing together in these habitats, and in this chapter we use the terms interchangeably.

There is general agreement about the origins of aquatic plants although relationships of many of the families are poorly understood.[5] They are clearly derived from terrestrial ancestors. Most possess such terrestrial features as nonmotile sperm and emergent flowers. In some groups, various terrestrial structures have degenerated (loss of secondary thickening), become simplified (leaf structure), or lost their function (stomates). Many primitive monocot families are aquatic, suggesting early adaptive radiation of monocots into aquatic habitats.[6] Aquatic members of otherwise terrestrial monocot, dicot, and pteridophyte families are probably of more recent origin and have arisen by convergent evolution from diverse backgrounds.[7]

Classification schemes for aquatic plants are based either on morphological and physiological features, e.g., types of foliage or inflorescence, or phytosociologic criteria based on life and growth form, or growth in relation to the water surface.[7,8] Conversely, wetlands and their characteristic vegetation are classified by flooding regime, substrate material, and life form.[9] The life form concept links wetland classification and the classification of aquatic plants.

Aquatic plants are divided into free floating and rooted forms. The rooted class is then subdivided into emergent, floating, and submerged classes. Hutchinson[8] recognized 26 subdivisions based on ecological growth forms that combined taxa with similar morphological adaptations to specific habitats. However, his classification omits woody vegetation characteristic in many seasonally flooded wetlands, including willows, mangroves, ericaceous shrubs, ashes, gums, cypress, and water oaks. These woody species have all evolved adaptations for persistence in at least seasonally waterlogged soils.

ADAPTATIONS OF PLANTS TO THE AQUATIC ENVIRONMENT

Wetland plants have evolved many structural and physiological adaptations for survival in water-dominated environments. A flooded anaerobic environment reduces oxygen available for respiration; light energy penetration for photosynthesis; and carbon dioxide and mineral nutrient availability for metabolism. Many useful reviews of the various ecological adaptations of aquatic and wetland plants are available.[7,8,10–15] Plants invading water-dominated environments did not evolve new and novel structures for dealing with that environment. Instead, existing structures were modified in evolving similar solutions to common problems among unrelated plant groups. The emergents are the least modified of the nonwoody wetland plants. The trend from emergent to floating to submerged plants reflects a gradual loss of many structural and functional terrestrial characteristics and modification of others. Among emergent plants, the monocots tend to have erect linear leaves that facilitate light penetration into the canopy. Dicots have erect leafy stems that also maximize reception of light for photosynthesis.

Formation of lacunae and/or aerenchyma tissue is a characteristic feature of nonwoody wetland plants. Lacunae function as nonliving support structures, reducing metabolic costs and providing for movement and storage of gases. Woody species rarely have lacunae, but many possess specialized structures to cope with periodic flooding. Lenticels, knees, adventitious roots, prop roots, and butt swellings are examples of the better-known adaptations that allow gas exchange with the atmosphere.

Adaptations of floating and submerged species reflect conservation of form and structure. Submerged plants have little need for elaborate support structure, and these tissues are absent or reduced. Instead, leaves are thin and pliable, with aerenchyma tissue and gas-filled lacunae providing buoyancy. Many of the adaptations found in shaded terrestrial leaves are also found in the submerged species, including a thin cuticle, chloroplasts in the epidermis, and thin leaves. These adaptations reflect the reduction in light intensity resulting from the relatively rapid extinction of light downward through the water column. Similarly, submerged species have maximized leaf surface-to-volume ratios. Their leaves are either long and thin or deeply dissected along the margins, or the leaf blades are separated into leaflets.

Despite diverse evolutionary lineages, floating leaved plants are morphologically similar. Few options exist for leaf development in a spatially restricted two-dimensional environment. They are adapted to existing at the air-water interface, where their distinct dorsoventral structure exposes the upper surface to air and the lower surface to water. Most species have circular leaves with entire margins that reduce tearing and a tough, leathery texture with a hydrophobic upper surface that prevents excessive wetting. Unlike most terrestrial plants, stomata are found on the upper surface. Their long, flexible petioles allow leaf blades to spread out on the water surface and reduce wave stress.

Among free-floating plants, water hyacinth (*Eichhornia crassipes*) leaves form supported rosettes, vegetative parts of duckweeds (*Lemna* spp.) are reduced in structure and float on the water surface, while the salvinids (e.g., *Salvinia*) have sessile leaves. Free-floating rosette plants have root systems that comprise 20–50% of the biomass. However, the salvinids and some lemnids have lost their roots, and nutrients are absorbed through modified leaves.

PRODUCTION AND GROWTH

Plants in wastewater systems have been viewed as storage compartments for nutrients where nutrient uptake is related to plant growth and production. Plants absorb nutrients and excrete or lose small amounts during the growing season but release a large percentage at senescence. Harvesting before senescence may permanently remove nutrients from the system. Alternatively, nutrients tied up in litter and, eventually, sediments represent semipermanent storage.

Productivity varies widely among wetland plants,[10,16–18] reflecting the availability of resources, environmental stress, and adaptations to their environment. Emergent plants, with long erect linear leaves, reduce self-shading while creating a high leaf area index and favorable microclimates for photosynthesis. Many emergents have high light saturation levels (e.g., *Typha* and *Sparganium*) and high temperature optima for photosynthesis. These plants also have high transpiration rates[19,20] with high stomatal water loss that increases with rising temperature and light intensity until photosynthesis stops.

Some species possess the C4 high-efficiency photosynthetic metabolic pathway for CO_2 fixation. Terrestrial plants with this photosynthetic pathway are found in hot, sunny areas where water is scarce. Water is rarely limiting in wetlands, but some emergent freshwater and salt marsh species also possess this pathway. It may have evolved in certain salt marsh species in response to metabolic drought from salinity stress and in some freshwater species in reducing transpiration to decrease the harmful effects of ferrous iron on roots while maintaining optimal rates of carbon fixation.[17,21,22]

The productivity of floating and free-floating aquatic plants is as high or higher than emergents.[10,16] In addition to common favorable environmental conditions, floating leaved species have less respiratory tissue and possess air-filled spaces which may enable them to utilize photorespired CO_2.

Production of submerged aquatic plants is generally low because of low light intensities under water and the low diffusion of CO_2 in water. Maximum photosynthesis usually occurs at light levels of 15–30% of full sunlight[23] although plants can photosynthesize at high temperatures.[24]

Submerged plants can utilize carbon dioxide from alternative sources. In alkaline environments, some species may utilize HCO_3^-,[25,26] although evidence for the general use of this ion is ambiguous.[27] Other adaptations by submerged plants to the restricted availability of carbon dioxide include (1) the CAM metabolic pathway for carbon fixation,[28] (2) refixation of respired CO_2,[29,30] and (3) the use of gaseous sediment carbon collected by roots and moved via the lacunae system to the leaves, where it is used in photosynthesis.

NUTRIENT UPTAKE

Few generalizations can be made about mineral uptake by wetland plants. Emergent plants utilize their roots to obtain sufficient nutrients from the interstitial water. Free-floating species have roots with numerous root hairs and can successfully obtain nutrients from the water column. Submerged plants use nutrients from both the water column and substrate.

It is unclear which source of nutrients is used by submerged aquatic plants.[31,32] The morphology and anatomy of submerged plants seems adapted for nutrient uptake from the water column. And, in fact, materials may pass through the leaves.[31] However, several studies[33-35] have demonstrated preferential uptake from the sediments. In water lilies (*Nuphar luteum*), phosphorus was preferentially taken up by roots, then submersed leaves, and then floating leaves.[36] Leaf uptake is probably related to relative nutrient concentration in the water and substrate.

Mineral element concentrations of aquatic plants differ among species at single sites and between sites for a single species.[37,38] Differences in nutrient composition may be environmentally induced or genetically determined. Cultivars of certain species exist that are adapted to specific nutritional environments. However, nutritional adaptations need not be advantageous in wastewater systems. Some wetland plants adapted to slow growth in nutritionally poor sites[39] cannot respond to increased nutrients by increasing growth.

Aquatic plants can accumulate nutrients against a concentration gradient, serving as indicators of nutrient availability in aquatic environments[8,37,40] and also of heavy metals and rare earth elements.[40-42] High K/Ca ratios occur in various wetland grasses and graminoid plants (including *Typha angustifolia*, *Sparganium erectum*, and *Glyceria maxima*).[40]

Many wetland plants use ammonia as a nitrogen source and are nitrogen limited (except see Barko[43]). Adequate amounts of ammonia in wastewater systems will alleviate this limitation, enhancing other nutrient uptake functions. Species of *Typha, Azolla,* and *Glyceria* support nitrogen-fixing microbes in their rhizosphere.[13] Without nitrogen limitation, these plants grow rapidly and are able to assimilate larger quantities of phosphorus. Some wetland species are also associated with mycorrhiza[44] implicated in nutrient uptake.

Generally, plants from nutrient-rich habitats accumulate more nutrients than plants typical of nutrient-poor habitats.[39] Plants in fertile habitats have high relative growth rates and increase growth with added nutrients, thereby increasing nutrient uptake. However, trade-offs are associated with life in high nutrient environments. Root absorptive capacities are sensitive to high photosynthetic rates and require high rates of root respiration that can be affected by environmental stress. Rapid growth also results in rapid leaf and root turnover with substantial nutrient loss from senescent tissue. In these plants, nutrient uptake and recovery efficiencies decrease and plants rely more on current uptake than nutrient recycling from belowground to aboveground tissues.[45]

ADAPTATIONS TO ANAEROBIC CONDITIONS

Wetland plants often grow in oxygen-poor substrates. Despite their ability for short-term anaerobic respiration, they grow best when oxygen is available for respiration. Most wetland plants possess an extensive internal lacunal system that may occupy up to 70% of the total plant volume.[10,12] This led Armstrong[12] to theorize that wetland plants can satisfy their oxygen requirements through oxygen transport in the lacunal system. Pressurized flow through the lacunae transports oxygen to the roots in yellow water lily (*Nuphar luteum*)[46,47] and similar systems may be widespread in other wetland plants.[48] In addition, gas flux from roots and rhizomes also carries carbon dioxide that can be fixed photosynthetically in the leaves. Dacey and Klug[49] also measured significant quantities of methane leaving water lilies. Methane generated through anaerobic decomposition in sediments moved into water lily roots and rhizomes and escaped to the atmosphere as part of the pressurized flow.

Crawford[50] emphasized the metabolic adaptations to anoxia. Even in well-aerated plants oxygen concentrations are critically low in root meristematic regions. Lower metabolic rates and anaerobic metabolites acting as electron acceptors are common.[52] However, anaerobic respiration can produce toxic levels of ethanol, although some wetland species can excrete ethanol and others accumulate nontoxic metabolic end products such as malate.

Few woody species survive in permanently flooded soil. Those that do survive have adaptations for circulating air to belowground structures, including gas flow through lenticels in the stem and other specialized structures as well as some metabolic adaptations.[53–55]

Response to waterlogging in sweet gum (*Nyssa sylvatica*) involves both metabolic and morphological adaptations.[56] At the onset of flooding, anaerobic metabolism increases to support root metabolism. Later, new root systems with aerenchyma are produced that support aerobic metabolism. Mendelsohm et al.[57] also demonstrated that a combination of metabolic and morphological adaptations to anaerobic conditions is possible. The presence of *Spartina alterniflora* response zones corresponding to differences in substrate reduction suggested that the degree of substrate reduction induced the type of response to anoxia.

Hook and Scholtens[58] proposed that flood tolerance and the degree to which different adaptations take effect is a function of both site factors and the type of flood water introduced. Soils with low organic matter content have lower oxygen demand than highly organic soils. Stagnant waters have lower oxygen concentrations than moving water. These considerations are important in wastewater treatment system operations.

FACTORS LIMITING PLANT GROWTH

What factors limit wetland plant growth? Water depth influences species distribution, although secondary factors or correlated changes may be mechanistically responsible. Increased water depth correlates with the onset of anaerobic conditions and reduced light availability. For submerged species, suspended sediments influence the quantity and quality of light and substrate composition.

Water flow rate can have profound effects on plant development as well as oxygen and nutrient availability and wetland substrates.[59,60] Variation in leaf form of wetland species with depth and velocity[7,61] is probably an adaptation to mechanical stress. Keough[62] demonstrated experimentally that the allocation of structural material in an emergent plant (*Scirpus validus*) varied in relation to water depth and velocity. Growth declined with increased depth and more biomass was allocated to roots as velocity increased in shallow water.

Nutrient availability is related to hydrology through renewal of nutrient-depleted waters, improved substrate aeration, and the water source.[63,64] Cypress tree (*Taxodium distichum*) production has been correlated to water flow, which increases nutrient renewal or aerates the substrate.[65] Nutrient depletion has also been associated with the growth of plants in dense submerged macrophyte beds.[66,67] Water movement can eliminate the boundary layer surrounding submerged plants and replenish nutrients needed for growth and photosynthesis.[68,69]

Increased flow rates are also implicated in the amelioration of the effect of toxic substances in the substrate.[54,70,71] Increased water flow raised oxygen levels and lowered the concentration of certain toxic metal forms in the peat substrate of several Canadian wetlands.[71]

Substrate effects on growth are tied to nutrient availability, but substrate composition may have other consequences. Sediment texture can affect the rooting depth of plants.[60] Highly organic soils can become quite anaerobic, and metals may shift to soluble toxic forms (e.g., iron and manganese).

Air and water temperature affects biochemical reactions and inhibits growth if thermal tolerances are exceeded. Many subtropical species, such as water hyacinth (*Eichhornia crassipes),* cannot tolerate low temperatures. Optimal growth of temperate submerged plants occurs at 28–32°C.[72] In general, higher temperatures within thermal tolerances promote increased production. However, warm water discharge to wetlands throughout the winter can lead to changes in species composition as perennial species deplete their carbohydrate stores to maintain higher respiration rates and die.[73]

Davis and Brinson[74] ranked submerged vegetation tolerance to ecosystem alteration. Species normally dominant in disturbed sites included *Ceratophy-*

llum demersum, *Najas guadalupensis*, *Potamogeton perfoliatus*, *Vallisneria americana*, and adventive species such as *Myriophyllum spicatum* and *Potamogeton crispus*.

Wetzel and Hough[75] portrayed hypothetical changes in primary productivity in aquatic plant communities along a gradient of increasing fertility. Submerged macrophytes do well at intermediate levels but decrease in abundance because of shading by algae and other epiphytes. Floating leaved plant productivity increases until they become crowded and occupy all available surface space. Emergent macrophyte productivity increases but occurrence in highly eutrophic waters may be limited by organic matter accumulation.

VEGETATION INTERACTIONS WITH WASTEWATER

Adaptations of wetland vegetation to water-dominated environments are the basis for their use in wastewater treatment systems. We use these species because they help transform wastewater constituents in such a way that various state and federal regulations for disposal or reuse are met. Our understanding of environmental effects on macrophytes and the effect wetland plants have on their environment are the keys to determining which types of vegetation to use in treatment systems.

Emergent wetlands have been proposed as sites for wastewater treatment because they appear to assimilate inorganic and organic constituents of wastewater. Plants with high growth rates and large standing crops can temporarily store various mineral nutrients, but long-term nutrient removal may be limited.

Emergent plants growing on organic soils incorporate only a small percentage of added nutrients into biomass.[76-79] However, growth of emergent vegetation on gravel substrates has led to significant reductions in substrate mineral nutrient concentrations.[80-83] Water hyacinth and other free floating wetland plants also sequester significant amounts of nutrients and heavy metals.[84-87]

Oxygen transported to belowground tissues of wetland plants can leak out of roots and oxidize the surrounding substrate.[12,88] Substrate oxidation supports aerobic microbial populations in the rhizosphere that modify nutrients, metallic ions, and trace organics.[89] Aerobic microbial metabolism also detoxifies substances potentially hazardous to plants. Metals such as iron and manganese are oxidized and immobilized.[88] Altered pH and oxidation-reduction potential affect trace and toxic metal solubility and uptake in salt and brackish marshes.[90,91] Increased oxidation of the substrate may result in decreased sulfide concentrations increasing metal solubility.[92]

Oxygen changes can also occur in the water column.[93-95] Oxygen deficiencies occur under mats of floating leaved wetland plants such as water hyacinth.[96] In general, floating leaved macrophytes do not oxygenate the water as well as submersed plants.[97]

Wetland vegetation can influence water movement.[98] Weiler[99] modeled significant flow reduction in beds of the submerged species *Myriophyllum spicatum*. Changes in plant density and life form can affect the ability of macrophytes to retard water flow, causing significant reductions in suspended solids.[100] Emergents can substantially lower the water level because of their high transpiration rates.[10]

SUMMARY AND RECOMMENDATIONS

Only a few taxa of wetland plants are used in wastewater treatment studies (Table 1). Sculthorpe[7] lists over 1000 species found entirely in aquatic families. If we include aquatic species in otherwise terrestrial families and woody species found in forested wetlands, less than 1% of the available taxa have actually been tried.

Many early wastewater treatment designs used emergent wetlands (both woody and herbaceous) for wastewater treatment (partial lists in Nichols[1] Guntenspergen and Stearns,[2] and Heliotis[101]). Analysis of these systems suggested that the vegetation acted as a temporary storage pool, with most pollutant transformations and sequestering processes occurring in the substrate.[1] Emergent plants are often grown in gravel beds to stimulate uptake and create suitable conditions for the oxidation of the substrate, thereby improving the ability of the system to treat wastewater.

Emergent and floating leaved species have been preferentially used in pilot studies of constructed wetlands. Floating leaved species have been used because of their high growth rates, large standing crops, and ability to strip nutrients directly from the water column.[84-87,102,103] Their roots provide sites for filtration and adsorption of suspended solids and the growth of microbial communities that sequester nutrients from the water column.

Potentially useful emergent species include many members of the cattail, reed, rush, sedge, and grass families. *Phalaris*, *Spartina*, *Carex*, and *Juncus* all have potentially high uptake and production rates. They are widespread, able to tolerate a range of environmental conditions, and can alter their environment in ways suitable for wastewater treatment. Rhizosphere processes are largely unknown but presumably have positive impacts (e.g., through oxygenation) and should demand further investigation.

Submerged aquatic plants do not appear to have attributes that would be useful in wastewater treatment. They have low production rates and many species are intolerant of eutrophic conditions and/or have detrimental interactions with algae in the water column. However, some species do possess large leaf surface areas, oxygenate the rhizosphere, and are adapted to eutrophic disturbed systems. Alone, submergents may be unsuitable, but they may have a role to play in conjunction with other species in a wastewater treatment system. However, adventive or exotic species that are known nuisances (e.g., *Myriophyllum spicatum*) should be avoided.

Our knowledge of wetland species autecology is incomplete with the available information scattered throughout the literature. Stephenson et al.[104] reviewed the uptake of specific substances by emergent, floating, and submerged plant species. Compilations of biological indicators of polluted waters or surveys of plants associated with heavy metals may also assist in selecting species. However, in addition to tolerance to wastewater, we must consider how wetland plants affect their environment beyond their influence on wastewater alteration. Although many aquatic plants have potential for affecting wastewater quality, certain species may be inappropriate: (1) nuisance plants well-adapted for use in wastewater treatment systems that may escape and cause serious problems in natural wetlands and (2) plants that produce undesirable environmental change.

Universal criteria to determine wetland species suitability for wastewater treatment are not possible because different facilities have different objectives and standards. Municipal wastewater treatment requires modification of organics and nutrients; storm water runoff carries heavy metals and refractory organic substances; and mine drainage has metals and low pH. One species or set of species will not be applicable for all cases. However, only a few species are used now, so it is reasonable to suspect that many other species may be useful. Cooperative screening studies on the ecology and physiology of candidate species should be implemented.

ACKNOWLEDGMENTS

We wish to thank H. H. Allen and D. Hammer for reviewing an earlier version of this chapter and J. R. Keough for helpful discussions. GRG acknowledges support from the National Science Foundation (BSR-8604556) during the preparation of this manuscript.

REFERENCES

1. Nichols, D. S. "Capacity of Natural Wetlands to Remove Nutrients from Wastewater," *J. Water Pollut. Control Fed.* 55:495–505 (1983).
2. Guntenspergen, G. R., and F. Stearns. "Ecological Limitations on Wetland Use for Wastewater Treatment," in *Wetland Values and Management,* B. Richardson, Ed. (St. Paul, MN: Water Planning Board, 1981), pp. 273–284.
3. Guntenspergen, G. R., and F. Stearns. "Ecological Perspectives on Wetland Systems," in *Ecological Considerations in Wetlands Treatment of Municipal Wastewaters,* P. J. Godfrey, E. R. Kaynor, S. Pelczarski, and J. Benforado, Eds. (New York: Van Nostrand Reinhold Company, 1985), pp. 69–97.
4. Penfound, W. T. "An Outline for Ecological Life Histories of Herbaceous Vascular Hydrophytes," *Ecology* 33:123–128 (1952).
5. Les, D. H. "The Origin and Affinities of the Ceratophyllaceae," *Taxon* 37:226–261 (1988).
6. Takhtajan, A. *Flowering Plants: Origin and Dispersal* (Edinburgh: Oliver and Boyd Ltd., 1969), p. 310.

7. Sculthorpe, C. D. *The Biology of Aquatic Vascular Plants* (London: Edward Arnold Ltd., 1969), p. 610.
8. Hutchinson, G. E. *Limnological Botany* (New York: Academic Press, Inc., 1975), p. 660.
9. Cowardin, L. M., V. Carter, F. C. Golet, and E. T. LaRoe. "Classification of Wetlands and Deepwater Habitats of the United States," FWS/OBS-79/31, Biological Services Program, Fish and Wildlife Service (1979).
10. Wetzel, R. G. *Limnology* (Philadelphia, PA: Saunders College Publishing, 1985), p. 762.
11. Riemer, D. N. *Introduction to Freshwater Vegetation* (Westport, CT: AVI Publishing Co., Inc., 1984), p. 207.
12. Armstrong, W. "Aeration in Higher Plants," *Adv. Bot. Res.* 7:226–332 (1979).
13. Bristow, J. M. "The Structure and Function of Roots in Aquatic Vascular Plants," in *Development and Function of Roots*, J. G. Torrey and D. T. Clarkson, Eds. (New York: Academic Press, Inc., 1975), pp. 221–236.
14. Hook, D. D., and R. M. M. Crawford. *Plant Life in Anaerobic Environments* (Ann Arbor, MI: Ann Arbor Science Publishers, Inc., 1978), p. 564.
15. Spence, D. H. N. "The Zonation of Plants in Freshwater Lakes," *Adv. Ecol. Res.* 12:37–126 (1982).
16. Westlake, D. F. "Comparisons of Plant Productivity," *Biol. Rev.* 38:385–425 (1963).
17. Garrard, L. A., and T. K. Van. "General Characteristics of Freshwater Vascular Plants," in *CRC Handbook of Biosolar Resources,* A. Mitsui and C. C. Black, Eds. (Boca Raton, FL: CRC Press, Inc., 1982), pp. 75–85.
18. Richardson, C. J. "Primary Productivity Values in Freshwater Wetlands," in *Wetland Functions and Values: The State of Our Understanding,* P. E. Greeson, J. R. Clark, and J. E. Clark, Eds. (Minneapolis, MN: American Water Resources Association, 1978), pp. 131–145.
19. Bernatowicz, S., S. Leszcznski, and S. Tyczynska. "The Influence of Transpiration by Emergent Plants on the Water Balance of Lakes," *Aquat. Bot.* 2:275–288 (1976).
20. Krolikowska, J. "The Transpiration of Helophytes," *Ekol. Polska.* 26:193–212 (1978).
21. Jones, M. B., and T. R. Milburn. "Photosynthesis in Papyrus (*Cyperus papyrus*)," *Photosynthetica* 12:197–199 (1978).
22. Jones, M. B., and F. M. Muthuri. "The Diurnal Course of Plant Water Potential, Stomatal Conductance, and Transpiration in Papyrus (*Cyperus papyrus*) Canopy," *Oecologia* 63:252–255 (1984).
23. Hough, R. A. "Photosynthesis, Respiration, and Organic Carbon Release in *Elodea canadensis* Michx.," *Aquat. Bot.* 7:1–11 (1979).
24. Titus, J. E., and M. S. Adams. "Co-existence and the Comparative Light Relations of the Submersed Macrophytes *Myriophyllum spicatum* L. and *Vallisneria americana* Michx.," *Oecologia* 40:273–286 (1979).
25. Lucas, W. J., M. T. Tyrie, and A. Petrov. "Characterization of Photosynthetic ^{14}C Assimilation by *Potamogeton lucens.*" *J. Exp. Bot.* 29:1409–1421 (1978).
26. Beer, S., and R. G. Wetzel. "Photosynthetic Carbon Metabolism in the Submerged Aquatic Angiosperm *Scirpus subterminalis.*" *Plant Sci. Lett.* 21:199–207 (1981).

27. Lucas, W. J. "Photosynthetic Assimilation of Exogenous HCO_3^- by Aquatic Plants," *Ann. Rev. Plant Physiol.* 34:71–104 (1983).
28. Boston, H. L., and M. S. Adams. "Evidence of Crassulacean Acid Metabolism in Two North American Isoetids," *Aquat. Bot.* 15:381–386 (1983).
29. Sondergaard, M., and K. Sand-Jensen. "Carbon Uptake by Leaves and Roots of *Littorella uniflora* (L.) Aschers.," *Aquat. Bot.* 6:1–12 (1979).
30. Sondergaard, M., and R. G. Wetzel. "Photorespiration and Internal Recycling of CO_2 in the Submerged Angiosperm *Scirpus subterminalis* Torr.," *Can. J. Bot.* 58:591–598 (1980).
31. Denny, P. "Solute Movement in Submerged Angiosperms," *Biol. Rev.* 55:65–92 (1980).
32. Denny, P. "Sites of Nutrient Absorption in Aquatic Macrophytes," *J. Ecol.* 60:819–829 (1972).
33. Carignan, R. "An Empirical Model to Estimate the Relative Importance of Roots in Phosphorus Uptake by Aquatic Macrophytes," *Can. J. Fish. Aquat. Sci.* 39:243–247 (1982).
34. Carignan, R., and J. Kalff. "Phosphorus Sources for Aquatic Weeds: Water or Sediments?" *Science* 207:987–989 (1980).
35. Barko, J. W., and R. M. Smart. "Mobilization of Sediment Phosphorus by Submersed Freshwater Macrophytes," *Freshwater Biol.* 10:229–238 (1980).
36. Twilley, R. R., M. M. Brinson, and G. J. Davis. "Phosphorus Absorption, Translocation, and Secretion in *Nuphar luteum,*" *Limnol. Oceanog.* 22:1022–1032 (1977).
37. Boyd, C. E. "Chemical Composition of Wetland Plants," in *Freshwater Wetlands: Ecological Processes and Management Potential,* R. E. Good, D. F. Whigham, and R. L. Simpson, Eds. (New York: Academic Press, Inc., 1978), pp. 155–168.
38. Guntenspergen, G. R. "The Influence of Nutrients on the Organization of Wetland Plant Communities," PhD thesis, University of Wisconsin-Milwaukee (1984).
39. Chapin, F. S. "The Mineral Nutrition of Wild Plants," *Ann. Rev. Ecol. Systematics* 11:233–260 (1980).
40. Dykyjova, D. "Selective Uptake of Mineral Ions and Their Concentration Factors in Aquatic Higher Plants," *Folia Geobot. Phytotax. Praha.* 14:267–325 (1979).
41. Cowgill, U. M. "Biogeochemistry of Rare-Earth Elements in Aquatic Macrophytes of Linsley Pond, North Branfort, Conn.," *Geochim. Cosmochim. Acta* 37:2329–2345 (1973).
42. Adams, F. S., H. Cole Jr., and L. B. Massie. "Element Constitution of Selected Aquatic Plants from Pennsylvania: Submersed and Floating Leaved Species and Rooted Emergent Species," *Environ. Pollut. (London)* 5:117–147 (1973).
43. Barko, J. W. "The Growth of *Myriophyllum spicatum* L. in Relation to Selected Characteristics of Sediment and Solution," *Aquat. Bot.* 15:91–103 (1983).
44. Sondergaard, M., and S. Laegaard. "Vesicular-Arbuscular Mycorrhiza in Some Aquatic Vascular Plants," *Nature* 268:232–233 (1977).
45. Shaver, G., and J. M. Mellilo. "Nutrient Budgets of Marsh Plants: Efficiency Concepts and Relation to Availability," *Ecology* 65:1491–1510 (1985).
46. Dacey, J. W. H. "Internal Winds in Water Lilies: An Adaptation to Life in Anaerobic Sediments," *Science* 210:1017–1019 (1980).
47. Dacey, J. W. H. "Pressurized Ventilation in Yellow Water Lily," *Ecology* 62:1137–1147 (1981).

48. Grosse, W., and J. Mevi-Schutz. "A Beneficial Gas Transport System in *Nymphoides peltata*," *Am. J. Bot.* 74:947–952 (1987).
49. Dacey, J. W. H., and M. J. Klug. "Methane Efflux from Lake Sediments Through Water Lilies," *Science* 203:1253–1255 (1979).
50. Crawford, R. M. M. "Metabolic Adaptations to Anoxia," in *Plant Life in Anaerobic Environments,* D. D. Hook and R. M. M. Crawford, Eds. (Ann Arbor, MI: Ann Arbor Science Publishers, Inc., 1978), pp. 119–136.
51. Crawford, R. M. M., and M. Baines. "Tolerance of Anoxia and Ethanol Metabolism in Tree Roots," *New Phytologist* 79:519–526 (1977).
52. Crawford, R. M. M. "Root Survival in Flooded Soils," in *Mires: Swamp, Bog, Fen, and Moor,* A. J. P. Gore, Ed. (Amsterdam: Elsevier Science Publishing Co., Inc., 1983), pp. 257–283.
53. Hook, D. D., C. L. Brown, and P. P. Kormanitz. "Lenticels and Water Root Development of Swamp Tupelo Under Various Flooding Conditions," *Botanical Gazette* 131:217–224 (1970).
54. Armstrong, W. "Root Aeration in the Wetland Condition," in *Plant Life in Anaerobic Environments,* D. D. Hook and R. M. M. Crawford, Eds. (Ann Arbor, MI: Ann Arbor Science Publishers, Inc., 1978), pp. 269–298.
55. Scholander, P. F., L. van Dam, and S. I. Scholander. "Gas Exchange in the Roots of Mangroves," *Am. J. Bot.* 42:92–98 (1955).
56. Keeley, J. "Population Differentiation Along a Flood Frequency Gradient: Physiological Adaptations to Flooding in *Nyssa svlvatica,*" *Ecol. Monog.* 49:89–108 (1979).
57. Mendelssohn, I. A., K. L. McKee, and W. H. Patrick. "Oxygen Deficiency in *Spartina alterniflora* Roots: Metabolic Adaptation to Anoxia," *Science* 214:439–441 (1981).
58. Hook, D. D., and J. R. Scholtens. "Adaptations and Flood Tolerance of Tree Species," in *Plant Life in Anaerobic Environments,* D. D. Hook and R. M. M. Crawford, Eds. (Ann Arbor, MI: Ann Arbor Science Publishers, Inc., 1978), pp. 299–332.
59. Dai, T. "Studies on the Ecological Importance of Water Flow in Wetlands," PhD thesis, University of Toronto (1971).
60. Haslam, S. M. *River Plants* (Cambridge: Cambridge University Press, 1978), p. 396.
61. Wallace, B., and A. M. Srb. *Adaptation* (Englewood Cliffs, NJ: Prentice-Hall, Inc., 1964).
62. Keough, J. R. "Response of *Scirpus validus* to the Physical Environment and Consideration of Its Role in a Great Lake Estuarine System," PhD thesis, University of Wisconsin-Milwaukee (1987).
63. Gosselink, J. G., and R. E. Turner. "The Role of Hydrology in Freshwater Wetlands," in *Freshwater Wetlands: Ecological Processes and Management Potential,* R. E. Good, D. F. Whigham, and R. L. Simpson, Eds. (New York: Academic Press, Inc., 1978), pp. 63–78.
64. Verry, E. S., and D. H. Boelter. "Peatland Hydrology," in *Wetland Functions and Values: The State of Our Understanding,* P. E. Greeson, J. R. Clark, and J. E. Clark, Eds. (Minneapolis, MN: American Water Resources Association, 1977), pp. 389–402.
65. Brown, S. L. "A Comparison of the Structure, Primary Productivity, and the

Transpiration of Cypress Ecosystems in Florida," *Ecol. Monog.* 51:403–427 (1981).
66. Van, T. K., W. T. Haller, and G. A. Bowes. "Comparison of the Photosynthetic Characteristics of Three Submersed Aquatic Plants," *Plant Physiol.* 58:761–768 (1976).
67. Bowes, G. A., S. Holaday, and W. T. Haller. "Seasonal Variation in the Biomass, Tuber Density, and Photosynthetic Metabolism of *Hydrilla* in Three Florida Lakes," *J. Aquat. Plant Management* 17:61–65 (1979).
68. Westlake, D. F. "Some Effects of Low Velocity Currents on the Metabolism of Aquatic Macrophytes," *J. Exp. Bot.* 18:187–285 (1967).
69. Madsen, T. V., and M. Sondergaard. "The Effects of Current Velocity on the Photosynthesis of *Callitriche stagnalis,*" *Aquat. Bot.* 15:187–193 (1983).
70. Wiegert, R. G., A. G. Chalmers, and P. F. Randerson. "Productivity Gradients in Salt Marshes: The Response of *Spartina alterniflora* to Experimentally Manipulated Soil Water Movement," *Oikos* 41:1–6 (1983).
71. Sparling, J. H. "Studies on the Relationship Between Water Movement and Water Chemistry in Mires," *Can. J. Bot.* 44:747–758 (1966).
72. Barko, J. W., D. G. Hardin, and M. S. Matthews. "Growth and Morphology of Submersed Macrophytes in Relation to Light and Temperature," *Can. J. Bot.* 60:877–887 (1982).
73. Bedford, B. "Alterations in the Growth and Phenology of Wetland Plants as Indicators of Environmental Change," in *Proceedings of the 4th Joint Conference on Sensing of Environmental Pollutants* (American Chemical Society, 1978), pp. 170–174.
74. Davis, G. J., and M. M. Brinson. "Responses of Submersed Vascular Plant Communities to Environmental Change," FWS/OBS79/33, Fish and Wildlife Service (1980).
75. Wetzel, R. G., and R. A. Hough. "Productivity and Role of Aquatic Macrophytes in Lakes: An Assessment," *Pol. Arch. Hydrobiol.* 20:9–19 (1973).
76. Chalmers, A. G. "The Effects of Fertilization on Nitrogen Distribution in a *Spartina alterniflora* Salt Marsh," *Est. Coastal Mar. Sci.* 8:327–337 (1979).
77. Tyler, G. "On the Effects of Phosphorus and Nitrogen Supplied to Baltic Shore Meadow Vegetation," *Botanisk Notiser* 120:433–447 (1967).
78. Sloey, W. E., F. L. Spangler, and C. W. Fetter, Jr. "Management of Freshwater Wetlands for Nutrient Assimilation," in *Freshwater Wetlands: Ecological Processes and Management Potential,* R. E. Good, D. F. Whigham, and R. L. Simpson, Eds. (New York: Academic Press, Inc., 1978), pp. 321–340.
79. Ewel, K. C., and H. T. Odum. "Cypress Domes: Nature's Tertiary Treatment Filter," in *Utilization of Municipal Sewage Effluent and Sludge on Forest and Disturbed Land,* W. E. Sopper and S. V. Kerr, Eds. (University Park: Pennsylvania State University Press, 1979), pp. 103–114.
80. Laksham, G. "An Ecosystem Approach to the Treatment of Wastewater," *J. Environ. Qual.* 8:353–361 (1979).
81. Bowmer, K. H. "Nutrient Removal from Effluents by an Artificial Wetland: Influence of Rhizosphere Aeration and Preferential Flow Studied Using Bromide and Dye Tracers," *Water Res.* 21:591–599 (1987).
82. Finlayson, C. M., and A. J. Chick. "Testing the Potential of Aquatic Plants to Treat Abattoir Effluents," *Water Res.* 17:415–422 (1983).
83. Gersberg, R. M., B. V. Elkins, and C. R. Goldman. "Use of Artificial Wetlands

to Remove Nitrogen from Wastewater," *J. Water Pollut. Control Fed.* 56:152–156 (1984).

84. Mishra, B. B., D. R. Nanda, and B. N. Misra. "Accumulation of Mercury by *Azolla* and Its Effect on Growth," *Bull. Environ. Contam. Toxicol.* 39:701–707 (1987).
85. Reddy, K. R., and W. F. DeBusk. "Nutrient Removal Potential of Selected Aquatic Macrophytes," *J. Environ. Qual.* 14:459–462 (1985).
86. Oron, G., A. De-Vegt, and D. Porath. "Nitrogen Removal and Conversion of Duckweed Grown on Wastewater," *Water Res.* 22:179–184 (1988).
87. Hauser, J. R. "Use of Water Hyacinth Aquatic Treatment Systems for Ammonia Control and Effluent Polishing," *J. Water Pollut. Control Fed.* 56:219–226 (1984).
88. Armstrong, W. "Waterlogged Soils," in *Environment and Plant Ecology,* J. R. Etherington, Ed. (Chichester: John Wiley & Sons, Inc., 1982), pp. 290–332.
89. Gersberg, R. M., B. V. Elkins, S. R. Lyon, and C. R. Goldman. "Role of Aquatic Plants in Wastewater Treatment by Artificial Wetlands," 20:363–368 (1986).
90. Gambrell, R. P., R. A. Khalid, M. G. Veroo, and W. H. Patrick. "Transformations of Heavy Metals and Plant Nutrients in Dredged Sediments as Affected by Oxidation-Reduction Potential and pH," Contract No. DACW39-74-C-0076, Army Corp of Engineers (1977).
91. Giblin, A. E. "Comparisons of the Processing of Elements by Ecosystems II: Metals," in *Ecological Considerations in Wetlands Treatment of Municipal Wastewaters,* P. J. Godfrey, E. R. Kaynor, S. Pelczarski, and J. Benforado, Eds. (New York: Van Nostrand Reinhold Company, 1985), pp. 158–179.
92. Giblin, A. E., I. Valiela, and J. M. Teal. "The Fate of Metals Introduced into a New England Salt Marsh," *Water Air Soil Poll.* 20:81–98 (1983).
93. Ondok, J. P., J. Pokorny, and J. Kvet. "Model of Diurnal Changes in Oxygen, Carbon Dioxide, and Bicarbonate Concentrations in a Stand of *Elodea canadensis* Michx.," *Aquat. Bot.* 19:293–305 (1984).
94. Carpenter, S. R., and A. Gasith. "Mechanical Cutting of Submersed Macrophytes: Immediate Effects on Littoral Water Chemistry and Metabolism," *Water Res.* 12:55–57 (1978).
95. Schreiner, S. P. "Effect of Water Hyacinth on the Physiochemistry of a South Georgia Pond," *J. Aquat. Plant Management* 18:9–12 (1980).
96. Rai, D. N., and J. D. Munshi. "The Influence of Thick Floating Vegetation (Water Hyacinth: *Eichhornia crassipes)* on the Physiochemical Environment of a Freshwater Wetland," *Hydrobiologia* 62:65–69 (1979).
97. Pokorny, J., and E. Rejmankova. "Oxygen Regime in a Fishpond with Duckweeds (Lemnaceae) and *Ceratophyllum,*" *Aquat. Bot.* 17:125–137 (1983).
98. Watson, D. "Hydraulic Effects of Aquatic Weeds in U.K. Rivers," *Regul. Rivers* 1:211–227 (1987).
99. Weiler, P. R. "Littoral-Pelagic Water Exchange in Lake Wingra, Wisconsin as Determined by a Circulation Model," Report 100, University of Wisconsin, Institute of Environmental Studies (1978).
100. Boto, K. G., and W. H. Patrick Jr. "Role of Wetlands in the Removal of Suspended Sediments," in *Wetland Functions and Values: The State of Our Understanding,* P. E. Greeson, J. R. Clark, and J. E. Clark, Eds. (Minneapolis, MN: American Water Resources Association, 1978), pp. 479–489.
101. Heliotis, F. D. "Wetland Systems for Wastewater Treatment: Operating Mecha-

nisms and Implications for Design," Report 117, University of Wisconsin, Institute of Environmental Studies (1982).

102. Steward, K. K. "Nutrient Removal Potentials of Various Aquatic Plants," *Hyacinth Control J.* 8:34–35 (1970).
103. Boyd, C. E. "Vascular Aquatic Plants for Mineral Nutrient Removal from Polluted Waters," *Econ. Bot.* 24:95–103 (1970).
104. Stephenson, M., G. Turner, P. Pope, J. Colt, A. Knight, and G. Tchobanoglous. "The Use and Potential of Aquatic Species for Wastewater Treatment. Appendix A. The Environmental Requirements of Aquatic Plants," Publication No. 65, California State Water Resources Control Board, Agreement No. 8-131-400-0 (1980).

CHAPTER 6

Wetlands Microbiology: Form, Function, Processes

Ralph J. Portier and Stephen J. Palmer

INTRODUCTION

The number of organic compounds introduced into the environment by humans has increased dramatically in recent years.[1] The environmental fate of xenobiotic (man-made) compounds—pesticides, polycyclic aromatic hydrocarbons, and domestic wastes—is an important issue. Disappearance, persistence, and/or partial transformation and potential hazardous effects are particular concerns. While many compounds are readily biodegradable, others persist in soil and water.

Recent research on the biochemistry and genetics of xenobiotic-degrading microorganisms, the newer biotechnology literature, and older literature on industrial microbiology describe processes in which microbial cultures play an important role.[2] Although some microbes can cause adverse effects, most adverse effects are controllable. Most species are benign, functioning in beneficial ways, and only a few species are pathogenic. Microorganisms have a substantial role in transformation of organic and inorganic substances critical to all life on earth, such as the transforming of free nitrogen molecules in the air for use by plants.

This chapter provides an overview of microbial processes of importance to constructed wetlands for wastewater treatment. Bacterial processes are the primary focus of discussion because more information is available. However, fungal and actinomycetous contributions, equally important, are also discussed. Information will be presented on microbial transformation processes, fate of anthropogenic organics, metals metabolism, and habitat for optimal microbial enzymology in a constructed wetland.

MICROBIAL FORMS

Introduction

Evolution of different life forms produced many large groups that can be clearly categorized as plants or animals, with well-established characteristics for each. Microorganisms developed differently, and classification as plant or animal is difficult—even the criteria for life have to be modified.[3] Most organisms in the major microbial groups are microscopic, lack differentiation of tissue, and live in interrelated groups. Bacteria, for example, include many sizes and shapes of unicellular microorganisms and are present in almost all natural environments, often in large numbers. Like plants, they have rigid cell walls, but, like animals, some are motile and require complex organic nutrients. Bacteria are usually not photosynthetic. Bacteria, molds, yeasts, viruses, and algae are assigned to the vegetable kingdom.

To chemically change an organic material, whether it be a pesticide or cellulose, microorganisms will utilize an enzymatic or biological chemical process to "bioconvert" a targeted substrate. This, of course, is the whole focus of a constructed wetland, namely, to optimize the contact of microbial species with a substrate, with the final objective being bioconversion to carbon dioxide, biomass, and water. Evidence of this result consists of reduced biological oxygen demand (BOD_5) or chemical oxygen demand (COD). Verification of microbial effect on a substrate, such as a pesticide, can further be quantified using gas and liquid chromatography.

Microbial species present in a constructed wetland can be considered an ever-more sophisticated hierarchy of metabolic systems, each bringing a unique enzymatic capability to the bioconversion process. The microorganisms present in a constructed wetland as result of numerous biological and nonbiological or abiotic conditions are the deciding factor in this biotechnological process.

Morphology of Bacteria

Bacteria are in the Protophyta division, class Schizomycetes, which is subdivided into 10 orders. The orders Eubacteriales and Pseudomonadales contain the largest number of species and include most bacteria important to man.

Bacteria are divided by shape into three conventional groups: cocci, bacilli, and spiral forms (Figure 1). *Cocci* are spherical, varying in size from 0.5 to 1.0 μm in diameter. Their arrangement depends on the order of successive cell division. If this is random, the cocci may grow in clusters and are called *staphylococci* (from the Greek word *staphylē*, "bunch of grapes.") When division takes place in the same plane and the daughter cells adhere to each other, chains called *streptococci* are formed. *Bacilli,* by far the most common, are shaped like rods or cylinders. They are about 1.0–10 μm long and 0.3–1.0 μm wide. The end of the rod appears to be rounded or square, and some bacilli

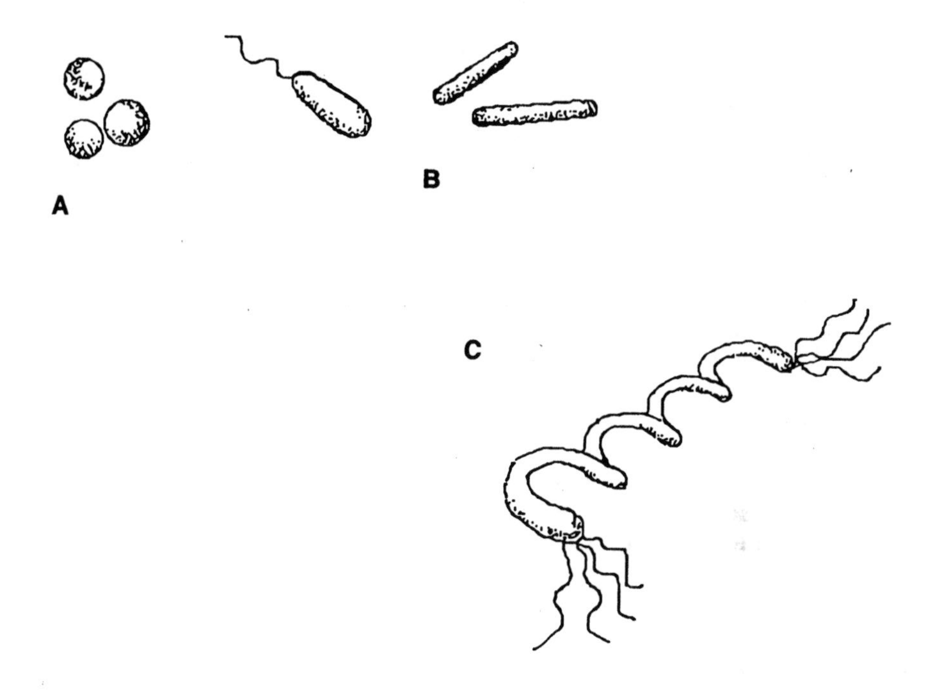

Figure 1. Bacterial morphology: (A) cocci form; (B) bacilli; (C) spiral form. (Modified from Poindexter.[4])

tend to form chains. Bacilli are motile; cocci are usually nonmotile. *Spiral forms* comprise a large variety of cylindrical bacteria. *Vibrios* are curved rods, *spirilla* are relatively rigid spirals, and *spirochetes* are spirals able to flex and wriggle. *Filamentous forms* of bacteria may be several hundred μm long but only about 1–2 μm in diameter. Bacterial shape is determined by heredity, but age and environmental conditions may change their morphology.

Structure of a Bacterial Cell

All kinds of living cells have some form of outer wall or membrane, cytoplasm, and nuclear material. The outer part of the bacterial cell is made up of three structures: slime layer or capsule, cell wall, and cytoplasmic membrane. The jellylike *slime layer* is the outside coating; when it becomes thick and firm, it is called a *capsule.* It is usually a polysaccharide (or a polypeptide) continually produced by metabolic activity of the cell with formation depending on the presence of carbohydrate. The slime layer, not an integral part of the cell, is influenced by the environment.[5] Capsular material confers *type specificity* on the organism; for example, the type of pneumonia depends on the molecular composition of the capsule. The *cell wall* limits the volume occupied by cytoplasm, providing a structural component to support high osmotic pressure caused by concentrations of cytoplasmic materials. The cell wall also has roles in cell division and in regulating the material passing between the internal and external environments.

The *cytoplasmic membrane,* inside the cell wall, is a semipermeable membrane composed mainly of proteins and lipids acting as part of the osmotic barrier between the external and internal environments of the cell. It regulates permeability of substances entering and leaving the cell and contains many oxidation-reduction enzyme systems concerned with energy metabolism.

The *cytoplasm* is the internal environment of the cell. It is a colloidal system containing salts, sugars, proteins, fats, carbohydrates, vitamins, granules, and other materials characteristic of a particular organism. The cytoplasm contains most of the enzymes necessary for metabolic processes of the cell and growth of the organism. Bacteria are *procaryotic* cells, that is, they do not have a true nucleus. The procaryotic cell has none of the specialized structures found in eucaryotic cells—no mitochondria for respiration, no endoplasmic reticulum as an extension of the cell membrane, no lysosomes with hydrolytic enzymes, and no Golgi apparatus to transport metabolic products.[6] The procaryotic nucleus has no membrane and does not undergo mitosis. The nuclear region is a weakly contrasting area that contains thin, fibrillar deoxyribonucleic acid (DNA), the genetic material. Sometimes more than one nuclear region is seen, but each probably contains only a single DNA molecule.

Bacteria are divided into two groups on the basis of a differential staining technique called the *gram stain.* They differ mainly in whether their cell walls retain the stain (gram-positive) or not (gram-negative). Gram-negative bacteria have short, filamentous fibers composed of protein called *pili* or *fimbriae* attached to their walls. Such bacteria tend to stick to each other because pili apparently are used for attachment to surfaces.

MICROBIAL FUNCTIONS

Nutritional Requirements

Microbes are classified in three major groups by types of material used as energy sources: (1) *chemoorganotrophs* use the energy of organic compounds; (2) *photoautotrophs* utilize radiant energy; and (3) *chemolithotrophs* oxidize inorganic molecules. Most bacteria are chemoorganotrophs. As with other life forms, bacteria require water, minerals, vitamins, and sources of carbon and nitrogen but in relatively smaller quantities; tap water will often meet their mineral needs. Necessary mineral ions include such trace elements as molybdenum, manganese, and cobalt. Tables 1 and 2 list the major and minor bioelements, their sources, and some of their functions in metabolism. The chief vitamin requirement for bacteria is B complex. Biological assay methods based on these requirements for specific vitamins or minerals by specific strains of bacteria have been developed.[8]

Table 1. The 10 Major Bioelements, Their Sources, and Some of Their Functions in Microorganisms; Adapted from *Bacterial Metabolism*

Element	Source	Function in Metabolism
C	Organic compounds, CO_2	Main constituents of cell material
O	O_2, H_2O, organic compounds, CO_2	
H	H_2, H_2O, organic compounds	
N	NH_4^+, HO_3^-, N_2, organic compounds	
S	SO_4^{2-}, HS^-, S^0, $S_2O_3^{2-}$	Constituent of cysteine, methionine thiamine pyrophosphate, coenzyme A, biotin, and lipoic acid
P	HPO_4^{2-}	Constituent of nucleic acids, phospholipids and nucleotides
K	K^+	Principal inorganic cation in the cell; cofactor of some enzymes
Mg	Mg^{2+}	Cofactor of many enzymes (e.g., kinases); present in cell walls, membranes, and phosphate esters
Ca	Ca^{2+}	Cofactor of enzymes; present in exoenzymes (amylases, proteases); Ca-dipicolinate is important component of endospores
Fe	Fe^{2+}, Fe^{3+}	Present in cytochromes, ferredoxins, and iron-sulfur proteins; cofactor of enzymes

Source: Adapted from Gottschalk.[7]

Table 2. Minor Bioelements, Their Sources, and Some of Their Functions in Microorganisms

Element	Source	Function in Metabolism
Zn	Zn^{2+}	Present in alcohol dehydrogenase, alkaline phosphatase, aldolase, RNA and DNA polymerase
Mn	Mn^{2+}	Present in bacterial superoxide dismutase; cofactor of some enzymes (PEP carboxykinase, recitrate synthase
Na	Na^+	Required by halophilic bacteria
Cl	Cl^-	
Mo	Mo_4^{2-}	Present in nitrate reductase, nitrogenase, and formate dehydrogenase
Se	SeO_3^{2-}	Present in glycine reductase and formate dehydrogenase
Co	Co^{2+}	Present in coenzyme B_{12}-containing enzymes (glutamate mutase, methylmalonyl-CoA mutase)
Cu	Cu^{2+}	Present in cytochrome oxidase and oxygenases
W	WO_4^{2-}	Present in some formate dehydrogenases
Ni	Ni^{2+}	Present in urease; required for autotrophic growth of hydrogen-oxidizing bacteria

Source: Adapted from Gottschalk.[7]

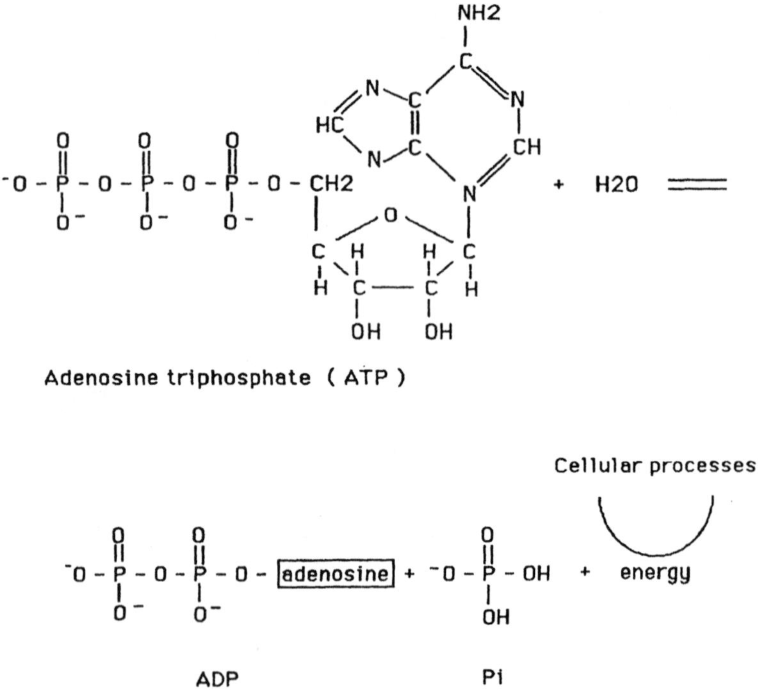

Adenosine triphosphate (ATP) + H2O

Cellular processes

ADP + Pi + energy

Figure 2. Reverse reaction between ATP and ADP. (Modified from Ucko.[9])

The source of carbon in synthetic media is usually glucose, but other carbohydrates can be used in a diagnostic test. Few species form lipases, or enzymes capable of hydrolyzing fats. Energy released by a catalyzed enzyme oxidation of carbon is accumulated in chemical bonds of adenosine-5-triphosphate (ATP) when formed from the addition of inorganic phosphate $H_2PO_4^-$ (P_i) to adenosine-5-diphosphate (ADP) during oxidative phosphorylation, a vital energy process required by all living cells. Cellular processes release energy of biological ATP by changing ATP back to ADP and inorganic phosphate in a reverse reaction (see Figure 2).

Formation of phosphate bonds requires energy. Energy stored in ATP is released when the bond connecting the last phosphate is broken. The principle involved when ATP absorbs energy given off during oxidation and transfers it to the different processes of the cell is called *energy coupling* and is applied to many metabolic reactions. The energy liberated in these metabolic reactions is directed primarily toward biosynthesis of cell materials (see Figure 3).

Chemical changes in the body of living organisms depend on *enzymes*, biological catalysts that increase reaction rate but are not used up in the process. Bacterial enzymes determine whether a bacterium can digest a complex material and use it for food. Enzymes are very large protein molecules

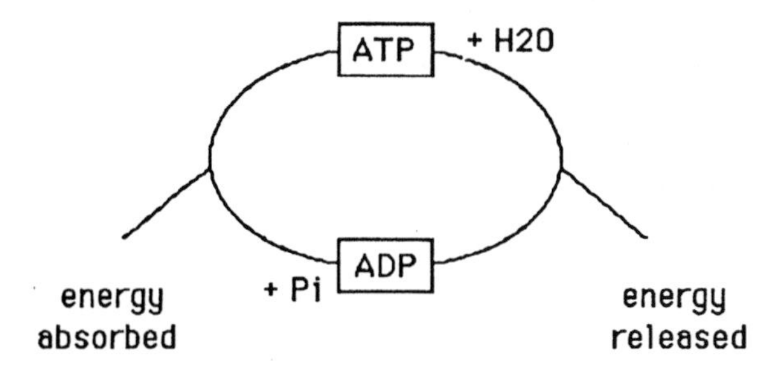

Figure 3. Relation between adenosine triphosphate (ATP) and adenosine diphosphate (ADP). (Modified from Ucko.[9])

that bind to the substrate, reacting chemically. Each enzyme has a special region onto which it binds (called an *active site)* and a specific activity for the substrate. Certain substances (known as *inhibitors)* prevent or slow down the action of enzymes. *Competitive inhibitors* are molecules similar to those of the substrate. They bind reversibly at the active site, stopping the enzyme from catalyzing the reaction.[9]

Molecules involved in a reaction must have energy, called the *activation energy.* Enzymes decrease the activation energy barrier of the reaction, resulting in more product in a shorter period of time. Some enzymes need an extra nonprotein to function, called a *cofactor.* If the cofactor is an organic molecule, it is called a *coenzyme.*

Habitat Requirements Within Constructed Wetlands

The use of O_2 as an electron acceptor is called *respiration,* measured by uptake of oxygen gas by the respiring organism. Many respiratory reactions are fundamental for almost all forms of life, including bacteria. If free oxygen enters the reaction, it is called *aerobic respiration.* Atmospheric oxygen functions as the final hydrogen acceptor in the series of oxidation and reduction reactions liberating energy from food in the metabolic process. When O_2 accepts electrons, it is reduced to H_2O. In constructed wetlands, this respiration requirement is necessary for significant turnover of introduced organic carbon. Since enzymes are proteins existing in living cells, certain environmental conditions, such as temperature, pH, and salt concentration, must be met for the enzyme to be active.

Temperature

One of the most important factors affecting microbial rate growth in constructed wetlands is environmental temperature. At some minimum temperature, growth will not occur because of deactivation of the enzyme-catalyzed system, and at the maximum, heat denaturalization will occur. Between these limits, an optimal temperature for microbial growth results in rapid increases in the activation rates of heat-sensitive cell components: enzymes, ribosomes, DNA, and membranes.[4] Most enzymes have optimal activity between 20°C and 30°C. In general, an increase in temperature produces increased molecular motion, promoting more rapid microbial growth.

Hydrogen Ion Concentration

The hydrogen ion concentration (acidity or alkalinity) of an effluent entering a wetland markedly affects enzyme activity. Some enzymes are active at pH values of 3–4, while others may be active at pH as high as 11 or 12. Most microorganisms prefer a neutral medium with a maximum activity in the range of pH 6–8. This is fortunate in that wetland systems are highly buffered at this neutral pH range.

Salt Concentration

Concentrations of sodium chloride increase bacterial growth by increasing osmotic pressure to an optimal point. However, if salt concentration is too high, high osmotic pressure inhibits bacterial growth.

Reproduction of Bacteria and Population Growth

Bacteria multiply by elongation of the cell, followed by division of the enlarged cell into two cells through *binary fission*. Although bacteria can vary in size, they retain their unicellular structure—the definition of *bacterial growth* is an increase in the number of individuals. Under favorable conditions in a constructed wetland, where generous surface areas exist for attachment and colonization, almost all microbial species reproduce very rapidly, and the population increases through several well-defined steps (Figure 4) in a predictable manner.

A period of adaptation to the new environment is called the *lag phase,* the duration of which varies from an hour to several days depending on the bacteria type; the culture age; the surface colonized, i.e., root rhizosphere, gravel, or sand; and available nutrients in the medium. This period is characterized by a lag only in multiplication, because cells are active metabolically (see Figure 4). Following the lag phase is the period of rapid reproduction or exponential growth, the *log phase.* Individual cells grow linearly with time, while the population of cells grow exponentially, doubling at each period of cell division. Constructed wetlands are designed to optimize microbial cell performance at this level. When rapid growth is halted by nutrient exhaustion, an oxygen deficiency, accumulation of toxic end products or bioaccumulation of metals, growth declines and the number of cells remains constant. The length of the *stationary phase* depends upon favorable conditions and specific microorganisms. Unless environmental conditions improve, i.e., improvement in dissolved oxygen or additional carbon in the form of effluent, the cells will

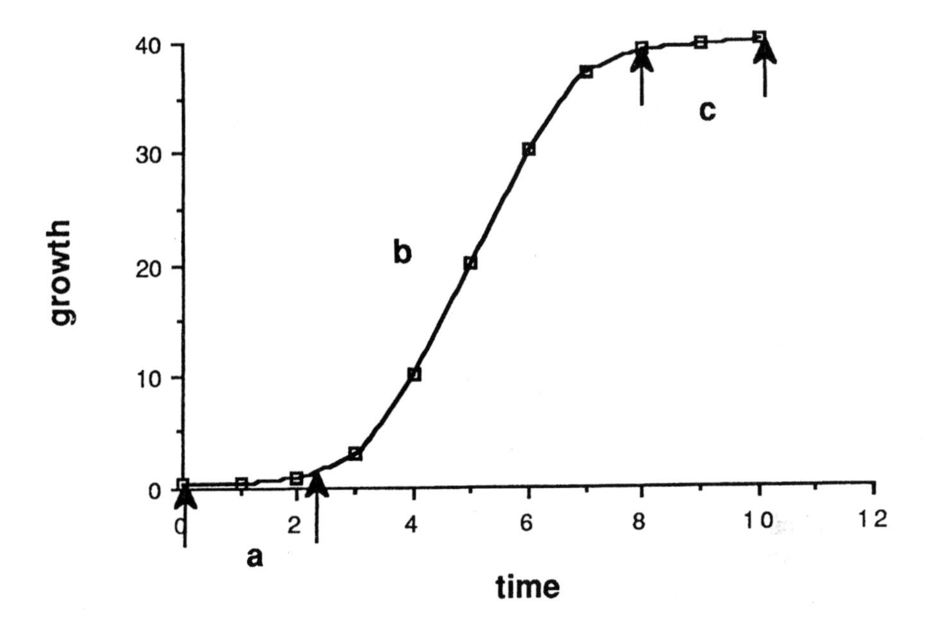

Figure 4. A common form of bacterial growth under favorable conditions: (a) lag phase; (b) log phase; (c) stationary phase.

eventually die. This *death phase* in old cultures often becomes exponential until no live cells remain.

Kinetics of Microbial Growth

When the logarithm of cell numbers is plotted against time, a straight line results. The rate of exponential growth is expressed as *generation time,* or doubling time, which is the time taken for the population to double.[10] Microbial populations can be maintained in the logarithmic phase by continually providing new supplies of energy, nutrients, and other requirements. Generation time depends on the type of organism, concentration of available nutrients, temperature, pH, and oxygen. In general, species multiply rapidly when provided with favorable conditions.[11] Thus, a primary operational concern in constructed wetlands management would be the maintenance and optimization of these factors to achieve acceptable wastewater treatment.

MICROBIAL PROCESSES

Geochemical Influences in Wetland Microbiology

Reaction conditions in a constructed wetland are mediated not only by the organic substrates introduced to the system but also by inorganic chemicals.

Many are indigenous to the constructed wetland habitat, such as iron and manganese, and play a fundamental role as electron acceptors in anaerobic respiration. Others are transitory and have no impact. Still others, such as mercury, can undergo methylation and result in biological accumulation and magnification problems. Thus, metals can be the major culprit in dysfunction of the constructed wetlands bioprocess and, ultimately, federal- or state-mandated closure of the wetland.

Metals Requirements and Metals Toxicity

Very low concentrations of certain metals are required by all microorganisms for normal cell functioning. These include potassium, magnesium, manganese, calcium, iron, cobalt, copper, zinc, and molybdenum. Copper, zinc, and molybdenum are constituents of specialized enzymes, and cobalt is found in vitamin B_{12} and its coenzymes. Magnesium, iron, manganese, calcium, and potassium are also enzyme cofactors.[12]

High levels of heavy metals in the environment are usually toxic to microorganisms. Some microorganisms may even be affected by low concentrations of particularly toxic metals. For the overall cell population, toxicity may be manifested as a drop in cell numbers due to cell death, bacteriostasis, or extension of the lag phase of the cell cycle.[13] Heavy metal toxicity may also cause altered cell morphology.[14]

In general, heavy metal toxicity is determined by degree of attraction to natural metal binding sites on and within the cell. The similarity in chemistry of some heavy metals to other elements required for cellular functioning may result in some being actively accumulated within the cell. The ability of a metal to enter the cell cytoplasm is a significant measure of its potential toxicity.[13] However, metal toxicity is also affected by nutritional and environmental factors. Nutritional state of the organism may alter toxicity, because cells in a nutrient-depleted environment are often more susceptible. Environmental factors influence heavy metal toxicity in several ways.

Microbes in constructed wetlands compete with cellulose and chitin as chelating surfaces for metals absorption, and metal-chelating materials such as clays, other ions, and pH affect the toxicity of heavy metals to microorganisms. Chelating agents affect toxicity by binding the metal. For example, kaolinite and montmorillonite clays, humic and fulvic acids, and proteins reduce heavy metal toxicity by binding the metal.[15] Synthetic chelating agents such as EDTA also reduce heavy metal toxicity in microorganisms.[13] Other cations, particularly those of similar ionic radii, can decrease toxicity due by competing for binding sites.[13,15-17] Low pH reduces metal toxicity[13,15] by ionic competition between hydrogen ions and metal ions.[18] High pH may enhance metal toxicity[19] due to less ionic competition. But for some metals, an increase in pH may *lower* toxicity because precipitation removes metals from solution.[16,17]

That these variables should be remembered in considering location and management of a constructed wetland cannot be emphasized too strongly. Many petroleum API landfarms in coastal Louisiana have been abandoned due to a lack of attention to geochemistry of metals in a wetland soil.

Resistance to Metal Toxicity

Microorganisms exposed to adverse environmental conditions soon produce strains, through genetic modification, capable of surviving a hostile condition. In many cases, the evolved mechanisms are highly specific. In bacteria, metal resistance is often plasmid-linked[14,20,21] and often associated with antibiotic resistance.[20] Resistance to toxic metals is achieved by (1) increasing impermeability of the cell and (2) biochemically transforming the metal. The former process protects the cell from toxic elements in its environment. The latter detoxifies the immediate cell environment by eliminating toxic metals or altering them to nontoxic forms.[14,21] Increased impermeability may be achieved nonspecifically by production of an outer protective layer around the cell, allowing metal binding at a distance from the cell wall and causing little damage.[17] A layer or matrix of extracellular polymer enhances cell tolerance of toxic metals by immobilizing them away from the cell wall, preventing metal from binding to functional groups on or within the cell. It also minimizes phagocytosis by amoebae or phagocytes, protects against bacteriophage and desiccation, and might act as a food reserve.[22] A protective polymer capsule may not have evolved in some bacteria, however.

Transformation of toxic metals into nontoxic forms may be achieved intracellularly but is more commonly done extracellularly. Alternatively, a toxic metal may be changed into a form unassimilable by the microorganism. Toxic metals may be oxidized, reduced, or methylated to produce less toxic compounds. Mercury resistance is often plasmid-linked via an enzyme that transforms mercury and organomercurials into volatile forms lost from the environment. Most heavy metals form insoluble sulfides, and sulfide production by the bacterium *Desulphovibrio desulphuricans*, the fungus *Poria vaillantii,* and some strains of the yeast *Saccharomyces cerevisiae* precipitates metals from solution.[15] Some fungi produce chelating agents that bind metal away from the cell. *Corrollus palustris* and others produce oxalic acid to enhance their copper tolerance.[15] Microorganisms enhance their tolerance of toxic metals through many mechanisms and, in so doing, immobilize, precipitate, and bind contaminants in treatment processes for polluted water.

Accumulation of Heavy Metals

Microorganisms remove heavy metals from solution by two general mechanisms: (1) metabolism-dependent uptake *into* the cell and (2) binding of metal ions to extracellular material (capsular polymer) or the cell wall.[22-24] Some potentially toxic divalent metal ions are micronutrients at low concentration, and microbes have active uptake systems to bind these ions. These divalent cation uptake systems tend to be specific but, in some cases, do transport other metals into the cell. The magnesium uptake system of *Escherichia coli* also accumulates Ni^{2+}, Co^{2+}, and Zn^{2+}. The Mg^{2+} transport system of *S. cerevisiae* takes up Co^{2+}, Mn^{2+}, Zn^{2+}, and Ni^{2+}. Generally, ion uptake systems are specific for ions of a certain ionic radius. Thus, monovalent cation uptake mechanisms exclude divalent metal ions or metal ions of a higher valency, including the toxic heavy metals. However, cesium (Cs) and its radioisotopes and Ti^{1+} have been taken into the cell via the potassium transport system.[24]

The term *biosorption,* describing the nonactive adsorption of heavy metal ions by microorganisms or biological polymers, is defined as "the non-directed, physical-chemical complexation reaction between dissolved metal species and charged cellular components, akin in many respects to ion exchange."[12] These processes occur as interactions between negatively charged ligands and metal ions and as ion-exchange or complex formation. The most likely components of microbial polymers capable of ion exchange are carboxyl groups, organic phosphate groups, and organic sulfate groups. Chelation or complex formation occurs on biopolymers where neutral divalent oxygen, sulfur atoms, or trivalent nitrogen atoms are present. Heavy metals are bound to extracellular polymers such as those of the bacteria *Zooglea ramigera* and *Klebsiella aerogenes.* Accumulation of metals also occurs with the extracellular polysaccharides of the algae *Mesotaenium kramstei* and *Mesotaenium caldariorum.*[25]

Implications of Anthropogenic Organics: Petroleum Aromatic Hydrocarbon Degradation as a Case Study

Introduction

Of the many possible organics introduced to a constructed wetland, the aromatic hydrocarbons pose a most difficult challenge, due to the complexity of the chemical mixture itself, plant toxicity, and limitations in anaerobic degradation. Microbial metabolism of hydrocarbons has been reported for several decades. Investigations as early as 1928 showed soil bacteria capable of decomposing certain aromatic compounds.[26] In 1941 microbial degradation of certain hydrocarbons was documented,[26] and in 1947 marine microorganisms experimentally degraded aromatic hydrocarbons.[27] Polyaromatic hydrocarbons (PAHs) were introduced into mixed cultures of marine bacteria adsorbed to ignited sand. Phenanthrene and anthracene were metabolized more rapidly than naphthalene, benz(a)anthracene, and dibenz(a,h)anthracene.[27] Hydrocarbons are ubiquitous. Because sources are natural as well as anthropogenic, they even occur in pristine areas. Many exhibit toxicity, mutagenicity, and carcinogenicity after metabolic activation, and may therefore be a hazard to the biota and to human health.[28]

Aliphatic hydrocarbons are straight- or branched-chain compounds naturally present in waxes and other constituents of plant tissues and in petroleum or petroleum products. Their transformations are significant in the terrestrial carbon cycle, and decomposition rate is markedly affected by length of the hydrocarbon chain.[29] Aromatic hydrocarbons have the benzene ring as the parent hydrocarbon. Several benzene rings joined together at two or more ring carbons form PAHs. Toxicity is determined by the arrangement and configuration of benzene rings. Hydrogens in aromatic hydrocarbons may or may not be substituted by a Cl (chloro); Br (bromo); I (iodo); NO_2 (nitro); NO (nitroso); and CN (cyano).[30]

Major environmental fate/transport mechanisms include (1) evaporation (volatilization); (2) photochemical oxidation; (3) sedimentation; and (4) microbial degradation. Of these, microbial degradation is most extensively studied and commercialized. It is a major mechanism for compound removal from sediments and terrestrial systems. A constructed wetland follows degradation pathways similar to those of marine sediments. Degradation of aromatic hydrocarbons by bacteria as well as fungi has been documented in numerous publications. Degradation processes are generally inversely proportional to the ring size of the respective PAH molecule. The lower-weight PAHs are rapidly degraded, but those with more than three condensed rings generally do not support microbial growth[31]—hence the effectiveness of creosote as a wood preservative.

Aromatic and Aliphatic Hydrocarbons in Constructed Wetlands

A microorganism's ability to grow in a given habitat is determined by its ability to utilize nutrients in its surroundings.[32,33] Energy sources available to constructed wetland heterotrophic microorganisms include cellulose, hemicellulose, lignin, starch, chitin, sugars, proteins, hydrocarbons, and various other compounds.[33] Numerous hydrocarbons or derivatives are naturally synthesized within this system, while others are added from pollution sources. Mineralization and formation by indigenous microflora are important in the general carbon cycle.[33] The three major types of microbial metabolism are fermentation, aerobic respiration, and anaerobic respiration.[33,34] Aerobic respiration is most important in efficiently transforming PAHs; very little anaerobic respiration has been reported. However, anaerobic biodegradation of PAHs has been observed where suitable electron acceptors were supplied.[31] Aerobic respiration initially involves incorporation of molecular oxygen in hydrocarbon molecules and conversion to more oxidized products. Energy produced during oxidation processes is partially used in the synthesis of cellular constituents.[30]

There is no agreement on the physical form under which hydrocarbons are metabolized. Most reported microbial hydrocarbon metabolism processes are intracellular oxidation processes.[31] Historically, investigations of PAH biodegradation measured the amount of CO_2 produced or the fractions of the toxicants (parent molecule) converted into CO_2. The need to investigate intermediates formed and the ratio of polar compounds to CO_2 was recently recognized because oxygenated polar compounds may be mutagenic and accumulation in the environment could be hazardous to living organisms. This is of particular concern for constructed wetlands wherein concentration effects may also result in adverse changes in the wetlands species diversity. Plants are very susceptible to phytotoxicity by elevated PAH concentrations.[35]

Bacteria in constructed wetlands are the dominant group involved in PAH degradation. The most widely occurring are *Pseudomonas*, *Myobacterium*, *Acinetobacter*, *Arthrobacter*, *Bacillus*, and *Nocardia*.[29] Bacteria can oxidize PAHs ranging from the size of benzene to benzo(a)pyrene but not PAHs more highly condensed.[30] Bacterial mechanisms for introduction of hydroxyl moieties into PAHs depend on alkyl substituents in the substrate.[35] The initial step of aromatic metabolism consists of substituent modification or removal on benzene rings and introduction of hydroxyl groups.[30] The first metabolites of unsubstituted PAHs created by bacteria are *cis*-dihydrodiols, formed by the incorporation of two atoms of molecular oxygen.[35] Fungi, in contrast, form *trans*-dihydrodiols.[36] Naphthalene and its alkylated homologs are among the most water-soluble and potentially toxic compounds in petroleum. Anthracene degradation and its derivatives are important because these compounds possess a structure found in other carcinogenic PAHs.[37] Degradation of anthracene reported for bacteria and fungi follows the general degradation pattern of other PAHs.

Many fungi cannot grow with PAHs as their sole source of carbon and energy, but they still have the ability to oxidize these compounds.[28,36] In the degradation process, fungi carry out reactions similar to those of mammals and are often used as model systems. Their enzyme systems for the oxidation of PAHs differ from those of bacteria (e.g., monooxygenases) but are similar to those of higher organisms. Thus, they are excellent monitoring species for ecosystem stress in wetlands.[37]

SUMMARY

Aquatic organisms, whether in a constructed wetland or in deep ocean microhabitats, are constantly faced with fluctuating environmental conditions. The most important parameter is the availability and quality of nutrients. Most aquatic environments are oligotrophic (starvation conditions), and their inhabitants must cope with uncertain, often unsuitable conditions for survival. Each species has a characteristic threshold for nutrient requirements which may be lower in organisms with high affinity but low specificity for nutrient uptake. Below threshold concentrations, onset of starvation survival is characterized by bacterial division without concurrent growth to produce ultramicrocells. The complexity and diversity of aquatic microorganisms and of their environments makes it virtually impossible for optimal conditions to exist for each organism. At any one time, most microbes are surrounded by waters lacking sufficient nutrients from which ATP can be produced; in other words, waters lacking sufficient energy-yielding substrates.

In this discussion, which has concentrated on heterotrophic bacteria, biodegradable organics are the energy-yielding substrates. When water currents deposit bacteria in an area of nutrient deficiency, the organisms must adapt or perish. As such, many microorganisms may become dormant until external conditions improve. Apparently, bacteria are very "patient" and can maintain this state for many, many years. Microbes must be considered living catalysts; their function is to convert DOC to POC for higher-order consumers. As the base of the food web, microbes facilitate the flow of organics through the system. By increasing efficiency, microbes affect the rate of carbon cycling through the system. But as catalysts, they are the central focus in biotreatment effect in a constructed wetlands system. And as living catalysts, they must be handled with some degree of patience.

REFERENCES

1. Pfaender, F. K., and G. W. Bartholomev. "Measurement of Aquatic Biodegradation Rates by Determining Heterotrophic Uptake of Radiolabelled Pollutants," *Appl. Environ. Microbiol.* 44(1):159–164 (1982).
2. Chibata I., T. Tosa, and T. Sato. "Immobilized Aspartase-Containing Microbial Cells: Preparation and Enzymatic Properties," *Appl. Microbiol.* 27:878–885 (1974).
3. Walter, W. G., R. A. McBee, and K. L. Temple. *Introduction to Microbiology* (New York: Litton Educational Publishing, Inc., 1973), pp. 1–7; 59–71; 121–141.
4. Poindexter, J. S. *Microbiology: An Introduction to Protists* (New York: Macmillan Publishing Co., Inc., 1971), pp. 23–66; 149–199.
5. Walter, W. G., and R. H. McBee. *General Microbiology,* 2nd ed. (New York: D. Van Nostrand Company, 1962), pp. 27–114; 145–180.
6. Wilkinson, J. F. *Introduction to Microbiology, Vol. 1,* 2nd ed. (Boston, MA: Blackwell Scientific Publications, Inc., 1975), pp. 11–91.
7. Gottschalk, G. *Bacterial Metabolism,* 2nd ed. (Heidelberg: Springer-Verlag, 1986).
8. Nester, E. W., C. E. Roberts, N. N. Pearsall, and B. J. McCarthy. *Microbiology,* 2nd ed. (New York: Holt, Rinehart & Winston, 1973), pp. 237–274.
9. Ucko, D. A. "Storage and Transfer of Energy," in *Living Chemistry: An Introduction for the Health Sciences,* E. Editor, Ed. (New York: Academic Press, Inc., 1977), pp. 388–419.
10. Caldwell, D. E., and J. R. Lawrence. "Growth Kinetics of *Pseudomonas fluorescens* Microcolonies Within the Hydrodynamic Boundary Layers of Surface Microenvironments," *Microb. Ecol.* 12:299–312 (1986).
11. Volk, W. A., and M. F. Wheeler. *Microbiology,* 3rd ed. (Philadelphia, PA: J. B. Lippincott Company, 1973), pp. 3–129.
12. Strandberg, G. W., S. E. Shumate, and J. R. Parrott. "Microbial Cells as Biosorbents for Heavy Metals: Accumulation of Uranium by *Saccharomyces cerevisiae* and *Pseudomonas aeruginosa,*" *Appl. Environ. Microbiol.* 41:237 (1981).
13. Bitton, G., and V. Freihofer. "Influence of Extracellular Polysaccharides on the Toxicity of Copper and Cadmium to *Klebsiella aerogenes,*" *Microb. Ecol.* 4:119–125 (1984).

14. Ehrlich, H. L. "How Microbes Cope with Heavy Metals, Arsenic and Antimony in Their Environment," in *Microbial Life in Extreme Environments,* D. J. Kushner, Ed. (New York: Academic Press, Inc., 1978), pp. 381–408.
15. Gadd, G. M., and A. J. Griffiths. "Microorganisms and Heavy Metal Toxicity," *Micro. Ecol.* 4:303–317 (1978).
16. Tuovinen, O. H., and D. P. Kelly. "Toxicity of Uranium to Growing Cultures and Tolerance Conferred by Mutation, Other Metal Cations and E.D.T.A.," *Arch. Microbiol.* 95:153–164 (1974a).
17. Tuovinen, O. H., and D. P. Kelly. "Influence of Uranium, Other Metal Ions and 2,4-Dinitrophenol on Ferrous Iron Oxidation and Carbon Dioxide Fixation," *Arch. Microbiol.* 95:165–180 (1974b).
18. Friis, N., and P. Myers-Keith. "Biosorption of Uranium and Lead by *Streptomyces longwoodensis,*" *Biotech. Bioeng.* 28:21–28 (1986).
19. Babich, H., and G. Stotzky. "Sensitivity of Various Bacteria, Including Actinomycetes and Fungi, to Cadmium and the Influence of pH on Sensitivity," *Appl. Environ. Microbiol.* 33:681–695 (1977).
20. Hardy, K. In *Bacterial Plasmids,* J. A. Cole and C. J. Knowles, Eds., Aspects of Microbiology Series #4 (London: Thomas Nelson and Sons, 1983), pp. 62–63; 71–72.
21. Iverson, W. P., and F. E. Brinckman. "Microbial Metabolism of Heavy Metals," in *Water Pollution Microbiology,* R. Mitchell, Ed. (New York: John Wiley & Sons, Inc., 1978), pp. 201–232.
22. Wilkinson, J. F. "Extracellular Bacterial Polysaccharides," *Bacteriol. Rev.* 22:46–72 (1957).
23. Shumate, S. E., and G. W. Strandberg. "Accumulation of Metals by Microbial Cells," in *Comprehensive Biotechnology,* C. W. Robinson and J. A. Howell, Eds. (Oxford: Pergamon Press, Inc., 1985), p. 235.
24. Kelly, D. P., P. R. Norris, and C. L. Brierly. "Microbiological Methods for Extraction and Recovery of Metals," in *Microbial Technology: Current State, Future Prospects,* A. T. Bull, C. Ellwood, and C. Ratledge, Eds. (New York: Cambridge University Press, 1979).
25. Mangi, J. I., and G. J. Schumacher. "Physiological Significance of Copper-Slime Interactions in *Mesotaenium* [Zygnematales; Chlorophyta]," *Am. Midl. Nature* 102:134–139 (1979).
26. Neff, J. M. *Polycyclic Aromatic Hydrocarbons in the Aquatic Environment. Sources, Fates and Biological Effects.* (London: Applied Science Publishers, Ltd., 1979), pp. 61–73; 102–149.
27. Sisler, F. D., and C. E. ZoBell. "Microbial Utilization of Carcinogenic Hydrocarbons," *Science* 106:521–522 (1947).
28. Cerniglia, C. E. "Microbial Transformation of Aromatic Hydrocarbons," in *Petroleum Microbiology*, R. M. Atlas, Ed. (New York: Macmillan Publishing Co., Inc., 1984), pp. 99–129.
29. Alexander, M. *Introduction to Soil Microbiology* (New York: John Wiley & Sons, Inc., 1977), pp. 115–225
30. Gibson, D. T. "Microbial Metabolism," in *Handbook of Environmental Chemistry, Vol. 2, Part A,* O. Hutzinger, Ed. (New York: Springer-Verlag, 1980), pp. 161–189; 231–244.
31. Zander, M. "Polycyclic Aromatic and Heteroaromatic Hydrocarbons," in *Hand-*

book of Environmental Chemistry, Vol. 3, Part A,* O. Hutzinger, Ed. (New York: Springer-Verlag, 1980), pp. 109–128.

32. Alexander, M. "Biodegradation of Chemicals of Environmental Concern," *Science* 211:132–139.
33. Mandelstam, J., and K. McQuillen. *Biochemistry of Bacterial Growth,* 3rd ed. (Oxford: Blackwell Scientific Publications, Ltd., 1982), pp. 142–158.
34. Hutzinger, O. *Handbook of Environmental Chemistry, Vol. 2, Part C,* (New York: Springer-Verlag, 1985), pp. 117–147.
35. Harvey, R. G. "Synthesis of Dihydrodiol and Diol Epoxide Metabolites of Carcinogenic Polycyclic Hydrocarbons," in *Polycyclic Hydrocarbons and Carcinogenesis,* R. G. Harvey, Ed. ACS Symposium Series 283, pp. 19–35 (1985).
36. Cerniglia, C. E. "Microbial Metabolism of 4-, 7-, 10-, Methylbenz[a]anthracenes," in *Polynuclear Aromatic Hydrocarbons: Formation, Metabolism and Measurement,* M. Cooke and A. J. Dennis, Eds. (Columbus, OH: Battelle Press, 1983), pp. 283–293.
37. London, S. A., C. R. Mantel, and J. D. Robinson. "Microbial Growth Effects of Petroleum and Shale-Derived Fuels," *Bull. Environ. Contam. Toxicol.* 32:602–612 (1984).

CHAPTER 7

Wetlands: The Lifeblood of Wildlife

J. Scott Feierabend

INTRODUCTION

This chapter provides a general overview of the types of wildlife that wetlands—both naturally occurring as well as constructed or man-made—can attract rather than an exhaustive treatment of the subject. Although the literature on wildlife associated with naturally occurring wetlands is considerable, little information is available for wildlife associated with constructed wetlands and even less for wetlands constructed for wastewater treatment.

It includes an introduction to the importance of wetland ecosystems as wildlife habitat, an overview of the types of wildlife associated with naturally occurring wetlands in the United States, and design considerations, issues, and research needs concerning wildlife utilization of constructed wetlands for wastewater treatment. It also examines how constructed wetlands might augment the nation's diminished wetlands inventory and provide additional habitat for wildlife.

WETLANDS AS WILDLIFE HABITAT

Two points bear mentioning before introducing this section. First, as used in this chapter, the term *wildlife* refers to invertebrate animals as well as vertebrate animals. Wildlife biologists oftentimes exclude "lower" life forms such as mollusks and arthropods from working definitions of wildlife, instead focusing narrowly on "high profile" species such as waterfowl and large carnivores. Although this distinction may at times be useful, because invertebrate animals provide critical links in most wetland food chains, it is essential they be included in any discussion of wetlands wildlife.

Second, wetlands created where none previously existed should be described as "constructed" rather than "artificial." Man-made wetlands—regardless of whether they provide one function or a dozen—are, indeed, real wetlands and not artificial.

Wetlands are described as providing both value and function. Importantly, these terms are not synonymous. *Function* refers to what a wetland does, regardless of interpretation of its worth. For example, a wetland may function by storing 25,000 m^3 of flood water, producing 100 mallard ducks/ha, or retaining 15 tons/ha/yr of sediment. Wetland value, on the other hand, is an interpretation of relative worth of a wetland function and can be high or low. For example, flood storage capacity of a wetland upstream from a town has high value to the town residents, yet the same wetland downstream might have low value to town residents because it affords them no flood protection. Although 15 or more functions have been described for wetlands,[1] those most frequently cited include water quality, flood conveyance, sediment control, barriers to waves and erosion, open space and aesthetic values, and wildlife habitat.

Wildlife habitat, in simplest terms, is the combination of food, water, and cover needed by a species to survive and reproduce. Odum[2] describes habitat as the place in which an organism lives or where one would go to find it, and Weller[3] defines it as the place where an organism finds food, shelter, protection from enemies, and resources for reproduction. I like to describe wildlife habitat as simply its kitchen, dining room, and bedroom—the requisites for life itself.

Wetlands are among the most vulnerable and most threatened habitats[4] of all our natural heritage. Of the 87 million ha of wetlands existing in the United States when the colonists arrived, less than half, or approximately 44 million ha remain.[5] Wholesale destruction of this natural resource is the result of generations of contempt and ambivalence toward wetlands, due in part to lack of appreciation and understanding of the important ecological and economic benefits they provide. Wetlands were believed wastelands, an impediment to progress, and something better drained, ditched, filled, and developed. Although misguided attitudes toward wetlands have changed somewhat, we continue to lose an estimated 150,000 to 225,000 ha of wetlands each year.

Wetlands represent a very small fraction of our total land area,[4] but they harbor an unusually large percentage of our nation's wildlife. For example, 900 species of wildlife in the United States require wetland habitats at some stage in their life cycle, with an even greater number using wetlands periodically.[6] Representatives from almost all avian groups use wetlands to some extent and one-third of North American bird species rely directly on wetlands for some resource.[7]

Stability is neither common nor desirable in wetland systems.[3] Unlike upland habitats, wetlands are dynamic, transitional, and dependent on natural perturbations. The most visible and significant perturbation is periodic inundation and drying. Changing water depths, either daily, seasonal, or annual, have an overbearing influence on plant species composition, structure, and distribution. Other influences, such as complex zones of water regimes, salt and temperature gradients, and tide and wave action produce wetlands vegetation that is generally stratified, much like forests. These factors combine to create a diversity and wealth of niches[3] that make wetlands important wildlife habitat.

Constantly varying environmental conditions of wetlands compel plants to adapt to stress and tolerate ever-changing biophysical conditions to establish, live, and reproduce. Like wetlands plants, animals inhabiting wetlands have evolved adaptations for surviving in these dynamic habitats. One common strategy—exemplified by birds—is mobility that allows daily, seasonal, and annual movement. Another common strategy is to use wetlands for only a part of a life cycle, as do most reptiles and many amphibians. Many wildlife species totally dependent on wetlands have developed mechanisms, such as dormancy, to survive periods of low water or times of no water at all. Several species of upland wildlife—primarily mammals and birds—also use wetland habitats extensively.[8]

WETLANDS WILDLIFE

The following is a brief overview of wildlife types occurring in wetlands. Because the literature is voluminous, this section is a general introduction and not a comprehensive or exhaustive treatment of the topic. Moreover, the information derives from research on natural or restored wetlands rather than from constructed systems. Nonetheless, because natural and constructed wetland habitats have similarities, cautious extension of these data to constructed systems is justified. For organizational purposes, this review is arranged by taxonomic group, beginning with invertebrates and concluding with mammals.

Invertebrates

Invertebrates are critical to energy dynamics and functions of wetlands and the foundation of wetland food chains. Some of the least conspicuous forms have important roles in converting plant energy into animal food chains. Despite this, invertebrates have been largely ignored, and far too little is known, especially regarding invertebrate habitat selection and niche segregation.[3,9] All freshwater wetlands contain protozoans and 25 species of sponges occur in freshwater wetlands of the United States. Several rare and unique groups of freshwater jellyfish and sponges have been documented, but the more abundant, ubiquitous invertebrates provide the principal source of food that fuels wetland food chains.

Most wetlands macroinvertebrates fall into four groups: annelid worms, mollusks, arthropods, and insects. Annelid worms are most frequently represented by the oligochaetes, which burrow into substrates or adhere to submersed aquatic vegetation. Densities of these worms can be very low—as in acidic bog lakes—or extremely high, as in California fens with 30,000 worms/m^2. Although we now recognize annelids, flatworms, leeches, earthworms, nematodes, and other worms as vital to wetlands food webs, far too little is known about the natural history and role of these animals.

The second category, mollusks, are a dominant group of animals found in wet meadows as well as in deepwater habitats. Mollusks include several genera of snails—both aquatic and terrestrial—clams, and mussels. Mollusks are primarily benthic or associated with aquatic vegetation and may have densities up to 40,000/m^2 (fingernail clams).[10] Mollusks are usually abundant and important food items for many vertebrate species, including ducks, fish, mink, otter, muskrat, raccoons, turtles, and salamanders.

Wetlands are favored habitats for diverse, abundant and oftentimes large arthropods, which are important foods of frogs, fish, toads, turtles, birds, and mammals. Most common are crustaceans—exemplified by crayfish and freshwater shrimp—the most abundant and widespread of all wetland invertebrates. Crayfish are particularly important food items for mink, raccoons, and predaceous fish.

Finally, 11 orders of insects are aquatic or semiaquatic, including such familiar examples as stoneflies, damselflies, dragonflies, mayflies, midges, mosquitoes, aquatic beetles, water striders, and springtails. As with most invertebrates, the larval stage—largely unnoticed by humans—provides the most important conduit of energy in wetlands systems. Insect aquatic larvae commonly occur on edges and surfaces of freshwater wetlands as well as in bottom soils and organic debris, providing abundant food for fish, frogs, ducks, shorebirds, and other invertebrates.

Although information on wetland invertebrates is limited, we recognize their crucial role in wetlands systems. For example, invertebrates are vital links in food chains supporting valuable animals such as songbirds and waterfowl. They not only process living and dead organic matter, channeling it to producer and detrital food chains, but they physically modify wetland habitats, enhancing their value for other wildlife species.[9] Major factors influencing abundance and diversity of wetland invertebrates appear to be wetland size, location relative to other wetlands, wetland setting, substrate, vegetation structure, water regime, water quality, competition, and predation. Significantly, these same factors influence abundance and diversity of poikilotherms, birds, and mammals in wetlands.

Poikilotherms (Cold-Blooded Vertebrates: Amphibians, Reptiles and Fish)

The conglomerate of wetland poikilotherms includes species totally dependent on wetlands, such as fish and amphibians, as well as those species requiring wetlands for only a limited period of their life cycle, such as reptiles.

Amphibians

Two groups of amphibians are endemic to the United States: anurans—the frogs and toads, and urodels—the salamanders. Although certain members of these groups may be largely terrestrial, there are few exceptions in the 190 species of North American amphibians that do not require wetlands for at least a part of their life cycle.

Frogs and toads are so ubiquitous that most every freshwater wetland harbors some species of frog and toad during either the breeding or the nonbreeding season. Some species, such as the leopard frog and the wood frog, are predominantly terrestrial, feeding mainly on upland insects. These species survive dry periods by taking refuge in wet depressions. Similarly, toads and wood frogs are terrestrial except when they move into wetlands during the breeding season.

In the warmer climes of the Southeast, bullfrogs are a dominant life form and a major predator, consuming other frogs as well as small ducklings. The green frog is also very closely tied to the marsh, with almost a third of its diet derived from aquatic organisms.

Salamanders occur at various levels in wetlands, but perhaps the most numerous and widespread species is the tiger salamander, the only salamander occurring in the dry Southwestern United States.[11] The tiger salamander is also best known because of the public attention in spring when hundreds—and sometimes thousands—move across highways from deepwater habitats to shallow water to breed.

Adult salamanders and frogs, as well as millions of egg masses and tadpoles, are important foods for birds such as ibises, egrets, and white pelicans; snakes and fish; small mammals such as otter, mink, and raccoons; and other frogs, toads, and salamanders. Species richness and diversity of frogs and salamanders vary with latitude and annual rainfall. Thus, these animals are abundant and well-represented in southern marshes but much less so in the north and west. For example, a study in Florida found 1600 salamanders and 3800 frogs and toads using a pond less than 30 m wide.[9]

Fish

Diversity of fish species, especially in dense marsh systems, has not been explored in depth, but these systems provide important shelter and food for fish. Major factors influencing a wetland's habitat value for fish include water quality—temperature and dissolved oxygen—water quantity, and cover—substrate and interspersion.

The freshwater wetlands fishes are dominated by forage species such as killifishes, shiners, mosquito fishes, and sunfishes. In fact, forage minnows account for a high proportion of the fish population as well as the fish biomass of freshwater wetlands. For example, three species comprise more than 75% of the population and more than 85% of the biomass in Florida's Everglades and Big Cypress Swamp. However, not all freshwater wetlands—notably small isolated or seasonally dry potholes and acidic bog lakes and fens—support fish.

Most larger fish species are transients of freshwater marshes, entering only diurnally to feed or seasonally to breed and spawn. Familiar species that move from adjacent lakes into marshes to breed include northern pike, walleye, black bullhead, yellow perch, and bluegill. Freshwater fish and their fry are important food items for wading birds such as storks and herons, amphibians and reptiles, and many aquatic and upland mammals.

Reptiles

Most reptile species in freshwater wetlands employ a reverse strategy to amphibians: they use wetlands for food, cover, and water but seek out drier areas and wetland fringes to reproduce. Three major groups of reptiles occur in freshwater wetlands of the United States: snakes, turtles, and alligators.[9] Of these, turtles are the most diverse and the most common. Familiar representatives include mud and musk turtles, softshell turtles, painted turtles, sliders, cooters, box turtles, and pond turtles. Some, snapping turtles, are omnivorous while others, softshells, are carnivorous or vegetarian. Dependence on water varies dramatically with various species of turtles. Some, snapping turtles and mud turtles, are truly aquatic, emerging from water only to lay their eggs. Others, box and wood turtles, are largely terrestrial, consuming fruit, berries, worms, and insects and entering the water only to hibernate in the muddy substrate.

Snakes are the second major reptile group inhabiting freshwater wetlands. The only snake common in northern marshes is the garter snake, occurring primarily in fringe wetlands and feeding on amphibians, eggs, and small birds. In the south, the cottonmouth moccasin is most frequently mentioned, but it is less important in numbers or biomass than water snakes found in freshwater wetlands throughout the United States. Other snakes common in freshwater wetlands include queen snakes, mud snakes, and swamp snakes. Many snakes are highly dependent on wetlands wildlife, preying heavily on crayfish, fish, salamanders, ducklings, earthworms, slugs, and snails.

The third category, alligators, is a dominant factor in some freshwater wetlands. Alligators from Texas and Oklahoma to North Carolina dig "holes" in the marsh for nesting, reproduction, and refuge in dry periods that provide critical habitat for other wildlife. Their predation on fish, birds, and mammals can be significant and is well documented. So, too, is man's historic predation on the American alligator for food and hide, which decimated populations until the species was listed as endangered. Through research and management, alligator populations have rebounded and now sustain a closely-regulated harvest in several Gulf Coast states.

Birds

Wetland birds are those species deriving any essential part of their existence from wetlands. Freshwater wetlands use by many species of ducks, geese, swans, coot, loons, pelicans, grebes, shorebirds, cranes, and others is well known. These birds come to mind when one mentions wetlands, and they have become "ambassadors" of wetlands. Because these birds are spectacular to observe, relatively easy to find, and have much broader appeal than amphibians or fish, they have historically and currently raised the public consciousness on the importance and plight of our nation's wetlands. One need simply reflect on the artwork, organizations, and revenues that pivot on these birds to understand their substantial roles.

Despite the attention, excitement, and broad constituency that revolves around some wetland birds, another important category is not so commonly associated with wetlands but very much a part of and dependent on these systems. These birds and their use of wetlands is more subtle and indirect than trumpeter swans or sandhill cranes. For example, the blackburnian warbler nests in upland spruce but flies miles to collect bog lichens for nest materials.[7] In fact, representatives from almost all avian groups use wetlands to some extent, and one-third of North American bird species depend directly on wetlands for some resource.

Birds are unique because of their mobility and ability to use disjunct wetlands. Many species are migratory, allowing them to use geographically disparate wetlands—many of which are seasonal—at varying times of the year. Some birds, such as the pied-billed grebe which can hardly walk on land and even builds a floating nest, are intimately tied to wetlands for survival. Others, such as flycatchers, are less dependent, only using the areas periodically for foraging or for nesting. Wetlands provide birds with food (tubers, invertebrates, seeds, and vertebrates); cover (nesting, hiding, and weather protection); and, obviously, water. Wetlands are strong attractors of birds not only because of abundant food resources but also because they provide excellent nesting and feeding sites protected from predators.

Examples of some representative wetlands birds bear mentioning. Waterfowl have been discussed. Loons and diving ducks are typically associated with larger, more sterile lakes, whereas grebes are associated with marshy areas. Of the herons, bitterns are probably most closely associated with freshwater wetlands, building solitary nests in reeds and rushes and feeding on frogs, fish, and small mammals. Herons, egrets, and ibises, on the other hand, nest colonially in wetlands, feeding on frogs, fish, crayfish, and other invertebrates. Occasionally, local colonies of these birds may conflict with humans. Rails are common residents of larger marshes. These rarely seen ground dwellers use the entire marsh, from the driest edges to areas of standing water. In addition to waterfowl, passerines, and shorebirds, freshwater wetlands are also home to many of our nation's raptors, or birds of prey. The more common include marsh hawks, bald eagles, and osprey, all using the abundant prey base—frogs, fish, birds, invertebrates, and small mammals— associated with freshwater wetlands.

Because so many species depend on wetlands, birds are one of our best bellwethers of the health of our wetlands resources. Said differently, as the birds fare, so do the wetlands. That the United States is entering what may be the worst year of waterfowl production on record should remind us all that unabated alteration and outright destruction of our nation's wetlands must no longer be condoned.

Mammals

Unlike many representatives of the above categories of wildlife, very few mammals are marsh specialists clearly adapted to water and hydrophytes and characterized as truly aquatic or wetland-dependent. Nevertheless, many mammals have individuals or populations that are wetland-dependent in certain areas and at certain times of the year. A primary reason wetland-dependent mammals may be underrepresented is that these animals, unlike birds, are nonmigratory and unable to escape and survive seasonal and annual dry periods. Despite fewer numbers of mammalian species associated with wetlands, they are extremely important to the maintenance and functioning of these systems.

Examples of mammalian species that are totally wetland-dependent are nutria, beaver, marsh rice rats, and swamp rabbits. The muskrat is perhaps the most characteristic example of a mammal inhabiting the broadest array of wetland types. Muskrats are distributed widely throughout North America, living in fresh and brackish water covering thousands of hectares as well as small roadside pools. Muskrats are dependent on emergent marsh vegetation, notably cattails, for food and shelter and can "eat out" a marsh if their population is too large. They can, however, shift to upland foods in times of stress. Surprisingly, cutting vegetation for lodging, storage, and nests influences the marsh more than food consumption.

The muskrat house, common in marshes nationwide, can reach 2 m in height and 5 m in diameter. These houses are obviously important to muskrats. They are also important as nesting and loafing areas for birds, living quarters for other animals, and germination sites for semiaquatic plants. Muskrat houses create microhabitats for smaller organisms such as spiders, mites, and insects and aquatic crustaceans, and resting and feeding areas for frogs, toads, turtles, garter snakes, water snakes, birds, mink, and raccoon. Interestingly, the warmth generated by vegetation decomposition in the house provides hospitable microclimates for poikilotherms and enhances early ice melting in northern marshes.

Partially wetland-dependent mammals include mink and raccoon. The meadow mouse is mainly a field dweller but swims well underwater and may live in over-water nests of diving ducks or in muskrat houses. The list of wetlands mammals would not be complete without bog lemmings, cotton mice, wood rats, swamp rabbits, and short-tailed shrews. Most of these species occupy wetlands and regularly use adjoining upland habitats.

Finally, wetlands are used by larger mammalian species such as coyote, fox, bobcat, white-tailed deer and moose that are more or less dependent on wetlands. Moose, for example, browse on rooted aquatic and submerged vegetation in northern bogs during the summer to supply their sodium requirements.

Other use of freshwater wetlands by large mammals is seasonal—such as winter protection, shelter, and feeding for white-tailed deer.

Threatened and Endangered Wildlife

In addition to abundant species of wetlands wildlife discussed above, wetlands also provide essential habitat for many threatened and endangered species. Williams and Dodd[4] reported that 16% (5 of 33) of endangered mammals, 31% (22 of 70) of threatened and endangered bird species, 31% (22 of 70) of endangered and threatened reptiles, and 54% (22 of 41) of threatened and endangered fishes are dependent on wetlands or found in freshwater wetland habitats during part of their life cycle. Wetlands obviously play a vital role in the maintenance of biological diversity.

CONSTRUCTED WETLANDS FOR WASTEWATER TREATMENT AS WILDLIFE HABITAT: CONSIDERATIONS, ISSUES, AND RESEARCH NEEDS

In certain circumstances, constructed wetlands can be a useful mechanism for treating wastewater—municipal and otherwise. Societal and ecological benefits accruing from this low level technology are extremely exciting. Wetlands constructed for the singular purpose of wastewater treatment can obviously yield benefits beyond simply discharging water that meets local, state, and federal water quality standards. For example, appropriate design and siting in the landscape may contribute to groundwater recharge or help moderate storm surges. From the perspective of a wildlife biologist, one of the most exciting derivatives of constructed wetlands may be their benefits as wildlife habitats. Although wildlife cannot be the primary purpose, proper planning, implementation, and maintenance can enormously enhance the value of constructed wetlands for wildlife. For example, establishing vegetation for wildlife food and cover should be an integral design element. Constructing wetlands that maximize edge, provide transition zones into uplands, and use existing wildlife corridors are examples of important design considerations. Planning should also incorporate control structures, public entry, and training facilities such as boardwalks and viewing platforms to enhance public enjoyment.

However, full benefit realization is dependent on related issues and research needs. First, what, if any, impact might wetlands constructed for wastewater treatment have on the short and long-term viability of wildlife attracted to these areas? Might we be innocently creating another Kesterson or Salton Sea National Wildlife Refuge by drawing animals into potentially hazardous and harmful environments? What might be the lethal and sublethal impacts of contaminants in water, vegetation, and soils on wildlife? Who has legal responsibility for potential impacts, especially in the case of migratory or threatened and endangered species? How can these problems be anticipated, avoided, and remedied? Obviously, we need to initiate long-term research to answer these and other questions on operating systems.

Another important issue is how or whether constructed wetlands for wastewater treatment will be managed. The dynamic fluctuating nature of natural wetlands makes these systems productive and important to wildlife. Can perturbations be included in designs of constructed wetlands, and can these systems be managed so as to replicate natural wetlands without impairing their function and utility for treating wastewater effectively? Managing change in constructed wetlands, as we do in many other wetlands, will significantly influence the wildlife value of these areas.

Public information and education programs should be considered in planning, design, and operating stages. Certainly, public information campaigns and outreach programs will be vital to acceptance and long-term success of this technology and should be a major element of any planning effort. Encouraging public visits to view the processes at work will enhance a project's acceptance, win over valuable allies, and garner broader public support for using constructed wetlands for wastewater treatment.

Part of this effort should develop support from local sportsmen's organizations; environmental groups; naturalist, birdwatching, and wildflower societies; and other associations and individuals interested in wetlands and wildlife. Constructed wetlands may prove to be excellent areas for hunting and birdwatching and other outdoor recreation. For example, a 1988 issue of *Birder's World,* a national birdwatching tabloid, has a five-page article on California's Arcata Marsh and Wildlife Sanctuary.[12] Entitled "Birding Hot Spots," the story describes the history of the town's wastewater treatment project and over 200 species of birds and thousands of waterfowl and shorebirds using the marsh. User surveys of visitors to the Arcata wetland rank walking and isolation as important values. Widening the support circles to include average citizens will improve acceptance of this important technology.

CONSTRUCTED WETLANDS TO AUGMENT OUR NATION'S WETLANDS BASE

We have already destroyed more than 50% of our nation's wetlands. Although increased attention may attenuate these losses, clearly it will be years, if not decades, before we will stop and perhaps reverse these losses. Will constructed wetlands for wastewater treatment play a role in augmenting our nation's wetlands inventory?

Because constructed wetlands are small, isolated, and slow to come "on line," it is unrealistic to suggest they will play a significant role in augmenting the nation's overall wetland inventory. Nonetheless, constructed wetlands can be significant in increasing local wetlands and in creating wetland habitats where none previously existed. These activities can have positive benefits for local wildlife populations by enhancing habitat richness in areas lacking wetlands. But perhaps the greatest potential of constructed wetlands for wastewater treatment is that we will not only be gaining wetlands, but doing so using low-impact technology to solve an important environmental problem.

In conclusion, constructed wetlands for wastewater treatment is a shining example of finding creative solutions to tough problems. This approach to a national—indeed an international—issue is bold, exciting, and visionary but not without risk. However, with determination and commitment of time, research, and capital resources, constructed wetlands for wastewater treatment can mature into one of those rare "win-win" situations. We owe it to ourselves, as well as to our children, to ensure that the technology succeeds so that we as well as our wildlife resources may benefit.

REFERENCES

1. Kusler, J. A. *Our National Wetland Heritage: A Protection Guidebook* (Washington, DC: Environmental Law Institute, 1983).
2. Odum, E. P. *Fundamentals of Ecology* (Philadelphia: W. B. Saunders Company, 1971).
3. Weller, M. W. "Wetland Habitats," in *Wetland Functions and Values: The State of Our Understanding,* P. E. Greeson, J. R. Clark, and J. E. Clark, Eds. (Minneapolis: American Water Resources Association, 1979), p. 210.
4. Williams, J. D., and C. K. Dodd. "Importance of Wetlands to Endangered and Threatened Species," in *Wetland Functions and Values: The State of Our Understanding,* P. E. Greeson, J. R. Clark, and J. E. Clark, Eds. (Minneapolis: American Water Resources Association, 1979), p. 565.
5. "Wetlands: Their Use and Regulation," Office of Technology Assessment OTA-0-206, U.S. Government Printing Office (1984).
6. Willard, D. E. "Support for Birds and Mammals," in National Symposium on Wetlands, Lake Buena Vista, FL, November 1978.
7. Kroodsma, D. E. "Habitat Values for Nongame Wetland Birds," in *Wetland Functions and Values: The State of Our Understanding,* P. E. Greeson, J. R. Clark, and J. E. Clark, Eds. (Minneapolis: American Water Resources Association, 1979), p. 320.
8. Schitoskey, F., Jr., and R. L. Linder. "Use of Wetlands by Upland Wildlife," in *Wetland Functions and Values: The State of Our Understanding,* P. E. Greeson, J. R. Clark, and J. E. Clark, Eds. (Minneapolis: American Water Resources Association, 1979), p. 307.
9. Clark, J. "Fresh Water Wetlands: Habitats for Aquatic Invertebrates, Amphibians, Reptiles, and Fish," in *Wetland Functions and Values: The State of Our Understanding,* P. E. Greeson, J. R. Clark, and J. E. Clark, Eds. (Minneapolis: American Water Resources Association, 1979), p. 330.
10. Gale, W. F. "Bottom Fauna of a Segment of Pool 19, Mississippi River, Near Fort Madison, Iowa, 1967–1968," *IA State Jour. Res.* 49:353–372 (1975).
11. Ohmart, R. D., and B. W. Anderson. "Wildlife Use Values of Wetlands in the Arid Southwestern United States," in *Wetland Functions and Values: The State of Our*

Understanding, P. E. Greeson, J. R. Clark, and J. E. Clark. Eds. (Minneapolis: American Water Resources Association, 1979), p. 278.
12. Higley, J. M. "Birding Hotspots," in *Birder's World* 2(3):36 (May-June 1988).

SECTION II

Case Histories

CHAPTER 8

Constructed Free Surface Wetlands to Treat and Receive Wastewater: Pilot Project to Full Scale

R. A. Gearheart, Frank Klopp, George Allen

INTRODUCTION

The City of Arcata, California has completed four years of pilot project studies with a constructed freshwater wetlands to polish secondary-treated wastewater.[1] The pilot project demonstrated to the California State Water Resources Control Board (SWRCB) that a constructed wetland can provide reliable tertiary treatment for municipal wastewater,[2-7] and the wetland effluent can enhance water quality in Humboldt Bay.[8-10] This chapter presents results and conclusions from the pilot studies and two years of full-scale wetland operation. In addition, wetland management and design criteria for wastewater treatment systems is discussed.[11]

Freshwater wetlands are typically formed by advanced eutrophication of lakes, ponds, river deltas, and shallow low areas with poor drainage. Because of continual inflow of mineral and detritus-laden water from streams and other sources, wetlands are seldom nutrient-limited and thus are highly productive. Constructed wetlands for processing wastewater are highly productive wetland systems.[12-15] Since nitrogen and phosphorus concentrations are high in raw and treated waste effluent, nutrient cycling is rapid in these systems.[16,17] Rapid nutrient turnover and high standing crop create considerable biomass that represents a harvestable source of fermentable and digestible products. Without harvest, accumulated detritus reduces the volume of surface water in the wetland volume, and organic material exerts pressure on the oxygen budget.[18,19] Management questions that arise from materials harvesting include effects on (1) water quality, (2) process kinetics, (3) wildlife, and (4) mosquito production.

EXPERIMENTAL APPROACH

In the first study,[1] 12 experimental wetlands received oxidation pond effluent from September 1980 to September 1982, after a one-year construction and startup period. Hydraulic loadings to experimental cells varied from 0.02 to 0.24 m^3/m^2/day (0.021–0.252 mgd/ac), representing proposed loadings to the operating scale wetlands, the Arcata Marsh and Wildlife Sanctuary. While many water quality constituents were monitored,[5] this chapter focuses on dissolved oxygen (DO), suspended solids (SS), biochemical oxygen demand (BOD_5), and coliforms results.

The Arcata Marsh Pilot Project has received oxidation pond effluent for over five years, and wetland effluent quality stabilized after four years (Figure 1). Vegetation harvesting appears necessary only on a long-term basis or for energy production or composting. Dry weight biomass production ranging[7] from 4.9 to 16,330 g/m^2 could augment the primary digesters or ethanol fermentation.

The same pilot wetlands were used as in the first study. Ten experimental cells, 6.1 m by 61 m and 1.22 m deep including the berm, could be operated at variable depth and variable hydraulic loading. Aquatic vegetation was principally bulrush *(Scirpus validus)*. Nonchlorinated oxidation pond effluent was pumped continuously to a stilling tower that delivered flow to three division boxes, each feeding the 10 cells. Design flow for each cell from the division boxes was maintained at 0.34 L/sec.

RESULTS OF THE FIRST PILOT PROJECT

Suspended solids were effectively removed to less than 10 mg/L at all hydraulic loading rates with no seasonal variations observed.[1] Average effluent concentration was 5.3 mg/L, representing 85% removal (Table 1). Average SS effluent values ranged from 4.0 to 9.4 mg/L. Daily influent SS values were highly variable, with six of eight study quarters above the 30 mg/L monthly standard. Within the tested loading range (0.02–0.24 m^3/m^2/day), the effectiveness of SS removal was not a function of hydraulic or solids loading (Figure 2) but appeared related to some minimal detention time. In six of eight study quarters, all cells produced constant effluent SS of less than 10 mg/L. These results indicate 85% removal of SS can be expected at hydraulic loadings up to 0.24 m^3/m^2/day (0.25 mgd/ac). After two years of accumulation, SS had not progressed more than 12 m (20% of length) in the highest loaded cells, and detrital solids had accumulated to 15 cm depth.

Average BOD_5 effluent values from the experimental wetlands ranged from 9.0–15.3 mg/L and averaged 13.3 mg/L (Table 1). The BOD_5 removal rate varied from 41% to 65% and averaged 56%. Influent BOD_5s from the oxidation pond were highly variable, similar to influent SSs. Lower hydraulic loading rates generally produced the highest removals. Seasonal effluent variations were significantly affected by vegetation type, biomass, and amount of open water. All cells responded to increased organic load from digester supernatant added to the oxidation pond by producing higher BOD_5 effluents during the third and fourth quarters. Fourteen of the quarterly average wetland effluents

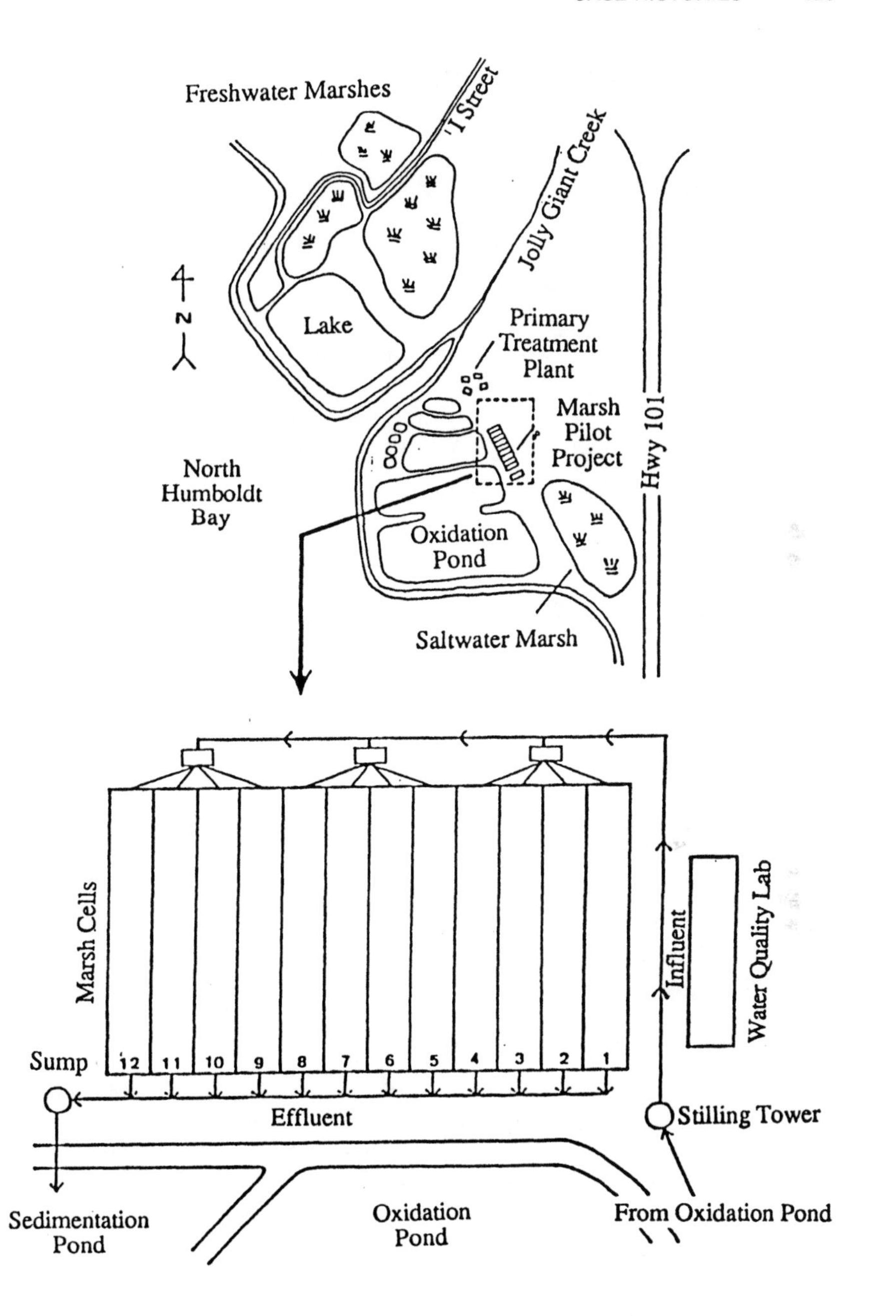

Figure 1. City of Arcata Marsh Pilot Project. Experimental cells and associated structures. (Modified from Gearheart et al.[1])

Table 1. Summary of BOD_5, SS, and Coliform Densities for the Entire Study Period for the 12 Experimental Wetland Cells

	Detention Times (hr)	Hydraulic Loading Rate ($m^3/m^2/day$)	BOD_5 (mg/L)	SS (mg/L)	Fecal Coliform (CFU/100 mL)
Influent			26	37	3183
Cell 1	52	0.24	11.2	6.8	317
Cell 2	38	0.24	14.1	4.3	272
Cell 3	65	0.19	13.3	4.7	419
Cell 4	37	0.22	12.7	5.6	549
Cell 5	88	0.12	14.0	4.3	493
Cell 6	59	0.11	10.7	4.0	345
Cell 7	106	0.11	13.3	7.3	785
Cell 8	58	0.11	15.3	7.2	713
Cell 9	160	0.07	11.9	9.4	318
Cell 10	90	0.07	12.6	4.9	367
Cell 11	183	0.06	9.4	5.7	288
Cell 12	132	0.06	9.0	4.3	421
Average all cells			11.4	5.3	440
Percent removal	—	—	56%	85%	86%

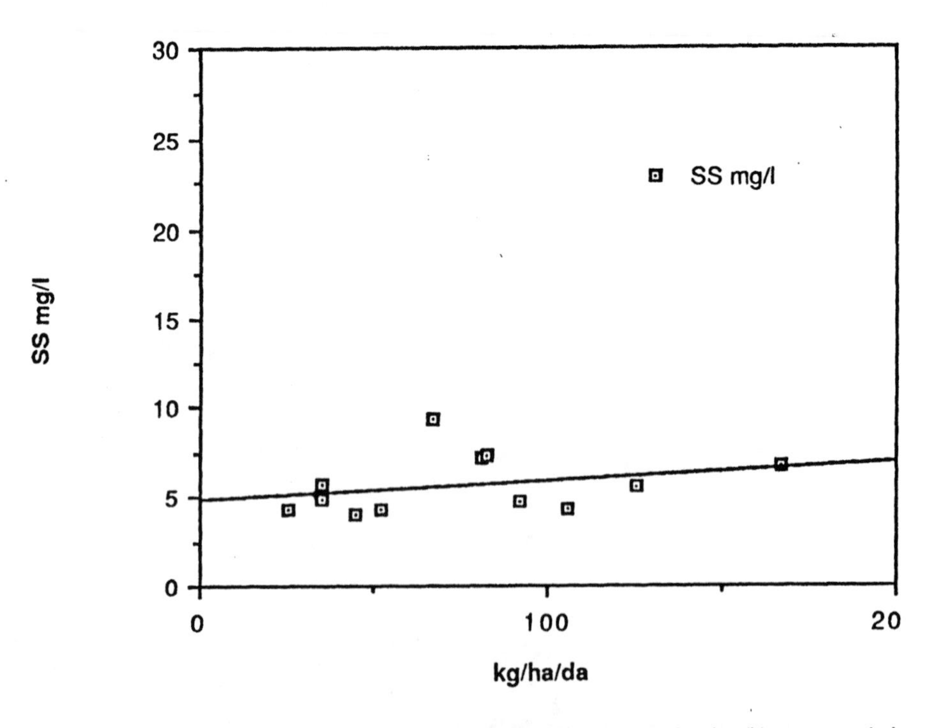

Figure 2. Relation of wetland effluent suspended solids concentration (mg/L) to suspended solids loading (kg/ha/da) based upon seasonal average values.

were less than 20 mg/L, and all were less than 30 mg/L. The lower loaded cells consistently produced effluent BOD_5s less than 20 mg/L.

Wetlands can effectively and consistently produce an effluent quality below SS and BOD_5 standards and can produce 10:10 BOD_5-SS effluents.[1]

RESULTS OF THE SECOND PILOT PROJECT

A second concern in some treatment wetlands is the important requirement for open water to support and enhance wildlife values. Specifically, the relation between inlet-outlet regions and open water macrophytes must be understood to maintain National Pollutant Discharge Elimination System (NPDES) effluent standards while operating and maintaining (harvesting) these systems.

This investigation[8] explored a range of operational and management strategies to minimize mosquito production; maximize DO levels in the effluent; maximize BOD_5, SS, and nutrient removal; and to treat primary wastewater to secondary standards. Objectives of this 1984–1986 study were to (1) determine the effect of vegetation harvesting on effluent water quality constituents, and (2) determine the kinetics of BOD_5 and fecal coliform removal in a wetland treatment system. A 60° V-notch weir controlled flows to each cell, and a 90° V-notch weir controlled depth and allowed effluent flow monitoring. Cell seepage was insignificant because heavy clay soil was used in cell construction.

To study wetland process kinetics, compartments were produced with intracell weirs in cells 6, 7, and 8 to create 2, 4, and 8 compartments, respectively (Figure 3). Flow from each compartment was constrained by three 90° V-notch weirs in the baffle. Cell 6 had one baffle, dividing the cell into two 30.5-m subbasins. Cell 7 had three baffles, dividing the cell into four 15.25-m subbasins. Cell 8 had seven baffles, creating eight internal basins. Cell 5, without baffles, was our control.

Average flow rate into each cell was 0.34 L/sec (5.5 gal/min) over the quarter. All 10 cells were operated at a weir elevation of 60 cm (2 ft), creating a theoretical liquid volume of 225 m^3 (59,800 gal). Baffled cells lost 2.54 cm of water elevation per baffle, resulting in elevation drops of 2.54 cm, 7.5 cm, and 17.5 cm in cells in 6, 7, and 8, respectively.

Based on the average flow of 0.34 L/sec., theoretical detention time in the cells was 7.5 days. Calibration tests prior to harvesting indicated that 30% of cell volume was plants and detritus. Theoretical detention time corrected for biological material but not short circuiting was five days for an unharvested cell and 6.1 days for a cell with 50% of the vegetated material removed.

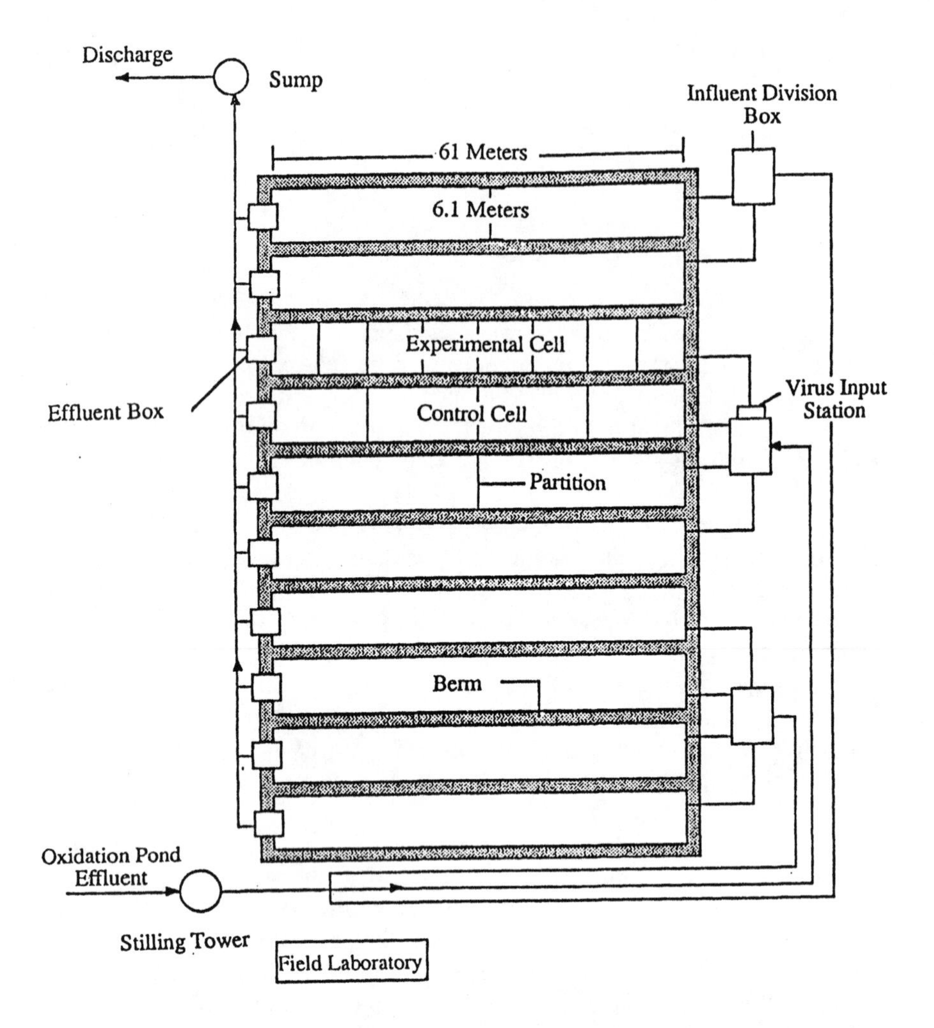

Figure 3. Physical layout of the 10 experimental marsh cells that comprise the pilot project, showing locations of the field laboratory, stilling tower, division boxes, partitions, effluent boxes, sump pump, and physical dimensions of the cells.

MANAGEMENT TECHNIQUES: HARVESTING

Vegetation harvesting was accomplished with a weed eater, machetes, and rakes. Cells 1, 2, 3, 4, and 10 tested vegetation harvest effects on effluent quality. Except for cells 3 and 5, all cells had 50% of the vegetation removed in various experimental configurations (Figure 4). Vegetation was harvested in November 1984 with hand-carried weed cutters, and vegetation was cut to within 6 cm of the bottom. Cut and loose vegetation (submergent and floating macrophytes) was removed from the wetland cell, but detrital banks of algal

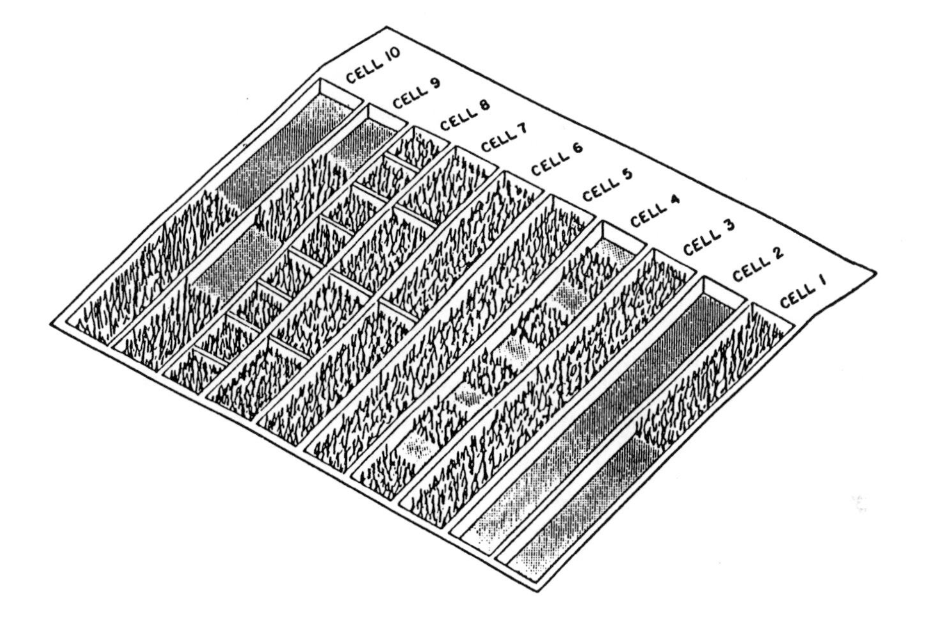

Figure 4. Second pilot project cell configuration with various vegetation harvesting patterns and intracell compartments.

and zooplankton solids were not removed. The manual harvesting rate was approximately 181 m^2/person/day.

EXPERIMENTAL WATER QUALITY CONDITIONS: CITY OF ARCATA'S OXIDATION POND EFFLUENT

Influent BOD_5 was highly variable due to a wastewater treatment plant upgrade during this period, resulting in primary and secondary digester solids being loaded into the oxidation pond. Monthly average effluent BOD_5 from the oxidation pond ranged from 31 to 91 mg/L, with a project period average of 50 mg/L (Tables 2 and 3). In summer, BOD_5 values were high (71, 68, and 91 mg/L for June, July, and August, respectively), and nonfilterable residue (SS) patterns from the oxidation pond were similar.

BIOCHEMICAL OXYGEN DEMAND

High-BOD_5 effluent provided an opportunity to observe the reaction of wetland treatment systems to high organic loads during the warmer months. Influent BOD_5 loadings varied from 15.75 to 67.5 kg/ha/day.

BOD_5 effluent values from the wetland cells increased from 10–12 mg/L in the spring to 18–25 mg/L during this summer period, not significantly differ-

Table 2. BOD_5, SS, and Dissolved Oxygen Effluent Levels Through Wetlands in Which a Different Harvesting Strategy Was Applied

		BOD_5		SS	
	DO	Effluent	% Removal	Effluent	% Removal
Influent	4.7	50.0		49.2	
Cell 1—50% harvested	0.7	19.4	61	11.5	76
Cell 1—100% harvested	1.5	18.2	63	15.3	69
Cell 3—not harvested	1.5	11.1	78	9.7	80
Cell 4—alternating 10% harvested	0.8	12.5	75	7.7	84
Cell 5—not harvested	0.72	12.6	75	8.7	82
Cell 6—two baffled compartments	0.7	14.6	71	10.0	79
Cell 7—four baffled compartments	1.1	11.1	78	10.2	79
Cell 8—eight baffled compartments	1.1	10.0	86	8.6	82
Cell 9—50% harvested alternating 1/4	1.5	18.4	63	13.8	72
Cell 10—50% harvested influent	1.4	10.3	79	13.6	72
Average	1.1	13.8	72.9	10.9	77.5

ent from values observed in the pilot project study. Breakthrough organic loading for the wetland cells had apparently not been reached. Generally, the segmented cells 6, 7, and 8 produced a lower BOD_5 effluent throughout the summer months. Cell 3, a nonharvested, nonsegmented control cell, also produced a consistent BOD_5 during the entire study period.

DISSOLVED OXYGEN

The DO in the oxidation pond had high average monthly values (7.0 mg/L or greater) during January and March, declined with winter storms in late January, and remained low in the summer (below 3.5 mg/L). DO levels were 1-2 mg/L during the last quarter, indicating a high level of decomposition and possible nitrification (Figure 5). DO levels decreased in the summer as the water temperature increased (14–16°C). Wetland cell influent from the oxidation pond ranged from 16°C to 19°C for the same period. The completely harvested wetland cell maintained a higher DO level during the spring, but by summer was similar to the other cells.

Winter conditions in wetlands minimize shading of periphyton and phytoplankton standing crop via senescent plant material and detrital debris in the water column. Differences in DO level in wetland cell effluents under various

Table 3. Arcata Marsh Wildlife Sanctuary Full-Scale Operations Monthly Average BOD_5 and SS

	BOD_5			SS		
	Pond Effluent (mg/L)	Wetland Effluent (mg/L)	% Removal	Pond Effluent (mg/L)	Wetland Effluent (mg/L)	% Removal
August 1986	16	11	32	22	14	37
September	22	13	41	22	13	41
October	41	12	71	53	7	83
November	37	20	38	45	27	40
December	48	19	61	38	29	25
January 1987	33	16	52	33	16	52
February	27	22	19	27	18	34
March	30	17	44	40	14	65
April	19	15	22	37	24	36
May	26	13	50	35	21	40
June	24	12	50	54	18	67
July	15	7	50	22	4	82
August	15	12	20	13	7	47
September	16	9	49	17	8	53
October	45	18	60	22	19	14
November	35	28	20	28	20	29
December	17	15	12	20	10	50
January 1988	17	12	30	21	10	53
February	16	14	13	26	11	58
March	29	13	56	21	8.5	60
April	12	5	59	20	3.5	82
May	25	4	86	29	1.5	95
Average	26	12	55	30	14	54
Variance	11	5.3	—	12	7.5	—
Violation of 30 mg/L standard	30%	0%	—	40%	0%	—

harvesting and baffling conditions (except for cell 3) were not significant. The DO level in cell 3—unharvested, with predominantly hardstem bulrush—was higher during the entire study period. Hardstem bulrush (*Scirpus* spp.) does not appear to contribute significant amounts of detritus, in contrast to cattail (*Typha*), which produces large amounts of detritus during fall dieback. In addition, this cell has historically received low organic loadings.

SUSPENDED SOLIDS

The SS values of the oxidation pond effluent were highly variable over the winter quarter. Warm weather in January produced conditions favorable for algal communities in the oxidation pond, and effluent response to varying influent values was observed in some cells. Overall SS concentration in effluents of 10 experimental cells was 9 mg/L. Certain cells produced average effluent SS concentrations of 4 mg/L over the winter quarter. Cell 4 has predominantly hardstem bulrushes.

The SS concentrations in the wetland cells were low, ranging from 2 to 18

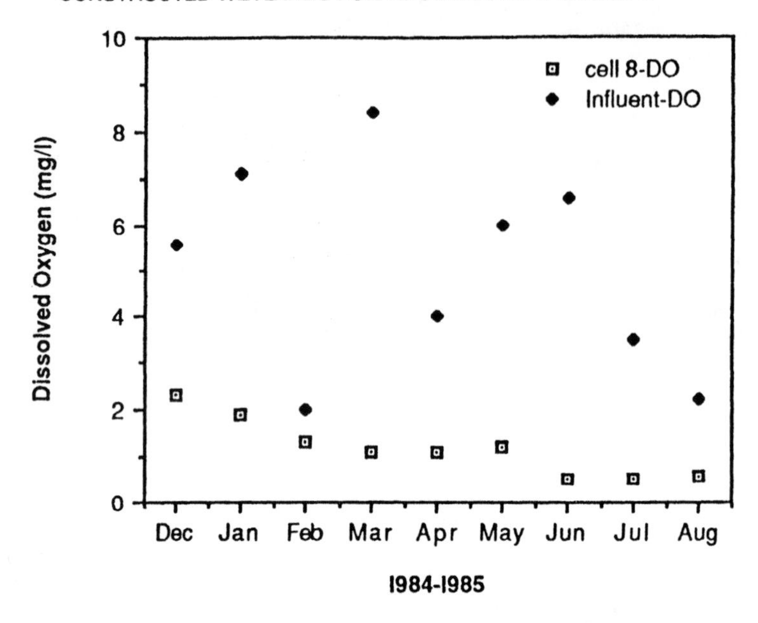

Figure 5. Comparison of dissolved oxygen levels in wetland pilot project cell 8 to the dissolved oxygen of the influent from an oxidation pond (cell 8, with eight internal cells, was 100% vegetated).

mg/L with averages of around 8 mg/L. The oxidation pond had unusually high concentrations of algae in the summer of 1988 due to the high organic loadings, high ammonia nitrogen, and clear skies. Since the wetland cells consistently produced less than 10 mg/L SS effluents, the percentage removed has significantly increased, averaging over 90% (Table 2).

Harvested cells produced higher SS concentrations than nonharvested cells. Cells harvested next to the effluent end produced higher SS values than those with vegetation at the effluent. Higher effluent SS occurred in those cells with open water areas near the effluent.

HYDROGEN ION

Average monthly pH of oxidation pond effluent varied from 6.9 to 7.6, with weekly values as high as 8.2 during late spring and early summer. Monthly average pH of wetland cells was 0.5-1.5 pH units lower than the oxidation pond effluent, ranging from 5.8 to 6.8. Characteristically, low and constant pH appeared related to decomposition of detritus, producing organic acid metabolites.

SUMMARY

Effluent BOD_5 values from both unharvested and baffled wetland cells were consistently low for the study period (Table 2). Cells with 50% or more harvest showed the poorest BOD_5 effluent. Having the last portion of a cell vegetated appeared important in maintaining low effluent BOD_5 and SS values. Effluent BOD_5 values were less than 20 mg/L for all test conditions and averaged 13.8 mg/L, representing 72% BOD_5 removal through wetland cells. SS removal was consistent and efficient. Harvested cells were least effective in removing SS. Average effluent SS value for the study period was 10.9 mg/L, a 78% SS removal rate. SS values had the least response to changes in influent values. Baffled cells had effluent values statistically similar to those of nonharvested cells.

REMOVAL KINETICS

An objective of the second study was to determine removal kinetics for lumped parameters such as BOD_5, SS, and fecal coliforms.[20] Wetland project cell 8, with eight internal cells, was used for collecting compartment samples every 8.3 linear meter through the cell. A composite sample was taken from each of three weirs on each baffle. Theoretical detention time in cell 8 was 144 hr, or 18 hr per compartment. Figure 6 shows the 12-month average BOD_5 remaining in each cell. A plot of the log of BOD_5 remaining against contact time in the wetland produced an excellent fit of the experimental data. Using the six-day detention time, a prediction model with a correlation coefficient of 0.96 covers the range of organic loadings commonly found from oxidation pond effluents and temperatures of 8°C to 17°C. The rate of BOD_5 removal through the wetland system for the seven-day contact period was 0.105 per day.

Autoflocculation and settling account for the high removal of suspended solids in the first intracell (Figure 7). Removal of suspended solids does not follow first-order kinetics past the first 16 m of the cell, and suspended solids were low through the remainder of the pilot wetland. Predation probably accounts for SS reduction in the second half of the cell.

A plot of the log of remaining fecal coliforms against detention time yields a regression equation with an excellent correlation coefficient (Figure 8). Six days of contact in a wetland system reduced fecal coliform levels approximately 2 log orders. The average oxidation pond fecal coliform level was 12.5 $\times 10^3$ cfu/100 mL, with an average wetland cell effluent of 316 cfu/100 mL. The log normal fit of the experimental data gave a correlation coefficient of 0.99, yielding a 0.29/day removal rate constant for reduction of fecal coliform density through a wetland system with emergent vegetation.

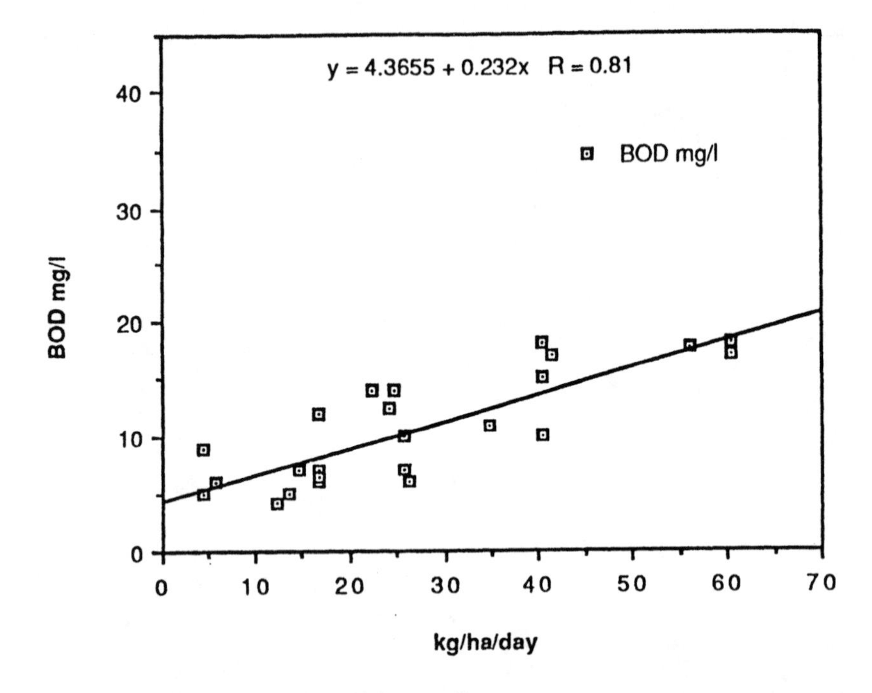

Figure 6. Wetland effluent BOD_5 concentration (mg/L) as a function of BOD_5 loading (kg/ha/da) seasonal loadings over the study periods (1982–1986).

FULL-SCALE OPERATIONS

The Arcata Wastewater Treatment plant upgrade, finished in June 1986, included installation of a new headworks, a retrofitted primary digester with cogeneration capability, cellularized oxidation ponds (2.3 ha of treatment wetlands) a chlorination/dechlorination facility and 13.6 ha of effluent polishing wetlands. These wetlands are part of the Arcata Marsh and Wildlife Sanctuary. The effluent polishing wetlands will serve as unit processes for SS and BOD_5 removal; a wildlife refuge; and a passive recreation facility.

Treatment wetlands were constructed last and were not planted until 1987. Vegetation density is inadequate, and design criteria were developed from the first wetland pilot project. The treatment wetlands will remove SS and associated BOD_5 prior to chlorination, allowing Arcata to meet its NPDES SS:BOD_5 standard of 30 mg/L:30 mg/L and reducing chlorine and sulfur dioxide operating costs.

These three wetlands receive treated effluent and BOD_5 and SS data for operations from August 1986 to May 1988 (Table 3). Oxidation pond BOD_5 effluents ranged from 16 to 48 mg/L, and average SS ranged from 13 to 53 mg/L. The 22-month BOD_5 and SS averages from the oxidation pond were 26 and 30 mg/L and Hauser Marsh effluent is blended with oxidation pond effluent if the flow is above 8706 m^3 (2.3 mgd) or is the total Arcata release at

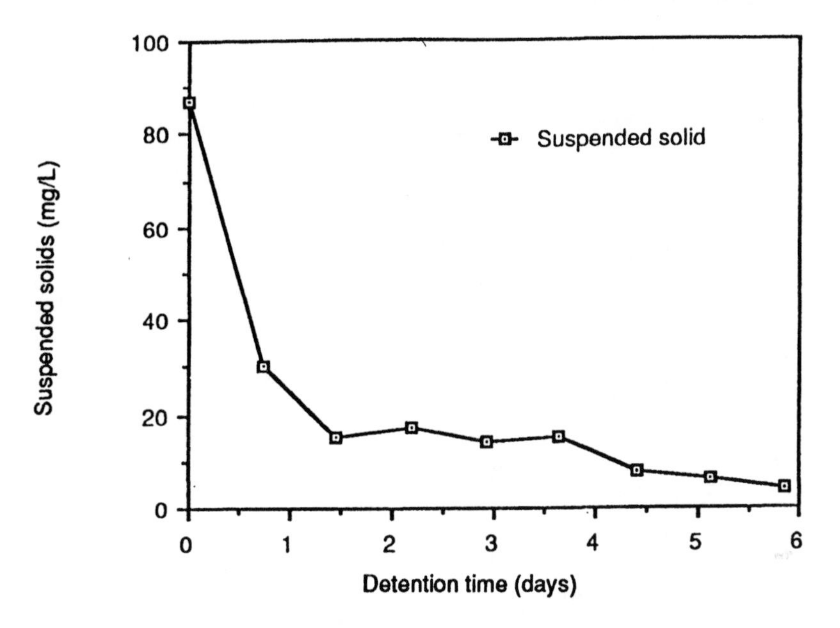

Figure 7. Suspended solids removal through each intracell compartment of cell 8, which has a total detention time of 5.9 days.

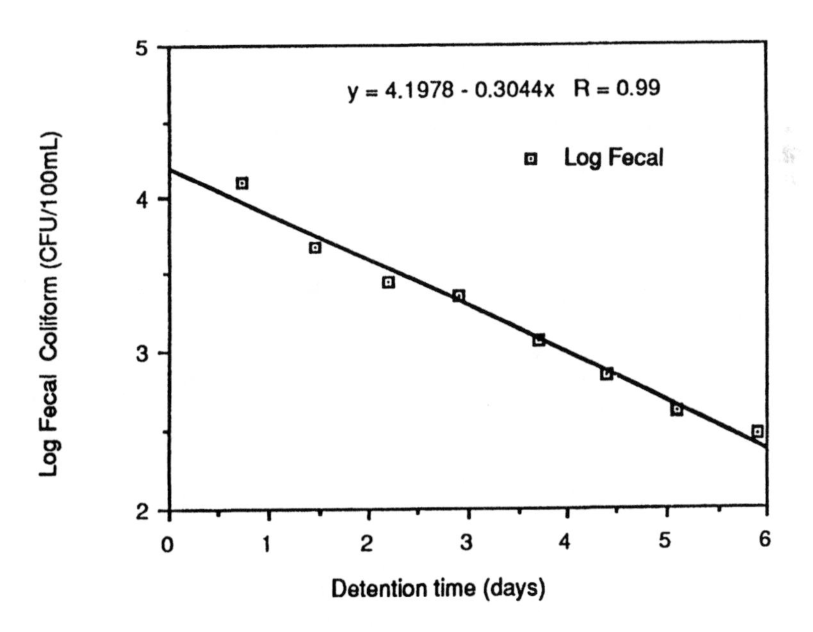

Figure 8. Fecal coliform removal through intracell compartments of cell 8, which has a total detention time of 5.9 days.

Table 4. Number of Potential SS and BOD_5 Violations from Oxidation Pond Wetland Pilot Project and Full Scale Wetland Effluent over Their Respective Study Periods

	Oxidation Pond (Effluent/N)	Pilot Project Wetland (Effluent/N)	Full-Scale AMWS (Effluent/N)
Daily (not to exceed 60 mg/L)			
SS	14/730	0/8760	—
BOD_5	7/730	0/8760	—
Weekly (not to exceed 45 mg/L)			
SS	30/104	0/1248	0/88
BOD_5	10/104	0/1248	0/88
Monthly (not to exceed 30 mg/L)			
SS	16/24	0/288	0/22
BOD_5	10/24	5/288	0/22

Note: N = number of samples.

flows below that level. Monthly averages ranged from 4 to 28 mg/L, with the higher values in late fall and winter and lower values in spring and summer. Monthly average SS ranged from 1.5 to 29 mg/L, and the higher SS values were in fall and winter with lower values in spring and summer. Overall BOD_5 removal was 54%, and overall SS removal was 53%.

Preliminary NPDES standards for Arcata for BOD_5 and SS are shown in Table 4 with the corresponding number of potential violations. The oxidation pond effluent would have violated the daily BOD_5 standard seven times and the daily SS standard 14 times over the two-year period. None of the 12 experimental wetlands with hydraulic loads ranging from 0.02 to 0.24 $m^3/m^2/$ day (0.021–0.252 mgd/ac) violated either the daily BOD_5 or the daily SS standard. The oxidation pond effluent exceeded the SS weekly standard 30 times and the BOD_5 weekly standard 10 times, representing a 28% and 9.5% noncompliance condition, respectively. Of 1248 weekly BOD_5 and SS values for the experimental wetlands collected, none were in violation of the weekly standard.

There were 16 monthly average influent SS values that exceeded the 30 mg/L SS standard (a 66% noncompliance), with no weekly average SS values from the experimental wetlands in excess of 30 mg/L. Five of 288 monthly wetland cell BOD_5 values exceeded 30 mg/L (a 1.7% noncompliance), but noncompliance values were no larger than 32 mg/L. Since the proposed hydraulic loading will be lower to the Arcata treatment wetlands, these results strongly suggest a potential 100% compliance for daily, weekly, and monthly standards.

The first 22 months were considered startup conditions because vegetation patterns and density were changing rapidly during initial wastewater loadings. Sago pondweed *(Potamogeton pectinatus)* plays a significant role in reducing light penetration and providing microbial attachment surfaces in the water column. Migrating waterfowl harvest most of it in the fall and winter, reducing water quality during those seasons. Emergent vegetation density is still inadequate to perform this function.

Table 5. Comparison of Pilot Project and Full-Scale Results for NPDES Parameters (Percent Change Through Wetlands)

	First Pilot Project, All Cells	Second Pilot Project, All Cells	Full-Scale Operation
BOD_5 (mg/L)	11.4	13.8	12
% change	–56	–73	–55
SS (mg/L)	5.3	10.8	14
% change	–85	–80	–54
Dissolved oxygen (mg/L)	1.5	1.1	5.0
% change	–73	–76	–27
pH	6.5	6.1	7.1
% change	29	–14	–6
Theoretical detention time/s (days)	1.5–30	5.4	8.5
1981 average	3.7		
1982 average	9.0		
Open water fraction	30–50%	0–100%	75%

Note: NPDES = National Pollutant Discharge Elimination System.

DISCUSSION

The results of two pilot projects and the first two years of full-scale operations are summarized in Table 5. The use of 375-m^2 pilot cells was effective in predicting the behavior of larger scaled wetland systems. BOD_5 values of the three systems were similar (12 mg/L), but SS values varied significantly (10 mg/L and 5 mg/L). The systems had significantly different proportions of open water area versus vegetated areas, which affected SS removal. Open water areas in constructed wetlands permit phytoplankton production, thereby contributing to the suspended solids level. If a canopied vegetative zone does not follow open water zones, suspended solids will increase.

DO values are also a function of the amount of open water. Natural reaeration plus photosynthetic contribution from phytoplankton populations produce higher levels. The pH of wetland systems are well buffered at or slightly below 7.0, with higher organic loading and vegetative cover decreasing the pH because of organic acids production by anaerobic breakdown of detritus. These studies and other studies[21,22] demonstrate the effectiveness and reliability of wetland systems in meeting secondary NPDES standards, although a biological similitude factor may be necessary when scaling up macrophyte pilot project results.

BOD_5 and fecal coliform removal can be effectively modeled by first-order kinetics. Effluent BOD_5 from a wetland is a function of organic loading rates, while SS effluent level is a function of some minimal detention time. Wetland treatment systems are effective and reliable treatment systems for treating oxidation pond effluent to a 30 mg/L BOD_5 and SS standard and can be designed and operated to produce 10 mg/L BOD and SS effluents.

REFERENCES

1. Gearheart, R. A., et al. "Volume 1—Final Report City of Arcata Marsh Pilot Project," City of Arcata Department of Public Works, April 1983.
2. Allen, G. H., and R. A. Gearheart. "Arcata Integrated Wastewater Treatment Reclamation and Salmon Ranching Project," paper presented at the winter meeting of The American Society of Agricultural Engineers, Chicago, Illinois (1978).
3. Boyd, R. L., S. E. Bayley, and J. Zoltek, Jr. "Removal of Nutrients from Treated Municipal Wastewater by Wetland Vegetation," *J. Water Pollut. Control Fed.* 49:780 (1977).
4. Cornwell, D. A., J. Zoltek, Jr., D. D. Patriuels, T. de S. Furmau, and J. I. Kim. "Nutrient Removal by Water Hyacinths," *J. Water Pollut. Control Fed.* 49(1):57–65 (1977).
5. Seidel, K. "Macrophytes and Water Purification," in *Biological Control of Water Pollution,* J. Tourbier and R. W. Pierson, Jr., Eds. (Philadelphia: University of Pennsylvania Press, 1976).
6. Wolverton, B. C. "Aquatic Plants for Wastewater Treatment: An Overview," in *Aquatic Plants for Water Treatment and Resource Recovery,* K. R. Reddy and W. H. Smith, Eds. (Orlando, FL: Magnolia Publishing Inc., 1987), pp. 3–15.
7. Wolverton, B. C., R. M. Barlow, and R. C. McDonald. "Application of Vascular Aquatic Plants for Pollution Removal, Energy, and Food Production in a Biological System," in *Biological Control of Water Pollution,* J. Tourbier and R. W. Pierson, Jr., Eds. (Philadelphia: University of Pennsylvania Press, 1976), pp. 141–150.
8. Cederquist, N. W., and W. M. Roche. "Reclamation and Reuse of Wastewater in the Suisun Marsh, California," in Proceedings, AWWA Water Reuse Symposium I, Washington, DC, March 25–30, 1979, pp. 685–702.
9. Ryther, J. H. "Preliminary Results with a Pilot Plant Waste Recycling-Marine Aquaculture System," paper presented at International Conference on the Renovation and Reuse of Wastewater Through Aquatic and Terrestrial Systems, Bellagio, Italy, July 15–21, 1975.
10. Valiela, I., S. Vince, and J. M. Teal. "Assimilation of Sewage by Wetlands," in *Estuarine Processes, Vol. I.,* M. Wiley, Ed. (New York: Academic Press, Inc., 1976), pp. 234–253.
11. Stephenson, M., G. Turner, P. Opoe, J. Colt, A. Knight, and G. Tchobanoglous. "The Use and Potential of Aquatic Species for Wastewater Treatment, Appendix A, The Environmental Requirements of Aquatic Plants," Publication No. 65, California State Water Resources Control Board, Sacramento, California (1980).
12. Stowell, R., R. Ludwig, J. Colt, G. Tchobanoglous. "Concepts in Aquatic Treatment System Design," *J. Environ. Eng. Div. ASCE* 107(EE5):16555–16569 (1981).
13. Tchobanoglous, G. "Aquatic Plant Systems for Wastewater Treatment: Engineering Considerations," in *Aquatic Plants for Water Treatment and Resource Recovery,* K. R. Reddy and W. H. Smith, Eds. (Orlando, FL: Magnolia Publishing Inc., 1987), pp. 27–48.
14. Watson, J. T., F. D. Diodato, and M. Launch. "Design and Performance of the Artificial Wetlands Wastewater Treatment Plant at Iselin, Pennsylvania," in *Aquatic Plants for Water Treatment and Resource Recovery,"* K. R. Reddy and W. H. Smith, Eds. (Orlando, FL: Magnolia Publishing Inc., 1987), pp. 263–271.

15. Knight, R. L. "Wetlands: An Alternative for Effluent Disposal, Treatment, and Reuse," *Florida Water Res. J.* (November-December 1985), pp. 6–9.
16. Kadlec, R. H. "Wetland Tertiary Treatment at Houghton Lake, Michigan," in *Aquaculture Systems for Wastewater Treatment: Seminar Proceedings and Engineering Assessment,* R. K. Bastian and S. C. Reed, Eds. U.S. EPA Report-430/9-80-006 (1979), pp. 101–139.
17. Nichols, D. S. "Capacity of Natural Wetlands to Remove Nutrients from Wastewater," *J. Water Pollut. Control Fed.* 55(5): (1983).
18. DeJong, J., T. Kok, and A. H. Koridon. "The Purification of Wastewater and Effluent Using Marsh Vegetations and Soils," Proceedings EWRS 5th Symposium on Aquatic Weeds (1985).
19. Williams, J. R., Jr. "The Relationship Between the Standing Crop of Macrophytes and the Nutrient Removal Efficiency," MS thesis, Humboldt State University, Arcata, CA (1985).
20. Gearheart, R. A., J. Williams, H. Holbrook, and M. Ives. "Wetland Speciation and Harvesting Effects on Effluent Quality; Final Report to State Water Resources Control Board," Project No. 3-154-500-0, Humboldt State University, Arcata, CA (1986).
21. Demgen, F. C. "An Overview of Four New Wastewater Wetlands Projects," in *Future of Water Reuse, Vol. 2, Proceedings Water Reuse Symposium III* (San Diego, CA: AWWA Research Foundation, 1985), pp. 579–595.
22. Hammer, D., and R. H. Kadlec. "Design Principles for Wetland Treatment Systems," U.S. EPA Report-600/S2-83-026 (May 1983).

CHAPTER 9

The Iselin Marsh Pond Meadow

Thomas E. Conway and John M. Murtha

Sewage management in Pennsylvania faces many problems. The Commonwealth of Pennsylvania has one of the highest number of individual sewage systems in the nation, serving one-third of Pennsylvania's population of 11 million. In 87% of the cases, soils are not suitable for conventional on-lot systems. In the early 1960s, a study conducted by Johns Hopkins University revealed over 65,000 visible malfunctions of on-lot disposal systems.[1]

To address these problems, the Commonwealth enacted legislation known as the Sewage Facilities Act of 1966. One of the intentions of this act was to prevent development of sewage problems by regulating new on-lot disposal systems and to resolve existing and future community problems through municipal sewage plans. The initial approaches favored large, centralized regional systems as compared to smaller package plants. The majority of the funding for the early projects came from the federal government. It became increasingly difficult for small communities to compete for federal funding and, by the late 1970s, nearly impossible. With approximately 3500 small communities unable to obtain funding, new approaches were needed.

RESEARCH AND DEMONSTRATION PROJECT

In 1978 Clifford E. Jones, then Secretary of the Department of Environmental Resources (DER), issued a directive initiating development and demonstration of affordable sewage treatment technologies and institutional schemes for small communities. During 1978–1979, the DER staff screened over 600 communities for a suitable research and demonstration site. Of the sites reviewed, 10 areas earned final consideration. Iselin, a small western Pennsylvania village in Young Township, Indiana County, was selected as the project site for the following reasons: (1) the area was considered a typical Pennsylvania small community in the mid-range of 10 to 1000 housing units; (2) community pride was evident, with most families undertaking renovation of homes originally owned by coal companies; (3) a management/operational agency was already in place; and (4) the site was geographically proximal to a research, monitoring, and consulting institution at the Indiana University of Pennsylvania (IUP).

Selecting a technology was based on several criteria. The project should have a low initial construction cost. Additionally, after completion of the construction phase, operation and maintenance (O&M) should be minimal in both man-hours and energy costs. Finally, the construction techniques should be simple so the technology could be applied in other rural communities using local contractors and municipal employees. This would also facilitate future expansion and modifications of the treatment process if indicated by data and recommendations from evaluation of this initial project.

Planning and design of the project addressed both the collection system and treatment facilities. Because of treatment and cost advantages, consideration was given to use of a septic effluent conveyance system (SEC). This could not be pursued at the time for legal and administrative reasons. The collection system selected was a shallow-placement, gravity-flow system utilizing 15-cm (6-in.) diameter and 20-cm (8-in.) diameter PVC pipe. The total service area of 104 homes was divided into two parts. Forty-two homes were placed on a community on-lot disposal system (COLDS). COLDS utilized a large septic tank with an absorption sand bed built on a coal refuse pile (bony dump). The remaining 62 homes were served by a marsh pond meadow system (MPM).

BUILDING THE ISELIN R&D PROJECT

Financing for the project was provided by the DER with funds from fines collected under the Clean Streams Law. This was Pennsylvania's first small community wastewater research and demonstration project and one of the few ongoing research projects in the nation that addresses this problem. A contractual agreement was made among four parties: the DER, which was responsible for planning, designing, financing, construction, inspection, and R&D; Indiana University of Pennsylvania (IUP), which served as consultants for design, monitoring, research, and evaluation; The Indiana County Municipal Services Authority (ICMSA), which handled contract and bid administration, design review, ownership, and O&M; and the Young Township Supervisors, who were responsible for the adoption of an official sewage plan revision and community support.

During the construction phase, which commenced in 1982, the DER Bureau of Operations stressed strict compliance to the terms of construction project specifications. To our knowledge, Iselin was the most thoroughly inspected community system in Pennsylvania's wastewater treatment history. Strict attention was paid to infiltration and inflow considerations because COLDS could be severely affected by any increase in extraneous flows and there was similar concern about the capabilities of the MPM.

Capital cost of the system was approximately $500,000, with breakdown of costs as follows:

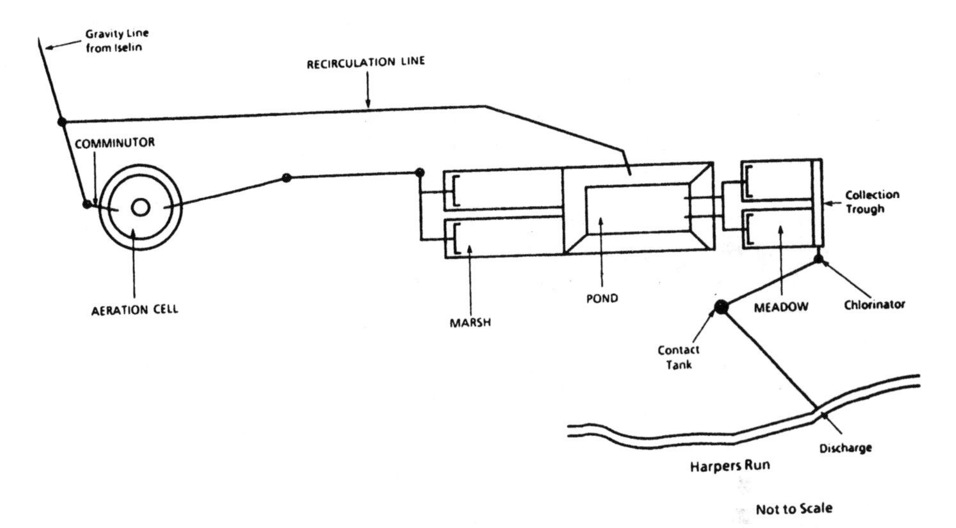

Figure 1. Ismelin marsh pond meadow schematic.

1. *Collection system*—3050 meters (10,000 feet) of 15-cm (6-in.) diameter and 20-cm (8-in.) diameter gravity sewer at $73.93 per meter ($22.50 per linear foot)—$225,500.
2. *COLDS*—total—$60,000 ($31,000 was for septic tank construction, with the remaining $29,000 for excavation, drain-field piping, and aggregate/sand materials).
3. *MPM* at $174,000 was competitive with package plants, but had lower O&M. This cost included a 20–30% cost for extensive cutting and filling of the project site.

THE ISELIN MARSH POND MEADOW

Design of the marsh pond meadow was based on pioneering work done by Maxwell M. Small of Brookhaven National Laboratory.[2] Final design considerations also incorporated experience gained from a short-lived (one year) experimental system installed in southeastern Pennsylvania to serve a new retirement center.[3]

The MPM system was designed to serve an average daily flow of 45.4 m^3/day (12,000 gpd) from 62 homes. It contains seven components (Figure 1), all of which are sealed from groundwater infiltration and exfiltration by either bentonite clay or Hypalon liners. The components are:

1. a small comminutor and bar screen with low energy requirements.
2. an aeration cell utilizing a 3730-watt (5 horsepower) floating surface aerator (retention time of 2.86 days).
3. a lateral-flow marsh planted with cattails *(Typha latifolia)* in a sand medium.

Dissolved minerals are removed by microbial digestion, plant metabolism, and physical filtration through sand. The rhizomes create an environment that encourages the growth of treatment bacteria; stabilizes the sand to prevent channelization; and reduces clogging and maintenance of the sand surface. The marsh is designed for an application rate of 467.5 m^3/ha/day (50,000 gal/ac/day). For operational purposes, duplicate marshes (11.5 m × 43 m)—each capable of treating one-half of the daily design flow—are provided. The growth/treatment substrate in the marsh consists of 41 cm (16 in.) of sand overlaid by 10 cm (4 in.) of 2B aggregate (12–64 mm). A slope of 2% along the length of the marsh is provided.

4. a facultative pond with plant life *(Lemna* sp., *Sagittaria* sp., *Nuphar* sp., and *Amacharis* sp.) to remove dissolved nutrients, and fish *(Cyprinus carpio, Ictalurus nebulosis,* and *Lepomis* sp.) to remove the plants. The pond is designed for 50% BOD_5 reduction under winter conditions. Thus, the 1000 m^3 (264,000 gal) capacity provides retention of approximately 22 days.
5. a meadow planted with reed canary grass *(Phalaris arundinacea)* to provide final filtration, treatment, and nutrient uptake. Like the marsh, this unit is divided into two identical units (11.5 m × 21.5 m), each capable of treating one-half of the daily design flow at an application rate of 935 m^3/ha/day (100,000 gal/ac/day).
6. a concrete trough that collects and discharges the treated effluent.
7. an erosion chlorinator and contact tank (120 minutes) for bacteriological control.

Start-up of the treatment system occurred in February 1983. Monitoring and sample collection were completed by two student interns from the Indiana University of Pennsylvania supervised and directed by the authors. Double samples were collected from each of the 15 collecting points from influent to final effluent. One sample was analyzed at the DER laboratory in Harrisburg and the second at IUP facilities using a Hach DR-EL/4 in addition to some wet chemistry using Standard Analytical Methods. Dissolved oxygen (DO), rainfall, flow rates, and weather conditions were also collected. Data analysis by Watson, Tennessee Valley Authority[4] indicated the system has the capability to exceed Pennsylvania's secondary discharge requirements (30 mg/L suspended solids, 30 mg/L BOD_5 and effective disinfection). This analysis showed average yearly reductions throughout the system for BOD_5 of 97% and suspended solids of 89%. Additionally, yearly fecal coliform reduction exceeds 99%. For these three parameters no significant difference between summer and winter values was noted.

Conversely, the effectiveness of the system to reduce ammonia nitrogen and total phosphorus varied between summer and winter conditions. Yearly ammonia nitrogen reduction averaged 77% but ranged from 93% (summer) to 54% (winter). Similarly, total phosphorus was reduced, on the average, by 82%. The range of reduction varied from 68% (winter) to 90% (summer).

Based on the data collected as well as operational experiences accumulated over the past five years,[5,6] several physical modifications to the system have been made. These include (1) installation of a collection pipe in the collection trough to eliminate the effect of extraneous organics and decaying animals in the trough on the final effluent; (2) installation of riprap faces on the pond banks to discourage muskrat damage (a trapping control program is a must); and (3) installation of surface baffles in the marsh to force lateral movement through the sand, increase exposed surface area for free oxygen exchange, reduce surface channelization, and encourage solids settling. These changes have, for the most part, been successful. Additional modifications are underway to increase treatment capabilities of the system and allow introduction of additional flows from COLDS. An anaerobic digester (two-compartment septic tank, retention of 1.5 days) will be installed in line well ahead of the aeration cell to begin the ammonification process while collecting and digesting solids. This should reduce loading on the aeration cell and the rest of the system (marsh, primarily), which should, in turn, encourage increased efficiency for winter operations. Second- and third-generation designs elsewhere in Pennsylvania now use the anerobic-aerobic concept successfully.

Experimentation with different emergent species, distribution, and combinations *(Typha latifolia; Phragmites australis,* etc.) continues at this site and others.

SUMMARY

Additional ongoing research and evaluation on these systems will be done before a final conclusion is reached. At present, the MPM system has the potential to meet the sewage needs of many of Pennsylvania's small communities. Low construction cost, low O&M, simple construction techniques, uncomplicated operational requirements, and system flexibility are some of the advantages. Currently, there are seven similar or modified systems in the planning, design, and/or construction stages in Pennsylvania. Additionally, other states and agencies have shown increased interest in the use of these systems.

REFERENCES

1. "Pennsylvania's Health," Report of a Study of Health Needs and Resources of Pennsylvania, Johns Hopkins University (May 1961).
2. Small, M. M. "Natural Sewage Recycling Systems," Brookhaven National Laboratory United States Energy Research and Development Administration Report EY-76-C-02-0016 (1977).
3. SME-Martin, Environmental Consultants. "Final Report: Marsh-Pond-Meadow Sewage Treatment Facility," Experimental Permit No. 4677452 (1980).
4. Watson, J. T., F. D. Diodato, and M. L. Lauch. "Design and Performance of the Artificial Wetlands Wastewater Treatment Plant at Iselin Pennsylvania," in *Aquatic Plants for Water Treatment and Resource Recovery,"* K. R. Reddy and W. H. Smith, Eds. (Orlando, FL: Magnolia Publishing Inc., 1987).

5. Lauch, M. L., and J. M. Murtha. Unpublished results and reports for Department of Environmental Resources, Commonwealth of Pennsylvania (1983–1987).
6. Conway, T. E. Personal communication (1985).

CHAPTER 10

Integrated Wastewater Treatment Using Artificial Wetlands: A Gravel Marsh Case Study

Richard M. Gersberg, Stephen R. Lyon, Robert Brenner, and Bert V. Elkins

INTRODUCTION

Wetland ecosystems have the ability to remove aquatic pollutants through a variety of physical, chemical, and biological processes occurring in the soil-water matrix and in the plant rhizosphere. There is now convincing evidence that natural wetland ecosystems may be used to treat wastewater effluents[1] and can perform integrated wastewater treatment by combining secondary as well as advanced wastewater treatment capabilities in a single system.[2] However, natural wetlands are usually unavailable at treatment sites and typically not amenable to controlled scientific experimentation. Studies of the wastewater treatment capability of artificial wetlands have suggested that it may be possible to derive benefits of wetland treatment in constructed marsh systems without endangering the function or extent of natural wetlands.

Gersberg et al.,[3-5] in a series of controlled experiments using artificial wetlands created by propagating cattails *(Typha* sp.), bulrushes *(Scirpus* sp.), and reeds *(Phragmites* sp.) in a gravel substrate, demonstrated removal of a wide variety of chemical and biological contaminants from municipal wastewaters. Gravel in these artificial systems affords high substrate permeability ($K = > 10^{-5}$ m/sec), thus assuring that much of the wastewater percolates through the well-developed rhizosphere (root zone), where microbial activity is high. Similar gravel-based systems planted with reeds or cattails have been described by Wolverton,[6] who found them capable of nearly 90% BOD_5 removal from domestic sewage. A recent cost estimate of a gravel marsh treating primary lagoon effluent at Benton, Kentucky yielded a unit cost value of \$0.18 per m^3/day.[7] The capability of artificial wetlands to perform integrated wastewater treatment using natural processes, with low energy input, and capital and operation and maintenance (O&M) expense, make them very attractive for use by small to medium-sized communities for meeting discharge limitations.[8]

A primary objective of this chapter is to present design and operational data on the use of artificial wetlands for performing secondary treatment of primary municipal wastewaters. A second objective is to describe the mechanisms by which these artificial wetlands may also remove nitrogen and total coliform bacteria from inflowing wastewaters.

MATERIALS AND METHODS

The artificial wetland bed used for testing of secondary wastewater treatment capability consisted of a plastic-lined (Hypalon, 30 mil) excavation containing emergent vegetation *(Scirpus* sp.) growing in gravel. The bed was 71 m long × 11.6 m wide × 0.76 m deep. Primary effluent from existing sedimentation tanks at the Santee Water Reclamation Facility (Padre Dam Municipal Water District, Santee, California) was used as influent to this wetland bed.

For testing of nitrogen and total coliform bacteria removal, we used marsh beds of similar design to that above, but smaller. Each bed was 18.5 m long × 3.5 m wide × 0.76 m deep. Nitrified effluent from the existing activated sludge (secondary) treatment plant at Santee, California was used as inflow for bacterial studies, while primary effluent was the inflow for our ammonia removal studies.

All water samples were analyzed according to EPA-approved methodology.[9] Ammonia-N was analyzed with an Orion ion-selective electrode. BOD_5 was measured using a five-day incubation at 20°C. Suspended solids (nonfilterable residue) were determined gravimetrically by drying at 103–105°C. Oxygen was measured with a YSI 54A oxygen meter calibrated weekly using the Winkler procedure. Total coliform bacteria were enumerated by the multiple-tube fermentation method (presumptive and confirmed tests).

Samples were collected weekly by pumping water from a standpipe reaching to the bottom at the effluent end of each bed. Valves at the inflow and outflow of each bed were used to regulate flow rates (application rates) and control water level.

RESULTS AND METHODS

Since secondary wastewater treatment accounts for about 75% of the total construction cost and 60–70% of the total O&M expense of a conventional secondary wastewater treatment facility,[10] then from a cost standpoint it would be most advantageous to use artificial wetlands for secondary treatment of primary wastewaters. Figure 1 shows the capacity of the artificial wetlands for BOD_5 removal when primary effluent was the sole source of inflow. At an application rate of about 5 cm/day (7.5–8 ha per 3785 m^3/day), the mean BOD_5 removal efficiency for the period from October 1983 to August 1984 was 90%, with the mean influent level of 113 mg/L reduced to 11 mg/L in the wetland effluent. On four sampling dates during late winter of 1984, BOD_5

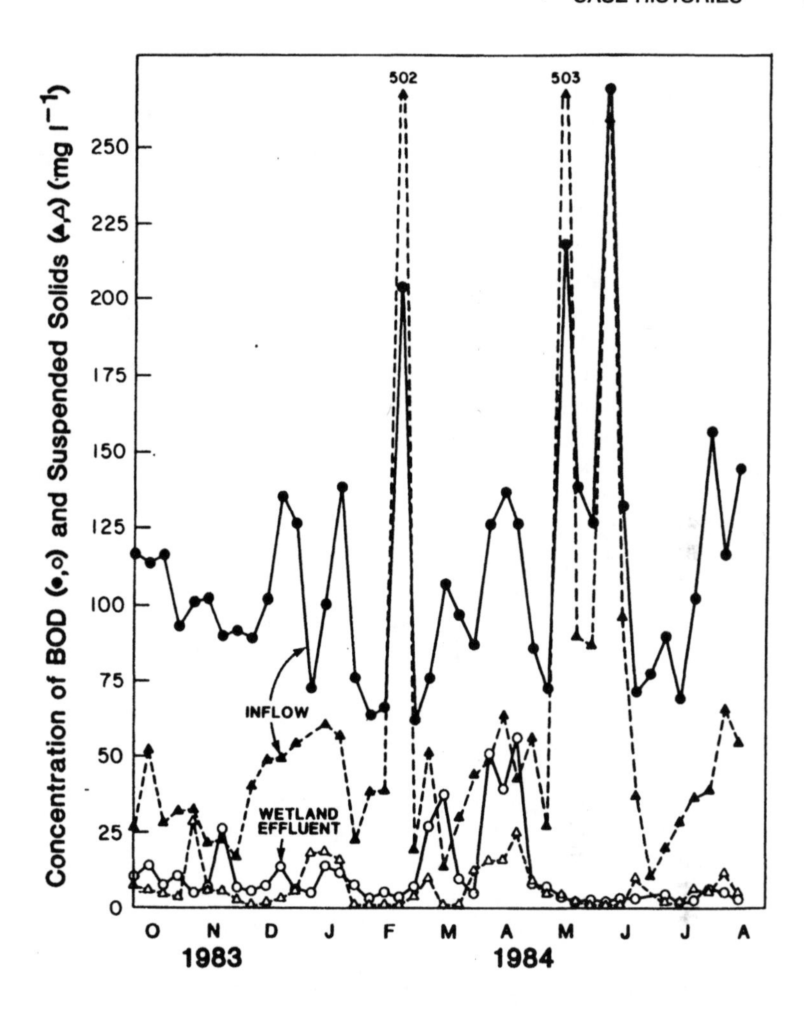

Figure 1. Level of BOD_5 and suspended solids in the inflow and the effluent of an artificial wetlands, at a primary wastewater application rate of 5 cm/day.

exceeded 30 mg/L but thereafter dropped to below 10 mg/L and remained at this low level for nearly four months until the end of the study.

Figure 1 also shows that at the primary wastewater application rate of 5 cm/day, the artificial wetlands removed approximately 90% of the influent suspended solids, with mean inflow level of about 70 mg/L reduced to a mean level of 6.8 mg/L in the wetland effluent. At no time throughout the period did suspended solids in the wetland effluent ever exceed 30 mg/L. At the hydraulic application rate of 5 cm/day, about 7.5–8 ha of constructed wetlands would be required to treat 3785 m^3 (1 mgd) of primary wastewaters to secondary treatment levels (<30 mg/L for BOD_5 and suspended solids). Capital cost is estimated to be \$1.7 million (ENR = 4200) as compared to \$2.5 million for a conventional secondary treatment facility (excluding the capital cost of primary sedimentation facilities). Estimated O&M expense (labor, energy cost, parts and supplies, and harvesting) for artificial wetlands would be about \$137 per 3785 m^3 treated, less than half the corresponding cost for conventional secondary treatment.[11]

The capability of wetlands to remove nitrogen through denitrification is well established.[12] Denitrification is an anaerobic respiration whereby nitrate (or nitrite) is used as the terminal electron acceptor for oxidation of organic compounds, and is ultimately reduced to the gaseous end products N_2O or N_2. This process is the most successful procedure for removal of nitrate from secondary effluents and agricultural return flows.[13,14]

Our studies of artificial wetlands at Santee have demonstrated sustained high rates of denitrification (>95% removal) at secondary wastewater application rates as high as 102 cm/day when methanol was added as an electron donor to drive denitrification.[15]

When nitrogen is in the form of ammonia (as in primary wastewaters), biological nitrogen removal can be accomplished by sequential nitrification-denitrification, whereby the ammonia is first oxidized (aerobically) to nitrate by nitrifying bacteria, which is then denitrified to N_2 gas in anaerobic microenvironments. Artificial wetlands with permeable soils can sustain relatively high rates of sequential nitrification-denitrification due to the alternating aerobic-anaerobic conditions formed at the soil-rhizosphere interfaces. At primary wastewater application rates of about 5 cm/day, total nitrogen removals above 80% were observed in both bulrush and reed artificial wetland beds.[3]

Under conditions where dissolved organic carbon is not limiting (as when primary wastewaters are applied), the factor most limiting nitrogen removal appears to be the supply of O_2 necessary to sustain nitrification. In this regard, ability of the aquatic plants to translocate O_2 from shoots to roots, establishing an oxidized rhizosphere where nitrification can proceed, is an important factor. Capacity for ammonia removal by vegetated versus unvegetated artificial wetland beds is shown in Figure 2. We measured 94% removal for bulrush wetlands, 78% for reeds, and 28% for cattails as compared with only 11% for an unvegetated bed.[3] Clearly, sequential nitrification-denitrification was impeded in the unvegetated bed as compared to reed and bulrush beds. Similarly, Hansen and Anderson[16] showed that potential nitrification rate in sediments from a reed swamp was three times higher than for sediments from deeper waters without plants. This evidence supports the hypothesis that nitrifying bacteria can be directly stimulated by oxidizing abilities of the plant rhizome. Hence, aquatic plants perform a function analogous to a compressor supplying air to an activated sludge tank.

Using data shown in Figure 2 for ammonia removal in unvegetated versus vegetated beds and similar data showing the excess BOD_5 removal in planted versus unplanted beds (96% and 81% for bulrushes and reeds, respectively, versus 69% in the unplanted bed),[3] we can calculate the minimum amount of

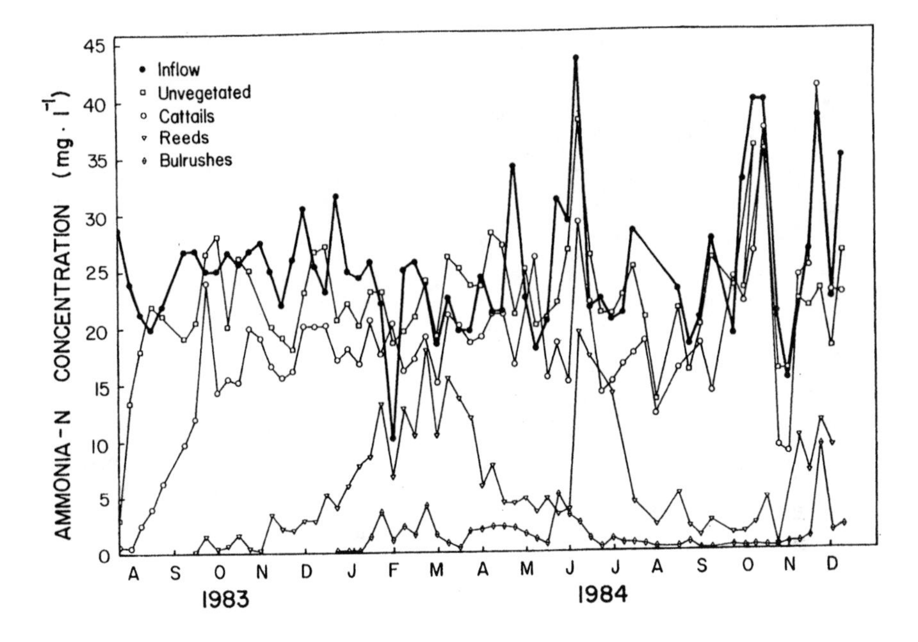

Figure 2. Level of ammonia nitrogen in the inflow and effluent of the vegetated and unvegetated beds, at a primary wastewater application rate of 5 cm/day.

O_2 that is introduced into the substrata due to plant translocation. For example, the excess level of nitrification in the bulrush bed as compared to the unvegetated bed accounted for the removal of 19.8 mg/L of NH_4-N. The oxygen demand of nitrification is 4.5 mg O_2 per mg N. Therefore, at least 89 mg/L of O_2 was needed for nitrification of this amount of ammonia. Similarly, excess oxygen supplied to satisfy aerobic respiration in the vegetated bed (as measured by BOD_5 removal) amounted to 31 mg/L. Therefore, a minimum amount of 120 mg/L of O_2 must have been added to the bulrush bed. Since each bed (area = 66 m^2) received about 3957 L/day of wastewater, then approximately 7.2 g of O_2 must have been supplied per m^2 of marsh surface. This amount is more than enough to reaerate the wastewater (each time to saturation equal to 8.6 mg/L O_2 at 23°) three times over each day. Obviously, higher aquatic plants can play an important role in translocating oxygen in tertiary treatment of wastewaters.

There is much less information available on the fate of biological indicators of pollution (such as total coliform bacteria) in wetlands. Artificial wetlands act as biofilters, offering a unique combination of physical, chemical, and biological factors that contribute to inactivation and removal of both pathogenic viruses and bacteria. In addition to filtration through the substrate and the attached biofilm, physical removal factors include sedimentation, aggregation, and inactivation by ultraviolet radiation. Chemical factors include oxida-

Table 1. Survival of Total Coliform Bacteria in a Vegetated (Bulrush) Wetland Bed with Secondary Wastewater Inflow

Date	Most Probable Number per 100 mL Inflow	Wetland Effluent	Percent Removal
11/27/84	3.5×10^4	3.5×10^2	99.0
12/03/84	3.4×10^4	3.3×10^2	99.1
12/07/84	9.2×10^4	9.2×10^2	99.0
1/14/85	5.4×10^4	4.6×10^2	99.1
1/21/85	9.2×10^4	3.3×10^2	99.6
1/28/85	9.2×10^4	3.3×10^2	99.6
2/04/85	5.4×10^4	1.7×10^3	99.7
2/11/85	4.9×10^4	1.3×10^2	99.7
2/18/85	1.3×10^4	4.9×10^2	96.2
2/24/85	4.6×10^4	3.3×10^2	99.3
3/04/85	2.7×10^4	3.3×10^2	98.8
3/12/85	4.6×10^4	2.7×10^2	99.4
3/18/85	4.9×10^4	3.3×10^2	99.3
3/27/85	4.9×10^4	1.6×10^4	67.4
3/30/85	3.5×10^5	3.3×10^2	99.9
4/08/85	5.4×10^5	2.3×10^2	99.9
4/23/85	3.3×10^5	1.7×10^3	94.8
4/30/85	1.7×10^4	2.4×10^3	85.8
5/07/85	7.9×10^4	2.2×10^2	99.7
5/13/85	7.9×10^4	1.7×10^3	97.8
5/20/85	4.6×10^4	3.3×10^2	99.2
5/28/85	3.3×10^4	2.7×10^2	99.2
6/03/85	2.3×10^4	3.3×10^2	98.6
6/11/85	4.9×10^4	2.4×10^3	95.1
6/17/85	7.9×10^4	3.5×10^3	95.6
6/25/85	3.3×10^4	1.4×10^2	99.6
		x (Mean) =	97.0%

Note: Application rate = 18 cm per day.

tion, exposure to biocides excreted by plants, and adsorption to organic matter and the biofilm. Biological removal mechanisms include antibiosis, ingestion by nematodes or protozoa, attack by lytic bacteria (or viruses), and natural cell die-off.[17]

Gersberg[5] showed that at a primary wastewater application rate of 5 cm/day mean total coliform bacteria level was reduced 99.1% by flow through a vegetated artificial wetland bed. We now present results on secondary effluent as the sole source of inflow. The capability of a bulrush bed for removal of total coliform bacteria from secondary wastewater (application rate of 18 cm/day) is shown in Table 1. For the study period from November 1984 through June 1985, mean total coliform removal efficiency was 97%.

In a study of artificial wetlands in Ontario, Canada, Palmateer et al.[18] determined a fecal coliform bacteria removal efficiency of approximately 90% when operated at a 6–7 day residence time, while Gearheart et al.[19] found total coliform removal efficiencies of 70–90% during winter and 0–50% during summer, at 1.5–7 day retention times of oxidation pond effluents in artificial marshes in Arcata, California. These removal efficiencies are somewhat lower than total coliform removal efficiencies we observed in the artificial bulrush wetlands at Santee at a 1.5-day hydraulic residence time. Perhaps the higher permeability of a gravel-based wetlands allows increased rhizosphere effects and substrate-biofilm interactions, leading to higher total coliform removal efficiencies than have previously been measured.

Ability of artificial wetlands to perform integrated wastewater treatment by removing bacterial indicators as well as BOD_5, suspended solids, and nitrogen makes them an attractive alternative for meeting treatment needs of small to medium-sized communities. Where land is available, these artificial wetlands offer a simple process design, low O&M expense, and wildlife enhancement value, and represent an innovative solution to cost-effective wastewater treatment.

ACKNOWLEDGMENTS

This study was supported by Grant B-54835 from the California Department of Water Resources and by Grant CR 807299-01-1 from the U.S. Environmental Protection Agency. The San Diego County Water Authority and the members of the San Diego Region Water Reclamation Agency also provided financial support. We thank the staff of the Santee Water Reclamation Facility (Padre Dam Municipal Water District) for space and facilities and V. Rosenbrook for manuscript preparation.

REFERENCES

1. Nichols, S. "Capacity of Natural Wetlands to Remove Nutrients from Wastewater," *J. Water Pollut. Control Fed.* 55:495–505 (1983).
2. Best, G. R. "Natural Wetlands—Southern Environment: Wastewater to Wetlands, Where Do We Go from Here?" in *Aquatic Plants for Water Treatment and Resource Recovery*, K. R. Reddy and W. H. Smith, Eds. (Orlando, FL: Magnolia Publishing Inc., 1987), p. 99.
3. Gersberg, R. M., B. V. Elkins, S. R. Lyon, and C. R. Goldman. "Role of Aquatic Plants in Wastewater Treatment by Artificial Wetlands," *Water Res.* 20:363–368 (1986).
4. Gersberg, R. M., S. R. Lyon, R. Brenner, and B. V. Elkins. "Fate of Viruses in Artificial Wetlands," *Appl. Environ. Microbiol.* 53:731–736 (1987).
5. Gersberg, R. M., R. Brenner, S. R. Lyon, and B. V. Elkins. "Survival of Bacteria and Viruses in Municipal Wastewaters Applied to Artificial Wetlands," in *Aquatic Plants for Water Treatment and Resource Recovery*, K. R. Reddy and W. H. Smith, Eds. (Orlando, FL: Magnolia Publishing Inc., 1987), p. 237.
6. Wolverton, B. C. "Aquatic Plants for Wastewater Treatment: An Overview," in *Aquatic Plants for Water Treatment and Resource Recovery*, K. R. Reddy and W. H. Smith, Eds. (Orlando, FL: Magnolia Publishing Inc., 1987), p. 3.
7. Steiner, G. R., J. T. Watson, D. A. Hammer, and D. F. Harker, Jr. "Municipal Wastewater Treatment with Artificial Wetlands—A TVA/Kentucky Demonstra-

tion," in *Aquatic Plants for Water Treatment and Resource Recovery*, K. R. Reddy and W. H. Smith, Eds. (Orlando, FL: Magnolia Publishing Inc., 1987), p. 923.

8. Watson, J. T., F. D. Diodato, and M. Lauch. "Design and Performance of the Artificial Wetlands Wastewater Treatment Plant at Iselin, Pennsylvania," in *Aquatic Plants for Water Treatment and Resource Recovery*, K. R. Reddy and W. H. Smith, Eds. (Orlando, FL: Magnolia Publishing Inc., 1987), p. 263.
9. "Methods for Chemical Analysis of Water and Wastes," U.S. Environmental Protection Agency Report 600/4-79-020 (1979).
10. "Construction Costs for Municipal Wastewater Treatment Plants," U.S. Environmental Protection Agency Report 430-9-80-003 (1980).
11. Gersberg, R. M., B. V. Elkins, and C. R. Goldman. "Wastewater Treatment by Artificial Wetlands," *Water Sci. Technol.* 17:443–450 (1984).
12. Good, B. J., and W. H. Patrick, Jr. "Root-Water-Sediment Interface Processes," in *Aquatic Plants for Water Treatment and Resource Recovery*, K. R. Reddy and W. H. Smith, Eds. (Orlando, FL: Magnolia Publishing Inc., 1987), p. 359.
13. Sharma, B., and R. C. Ahler. "Nitrification and Nitrogen Removal," *Water Res.* 11:897–925 (1977).
14. Brown, F. L. "The Occurrence and Removal of Nitrogen in Subsurface Agricultural Drainage from the San Joaquin Valley, CA," *Water Res.* 9:529–546 (1975).
15. Gersberg, R. M., B. V. Elkins, and C. R. Goldman. "Nitrogen Removal in Artificial Wetlands," *Water Res.* 17:1009–1014 (1983).
16. Hansen, J. I., and F. O. Andersen. "Effects of *Phragmites australis* Roots and Rhizomes on Redox Potentials, Nitrification, and Bacterial Numbers in the Sediment," in *Proceedings of the 9th Nordic Symposium on Sediments* (Norr Malmo, Sweden: Scripta Limnologica, 1981).
17. Mitchell, R., and C. Chamberlin. "Factors Influencing the Survival of Enteric Microorganisms in the Sea: An Overview," in *Proceedings of the International Symposium on Discharge of Sewage from Sea Outfalls*, London, 1974, paper number 25.
18. Palmateer, G. A., W. L. Kutas, M. J. Walsh, and J. E. Koellner. "Recovery of Pathogenic and Indicator Bacteria from Wastewater Following Artificial Wetland Treatments of Domestic Sewage in Ontario," Abstracts of the 85th Annual Meeting of the American Society for Microbiology, Las Vegas, Nevada, March 3–7, 1985.
19. Gearheart, R. A., S. Wilbur, J. Williams, D. Hull, N. Hoelper, K. Wells, S. Sandberg, D. Salinger, D. Hendrix, C. Holm, L. Dillon, J. Morita, P. Grieshaber, N. Lerner, and B. Finney. "City of Arcata—Marsh Pilot Project," Second Annual Progress Report. Project No. C-06-22770 State Water Resources Control Board, Sacramento, CA (1982).

CHAPTER 11

Sewage Treatment by Reed Bed Systems: The Present Situation in the United Kingdom

P. F. Cooper and John A. Hobson

INTRODUCTION

In July 1985 a group from the United Kingdom's Water Authorities and WRc visited the Federal Republic of Germany to see Root Zone Method (RZM) treatment systems designed by Kickuth (Gesamthochschule Kassel, Universität des Landes Hessen). The group agreed that the process had potential for sewage treatment for small rural situations, but it was clear that there were several areas of uncertainty. To make rapid progress and prevent duplication, a group was formed to coordinate research and development of the technique in October 1985 under the aegis of the Water Authorities' Association (WAA) with WRc acting as secretariat. Group objectives were to define work needed and conduct research and development to produce a design and operations manual for reed bed treatment systems (RBTS) by 1990.

A number of international contacts were made to exchange experience and data on RBTS systems, and in October 1986 the European Emergent Hydrophyte Treatment Systems (EHTS) Group was formed. It has recently been accepted as a European Community (EC) Expert Contact Group.

The field was reviewed in a paper[1] to the Orlando Conference in July 1986. This chapter will describe principles behind RBTS and outline progress made to December 1987 (halfway through the 4-year R&D program).

DESCRIPTION OF REED BED TREATMENT SYSTEMS

Reed bed treatment systems (RBTS) are used here to describe methods of sewage treatment in an artificial (constructed) wetland containing common reeds *(Phragmites australis)*. It depends upon the flow of sewage through a bed of either soil or gravel in which reeds are growing (Figure 1).

Some of the data and figures in this chapter have been used in a paper published in the *Journal of the Institution of Water and Environmental Management* in the United Kingdom.

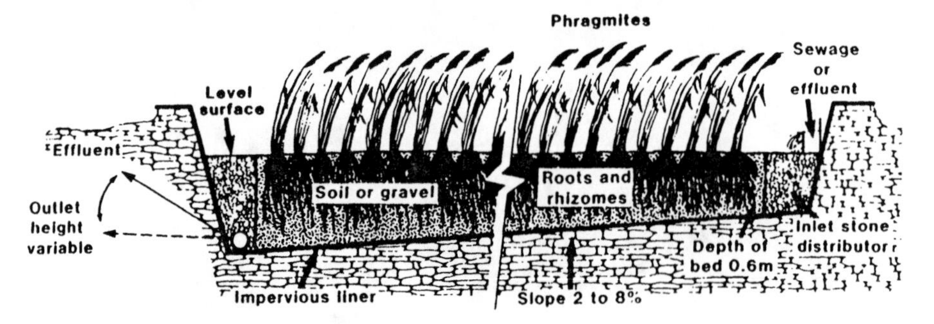

Figure 1. Typical arrangement for a reed bed treatment system.

Key process features are (1) rhizomes of reeds grow vertically and horizontally, opening up the soil to provide a "hydraulic pathway"; (2) wastewater is treated by bacterial activity (aerobic treatment takes place in the rhizosphere with anoxic/anaerobic treatment in the surrounding soil); (3) atmospheric oxygen passes to the rhizosphere via leaves and stems, then to the hollow rhizomes, and out through the reed roots; and (4) suspended solids in the sewage are aerobically composted in an aboveground layer of straw formed from dead leaves and stems.

The following advantages have been claimed for RBTS:[1] (1) low capital costs; (2) simple construction involving no mechanical or electrical equipment; (3) low maintenance cost; (4) robust process able to withstand a wide range of operating conditions; (5) consistent effluent quality; and (6) environmentally acceptable with potential for wildlife conservation.

Land area needed for treatment of screened, degritted sewage is estimated to be 2–5 m^2/pe* to achieve an effluent with less than 20 mg BOD_5/L on a 95% basis. The method can also be used for tertiary treatment to improve poor quality effluents.

Capital costs of installing a RBTS vary with (1) ease or difficulty of working with existing soil on the site, (2) size, and (3) degree of sophistication used. On the basis of the plants built so far, costs range from £50-£150/pe.

RBT SYSTEMS IN THE UNITED KINGDOM AND EUROPE

United Kingdom. The first beds in the United Kingdom were built at Acle (Anglian Water) and St. Paul's Walden (Thames Water) in October 1985. Since that time, another 22 systems have been built in the United Kingdom. Table 1 lists these plants with design details, and performance of some of these systems is discussed later.

Denmark. More than 100 systems have been built (the exact number is uncertain and may be up to 200). Many Danish beds have problems—mainly

*pe = population equivalent based on 56 g BOD_5/person/day (and 170 L/person/day).

Table 1. Reed Bed Treatment Systems Completed in the United Kingdom in October 1987

Site	Feed	Flow (m^3/day)	Population Equivalent	Length (m)	Width (m)	Slope (%)	Planting Time	Comments/Special Features
Acle	Screened crude sewage	224	1260	50	35	2	Oct 1985	Two beds
Kirmington	Screened crude sewage	120	550			3	Oct 1986	Three beds, different sizes
Chalton	Screened crude sewge		200	30	40	2	May 1987	
East Haddon	Screened crude sewage	180	550	33	15	2	April 1987	Four beds
Holtby	Screened crude sewage	30	130	18	34	5	June 1986	Designed by OIS; German reeds
Nun Monkton	Crude sewage/settled sewage	36	160	20	40	1.5	April 1987	Limestone gravel
Westow	Settled sewage	28	220	27	15		April 1987	Designed by OIS; two beds; 50% German reeds; 50% UK reeds
St. Paul's Walden	Crude sewage	11	50	20	10	2	Nov 1985	Sealed with clay
Bracknell	Sludge	N.A.	N.A.	4	4	N.A.	May 1987	
Standon	Sludge	N.A.	N.A.			N.A.	May 1987	Sand/gravel
Windsor	Crude sewage/sludge					1,2, & 3	May 1987	Four beds, different sizes
Stormy Down	Screened crude sewage	28	200	13.5	18.5	1.5	Oct 1986	
Kingstone and	Settled sewage/	215	1200	16	40	1.75	May 1987	Two beds
Madley	Screened crude sewage			16	30			Two beds
Middleton	Humus tank effluent	40		15	30	5.3	May 1987	Recovered sand/gravel
Little Stretton	Septic tank effluent	10	60	10	2	2.0	July 1987	Sewer dike; eight small beds built in series down side of a hill; 12-mm gravel

Table 1. (contd.)

Site	Feed	Flow (m^3/day)	Population Equivalent	Length (m)	Width (m)	Slope (%)	Planting Time	Comments/Special Features
Marnhull	Screened crude sewage/ settled sewage	85	375	21	35	4	July 1986	Two beds
Gravesend	Crude sewage/settled sewage/sludge liquor	100	1000	35	35	2	April/June 1986	Three gravel beds
Lustleigh	Crude sewage	180	470	16	46	3.5	Sept 1987	Two beds, level surface
West Buckland	Final effluent	33	180	16	33	2.0	March 1987	7-mm gravel, level surface
Castleroe	Settled sewage	5 5	25 25	8 8	3.6 3.6	2.4	March 1986 March 1987	Two soil beds Two gravel beds
Rugeley	Septic tank effluent Limestone gravel	3.9		18.6	5	1.5	Nov 1986	Pig farm waste Level surface
Audlem	Crude sewage			6	2.8	2	April 1987	10 pilot beds to test media and planting techniques
Valleyfield	Screened crude sewage	17	60	4.5	10	2	April 1987	Four beds: three waste products, one gravel

N.A. = not applicable—sludge drying beds.

overland flow because of inadequate soil selection. The Botanical Institute at the University of Århus has surveyed 14 of the systems.[2]

West Germany. It is believed 188 plants exist in the Federal Republic of Germany.

Austria. There are several root zone–type systems in Austria and a number of Max-Planck Institute Process (MPIP) plants.[1] The most comprehensive study of an RBTS has been done at Mannersdorf, where the Universität für Bodenkultur, Wien and the Universität Wien have cooperated with the Government of Lower Austria in monitoring a plant built in June 1983.[3]

France. A number of RBTS based on the MPIP Process are being evaluated in France by CEMAGREF (Centre Nationale du Machinisme Agricole du Genie Rural, des Eaux et des Forets).

Belgium. Work using *Typha* is in progress at the Fondation Universitaire Luxembourgeoise at Arlon.

Luxembourg. Work is planned using RBTS.

Holland. Much work has been done on artificial wetlands in Holland, especially with the Lelystad Process.[1] We hope to involve Holland in the EC group next year.

Data and information from all these countries are compiled by the European Community EHTS group. Informal links for exchange of information and experience have been made with research groups in the United States (Tennessee Valley Authority), Australia, and New Zealand.

DESIGN AND CONSTRUCTION DETAILS

Sizing of Beds

At present, the basis for sizing RBTS in the United Kingdom is the equations given by Kickuth on the July 1985 visit.[1] One research objective is to refine these equations using the performance results from the RBTS in the United Kingdom and Europe.

For crude or settled sewage, the surface area is calculated from

$$A_h = 5.2\, Q_d\, (\ln C_o - \ln C_t) \tag{1}$$

where A_h = surface area of bed, m^2
Q_d = daily average flow rate of sewage, m^3/d
C_o = daily average BOD_5 of the feed, mg/L
C_t = required average BOD_5 of the effluent, mg/L

For domestic sewage this tends to produce an area of 2.2 m^2/pe. The 1985 Mannersdorf report[2] indicates that this value may be optimistic and that area needed may be 3–4 m^2/pe.

Most U.K. plants have been designed with 3-5 m^2/pe or built in low-risk situations where RBTS effluent quality is not critical. Some beds have been built as tertiary systems until mature, when they will be switched to treating settled or screened sewage.

The cross-sectional area of the bed (A_c) has been established from a statement of Darcy's law:

$$A_c = \frac{Q_s}{k_f \frac{\partial H}{\partial s}} \qquad (2)$$

where Q_s = average flow rate of sewage, m^3/s
k_f = hydraulic conductivity of fully developed bed, m/s
$\partial H/\partial s$ = slope of the bed, m/m

Kickuth stated that k_f would reach 3×10^{-3} m/s in fully developed RZM within 3–5 years. But doubt has been raised from many quarters since this value is much higher than expected from soils at the start. The Mannersdorf plant in Austria has been in operation for four growing seasons, and hydraulic conductivity has changed little ($1–3 \times 10^{-5}$ m/s in 1987). WRc[4] have measured hydraulic conductivity in an old, natural reed bed at 5×10^{-6} m/s. Bucksteeg[5,6] also doubts the values used by Kickuth. Using design values less than 10^{-4} m/s results in short, unacceptably wide bed designs. Most U.K. beds have been designed at values between 3×10^{-3} and 10^{-4} m/s, and some overland flow is expected. With gravel it is possible to achieve an hydraulic conductivity of 10^{-3} m/s or more and a number of the U.K. beds have been built with gravel or a waste product with similar hydraulic conductivity.

Kickuth recommended limiting *horizontal velocity* to a maximum of 10^{-4} m/s, irrespective of the slope, to avoid disrupting the mosaic of aerobic, anoxic/anaerobic zones in the bed, and to prevent erosion and surface channeling. To date, this advice has been followed in the U.K. designs.

Depth

Most U.K. beds have bed depths of 0.6 m based on (1) the statement that beyond 0.6 m roots start to weaken and that (2) thinner beds may suffer from freezing. Three beds (one at Nun Monkton and two at Castleroe) have recently been built 0.3 m deep. Most beds have at least 0.5 m head over the bed surface to the top of bund walls to accommodate anticipated accumulation of 25-mm/yr and a 20-year life before the front end of the bed needs skimming.

Slope

Kickuth stated that beds could be built with slopes of 2–8%.[7] The Holtby bed was built by Oceans International Services (OIS, Kickuth's U.K. licensees) with a 5% slope on the surface and the base. OIS also designed the Marnhull beds with a 4% slope on the surface and base. Neither bed could be flooded, and severe weed problems were experienced at Holtby, Marnhull, and Acle. A level surface will allow flooding for weed control. Current advice in the United Kingdom is to design beds with minimum slope on the base needed to pass water through the bed (calculated from Darcy's law [Equation 2]) and to use a level surface. Hydraulic conductivity (i.e., the medium used in the bed) has substantial influence on calculations of base slope.

Soil or Growth Medium

As earlier indicated, this is the greatest area of uncertainty at present. Except Othfresen (iron-mining waste),[7] Kickuth-designed plants in Germany and Denmark were built with soil. Several have problems with overland flow because of the low k_f and have suffered from erosion and poor reed growth. Beds in Denmark and San Diego, California[8] have used gravel. Gravel should allow flow-through of water initially, and if solids cause blockage, roots and rhizome growth may counteract this and open up the bed. There are 14 sewage treatment beds built with gravel and others built with granular mining or industrial waste in the United Kingdom. Silica gravel beds have little potential for removing phosphate, but phosphate removal is not a major objective in the United Kingdom. Two beds were built with limestone gravel.

Plants and Planting

The common reed *(Phragmites australis)* is used in U.K. systems at recommended planting densities of 2–4 rhizomes or stems/m^2. Plantlets, rhizome sections, and clumps of plants have been used. Until recently, clumps (20-cm square sections) from existing reed beds were favored with small plantlets as second choice, and beds have been successfully grown from rhizome sections. The Institute of Terrestrial Ecology (ITE) has studied propagation under WRc contract. Seed propagation was believed very difficult because British seed was considered infertile. However, the ITE research has evolved a method of growing *Phragmites* from British seed with near 100% success. Based on experience with other grasses, ITE researchers believe stronger, denser reed beds may be developed from individual seedlings (grown from seed) or rhizome fragments rather than from clumps.

Weed Control

Weed problems in several U.K. reed beds stem from excessive proliferation in spring and early summer, and from faster-growing weeds shading *Phragmites* shoots and retarding growth. Long-term effects are unclear but (1) the most successful weed control method tested so far is flooding with water/effluent/sewage (a level surface enhances complete flooding); and (2) reed beds grown in gravel lack the seed bank present in soils. If weeds start in a gravel bed from windblown seeds, hand weeding is easier because root attachments are less secure in gravel than in a soil bed.

Where flooding was used to control weeds on a sloping surface bed at Kingstone, new reed shoots grew well in the middle where the water was less than 30 cm deep. At the outlet with water deeper than 30 cm, reed shoot development was inhibited, and at the inlet the soil could not be flooded and weeds proliferated. As a result, we recommend a level bed surface and water depth of 5–15 cm.

Weeds were a serious problem at Acle in the summers of 1986 and 1987. Herbicides were effective in killing weeds but dead plants collapsing onto reed shootlets retarded their growth. In spring 1987 neither weed killers nor terracing to hold water and drown weeds appeared successful, but in late summer the reeds started to spread and may have overcome the weeds.

Feed Distribution

Most U.K. beds have a 0.5-m wide inlet distributor zone filled with 60–100 mm stones. In several cases, a castellated weir distributes the feed along the inlet zone, but it is costly and more complicated than necessary. In some recent beds, a simple pipe with a series of adjustable T-pieces has been used. Gravel beds may not require a specific inlet zone.

Outlet Collector

Outlet collector zones tend to be 0.5-m wide zones of stones similar to the inlet zones with a slotted pipe at the base. In all cases, provision for raising or lowering the bed water table with a control at the outlet pipe is included.

Liner

In most cases, a plastic liner has been used. This has either been HDPE of the Schlegel type (2 mm thick) or lower density polyethylene of the Monarflex type (0.5–1.0 mm thick). One bed (St. Paul's Walden) is sealed with puddled clay that was present on the site. Some of the beds are constructed in disused concrete-walled sludge-drying beds.

PERFORMANCE OF THE FULL-SCALE BEDS

A database for bed design and performance has been established at WRc Stevenage. Since most beds were planted in 1987 and only three beds have gone through two growing seasons, limited long-term performance data are available. However, performance summaries for seven beds with the most comprehensive data are given in Table 2 and in Figures 2 to 5.

Table 2. Summary of Performance Data from United Kingdom Reed Bed Treatment Systems

Gravesend beds (447 days—October 26, 1986 to January 18, 1988)

	BOD_5		SS		NH_3-N		o-PO_4-P	
	Inf	Eff	Inf	Eff	Inf	Eff	Inf	Eff
Bed 1								
Minimum	36	16	35	4	14.0	14.0	3.70	0.30
Maximum	510	190	400	350	73.0	71.0	25.00	15.00
Average	237	84	131	64	43.0	37.0	12.60	7.20
Bed 2								
Minimum	36	35	34	15	14.0	21.0	3.70	2.70
Maximum	510	210	400	220	73.0	54.0	25.00	15.00
Average	237	90	131	58	43.0	36.0	12.60	7.50
Bed 3								
Minimum	36	7	35	10	14.0	17.0	3.70	0.30
Maximum	510	135	400	110	73.0	49.0	25.00	11.00
Average	237	63	131	38	43.0	36.0	12.60	5.00

Holtby bed (602 days—June 4, 1986 to January 27, 1988)

	BOD_5		SS		NH_3-N		o-PO_4-P	
	Inf	Eff	Inf	Eff	Inf	Eff	Inf	Eff
Minimum	25	12	44	6	6.3	12.3	2.96	0.03
Maximum	685	109	972	72	70.4	51.5	37.30	13.30
Average	233	49	163	24	34.9	32.9	7.79	6.82

Marnhull beds (265 days—April 23, 1987 to January 13, 1988)

	BOD_5			SS			NH_3-N			Flow	
	Inf	Eff 1	Eff 2	Inf	Eff 1	Eff 2	Inf	Eff 1	Eff 2	Bed 1	Bed 2
Minimum	8	5	4	15	8	8	8.5	9.7	9.3	27.6	40.6
Maximum	160	33	42	170	47	52	41.4	38.2	40.6	40.6	59.2
Average	87	13	17	74	23	20	28.9	26.4	28.2	32.8	50.7

Middleton reed bed (241 days—June 1, 1987 to January 27, 1988)

	BOD_5		SS		NH_3-N		Oxd N		o-PO_4-P	
	Inf	Eff	Inf	Eff	Inf	Eff	Inf	Eff	Inf	Eff
Minimum	4.7	0.4	11	1	0.3	0.2	11.1	5.1	3.4	3.8
Maximum	29.7	8.2	86	35	10.0	4.2	60.5	46.0	17.4	13.3
Average	11.0	3.0	30	8	2.8	1.5	29.5	21.6	10.8	7.6

Table 2. (contd.)

Castleroe reed beds
(Start dates: old beds, March 25, 1986; new beds, March 26, 1986. Data to January 11, 1988)

	BOD_5					SS				
	Inf	New LB	New RB	Old LB	Old RB	Inf	New LB	New RB	Old LB	Old RB
Minimum	68	2	26	3	16	46	9	4	6	11
Maximum	300	112	156	89	110	128	86	67	60	98
Average	157	53	70	37	60	73	30	29	33	37

	NH_3-N					o-PO_4-P				
	Inf	New LB	New RB	Old LB	Old RB	Inf	New LB	New RB	Old LB	Old RB
Minimum	9.0	0.0	0.0	4.2	9.6	1.5	0.0	1.4	0.7	1.1
Maximum	25.9	24.7	26.0	21.3	26.0	9.8	7.1	9.1	5.5	9.6
Average	19.1	14.5	17.3	13.9	18.2	5.0	4.0	4.8	2.5	5.6

Valleyfield reed beds (305 days—May 1, 1987 to March 1, 1988)

	BOD_5					SS					NH_3-N				
	Inf	Eff 1	Eff 2	Eff 3	Eff 4	Inf	Eff 1	Eff 2	Eff 3	Eff 4	Inf	Eff 1	Eff 2	Eff 3	Eff 4
Minimum	72	13	6	6	25	109	8	2	15	8	8	12	11	11	6
Maximum	334	84	101	57	96	252	62	34	49	29	46	35	30	26	29
Average	207	49	43	30	51	167	26	17	26	17	27	24	20	20	22

	o-PO_4-P					Flow (m^3/day)			
	Inf	Eff 1	Eff 2	Eff 3	Eff 4	Bed 1	Bed 2	Bed 3	Bed 4
Mimimum	4.41	0.76	0.49	0.10	0.10	0.00	1.44	0.00	1.57
Maximum	20.00	7.20	4.90	2.40	5.91	9.09	4.32	8.64	12.34
Average	10.49	3.64	1.71	0.93	3.14	4.84	2.81	4.47	4.54

Little Stretton sewer dike—before installation of reed bed (60 days—May 1, 1987 to June 30, 1987)

	BOD_5		SS		NH_3-N	
	Inf	Eff	Inf	Eff	Inf	Eff
Miminum	25.5	9.3	19	10	3.4	2.1
Maximum	3042.0	311.0	1550	343	11.1	19.1
Average	464.1	60.4	417	83	7.2	6.8

Little Stretton—after installation of reed bed (179 days—July 3, 1987 to December 19, 1987)

	BOD_5		SS		NH_3-N		Flow
	Inf	Eff	Inf	Eff	Inf	Eff	(m^3/d)
Minimum	14.2	2.0	22	4	1.9	3.7	14.3
Maximum	426.0	183.0	870	115	21.0	27.6	115.7
Average	140.9	32.9	127	20	9.8	10.2	41.6

Note: Inf = influent; eff = effluent. Values except flows in mg/L; flows in m^3/day.

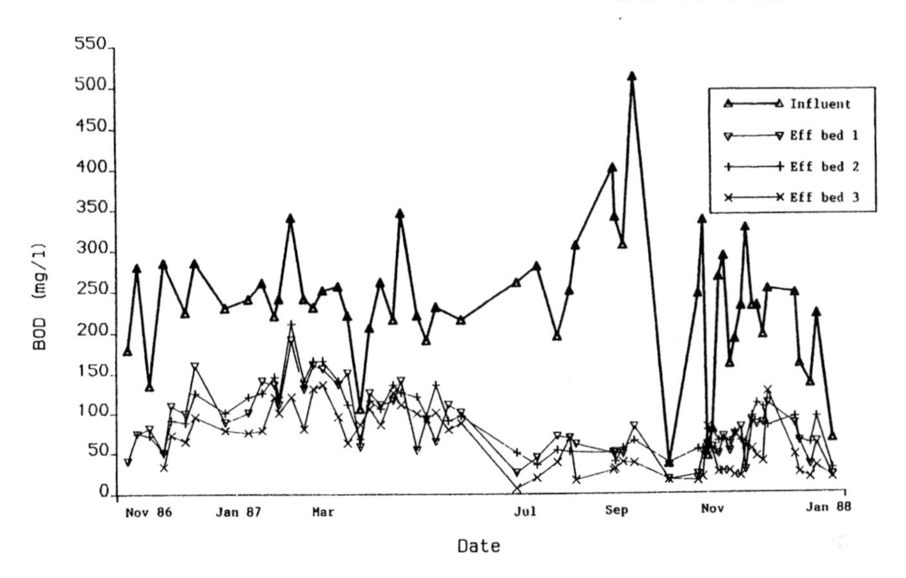

Figure 2. Gravesend reed beds—BOD_5.

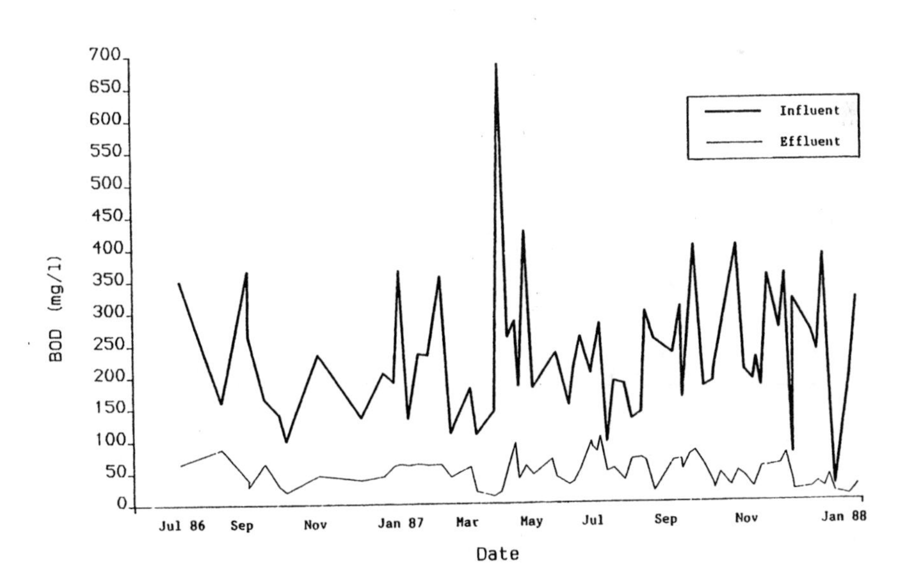

Figure 3. Holtby reed bed—BOD_5.

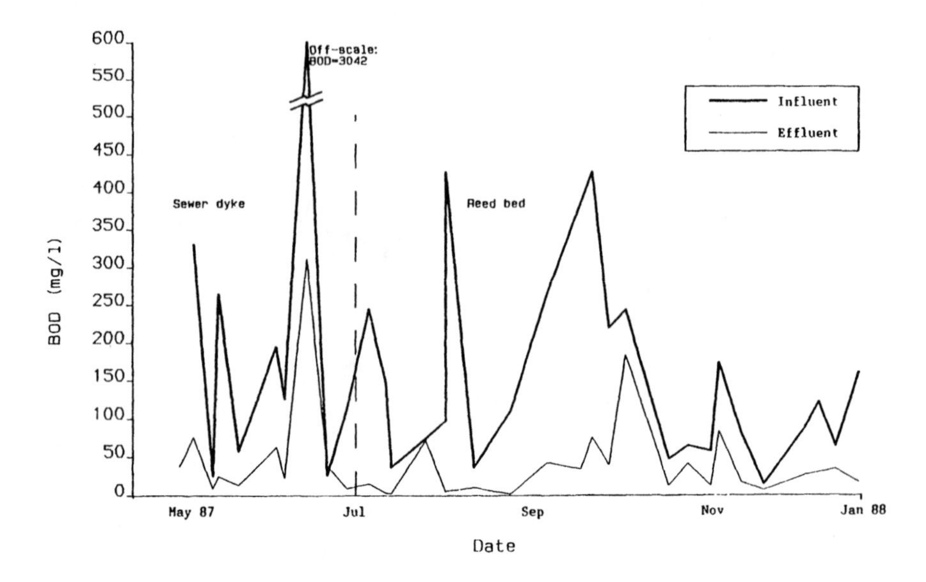

Figure 4. Little Stretton reed bed—BOD_5.

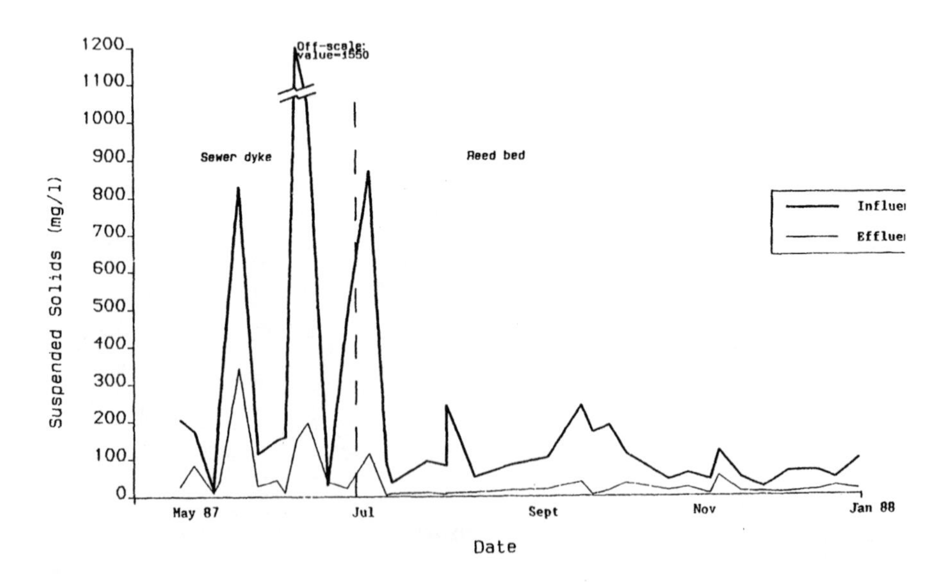

Figure 5. Little Stretton reed bed—suspended solids values.

Gravesend Reed Beds (Southern Water)

Gravesend was the first gravel media system. The beds have a well-developed covering of reeds. All three beds were started with settled sewage as the feed until beds matured. The loading rate is 3.7 m^2/pe. Target effluent quality for settled sewage treatment is 120 mg BOD_5/L. On average this has been achieved, but system performance is expected to improve with maturity.

The Gravesend beds produced an improved quality effluent from mid-June 1987 (Table 2, Figure 2). The reason is unclear but may be associated with gradual maturation in its second growing season. Throughout the operational period, Bed 3 has performed the best.

Marnhull Reed Beds (Wessex Water)

Marnhull reed beds have local soil and a design loading rate of 3.9 m^2/pe for a flow of 85 m^3/day of screened crude or settled sewage. At present, two beds are fed with settled sewage at average flows of 31 and 51 m^3/day. Major weed problems have occurred at Marnhull (a 4% surface slope precludes flooding), but effluent quality has been good averaging 13 and 17 mg BOD_5/L after two growing seasons. However the current loading rate of 11.6 m^2/pe on Bed 1 and 7.5 m^2/pe on Bed 2 are lower than design because settled sewage is weak and low flows are used for gradual start-up.

Holtby Reed Bed (Yorkshire Water)

The Holtby Reed Bed[9] was designed and constructed by OIS as the only form of treatment for a small village. The bed has a gravel layer (and valving system) below local soil medium that passes part of the flow through the base of the bed rather than through the soil (an OIS development). Design effluent standard is 20 mg BOD_5/L, 30 mg SS/L as a mean, and the loading rate is a conservative 4.7 m^2/pe. The bed has had severe weed problems that cannot be removed by flooding because of the 5% slope. However, treatment performance is encouraging (Table 2 and Figure 3) even though the bed is not fully developed. Effluent quality approaches desired levels for SS but is high for BOD_5 and little phosphorus is removed.

Castleroe Reed Beds (Department of the Environment, Northern Ireland)

The Castleroe Reed Beds are small experimental beds built in old concrete-walled sludge-drying beds. Some data in Table 2 are from the two soil beds planted in April 1986 ("old" beds). Performance has been disappointing because of the difficulty in passing the sewage through the soil and overland flow was evident. As a result, the second set of two beds was built in gravel in March 1987 ("new" beds).

Middleton Reed Bed (Severn-Trent Water)

The bed is designed for initial use as a tertiary treatment unit to prevent noncompliance with the effluent consent for a biological filter works. A proportion of screened crude sewage will be diverted to the bed when it is fully developed. At present, it takes humus tank effluent with a feed design value of 20 mg BOD_5/L. The bed was planted in mid-May 1987 using short rhizome sections with shoots (3/m^2) from a former sludge lagoon. Performance has been very good from the start, producing an effluent not exceeding 8 mg BOD_5/L and averaging 4 mg BOD_5/L and 8 mg SS/L (Table 2). The oxidized nitrogen concentrations in and out indicate some denitrification is taking place. The bed was drained down and not fed from August 14 to October 1, 1987, to encourage root and rhizome penetration.

Valleyfield Reed Beds (Fife Regional Council)

The four Valleyfield reed beds are pilot units to investigate four alternative media prior to construction of a reed bed to treat flow from a large village. The media are fine pulverized fuel ash (PFA, i.e., burnt coal waste from a power station, Bed 1), coarse PFA (Bed 2), pea gravel (Bed 4), and rejected overburden from a sand/gravel pit (Bed 3). Each pilot bed is fed screened crude sewage at a loading equivalent to 3.0 m^2/pe.

Reeds have grown well and have reached 2 m high after the first growing season. Effluents from the new gravel beds (Beds 3 and 4) are impressive; they have not reached but are approaching design quality of 20 mg BOD_5/L (mean value), and phosphorus removal is better than in other UK beds. The two PFA beds have suffered from surface flows because of lower hydraulic conductivity—8×10^{-5} m/s in PFA beds compared with 1×10^{-3} m/s for quarry rejects and 8×10^{-1} m/s for pea gravel.

Little Stretton Reed Beds (Severn-Trent Water)

The Little Stretton reed bed system is unusual because it is a series of eight small gravel beds built in terraces down the side of a hill. They replaced a sewer dike that took the sewage from a small village into a stream. Design loading is 2.7 m^2/pe with allowance for dairy farm floor washings occasionally mixed with the domestic sewage flow. The media is 12-mm gravel with *Phragmites* plantlets grown from seed by the ITE. Plantlets grew well after planting in July 1987 with nearly 100% survival.

Data in Table 2 and Figures 4 and 5 represent conditions before and after reed bed construction. Some treatment was taking place in the sewer ditch, but effluent quality has improved with reed bed installation. Although the effluent only has a qualitative consent, effluent quality is at least as good as the design target—but this could be purely physical filtration. Influent peaks (probably occasional flows from the farm) and the moderating effect of the reed bed on effluent quality are noteworthy.

The bed was designed for 60 pe on BOD_5 and flow (56 g BOD_5/pe and 170 L/pe) with 40 villagers and an allowance of 20 pe for washings from the farm. Design flow was 10 m^3/day, but monthly average loading ranged from 61 to 154 pe for BOD_5 and from 167 to 598 pe on a flow basis since August 1987. The increase is due partly to field runoff but more so to increased flows from the farm. Despite loading over design levels (down from 2.7 m^2/pe to 1.0 m^2/pe in September) treatment achieved was impressive with an immature bed. Effluent quality is close to that predicted by Kickuth's first equation.

SMALL-SCALE TESTS IN MOBILE TEST TANKS

Although we felt that most information on RBTSs would come from full-scale beds, we decided to test the process on a smaller scale. Small scale RBTSs could serve as effluent treatability units. Also, small tanks filled with a section from an existing reed bed might produce an "instant" RBTS. Mobile test tanks (Figure 6) have been given to two Water Authorities to conduct treatability tests (one on landfill leachate), and four test tanks are used at WRc's Little Marlow site and six at Stevenage. At Stevenage, Bed 1, filled with several clumps of mature reeds and packed with sieved soil, started operation at the end of June 1986. The remaining beds (2 through 6) were prepared by mixing rhizomes with the medium, and operation started the end of August 1987. Beds 1 and 2 contained soil, Beds 3 and 4 gravel, and Beds 5 and 6 Pozzolanic Lytag (synthetic spheres made from fly ash). Bed 1 was fed with 48 L/day settled sewage from July 1986 and reduced to 24 L/day in November 1986 (the latter is equivalent to 3–3.5 m^2/pe). Beds 2–6 were fed with settled sewage at 24 L/day from August 1987.

RESULTS AND DISCUSSION

Performance results are shown in Tables 3 and 4 and Figure 7. On day 60, the Bed 1 feed arrangements were changed to increase BOD_5 concentration to a typical value of 200 mg/L, and performance of the bed fell dramatically. On day 126, the feed rate was halved to 24 L/day, and effluent BOD_5 values showed some improvement. However, at day 200 the bed froze and was moved indoors for two weeks and effluent quality deteriorated. Since then, effluent BOD_5 values have steadily fallen over a year to 20 mg/L. The results could demonstrate that the units will acclimatize (as the bacterial populations develop) to give good treatment with settled sewage at a loading of 3.5 m^2/pe but that treatment is damaged by freezing. However, treatment may have been damaged by draining on day 30 and to some extent on day 200 when the bed was moved. Manipulating water levels for weed control and to encourage rhizome penetration is likely to be desirable, but if this could harm treatment, water level fluctuations should be used sparingly if at all.

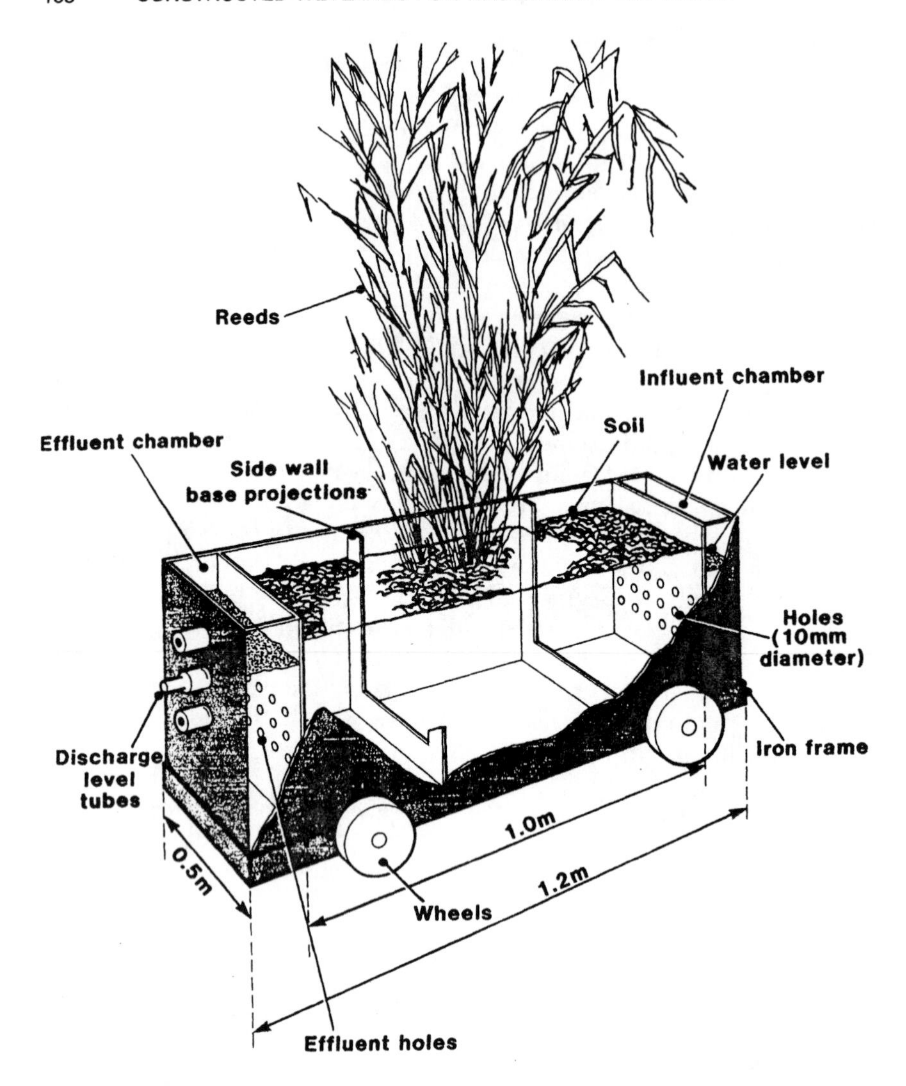

Figure 6. WRc mobile test tank.

The two soil beds and one gravel bed perform much better at BOD_5 removal than the other gravel bed and the two Lytag beds. Lack of correlation between BOD_5 and SS effluent values is also interesting. The two Lytag beds and Bed 4 (gravel) are better at removing solids (probably by filtration) than at removing BOD_5. The new soil bed (Bed 2), has high SS values and low BOD_5 values consistent with inorganic soil particles washing from the bed.

All beds removed significant amounts of ammonia during the first three weeks and then, with the exception of soil Bed 2, gradually lost this ability. Bed

Table 3. Performance Data for Stevenage Mobile Bed I

	Flow	Loading	BOD_5 (mg/L)		SS (mg/L)		NH_3-N (mg/L)		o-PO_4-P (mg/L)	
	(L/day)	Rate[a] (m^2/pe)	Inf	Eff	Inf	Eff	Inf	Eff	Inf	Eff
July 1986	48	6.3	92	32			47	28	8.2	4.9
August	48	6.3	93	30			44	33	7.1	6.2
September	48	3.3	175	58	107	71	46	42	8.3	9.9
October	48	3.7	157	92	107	47	45	44	7.1	10.4
November	24	6.7	175	59	118	32	43	45	7.5	9.9
December	24	5.0	235	49	130	38	47	45	8.5	10.7
January 1987	24	6.3	186	91	145	34	44	51	9.2	12.0
February	24	5.7	204	85	124	26	42	51	8.0	12.7
March	24	6.7	175	108	107	32	36	48	7.3	9.7
April	24	4.4	264	46	156	39	44	44	10.1	12.4
May	24	5.5	212	58	134	32	43	42	11.7	12.2
June	24	7.9	148	63	121	38	39	41	7.9	9.2
July	24	6.0	196	44	134	45	37	33	9.0	9.5
August	24	5.7	205	40	163	26	39	32	10.0	10.3
September	24	6.4	181	24	103	14	38	26	8.4	7.9
October	24	6.9	168	16	117	12	39	21	8.3	6.4
November	24	6.2	188	16	114	13	38	21	7.7	6.3
December	24	5.4	218	33	121	16	42	31	8.4	9.2
January 1988	24	8.2	142	36	130	14	37	32	9.1	9.2
February (to 15th)	24				132	11	35	29	7.4	7.5

Note: 594 days—July 1986 to February 1988. Values are monthly averages based on composite samples analyses. (Until May 29, 1987: 3 day/week for BOD_5; 5 day/week for all other parameters. After May 29, 1987: 2 day/week for BOD_5; 5 day/week for SS; 3 day/week for all other parameters.)

[a]Based on 56 g BOD_5/pe/day.

Table 4. Performance Date for All the Stevenage Mobile Beds

	Inf	Eff 1[a]	Eff 2[b]	Eff 3[c]	Eff 4[d]	Eff 5[e]	Eff 6[f]
BOD_5	177	24	22	64	66	79	88
SS	118	13	35	27	23	22	24
COD	375	79	91	142	151	159	170
NH_3-N	38	26	20	26	31	29	30
o-PO_4-P	8.2	7.7	5.0	7.4	8.0	7.2	7.4

Note: Average values for the period August 28, 1987 to February 15, 1988 (BOD_5 values to January 29, 1988). All results in units of mg/L. For all beds, flow rate = 24 L/day; loading rate = 6.4 m^2/pe (8.7 g BOD_5/m^2/day)

[a]Bed 1 Mature soil bed—started June 1986.
[b]Bed 2 New soil bed—started August 1987.
[c]Bed 3 Gravel bed 1—started August 1987.
[d]Bed 4 Gravel bed 2—started August 1988.
[e]Bed 5 Lytag bed 1—started August 1988.
[f]Bed 6 Lytag bed 2—started August 1988.

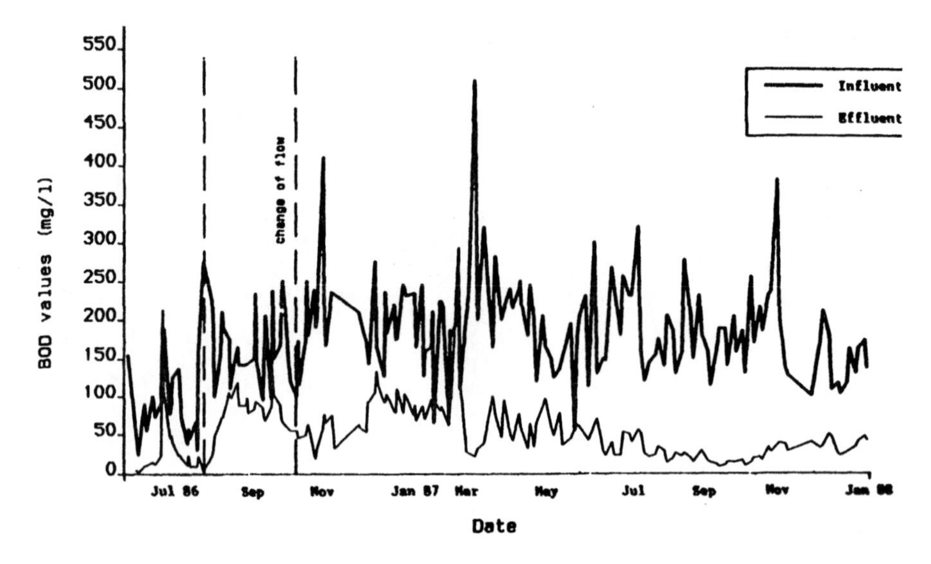

Figure 7. Stevenage mobile reed bed—BOD_5 values.

1 is only removing ammonia consistent with normal nutrient uptake of carbonaceous bacteria, and all the beds are likely to settle at this level.

CONCLUSIONS

The small beds are giving useful information, but how representative is unknown until more results from larger-scale reed beds are analyzed. Results from Bed 1 suggest that a year of constant flow operation is inadequate to reach a steady state. Although Bed 1 results appear to be improving with time, seasonal variation in treatment cannot be discounted. Also, effluent quality appears to deteriorate slightly in winter (December to March) as in full-scale beds (Table 3 and Figure 7). Moving and/or draining beds may adversely effect treatment.

CONCLUDING REMARKS

Data from most U.K. reed bed systems with full sampling and analysis are encouraging, particularly considering that none are near maturity (considered to take 3–5 years), but it is early to draw conclusions. Reeds grow well in gravel—those systems have performed well from the start—but most treatment at this stage is likely due to physical filtration. Some systems built in soil have suffered from (1) overland flow and (2) excessive weed growth. However, performance of the soil beds at Holtby, Marnhull, and Middleton is quite encouraging. The roles of soil and the litter or humus layer must be evaluated for soil systems. Evaluation of data from 24 sites over the next three years may provide answers to current questions.

ACKNOWLEDGMENTS

Data reproduced in this chapter (with the exception of small-scale mobile trial units) has been supplied by Water Authority staff as part of the WAA-RBTS coordinated program. The authors are particularly grateful to the following for supplying the data used here: John Christian (Southern Water); Fred Psyk, Peter Brewer, and Dr. Alan Frake (Wessex Water); Dr. Liz Chalk and Gordon Wheale (Yorkshire Water); Dr. Bill Storey (DOE, Northern Ireland); Ben Green, Paul Griffin, and Phil Harding (Severn Trent Water); Ron Dickson (Fife Regional Council); and Colin Bayes (WRc Scottish Office). This chapter was produced as part of the WAA Reed Bed Treatment System Coordinating Group's Programme but the opinions expressed are entirely those of the authors.

REFERENCES

1. Cooper, P. F., and A. G. Boon. "The Use of *Phragmites* for Wastewater Treatment by the Root Zone Wetland: The UK Approach," in *Aquatic Plants for Water Treatment and Resource Recovery,* K. R. Reddy and W. H. Smith, Eds. (Orlando, FL: Magnolia Publishing Inc., 1987), pp. 153–174.
2. Brix, H., and H-H. Schierup. "Root Zone Systems. Operational Experience of 14 Danish Systems in the Initial Phase," University of Århus, Denmark (1986).
3. "Pflanzenklr—Anlage Mannersdorf/L, Jahresbericht 1985," Universität für Bodenkultur/Amt der NO Landesregierung, Wien, Austria (1986).
4. Clark, L. "Determination of the Hydraulic Conductivity of a Reed Bed in the River Ant, How Hill, Norfolk," WRc Report PRS 1470-M, Medmenham, UK (1987).
5. Bucksteeg, K. "Discussion of Treatment of Wastewater in the Rhizome Sphere of Wetland Plants—The Root-Zone Method—H Brix," *Water Sci. Technol.* 19:1063–1064 (1987).
6. Bucksteeg, K. "Sewage Treatment in Helophyte Beds. First Experiences with a New Technique," paper presented at the 13th IAWPRC Conference Post-Conference Seminar on the Use of Macrophytes in Water Pollution Control, Piracicaba, Brazil, August 24–28, 1986.
7. Boon, A. G. "Report on a Visit by Members and Staff of WRc to Germany (FRG) to Investigate the Root Zone Method for Treatment of Wastewater," WRc Report 376-S/1, Stevenage, UK (September 1985).
8. Gersberg, R. M., B. V. Elkins, and C. R. Goldman. "Wastewater Treatment by Artificial Wetlands," *Water Sci. Technol.* 17:443–450 (1984).
9. Chalk, E., and G. Wheale. "The Root Zone Process for Sewage Treatment at the Holtby Works, Yorkshire Water," paper presented to the Institution of Water and Environmental Management, Scientific Section Meeting at Matlock, UK, October 8, 1987.

CHAPTER 12

Aquatic Plant/Microbial Filters for Treating Septic Tank Effluent

B. C. Wolverton

INTRODUCTION

Septic tank systems that serve 25% of the U.S. population are installed in a variety of soil types and temperature ranges throughout the country.

A primary concern of many septic tank owners is odor, often emitted from house roof vents directly connected to the septic tank. These vents not only produce foul odors but also allow flies and other insects to breed in the tank, causing nuisance and potential health problems. To avoid this problem, the inlet tee inside the septic tank should be capped or the tank constructed using an ell. Open tees should be installed only at the effluent discharge point inside the tank to allow trapped gases to escape into the leach field or aquatic plant/microbial filter. Odors are also created when leach field clogging causes surface pooling of septic tank effluent before complete treatment has occurred.

Although some reduction in fecal coliform bacteria may take place inside the septic tank, removal of pathogenic bacteria and viruses is the concern of health officials. In general, pathogenic microorganisms are quite host-specific and do not survive long apart from the host. Because viruses are charged particles and respond to flocculants, most become attached to septic tank solids and remain in the tank sludge.[1]

Problems with septic tank systems are not normally associated with properly installed, sealed tanks, but with the leach field component of the system. To make the septic tank system more versatile and acceptable in most climates and soil conditions, an aquatic plant/microbial filter can be used to replace the leach field.[2] The use of natural biological processes for treating various types of wastewater has been developed by NASA at the John C. Stennis Space Center (SSC) during the past 15 years.[3-9] An aquatic plant filter constructed of washed gravel receives the partially treated, odorous discharge from the septic tank and continues the treatment process. The filter must be maintained at a depth such that the anaerobic septic tank effluent is converted to aerobic conditions and maintained aerobic throughout the filter. Once aerobic condi-

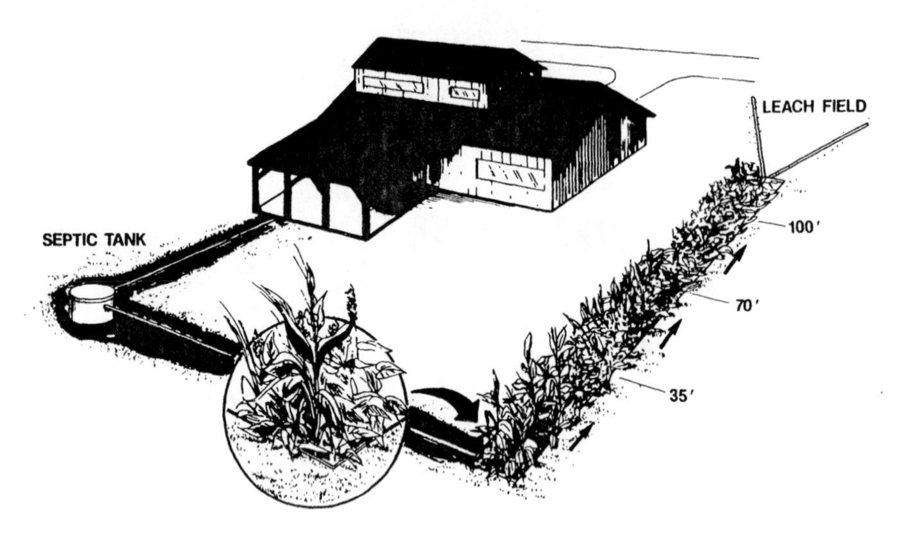

Figure 1. Artificial marsh system for treating discharge from septic tanks.

tions are achieved, odor removal and water clearing will begin. At this point in the rock/plant filter, protozoa begin to grow. These large microorganisms feed on bacteria and other microscopic life and are essential in the natural processes used to remove coliforms and other bacteria. Soil conditions determine whether the filter should be lined with a layer of clay or a plastic sheet liner. Purity of the filter effluent is determined by the length and depth of the filter in addition to the retention time.

There are more than 10 single-home septic tank/rock/plant sewage treatment systems in operation in the Picayune, Hattiesburg, and Philadelphia, Mississippi areas. The Philadelphia area systems are on the Choctaw Indian Reservation. These single-home units are designed as shown in Figure 1. Approximately 400 ft^2 (37.2 m^2) of surface area is recommended for the single-home rock/plant filter, and approximately 20 ft (6 m) of 4-in. (10.2 cm) perforated leach field tubing is used to disperse the highly treated rock/plant filter effluent beneath the soil.

The single-home system used to obtain the design parameters is located in Picayune, Mississippi. A 70-foot (21.3 m) section of rock/plant filter was monitored for several years with average data shown in Table 1.

A septic tank/rock/plant sewage treatment system for treating wastewater from a radio station at Hattiesburg, Mississippi began operation in May 1988. A 12,000-gpd (45.4 m^3) septic tank/rock/plant system is also in operation for a mobile home park at Pearlington, Mississippi, on the Gulf Coast (Figure 2). Over 20 single-home septic tank/rock/plant sewage treatment systems are operational in Louisiana, with an additional 20 approved for installation by the Louisiana Health Department.

Table 1. Single Home Wastewater Treatment System Using a Rock/Plant Filter to Treat Septic Tank Effluent

Parameter	Filter Influent Septic Tank Effluent	After 35 Ft (10.7 m)	After 70 Ft (21.3 m)
BOD_5, mg/L	100	32	10
NH_3-N, mg/L	28	24	7
Fecal coliform colonies/100 mL	600,000	72,000	19,000

Note: Data from a 3.28 ft (1 m) wide × 70 ft (21.3 m) long filter with a 12 in. (0.3 m) wastewater depth containing 4–6 in. (10–15 cm) of gravel on top. Elephant ears and calla lilies were the dominant plants in this system. From this data, a 4 ft (1.2 m) wide × 100 ft (30.5 m) long filter system is recommended for achieving tertiary level treatment of septic tank effluent from single homes, 2–3 people per home.

SCIENTIFIC BASIS FOR USING AQUATIC PLANTS IN WASTEWATER TREATMENT

Biologically, the aquatic plant systems are far more diverse than present-day mechanical treatment systems. Oxidation ditches and other types of extended aeration treatment systems use energy-intensive mechanical aerators to supply large amounts of oxygen for growing aerobic microorganisms that treat the wastewater.

The scientific basis for waste treatment in a vascular aquatic plant system is the cooperative growth of plants and microorganisms associated with the plants. Much of the treatment process for degradation of organics is attributed to microorganisms living on and around plant root systems.

Once microorganisms are established on aquatic plant roots, they form a symbiotic relationship with higher plants. This relationship normally produces a synergistic effect, resulting in increased degradation rates and removal of organic chemicals from the wastewater surrounding the plant root systems. Products of microbial degradation of the organics are absorbed and utilized as a food source by the plants along with nitrogen, phosphorus, potassium, and other minerals. Microorganisms also use metabolites released through plant roots as a food source. Each component using waste products of the other sustains a favorable reaction for rapid removal of organics from wastewater. Electrical charges associated with aquatic plant root hairs also react with opposite charges on colloidal particles such as suspended solids, causing them to adhere to plant roots where they are removed from the wastewater stream and slowly digested and assimilated by the plant and microorganisms. Aquatic plants have the ability to translocate O_2 from the upper leaf areas into the roots, producing an aerobic zone around the roots that is desirable in domestic sewage treatment.

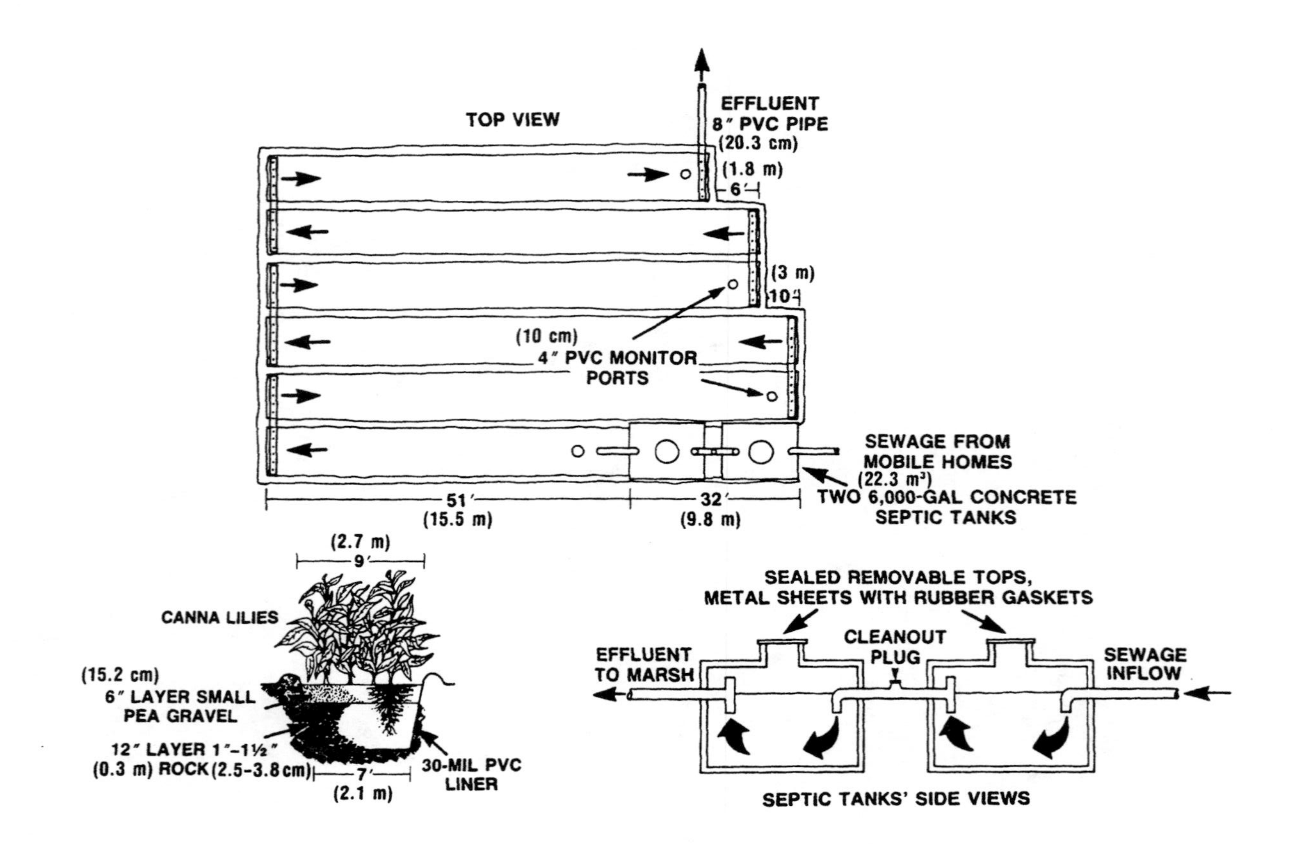

Figure 2. Septic tank/rock/plant wastewater treatment system for Sunrise Haven Mobile Home Park, Pearlington, Mississippi.

TREATMENT EFFECTS ON THE SEPTIC TANK/ROCK/PLANT MARSH TREATMENT SYSTEM

Studies conducted in the states of Washington and Alaska and in Canada have indicated that septic tank systems perform satisfactorily during the winter months in these cold climates. Experiments at Fairbanks and Anchorage, Alaska demonstrated that heat provided to the septic tank by residential wastewater was a significant factor in maintaining the disposal system at an operable temperature.[10] Cold-tolerant plants, bulrushes *(Scirpus),* and cattails *(Typha)* must be used when installing septic tank/rock/plant wastewater treatment systems in cold climates. Studies in Anchorage also demonstrated the better insulating properties of concrete tanks over steel tanks.

Temperature studies in Washington and Wisconsin, where the septic tank mound system is used, indicate the rock/plant system could replace the mound system at less cost by eliminating the pumping chamber used in most mound systems. The mound system is essentially an elevated soil absorption system.[1]

SUMMARY

Our studies indicate that septic tank effluent from single homes can be treated to advanced secondary levels or better by using a 400-ft^2 (37.2 m^2) washed gravel filter. The filter should be 1.0–1.5 ft (0.3–0.46 m) deep with the wastewater level maintained 6 in. (15.2 cm) below the rock surface. When aesthetic plants—calla lily *(Zantedeschia aethiopica),* canna lily *(Canna flaccida),* elephant ears *(Colocasia esculenta),* yellow iris *(Iris pseudacorus),* and ginger lily *(Hedychium coronarium)*—are planted in the rock filter, their roots penetrate into the wastewater level, adding oxygen and increasing biological activity. If a point source discharge is undesirable, approximately 20 ft (6 m) of 4-in. (10.3-cm) perforated leach field tubing should be used to disperse the treated rock/plant filter effluent beneath the soil.

Large septic tanks used in small towns and communities in lieu of open sewage lagoons have many advantages over open lagoon systems. Underground tanks can be installed in different locations throughout the collection area, taking advantage of the land elevations. All tanks can be connected to drain pipes carrying the effluent to one or more rock/plant filters for treatment. Size of the rock/plant filter system will be dictated by the volume of septic tank effluent and the level of treatment desired. When these systems are properly installed, sewage will not have open-air exposure before treatment has been accomplished.

REFERENCES

1. Canter, L. W., and R. C. Knox. *Septic Tank System Effects on Ground Water Quality* (Chelsea, MI: Lewis Publishers, Inc., 1985), pp. 38–43.

2. Wolverton, B. C., R. C. McDonald, C. C. Myrick, and K. M. Johnson. "Upgrading Septic Tanks Using Microbial/Plant Filters," *J. MS Acad. Sci.* 29:19–25 (1984).
3. Wolverton, B. C. "Hybrid Wastewater Treatment System Using Anaerobic Microorganisms and Reed *(Phragmites communis),*" *Econ. Bot.* 36(4):373–380 (1982).
4. Wolverton, B. C., R. C. McDonald, and W. R. Duffer. "Microorganisms and Higher Plants for Wastewater Treatment," *J. Environ. Qual.* 12(2):236–242 (1983).
5. Wolverton, B. C. "Artificial Marshes for Wastewater Treatment," in *Aquatic Plants for Wastewater Treatment and Resource Recovery*, K. R. Reddy and W. H. Smith, Eds. (Orlando, FL: Magnolia Publishing Inc., 1987), pp. 141–152.
6. Wolverton, B. C. "Natural Systems for Wastewater Treatment and Water Reuse for Space and Earthly Applications," in *Proceedings of American Water Works Association Research Foundation, Water Reuse Symposium IV* (Denver, CO: 1987), pp. 729–741.
7. Wolverton, B. C. "Aquatic Plants for Wastewater Treatment: An Overview," in *Aquatic Plants for Wastewater Treatment and Resource Recovery*, K. R. Reddy and W. H. Smith, Eds. (Orlando, FL: Magnolia Publishing Inc., 1987), pp. 3–15.
8. Wolverton, B. C., and R. C. McDonald. "Natural Processes for Treatment of Organic Chemical Waste," *Environ. Prof.* 3:99–104 (1981).
9. Wolverton, B. C., and R. C. McDonald-McCaleb. "Biotransformation of Priority Pollutants Using Biofilms and Vascular Plants," *J. MS Acad. Sci.* 31:79–89 (1986).
10. Viraraghavan, T. "Temperature Effects on Onsite Wastewater Treatment and Disposal Systems," *J. Environ. Health* 48(1):10–13 (1985).

CHAPTER 13

13a Creation and Management of Wetlands Using Municipal Wastewater in Northern Arizona: A Status Report

Mel Wilhelm, Sam R. Lawry, and Douglas D. Hardy

INTRODUCTION

Wetlands are a scarce habitat type in Arizona. Most historically occurred along the Colorado and Gila rivers, now highly modified or dewatered by human development. Some seasonal or yearlong wetlands occurred in higher elevation natural lakes. Appropriation of water for agricultural, industrial, and municipal uses has decreased wetland habitats, and increased water demands by Arizona's growing population is expected to continue. However, a new factor, municipal wastewater used to create new wetlands in northeastern Arizona, may offset natural wetland losses.

PROJECT AREA

The Apache-Sitgreaves National Forests, U.S. Department of Agriculture, are situated in the northeastern portion of Arizona along a major escarpment, the Mogollon Rim. These two forests comprise some 810,931 ha, with ponderosa pine *(Pinus ponderosa)* and pinyon juniper *(Pinus edulis-Juniperus* spp.) the two major habitat types and elevations varying from 1370 to 3050 m. Forest staff recently prepared a 10-year management plan including added emphasis to wildlife and recreation use.[1] The potential for wetland habitat improvements was first documented in a management program report[2] that included the following: "The objective is, through a 20-year comprehensive program, to improve the quality of the wetland ecosystem on the Apache-Sitgreaves National Forests. The paucity of adequate nesting and rearing cover is the primary factor limiting reproduction and survival of wetland faunal species on the Forests." Through proper management, limiting factors may be alleviated by manipulating vegetative conditions to create, maintain and improve the wetland ecosystem. This objective can be accomplished by peripheral fencing of critical nesting areas, construction of suitable nesting islands, augmentation of existing food and cover species by planting, and fabrication of artificial nesting structures. In addition, the potential exists in the Show Low area to create substantially more wetland habitat by utilizing treated effluent to flood selected areas.

The Wetlands Management Program estimated 1976 waterfowl production at 400 young per year and potential after implementation of the plan up to 58,800 per year.

The Wetlands Management Program was partially implemented with natural wetlands improved and new wetlands created using municipal effluent. From a city standpoint, these marshes act as large evaporation ponds for wastewater disposal. From a wildlife standpoint, they offer opportunities to design optimum habitat.

EFFLUENT MARSHES PHASE I: PINTAIL LAKE

Prior to 1979, the City of Show Low, Arizona disposed of sewage effluent in a volcanic depression north of the city limits. The effluent received secondary treatment and chlorination and posed no public health hazard, but would lead to nutrient enrichment and pollution if released into live watercourses. Tertiary treatment was deemed too expensive. Projected population growth pointed to the need for additional effluent storage areas.

Upon receipt of city application to dispose of surplus effluent in dry lake beds, the Lakeside Ranger District (ASNF) began discussions with Arizona Game and Fish (AG&F) Department personnel (Region 1) on benefits and feasibility of developing effluent-fed waterfowl areas.

Various sites on National Forest System lands north of Show Low were evaluated. Some sites were rejected because of cost or proximity to the local airport. Pintail Lake was selected because of size, secluded location, and proximity to an existing pipeline. Pintail Lake is a natural basin in a pinyon-juniper woodland, 5 km north of Show Low. An agreement was reached in late 1977 between the city and ASNF in which the Forest Service would fund and construct, on National Forest System lands, depressions and dikes to form a marsh. The city, with U.S. Environmental Protection Agency (EPA) grant fund assistance, installed pipeline and valves and agreed to supply effluent to maintain optimum pond water levels. Total project cost was $146,750—$91,750 from the City and EPA and $55,000 from the Forest Service. Forest Service funding was budgeted for the Comprehensive Wetlands Management Program for the Apache-Sitgreaves National Forests.

In December 1977, the AG&F Commission in a cooperative agreement with ASNF, accepted maintenance and operation responsibilities for the marsh system for 20 years. Marsh construction, including 14 islands, was completed by spring of 1979, and effluent inflow began that fall.[3]

In 1979, ASNF contracted with the University of Arizona for an evaluation

Table 1. Number of Nests/ha Found at Pintail Lake, Arizona

Species	1979	1980	1981	1982	Total
Mallard (*Anas platyrhynchos*)		10	57	143	210
Northern pintail (*A. acuta*)		3	23	24	50
Gadwall (*A. strepera*)		7	18	53	78
Northern shoveler (*A. clypeata*)				2	2
Cinnamon teal (*A. cyanoptera*)		15	52	122	189
Green-winged teal (*A. crecca*)	3	4	26	23	56
Redhead (*Aythya americana*)			2	3	5
Ruddy duck (*Oxyura jamaicensis*)		4	15	10	29
Totals	3	43	193	380	619

of habitat improvements accomplished under the Wetlands Management Program, including Pintail Lake. The marsh proposal for Pintail Lake was questioned by some waterfowl authorities in Arizona on effectiveness of islands for Arizona's "desert ducks" and if optimum habitats would result in increased waterfowl production. After three years of data collection on Pintail Lake, Piest[4] stated, "The response of breeding waterfowl to these conditions has been dramatic. I estimated that 1544 ducklings, or 76.4 ducklings/ha (30.93/ac) were produced in 1981. This was only the second season after the lake had begun to receive sewage effluent. These totals far surpassed those from any other wetland that I studied. The density and success rate that I observed for nests on the islands in 1981 were 163.95 nests/ha (66.38/ac) and 96.8%, respectively. During my review of the published literature, I found only one study that documented a higher density of nests of a wild nesting population of ducks, and I found no studies that documented a higher rate of nest success."

Man-made islands were preferred nesting sites at Pintail Lake, supporting 90% of all nests surveyed in 1981. In 1982, nest density increased 97% over 1981. Tables 1 and 2 summarize nests found and ducklings produced for four years at Pintail Lake.[5]

Budget reductions later reduced intensive monitoring, but AG&F reported that nesting intensities and ducklings produced have declined since 1982. Effluent quality decline from the City's overtaxed system may have caused the decline in waterfowl production.[6]

Pintail Lake has become a local attraction for consumptive and nonconsumptive users; hunter use and bird watching are roughly equal, and the town council and mayor have expressed pride in this development.

In 1976, Pinetop and Lakeside communities needed to dispose of additional effluent, and plans were approved for another marshland (Jacques) north of Lakeside in the pinyon-juniper habitats at 2048 m elevation on National Forest

Table 2. Ducklings Hatched at Pintail Lake, Arizona

Species	1979	1980	1981	1982	Total
Mallard (*Anas platyrhynchos*)		93	521	1188	1802
Northern pintail (*A. acuta*)		27	150	158	335
Gadwall (*A. strepera*)		49	145	316	510
Cinnamon teal (*A. cyanoptera*)		112	170	884	1166
Green-winged teal (*A. crecca*)	9	26	174	162	371
Redhead (*Aythya americana*)			14	27	41
Ruddy duck (*Oxyura jamaicensis*)		32	114	82	228
Unidentified				2	2
Totals	9	339	1588	2819	4755

System lands. The terrain was relatively flat; natural depressions were not available as at Pintail Lake. Dikes created seven interconnecting ponds with a combined surface area of 38 ha, and 18 nesting islands were built within the ponds. Total construction costs were \$286,600—\$265,600 provided by the communities and \$21,000 from Forest Service funds. Jacques Marsh did not receive significant effluent until 1984, but natural and artifically seeded vegetation supported some waterfowl nesting in 1985. By 1986, 1540 m3/day of effluent were entering Jacques Marsh, filling three ponds totaling 18 ha.

To date, 10 of 16 introduced hydrophytes have become established, providing important nesting, feeding, and cover areas. Fall, single-day waterfowl surveys conducted by the AG&F in 1986 and 1987 revealed 1300 and 1500 birds, respectively. Spring, single-day waterfowl surveys for the same years revealed 950 and 1900 birds, respectively. In spring 1987, nesting density was 104 nests/ha (42.1/ac); most were mallard (*Anas platyrhynchos*) and cinnamon teal (*Anas cyanoptera*), and 95% were on nesting islands. Effluent quality at Jacques is expected to remain high, and by 1990 the marsh is expected to reach capacity effluent inflows and produce 450 nests and 3000 ducklings annually. Waterfowl productivity for Jacques is expected to exceed Pintail Lake because of better effluent quality, larger size, and higher elevation.

Jacques Marsh also attracted Rocky Mountain elk *(Cervus canadensis)* during winter that by spring had grazed the entire 53 ha fenced to exclude domestic livestock. As temperatures warm in April and May, elk leave for higher elevation summer range. Elk grazing in this fashion seems compatible with waterfowl use, because adequate regrowth occurs in time for nesting. Jacques Marsh attracts an estimated 100 hunter-days a year and is becoming a popular bird watching area.

Success of Pintail and Jacques Marsh prompted interest by Springerville, Arizona, and a cooperative effort between the city and AG&F built another effluent marsh on AG&F land (Springerville Marsh). Completed in 1983, it consists of five ponds with 15 islands totaling 36 ha in a grassland habitat at 2116 m elevation. The system receives 385 m3/day of effluent, which only effectively supplies one pond. Some waterfowl nesting has occurred but no surveys have been conducted. Introduced submergent and emergent plants should increase nesting activities as at Jacques Marsh. Springerville Marsh is closed to hunting due to its proximity to the town and should receive excellent waterfowl use.

EFFLUENT MARSHES PHASE II

At present, the marsh disposal systems for Show Low and Pinetop-Lakeside are being expanded to handle larger volumes of treated effluent. These newer marshes are referred to as Phase II and are designed to alleviate some problems encountered in Phase I. Problems encountered, and changes in designs and/or management are discussed below:

1. Maintaining marsh levels during nesting seasons is difficult if effluent quantities vary substantially. Islands do not protect nesting ducks when shallow water levels permit coyotes *(Canis latrans)* to wade to them. Marsh vegetation also suffers from lack of water at critical times. Phase II marshes have requirements for effluent storage during winter months and release during nesting seasons to maintain water levels and vegetation.
2. Water quality is crucial; clarity allowing sunlight penetration seems to be important. Factors inhibiting aquatic insect production may cause declines in waterfowl productivity as at Pintail Lake. The City of Show Low's system was upgraded and better effluent quality is occurring. Other improvements incorporated include a storage facility at Telephone Lake, open channel flow for 0.6 km, and a flow-through riparian area to improve water quality before it enters the marsh system.
3. Operating flexibility, including ability to fluctuate water levels; closed basins with no outlets; and lack of ability to spread water on adjoining areas for forage enhancement caused problems. Phase II marshes are designed as flow-through with water release. New head works incorporate this and water boards and head gates are independent. Marsh dikes in the lower portions of a natural drainage basin spread effluent anywhere throughout the basin without releasing effluent outside the system.
4. Adequate vegetation to control erosion of islands and dikes has been a problem. Northern Arizona experiences windy spring conditions. Better control of water levels to favor shoreline vegetation and storage facilities will improve vegetation establishment. Invasion of undesirable plants—salt cedar *(Tamarix pentadra)* and cocklebur *(Xanthium saccharatum)*—can also be discouraged by water level control.
5. Coyote predation is the main cause of nest destruction, and shoreline vegetation is important in reducing predation. Thick bands of shoreline vegetation (including island shorelines) may reduce island visibility. Net wire fences were installed to reduce access to dikes used for nesting.
6. Marsh area fencing has reduced forage available for livestock grazing, and

livestock permittees are concerned over potential forage losses for wildlife benefits. Provisions have been made to irrigate areas outside the marshes to enhance forage for livestock and benefit livestock permittees.

RESULTS AND DISCUSSION

Municipal effluent used to create marshlands in Northern Arizona has received substantial interest, because it provides economical disposal and benefits for wildlife. Marsh disposal for Show Low's secondary-treated effluent cost 32% less than the lowest nonmarsh alternative, and annual operating costs were 47% less with the marsh.[7] Unless regulations for discharging effluent into natural stream courses are relaxed, use of marshes as disposal areas and evaporation ponds will continue to be economically attractive. Effluent marshes have been an obvious success for wildlife habitat. Articles in national publications have encouraged program expansion, and local groups use the areas for observing wildlife.

FUTURE

Opportunities to expand these developments will continue because populations of most area communities are expanding rapidly. Effluent production rates are projected to increase from 2200 to 7570 m3/day for Show Low and from 1514 to 9463 m3/day for Pinetop-Lakeside by the year 2025. Show Low recently expanded its wastewater facilities with a new 20-ha marsh on National Forest System lands 5 km west of Pintail Lake. The addition includes effluent storage in winter, with a riparian growth area and forage enhancement for wildlife and livestock. Pinetop and Lakeside developments recently constructed include a 16-ha storage facility, flexible water distribution and marsh water level control, and an overland irrigation area for elk forage. Evaluation must continue to improve management and develop early warning methods for any adverse impacts, but marshes from wastewater effluents have much potential.

REFERENCES

1. "Proposed Apache-Sitgreaves National Forests' Plan," USDA Forest Service, Southwestern Region (1986).
2. McKibben, J. "Wetlands Management Program and Environmental Analysis Report," Apache-Sitgreaves National Forest, Springerville, AZ (1976).
3. "The Allen Severson Memorial Wildlife Area Management Plan, 1980–85," Arizona Game and Fish Department, Region 1, Pintail Lake (1980).
4. Piest, L. "Evaluation of Waterfowl Habitat Improvements on the Apache-Sitgreaves National Forest, Arizona," (1981).
5. Piest, L., Arizona Cooperative Wildlife Research Unit, University of Arizona, Tucson. Personal communication (1983).

6. O'Neil, J., Region 1 Big Game Supervisor, Arizona Game and Fish Department, Pinetop. Personal communication (1986).
7. "Environmental Assessment City of Show Low, Arizona, Wastewater Treatment and Disposal System Expansion," Morrison-Knudsen Engineers, Inc. (1985).

13b Land Treatment of Municipal Wastewater on Mississippi Sandhill Crane National Wildlife Refuge for Wetlands/Crane Habitat Enhancement: A Status Report

Joe W. Hardy

INTRODUCTION

The endangered Mississippi sandhill crane *(Grus canadensis pulla)* occupies a designated critical habitat area of 10,500 ha in southeast Mississippi.[1] These lands contain most breeding and roosting habitats essential to the current crane population, but some winter feeding areas are excluded. A 7700-ha national wildlife refuge established in 1975 falls within the critical habitat boundaries. Major historical habitat changes—timber production emphasis, accelerated urban and commercial development, and repeal of the free-range law from the mid-1950s to mid-1970s—caused a steady decline in the crane population.[2] Survival of this subspecies can be linked to infertile, acidic soils of the area and associated native plant communities. These poorly drained soils support pine forest, savannas, and swamp plant communities. Of greatest importance to the cranes are the savannas. These wet prairies support scattered longleaf pine *(Pinus palustris)* and pond cypress *(Taxodium* spp.). Principal ground covers are wiregrass *(Aristida* spp.) and bluestems *(Andropogon* spp.), along with pitcher plants *(Sarracenia alota),* panic grasses *(Panicum* spp.), rushes, sedges, and a variety of annuals. Savannas, used by cranes for nesting, feeding, and roosting, originally occupied nearly 60% of the crane habitat. During the 1950s, paper companies acquired large tracts of land, which were ditched and planted to thick stands of slash pine *(Pinus elliotti).* Fire suppression and lowered water tables from drainage further reduced the habitat suitable for cranes. Residential and commercial development along with road construction, including Interstate 10, through prime crane habitat have added to habitat losses.

A primary objective for recovery and survival of the crane is to restore desirable habitats, including plant communities and water regimes. Given the federal emphasis of the early 1980s on "good neighbor" policy and community cooperation, discussions between the area's growing suburban community and the refuge relative to new technologies for wastewater treatment were just a matter of course. Assuming adverse impacts to the crane could be avoided, the local municipalities could satisfy new Clean Water Act requirements while providing an endangered species with enhanced habitat conditions.[3]

PROJECT AREA

The Ocean Springs Regional Land Treatment site is located 9.7 km north of Ocean Springs in Jackson County, Mississippi along the northern edge of the Mississippi Sandhill Crane National Wildlife Refuge. The lower Gulf of Mexico Coastal Plain site has land elevations ranging from 11.3 m above mean sea level (msl) along the southern portions to 24.1 m in the northeast corner. Topography ranges from nearly level (0–2%) to sloping (5–8%), and drainage is through the refuge southeast to Old Fort Bayou and southwest 50 km to the Gulf of Mexico.[4]

Land use of the area in and adjacent to the land treatment site is agronomic and silvicultural. Agricultural use includes pastures, small truck crops, and general farming operations. Silvicultural activities are slash pine plantations for pulpwood. Several residential areas are scattered throughout the area. Climate is characterized as humid and subtropical, with hot summers and alternately warm and cold winters. Annual rainfall ranges from 152 to 178 cm, mostly occurring from June to September.

Soils on site are primarily loamy sands with some sandy clays and clays as subsoils. Well and moderately well drained soils are found on gently sloping narrow ridges between drainage ways. Poorly drained soils are found in the broad, nearly level coastal flatwoods. Poorly drained soils and swampland types are found in bayheads and adjacent to drainageways. A gray clay to sandy clay substratum occurs at depths of 1.5–2.5 m, restricting vertical water movement through the profile and locally perching a seasonally high water table.

The land treatment study area has a mosaic of biotic communities developed in response to variable, mostly man-influenced, environmental factors, particularly hydroperiod. Relatively distinct plant communities include pineland, mixed bayheads and pinelands, savannas, cleared areas, pine-savanna transitional, and bayheads. The pineland community, largely plantation, occupied 78% of the study area with some longleaf pine *(Pinus palustris)* dominating drier areas, and slash pine *(Pinus elliotti)* dominating wetter areas. Most of the study area is planted slash pine. A thick understory of shrubs and herbaceous vegetation may or may not be present depending on the history of burning and pruning as well as soil moisture conditions. The savanna community is found in wetter areas and has an abundance of herbaceous species with few scattered trees. Periodic ground fires and a high groundwater table are required to prevent invasion of this community by pines or water-tolerant hardwoods. Bayhead and branch-bottom communities representing 12% of the area are found primarily along Perigal Bayou. No threatened or endangered plant species are known to be present and a wide variety of resident mammals, songbirds, and raptors are common to the area.

The study area is critical habitat for the Mississippi sandhill crane, an endangered subspecies of the sandhill crane. In 1980, only 40 of these birds existed, and the current population of 50 includes some captive reared-and-released cranes. Preferred nesting and feeding habitats of the cranes are the savanna and transitional pineland/savanna communities discussed earlier.

PROJECT OBJECTIVES AND LAYOUT

Following lengthy feasibility reviews, coordination with concerned agencies and groups, and evaluation of environmental and endangered species impacts, in 1983 the U.S. Fish and Wildlife Service ("Service") and the Mississippi Gulf Coast Regional Wastewater Authority ("Authority") signed a memorandum of understanding (MOU) to allow land treatment of primary-treated effluent on the Mississippi Sandhill Crane National Wildlife Refuge. The Authority was responsible for construction and operation of the facility. Service responsibility was limited to providing 100 ha of refuge land and technical input during construction and operation on potential adverse impacts to cranes, their habitat, and refuge wetlands. The 20-year agreement has provisions for extension and contingency plans, and for immediate termination if adverse impacts on cranes or their habitat were identified and problems were not corrected.

An amendment to the original MOU signed in 1986 granted the Service access and management control of 235 ha owned and used by the Authority for land treatment of primary wastewater. Land management authority granted to the Service for crane breeding and feeding habitat and trespass control may not conflict with interests of third parties or normal operation, maintenance, or expansion of the Authority's facility.

With the 1986 amendment, the refuge added 335 ha of potentially valuable, short-supply crane habitat, but risks of adverse impact to cranes were involved with field application of wastewater technology. However, project success would enhance achievements of Clean Water Act mandates and area environmental quality.

The Authority contracted with CH2M HILL for design and oversight of construction and initial operation. The project has two major components: (1) the lagoon system, including the facultative lagoon/storage facilities and screening and distribution pump station facilities; and (2) the land treatment system, including the distribution and sprinkler facilities and the underdrain and wetland cell facilities.[4]

The lagoon system is located adjacent to refuge property. Proper functioning of this system provides adequate pretreatment, screening, and distribution for the land treatment stage. The lagoon system was designed (depths, bottom slopes) to discourage use by cranes.

The land treatment system is important to the refuge and crane management because any crane benefits will come from this activity. Conversely, land treatment and water management problems could negatively affect crane habitats. Treated effluent is pumped to the land treatment site via a 76.2-cm diameter ductile iron or concrete pipe. The 150-ha site has 421 permanent gun-type sprinklers spaced on 45.75-m centers on PVC lateral feed lines. With 3.45 × 105 Pa nozzle discharge pressure, the effluent application rate of 0.76 cm/hr should result in zero surface runoff. The site has coastal bermuda grass *(Cynodon dactylon)* with fall overseeding of annual ryegrass *(Lolium temulentum)* for a potential nitrogen removal capacity of 80–100 kg/ha/yr. The March through October growing season will produce five to seven hay cuttings per year, and 4 ha of corn will be planted along field margins for cranes.

The underdrain and wetland cell portion of the treatment system is of primary concern because soil moisture/water changes on 400–500 ha may have negative or positive impacts on crane habitats. The underdrain system will maintain unsaturated soil to 1 m depth within the treatment area for percolate treatment and crop growth. The system consists of 10–61 cm diameter perforated, corrugated polyethylene tubing spaced on 15-m centers 2 m below grade. A gravel filter of 7.62 cm is located around underdrain piping to minimize filter plugging.

Three wetland cells located 500–1500 m downgrade were constructed to receive underdrain percolate from the treatment site. One wetland cell was constructed by placing a vertical, impervious synthetic barrier into the subsurface clay layer, causing groundwater above the clay to surface. Two other wetland cells were constructed without the barriers. Water level control in these wetland cells should provide favorable feeding, roosting, and nesting crane habitat.

The MOU established a monitoring program covering soils, crops, biota, groundwater, and land treatment system effectiveness. The program's selected parameters and sampling frequency are based in part on Service and U.S. EPA recommendations. The Authority will monitor soils for pH, EC (electrical conductivity) of saturation extract, available nitrogen, phosphorus, plant tissues from the hay crop, biota of pine woodlands and savanna areas downgrade from spray fields, and biotic change in receiving wetland cells. In addition, water depth, specific conductance, total phosphorus, and nitrate/nitrite will be sampled in eight shallow-aquifer monitoring wells.

Spray field operation trials began in January 1986 and became fully operational in October 1987. Several problem areas as well as successes have been noted. Considerable surface runoff of primary-treated effluent from one section of the site has contaminated and damaged 4–6 km of refuge service roads and ditches. Unseasonably high soil saturation levels have curtailed timber harvests. Although normal rainfall restricts woodlands access during rainy periods and causes some roadway erosion, heavy flows from spray field runoff have contributed to these problems. Wetland cells have operated at less than 25% of predicted efficiency. Major structural erosion problems have plagued all three cells. Little percolate has reached the cells. The grassland sod in the treatment area may require three years to develop full percolate capacity.

However, clearing heavy slash-pine forest and the haying operation on the treatment area have resulted in increased crane use. Since 1985, spray fields have become a major feeding area for birds, especially during winter, and mowed hay fields should increasingly provide excellent feeding areas for young cranes. One pair and their chick used the fields for 14 weeks in 1987.

The success of this cooperative effort in using municipal wastewater for wetlands and sandhill crane habitat enhancement will require several years of evaluation. The MOU between the Wastewater Authority and U.S. Fish and Wildlife Service provides an adequate vehicle to correct problems and enhance the goals of both agencies. Close adherence to the agreement, including monitoring, operational adjustments, system modifications, and implementation of the contingency plans if necessary are important to the final outcome.

REFERENCES

1. "Recovery Plan for the Mississippi Sandhill Crane (*Grus canadensis pulla*)," USDI Fish and Wildlife Service, Region IV (1984), p. 56.
2. "Mississippi Sandhill Crane National Wildlife Refuge Operation Plan, FY85-90," USDI Fish and Wildlife Service, Region IV (1984), p. 18.
3. Valentine, J. M., Jr. "Report on the Ocean Springs Wastewater Facilities Plan and the Relationship of the Project to the Mississippi Sandhill Crane and Its Environment, Parts I and II," LaFayette, LA (1983), p. 28.
4. CH2M HILL Engineers, Planners, Economists, and Scientists, Inc. "Ocean Springs Regional Land Treatment System, Jackson County, Mississippi, Volume 3—Conceptual Documents and Technical Memoranda," Gainesville, FL (1983), p. 190.

CHAPTER 14

Waste Treatment for Confined Swine with an Integrated Artificial Wetland and Aquaculture System

J. J. Maddox and J. B. Kingsley

INTRODUCTION

Many biological and agricultural approaches have been used in recovering plant nutrients from livestock wastes and to treat wastewater from livestock facilities.[1,2] Aquatic reclamation systems using livestock manures or municipal wastewater include phytoplankton and bacterial communities as the basis of the food chain. Productivity can be high and results in net increases of dry matter from photosynthesis of algae and higher plant life.[3] Mechanically removing algae in waste treatment systems can be expensive and is usually done on a large scale to accomplish wastewater treatment not for resource recovery. Problems with maintaining high-density algae and bacterial cultures in temperate climates, culture quality control, social constraints, and lack of profit incentives are strong deterrents to systems that depend on mechanical methods to remove suspended solids.

The livestock waste treatment described in this chapter uses fish culture and sand bed filtration in the artificial wetlands to circumvent problems of algae harvesting. The recommended fish are filter-feeders that graze on single-cell algae. Sand beds are planted with Chinese water chestnuts *(Eleocharis dulcis)* in the artificial wetlands to receive suspended solids and dissolved plant nutrients.

Livestock Waste

Swine production is second only to poultry in terms of animal numbers in the Tennessee Valley states and is the third most abundant source of livestock manures.[4] Livestock production systems with automated or daily removal of manure, predictable quantities, and a reliable quality have the best potential for applying the integrated approach to waste treatment described in this chapter. Other types of livestock waste can be treated with aquaculture and artificial wetlands but may be more labor-intensive. Waste treatment on a farm must meet two requirements: (1) protecting the environment and (2) maintaining economic viability of the farm enterprise.

Poultry broiler production uses mechanical methods for removal of the poultry litter from the broiler house on an intermittent basis. Storage and waste handling problems make its direct use in artificial wetlands more difficult.[5] Poultry layer waste from egg production facilities are a more suitable source of treatable manure than broiler litter using artificial wetlands as described herein.

Dairy waste contains straw and large volumes of wash water containing pesticides for cleaning and disinfecting purposes. Volume and quality of this water varies depending on management practices.

Swine production uses waste flushing and holding methods easily adapted to an integrated aquaculture and artificial wetland approach. Swine finishing facilities can be designed for complete confinement, environmental control, water flushing, and lagoon waste stabilization. Swine finishing (market hogs) requires a nutritionally balanced feed lacking bulky straw and cellulosic fiber content found in other types of livestock waste.

Manure generated from a 1000-head swine finishing facility can produce 5000 kg of nitrogen, 7700 kg of P_2O_5, and 8200 kg of K_2O annually.[6] This implies a fertilizer value of $7700 (N @ $0.38/kg; P_2O_5 @ $0.49/kg; and K_2O @ $0.25/kg) for the N-P-K value of the manure, offering great potential for economic resource recovery to offset waste treatment cost.

Fish

Growing fish in waters enriched with livestock manures originated with the Chinese several thousand years ago,[7] including "polyculture" with several fish species of differing food habits to more efficiently use the existing fish food.[8]

Several of the Chinese carps and tropical cichlids (tilapia) are suitable for polyculture combinations. The silver carp *(Hypophtholmichthys molitrix)*, bighead carp *(Aristichthys nobilis)*, and several tilapias have special anatomical structures to filter fine particulate matter from the water. Silver carp feed on phytoplankton in the size range 8 to 100 μm, with most organisms in the range of 17 to 50 μm. Bighead carp feed on zooplankton, phytoplankton, and detritus in the size range 17 to 3000 μm, with most in the range of 50 to 100 μm.[9] Cichlids of the genus *Sarotherodon* (formerly *Tilapia)* are efficient filter-feeders, grazing on benthic invertebrates and detritus.[10]

Water Chestnuts

The Chinese water chestnut *(Eleocharis dulcis)* is an anchor-rooted, emergent species first introduced into the United States in the 1940s.[11,12] The plant is grasslike and grows to a height of 1.0 to 1.5 m (3.5 to 5.0 ft). It requires 150–220 days to mature, needs 250–300 kg/ha of nitrogen, and produces a high starch, sweet chestnut (corm) with a retail value of $5–7/kg ($2–3/lb).

Corm sizes vary with the variety or plant selection but are marketable when 2.5 cm (1 in.) or greater in diameter. Fertilizer requirements, plant disease, weed control, mechanical harvesting, problems of storage, and food processing, including peeling, have been studied and reviewed for commercial development in the United States.[13]

In 1979, a production test facility at Muscle Shoals, Alabama was used to establish (1) the ratio of fish culture area to plant growing area in artificial wetlands, (2) effluent water quality, and (3) predicted plant yields as a function of irrigation rates. We report here on the second year's experience with a small-scale demonstration facility that integrated livestock production with fish aquaculture and artificial wetlands.

MATERIALS AND METHODS

The experimental schematic for the test facility is shown in Figure 1. A serial water flow was maintained from a groundwater source through a swine manure-fertilized fish culture tank and onto sand filtration beds where water chestnuts were grown.

Fish Culture

Test tanks were 10.5 m^2 (0.001 ha), 76 cm deep, lined with Hypalon, and equipped with PVC plumbing. *Tilapia nilotica,* silver carp, and bighead carp were stocked at 35,000 fish/ha (600–650 kg/ha). All-male tilapia were stocked to minimize food competition resulting from stocking a tilapia breeding population. Well water was passed through to exchange the system completely in 10 days (10 RT, retention time of 10 days). The fertility rate was equivalent to 70 kg/ha of dry swine manure per day, equivalent to 180–200 hogs/ha (73–81 hogs/ac) weighing an average of 61.4 kg (135 lb), applied directly to fish test units from a flushing gutter swine facility (Table 1).

Fish were grown in polyculture for 158 days and harvested on October 23, before water temperatures became lethal ($<10°C$). A polyculture of silver carp and bighead carp was restocked, and testing continued until April 22 (Table 2).

Artificial Wetlands

Fish wastewater effluent (10 RT) was applied to three test sizes of water chestnut sand filtration beds (1.8 m^2, 2.7 m^2, and 4.5 m^2) with irrigation rates equivalent to 60, 100, and 150 L/m^2/day (Figure 1). Sand filtration beds were lined with two layers of 6-mil greenhouse clear plastic equipped with bottom drain and level control standpipes and filled with 30.5 cm (12 in.) of masonry sand. Water depth was maintained at 5.1 cm above the sand, and water detention times were 2.3, 3.5, and 5 days for irrigation rates corresponding to 150,

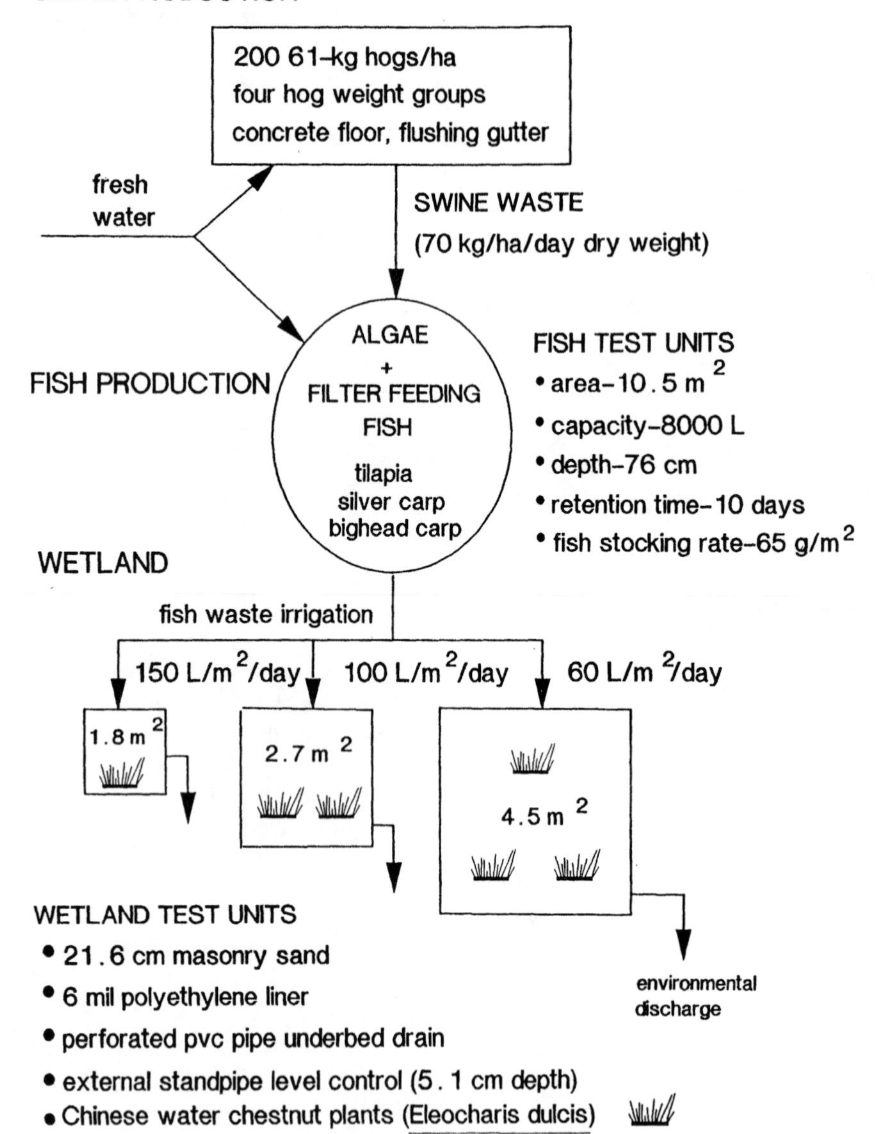

Figure 1. Schematic of the test facility for integrated constructed wetlands and aquacultural system for livestock waste treatment.

Table 1. Characteristics of Swine Manure and Water Quality of Influent and Effluent in the Production Test (Spring/summer/fall, May-December 1980)

			Effluent (mg/L)			
				Wetlands Irrigation, (L/m²/day)		
Parameter	Manure (mg/L)	Water Source[a] (mg/L)	Fish	150	100	60
BOD	1953	6.9	19	5	5	4
COD	4559	5.0	80	26	25	23
SS	2567	2.4	53	12	8	9
TS	4044	453	581	532	544	514
FC	7×10^6	1733	5900	450	650	750
TKN	355	1.0	3.0	1.1	1.0	0.8
TP	124	0.25	1.3	0.6	0.4	0.3
K	198	5.5	8.4	7.5	7.2	6.7
pH		7.2	8.5	6.9	6.8	6.8

Note: BOD = biological oxygen demand; COD = chemical oxygen demand; SS = suspended solids; TS = total solids; FC = fecal coliforms (colonies/100 mL); TKN = total kjeldahl nitrogen; TP = total phosphorus; K = total potassium.
[a]Groundwater supply.

100, and 60 L/m²/day, respectively. Water chestnuts were grown in a hothouse for 59 days before planting in test units on May 11. They grew 188 days in the test until killed by a frost on October 14 and were harvested on December 12 (Table 3).

During this plant postharvest test period—December 12 to April 22, the winter/spring season (Table 2)—the sand filtration beds were reconditioned. Wastewater was applied continually to the sand filtration beds through the winter/spring season at similar rates.

RESULTS AND DISCUSSION

Net fish production after 160 days of culture was 4300 kg/ha. Flowing water systems produced 31% more when compared with static water trials during this same period.[3] Tilapia averaged 184 g each (0.4 lb) after initially stocking at 20–30 g each. Previous fish production tests had indicated that a 10% daily replacement rate (10 RT) was better than 5 RT or 15 RT.[14]

Yields of water chestnuts and hay were correlated with irrigation rate and compared favorably with results obtained in 1978 (Table 3). In this test, irrigation rates were a function of water chestnut area. Ratios of fish pond surface area to chestnut system surface area were 2.29, 1.32, and 0.77 for the irrigation rates of 150, 100, and 60 L/m²/day, respectively (Figure 1). Plant response to irrigation rate did not depend upon fish pond wastewater retention time but was linearly correlated with irrigation rates ($R^2 > 90\%$, intercept = −5.819, and slope = 0.351). Also, higher flow rates produced more water chestnuts near the sand bed surface and did not significantly influence yields at 7.6–22.9 cm (3–9 in.) deep. Nearly 70% of the water chestnuts were produced at the

Table 2. Seasonal Averages of Effluent Water Quality from Production Test with Fish and Water Chestnuts (Winter/spring, January-April 1980)

		Effluent (mg/L)			
			Wetlands Irrigation, ($L/m^2/day$)		
Effluent Variable	Water[a] Source (mg/L)	Fish	150	100	60
BOD	1.45	22.2	19.7	16.5	13.2
COD	7.00	61.7	59.7	53.2	47.1
SS	1.57	30.0	13.0	10.4	13.3
TS	380	386	362	379	359
FC[b]	357	2461	1062	433	561
TKN	0.09	3.29	3.53	3.42	2.70
NH_3 + NH_4^+–N	0.02	0.73	1.92	1.83	1.28
ORGN	0.07	2.56	1.75	1.59	1.37
TP	0.01	1.31	1.41	1.57	1.08
TOC	61.3	171	113	95.7	122
Ca	39.4	39.1	41.1	42.9	40.5
K	4.92	7.26	8.01	8.21	7.74
pH	7.01	8.36	7.27	7.47	7.54
T (°C)	8.5	8.6	8.6	8.6	8.6

[a]Groundwater.
[b]Colonies/100 mL.

depth from the surface to 7.6 cm (0–3 in.) at the highest irrigation rate, and this flow rate produced the equivalent of 39.5 MT/ha (17.6 ton/ac).

Irrigation rate onto sand infiltration beds may be more limited by management criteria than plant yield or water quality. Higher irrigation rates resulted in reduced water infiltration rates near the end of the growing season. The effect was to increase standing water depth above the sand surface, requiring

Table 3. Yields of Water Chestnuts and Hay in Response to Daily Irrigation with Wastewater from a Fish Pond (1979 Colbert County, Alabama)

		Yield (MT/ha) Wetlands Irrigation ($L/m^2/day$)		
Soil Depth (cm)		150	100	60
			Water Chestnuts	
Top:[a]	0– 7.6	27.0 a	12.4 b	3.7 c
Mid:	7.6–15.2	11.5 a	8.6 a	12.6 a
Bottom:	15.2–22.9	1.0 a	2.0 a	2.2 a
Total[b]		39.5	23.0	18.5
			Dry Hay	
		15.7	5.3	4.3

Note: All means within a row followed by the same letter are not significantly different ($p < 0.05$) according to Duncan's New Multiple Range Test.
[a]R^2 Top = 0.942; Intercept = –12.565, Slope = 0.282
[b]R^2 Total = 0.922; Intercept = – 5.819, Slope = 0.351

10–20% higher dikes, and water depths greater than 25 cm (10 in.) to depress plant yields.

Water Quality

Characteristics for the spring/summer/fall season are compared in Table 1 for effluents from the fish and water chestnut production systems. Effective removal rates achieved were biological oxygen demand (BOD_5) = 75%; chemical oxygen demand (COD) = 69%; suspended solids (SS) = 82%; fecal coliforms (FC) = 74%; total Kjeldahl nitrogen (TKN) = 66%; and total phosphorus (TP) = 66% compared to the fish wastewater source. Effluent total solids (TS) from the artificial wetland were not changed appreciably, but the discharge of TS was mostly due to high dissolved solids content in the water supply.

During winter/spring most water quality variables tested had higher values in the effluent water than in the water source (groundwater) (Table 2). The wetland system did not change the TS, Ca, or temperature. Wastewater treatment equivalent to secondary standards for municipal waste treatment could be achieved during the winter period. Biological oxygen demand and SS averaged less than 20 mg/L and only occasionally were greater than 30 mg/L during the plant dormant season. Fecal coliforms averaged 20% higher in the effluent during the winter but lower than the water supply during the spring/summer/fall season. However, analytical sensitivity to FC analyses was not greater than 100 colonies/100 mL during this period. Total ammonia nitrogen (NH_3 + NH_4^+-N) averaged less than 2 mg/L. Due to low temperature (8.6°C) and pH values of less than 7.5, free ammonia averaged less than the recommended maximum level for discharge criteria—0.02 mg/L NH_3.[15]

The fish pools must be drained to harvest the fish by seining, and resulting discharge of accumulated organic constituents could pollute the receiving stream. Dramatic reductions in effluent BOD_5, COD, SS, and TKN were observed immediately after fish harvest, and the fish pools were refilled. Adding swine manure into the fish pools after refilling resulted in an accumulative increase of some pollutants during the winter period.

Based on these results, there is justification for the water quality improvement due to plant growth. However, water treatment improvement achieved by lower irrigation rates is probably not justified on the basis of reduced yield. Water chestnut yields were reduced by 42% and 53% in the lower irrigation rates, while BOD_5, SS, and TKN were only reduced by 25%, 30%, and 27% in the effluent concentrations. Reduction in irrigation rate did not generally result in a proportional improvement in water quality.

Nitrogen removal was 86% and phosphorus removal was 90% from both the aquaculture and the artificial wetland system. The plants used about 15% of the total nitrogen applied to the system and about 10% of the phosphorus. Best growth was at the higher irrigation rate (150 L/mg/day). The plants used 10.9% of total phosphorus from the higher irrigation rate and 11.0% of

Table 4. Comparison of Effluent from Sand Filtration Beds with and without Water Chestnuts

	% Difference[a]		
Treatment	5 RT[b]	10 RT	15 RT
Total N	−38	−62	−61
Total P	−44	−74	−75
K	−44	−42	−34
TSS	−67	−81	−76
BOD	−27	−67	−85
Algae	−27	−25	0
Fe	+55	+343	+267
Ca	+42	+42	+43

[a]% Difference = concentration of plant discharge − concentration of discharge without plants ÷ concentration of discharge without plants × 100.
− = decrease with plants;
+ = increase with plants.
[b]RT is the retention time in days; thus, 5 RT means a 5-day water retention time in the tanks discharging to the sand beds.[14]

phosphorus from the fish waste, but removal was less predictable at rates tested. Nearly all phosphorus harvested from the system was in the plant tissue. Most phosphorus remained in the system as precipitates of calcium and other cations, increasing the apparent treatment effectiveness for phosphorus.

The sand filtration beds contain microorganisms that decompose much of the organic fish waste and convert unavailable plant nutrients into available plant fertilizer nutrients. The plants substantially increased the treatment effectiveness of the system (Table 4).[14] Total suspended solids and total ammonia nitrogen were greatly reduced by adding plants to the filtration system.

Effluent Fe and Ca concentrations were increased by adding plants to the system. Presumably, the aquatic environment in the sand beds solubilized Ca and Fe allowing leaching. The aquatic environment also buffers effluent pH from an alkaline level (pH 8–9) to near pH 7.

The pumping cost for this type of system is an important factor in normal farm situations. However, recirculation of discharge water would reduce the water requirements. All make-up water could be made available through the swine facility for drinking and waste flushing purposes. Another adaptation would grow water chestnuts and fish together in shallow areas of a common pond. Other designs may reduce the manure fertilization rate, resulting in increases in land requirements to treat livestock waste with this integrated approach.

Anaerobic digester waste from methane generation would be a suitable fertilizer for artificial wetlands. Reduced oxygen demand as a result of this type of pretreatment would permit more waste loading and reduce the land area required to recover and treat a given amount of swine waste.[13]

Economic analysis of this type of system must be done on a site basis. Biomass harvesting is recommended to provide resource recovery via the production of marketable fish, water chestnuts, and hay. Both the yields and the markets for these products may be constrained by factors associated with siting criteria and economics. Generally, these systems are cost effective compared with alternative waste treatment methods that give comparable water quality results.[16] However, these waste treatment methods are not usually practiced, and farm experience with this type system is lacking. Acceptance of any new enterprise on the farm partly depends on the demand it places on farm materials, land, and labor, which are managed for a profit. If any new activity, such as waste treatment, does not reduce costs and/or increase income, it is unlikely to gain acceptance.

REFERENCES

1. Jewell, W. J., and M. S. Switzenbaum. "Waste Characteristics and Impacts," *J. Water Pollut. Control Fed.* 50:1245–1258 (1978).
2. Elliot, L. F., and F. J. Stevenson, Eds. "Soils for Management of Organic Wastes and Waste Waters," American Society of Agronomy, Crop Science Society of America, and Soil Science Society of America, Madison, WI (1977).
3. Behrends, L. L., J. B. Kingsley, J. J. Maddox. and E. L. Waddell, Jr. "Fish Production and Community Metabolism in an Organically Fertilized Fish Pond," *J. World Mariculture Soc.* 14:510–522 (1983).
4. Van Dyne, D. L., and C. B. Gilberston. "Estimating U.S. Livestock and Poultry Manure and Nutrient Production," U.S. Department of Agriculture, Economics Statistics, and Cooperative Service, ESCS-12, NTIS, Springfield, VA (1978).
5. Martin, J. B., B. G. Ruffin, J. O. Donald, and L. L. Behrends. "Application of New Technologies to Livestock Waste Management," in *Proceedings of Conference Perspectives on Nonpoint Source Pollution.* (Washington, DC: USEPA, Office of Water Regulations and Standards, 1985), pp. 218–221.
6. Hrubant, G. R., R. A. Rhodes, and J. H. Stoneker. "Specific Composition of Representative Feedlot Wastes: A Chemical and Microbial Profile," U.S. Department of Agriculture, Science, and Education Administration, North Central Region, Peoria, IL (1978).
7. Bardach, J. E., J. H. Ryther, and W. O. McLarney. *Aquaculture* (New York: Wiley Interscience, 1972).
8. Tang, Y. A. "Evaluation of Balance Between Fishes and Available Fish Foods in Multispecies Fish Culture Ponds in Taiwan," *Trans. Am. Fish. Soc.* 99(4):708–718 (1970).
9. Cremer, M. C., and R. O. Smitherman. "Food Habits of Silver and Bighead Carp in Ponds and Cages," *Aquaculture* 20:57–64 (1975).
10. Behrends, L. L., J. B. Kingsley, J. J. Maddox, and E. L. Waddell, Jr. "Integrated Agriculture/Aquaculture Systems," paper no. 5031, American Society of Agricultural Engineers, St. Joseph, MI (1982).
11. Hodge, W. H. "Chinese Water Chestnut or Matai – A Paddy Crop of China," *Econ. Bot.* 10:49–55 (1956).
12. Hodge, W. H., and D. A. Bisset. "The Chinese Water Chestnut," Circular 956, United States Department of Agriculture (1955).
13. Maddox, J. J., L. L. Behrends, D. W. Burch, J. B. Kingsley, and E. L. Waddell, Jr. "Optimization of Biological Recycling of Plant Nutrients in Livestock Wastes

by Utilizing Waste Heat from Cooling Towers," EPA-600/7-82-041, NTIS, Springfield, VA (1982).
14. Maddox, J. J., L. L. Behrends, R. S. Pile, and J. C. Roetheli. "Waste Treatment for Confined Swine by Aquaculture," paper no. 79-4077, American Society of Agricultural Engineers, St. Joseph, MI (1978).
15. "Quality Criteria for Water Quality," U.S. EPA (1975), pp. 10-13.
16. Martin, J. B., and C. E. Madewell. "Environmental and Economic Aspects of Recycling Livestock Wastes," *Southern J. Agric. Econ.* 3:137-142 (1971).

CHAPTER 15

Treatment of Acid Drainage with a Constructed Wetland at the Tennessee Valley Authority 950 Coal Mine

Gregory A. Brodie, Donald A. Hammer, and David A. Tomljanovich

BACKGROUND

In 1975, the Tennessee Valley Authority (TVA) opened its 225-ha 950 coal mine near Flat Rock, in Jackson County, Alabama. Surface mining continued until 1979, when reclamation was initiated. Impoundment 3, a sediment basin, received acid mine drainage and was treated with sodium hydroxide to meet effluent permit limitations (Table 1). Impoundment 3 was constructed in 1975 as a 250-m^3 sediment control pond, about 30 m from Kash Creek, a third order tributary to the Tennessee River in the head of a hollow created by mining. The drainage basin for Impoundment 3 consisted of 21 ha of undisturbed forest and 6 ha of reclaimed land.

A drainage diversion constructed in 1984 drained a mine highwall pond and diverted storm runoff 200 m upstream of Impoundment 3. This diversion discharged only after precipitation and snowmelt and reduced the drainage basin for Impoundment 3 from 27 ha to 0.8 ha but had no effect on the quantity of seepage entering Impoundment 3. Results from ground water monitoring near Impoundment 3 did not show correlation among the highwall pond, precipitation, and the acid seepage.

Between 1977 and 1986, TVA spent nearly $250,000 on chemical treatment, pond maintenance, monitoring, and administrative activities at Impoundment 3. Sodium hydroxide was used at a rate of approximately 3500 L/month, and pond cleaning was conducted annually, but noncomplying discharges were still frequent.

In February 1984, TVA received a Notice of Violation from the Alabama Department of Environmental Management (ADEM) for chronic effluent discharge violations at Impoundment 3. Immediate actions taken included dredging and deepening the basin, increasing frequency of monitoring and pH adjustment, and installing wood flow baffles to increase the flow path and

Table 1. Alabama Coal Mining Operation Effluent Limitations

Parameter	Daily Maximum	Monthly Average
pH[a]	6.0–9.0	
Total Fe	6.0	3.0
Total Mn	4.0[b]	2.0
Total suspended solids	70.0	35.0

Note: All units mg/L except pH (S.U.).
[a]pH was limited to a maximum of 10.5 s.u. to precipitate Mn at Impoundment 3.
[b]Mn limitations do not apply if drainage, prior to any treatment, contains less than 10.0 mg/L total Fe and pH = 6.0–9.0 s.u.

retention in the pond. These actions only slightly increased TVA's ability to maintain compliance at Impoundment 3.

Major problems associated with achieving compliance at Impoundment 3 included (1) inadequate retention/sedimentation capacity in the basin; (2) variable flows, influent water chemistry, and climatic conditions that influenced the wastewater treatability; and (3) vandalism of chemical treatment equipment.

TVA constructed its first wetland to treat acid drainage emanating from a coal preparation plant slurry disposal area in June 1985.[1,2] The success of this wetland led to an aggressive TVA program of wetlands construction and research for treating acid drainage at coal mines and coal-fired steam plants. Because Impoundment 3 had acceptable characteristics (moderate water quality, adequate siting characteristics, and suitable geology and hydrology) for constructed wetlands treatment of acid drainage, construction was initiated in summer 1986.

INITIAL SITE ASSESSMENT

Predesign assessments included evaluations of water chemistry analyses, hydrology, geology, permitting and regulatory requirements, surveys for cultural resources and threatened or endangered species, baseline stream monitoring, and land use, ownership, and availability.

Raw seepage was sampled and analyzed using standard methodologies for pH, dissolved oxygen, total and dissolved Fe and Mn, Al, SO_4, and total suspended solids.[3] Site geology— interpreted from mine maps, field observations, and drill hole logs—consisted of about 10 m of shaley, silty sandstone mine spoil overlying sandstone of the Pottsville formation.

Maximum expected flows were evaluated using historical data and computational methods. Average flow from Impoundment 3 was 87 L/min and ranged from 0.0–380 L/min. Maximum predicted runoff from the 10-year, 24-hour precipitation event, by various methodologies,[4] ranged from 600 to 1000 L/min. Total design impounded area of 0.1 ha was calculated using 380 L/min as the maximum expected flow and EPA's methodology for sediment basin design.[5] Design area was doubled to 0.2 ha at 0.3–0.5 m depth, but shallow bedrock limited the final size to 0.13 ha.

Initial and final designs were closely coordinated and approved by the ADEM, and an engineering report was submitted with a request to modify the existing National Pollutant Discharge Elimination System permit. Other regulatory considerations included reviews for threatened or endangered species and critical habitat (Endangered Species Act); floodplains and wetlands (Executive Orders 11988 and 11990); surveys for cultural resources (National Historic Preservation Act); and environmental review (National Environmental Policy Act).

Baseline stream monitoring included limited water chemistry analyses in Kash Creek upstream and downstream of Impoundment 3, and surveys for aquatic macroinvertebrates.

Land use at the time of construction was forested reclaimed land. All mining permits had expired and reclamation bonds released, and the Alabama Surface Mining Commission concluded that wetland construction was not associated with mining and no mining permit or postmining land use modification would be necessary. Approval from landowners to construct the wetland was obtained, and long-term access and control of the Impoundment 3 site is in negotiation.

DESIGN AND CONSTRUCTION

Figure 1 shows the Impoundment 3 wetland as constructed. Actual construction required modifying the first spillway and reducing the size of the third cell from design specifications.

Construction began September 5, 1986, by placement of a silt fence along Kash Creek and clearing and grubbing 0.3 ha for cells 2 and 3. The existing cell 1 was pumped down and sludge hydraulically removed and disposed of at the nearby Fabius Coal Preparation Plant gob disposal area. The discharge pipe and wood baffles were removed from the existing pond, and the dike was regraded. Pumping from cell 1 continued throughout the operation.

Cell 1 was filled and regraded to a uniform flat bottom using borrow from a sandstone spoil bank. Cells 2 and 3 were graded uniformly, and a dike parallel to Kash Creek was constructed using compacted material from cells 2 and 3, which consisted of alluvially derived Philo-Atkins silt loams.[6] Dikes were sloped about 2:1, and outslopes were lined with erosion control fabric. Two spillways consisted of channels lined with erosion fabric planted with aquatic vegetation; the final spillway was riprapped and fitted with a discharge pipe for monitoring flow. Freeboard was 0.5 m, while pond depths averaged between 0.3 m and 0.5 m.

Disturbed areas were seeded, mulched, and fertilized with a mixture of grasses. Ponds were planted with 16,000 cattails *(Typha latifolia)* and woolgrass *(Scirpus cyperinus)*, fertilized with 12–12–12 at 450 kg/ha, and limed at 9000 kg/ha.

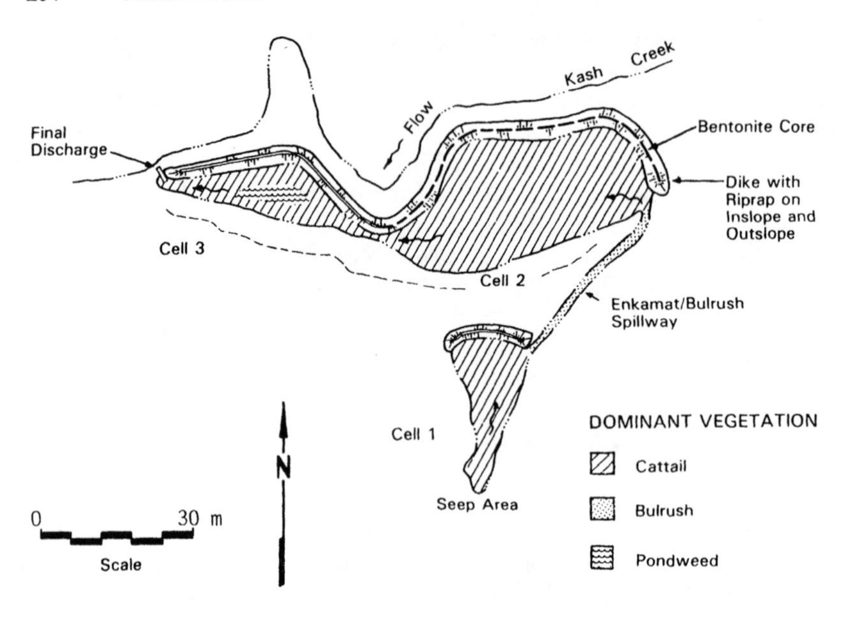

Figure 1. Schematic of as-constructed Impoundment 3 wetland.

Construction was completed on October 31, 1986, and the ponds allowed to fill. The first discharge from the new wetland occurred on November 20.

COSTS

Total costs of design, construction, and operation of the Impoundment 3 wetland and a comparison with prewetlands expenditures are shown in Table 2. From 1975 to 1985, TVA spent nearly $500,000 on chemical treatment, land reclamation, and engineering attempts aimed at mitigating the Impoundment 3 discharge, even though noncomplying discharges were a chronic occurrence (from 1984 to 1986, 67% of discharge samples were out of compliance). During the period 1984–1986, TVA averaged $28,500 annually on the treatment system at Impoundment 3. In 1986, the wetland was constructed for $41,200, with annual operating and monitoring costs of less than $3700.

RESULTS

Figures 2–5 show Impoundment 3 wetlands discharge data from November 20, 1986, to June 23, 1988, for pH, total Fe, total Mn, and total suspended solids. Since initial system operation, effluent water quality has met permit limits. Generally, the wetland has increased the pH from 6.1 to 6.9 and

Table 2. Comparative Costs of Impoundment 3 Wetland with Conventional Treatment System

ITEM	COST
Construction (dozer and backhoe)	$ 5,200
TVA construction equipment (pumps, air compressor, tank trucks)	4,700
Materials and supplies (gravel, clay, mulches, erosion control blanket, seed, fertilizer)	2,800
Labor	18,200
Supervision and administration	5,300
Design and site evaluation	5,000
Total wetland capital cost	$41,200
Annual wetlands operation (discharge monitoring, reporting, minor repairs)	$ 3,700
Annual prewetlands costs	
Monitoring, reporting	$ 4,500
Chemicals (NaOH)	12,000
Pond maintenance	10,000
Miscellaneous	2,000
Total prewetlands annual cost	$28,500

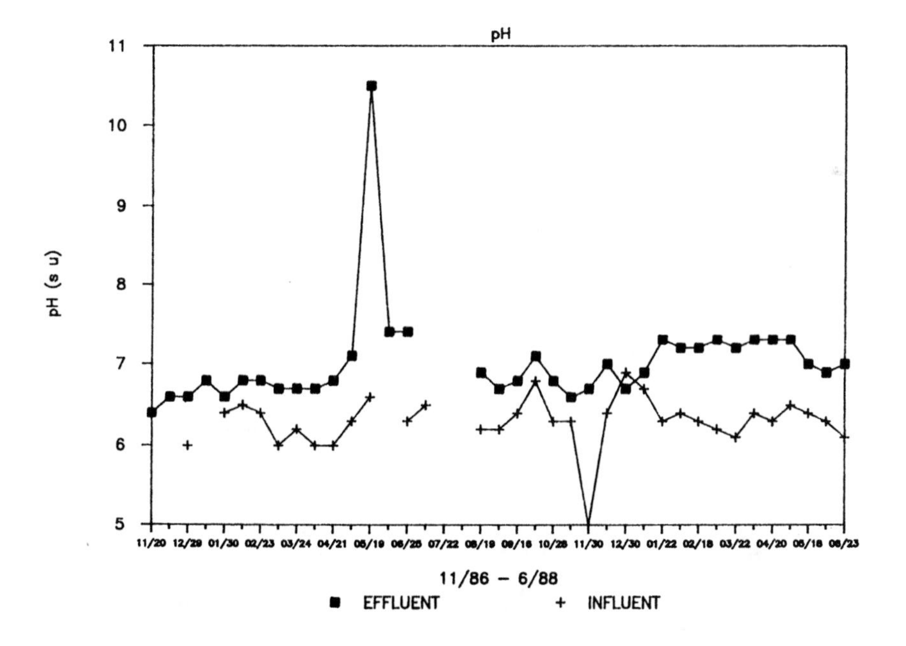

Figure 2. Influent/effluent pH of Impoundment 3 wetland.

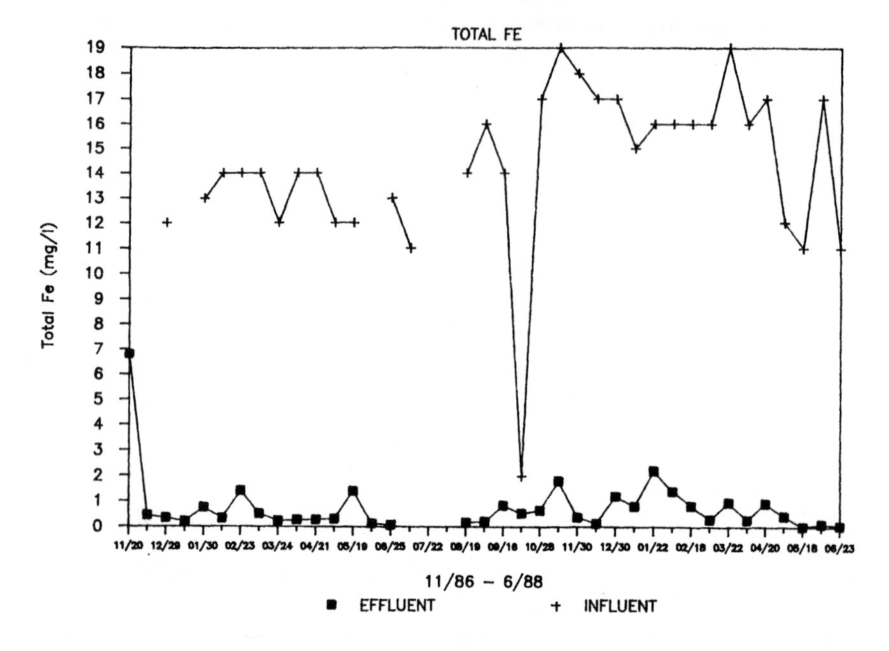

Figure 3. Influent/effluent total Fe of Impoundment 3 wetland.

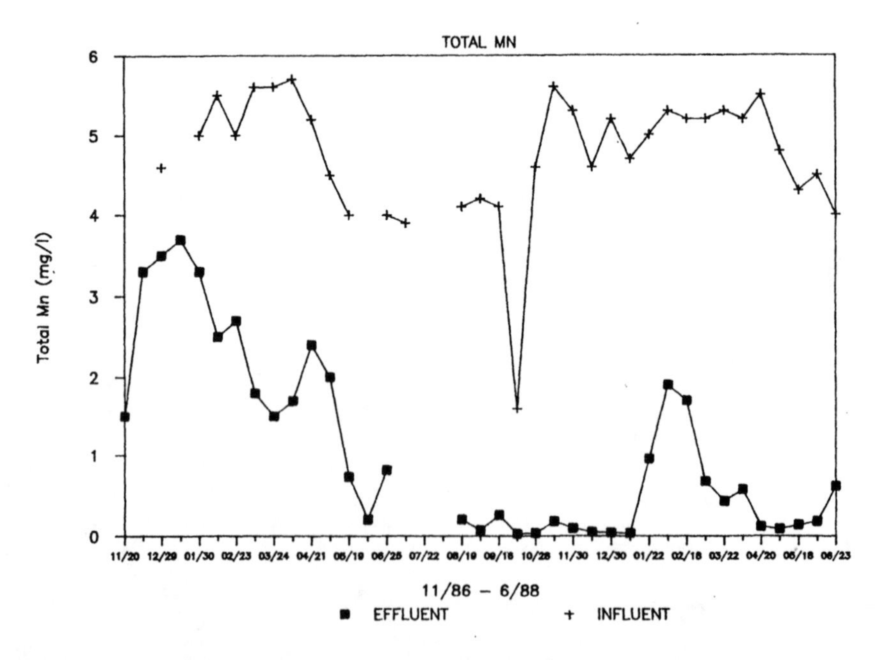

Figure 4. Influent/effluent total Mn of Impoundment 3 wetland.

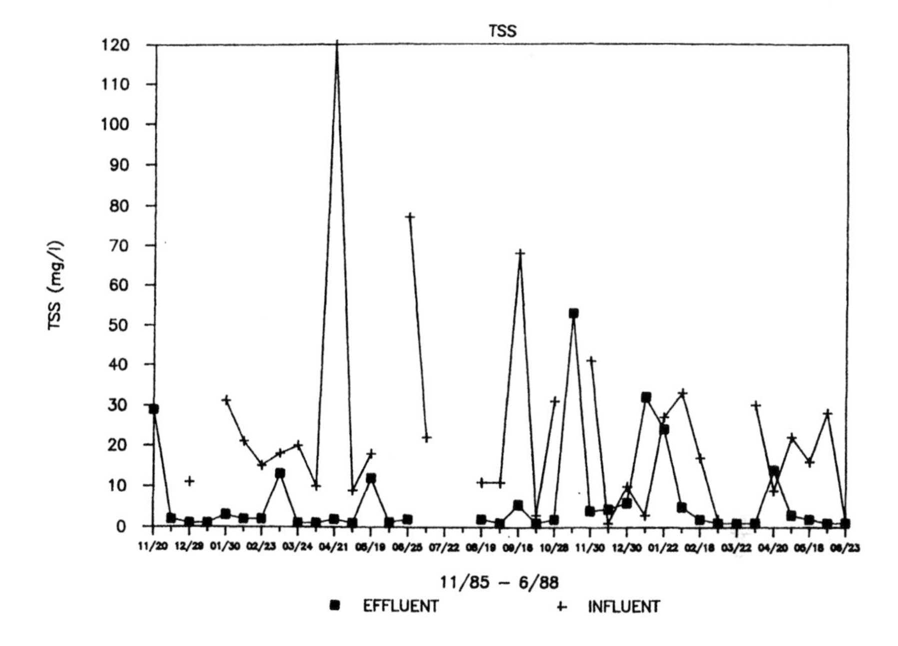

Figure 5. Influent/effluent total suspended solids of Impoundment 3 wetland.

reduced total Fe from 14.3 mg/L to 0.8 mg/L, total Mn from 4.8 mg/L to 1.1 mg/L, and total suspended solids from 24 mg/L to 7 mg/L.

Table 3 lists macroinvertebrates present downstream in Kash creek before and after wetland construction and those taxa present within the wetland six months after construction. Before wetland construction, only two taxa were present in Kash Creek, due to poor and variable water quality (the perennial headwater is the Impoundment 3 seepage). Only six months after wetland construction, 19 taxa, including four dominant species, were present in Kash Creek. Within Impoundment 3, 32 taxa were collected six months after construction.

Only broad-leafed cattail *(Typha latifolia)* and woolgrass *(Scirpus cyperinus)* were planted in the Impoundment 3 wetland, but 20 vegetative taxa were present on September 9, 1987, 13 months after construction, with three dominant species: *Typha; Scirpus;* and a rush, *Juncus acuminatus* (see following list).

Acer rubrum
Bidens frondosa
Bidens polylepis
Chara sp.
Echinochloa crusgalli
Eleocharis quadrangulata
Erechtites sp.
Eupatorium perfoliatum
Hypericum mutilum
Juncus acuminatus
Juncus diffusissimus
Juncus effusus
Juncus validus
Ludwigia alternifolia
Ludwigia palustris
Panicum dichotomiflorum
Pluchea camphorata
Salix nigra
Scirpus cyperinus
Typha latifolia

Table 3. Macroinvertebrates Present in 100 m of Kash Creek Below Discharge of Impoundment 3 Wetland Before and After Construction

	Sample Station		
	Kash Creek		Impoundment 3 Wetland
Taxa	June 1986[a]	June 1987	June 1987
Ablabesmyia sp.		X	
Anax sp.		X	X
Baetis sp.		X[b]	X
Berosus sp.			X
Ceratopogonidae		X	X
Chironomidae			X
Chironomis sp.		X	X
Coenagrionidae			X
Coleoptera		X	X
Collembola			X
Conchapelopia sp.			X
Corixidae		X	
Corydalus cornutus	X		
Cryptochironomus sp.		X	X
Culicidae			X
Dicrotendipes sp.			X
Dytiscidae			X
Enallagma sp.			X
Gerridae		X	X
Hydatophylax sp.		X	
Ischnura sp.			X
Larsia sp.			X
Libellulidae		X	
Lymnaeidae			X
Microtendipes sp.		X	
Notonecta sp.			X
Oligochaeta			X
Paramerina sp.			X
Plathemis lydia			X
Polypedilum sp.			X
Procladius sp.		X	X
Pycnopsyche sp.	X		
Sialis sp.		X[b]	
Sphaerium sp.		X	
Stenochironomus sp.		X	
Stictochironomus sp.		X[b]	X
Sympetrum sp.			X
Tabanidae			X
Tanytarsus sp		X[b]	
Tipulidae		X	X
Tribelos sp.			X
Tropisternus sp.			X

[a]Sample taken before wetland was constructed.
[b]Dominant taxa (>12 individuals per kick net sample).

Muskrats burrowing into the inner slope of cell 2 caused dike failure and bypass of wetlands water into Kash Creek in August 1987. The dike centerline was excavated to bedrock (2.5 m), backfilled with bentonite, revegetated, and the dike slopes lined with coarse riprap 0.7 m below and 1.0 m above waterline to prevent further muskrat excavation.

SUMMARY AND CONCLUSIONS

Installation of the Impoundment 3 wetland has demonstrated that this technology can be an environmentally effective and cost-beneficial alternative to conventional treatment of mine drainage. Experience at Impoundment 3 and other sites at Fabius[1,2] suggests that wetlands may be a long-term, self-maintaining treatment system capable of producing high-quality water from moderately polluted inflows.

Sizing of the Impoundment 3 wetland was determined with maximum expected flow techniques; the as-built size treatment area was somewhat smaller, resulting in a hydraulic loading of about 0.7 $L/m^2/min$, and a chemical loading of 1.1 $m^2/mg/min$ for Fe and 2.8 $m^2/mg/min$ for Mn. These loading rates are below average when compared to other wetland treatment installations by TVA but above previous recommendations[1] and have been adequate for the Impoundment 3 seep water.

REFERENCES

1. Brodie, G. A., D. A. Hammer, and D. A. Tomljanovich. "Constructed Wetlands for Acid Drainage Control in the Tennessee Valley," in *Mine Drainage and Reclamation,* Bureau of Mines Information Circular IC 9183, Pittsburgh (1988).
2. Brodie, G. A., D. A. Hammer, and D. A. Tomljanovich. "Treatment of Acid Drainage from Coal Facilities with Man-Made Wetlands," in *Aquatic Plants for Water Treatment and Resource Recovery*, K. R. Reddy and W. H. Smith, Eds. (Orlando, FL: Magnolia Publishing Inc., 1987), pp. 903–912.
3. "Methods for Chemical Analysis of Water and Wastewater," EPA-600/4-79-020, U.S. Environmental Protection Agency, Cincinnati, OH (1979).
4. Lyle, E. S. "Surface Mine Reclamation Manual," (New York: Elsevier Science Publishing Co., Inc., 1987).
5. "Erosion and Sediment Control," EPA-625/3-76-006, U.S. Environmental Protection Agency (1976).
6. "Soil Survey—Jackson County, Alabama," Soil Conservation Service, Series 1941, No. 8, U.S. Department of Agriculture (1954).

CHAPTER 16

Constructed Wetlands for Treatment of Ash Pond Seepage

Gregory A. Brodie, Donald A. Hammer, David A. Tomljanovich

INTRODUCTION

Coal processing and transporting and coal ash storage frequently results in acid drainage similar to seepage from surface and underground mine areas. Ash storage pond seepage has concentrations of metallic ions similar to acid mine drainage (AMD), but the aggregate flow from many seeps along one ash pond dike may be orders of magnitude greater than individual mine drainage seeps. Typically the Tennessee Valley Authority (TVA) ash pond seeps have pHs of 3–6, total iron (Fe) of 100–200 mg/L and total manganese (Mn) of 5–10 mg/L, but Fe may be 300–400 mg/L and Mn may be 70–80 mg/L and flows range from 500–2000 L/min. Ash pond seeps may also contain other contaminants such as Se, Be, Al, and heavy metals.

Since water chemistry is similar and constructed wetlands have successfully treated acid mine drainage at a number of TVA sites,[13] we decided to explore constructed wetlands for treatment of ash pond seepage at three TVA coal-fired generating plants.

METHODS

Widows Creek Steam Plant is located along Guntersville Reservoir in Jackson County, Alabama. This 2000-megawatt station has 129 ha of retired and 39 ha of active ash pond storage within the 399-ha site. Seepage (WIF 018) emanated from the toe of a reclaimed ash pond dike and flowed around the coal pile runoff pond and into Guntersville Reservoir. Previous attempts to remedy discharge quality using crushed limestone dikes failed because of Fe coating. A second seep area (WIF 019) occurred along the dike of an active ash pond and flowed into Widows Creek.

In April 1986, the discharge channel at WIF 018 was widened and dikes and spillways built to create a 0.5-ha wetlands (Figure 1). Cattail *(Typha latifolia)* and rush *(Juncus effusus)* were planted in cell 1 with cattail and a bulrush

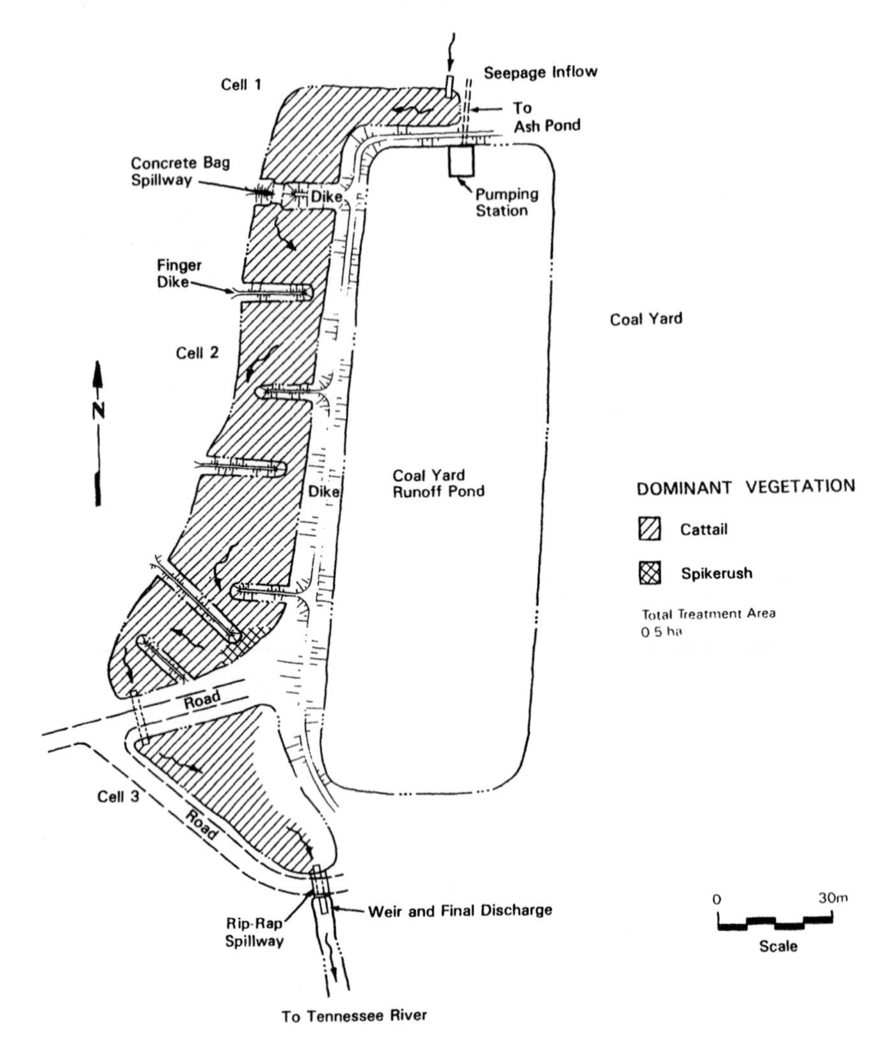

Figure 1. Plan view of Widows Creek Steam Plant (018) acid drainage treatment wetland.

(Scirpus cyperinus) and a few *Iris versicolor* and *Eleocharis quadrangulata* in cells 2 and 3. Fertilizer (Nutra Nuggets®) was applied after planting at 6.75 MT/ha (3 tons/ac).

An unusually heavy infestation of armyworms *(Simyra henrici)* was first noted in mid-August 1986. Lorsban® was aerially applied then and again on October 10. Similar treatments were necessary in July and September 1987 before the outbreak was brought under control. Cause of the infestation was unknown, but rapid armyworm population increases and severe defoliation of cattail necessitated chemical control during the first two summers of operation.[4]

The low-lying seepage area at WIF 019 had fostered a natural stand of cattail and bulrush. In May 1986, three earthen dikes and spillways were built to expand the area of the existing wetlands to 2.5 ha, and the discharge sampling point was relocated 300 m downstream. Previously unknown low pH seeps were encompassed at the lower end of the wetlands treatment area. This wetland was impounded during expansion of the active ash storage area in 1987 because of steam plant operational requirements.

Kingston Steam Plant is located along Watts Bar Reservoir in Roane County, Tennessee. This 1700-megawatt station has 14 ha of active ash storage ponds and 91 ha of retired ash ponds within the 330-ha site. Acid seepage through the ash dike of the sluice canal carrying bottom ash to the active storage area collects in a channel below the dike (KIF 006). Additional seepage emerging beyond the channel had created a small natural stand of cattail and bulrush that was enlarged by grading and dike construction in August and September 1986 to create 0.9 ha of treatment area in three cells (Figure 2). A flashboard riser on a 38-cm diameter culvert provides water depth adjustment in the collection ditch and controls inflow to cell 1; a similar water control provides adjustments in the final cell. Spillways can be modified with concrete bags to adjust water levels in cells 1 and 2, and rectangular flow monitoring weirs were later constructed at each spillway.

Planting was initiated after flooding the cells in mid-September and completed two months later. A hydroseeder applied 9.1 MT of lime and 318 kg of 12–24–24 fertilizer in early May 1987. Since only one-third to one-half of the original plants survived, bare areas were replanted and fertilized in July 1987.

In August and September 1987, a 1200-m^2 section of cell 3 was reexcavated, layered with crushed limestone (-60 mesh) to 20 cm and topped with 18 cm of spent mushroom compost in an unsuccessful attempt to elevate discharge pH.[5] This area was replanted in cattail, and 90 kg of phosphate and 90 kg of 17–19–19 fertilizer were broadcast throughout the cell. The limestone/compost bed and other bare spots were replanted with cattail in May 1988.

Colbert Steam Plant is located on Pickwick Reservoir in Colbert County, Alabama. Moderately acidic ash pond seepage (COF 013) had little Fe but total Mn levels exceeded discharge permit limits. In July 1987, TVA constructed four earthen dikes to increase the size of a natural wetland and enlarge the treatment area to 1 ha.

We are aware of one other example of constructed wetlands for ash pond seep treatment. Pennsylvania Electric Company (Penelec) completed a wetlands treatment system for ash pond seepage at the Homer City generating station in September 1988.[6] This system has an average flow of 110 L/min with little Fe but total Mn concentrations of 80 mg/L. Total treatment area is 5852 m^2 for a loading rate of 0.67 m^2/mg Mn/min. Penelec will construct a second system for ash pond seepage at their Iselin station in 1989.[6]

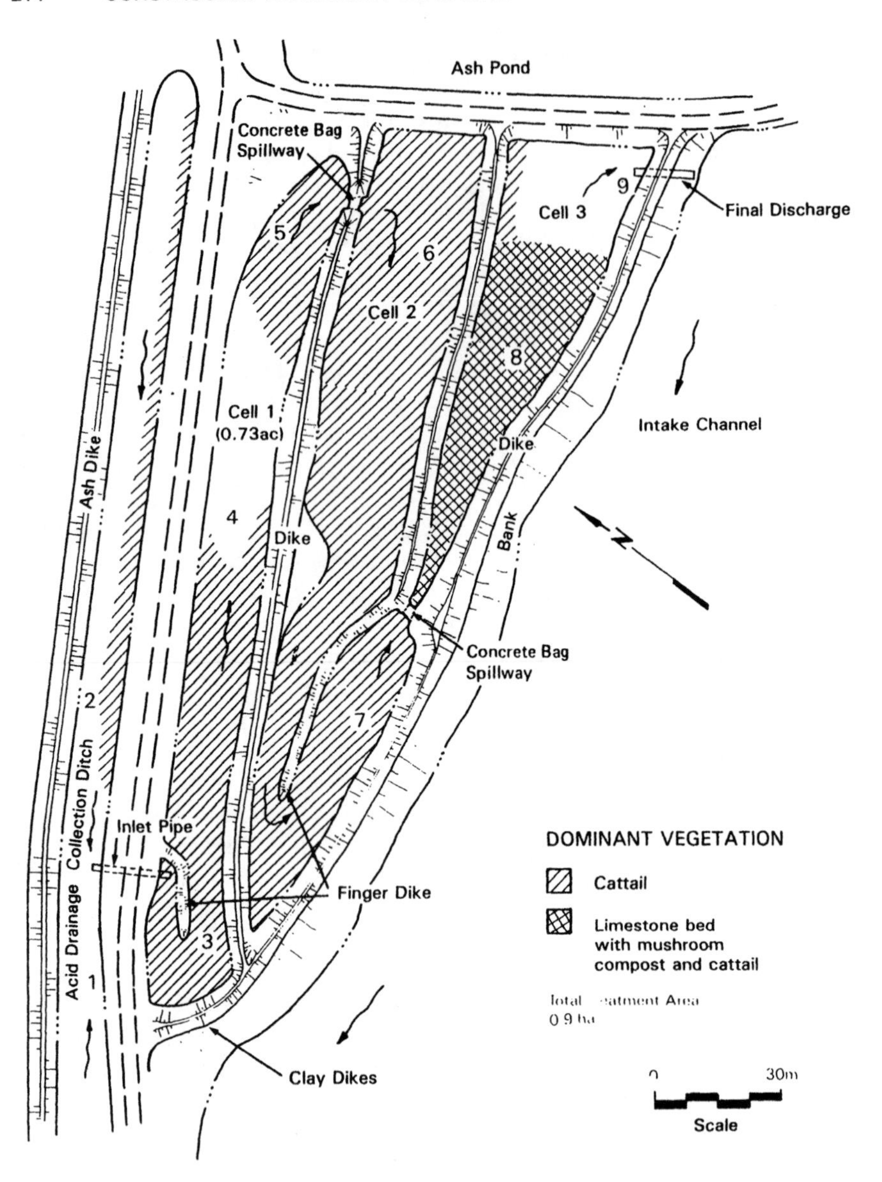

Figure 2. Plan view of Kingston Steam Plant (006) acid drainage treatment wetlands. Numbers depict sample sites in Table 3.

Table 1. Water Quality Parameters in Wetlands Discharge Samples at Widows Creek 018

	pH			Fe			Mn			TSS
	Min	Ave	Max	Min	Ave	Max	Min	Ave	Max	Ave
1984 n = 36 5/9–12/27	4.3	6.1	6.7	0.2	196.9	472	0.1	6.9	13	26.6
1985 n = 27 1/2–7/3	3.3	5.8	6.7	21	176.2	328	1.4	7.2	13.5	
1986–87 n = 37 5/7–1/14	2.6	3.9	6.8	0.3	6.3	9.1	0.7	6.4	19.2	
1987 n = 48 1/21–12/31	3.4	7.5	8.9	0.2	2.1	9.6	0.2	1.5	6.6	
1988 n = 18 1/7–5/16	3.2	7.5	9.0	0.2	2.4	8.5	0.1	1.1	4.9	

Note: Drip feed NaOH added in January 1987. Values in mg/L except pH.

RESULTS

Widows Creek

Water quality sampling results before and after wetlands treatment system construction and operation at Widows Creek 018 are shown in Table 1. Seepage contained an average of 186 mg/L total Fe but varied from less than 1 mg/L to over 470 mg/L. Total Mn levels were low, averaging 7.1 mg/L with extremes of 0.1 to 13.5 mg/L. Average pH was 6.0, with extremes of 3.3 and 6.7. The wetlands system removed over 97% of the Fe but only 9% of the Mn, and average discharge pH decreased 2.1 s.u. during the first eight months of operation. Average flow rate is 170 L/min with 186 mg/L Fe, yielding a treatment area of 0.2 m^2/mg Fe/min. In January 1987 a drip feed NaOH system was installed at the inlet to cell 3, and discharges have met compliance standards to date.

Kingston

Seeps at Kingston (KIF 006) had total Fe concentrations of 40–45 mg/L, total Mn concentrations of 4–5 mg/L and pH of 2.6–3.0, although small seeps were indistinct and difficult to sample within the natural marsh. Most seepage in the channel had variable metal concentrations and pH of 5.5 with an average flow of 1370 L/min. Since wetlands system construction in late summer 1986, influent samples have had up to 280 mg/L total Fe and 7 mg/L total Mn with pH varying from 4.4 to 6.3 (Table 2). Average flow was 1370 L/min with 170 mg/L Fe for a loading rate of 0.06 m^2/mg Fe/min.

Pollutant removal has been much less efficient at Kingston with reduction of 85% of total Fe but little Mn removal, and pH has consistently dropped 3 s.u. from the influent to the proposed discharge point. Final wetlands effluent will be pumped to the ash pond for treatment before discharge until the wetlands system performance improves or operation is modified. Recent samples

Table 2. Water Quality Parameters for Inflows and Discharge from Kingston 006 Wetlands

	Influent				Effluent			
	pH	Fe	Mn	TSS	pH	Fe	Mn	TSS
1985								
4/25					2.6	7.5		207
5/2					3.0	40		684
5/8					2.7	45		4
1987								
11/19		145	4.0	18.4		49	5.4	5
12/15		280	9.4	60.4	3	7	4.0	1.2
1987–88								
11/17–1/11	4.2	153	4.9	39.4	3.1	29.8	4.4	3.1
1988								
1/25–9/19	5.3 (2.9–6.3)	52	3.3		2.9 (2.7–3.1)			
n =	29	16	16		29			

Note: Values in mg/L except pH.

upstream and within the wetlands were not helpful in understanding the marked decrease in pH (Figure 2 and Table 3).

Colbert

Colbert 013 effluent had good pH (6.6), low total Fe (<1 mg/L) but high levels of total Mn (22.5 mg/L) immediately after construction of the wetlands system (Table 4). Sodium hydroxide treatment was continued in the final two cells of the system to reduce Mn to permit limits. However, within system Mn levels have consistently dropped. High initial levels probably resulted from construction disturbance and reflooding a natural wetland area containing considerable quantities of previously deposited Mn. Average influent flow was 496 L/min with 7.0 mg/L Mn yielding a treatment area of 1.3 m^2/mg Mn/min. Influent Fe was less than 1.0 mg/L.

COSTS

Construction costs for three treatment systems ranged from \$6.98 to \$14.21/m^2 of treatment area (Table 5), whereas typical mine site costs range from \$3.58 to \$14.12/m^2 of treatment area.[3] Unfortunately, costs for Widows Creek 018 and 019 were combined by the accounting system.

DISCUSSION

Although topography and land availability often restrict wetland system design at mining sites, these factors severely constrain size and location of treatment systems at steam plant sites. In addition, previous construction for steam plant operational requirements disturbed or deposited materials of

Table 3. Water Quality Parameters at Collection Sites Shown in Figure 2 Within the Kingston 006 Wetlands

							Mn				Fe							
	pH		DO		SO_4		Tot		Dis		Tot		Dis		$CaCO_3$		NFR	
Site	1	2	1	2	1	2	1	2	1	2	1	2	1	2	1	2	1	2
1	6.1	5.6	4.7	0.2	340	900	1.8	3.3	1.5	1.8	62	100	52	48	320	140	25	56
2	5.9	5.5	5.1	<0.1	290	350	1.4	6.5	1.2	5.8	42	160	38	140	270	250	21	40
3	6.0	5.7	4.4	<0.1	520	—	3.2	1.6	2.9	3.4	130	—	82	86	481	70	96	57
4	3.7	6.0	3.3	<0.1	560	560	3.4	3.4	3.1	3.1	71	79	64	74	469	140	46	16
5	3.4	5.2	2.2	<0.1	580	540	3.3	3.4	3.3	3.2	68	73	62	68	469	140	12	34
6	3.0	3.2	1.0	<0.1	550	550	3.7	3.6	3.3	3.4	42	46	35	42	502	140	1	9
7	2.9	3.0	1.6	6.6	580	610	3.7	4.5	3.5	4.0	17	13	16	12	477	150	5	5
8	2.9	3.1	4.0	4.1	580	610	3.6	3.7	3.7	3.6	15	17	14	17	498	140	1	2
9	2.9	—	3.6	—	590	610	3.8	3.7	3.5	3.7	17	16	13	15	527	140	2	<1

Note: Samples collected on 28 June and 29 August 1988. Values in mg/L except pH.

Table 4. Water Quality Parameters in Discharge Samples at Colbert 013

		pH			Fe			Mn			Flow
		Min	Ave	Max	Min	Ave	Max	Min	Ave	Max	(m^3/day)
1986											
10/23–10/28	n=3								6.2		
11/5–11/19	n=2								6.3		
1987											
3/23									5.6		
4/9–4/23	n=2								6.4		
5/7–5/19	n=2								8.9		
6/2–6/30	n=3								9.0		
7/13–9/24	n=10	6.2	6.7	6.9	0.4	0.7	1.5	8.3	13.6	22.5	423.4
10/1–12/23	n=11	6.2	6.5	6.9	0.3	0.6	0.8	2.5	6.6	8.4	498.9
1988											
1/6–4/27	n=9	8.0	8.3	8.6	0.3	0.4	0.6	0.8	1.1	1.5	567
8/30			8.2			0.4			0.8		

Note: Wetlands constructed in July 1987 and NaOH treatment initiated 3 Dec 1987. Values in mg/L except pH.

unknown origin at the construction sites. Proximity to coal yards, active ash ponds, or other plant activities caused additional, previously unknown seepage to enter the wetland systems at midpoint or in the lower portion of the treatment areas.

Limited land availability for wetlands treatment systems directly resulted in construction of systems at Kingston 006 and Widows Creek 018 that lacked sufficient treatment area for the loading rate. Treatment area at Kingston and Widows Creek was 0.06 and 0.2 m^2/mg Fe/min, respectively, whereas our successful wetlands treatment systems on mine areas have had more than 0.7 m^2/mg Fe/min.[3] Colbert 013 wetlands system had 45.6 m^2/mg Fe/min and 1.3 m^2/mg Mn/min, and discharge Mn has been high. However, treatment efficiency at Colbert is recovering from construction disturbance and may provide discharge limit treatment in the near future. In addition, high length-to-width ratios of wetland cells at Kingston may have reduced treatment efficiencies. Potential remedies under investigation include the possible lack of Fe-Mn coprecipitation at Colbert 013, alkalinity buffering and sulfate reduction mechanisms for Widows Creek 018 and Kingston 006, and flow reductions or increased treatment area at Widows Creek and Kingston.

Table 5. Construction Costs for Wetland Treatment System at Three Ash Pond/Seep Areas

System	Area (ha)	% Equip	% Labor	% Overhead	Total (000's)	\$/m^2
COF 013	0.9	32.4	30.4	37.2	7.9	1.16
WIF 018 & 019	2.9				209.0	6.98
KIF 006	0.9	28.6	43.3	28.1	131.7	14.21

REFERENCES

1. Brodie, G. A., D. A. Hammer, and D. A. Tomljanovich. "Treatment of Acidic Drainage from Coal Facilities with Manmade Wetlands," in *Aquatic Plants for Water Treatment and Resource Recovery*, K. R. Reddy and W. H. Smith, Eds. (Orlando, FL: Magnolia Publishing Inc., 1987), pp. 903–912.
2. Brodie, G. A., D. A. Hammer, and D. A. Tomljanovich. "Constructed Wetlands for Acid Drainage Control in the Tennessee Valley," in *Proceedings Wetlands—Increasing Our Wetlands Resources,* J. Zelazy and J. S. Feierabend, Eds. (Washington: National Wildlife Federation, 1987), pp. 173–180.
3. Brodie, G. A., D. A. Hammer, and D. A. Tomljanovich. "Constructed Wetlands for Acid Drainage Control in the Tennessee Valley," in *Proceedings of the Mine Drainage and Surface Mine Reclamation Control Conference,* Bureau of Mines Information Circular 9183, Pittsburgh (1988), pp. 325–331.
4. Snoddy, E.L., G. A. Brodie, D. A. Hammer, and D. A. Tomljanovich, "Control of the Armyworm, *Simyra henrici* (Lepidoptera: Noctuidae), on Cattail Plantings in Acid Drainage Treatment Wetlands at Widows Creek Steam-Electric Plant," Chapter 42i, this volume.
5. Pesevento, B., Fredonia, PA. Personal communication (1987).
6. Greco, J., Pennsylvania Electric Company, Johnstown, PA. Personal communication (1987 and 1988).

CHAPTER 17

Use of Wetlands for Treatment of Environmental Problems in Mining: Non-Coal-Mining Applications

Thomas R. Wildeman and Leslie S. Laudon

INTRODUCTION

Wetland treatment of environmental problems resulting from coal mining far outnumber non-coal-mining applications. Instead of reviewing existing technologies, this chapter will review the chemistry of metal mine drainages and differences from coal mine drainages; analyze the geochemistry of metals removal within wetlands; and summarize the results in the few pioneer examples.

Throughout, the argument will be made that effluent from a base- or precious-metal mining operation containing abundant pyrite will be most difficult for wetland system application. Drainage from the Big Five Tunnel in Idaho Springs, Colorado is an example. A pilot system testing wetland effectiveness on treatment of Big Five effluent will be described.

THE CHEMISTRY OF METAL MINE DRAINAGE

In coal mining areas, pyrite is the mineral responsible for acid mine drainage problems, and it causes most problems in metal mining situations. Understanding how pyrite weathers is essential to understanding the differences between metal mining and coal mining pollution problems. Stumm and Morgan[1] review the chemistry of pyrite weathering (the following description is summarized from their text). The overall stoichiometric reactions are:

$$FeS_2(s) + (7/2)O_2 + H_2O \rightarrow Fe^{2+} + 2SO_4^{2-} + 2H^+ \quad (1)$$

$$Fe^{2+} + (1/4)O_2 + H^+ \rightarrow Fe^{3+} + (1/2)H_2O \quad (2)$$

$$Fe^{3+} + 3H_2O \rightarrow Fe(OH)_3(s) + 3H^+ \quad (3)$$

$$FeS_2 + 14Fe^{3+} + 8H_2O \rightarrow 15Fe^{2+} + 2SO_4^{2-} + 16H^+ \quad (4)$$

Key features of the stoichiometry and mechanism are:

1. Weathering is by oxidation. Since pyrite formation only occurs in a reducing environment, oxygen gas from outside the deposit is the ultimate oxidant.
2. Hydrogen ions are produced by oxidation. For every mole of pyrite oxidized, two moles of H^+ are produced by sulfur oxidation (Reaction 1), and two moles of H^+ are produced upon the precipitation of ferric hydroxide (Reactions 2 and 3).
3. Since ferric hydroxide is so insoluble, pyrite oxidation is among the most acid producing of all weathering reactions.
4. The slow step in the mechanism is oxidation in solution of Fe(II) to Fe(III). Sulfur oxidation is relatively rapid.
5. Once the weathering has produced Fe(III), this species can rapidly oxidize pyrite, as shown in Reaction 4. Therefore, Fe(III) cannot persist in the presence of pyritic minerals.
6. Microorganisms can significantly catalyze the rate of pyrite oxidation, especially when they mediate the oxidation of Fe(II) to Fe(III).

Recent studies of the weathering of pyrite to form acid mine drainage have added some refinements to the mechanism.[2,3] Reaction 4 is a major cause of sulfide oxidation, and this reaction does not use molecular oxygen directly. Therefore, flooding mine workings to eliminate air-pyrite contact may not stop pyrite weathering. Weathering could continue by bacterial mediation of Reaction 1 and by Reaction 4, particularly if iron-oxidizing bacteria are present. In addition, where pyrrhotite is present along with pyrite, production of acid drainage is apparently more widespread.[4]

The concept of congruent and incongruent reactions is important to pyrite weathering and to reactions that form other constituents in acid mine drainage. To demonstrate incongruence, consider manganese in coal, which exists as rhodochrosite, $Mn(CO_3)$.[5] Below pH 4, the $Mn(CO_3)$ will react with H^+ according to the reaction:

$$Mn(CO_3)(s) + 2H^+ \rightarrow Mn^{2+} + H_2O + CO_2(g) \tag{5}$$

Being slightly soluble in water, CO_2 gas can escape, and if it does, $Mn(CO_3)$ cannot be reprecipitated in an acidic solution. With incongruent weathering of minerals, some other reaction or severe altering of solution conditions is necessary to cause reprecipitation of the reaction products. Reaction 5 is the basis for how Mn exists in mine drainage as Mn^{2+}. Reactions 1 through 4 show that Fe(II) and SO_4^{2-} in mine drainage cannot be changed back to pyrite through the reversal of a simple reaction. Other sulfide minerals can weather by incongruent reactions. For example, Fe^{3+} can oxidize sphalerite (ZnS) in the same way as in the dissolution of pyrite:

$$8Fe^{3+} + ZnS(s) + 4H_2O \rightarrow 8Fe^{2+} + SO_4^{2-} + 8H^+ + Zn^{2+} \tag{6}$$

In contact with acid mine drainage, ZnS and other sulfides will weather in a manner that cannot be easily reversed.

Table 1. Dissolved Constituents in Waters Associated with Mine Drainages

	Central City District				
Constituent	**Central Zone**	**Intermediate Zone**	**Peripheral Zone**	**Big Five Tunnel**	**Coal Mine Drainage**
SiO_2	50–70	40–70	20–40	40	90
Al	25–100			18	50
Fe	200–700	2–170	0.5–4	50	50–300
Mg	150–260	80–120	30–100	150	80
Ca	240–360	140–300	40–300	370	200
Mn	90–120	20–40	1.0–5.0	32	20–300
Cu	6–60	0.1–5	<0.01–0.11	1.6	
Zn	60–400	7–100	0.3–8	10	
Cd	0.2–3	<0.01–0.3	<0.01–0.04	0.03	
Pb	0.1–0.5	<0.01–0.2	<0.01–0.06	0.01	
Na	10–23	14–27	6–25	46	
K	1.4–3.5	4.8–7.6	3.0–5.1	9.2	
As	0.2–7	<0.001–0.01	<0.001	0.02	
SO_4^{2-}	2300–4000	900–1300	240–800	2100	20–2000
pH	2.1–2.7	4.0–6.0	5.4–6.9	2.6	3.0–5.5
	6	6	6	7	8, 9, 10

Source: Central City District data from Wildeman.[6] Big Five Tunnel data from Camp Dresser McKee Inc.[7] Coal mine drainage data from Girts and Kleinmann,[8] Barton,[9] and Hammer.[10]
Note: Except for pH, all concentrations are in mg/L.

Table 1 lists the chemistry of some acid mine drainages. Fe, Mn, and SO_4^{2-} dominate the constituents in coal mine drainage; reactions 1–5 explain their presence. In drainages from metal mines, Cu, Zn, Cd, Pb, and As as well as Fe, Mn, and SO_4^{2-} are present in amounts detrimental to the environment. Reaction 6 explains the presence in solution of these base metal cations. Groundwater hydrology, fluctuations in rainfall, and the manner of ore deposition, can also affect mine drainage chemistry.[11] However, reactions 1–6 are basic to the system, and the other factors cause secondary changes in the rate and extent of these reactions.

METAL MINE RECLAMATION

Wetlands have been effective for treating acid drainage problems from coal mines.[8] Assessing the potential of wetlands treatment in other mining situations raises two questions: Are the effluent problems limited to deposits with abundant pyrite, or are there other factors? If pyrite is present, is the situation that different from coal mine drainage?

In mining situations without pyrite, water pollution is significantly reduced. Molybdenum mining along the continental divide in Colorado is an example. The Climax, Urad, and Henderson operations are the largest molybdenum mining and milling operations in the world.[12] The ore is present as molybdenite, MoS_2, and acid mine drainage is not a problem.[12,13]

Pyrite is present in the Central City Mining District in Colorado, a typical example of a zoned hydrothermal deposit of gold and base metal ores.[11,14,15]

Pyrite is most abundant in the high temperature Central Zone, and only a minor mineral in the lower temperature Peripheral Zone. Cd, Zn, and Pb concentrations are lowest in Peripheral Zone waters even though the minerals of these metals are in highest abundance in that zone (Table 1). Fe(III) and H^+ in groundwater catalyze the dissolution of base metal sulfides to such an extent that they become important constituents in the drainage from a metal mine even though the base metals may not be abundant in the deposit.

With pyrite present, almost every type of heavy metal contaminant may be present in acid drainage from metal mine operations. Table 2 lists some examples. Contaminants such as Pb, Cd, and As are harmful at concentrations far lower than Mn and Fe. Zn and Cu are harmful to aquatic life at concentrations far less than the drinking water standards.[13] In a wetland, trace elements such as Cu, B, and Hg may kill plants long before the pH, Fe, or Mn would be harmful.[29]

Although generation is from pyrite in both cases, acid drainage from metal mines presents a more severe problem than most coal mine drainages because priority pollutants such as As, Cd, Pb, Hg, Cu, and Zn may be present in hazardous concentrations. Designing a wetland to treat metal mine effluent has to include whether these pollutants will be toxic to flora and fauna and whether the drainage from the wetland will still exceed aquatic standards. Also, removal processes for priority pollutants have to be such that release in the future will not occur. Reviewing contaminant removal processes in wetlands should assist in designing for these circumstances.

GEOCHEMISTRY OF WETLANDS REMOVAL

Low-cost immobilization of pollutants for long time periods is the goal of using wetlands for mine drainage treatment. Among the possible removal mechanisms are:

1. filtering suspended and colloidal material from water
2. uptake of contaminants into roots and leaves of live plants
3. adsorption or exchange of contaminants onto soil materials, live plant materials, dead plant materials, or algal materials
4. precipitation and neutralization through the generation of NH_3 and HCO_3^- by bacterial decay of biologic material
5. precipitation of metals in the oxidizing and reducing zones catalyzed by bacterial activity

Although removal processes have been studied for years,[30] only recently have applications to actual wetlands been reported.[2,4,23,25,31-33] The results do not identify one process as dominant. The following analysis of metals removal geochemistry in a wetland presents the hypothesis that mechanism 5 may be dominant. This hypothesis is based on what happens to a wetland over geologic time,[1,30,34,35] on recent wetland studies[31,32] and on recent experience at the Big Five Tunnel site.[27]

Table 2. Descriptions of Non-Coal-Mining Operations Using Wetlands Treatment

Site	Location	Ore	Pyrite	Treatment	Comments	Reference
Buick Mine	Viburnum, MO	Pb, Zn	Unknown	Secondary, water from tailings lake	Algae in stream meanders removing metals.	16, 17
Zinc Corp. of Amer.	Balmat, NY	Zn	Low content	Secondary, water from mine polishing pond	Marsh removes Zn from 1. to 0.1 ppm.	18, 19
Pennsylvannia Mine	Peru Creek, CO	Au, Ag	Unknown	Primary, abandoned mine drainage	Natural wetland used; project stopped because drainage leached metals from wetland.	20, 21
St. Kevins	Leadville, CO	Ag	Present	Primary, abandoned mine drainage	Natural wetland cleans typical metal drainage.	22
Northern Ontario	Canada	Au, Ag, U, Zn	Present	Primary, abandoned mine drainage	Promotion of algal and cattail growth and increase of surface area for metal removal.	2, 4, 37
Danka Mine	Minnesota	Taconite	Absent	Primary, mine water and stockpile runoff	Natural wetland removes Cu and Ni.	23, 24
Swamp Gulch	Montana	Ag, Pb, & Zn	Present	Primary, abandoned mine seepage	Natural wetland successfully removes Pb and Fe.	25, 26
Big Five Tunnel	Idaho Spgs, CO	Au, Ag	Present	Primary, abandoned mine drainage	Constructed pilot wetland tests substrates and removes Fe and Cu.	11, 27, 28,

If a wetland were buried, upon diagenesis, it would eventually become a bog deposit, coal, or black shale.[30,34] Reviewing metals occurrence in these sediment types that have undergone early diagenesis may identify forms with long-term stability. Mineral forms of manganese, iron, and other base metals in these sediments represent the most thermodynamically stable phases of these elements. In sediments formed by chemical precipitation, the stable iron minerals are hematite (Fe_2O_3), pyrite, or siderite ($FeCO_3$); stable manganese minerals are pyrolusite (MnO_2) and rhodochrosite.[1,30,34,35] Trace elements such as Co, Ni, Cu, Zn, Ag, Cd, Au, Hg, and U occur as sulfides, oxides, and carbonates. The same is true in lignite and coal deposits. With the possible exceptions of V and Ni, metals are not retained by the organic fraction in organic-rich reducing sediments.[1,5,34]

Organic material in a wetland system may only play a minor role in long-term storage. Since sulfides, oxides, and carbonates are the most stable form of trace element precipitates, then immobile organic forms of these elements are intermediate products that will eventually undergo diagenesis to inorganic forms. Therefore, wetland system optimization should concentrate on forming inorganic precipitates and use organic components to promote their formation. In other words, optimize the last two possibilities listed above to reverse Reactions 1–6 listed earlier.

Generally, microorganisms survive in nature by catalyzing chemical reactions that release energy to the organism.[27,36] In aerobic zones, these bacteria promote the oxidation of iron and manganese to more insoluble states. In anaerobic zones, sulfate-reducing bacteria promote the formation of H_2S. Throughout a wetland, bacterial mediation of organic decay will generate NH_3 and HCO_3^-, which will raise the pH and cause hydroxide precipitation. Wetland system design should include plant species that survive and produce large amounts of biomass to support microorganism growth. Metal uptake by plants is less important. An organic substrate should (1) have an anaerobic zone so H_2S is produced, (2) promote plant growth, (3) promote growth of bacteria that increase the pH, and (4) conduct drainage waters to sites of bacterial activity.

This list omits the ability of the organic substrate to adsorb or complex metals. Peat or other sources of humic acids may not be as important for complexing metals as for maintaining a suitable ecosystem. If organic complexing of metals is important, the reason may be to retain the metals in labile forms that can be used directly by bacteria or the products of bacterial growth.

Finally, precipitation of iron and manganese oxyhydroxides in the aerobic zone not only removes primary pollutants as in coal mine drainages, but these precipitates may also remove significant amounts of the trace elements through adsorption.[1,30,34] Manganese oxyhydroxide is far more important than the iron oxyhydroxide in this process,[1,34] so raising the pH is important.

With the above summaries, Table 2 can guide potential uses for wetlands for environmental problems associated with drainages from non-coal-mining situations. In addition, wetlands could possibly be used to control problems caused by tailings (Table 2). If a wetland was built over a tailings pile and anoxic wetland groundwater percolated through the pile, oxygen would not contact pyrite and Reaction 1 would never occur.[4] Also, the low oxidizing capacity of the water would maintain Fe(II), and Reaction 4 would not occur.

EXAMPLES OF SPECIFIC SYSTEMS

Existing metal mine drainage sites using wetlands can be divided into two types of treatment: (1) primary, where the wetland is the only type of treatment in use and (2) secondary, where the wetland is used as a polishing step after initial treatment. Treatment sites in several areas have been proposed but are not operational. Brief descriptions of operational sites are provided in Table 2. Those in Ontario, Minnesota, and Montana are the best documented and are described below along with the Big Five Tunnel site.

Systems in Ontario, Minnesota, and Montana

Several sites in Ontario are being studied to determine the feasibility of biological polishing and ecological engineering for treating wastes from copper-zinc mines.[4,37] In the Red Lake area, a pilot system incorporating various methods of treatment was initiated. Contaminated subsurface water was intercepted and treated in a polishing ditch and pond that provide additional surface area for precipitation of iron hydroxide. In a pond receiving seepage from tailings, cattail growth was promoted. Maintaining reducing conditions for the seepage prevents further release of precipitate by the algae, the polishing agent. Algal growth in a tailings pond was stimulated by adding material that increased surface area for growth.

At a tailings site in Sudbury, Ontario, the mechanism of oxidation of pyrite and pyrrhotite and factors controlling metal removal in biological polishing processes are being studied. Results indicate precipitation of oxides is the main metal removal mechanism and metal uptake by algae is secondary.[2] Other experiments with algae conducted in tailings ponds at several sites indicate that algae concentrate uranium, zinc, copper, nickel, and radium-226 under alkaline conditions.[2]

At the Danka Mine, LTV Steel Mining Company's open pit taconite mine in northeastern Minnesota, a naturally occurring white cedar wetland receives contaminated drainage from mine dewatering and stockpile runoff. The major source of metals is stockpile drainage that averages 18 mg/L nickel and 0.62 mg/L copper and has flow rates ranging from zero to 16 L/sec. During a year-long study, overall nickel and copper removal in the peatland was 84% and 92%, respectively, with peat uptake accounting for most of the metal removal.[23,24]

Laboratory experiments on substrate samples taken from a constructed wet-

Table 3. Concentrations in mg/L of Constituents and Percent Metal Reduction of Waters at the Big Five Tunnel Pilot Project

	Al	% Red.	Cu	% Red.	Fe	% Red.	Mn	% Red.	SO_4^{2-}	pH
December 11, 1987										
Mine drainage	5.8		1.02		32		34		1750	2.8
Output A	2.7	53	0.44	57	18	45	27	21	1560	4.6
Output B	5.0	14	0.89	12	24	26	33	1	1430	3.1
Output C	5.0	14	0.91	10	22	32	34	0	1520	3.3
February 13, 1988										
Mine drainage	5.9		0.89		28		28		1750	3.3
Output A	2.7	54	0.14	84	18	36	27	4	1690	4.7
Output B	5.6	5	0.92	0	28	0	31	0	1780	3.4
Output C	5.7	3	0.92	0	28	0	29	0	1700	3.4
May 4, 1988										
Mine drainage	5.7		0.93		32		30		1760	3.0
Output A	4.1	28	0.14	85	21	34	26	13	1720	3.7
Output B	5.1	10	0.71	23	19	40	26	13	1550	3.1
Output C	5.9	0	0.80	14	24	20	28	7	1600	3.1
May 31, 1988										
Mine drainage	4.8		0.75		44		25		1500	3.0
Output A	3.0	38	0.03	96	27	39	25	0	1330	4.3
Output B	4.7	3	0.64	15	17	61	25	0	1570	3.0
Output C	4.8	0	0.68	9	21	52	25	0	1220	3.0

land near Sand Coulee, Montana showed little remediation potential in the constructed wetland when compared to a natural wetland that had been receiving acid mine drainage from an abandoned lead-zinc mine. In situ study of the natural wetland found removal efficiencies from metals to be 70% for iron, 14% for copper, 5.8% for zinc, 0.7% for manganese, and 0.3% for cadmium. The constructed wetland substrates did not effectively remove iron or acidity, due to reduced microbial populations.[25,26]

Big Five Tunnel, Idaho Springs, Colorado

The Big Five Tunnel site was constructed as a pilot treatment system with all inputs (other than precipitation) and outputs controlled.[28] Drainage flow and chemistry is nearly constant throughout the year.[11] Three cells of 19 m^2 each were made at the site: Cell A was filled with mushroom compost; Cell B was filled with a mixture of one-third peat, one-third aged manure, and one-third decomposed wood products; Cell C was filled with 10–15 cm of limestone rock (5-cm pieces) and covered with the same organic mixture as Cell B.[28]

Based on four sets of samples, the section with mushroom compost (Cell A) has the highest metal removal efficiency and effluent pH (Table 3). Metal reduction ranges from almost none for Mn to complete removal for Pb and Cu in Cell A. In addition, (1) iron hydroxide precipitate is present in all sections; (2) iron oxidizing bacteria not present in the original materials are present in the substrates;[27] (3) although sulfate in the water is not significantly reduced, sulfate-reducing bacteria are present even during cold weather;[27] (4) algae were present throughout winter and spring, with the largest concentration in Cell A; and (5) September transplanting did not allow plants to become established; however, all species were vigorously growing in the spring.

The winter was colder than average and the plants were quite dormant. Nevertheless, substantial metal removal was occurring after only two months and has been continuing throughout the spring.

REFERENCES

1. Stumm, W., and J. J. Morgan. *Aquatic Chemistry,* 2nd ed. (New York: John Wiley & Sons, Inc., 1981), p. 780.
2. Van Everdingen, R. O., and H. R. Krouse. "Interpretation of Isotopic Compositions of Dissolved Sulfates in Acid Mine Drainage," in *Mine Drainage and Surface Mine Reclamation,* U.S. Bureau of Mines Information Circular 9183 (1988), pp. 147–156.
3. Taylor, B. E., M. C. Wheeler, and D. K. Nordstrom. "Stable Isotope Geochemistry of Acid Mine Drainage: Experimental Oxidation of Pyrite," *Geochim. Cosmochim. Acta* 48:2669–2678 (1984).
4. Kalin, M. "Ecological Engineering and Biological Polishing: Methods to Economize Waste Management," paper presented at the Canadian Mineral Processors Annual Operators Conference, Ottawa, Ontario, 1988.
5. Valkovic, V. *Trace Elements in Coal, Vol. 1* (Boca Raton, FL: CRC Press, Inc., 1983) p. 210.
6. Wildeman, T. R. "A Water Handbook of Metal Mining Operations," Report 113, Colorado Water Resources Research Institute, Fort Collins, CO (1981), p. 84.
7. Camp Dresser and McKee Inc. "Remedial Investigation Report Clear Creek/ Central City Site," EPA Contract No. 68-01-6939, Denver, CO (1987).
8. Girts, M. A., and R. L. P. Kleinmann. "Constructed Wetlands for Treatment of Acid Mine Drainage: A Preliminary Review," in *National Symposium on Mining Hydrology, Sedimentology, and Reclamation* (Lexington: University of Kentucky Press, 1986), pp. 165–171.
9. Barton, P. "The Acid Mine Drainage," in *Sulfur in the Environment Part II: Ecological Impacts,* J. O. Nriagu, Ed. (New York: John Wiley & Sons, Inc., 1978), pp. 314–358.
10. Hammer, D. A. Tennessee Valley Authority. Personal communication (1988).
11. Wildeman, T. R. "Chemistry of the Argo Tunnel," *Quart. Colorado School of Mines* 78(4):31–37 (1983).
12. Chappell, W. R., and K. K. Peterson. *Molybdenum in the Environment, Vol. 2* (New York: Marcel Dekker Inc., 1977), pp. 317–812.
13. Wentz, D. A. "Effect of Mine Drainage on the Quality of Streams in Colorado," Colorado Water Resources Circular No. 21, Colorado Water Conservation Board, Denver, CO (1971), p. 117.
14. Wildeman, T. R., D. L. Cain, and A. J. Ramirez. "The Relation Between Water Chemistry and Mineral Zonation in the Central City Mining District, Colorado," in *National Symposium of Water Resources Problems Related to Mining,* Proceed-

ings no. 18 (Bethesda, MD: American Water Resources Association, 1974), pp. 219–229.

15. Sims, P. K., M. A. Drake, and Tooker. "Geology and Ore Deposits of the Central City Mining District, Gilpin County, Colorado," U.S. Geological Survey Professional Paper 359, (1963), p. 231.
16. Wixon, B. G., and B. E. Davies. "Frontier Technology for Environmental Cleanup," in *Frontier Technology in Mineral Processing*, J. F. Spisak and G. V. Jergensen II, Eds. (New York: American Institute of Mining, Metallurgy, and Petroleum Engineering, 1985), pp. 33–42.
17. Kearney, W. M., Environmental Control Engineer, The Doe Run Co., Southeast Missouri Mining Division, Viburnum, MO. Personal communication.
18. Randall International Ltd. "Water Management and Treatment for Mining and Metallurgical Operations," Vol. 3. (1985), pp. 1651–1668.
19. Kreider, J. E., Zinc Corporation of America, Mining Division, Balmat, NY. Personal communication.
20. Emerick, J. C., W. W. Huskie, and D. J. Cooper. "Treatment of Discharge from a High Elevation Metal Mine in the Colorado Rockies Using an Existing Wetland," in *Mine Drainage and Surface Mine Reclamation*, U.S. Bureau of Mines Information Circular 9183 (1988), pp. 345–351.
21. Emerick, J., Environmental Sciences and Ecological Engineering, Colorado School of Mines, Golden, CO. Personal communication.
22. Walton-Day, K., Denver Federal Center, Lakewood, CO. Personal communication.
23. Eger, P., and K. Lapakko. "Nickel and Copper Removal from Mine Drainage by a Natural Wetland," in *Mine Drainage and Surface Mine Reclamation,* U.S. Bureau of Mines Information Circular 9183 (1988), pp. 301–309.
24. Eger, P., or K. Lapakko, Minnesota Department of Natural Resources, Division of Minerals, St. Paul, MN. Personal communication.
25. Dollhopf, D. J., and others. *Hydrochemical, Vegetational and Microbiological Effects of a Natural and a Constructed Wetland on the Control of Acid Mine Drainage*, Reclamation Research Publication 88–04, Montana Department of State Lands, Abandoned Mine Reclamation Bureau (1988), p. 213.
26. Hiel, M., Montana Department of State Lands, Capitol Station, Helena, MT. Personal communication.
27. Batal, W., L. S. Laudon, T. R. Wildeman, and N. Mohdnoordin. "Bacteriological Tests from the Constructed Wetland of the Big Five Tunnel, Idaho Springs, Colorado," Chapter 38h, this volume.
28. Howard, E. A., J. C. Emerick, and T. R. Wildeman. "Design and Construction of a Research Site for Passive Mine Drainage Treatment in Idaho Springs, Colorado," Chapter 42b, this volume.
29. Brooks, R. R. *Biological Methods of Prospecting for Minerals* (New York: Wiley-Interscience, 1983), p. 322.
30. Mason, B., and C. B. Moore. *Principles of Geochemistry,* 4th ed. (New York: John Wiley & Sons, Inc., 1982), p. 244.
31. Wieder, R. K. "Determining the Capacity for Metal Retention in Man-Made Wetlands Constructed for Treatment of Coal Mine Drainage," in *Mine Drainage and Surface Mine Reclamation,* U.S. Bureau of Mines Information Circular 9183 (1988), pp. 375–391.
32. Hedin, R. S., D. M. Hyman, and R. W. Hammack. "Implication of Sulfate-

Reduction and Pyrite Formation Processes for Water Quality in a Constructed Wetland: Preliminary Observation," in *Mine Drainage and Surface Mine Reclamation,* U.S. Bureau of Mines Information Circular 9183 (1988), pp. 382–388.
33. Brodie, G. A., D. A. Hammer, and D. A. Tomljanovich. "An Evaluation of Substrate Types in Constructed Wetlands Acid Drainage Treatment Systems," in *Mine Drainage and Surface Mine Reclamation,* U.S. Bureau of Mines Information Circular 9183 (1988), pp. 389–398.
34. Maynard, J. B. *Geochemistry of Sedimentary Ore Deposits,* (New York: Springer-Verlag, 1983), p. 305.
35. Lindsay, W. L. *Chemical Equilibria in Soils* (New York: John Wiley & Sons, Inc., 1979), p. 449.
36. Manahan, S. E. *Environmental Chemistry,* 4th ed. (Boston, MA: Willard Grant Press, 1984), p. 612.
37. Kalin, M., Boojum Research Limited, Toronto, Ontario. Personal communication.

CHAPTER 18

Constructed Wetlands for Wastewater Treatment at Amoco Oil Company's Mandan, North Dakota Refinery

Donald K. Litchfield and Darryl D. Schatz

INTRODUCTION

To comply with stringent standards of the National Pollutant Discharge Elimination System (NPDES) promulgated by the EPA in the early 1970s, a study was undertaken by Amoco's Mandan, North Dakota refinery to evaluate alternate wastewater treatment systems. This evaluation reviewed carbon adsorption, activated sludge, and expanding the refinery's existing biooxidation system. The decision to expand the refinery's existing biooxidation system was based on availability of open land with natural drainage, past success of the existing system, and economics. Expanding the existing biooxidation system cost $250,000 vs costs of alternate systems ranging from $1 million to $3 million.

The Mandan Refinery covers 122 ha on a 389-ha tract of land north of the city of Mandan, North Dakota, along the west bank of the Missouri River. The 267 ha surrounding the refinery are dedicated to wastewater treatment and wildlife management. The refinery has the capacity to process 7592 metric tons of crude oil per day, requiring 5.7 million L/d of water, which is pumped from the Missouri River.

The Mandan Refinery's original wastewater treatment facilities consisted of an American Petroleum Institute (API) oil-water separator as primary treatment and an aerated biooxidation lagoon as secondary treatment. This system offered limited flexibility. If a wastewater quality problem occurred from a process upset, wastewater could not be held or diverted while the problem was corrected. In addition, heavy rains and/or snowmelt reduced residence time significantly. The additional biooxidation ponds (average depth 1.2–1.8 m) were established by building dams across natural drainage channels. Dams are earth-filled, with clay keyways, overflow structures, and spillways, and construction was based on North Dakota Water Commission and local engineering guidelines. Upstream surfaces are riprapped for wave protection, and water levels are controlled by siphon lines, drain lines, and a series of culverts with slip gates. Cattail, bulrush, and a mix of other wetland plants have naturally invaded the lower ponds.

SYSTEM DESIGN AND OPERATION

Primary treatment of wastewater from the refinery process units is accomplished in a conventional API separator in which oil and other contaminants are separated and recovered. Water is then discharged into a 6-ha lagoon for initial secondary treatment and pumped to a high point for distribution among several routes through a series of cascading ponds and ditches before discharge to the Missouri River. Figure 1 is a schematic of the wastewater treating system, indicating the general flow route through the system.

Normally water flow is through six of the 11 ponds, with a treatment area of 16.6 ha. Five ponds (19.1 ha) are dedicated to wildlife management, providing diversion or holding capacity during heavy rain or snowmelt or an unexpected contaminant problem.

WASTEWATER QUALITY

Wastewater is routinely sampled at the API separator effluent (start of the biooxidation system), lagoon effluent (start of the cascade ponding system), and the final discharge point (Dam 4 discharge).

Sample analyses indicate reductions in all parameter concentrations as the wastewater flows through the treatment system (Table 1). Hexavalent chromium, total chromium, and total suspended solids are not analyzed in the API separator effluent because of the oily nature of this water. The remaining parameter concentrations are reduced by 90–100% between the API separator effluent and the final discharge point. Reductions in parameter concentrations of 36–99.9% are obtained in the primary lagoon. Concentrations are further reduced by 70–100% in the cascade ponding system which, in effect, acts as a polishing system. Average discharge concentrations to the Missouri River are well below NPDES limits; however, there are occasions when a daily limit can be exceeded. This usually occurs during an extremely heavy rainfall or snowmelt, when surface runoff from surrounding drainage flushes out the lagoon and cascade ponding system.

Although average parameter concentrations in effluent water from the primary lagoon are well below permit limits (Table 1), daily effluent limits would have been exceeded 32 times in 1987 had the lagoon discharge gone directly to the river instead of through the cascade ponding system. Specifically, TSS would have been exceeded 23 times, oil and grease seven times, pH once and phenols once. In addition, at least once or twice a year, some operating upset increases the lagoon concentration of one or more parameters above normal, requiring longer holding to complete treatment.

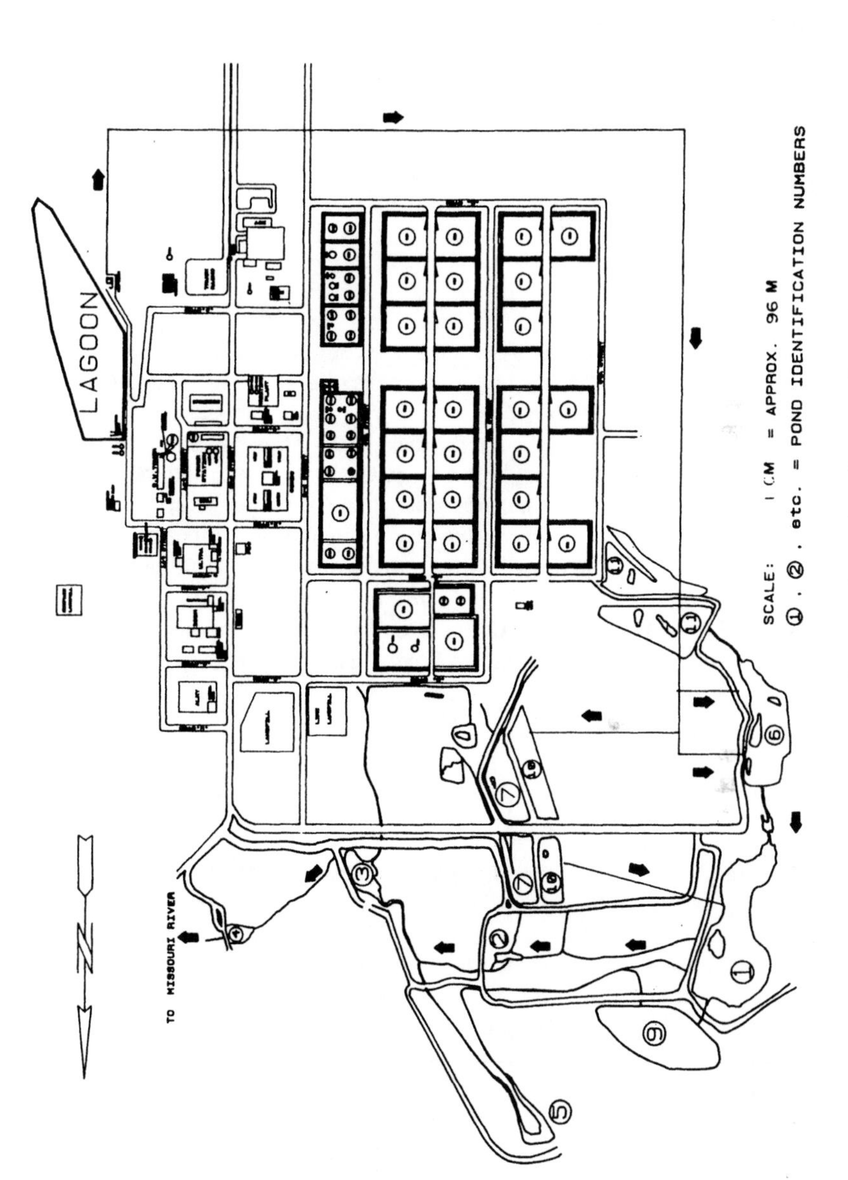

Figure 1. Cascade ponding system. Flow pattern from start (lagoon) to discharge (Dam 4).

Table 1. Wastewater Quality (1987)

		Discharge Points		
NPDES Parameter	NPDES Limits (kg/day)	API Sep. (kg/day)	Lagoon (kg/day)	Dam 4 (kg/day)
BOD	197.7	603.5	79.4	12.4
COD	1477.6	1226.5	346.7	101.0
NH_3-N	131.8	26.3	16.9	2.6
Sulfides	1.3	194.3	0.2	ND
Phenols	1.5	6.1	0.2	0.01
Oil-Grease	59.9	49.1	21.4	1.0
Hex CR	0.24	—	0.01	0.00
Total CR	3.00	—	0.72	0.18
TSS	137.8	—	106.1	11.7

Note: ND = not detectable.

In 1987, discharge from the API separator, lagoon, and Dam 4 averaged 2,411,000; 2,650,000; and 2,542,000 L/d, respectively. The primary lagoon has influent from stormwater runoff and discharges from surface impoundments in addition to the API separator. As the wastewater flows through the cascade ponding system, its volume is affected by evaporation, percolation, stormwater runoff, and influent from natural springs in the drainage area.

In 1979, a limited survey was conducted on chromium, copper, iron, nickel, lead, and antimony reductions through the wastewater treatment system. Besides a general reduction, the samples indicated most of the reduction occurred in a 0.8 km earthen canal between the primary lagoon and the first pond in the system. This canal had a heavy growth of vegetation, including reeds and bulrushes.

WILDLIFE HABITAT

Superior wastewater quality and cost-effectiveness are only two advantages of the refinery's artificial wetlands. Establishing the cascade ponding system has created a natural habitat for wildlife. Shortly after the development of the ponding system, a sharp increase in the number of wildlife species inhabiting the area was noticed. In conjunction with various state and federal wildlife officials, refinery staff encouraged and expanded this propagation of wildlife and waterfowl by creating nesting islands within the ponds and planting 50,000 trees, including 30,000 fruit-bearing trees and shrubs. One hundred eighty-four species of plant life have been identified in this area, including numerous wetland plants and grasses such as reed canary grass, wild rice, musk grass, cattails, and bulrushes. In addition, hectares of alfalfa, millet, flax, and corn were planted. Feeders and houses for wood ducks, swallows, and purple martins were installed and ponds were stocked with rainbow trout, bass, and bluegill. Rainbow trout up to 2.3 kg have been taken from these ponds for necropsy analysis. Results from annual necropsies have been normal.

Some of the most popular residents are giant Canada geese, which the North

Dakota Game and Fish Department is restoring throughout the state. Since 1977, when state officials released 35 yearlings at the refinery, 581 goslings have fledged at the site. More than 400 have been banded and released at other locations in North Dakota. In addition, 191 species of birds have been observed, and 51 species nest in the area. In addition to geese, numerous other wildlife such as pheasants, grouse, partridge, deer, fox, badger, skunk, and raccoons also inhabit this area.

CONCLUSION

In the past five years, only three excursions from the refinery's NPDES permit limits have occurred. These events happened during periods of heavy rainfall and/or snowmelt when so much water flowed through the cascade ponding system that kg/day limits were exceeded for either TSS or BOD_5.

In addition, the refinery has received many awards during the past decade for the development of this wastewater treatment/wetlands area. These include the 1980 Citizen Participation Award from the Environmental Protection Agency for care of this planet and protection of its life. The latest award was the 1986 Blue Heron Award from the National Wildlife Federation for use of a wastewater treatment wetlands as a wildlife sanctuary. This system demonstrates that industrial and environmental concerns can exist harmoniously.

CHAPTER 19

Utilization of Artificial Marshes for Treatment of Pulp Mill Effluents

Rudolph N. Thut

INTRODUCTION

Natural and artificial marsh systems have been used to successfully treat both primary and secondary municipal wastewaters. In general, the systems have been effective in removing BOD_5, TSS, and nitrogen from wastewaters.[1-3] More limited data also suggest that COD and bacteria are substantially reduced.[4] Results with phosphorus have been mixed; in some studies, phosphorus removal has been only marginally successful.[5]

Studies testing the efficacy of marsh systems on industrial effluents or industrial effluent components have been limited. Wolverton and McDonald[6] demonstrated that 60–90% of phenol and m-cresol could be removed by artificial marshes containing either cattail *(Typha* sp.) or reed *(Phragmites australis)* at a retention time of 24 hours. Allender[7] tested the removal effectiveness of a variety of aquatic plants native to Australia on pulp and paper mill effluent. The experiments were conducted under static conditions over a period of weeks. The plants proved effective in removing several pollutants of interest: lignosulfonates, foaming propensity, color, BOD_5, and TSS.

The Weyerhaeuser Company initiated preliminary studies in 1983 and 1984 to determine the potential of artificial marshes to treat pulp mill effluents at the National Space Technology Laboratories in Mississippi. Two studies were conducted with secondary effluent from a bleached kraft mill in a pilot scale anaerobic-filter reed treatment system.[8] The first study was a 24-hr static test; the second was a 96-hr flow-through test. The system was very effective for removal of nitrogen (organic-N, ammonia, and nitrate) and also effective for removal of phosphorus, total organic carbon, and color (although color treatment efficiency declined toward the close of the 96-hr study). These encouraging results led to a more rigorous, long-term study which is the subject of this chapter.

Table 1. Experimental Design

	6-hr Retention	24-hr Retention
No plants	X	X
Cordgrass[a]	X	X
Cattail	X	X
Reed	X	X

[a]Later in the experiment, the retention times were changed from 6 hours to 15 hours and from 24 hours to 15 hours for this plant type.

EXPERIMENTAL DESIGN

Eight artificial marshes were constructed in galvanized steel watering troughs with dimensions of 1.2 m × 3.6 m × 0.6 m deep. The interiors were painted with an epoxy paint to prevent leaching of zinc. A 0.4-m deep rock substrate was placed in the troughs. The substrate was marl (10–20 mm), a locally abundant rock composed principally of calcium carbonate and clay. Marl is relatively inexpensive and has good weight-supporting characteristics. If a full-scale artificial marsh were constructed, the substrate must periodically support vehicles needed to harvest plants or for other maintenance activities. Each substrate occupied a volume of 1.8 m^3. The interstitial (void) volume was 0.8 m^3. Six perforated stand pipes were placed in each marsh for sampling at different points and at different depths in the substrate.

The initial phase of the experiment determined the impact of key variables on pollutant removal efficiency. The independent variables considered were presence/absence of plants, plant type, and retention time (Table 1). Three plant species were tested: cattail *(Typha latifolia)*, reed *(Phragmites australis)*, and cordgrass *(Spartina cynosuroides)*. Cattail and reed were collected from the pond margins of a pulp mill treatment system. Cordgrass (also known locally as sawgrass) was collected from a nearby river. The plants (0.5 m high) and roots or rhizomes were harvested in the spring and immediately transplanted to the artificial marshes. In each marsh, 30 plants were placed at 0.3-m intervals. By the end of the first growing season, roots and rhizomes had sent out many more shoots, and the height of the plants approached 3 m.

Secondary-treated effluent from a bleached kraft mill was pumped to the eight artificial marshes. Flows were adjusted with ball valves to achieve nominal retention times of 6 and 24 hr. The flow rates were 3.2 m^3/d and 0.8 m^3/d, respectively. Treated effluent entered the troughs through a t-shaped tube that dispersed the liquid in two streams (left and right) onto the top of the substrate. The liquid was withdrawn on the opposite end of the trough near the bottom. Depth of the liquid, regulated by adjustments to a vertical tube at the outlet, was maintained so that the surface of the substrate looked wet but no standing water was visible.

The experiment was conducted over a period of three years. Most of the study of the independent variables was done in the first year. The long-term performance of the pilot systems was assessed in the second and third years. Four of the eight artificial marshes were left unaltered throughout the experi-

Table 2. Percent Removal Efficiency of Artificial Marshes for Polishing Pulp Mill Wastewater

	Control		Cordgrass		Cattail		Reed	
	6-hr	24-hr	6-hr	24-hr	6-hr	24-hr	6-hr	24-hr
TSS	39	46	35	42	48	58	26	47
BOD_5	21	45	39	44	36	49	32	36
Ammonia	9	24	40	75	32	79	25	79
Organic-N	31	38	33	36	21	33	28	27
Phosphorus	3	8	16	18	16	41	11	26

Note: Data from April 29 to August 20 of the first year; retention times changed on August 20.

ment, although the retention time was adjusted to 15 hr. The other four were reconstructed in various ways in the second year to test the effects of different substrate depths and configurations. Results of the latter studies will be presented elsewhere.

Influent and effluent samples were collected at weekly or biweekly intervals in the first year and at monthly intervals in the second and third years. Upon collection, the samples were chilled or preserved with mercuric chloride (nutrients). Water quality characteristics determined were pH, color, turbidity, total suspended solids, conductivity, temperature, dissolved oxygen, BOD_5, ammonia, total Kjeldahl nitrogen, nitrate, and total phosphorus.

Late in the year, the aboveground portion of the plants died (cattail in October and reed and cordgrass in December), and all this vegetation was harvested. Four substrate samples were collected in each marsh and the roots harvested. Plant material was weighed and a sample of it dried and weighed again. Nitrogen and phosphorus analyses were conducted on this dried material.

RESULTS

Some parameters did not change significantly in transit through the marsh systems (pH, temperature, conductivity, and color). The lack of an effect on conductivity indicates that water loss via evaporation and transpiration was not important. The lack of color removal was a disappointment. The earlier studies in Mississippi using the same effluent had led us to believe the marsh systems would remove color.

The concentrations of the remaining parameters were substantially reduced (Table 2). The reactors were more effective at retention times of 24 hours than at retention times of 6 hours. However, because four times more effluent flowed into the 6-hr reactors, total mass of pollutants removed was higher at the shorter retention times.

Presence of plants had no substantial effect on removal efficiencies for most pollutants. However, the plants did add to the marsh system effectiveness in removing ammonia and phosphorus. Removal mechanisms for the other pollutants were probably filtration and anaerobic breakdown. Dissolved oxygen

Table 3. Percent Removal Efficiency as a Function of Retention Time

	Retention Time		
	6-hr	15-hr	24-hr
TSS	52–55	60–62	62–68
BOD_5	32–39	37–40	41
Ammonia	11–31	59–64	52–88
Organic-N	16–19	24–29	24–27
Phosphorus	9–14	18–19	19–31

and oxidation reduction potential readings from within the rock substrate indicated an anaerobic, strongly reducing environment. From chemical analyses of plant tissue and suspended solids, it was possible to assess the relative importance of various removal mechanisms. Approximately 80% of the phosphorus removed from the effluent was due to plant uptake, with the remainder due to filtration of particulate phosphorus by the gravel matrix. In the case of nitrogen, 45% was removed by plant uptake and 10% by filtration. The remaining 45% was apparently removed by denitrification.

Differences in removal efficiency among the three types of plants were not striking. In general, cattail appeared to be somewhat superior to the other two. Data in Table 2 are representative of the period up to August 20, when the retention time in some reactors was changed. Until senescence in the fall, results from the six unaltered reactors were similar to results gathered prior to August 20.

In general, the marsh systems provided further treatment of an already high-quality effluent. Before treatment by the marshes, pollutant levels were TSS, 5–8 mg/L; BOD_5, 10–15 mg/L; ammonia, 1.5–3.0 mg/L; organic-N, 2.5–3.5 mg/L; and phosphorus, 0.5–0.8 mg/L. By applying the removal efficiencies from Table 2, pollutant levels after treatment can be estimated.

In the latter half of the first year, retention times in the two cordgrass reactors were changed to 15 hours. Because capital and operating costs would be a function of the areal extent of a marsh system, an intermediate retention time was tested to see if it would be as effective or almost as effective as 24-hr retention. In Table 3, the cordgrass reactors at 15 hours are compared with the other reactors with plants at 6 and 24 hours. As noted above, the differences among the three plants had been determined to be rather small. Predictably, the removal efficiencies at 15 hours were intermediate to efficiencies at 6 and 24 hours. However, the differences between 15 and 24 hours were slight. The lesser retention time could be used with only a modest penalty in effectiveness.

Removal efficiencies did not change substantially over the three-year study period (Table 4) but may have been slightly greater for ammonia and phosphorus in the second year as compared to the first year. Plants and root systems were more developed in the second year and would be expected to process ammonia and phosphorus more effectively. A concern about these systems is the potential of occlusion of the interstitial spaces due to plant or

Table 4. Percent Removal Efficiency in the First Three Years of Operation

	Year 1	Year 2	Year 3
TSS	55	61	46
BOD_5	28	27	32
Ammonia	42	80	70
Organic-N	33	32	35
Phosphorus	14	21	19

Note: Data from Reed marshes at 15-hour retention time.

other growth, thereby reducing retention time if influent flow was maintained at a constant level. Interstitial volume, measured at the end of each growing season, declined by 10% each year. This decline appeared to be due largely to growth of the root systems; however, interstitial bacterial growth was present and may have accounted for part of this reduction. In spite of this, removal efficiencies in the third year were not substantially or consistently lower than in the first two years of operation.

DISCUSSION AND CONCLUSIONS

Based on a three-year study of pilot scale facilities, artificial marsh systems proved successful in substantially reducing concentrations of several pollutants in a pulp mill effluent (TSS, BOD_5, ammonia, and organic nitrogen). The systems were only moderately effective in reducing phosphorus levels and were completely ineffective in reducing color levels.

The presence of plants was an important factor in ammonia and phosphorus removal; however, the plants did not contribute an increment of treatment for the remaining pollutants. The three plant species tested (cordgrass, cattail, and reed) thrived in pulp mill secondary effluent. Untreated black liquor was added over a period of two weeks to the marshes (simulating a serious spill) without apparent effect on the plants. Marl, chosen as the gravel matrix because of its low cost, proved to be an acceptable matrix in all regards. The three plant species tested were nearly equal in removal effectiveness. Cordgrass forms massive root systems that could impede water movement through the rock matrix and probably would have to be thinned periodically. The cattail died back much earlier than the other two species. If operation in November and December was important, cordgrass or reed would offer an advantage.

Removal efficiencies were a function of retention time in the range of 6–24 hr. However, the increment of improvement between 15 and 24 hours was slight. A facility sized for about 15 hours of treatment would probably be adequate for most purposes. At 15 hours, a full-scale artificial marsh at a typical pulp mill in the United States would be 20–40 ha.

A key unknown in assessing the practicality of marsh systems is long-term performance. Substrate interstices could become so occluded with roots and bacterial mass that retention time and treatment efficiency would be significantly reduced, although the increased biomass might offset this effect. No such impact on performance was noted in the three years of this study.

REFERENCES

1. Pope, P. R. "Wastewater Treatment by Rooted Aquatic Plants in Sand and Gravel Trenches," USEPA Project Summary, EPA-600/S2-81-091 (1981), pp. 1-6.
2. Gersberg, R. M., B. V. Elkins, and C. R. Goldman. "Nitrogen Removal in Artificial Wetlands," *Water Res.* 17:1009-1014 (1983).
3. Gersberg, R. M., B. V. Elkins, and C. R. Goldman. "Wastewater Treatment by Artificial Wetlands," *Water Sci. Technol.* 17:443-450 (1984).
4. DeJong, J. "The Purification of Wastewater with the Aid of Rush or Reed Ponds," in *Biological Control of Water Pollution,* J. Tourbier and R. W. Pierson, Jr., Eds. (Philadelphia: University of Pennsylvania Press, 1976), pp. 133-139.
5. Wolverton, B. C., R. C. McDonald, and W. R. Duffer. "Microorganisms and Higher Plants for Wastewater Treatment," *J. Environ. Qual.* 12:236-242 (1983).
6. Wolverton, B. C., and R. C. McDonald. "Natural Processes for Treatment of Organic Chemical Waste," *Environ. Prof.* 3:99-104 (1981).
7. Allender, B. M. "Water Quality Improvement of Pulp and Paper Mill Effluents by Aquatic Plants," *Appita* 37:303-306 (1984).
8. Wolverton, B. C. "Hybrid Wastewater Treatment System Using Anaerobic Microorganisms and Reed," *Econ. Bot.* 36:373-380 (1982).

CHAPTER 20

Use of Artificial Wetlands for Treatment of Municipal Solid Waste Landfill Leachate

Nancy M. Trautmann, John H. Martin, Jr., Keith S. Porter, and Kenneth C. Hawk, Jr.

INTRODUCTION

New York State contains over 250 municipal sanitary landfills. Although collection and treatment of leachate from these facilities are required by the New York State Department of Environmental Conservation (NYSDEC), many landfills currently do not comply with regulations. Leachate treatment at municipal wastewater treatment facilities is one option for meeting water quality standards. This alternative can be expensive and energy-intensive, particularly for small landfills in rural areas because of the need for leachate transport. On-site treatment eliminates transportation costs, but treatment costs still can be substantial if the common approach of employing aerated lagoons is used. One possible method to reduce cost and energy requirements is to treat the leachate on-site using artificially constructed wetlands. This chapter describes a proposed study to evaluate the feasibility of this approach at the municipal sanitary landfill in Fenton, Broome County, New York.

The Town of Fenton disposes of solid waste at a 3.6-hectare site, operated for 20 years as an open dump and for the past nine years as a sanitary landfill. Only household wastes are accepted for disposal. The Fenton landfill has operated without a permit since 1978 and is under closure order by the NYSDEC. Negotiations with the NYSDEC are underway to develop a mutually agreeable timetable for closure. Leachate from precipitation percolating through the refuse flows through a series of five filter ponds (Figure 1) that provide partial treatment and then into a nearby stream.

Concentrations of ammonia, iron, manganese, nitrate, and benzene in the Fenton landfill leachate exceed discharge limits specified by New York State Water Quality Regulations (Table 1), and the Town of Fenton was told it must transport all leachate to a wastewater treatment facility. To avoid such a costly and energy-intensive measure, the town has worked with Hawk Engineering and with scientists and engineers at Cornell University and Ithaca College to

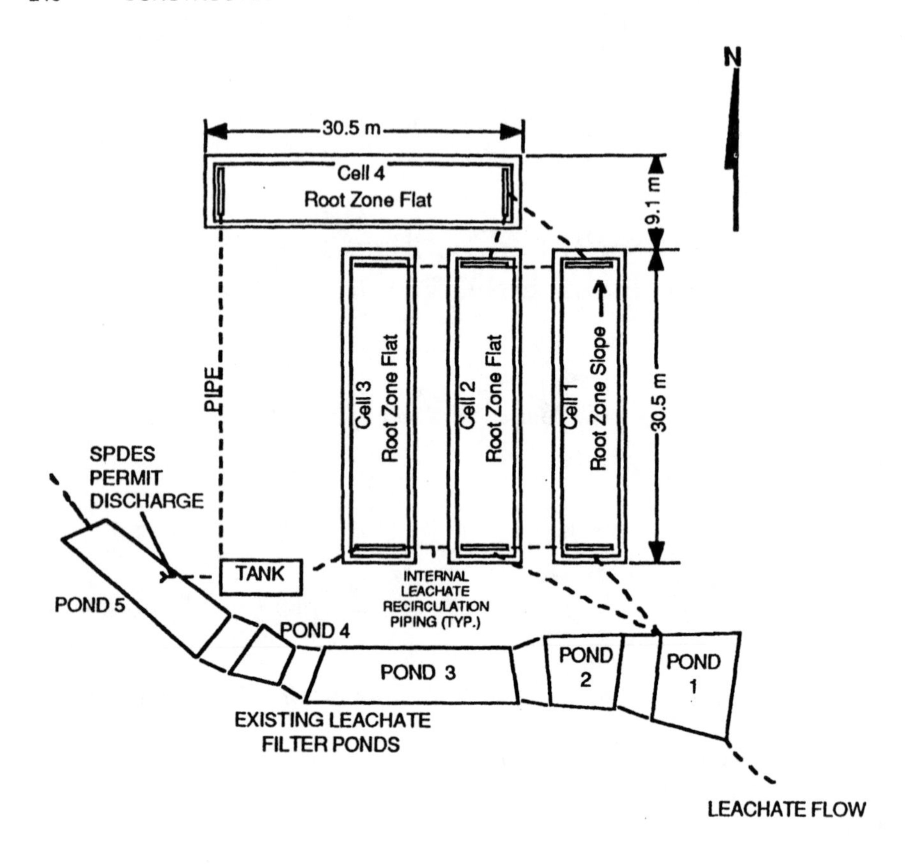

Figure 1. Site plan for the prototype landfill leachate treatment system.

design and construct a system using the purification capabilities of wetland plants for on-site treatment. This on-site system is projected to save the town over $21,000 per year and to reduce energy consumption by 25,000 kwh annually compared with the alternative of transporting 3.8 m^3/day of leachate 74 km to a wastewater treatment plant for disposal.[1]

Table 1. Characteristics of Fenton, New York Landfill Leachate for 1986, mg/L

Parameter	Average Concentration	Maximum Concentration	Discharge Limits
Ammonia-N	11	29	20
Chlorides	183	310	250
Iron	3.2	8.2	0.3
Manganese	4.2[a]	4.2[a]	0.3
Nitrite	0.03	0.05	0.02
Benzene	14[a]	14[a]	—[b]

[a]Results from one sample taken on 7/31/86.
[b]Below detection limit.

Due to the experimental nature of this approach for landfill leachate treatment, assistance has been requested from the New York State Energy Research and Development Authority (NYSERDA), which has agreed in principle to provide partial financial support. A two-year evaluation will begin after a contractual agreement has been approved by the Town of Fenton, Cornell University, and NYSERDA.

ARTIFICIAL WETLANDS FOR WASTEWATER TREATMENT

In recent years, artificially constructed beds of wetland plants have received much attention in Europe for treatment of domestic wastewater, especially from rural areas or sites with seasonally fluctuating loads.[2-4] A bed is excavated, lined with an impermeable layer, filled with gravel or soil, and planted with wetland plants. Usually reeds *(Phragmites* sp.) are used, but rushes *(Scirpus* sp.), cattails *(Typha* sp.), and sedges *(Carex* sp.) also are common.

Although the capability of wetlands to treat wastewater is widely recognized,[5,6] the root-zone systems described here differ from natural and conventionally constructed wetlands in that they rely on percolation of wastewater through the soil rather than over the surface. Theoretically, wastewater flows horizontally through the root zone, where plant roots supply oxygen and channels for wastewater flow.[7] Solids are aerobically decomposed in the layer of plant litter at the soil surface. Pathogens are filtered out of the wastewater by the soil, and soluble organic matter is removed by soil microorganisms as in other land treatment systems. Wetland plants, although essential to the treatment process, are not thought to play a significant role in removing organics and nutrients from wastewater. Rather, their root structure theoretically maintains or increases soil hydraulic conductivity and supplies oxygen to soil microorganisms. In this environment, soil microorganisms can oxidize organic matter and nitrify ammonia nitrogen. Phosphorus is removed through oxidization to phosphate, which precipitates and adsorbs to soil particles. Heavy metal removal has received little attention to date because most existing systems treat domestic wastewaters with low metal concentrations, but removal by plant uptake and adsorption has been documented.[8,9]

PROJECT DESIGN

Employing concepts introduced in Europe for wastewater treatment using artificially constructed beds of wetland plants, an on-site leachate treatment system was designed for the Town of Fenton landfill. Leachate will be channeled into a series of four beds 3.7 m × 30.5 m (Figure 1) lined with a 60-mil geomembrane and a 61-cm clay liner to prevent leachate migration to the underlying aquifer (Figure 2). The first two beds will be used for pretreatment via overland flow and will contain a 61-cm layer of soil vegetated with reed

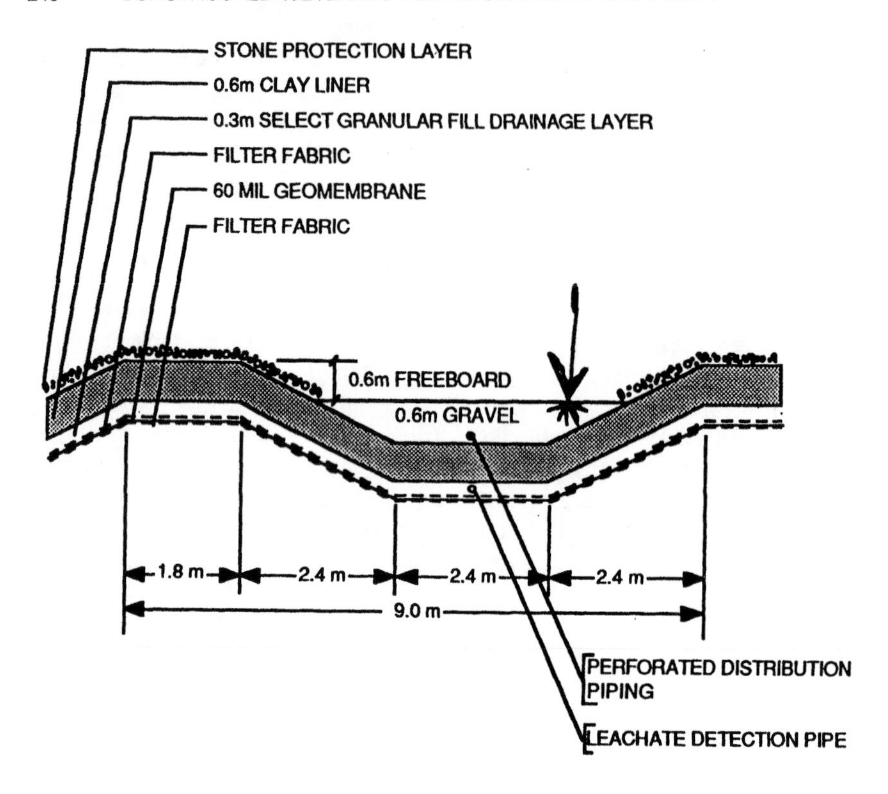

Figure 2. Root-zone bed cross section.

canary grass *(Phalaris arundinacea)*. Effluent from the pretreatment beds will be split between two parallel root-zone beds containing a 61-cm layer of coarse gravel. One bed will be planted with cattails *(Typha* sp.); the other will remain unplanted as a control.

Landfill leachate may be a problematical wastewater for root-zone treatment because of high levels of dissolved iron and manganese. When reduced forms of these metals are oxidized, precipitation results in reduction and, ultimately, blockage of flow through the root zone. Thus, the first two beds in Fenton were designed to remove much of the iron and manganese by precipitation before the leachate enters the third and fourth beds. A second pretreatment objective is to remove a substantial part of the readily oxidizable organic matter and to volatilize ammonia nitrogen or oxidize it microbially to nitrate nitrogen. Volatilization of benzene also may occur.

In the third and fourth beds, denitrification of nitrate, decomposition of dissolved organic matter, and adsorption and precipitation of phosphorus and metal ions should continue as the leachate moves horizontally through the gravel media. The unplanted control bed will provide the opportunity to characterize the significance of the wetland plants in these processes.

The four beds were designed larger than needed to treat the volume of leachate generated at the Fenton landfill so that removal rates for various constituents could be determined and design guidelines established for leachate treatment at other sites.

DATA COLLECTION

To evaluate the effectiveness of the Fenton leachate treatment system, influent and effluent samples, as well as samples from intermediate points in each of the four beds, will be collected weekly. Samples will be analyzed to determine changes in the concentrations of the parameters listed below:

Weekly:

biochemical oxygen demand
chemical oxygen demand
total organic carbon
total Kjeldahl nitrogen
ammonia nitrogen
nitrite + nitrate nitrogen
phosphorus
iron
manganese
pH
temperature

Monthly:

cadmium
chromium
copper
lead
nickel
zinc
calcium
potassium
magnesium
toxic organics

Concentrations of metals will be determined monthly. Since the fate of toxic organic compounds commonly present in landfill leachate is of interest, influent and effluent samples from each bed will be analyzed monthly. For the first three months of system operation, these analyses will be done for all priority pollutants. Once the toxic organic compounds present in significant concentrations are identified, subsequent monthly analyses will be limited to those compounds.

In addition to leachate sample analyses, a series of soil samples from both pretreatment beds will be analyzed routinely to determine changes in concentrations of metals and nutrient and toxic organics that have been routinely detected in the untreated landfill leachate. Before the initial system operation and every six months thereafter, samples of the soil layers at 0–5 cm, 5–15 cm, and 15–30 cm will be analyzed. Each root-zone bed will be sampled similarly to determine fate of metals, nutrients, and toxic organics.

LABORATORY EXPERIMENTS

In conjunction with work at the landfill site, laboratory experiments will be conducted to investigate wetland plant effects on hydraulic conductivity and oxygen content of the root zone. Objectives of laboratory experiments are to:

1. quantify differences in the ability of selected wetland plant species to supply oxygen to the root zone
2. measure iron oxide formation in the root zone and relate it to oxygen transport in the roots
3. determine whether rhizome systems of selected wetland plants increase the hydraulic conductivity within the root zone

Initial experiments on changes in soil hydraulic conductivity have been conducted. In a 7-month study using *Typha glauca* Gadr and *Scirpus acutus* Muhl planted in boxes of sand and grown in a greenhouse, saturated hydraulic conductivity decreased over time in both vegetated and nonvegetated systems, with a significantly greater decrease in the vegetated systems.[10] Saturated hydraulic conductivity declined by 41% in control systems, probably because of sand settling, and by 55% in vegetated systems. Ammonia nitrogen and nitrate nitrogen were supplied in ranges of rates comparable to municipal wastewater. Systems with vegetation reduced ammonia nitrogen and nitrate nitrogen concentrations significantly better than those without vegetation. *Typha* and *Scirpus* showed comparable rates of nitrogen removal.

SUMMARY

A potentially cost-effective and energy-efficient prototype system for on-site treatment of landfill leachate has been designed for the Town of Fenton, New York. Leachate will be pretreated using overland flow followed by horizontal flow through the root zone in a bed of wetland plants. In the absence of previous experience in treating landfill leachate with this technique, its efficacy is not known. Overland flow theoretically will remove much of the dissolved iron and manganese, readily oxidizable organic matter, ammonia nitrogen, and volatile organics such as benzene. Additional removal of nitrogen and organic matter as well as removal of phosphorus and metal ions will occur in the root-zone bed, producing an effluent that can be directly discharged to surface waters. Through a combination of laboratory studies and evaluation of the prototype system performance, the feasibility of this approach will be determined.

REFERENCES

1. Hulbert, E. H., D. R. Bouldin, J. M. Bernard, K. S. Porter, and T. E. Lauve. "Innovative Technologies for Landfill Leachate Management," proposal submitted to the N.Y.S. Energy Research and Development Authority, Hawk Engineering, p.c., Binghamton, NY, (1987), p. 3.
2. Brix, H., and H.-H. Schierup. "Root-Zone Systems. Operational Experience of 14 Danish Systems in the Initial Phase," report to the Environmental Protection Board (Botanical Department, Århus University, 8240 Risskov, Denmark, 1986).

3. Bucksteeg, K. "Sewage Treatment in Helophyte Beds—First Experiences with a New Treatment Procedure," *Water Sci. Technol.* 19(10):1–10 (1987).
4. Cooper, P. F., and J. A. Hobson. "State of Knowledge on Reed Bed Treatment Systems," Report No. 581-S, Water Research Centre, Hertfordshire, Great Britain (1987).
5. Nichols, D. S. "Capacity of Natural Wetlands to Remove Nutrients from Wastewater," *J. Water Pollut. Control Fed.* 55(5):495–505 (1983).
6. Gersberg, R. M., B. V. Elkins, and C. R. Goldman. "Wastewater Treatment by Artificial Wetlands," *Water Sci. Technol.* 17:443–449 (1984).
7. Kickuth, R. "Abwasserreinigung in Mosaikmatrizen aus Anaeroben und Aeroben Teilbezirken," Grundlagen der Abwasserreinigung. *Gwf-Schriftenreihe Wasser/Abwasser* 19:639–665 (1981).
8. Gersberg, R. M., S. R. Lyon, B. V. Elkins, and C. R. Goldman. "Removal of Heavy Metals by Artificial Wetlands," in *Future of Water Reuse, Vol. 2* (Denver, CO: American Water Works Association, 1984), pp. 639–648.
9. Universität für Bodenkultur. "Pflanzenklaeranlage Mannersdorf/C. Jahresbericht 1986 (annual report for the reed bed system in Mannersdorf, Austria)," Vienna, Austria (1986), pp. 139–142.
10. McIntyre, B. D. "A Root-Zone Method for Wastewater Treatment: Hydraulic Conductivity and Nutrient Loading Investigations," unpublished master's thesis, Department of Agronomy, Cornell University, Ithaca, NY (January 1989).

CHAPTER 21

Use of Wetlands for Urban Stormwater Management

Eric H. Livingston

INTRODUCTION

Wetlands are an essential part of nature's stormwater management system. Important wetland functions include conveyance and storage of stormwater, which dampens effects of flooding; reduction of flood flows and velocity of stormwater, which reduces erosion and increases sedimentation; and modification of pollutants typically carried in stormwater. Accordingly, there is a great amount of interest in the incorporation of natural wetlands (especially wetlands that have been previously drained) or constructed wetlands into stormwater management systems. This concept provides an opportunity to use, not abuse, one of nature's systems to mitigate effects of runoff associated with urbanization. In addition, by using wetlands for stormwater management, drained wetlands can be revitalized, and landowners and developers have an incentive to preserve or restore wetlands.

Unfortunately, the use of wetlands for stormwater management involves a large degree of uncertainty.[1] Little scientific information is available concerning the short- or long-term effects on wetlands, their natural functions, or associated fauna from the addition of stormwater. Use of wetlands for urban stormwater management should not be considered a panacea to stormwater problems. Much remains to be learned. This chapter will review the current state of the art and discuss the design and performance standards used for wetland stormwater treatment systems.

STORMWATER MANAGEMENT

In an undeveloped area, stormwater runoff is managed through the natural hydrologic cycle that effectively accommodates even severe rainfalls. Unfortunately, land use changes associated with urbanization alter the natural hydrology by changing peak flow characteristics, total runoff, water quality, and the hydrologic amenities.[2] These detrimentally affect downstream areas and receiving waters, creating a need for stormwater management systems.

Objectives of stormwater management are to (1) provide surface drainage, (2) control floods, (3) control erosion and sedimentation, (4) reduce pollutants in runoff, and (5) provide aesthetic amenities, including open space, recreation, and waterfront property. To achieve these goals, the volume, rate, timing, and pollutant load of runoff after development must be similar to predevelopment conditions. A stormwater management system must be an integral part of the site planning process for every project. Natural site attributes (soils, geology, slope, water table) will influence the type and configuration of the stormwater system. For example, sandy soils accommodate infiltration practices, while natural low areas and isolated wetlands offer opportunities for detention/wetland treatment. A stormwater management system should be viewed as a treatment train in which individual best management practices (BMPs) are the cars. The more BMPs included in the system, the better the performance of the train.

Stormwater pollution control has received increasing attention since the late 1970s, when the Federal Section 208 program funded numerous studies on water quality effects of runoff. Pollutant types and amounts in stormwater vary widely with land use and rainfall characteristics.[3] In addition, (1) higher pollutant concentrations are associated with more intensive development and greater imperviousness; (2) construction erosion and sedimentation can result in high loadings of suspended solids; and (3) stormwater pollutant levels are comparable to secondary-treated wastewater effluent.

Most water quality effects result from the "first flush." In the early stages of a storm, accumulated pollutants in the watershed, especially on impervious surfaces such as streets or parking lots, are flushed clean by rainfall and resulting runoff. In Florida the first flush equates to the first 2.5 cm of runoff, which carries 90% of the pollution load from a storm event.[4] Treatment of that runoff will help minimize water quality effects of stormwater.

WETLANDS AS STORMWATER TREATMENT SYSTEMS

Considerable attention has been devoted to use of wetlands for wastewater treatment and nutrient assimilation. However, stormwater hydrology and pollutants differ from partially treated wastewater typically placed into wetlands for additional pollutant removal. For example, nutrient concentrations in wastewater and stormwater may differ by a factor of 10 or more and treatment efficiencies of wastewater wetlands may not be applicable to stormwater treatment.

Investigations of wetlands for treatment of stormwater are limited and have dissimilar wetland types and stormwater characteristics.[5-10] Limited results indicate that (1) nutrient removal varies widely in stormwater treatment wetlands; (2) flow and seasonal factors influence pollutant removal capabilities;

and (3) removal is consistently better for BOD_5, suspended solids, and heavy metals.[7,8]

Review papers on wetlands for stormwater treatment summarize treatment efficiency and performance of wetland stormwater systems.[5-8] Due to variations in stormwater characteristics and poor understanding of wetland processes that remove pollutants, treatment efficiency predictions of a wetland stormwater system are not possible. Essentially, a wetland system must be designed as a BMP treatment train that accentuates the numerous assimilation mechanisms that are present in each type of wetland to maximize stormwater pollutant removal.

The type of wetland (peatland, cypress dome, marsh meadow) will influence the wetland's suitability as a component of a stormwater system, the final system design and system effectiveness.

Pretreatment

Pretreatment removes heavy sediment loads and other pollutants such as hydrocarbons that can damage the wetland. Pretreatment also attenuates stormwater volumes and peak discharge rates to maintain the wetland hydroperiod and to reduce scour and erosion. Erosion and sediment control practices during construction are essential to prevent sedimentation of the wetland.

Hydrology

Relations between hydrology and wetland ecosystem characteristics must be included in the design to ensure long-term effectiveness. The source of water, velocity, volume, renewal rate, and frequency of inundation influence the chemical and physical properties of wetland substrates which, in turn, influence species diversity and abundance, primary productivity, organic deposition and flux, and nutrient cycling.[7] Hydrology also influences sedimentation, aeration, biological transformation, and soil adsorption processes. Critical factors that must be evaluated include velocity and flow rate, water depth and fluctuation, detention time, circulation and distribution patterns, seasonal and climatic influences, groundwater conditions, and soil permeability.

After gaining an understanding of the hydrologic factors ongoing in the proposed treatment wetland, the information can be used to design the system. Establishing wetland hydroperiod is of primary importance because this determines the form, nature, and function of the wetland. Hydroperiod is the depth and duration of inundation measured over an annual wet or dry cycle.[11] Acceptable high and low water elevations will determine the stormwater treatment volume capacity of the wetland, the discharge structure, and bleed-down orifice elevations. Water depth and inundation period can change the plant community, with beneficial or detrimental effects on the wetland or stormwater pollutant removal.

Vegetation

Wetland vegetation is a function of climate, hydrology, and nutrient availability. In a natural wetland, the vegetation is a factor in the selection of loading rates and expected pollutant removal. In a constructed wetland, climate, hydrology, and pollutant response influence selection of plant species. In either case, a mixture of floating, emergent, and submerged plants is desirable, especially those with dense submerged stems and leaves or floating root mats that increase filtration and provide sites for microorganisms.

Wetland plants have specific tolerances to levels and types of pollutants. Polluted stormwater represents increased nutrients, which may change the plant community. Since new dominants reflect more efficient use of added nutrients or are more tolerant to pollutants, the plant changes should benefit pollutant removal.[7]

Maintenance

Though little information is available, maintenance is important because accumulation of organic matter and sediment can alter the pollutant removal effectiveness of the system by decreasing storage volume and changing the substrate. Likely maintenance will include sediment accumulation removal, conveyance system and discharge structure repair, and vegetation harvesting or burning.

MARYLAND CREATED WETLAND DESIGN STANDARDS

In 1982, Maryland legislation required development of stormwater management regulations to ensure new development stormwater was treated to reduce the pollutant load discharged to receiving waters. As part of its program, the state developed the following guidelines for the construction of shallow wetland stormwater systems:[12]

1. A permanent pool is essential to establish and maintain a shallow wetland. Water inflow from storms, base flow, and groundwater must be greater than water outflow via infiltration and discharge. If the basin is above the water table and the infiltration rate is high, then a clay or synthetic liner should be used.
2. Wetland surface area should be maximized for pollution control and wildlife benefits. A detention time of 24 hours for the one-year storm enhances pollutant removal and provides the storage volume recovery between storms. If extended detention is not possible, then the wetland surface area should be a minimum of 3% of the contributing drainage area.
3. Approximately 75% of the wetland should have depths less than 30 cm for emergent aquatic vegetation, with 25% of this over 15 cm deep and 50% under 15 cm deep. The remaining 25% should be 1 m deep, where submerged aquatic vegetation will thrive and the outlet structure placed.
4. Sediment forebays, 1 m deep and 10% of pool volume, are located at all

inflow points along with energy dissipation devices to lessen scour and damage to plants.

5. A length-to-width ratio of 2:1 will reduce short circuiting and maximize the flow path, but if inlet and outlet structures are close, baffles, islands or peninsulas will increase the flow path.
6. The outlet structure must maintain the wetland's permanent pool, detain the treatment volume, and produce slow and reliable discharges. The discharge orifice is set at the permanent pool elevation and sized to meet desired discharge rates. Wire mesh protects the small-diameter orifice from plant material or debris blockage.
7. Wetland soils with a pool of propagules and seeds are preferred and may have to be imported. If available, wet soils from an existing basin or flow paths should be stockpiled and spread over the basin after excavation. At least 10 cm of soil is needed to anchor plants securely.
8. Part of the basin should be planted to enhance short- and long-term development and reduce success of undesirable plants, especially aggressive volunteers such as *Typha* spp. (cattail) and *Phragmites australis* (common reed). Lists of primary (aggressive growers) and secondary species and their planting recommendations and a list of unacceptable species follow:

Primary species (use two)	Maximum depth (cm)
Sagittaria latifolia	30
Scirpus americanus	15
Scirpus validus	30
Secondary species (use three)	
Acorus calamus	8
Cephalanthus occidentalis	60
Hibiscus moscheutos	8
Hibiscus laevis	8
Leersia oryzoides	8
Nuphar luteum	150
Peltandra virginica	30
Pondederia cordata	30
Saururus cernuus	15
Unacceptable species	
Phragmites australis	
Typha latifolia	
Typha angustifolia	

For primary species:
1. Cover 30% of the shallow zone on 3 ft spacing.
2. Plant in four monospecific areas with suitable depth conditions.
3. Distribute an additional 40 clumps per acre of each species over entire wetland.

For secondary species:
1. Plant 50 individuals per acre.
2. For each species, plant 10 clumps of five individuals each

as close to the edge of the pond as possible.
3. Space clumps as far apart as possible, but no need to segregate species.

9. Five or more wetland species (two primary, three secondary) are needed to match plant requirements to variations in soil type, depth, and water circulation and to promote some active growth for nutrient uptake throughout the growing season.
10. Site preparation for planting the wetland plants requires that the substrate be soft enough to permit easy insertion of the plants. If planting is to be done before the wetland is flooded, no more than 24 hours (for bare root plants) to 72 hours (for peat-potted plants) should elapse between planting and flooding. Use of nursery-grown plants, plants transplanted from roadside ditches or other wet areas (where permissible), or transplanted dormant underground plant parts are recommended means of plant establishment. The growth form of the propagule will determine the suitable planting time during the year.
11. Functional lifetime, especially for nutrient and metal removal, of wetland basins is uncertain. Solids will accumulate and need removal. Outlet structures can be modified to raise permanent pool elevation when solids accumulate (until storage volume requirements are compromised, when excavation will be needed).

FLORIDA VEGETATED STORMWATER SYSTEMS

Stormwater has become the major source of pollutant loading to receiving waters in Florida. In February 1982, the Florida Stormwater Rule was implemented by the Florida Department of Environmental Regulation, requiring all newly constructed stormwater discharges to use appropriate BMPs to treat the first flush of runoff.[13] Vegetated systems, wet detention, or wetlands, are commonly used BMPs.

Wetland Systems

The 1984 Wetlands Protection Act authorizes the use of two types of natural wetlands for stormwater management in Florida: those wetlands that are connected to other state waters by (1) an artificial watercourse, or (2) by an intermittent watercourse that flows only when groundwater is at or above the land surface. In addition to providing natural stormwater management, the legislation provides an incentive to developers to use wetlands rather than destroy them and enhances restoration of ditched and drained wetlands by routing stormwater into them. Recognition of potential negative effects was included by requiring that such use shall protect the ecological values of natural wetlands.

Because using wetlands for stormwater treatment is untested, the department adopted design and performance standards for wetland systems that include monitoring to evaluate their effectiveness. Current standards will likely be modified with experience and analysis of monitoring data. Design and performance standards for wetland stormwater management systems are:

1. The wetland treatment facility is part of a comprehensive stormwater management system that uses wetlands in combination with other BMPs to treat runoff from the first 2.5 cm of rainfall; or, as an option for projects with drainage areas under 40 ha, treats the first 1.2 cm of runoff. Facilities which directly discharge to Outstanding Florida Waters must treat an additional 50% runoff volume.
2. Pretreatment swales or lakes are used to reduce sediments, oils, greases, and heavy metals, and to attenuate stormwater volumes and peak discharges so that the wetland's hydroperiod is not adversely altered.
3. Use shall not adversely affect the wetland by disrupting the normal range of water level fluctuation as it existed prior to construction of the wetland stormwater system. Normal low pool and high water elevations are established using existing biological or hydrological indicators in the wetland such as water level data, lichen and moss lines, adventitious root formation, epiphytic plant colonies, root tussocks and hammocks, emergent plant community zonation, hydric soils, the landward extent of wetland vegetation, and recent water stain lines and cast rack and debris lines.
4. Design features of the system shall maximize stormwater residence time, enhancing contact with wetland sediment, vegetation, and microorganisms. Stormwater must be discharged into the wetland via sheet flow so that channelized flow is minimized. The outfall structure must be designed to bleed down the treatment volume in no less than 120 hours with no more than one-half of the volume discharged within the first 60 hours.
5. Erosion and sediment controls must be used during construction and operation to minimize sedimentation of the wetland.

Wet Detention System Design

In addition to natural wetland stormwater treatment systems, many constructed wetland systems have been built in Florida since implementation of the Stormwater Rule. The most common type is a wet detention system consisting of a permanent water pool, a temporary stormwater storage area above the permanent pool and a littoral zone planted with native aquatic plants.

Wet detention systems provide a high level of pollutant removal for many constituents, particularly nutrients and metals.[3] Pollutant removal efficiency depends on detention time, amount of littoral zone, aquatic vegetation establishment, pond geometry, pond depth, area ratio, volume ratio, and the incorporation of other BMPs with the detention system into a "treatment train."[14]

Besides flood protection, peak discharge rate attenuation, and water quality protection, wet detention systems can also provide a source of fill to the developer, premium "lakefront" property, open space, recreational area, and aesthetic enhancement.

Current design and performance standards for wet detention systems in Florida are:

1. Treatment volume varies from a minimum of 2.5 cm of runoff up to 2.5 times the percent of impervious area. Treatment volume is slowly discharged in not less than 120 hr, with no more than half the volume discharged within the first 60 hr following a storm. Permanent pool volume should provide a residence time of at least 14 days to promote biological uptake of nutrients.[15]
2. At least 30% of the surface area should be a shallow, gently sloped (6:1) littoral zone with a 15-cm layer of wetland soil to encourage native aquatic vegetation. At least one-third of the littoral area should be planted with a variety of aquatic plants suitable for the depth conditions. Within one year, 80% of the littoral area must have desirable plant coverage and adequate maintenance to remove cattails.
3. The littoral zone should be concentrated near the discharge structure, located on shallow sills separating in-line tandem ponds, and near inlet areas to promote filtering before the runoff enters the main impoundment.
4. Permanent pool should have a mean depth of 2–3 m with a maximum depth of 6 m.
5. Pond geometry should assure a minimum 2:1 length-to-width ratio, with inlet and outlets widely separated and baffles, islands, or peninsulas as needed to increase the length-width ratio and flow path.
6. The discharge structure should include a skimmer or other device to prevent oils and greases from leaving the system.
7. Erosion and sediment control practices must retain sediment on-site during construction. A sediment sump or forebay should be included at inflows. Pretreatment, the treatment train concept, swale conveyances, small off-line landscape retention areas, and a perimeter swale and berm system will reduce sediment, oil, grease, and nutrient loadings to the system.

Wet Detention Regional Systems

Reducing pollutant loads from flood protection systems, which rapidly convey polluted stormwater to receiving waters, is a major problem. Regional wet detention systems serving large drainage areas with multiple property owners will be heavily relied upon to reduce this pollutant source. Advantages of regional facilities include stormwater management from existing and future land uses; economies of scale in construction, operation, and maintenance; and enhanced recreation and open space opportunities. Disadvantages include the need for advance planning and funding to locate and construct systems, and institutional problems in administering a master stormwater planning approach.

A regional stormwater system was constructed in Tallahassee in 1983 to reduce pollutant loads to Megginnis Arm and Lake Jackson.[16] The 900-ha watershed consists of a rapidly developing area with commercial, highway, and residential land uses. The stormwater system includes an 11.5-ha detention pond, a 1.8-ha intermittent underdrain sand filter, and a 2.5-ha artificial marsh. The three-section marsh has 45 cm average depth except for a 2-m pool near the outfall. Individual marsh cells were planted with *Cladium jamaicense* (sawgrass), *Scirpus* (bulrush), and *Pontedaria* (pickerelweed), respectively.

Under normal operating conditions, the system removes 95% of the suspended solid load by settling and filtration. Total nitrogen (75%); ammonia (37%); nitrate (70%); nitrite (75%); total phosphorus (90%); unfiltered phosphorus (53%); and filtered phosphorus (78%) removals were documented.[17] Ammonia- and nitrate-oxidizing bacteria on the filter's rocks removed much of the nitrogen and may assist in phosphorus removal. Marsh pollutant removal is seasonal, with good removal of nutrients and dissolved substances during the growing season (30–60%) but net export during winter from plant dieback and decay.

Recently, the City of Orlando initiated the Southeast Lakes Project, a watershed-wide program to modify existing stormwater systems to reduce stormwater pollutant loadings to the city's lakes. Regional and wetland systems are emphasized, and urban wetlands will be created or restored throughout the watershed to help reduce stormwater pollution. The recently constructed Lake Greenwood Urban Wetland includes a series of BMPs within a circuitous wet detention lake system with wetland vegetation planted in littoral zones and grown in cages for harvesting aquatic plants.

CONCLUSION

Wetlands have great potential to help solve stormwater management problems. However, more information is needed to ascertain possible effects on wetlands and their fauna from addition of untreated stormwater. Little is known about the potential for bioaccumulation of heavy metals or other toxics typical of stormwater. Monitoring of wetland stormwater systems is also essential to determine relations between design variables and pollutant removal efficiency.

REFERENCES

1. Lakatos, D. F., and L. J. McNemar. "Storm Quality Management Using Wetlands," paper presented at Freshwater Wetlands and Wildlife Symposium, Charleston, SC, March 24–27, 1986.
2. Leopold, L. B. "Hydrology for Urban Planning—A Guidebook on the Hydrologic Effect of Urban Land Use," Circular 554, USGS, Washington, DC (1968).
3. "Results of the National Urban Runoff Program (NURP)," U.S. EPA (December 1983).
4. Miller, R. A. "Percentage Entrainment of Constituent Loads in Urban Runoff, South Florida," USGS WRI 84-4319 (1985).
5. Stockdale, E. C. "The Use of Wetlands for Stormwater Management and Nonpoint Pollution Control: A Review of the Literature," report submitted to the Washington State Department of Ecology (October 1986).
6. Stockdale, E. C. "Viability of Freshwater Wetlands for Urban Surface Water Management and Nonpoint Pollution Control: An Annotated Bibliography," report submitted to the Washington State Department of Ecology (July 1986).

7. Chan, E., et al. "The Use of Wetlands for Water Pollution Control," Municipal Environmental Research Laboratory, U.S. EPA Report 600/2-82-036 (1982).
8. Harper, H. H., et al. "Stormwater Treatment by Natural Systems," report submitted to the Florida Department of Environmental Regulation (December 1986).
9. Silverman, G. S. "Seasonal Freshwater Wetlands Development and Potential for Urban Runoff Treatment in the San Francisco Bay Area," PhD dissertation, UCLA, Los Angeles, CA (1983).
10. Kutash, W. "Effectiveness of Wetland Stormwater Treatment—Some Examples," in *Stormwater Management: An Update*, University of Central Florida Environmental Systems Engineering Institute Publication 85-1, Orlando, FL (July 1985), pp. 145-157.
11. Knight, R. L., and J. S. Bays. "Florida Effluent Wetlands—Hydroecology," WTRDS No. 2, CH2M Hill, Gainesville, FL (1986).
12. "Guidelines for Constructing Wetland Stormwater Basins," Maryland Department of Natural Resources, Water Resources Administration, Annapolis, MD (March 1987).
13. Livingston, E. H. "Stormwater Regulatory Program in Florida," in *Urban Runoff Quality—Impact and Quality Enhancement Technology*, proceedings of an Engineering Foundation Conference, Henniker, NH, June 23-27, 1986, pp. 249-256.
14. Livingston, E. H., et al. *The Florida Development Manual: A Guide to Sound Land and Water Management*, Florida Department of Environmental Regulation, Tallahassee, FL (June 1988).
15. Hartigan, J. P. "Regional BMP Master Plans," in *Urban Runoff Quality—Impact and Quality Enhancement Technology*, proceedings for an Engineering Foundation Conference, Henniker, NH, June 23-27, 1986, pp. 351-365.
16. Esry, D. H., and J. E. Bowman. "Final Construction Report: Lake Jackson Clean Lakes Restoration Project," submitted to Florida Department of Environmental Regulation (June 1984).
17. Tuovila, B. J., et al. "An Evaluation of the Lake Jackson Filter System and Artificial Marsh on Nutrient and Particulate Removal from Stormwater Runoff," in *Aquatic Plants for Water Treatment and Resource Recovery*, K. R. Reddy and W. H. Smith, Eds. (Orlando, FL: Magnolia Publishing Inc., 1987), pp. 271-278.

SECTION III

Design, Construction and Operation

CHAPTER 22

Use of Wetlands for Municipal Wastewater Treatment and Disposal—Regulatory Issues and EPA Policies

Robert K. Bastian, Peter E. Shanaghan, and Brian P. Thompson

Wetlands were once regarded as wasted land. It is now clear they provide irreplaceable benefits to people and the environment. Wetlands provide natural flood prevention and pollutant filtering systems and contribute significantly to groundwater recharge. Many sport fish, migratory waterfowl, furbearers, and other valuable wildlife live and breed in wetlands.

Freshwater, brackish, and saltwater wetlands have inadvertently served as natural water treatment systems for centuries. Because of their transitional position between terrestrial and aquatic ecosystems, some wetlands have been subjected to wastewater discharges from both municipal and industrial sources. Wetlands have also received agricultural and surface mine runoff, irrigation return flows, urban stormwater discharges, leachates, and other sources of water pollution. Effects of these discharges on different wetlands have been varied.

In the past few decades, planned use of wetlands for wastewater treatment and other water quality objectives has been studied and implemented. The functional role of natural wetlands in water quality improvements has been a compelling argument for wetland preservation and, in some cases, for wetland creation.[1,2] Studies over the years have shown that wetlands are able to provide high levels of wastewater treatment.[3-8] However, concern has been expressed over (1) possible harmful effects of toxic materials and pathogens in wastewaters and (2) long-term degradation of wetlands due to the additional nutrient and hydraulic loadings from wastewater discharges.[9-11]

Due to these concerns and other factors, there has been considerable interest in using constructed (or artificial) wetlands for wastewater treatment. Constructed wetlands are planned systems designed and constructed to employ wetland vegetation to assist in treating wastewater in a more controlled environment than occurs in natural wetlands.

As presented in a recent U.S. EPA report,[12] this chapter reviews the extent and circumstances of this practice and summarizes the regulatory issues and EPA policies involved.

WETLAND TREATMENT SYSTEMS

Natural Wetlands

The term "wetlands" is a relatively new expression, encompassing what for years have simply been referred to as marshes, swamps, and bogs. Wetlands occur in a wide range of physical settings at the interface of terrestrial and aquatic ecosystems.

"Wetlands" is defined by federal regulatory agencies as "those areas that are inundated or saturated by surface or ground water at a frequency and duration sufficient to support, and that under normal circumstances do support, a prevalence of vegetation typically adapted for life in saturated soil conditions." They are vegetated systems, ranging from marshes to forested swamps. Wetlands occur in a wide range of natural settings and encompass a diversity of ecosystem types while exhibiting a wide array of primary functions and values such as providing wildlife habitat, groundwater recharge, flood control, water quality enhancement, and recreational opportunities. For regulatory purposes, almost all natural wetlands are considered waters of the United States.

Natural wetlands appear to perform all of the biochemical transformations of wastewater constituents that take place in conventional wastewater treatment plants, septic tanks, drain fields, and other forms of land treatment. Submerged and emergent plants, associated microorganisms, and wetland soils are responsible for the majority of the treatment effected by the wetland.[13-15]

Use of natural wetland treatment systems is limited to providing further treatment of secondary effluent* to meet downstream water quality standards. (Any applicable water quality standards for the wetland itself must be met near the point of discharge to the wetland.) Usually the objective is to reduce concentrations of biochemical oxygen demand (BOD_5), suspended solids (SS), and nutrients such as ammonia, other forms of nitrogen, and phosphorus in secondary effluent. Most natural wetlands can effectively remove BOD_5, SS, and nitrogen from secondary effluents. However, phosphorus removal capability varies among individual wetlands and depends largely on site-specific factors, especially soil type.[16-20]

Other pollutants of concern may also be removed in a natural wetland. Removal and die-off rates of pathogens from wastewater discharged into wetlands have been reported as "high" in some places but "variable" in others. However, variable numbers of coliform bacteria and salmonella (routinely used as indicators of human pathogens but also produced by wildlife) complicate monitoring for human pathogens in wetlands receiving wastewater dis-

*Wastewater that has received secondary or equivalent-to-secondary treatment.

charges. Levels of many inorganic and organic compounds present in wastewater are greatly reduced by passage through wetlands. Initially, heavy metals appear to be removed by sorption to wetland soils and sediments and microbial modification, although long-term studies are needed to determine exact cycling and removal mechanisms. Many organic compounds are degraded by microbial organisms associated with wetlands soil, sediment, and vegetation.

While it appears many wetlands have some capacity for improving water quality of wastewater, runoff, or industrial discharges, some wetlands are clearly not appropriate for continuous day in, day out use as a part of a wastewater disposal or treatment system. Potentially altering biotic communities of natural wetlands by wastewater additions is of great concern to EPA and groups interested in preserving existing wetlands.

Major costs and energy requirements of a natural wetlands treatment system include preapplication treatment, land costs, minor earthwork, and the wastewater distribution system. In addition to monitoring to assure maintenance of wetland biota, operational problems that may need to be considered include the potential for increased breeding of mosquitoes or flies, odor development, and maintenance of flow control structures.

Proposed physical modification of a natural wetland to allow wastewater application requires a permit from the Army Corps of Engineers (under section 404 of the Clean Water Act) and review under the National Environmental Policy Act.

Constructed Wetlands

Use of constructed wetlands for wastewater treatment takes advantage of the same principles in a natural system, within a more controlled environment. Small-scale wetlands have been created expressly for wastewater treatment purposes, while some large-scale systems have been developed involving multiple use objectives, such as using treated wastewater effluent as a water source for creating or restoring marshes for wildlife use and environmental enhancement.[21]

Constructed wetlands treatment systems can be established almost anywhere (including on lands with limited alternative uses), especially where wastewater treatment is the only function sought. They can be built in natural settings, or they may entail extensive earthmoving, construction of impermeable barriers, or building containments such as tanks or trenches. Wetland vegetation has been established on substrates ranging from gravel or mine spoils to clay or peat. Some systems recycle at least a portion of the treated wastewater by recharge of underlying groundwater. Others act as flow-through systems, discharging the final effluent to surface waters.[22,23]

Constructed wetlands have diverse applications across the country and around the world. They can be designed to accomplish a variety of treatment objectives. Influent to constructed wetlands treatment systems ranges from raw wastewater to secondary effluent.

The many advantages of constructed wetlands include site location flexibility, optimal size for anticipated waste load, potential to treat more wastewater in smaller areas than with natural wetlands, less rigorous preapplication treatment, and no alteration of natural wetlands. Disadvantages of using constructed wetlands for treatment relative to natural systems include the cost and availability of suitable land and the cost of gravel or other fill, site grading, liners, and so on. In addition, the sites are usable for wastewater treatment during the construction period, and reduced performance can be expected during the vegetation establishment period. Other possible constraints are costs of plant harvesting and disposal, if required, and the fact that artificial wetlands, like natural wetlands, provide breeding habitat for nuisance insects or disease vectors and may generate odors. When toxics are a significant component of the wastewater to be treated, pretreatment may need to be provided to avoid problems of bioaccumulation in wildlife (particularly waterfowl) attracted to the site.

NUMBER AND TYPES OF WETLAND TREATMENT SYSTEMS

Under appropriate conditions, both natural and constructed treatment systems have achieved high removal efficiencies for BOD_5, SS, nutrients, heavy metals, trace organic compounds, and pathogens from municipal wastewater. In limited situations, natural wetland treatment systems have provided small communities with alternatives to costly and complex advanced treatment facilities. More than 25 wetland-related municipal wastewater treatment projects have been at least partially funded by EPA construction grant funds in essentially all regions of the country (Figure 1).

However, rather than a single concept of wetland wastewater treatment, a number of approaches combine wastewater and wetlands as a part of water quality management projects. These include natural wetlands for wastewater disposal; wetland enhancement, restoration, or creation; natural wetlands for wastewater treatment or reclamation; and constructed wetlands for wastewater treatment.

Natural Wetlands for Wastewater Disposal

In many areas, natural wetlands serve as receiving waters for permitted discharges of treated wastewater. More than 400 such discharges exist in the Southeast[24,25] and another 100 occur in the Great Lakes States.[26] In addition, runoff from both rural and urban areas in many parts of the country receives considerable "treatment" as it passes through natural wetlands prior to entry into groundwater, estuaries, streams, and lakes. However, any use of natural wetlands for treatment purposes requires extensive preproject review to ensure the wetland ecosystem is not unacceptably altered.

Figure 1. Location of some known wetland wastewater treatment projects in the United States and Canada.

Wetland Enhancement, Restoration, or Creation

In arid regions, wastewater effluents are often used to create, maintain, restore, or enhance wetlands. In some cases, wastewater effluent is the sole or major water source for valuable wetland habitat (e.g., the Mt. View Sanitary District project near Martinez, CA; the Bitter Lake National Wildlife Refuge near Roswell, NM: and the Apache-Sitgreaves National Forests project near Phoenix, AZ). Considerable opportunity exists for treating and using wastewater effluents, drainage water, runoff, or other sources to enhance or restore existing wetlands stressed from lack of adequate water or other reasons. However, serious environmental problems have occurred where proper management practices were not observed (e.g., the Kesterson National Wildlife Refuge in Merced County, CA, which received extremely high levels of selenium in the irrigation drainage water from the western San Joaquin Valley).

Natural Wetlands for Wastewater Treatment or Reclamation

Engineered treatment systems in Michigan, Wisconsin, Florida, Oregon, and elsewhere effectively use the capabilities of natural wetlands to provide part of the treatment in a manner that minimizes ecological disturbance. Typically, the only construction activity in the wetland is installation of (1) a distribution system for dispersed application of secondary effluent and (2) monitoring stations. Such systems require pretreatment of the wastewater and are usually used to provide for nutrient removal or high-level effluent polishing to protect downstream water quality.

Constructed Wetlands for Wastewater Treatment

Constructed systems with wetland vegetation for treating municipal and industrial wastewater, urban stormwater, agricultural runoff, acid mine drainage, leachates, and other sources of contaminated water show great promise due to greater opportunity for process control and less chance for causing adverse environmental effects. In some cases, wetland systems treat high loadings of wastewater in as small an area as possible (e.g., Santee and Gustine, CA; Collins, MS; Benton, KY; Listowel, Ontario). In other cases, systems have been built to simulate natural wetlands, receiving low loadings of pretreated effluents for further polishing while enhancing wildlife habitat and other wetland values (e.g., Arcata, CA; Incline Village, NV; Harriman, NY).

DISCHARGE STANDARDS AND PERMIT REQUIREMENTS

The Clean Water Act, together with EPA's implementing regulations, governs discharges of wastewater to waters of the United States, including any wetlands considered to be "waters of the United States" (40 CFR Part 122.2). Municipal discharges to wetlands considered waters of the United States must meet minimum technology requirements and conform with state water quality standards. The exact level of treatment required for any discharge is specified in its National Pollutant Discharge Elimination System (NPDES) permit issued by EPA or an authorized state. Projects that may result in any filling or other alteration of wetlands will generally need to obtain a Section 404 permit and address other federal requirements.

When Are Wetlands "Waters of the United States"?

The definition of "waters of the United States" under 40 CFR Part 122.2 includes most natural wetlands. Wetlands adjacent to other waters of the United States (other than wetlands) are automatically waters of the United States. In addition, other wetlands (often called isolated wetlands) are waters of the United States if their use, degradation, or destruction would or could affect interstate or foreign commerce. Some examples of an adequate interstate commerce connection are (1) if the wetlands are or could be used by interstate or foreign travelers for recreational or other purposes, (2) contain or could contain fish or shellfish that could be sold in interstate or foreign commerce, or (3) are or could be used for industrial purposes by industries in interstate commerce. In addition, isolated wetlands that are or could be utilized by migratory waterfowl are regulated as waters of the United States.

Constructed wetlands designed, built, and operated as wastewater treatment systems are, in general, excluded from the definition of "waters of the United States." However, under certain circumstances, constructed multiple-purpose wetlands designed, built, and operated to provide (in addition to wastewater treatment) functions and values similar to those provided by natural wetlands may be considered waters of the United States and, as such, are subject to the same protection and restrictions as use of natural wetlands.

Minimum Technology Requirements

All municipal wastewater treatment systems, except for certain ocean discharges and aquaculture systems, must achieve the degree of effluent reduction attainable through application of secondary treatment prior to discharge to waters of the United States (including almost all natural wetlands). "Secondary treatment" is most often defined as "attaining an average effluent quality for both five-day Biochemical Oxygen Demand (BOD_5) and Suspended Solids (SS) of 30 milligrams per liter (mg/L) in a period of 30 consecutive days, an average effluent quality of 45 mg/L for the same pollutants in a period of 7 consecutive days and 85 percent removal of the same pollutants in a period of 30 consecutive days." Although this definition is based upon the performance of a large number of properly operating wastewater treatment plants of various types, EPA's secondary treatment regulation does not require the use of any of these specific technologies to achieve the effluent limits, only that the limits be achieved.

The Clean Water Act and its implementing regulations do provide for relaxing some aspects of the definition of "secondary treatment" under certain circumstances. The most notable case is when waste stabilization ponds (WSP) and trickling filters (TF) are used as the principal biological treatment process. When this is so, the facility may be considered as achieving treatment equivalent to secondary treatment if it achieves average BOD_5 and SS of 45 mg/L during a 30-day period, average BOD_5 and SS of 65 mg/L during a seven-day period, and 65% removal of these pollutants. Where WSPs are used, the state may further adjust upward the SS limit for treatment equivalent to secondary treatment to reflect the effluent quality achieved within a specific geographical area. This provision allows differing geographical/climatic conditions that affect WSP performance to be taken into account.

Permit limits for WSPs or TFs may only be adjusted where violation of water quality standards will not result. In attempting to understand how an average SS level in excess of 45 mg/L could be considered equivalent to secondary treatment (average SS of 30 mg/L), it is important to remember that the SS contained in WSP effluent are largely the result of the biological treatment process occurring in the pond and are quite different from the SS in raw wastewater. WSP effluent SS are for the most part algae, which occur widely in natural water bodies. Thus, depending upon the water body, a WSP effluent high in SS may not pose an undue burden.

Water Quality Standards

In addition to meeting minimum technology requirements, discharges to waters of the United States must comply with applicable state water quality standards. Very few states have established separate water quality standards for wetlands, and EPA has not yet developed water quality criteria specifically for wetlands. An internal EPA task force recently concluded that the lack of EPA water quality criteria for wetlands (and the resulting absence of state water quality standards for wetlands) is one of the most serious impediments to a consistent national policy on use of wetlands for wastewater treatment or discharge.

EPA is looking into the feasibility of developing water quality criteria and numerical or narrative biological quality maintenance criteria for wetlands which would serve as the basis for establishing separate state water quality standards for wetlands. Some activity has been initiated in this area. For example, EPA has prepared a research plan for developing water quality standards for wetlands.[27] EPA is exploring the feasibility of a twofold approach to water quality standards for wetlands, addressing both chemical water quality and biological integrity. Chemical water quality standards would aim to protect wetlands and the fish and wildlife that use them from water pollution. Standards for biological integrity would address effects from activities that physically alter wetlands so that ecosystem processes are impaired (e.g., discharges of dredge and fill material).

A related project is planned to improve state programs under Section 401 of the Clean Water Act for certifying that discharges into wetlands and other aquatic sites comply with existing water quality standards. The programs of states that have particularly effective or innovative approaches to their responsibilities under Section 401 will be reviewed in detail. Subsequently, other states will be encouraged to adopt and build upon the best approaches. Both efforts have significant potential for improving the level of protection for wetlands habitat.

In lieu of separate wetland standards, some states have often applied water quality standards for adjacent streams or lakes to wetlands. These standards are often inappropriate for wetlands because wetlands are vastly different ecosystems. Research has shown that wetlands can provide additional pollutant removal from secondary wastewater prior to its entering a stream or lake without significant impact on the wetland itself. Thus many wetlands can, without harm, accept higher loadings of nutrients and SS from municipal wastewater than can many streams or lakes. This, of course, assumes that contaminated materials are not contained in the SS at levels that would be harmful to fish and wildlife or would pose threats to human health. Water quality standards for wetlands need to reflect these aspects.

In a number of cases, advanced polishing by natural wetlands has been recognized in state-issued NPDES permits. These permits call for the wastewater entering the wetland to meet secondary treatment limits and water quality standards for the wetland, while water flowing out of the wetland into an adjacent stream must meet more stringent water quality standards for the adjacent stream.

In some cases, permits contain seasonal variations in discharge requirements. For example, a higher level of nutrient removal may be required during the summer than during the winter. Natural wetlands may be well suited to meeting such seasonal requirements because their peak pollutant removal and transformation capacity occurs during the summer growing season.

The State of Florida, in 1986, established complex standards for the use of wetlands for treatment. Florida's wetland standards include design criteria and regulation at three levels: effluent limits; standards to be met within the treatment wetland; and standards for discharge from the wetland to downstream water bodies. The Florida standards contain traditional physical and chemical parameters as well new "wetland biological quality" standards. Thus the standards recognize and allow wetland treatment capacity to be used while, at the same time, protecting the unique values and functions of wetlands and the water quality standards of the receiving waters.

However, Florida is the only state with such standards. In most cases, the NPDES permitting authority must review the use of wetlands systems to achieve downstream water quality standards on a case-by-case basis. This situation has led to different findings in similar situations, where questions have been raised concerning the use of natural wetlands for providing advanced treatment of secondary effluents.

Section 404 Permit Requirement

Generally proposals to use natural wetlands for wastewater treatment involve some alteration of the wetland, such as building dikes. It is generally necessary to obtain a permit for the discharge of dredged or fill material from the Army Corps of Engineers (or appropriate state agency) under Section 404 of the Clean Water Act before such construction will be allowed.

Regulations established by EPA under Section 404(b)(1) are primarily intended to protect existing wetland values. The corps or state review for the Section 404 permit will require determination that the effects of dredged or fill material on the wetland do not constitute significant degradation; that there are no practicable, environmentally preferable alternatives; that unavoidable effects have been minimized; and that unavoidable effects are mitigated through practicable compensatory actions. An additional determination by the corps also requires that the proposed wetland alteration is in the public interest. The proposed modification will also require a review under the National Environmental Policy Act (NEPA), consideration of other applicable federal laws and executive orders (such as the Endangered Species Act), and any applicable state laws governing wetland filling or other alteration.

EPA's APPROACH

Natural Wetlands

EPA regulates wastewater discharges to natural wetlands considered waters of the United States through the Clean Water Act NPDES permit program. Municipal discharges to such wetlands must meet minimum technology-based requirements and conform with all applicable state water quality standards, including the "free froms" and other narrative standards. In those cases where a natural wetland will be used either on a seasonal or year-round basis for effluent polishing, the NPDES permit must contain secondary (or equivalent to secondary) effluent limits, plus any additional limits and/or monitoring requirements necessary to protect the wetlands and adjacent and downstream water quality standards. These water quality–based limits must be implemented in the NPDES permit at the point of discharge into the wetlands (i.e., at end-of-pipe). Limits for adjacent or downstream water bodies based on end-of-pipe water quality may be based on the degree of effluent polishing provided by the wetlands. In the current absence of separate water quality criteria or state water quality standards for wetlands, EPA considers a conservative case-by-case approach, often combined with pilot testing, to be most appropriate for evaluating the use of natural wetlands for municipal wastewater discharge and treatment on a seasonal or year-round basis.

A recent internal EPA task force concluded that natural wetlands should be viewed primarily as protected water bodies, and that, in the absence of water quality criteria for wetlands, it is not possible to broadly identify conditions where they could be safely regarded as part of the "treatment system." Therefore, the agency continues to review requests for treatment systems involving discharges (treated to secondary or equivalent-to-secondary levels) to natural wetlands using a conservative, case-by-case approach. For example, EPA's Atlanta regional office has prepared "Freshwater Wetlands for Wastewater Management Handbook" (EPA 1985), which provides guidance for such case-by-case evaluation.

When such practices are allowed, a comprehensive monitoring system must be in place. At a minimum, the monitoring program should be designed to help avoid harmful accumulations of toxic materials present in trace amounts in the wastewater and to detect changes in the plant and animal communities due to changes in water flow and characteristics caused by the wastewater discharge.

Constructed Wetlands

EPA encourages the use of constructed (artificial) wetland systems through the innovative and alternative technology provisions of its construction grants program. Constructed wetland treatment systems can often be an environmentally acceptable, cost-effective treatment option, particularly for small communities. This technology can also help expand wetland-type habitats, although these systems rarely achieve the biological complexity of natural wetlands, and their ecological values are correspondingly less than natural systems.

Current EPA regulatory and construction grants policies create fewer problems for consideration of constructed wetlands as a treatment option. When constructed wetlands systems are designed, built, and operated for the purpose of wastewater treatment, the constructed wetlands treatment systems are not considered waters of the United States (40 CFR Part 122.2). As a result, some constructed wetland systems are used to treat primary effluents to meet secondary or advanced treatment requirements. Operational controls can be designed to closely regulate flow, application rate, and detention time to meet desired seasonal variations in operation and treatment needs. Plant species can be selected and utilized on various bases, such as nutrient uptake efficiency, ease of culturing or harvesting, value as biomass, and so on.

Many constructed wetlands are designed, built, and operated to provide, in addition to wastewater treatment, functions and values similar to those of natural wetlands. Under certain circumstances such constructed, multiple-purpose wetlands may be considered waters of the United States and, as such, subject to the same protection and restrictions on use as natural wetlands. This determination must be made on a case-by-case basis and may consider factors such as size and degree of isolation of the constructed wetlands. Where such constructed, multiple-use wetlands are found to function as waters of the United States, municipal discharges to such systems must be limited to secondary or equivalent-to-secondary effluents and any more stringent requirements necessary to meet applicable water quality standards.

Construction Grants Eligibility

As with other land treatment systems, land purchases as well as design and construction expenses for constructed wetland treatment systems can be eligible for funding under the construction grants program. Eligibility of natural wetlands for funding, however, remains a more complicated issue due to the case-by-case approach necessary in determining eligibility. Still, under certain limited circumstances natural wetlands may be eligible for grant funding. Such projects must involve the wetlands in meeting more stringent downstream water quality requirements and must be found to be both cost effective and environmentally sound. In addition, the facility must have a current NPDES permit that reflects minimum treatment technology and state water quality requirements into and out of the wetlands. As noted above, these proposals must be reviewed on a case-by-case basis and may require a Section 404 permit.

Future Actions by EPA

EPA is taking actions to help improve the basis for project design and resolve questions regarding the extent to which natural wetlands may be used to help treat municipal wastewater. One of the most important of these is EPA's exploration of water quality criteria for wetlands. The first step is to examine the current scientific data and possible approaches and to develop a research plan to fill in the missing information. A key component to developing appropriate criteria is additional information on the response of wetland ecosystems to various types and rates of discharges. In addition, EPA will continue to encourage and, to the extent possible, participate in monitoring current wetland wastewater treatment sites to determine (1) the fate of toxics and their effect on wildlife using the system and (2) the differences in ecological functions of constructed versus natural wetlands. Currently, use of constructed rather than natural wetlands is generally preferred by EPA when projects for wastewater treatment are proposed.

REFERENCES

1. Odum, E. "The Value of Wetlands: A Hierarchical Approach," in *Wetland Functions and Values: The State of Our Understanding,* P. E. Greeson, J. R. Clark, and J. E. Clark, Eds. (Minneapolis, MN: American Water Works Association, 1978), pp. 16–25.
2. Horwitz, E. L. "Our Nation's Wetlands, An Interagency Task Force Report Coordinated by the President's Council on Environmental Quality," U.S. Government Printing Office, Washington, DC (1978).
3. Kadlec, R. H. "Using Wetlands to Mitigate the Water Quality of Sewage Effluent," in *Proceedings of the Symposium on Monitoring, Modeling, and Mediating Water Quality,* S. J. Nix and P. E. Black, Eds. (Bethesda, MD: American Water Resources Association, 1987), pp. 415–427.
4. Ewel, K. C., and H. T. Odum. *Cypress Swamps* (Gainesville: University Presses of Florida, 1984).
5. Knight, R. L., T. W. McKim, and H. R. Kohl. "Performance of a Natural Wetland Treatment System for Wastewater Management," *J. Water Pollut. Control Fed.* 59:746–754 (1987).
6. Gearheart, R. A., B. A. Finney, S. Wilbur, J. Williams, and D. Hull. "The Use of Wetland Treatment Processes in Water Reuse," in *Future of Reuse, Water Reuse Symposium III, Vol. 2* (Denver, CO: American Water Works Association Research Foundation, 1985), pp. 577–1169.
7. Reddy, K. R., and W. H. Smith, Eds. *Aquatic Plants for Water Treatment and Resource Recovery* (Orlando, FL: Magnolia Publishing Inc., 1987).
8. Allen, G. H., R. A. Gearheart, and M. Higley. "Renewable Natural Resources Values Enhanced by Use of Treated Wastewaters Within Arcata Treatment and

Disposal System," in *Implementing Water Reuse, Water Reuse Symposium IV* (American Water Works Association Research Foundation, Denver, CO, 1987), pp. 1223–1245.

9. Godfrey, P. J., E. R. Kaynor, S. Pelczarski, and J. Benforado, Eds. *Ecological Considerations in Wetlands Treatment of Municipal Wastewaters* (New York: Van Nostrand Reinhold Company, 1985).
10. "The Ecological Impacts of Wastewater on Wetlands, An Annotated Bibliography," EPA Report 905/3-84-002, U.S. EPA and Fish and Wildlife Service, Washington, DC (1984).
11. Mudroch, A., and J. A. Capobianco. "Effects of Treated Effluent on a Natural Marsh," *J. Water Pollut. Control Fed.* 51(9):2243–2256 (1979).
12. "Report on the Use of Wetlands for Municipal Wastewater Treatment and Disposal," EPA Report 430/09-88-005, U.S. EPA Office of Municipal Pollution Control, Washington, DC (1987).
13. Bastian, R. K., and J. Benforado. "Water Quality Functions of Wetlands: Natural and Manmade Systems," in *Proceedings of the International Symposium on Ecology and Management of Wetlands, Vol. 1, Ecology of Wetlands,* D. Hook, Ed. (Beckenham, Kent, U.K.: Croom Helm Ltd., 1988), pp. 87–97.
14. Brinson, M. M., and F. R. Westall. "Application of Wastewater to Wetlands," Report No. 5, Water Research Institute, University of North Carolina, Raleigh, NC (1983).
15. Kadlec, R. H., and J. A. Kadlec. "Wetlands and Water Quality," in *Wetland Functions and Values: The State of Our Understanding* (Bethesda, MD: American Water Resources Association, 1979), pp. 436–456.
16. Richardson, C. J. "Mechanisms Controlling Phosphorus Retention Capability of Freshwater Wetlands," *Science* 228:1424–1427 (1985).
17. Nichols, D. S. "Capacity of Natural Wetlands to Remove Nutrients from Wastewater," *J. Water Pollut. Control Fed.* 55(5):495–505 (1983).
18. Nixon, S. W., and V. Lee. "Wetlands and Water Quality: A Regional Review of Recent Research in the U.S. on the Role of Freshwater and Saltwater Wetlands as Sources, Sinks, and Transformers of N, P and Heavy Metals," prepared by the University of Rhode Island for the U.S. Army Corps of Engineers, WES, Vicksburg, MS (1986 draft).
19. Gersberg, R. M., B. V. Elkins, and C. R. Goldman. "Use of Artificial Wetlands to Remove Nitrogen from Wastewater," *J. Water Pollut. Control Fed.* 56:152–156 (1984).
20. Hammer, D. E., and R. H. Kadlec. "Design Principles for Wetland Treatment Systems," EPA Report 600/2-83-026, U.S. EPA Office of Research and Development, Ada, OK (1983).
21. Reed, S. C., and R. K. Bastian. "Wetlands for Wastewater Treatment: An Engineering Perspective," in *Ecological Considerations in Wetlands Treatment of Municipal Wastewater,* P. J. Godfrey, et al., Eds. (New York: Van Nostrand Reinhold Company, 1985), pp. 444–450.
22. Reed, S. C., E. J. Middlebrooks, and R. W. Crites. *Natural Systems for Waste Management and Treatment* (New York: McGraw-Hill Book Company, 1987).
23. "Constructed Wetlands and Aquatic Plant Systems for Municipal Wastewater Treatment Process Design Manual," EPA Report 625/1-88/022, U.S. EPA Center for Environmental Information, Cincinnati, OH (1988).

24. "Saltwater Wetlands for Wastewater Management Environmental Assessment," EPA Report 904/10-84-128, U.S. EPA Region V, Atlanta, GA (1984).
25. "Freshwater Wetlands for Wastewater Management Environmental Assessment Handbook," EPA Report 904/9-85-135, U.S. EPA Region IV, Atlanta, GA (1985).
26. "The Effects of Wastewater Treatment Facilities on Wetlands in the Midwest," EPA Report 905/3-83-002, U.S. EPA Region V, Chicago, IL (1983).
27. Zedler, J. B., and M. E. Kentula. "Wetlands Research Plan, November 1985," U.S. EPA ERL-Corvallis, OR (1985).

CHAPTER 23

States' Activities, Attitudes and Policies Concerning Constructed Wetlands for Wastewater Treatment

Robert L. Slayden, Jr., and Larry N. Schwartz

INTRODUCTION

This chapter presents a cross section of current activities, attitudes, and policies of individual states concerning constructed wetlands for wastewater treatment. In most cases, states were chosen because of previous contact with the Tennessee Valley Authority (TVA). The information presented, a synopsis of a telephone survey of state regulatory agencies in May 1988 (or a review of their published papers), will be useful to developers, engineers, and other interested states. We will begin with Tennessee's perspective and then review the surveyed states in alphabetical order.

In Tennessee, the number one problem with treatment systems for small flows is unreliability. Operation of activated sludge package plants is extremely difficult and rarely are qualified personnel available for proper operation. A 1986 survey of package-activated sludge sewage treatment plants in Tennessee showed 50% of the plants in obvious noncompliance. Our experience indicates that if actual samples were taken on those visited, the violation total would approach 80%. Therefore, in Tennessee we have searched for more passive, "operator proof" technologies. The recirculating sand filter is one passive system that we have investigated and are now promoting, and Tennessee is examining constructed wetlands within this context. Will it be another alternative that we can promote in place of package-activated sludge plants for small flows?

ALABAMA

The Alabama Department of Environmental Management (ADEM) has considerable experience with natural systems for wastewater treatment and, in its research effort, has built a good foundation of knowledge concerning artificial wetlands. The state has contracted with Auburn University and TVA to do research and a demonstration project. ADEM prepared a report on artificial wetlands and distributed it to noncomplying municipalities to encourage community interest in the approach and has participated in several seminars on artificial wetlands.

ADEM is monitoring the performance of four constructed wetlands. Three full-scale projects are (1) Degussa, Inc.; (2) TVA-Fabius Mine/Washer; (3) Vredenburgh; and (4) Hurtsboro, a pilot-scale project on-line. Alabama favors the surface flow type system, using a combination marsh and hardwood species. It is acknowledged that variations of the marsh meadow and rock-reed filter may have applications in Alabama once there is greater experience with these types. Alabama has preliminary design criteria, applicable to secondary influent, which recommends a loading rate of 20–250 m^3/day/ha (13,000–164,000 gpd/ac) for a marsh system, depending on influent strength and desired effluent quality.[1,2]

The Alabama Department of Public Health regulates two constructed wetlands: Vredenburgh and Snow Hill School. Vredenburgh is a classic example of a poor rural community with either failing septic systems or no facilities and seriously in need of a sewerage system. The wastewater from about 100 families passes through two open cells and then through a rock-reed filter. The Vredenburgh system has operated for one year. The Snow Hill site, under construction, will take care of a failing septic system. The design flow regime consists of septic tank, rock-reed filter, then two cells of surface flow type wetlands. The Alabama Department of Public Health is very interested in the use of constructed artificial wetlands to replace failing septic tank/drain field disposal systems.

Both Alabama agencies have open minds about the subject of artificial wetlands. The engineer proposing to build a system in Alabama is likely to meet with knowledgeable regulatory personnel.[1-3]

CALIFORNIA

The often-referenced Santee, California research project creates the impression the state has many systems in place, but actual experience is somewhat limited. The Santee research project was the only subsurface flow system, but it is no longer on line. California has two operational constructed wetlands (at Gustine and Arcata, both small towns). Gustine became operational in 1987. Because plants at Gustine have not become established as quickly as anticipated and it has not met the suspended solids permit limit, the state has asked for a remedial plan. Arcata is in its second or third year full scale, but it was operated as a pilot for two years prior to that. Both Gustine and Arcata are surface flow systems.

The Department of Water Resources experimented with wetlands in conjunction with the Los Banos Desalinization project. Wetlands were used as pretreatment prior to reverse osmosis of saline irrigation water. The project was piloted at 3785 m^3/day (1 mgd) but was shut down because drain water from the reverse osmosis contained high amounts of selenium.

California has many water hyacinth (*Eichhornia crassipes*) type systems. State officials mentioned problems with rock media in a San Diego prepilot water hyacinth system that incorporates a rock filter after water hyacinth to remove suspended solids from algae production. After the 3.8–6.4 cm (1.5–2.5 in.) rock plugged, 15.2–20.3 cm (6–8 in.) rock was tried but it plugged also. Plugging was caused by a slime growth feeding on high sulfates in the wastewater stream.

Although construction of artificial wetlands is static in California, state officials hope for additional demonstration projects in the future.[4] The engineering community is cautious about wetlands, but the state indicated its major concern was the possibility of mosquito problems. Low maintenance is a major advantage, and constructed wetlands are likely to have most use for wastewater treatment in small communities.

FLORIDA

The Florida Department of Environmental Regulation concentrates on use of natural wetlands rather than created wetlands for wastewater treatment. State legislation in 1984 provided for use of existing wetlands and flow-altered wetlands for final treatment of secondary wastewaters. Regulations are designed to protect, and possibly enhance, wetlands and dictate the degree of treatment for effluent from the wetland.

Constructed wetlands are basically exempted from these criteria. Higher strength waters can be discharged to constructed wetlands than to natural wetlands. Constructed wetland systems have no minimum standards for biota or benthic life, but man-made wetlands must maintain the minimum criteria for surface waters, and the Class III heavy metals standards apply. Secondary treatment, disinfection, and pH controls are required. Where the system could affect groundwater, groundwater standards apply.

KENTUCKY

The Kentucky Division of Water oversees three artificial wetlands: Benton, Hardin, and Pembroke, small municipalities in western Kentucky. All have been built with TVA designs and funded by TVA and other funding sources. The Kentucky Construction Grant branch has issued a field test grant to TVA to do a three-year testing program at Benton and Hardin. None of the data has been published yet.

The Benton wetland is the oldest, becoming operational in fall 1986. It was constructed to free the town from a sewer moratorium, and much of the work was done with city crews and town volunteers. Benton's flow design uses a stabilization pond and three parallel wetlands, two with natural soil media and one with gravel media. The gravel appears to be working the best. Discharge limitations for ammonia are 4 mg/L in summer and 10 mg/L in winter. Benton has had problems establishing vegetation (50% coverage at the time of survey) and has discharged a year-round ammonia value of approximately 9 mg/L. The regulatory agency acknowledges that some give-and-take will be necessary during the establishment of the plants in a natural system. Benton, with TVA's help, has been doing everything possible to get thorough plant coverage during 1988.

Hardin is under construction and will use an old contact stabilization wastewater plant as preliminary treatment followed by two parallel gravel-media cells. Hardin is unique in that storm flows will be diverted into a diked bottomland hardwood area and later recirculated through the treatment system.

Pembroke has two parallel systems of the classic marsh-pond-meadow type. It is designed to treat approximately 378 m^3/day (100,000 gpd) and utilizes a 2-acre marsh as the wetland treatment.

Kentucky officials are confident about the concept of constructed wetlands for wastewater treatment, but admit they have not had time to make judgments. The existing facilities at Hardin and Pembroke were so bad it was easy to take the attitude: "Anything is an improvement over this." Kentucky regulators have reasonable experience with constructed wetlands. They know what the benefits are as well as the pitfalls.[5,6]

MISSISSIPPI

In Mississippi, the stream control authority is the Bureau of Pollution Control. Only one constructed wetland is on-line (at Collins, Mississippi, in operation about one year). Getting the plants started was difficult and caused delays. Three other constructed wetlands are planned. All are subsurface flow, rock-reed type rather than surface flow type. At Piciune, a rock-reed wetlands type corrects a failing septic tank/drain field system for a mobile home park with flows of 189 m^3/day (60,000 gpd).[7]

MISSOURI

The state water pollution regulatory agency in Missouri is the Department of Natural Resources, Division of Environmental Quality. One small artificial wetland system (submerged bed type) was constructed in 1986 to serve a mobile home park. Missouri could not cite any bad experiences with the constructed wetlands but presently regard these systems as experimental and riskier than alternative systems. More acceptable design data are awaited before use in Missouri is endorsed. If an artificial wetland system were approved for a small town, it would receive funding for field sampling. Missouri has the authority to issue interim limits in its permits.[8]

NORTH CAROLINA

North Carolina officials reported that no constructed wetlands exist in their state, but considerable research is underway in North Carolina on natural wetland systems for wastewater treatment.[9] During some preliminary discussions, a group of developers at Highlands, North Carolina expressed interest in a form of constructed wetlands. North Carolina requires references from the literature about operating systems before approving a demonstration project. The state is interested and receptive to new ideas.

PENNSYLVANIA

In Pennsylvania, the Department of Environmental Resources' Division of Permits and Compliance appears to have a broad knowledge of constructed artificial wetlands. There are 10 in the state, with examples of systems for acid mine drainage and small municipalities as well as two private domestic sources. Experience dates back to 1979, when research was done by Dr. Fred Brenner of Grove City on the use of artificial wetlands for treatment of acid mine drainage. Since then, the department has been actively involved with each project, experimenting with new ideas to improve the existing technology.

Pennsylvania started out with the marsh-pond-meadow approach then added an aerated cell before the marsh. It experimented with over-and-under baffles to bring anaerobic water to the surface to improve ammonia removal, but concluded that the pond was more of a liability than an asset and will leave it out of the next system approved. The constructed marsh uses 0.61 meters (2 ft) of sand covered by 15.2 cm (6 in.) of 1.27–5.1 cm (0.5–2 in.) stone and can operate with the media exposed or completely submerged. The importance of good preliminary treatment to avoid plugging is emphasized.

All of the Pennsylvania systems are 113.6 m^3/day (30,000 gpd) or less. One system treating truck stop wastewater uses a septic tank followed by an aerated pond, then followed by the artificial wetland. Most systems must meet a 1 mg/L summer and 3 mg/L winter ammonia limit. The state has design recommendations that are continually updated and modified. Operators are required to hold the equivalent of a package plant license for these small systems.

Wetlands are issued a National Pollutant Discharge Elimination System (NPDES) permit and an experimental permit under the system authorized by state law. If the system does not meet standards within one year, the permittee is required to file a report on plans of action to bring the system into compliance. This approach provides an extended startup period to allow for establishment of plants in artificial wetland systems.[10]

SOUTH DAKOTA

The Department of Natural Resources, Division of Water Resource Management has three existing wetlands (surface flow type) and 20 to 30 in some stage of planning. All have been proposed for upgrading lagoon effluent to meet secondary limitations. Because artificial wetlands are considered alternative treatment (they receive 75% funding through the construction grants program) and appear to be cheaper to build, there is tremendous artificial wetland activity in South Dakota. Since most treatment in South Dakota is with lagoons, officials in that state believe constructed artificial wetlands have a bright future in expanding or upgrading existing facilities.[11]

The state's experience is recent, with no activity before 1985. The wetlands are completely anaerobic during four months of ice cover, and winter storage of 150 days is required of each system. Three systems are total retention systems utilizing evaporation. The state has regulatory flexibility to allow the systems an extended startup period, and there are informal guidelines for design that will be revised soon.

TENNESSEE

Representatives of the Division of Water Pollution Control in Tennessee are enthusiastic about the potential for use of constructed wetlands. They believe wetlands may be used for small flows in place of unreliable mechanical plants.

Tennessee presently has two constructed wetlands. One operating as the polishing stage after an anaerobic fixed-film treatment process has been in place for over a year. It is a private system serving a small condominium development. The second, larger system serving a retirement development is constructed and is being planted. Both systems were constructed on the contour in hilly terrain. A third system approved for a state park will be a polishing stage following an anaerobic fixed-film treatment system. All Tennessee systems are subsurface flow type. The TVA has assisted with the design of each of these.

Although State of Tennessee staff are becoming more familiar with loading rates and other design criteria, much reliance upon the expertise of TVA personnel has been used in the approval process of these three systems. No design criteria have been published, but staff are familiar with current literature. Statewide field personnel are directed to suggest artificial wetlands as an alternative when conventional activated-sludge package plants are turned down because of unreliability. State officials feel that uncertainties about artificial wetlands are outweighed by the relative reliability of passive natural systems as compared to mechanical wastewater treatment systems.

State officials are receptive to municipal-scale wetland projects and have had preliminary discussions with Eagleville, an unsewered community. Consulting engineers are concerned about system reliability in meeting design criteria during startup, when plants are not well established. The state has agreed to issue relaxed ammonia limitations in a three-year permit to accommodate expected startup performance of the plant. In addition, because of the project's demonstration nature, much flexibility will be allowed during an extended startup period.

An artificial wetland system has been approved for correcting a failed septic tank system. This demonstration will treat wastewater with a fixed-film anaerobic package system followed by a constructed wetland. The effluent will discharge into an expanded drainfield that is still likely to outcrop, but water quality should be good at that point. State officials are hopeful this demonstration will provide another alternative to the Wisconsin Mound type system for correcting failed septic fields. Artificial wetland systems will not be approved for new construction on a single-family basis.

TEXAS

The Texas Water Commission reported that one constructed wetland pilot plant was in the engineering stages for Johnson City, Texas. The project was approved to examine alternatives for upgrading an existing municipal wastewater treatment plant.[12]

One Texas consultant designed an artificial wetland as a pretreatment step before a sand filter, with the effluent reused as irrigation water. The design was approved by Texas only after a considerable effort on the engineer's part. The project was never constructed for reasons unrelated to its design.[13]

VIRGINIA

The Virginia State Water Control Board has only one small demonstration project, treating wastewater seepage from acid mine tailings.[14]

CONCLUSION

This survey shows a wide spectrum of activities and attitudes concerning constructed wetlands. Few states have hard-and-fast policies or criteria on this technology. Everyone agrees there is still much to be learned about the optimization of the process. Some states are more willing to approve presently imperfect examples in an effort to further the knowledge on the system. Most contacted states have several examples of wetlands systems, and South Dakota is planning to build 20 to 30. When states are asked if they are promoting constructed wetlands, some states reply that promotion of certain types of systems is not their responsibility—that they only review what is submitted.

We submit that it *is* our responsibility to encourage use of more reliable systems. If state regulators continue to blindly approve whatever is submitted, we do nothing but perpetuate an already deplorable situation. We may not know everything there is to know about constructed wetlands, but we know enough to be sure they are more reliable than most of the alternatives in the 114–568 m^3/day (30,000–150,000 gpd) range. Look for places to try constructed wetlands. If we can get a few operating systems in, the word will spread quickly. The constructed wetland may just prove to be the best wastewater treatment alternative to come along in a long time.

REFERENCES

1. Horn, C. R., Alabama Department of Environmental Management. Personal communication (May 1988).
2. "Natural Treatment Systems for Upgrading Secondary Municipal Wastewater Treatment Facilities," Alabama Department of Environmental Management (December 1987).
3. Robertson, S., Alabama Department of Public Health. Personal communication (May 1988).
4. Martin, C., California Water Resource Control Board. Personal communication (May 1988).
5. Gatewood, B., Kentucky Division of Water. Personal communication (May 1988).
6. Ryker, R., Kentucky Division of Water, Western Section. Personal communication (May 1988).
7. Reed, R., Mississippi Department of Natural Resources. Personal communication (May 1988).
8. Markus, H., Missouri Department of Natural Resources. Personal communication (May 1988).
9. Dodd, R., North Carolina Department of Natural Resources and Community Development. Personal communication (May 1988).
10. Murtha, J., Pennsylvania Department of Environmental Resources. Personal communication (May 1988).
11. Henrickson, P., South Dakota Department of Natural Resources. Personal communication (May 1988).
12. McGinley, A., Texas Water Commission. Personal communication (June 1988).
13. Venhuisen, D. Personal communication (May, June 1988).
14. Sizemore, D., Kentucky Water Control Board. Personal communication (May 1988).

CHAPTER 24

Human Perception of Utilization of Wetlands for Waste Assimilation, or How Do You Make a Silk Purse Out of a Sow's Ear?

Richard C. Smardon

INTRODUCTION

This chapter will (1) introduce human perception of wetlands from a historical perspective; (2) review the literature on how people perceive environmental quality in relation to odor, water quality, and wetland quality; and (3) outline a data gathering framework to assess public perceptions on the role of wetlands in water quality enhancement.

HISTORICAL PERCEPTION OF WETLANDS

A negative attitude toward wetlands can be traced, in many cases, to our European ancestors. The "bog-swamp" mythology developed in Europe, where people thought pixies, heathens, and strange mythical creatures lived in wetlands,[1] was brought across the Atlantic and intensified with more stories of wolves, wild dogs, crop-destroying hordes of crows, and quicksand.[2] Attitudes toward wetlands changed from fear to indifference after the industrial revolution. Wetlands were commonly believed to be wastelands good for nothing except causing disease and producing mosquitoes.[3] In fact, an English scholar has noted that "wetlands have been significant mainly at half-conscious or subconscious levels of culture," and that "the psychology of wetland experiences has been such as to render them rather easily repressible."[4]

During the last few decades, centuries of bog-swamp mythology, ignorance, prejudice, and trouble associated with wetlands have been slightly reversed by the efforts of enlightened environmentalists. So you say, you are going to use wetlands for treating wastewater! Sounds like trouble if you are going to try to be rational with the American cultural psyche.

Table 1. Perceived Characteristics of Eutrophic Lakes in New York

	Large Lakes (%)	Small Lakes (%)
Growth of algae, plants, or scum	55	50
Not very clear, muddy	36	44
Build-up of shoreline growth	25	35
Strange odors	29	37
Dead fish	45	35
Strange colors	18	23
Film of gasoline or oil	8	23
No objections	9	5

Source: Kooyoomjian and Clesceri.[9]

RESEARCH ON PERCEIVED ENVIRONMENTAL QUALITY

A literature review provides some useful information as to how people will react to sensory environmental quality aspects of wetlands loaded with wastewater. We know that human perception of air quality or odor is directly related to seasonal behavior rather than actual pollution levels, e.g., Toronto residents indicated that summer was the most polluted, despite physical evidence to the contrary.[5] We also know that human recreational users of wetlands are liable to be particularly sensitive to levels of sulfur dioxide produced naturally in wetlands and which may be increased by wastewater loadings. Sulfurous odors may be noticeable particularly in the summer months unless management steps, such as water level drawdowns, are taken to modify soil/mulch decomposition processes.

When respondents are asked about their perception of water quality or, rather, water pollution, visual evidence is frequently cited. Green scum and algae and murky dark water are mostly readily mentioned,[6] followed by sewage; chemicals; debris; garbage; dead fish, birds, and other animals; oil; foam; filth; dirty, murky, scummy water; and characteristics such as discoloration, mud, and slime.[7] Outdoor recreation participants, naturally enough, are more apt to be much more critical about water quality—boaters are most tolerant and swimmers least.[8] Probably the most appropriate study in this regard is by Kooyoomjian and Clesceri[9] at the Rensselaer Fresh Water Institute of Lake George, New York, which asked respondents to comment on attributes of water pollution in eutrophic and nutrient-poor lakes (Table 1).

Eutrophic lakes were perceived to have several attributes of a polluted state by large portions of respondents. Similar characteristics were noted in nutrient-poor lakes by far fewer observers. In separate questioning of recreationists and cottage owners, "No objections" was cited by recreationists most frequently and by cottage owners least frequently. Recreationists did complain more than other groups about unclear and muddy water, strange colors, and floating objects. Fishermen complained more than others about films and oils and dead fish on the surface. Cottage owners had more complaints than other groups concerning strange odors, algae, and irritation to eyes or skin caused by water.

Research results indicate that untrained observers have little trouble in recognizing a nutrient-rich situation.[10] It is unknown, however, how perceptions would change with a given change in the degree of eutrophy, because only two conditions (nutrient-rich and nutrient-poor) were defined. It would be challenging to solicit differences in perceived water quality of naturally nutrient-rich wetlands versus wetlands additionally loaded with wastewater effluent. One would hypothesize that it would be much easier to solicit perceived differences in water quality in nutrient-poor wetlands with increasing levels of wastewater loadings. These studies suggest that it is critical to understand the patterns of human use of proposed treated wetlands.

The perceived risk of increasing habitat for disease vectors is also an important issue. In much of the United States, wetland habitats are important for the mosquito host of equine encephalitis. Marshes and swamps are known habitats for other disease-carrying species of mosquitoes and are often sprayed with pesticides as a preventive measure. Individuals will justifiably ask whether wetlands created for wastewater treatment will host disease-carrying vectors.

These public perception issues should be addressed in the early planning stages of any project utilizing wetlands for wastewater treatment. Public perceptions or misperceptions can suddenly arise at inappropriate times for project planners and managers.

PUBLIC USE AND PERCEPTION RESEARCH FRAMEWORK

To address a similar situation, I developed a public use research program for the Des Plaines River Wetlands Demonstration Project in Lake County, Illinois.[11]

The Des Plaines Project entails reconstruction of wetlands on a 182-ha site along a 5-km stretch of the Upper Des Plaines River. The site is in Lake County, near the town of Wadsworth, Illinois, approximately 57 km north of Chicago. The 55,000-ha watershed is 80% agricultural and 20% urban. Nonpoint source contaminants from both urban and agricultural land uses affect the river.

A broadened and naturally braided channel will be created by regrading the streambed to eliminate existing levees and reduce channelization. Experimental wetlands will be constructed on the floodplain adjacent to the newly created river channel. A portion of the river's flow pumped to these wetlands will be subjected to different experimental conditions and allowed to return to the channel. Duration and extent of land and water contact will be controlled by pumps and sluice gates and varied to suit research purposes. Three abandoned quarry lakes will be graded to create littoral zones conducive to growth and reproduction of aquatic vegetation and animals and will be physically connected to the river to create backwater areas. The river will flow through the third quarry lake to create a sediment trap for suspended solids. Native plant

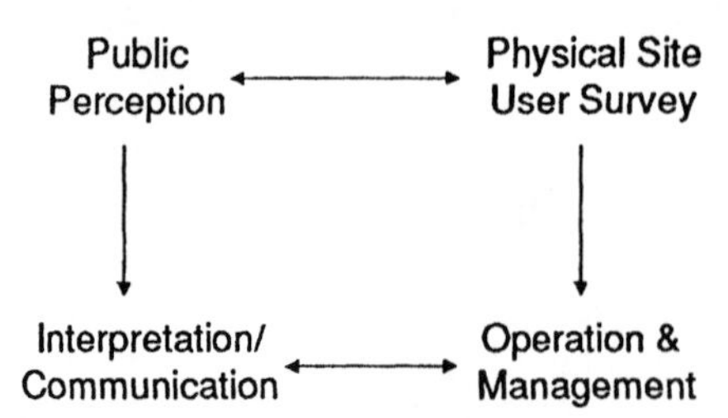

Figure 1. Research components of the Public Use Research Program at the Des Plaines River Wetlands Demonstration Project in Lake County, Illinois.

communities will be reintroduced through seeding and planting to create and manage fish and wildlife habitat for all native species and particular species of certain research programs.

A number of economic, ecological, aesthetic, and recreational benefits are expected to result from the project. Benefits will include water quality improvement through decrease of turbidity, increase of dissolved oxygen, achievement of nutrient balances, and control of algal blooms. Increased flood storage and reduced peak flows at the project site will diminish flood stage and consequent flood damage in downstream residential areas. Valuable fish and wildlife habitats restored through creation of native wetland, prairie, and associated plant communities planted and maintained on-site will support a wide variety of faunal species, including some which are rare or absent in northern Illinois. Aesthetic and recreational opportunities in the restored wetlands will provide benefits to individuals who choose to use the opportunities. Finally, the most important benefit of the project will derive from its demonstration and research activities, providing a role model for wetland restoration. The project will provide new information on how and at what cost natural wetland processes can be used to manage our water resources, which will be applicable to thousands of miles of degraded riverine systems in industrial regions (where wildlife habitat, nonpoint source pollution abatement, and flood storage are scarce and in demand).

Four important components of the Public Use Research Program at the Des Plaines Project or a wetland site utilized for wastewater treatment include (1) public perception; (2) physical site user survey; (3) interpretation/communication; and (4) operation and management. These research components are interlinked as illustrated in Figure 1. The *public perception* research element is a direct input to *interpretation/communication,* and a *user survey* is a direct input to *operation and management.* Each of these elements is presented below as a set of research questions followed by a methodological approach and projected analyses.

Public Perception

The relation between the potential local and regional user population and the project site is addressed by two sets of significant questions. (1) The first set concerns concepts, attitudes, and knowledge of landscape change, wetlands in general, water pollution control, and ecology. (2) The second set addresses energy costs and trade-offs on a system-wide basis.

Concepts, Attitudes, and Knowledge

The first question concerns the concept of landscape change over time. It is difficult for people to perceive how the altered landscape will appear or to anticipate its change over time. While it is easy to conceptualize discrete "before" and "after" conditions, means for understanding how the population at large conceives landscape change need to be developed because landscape change is continuous. For example, Palmer[12,13] is testing reactions of different population groups to visual representations of common forest harvesting techniques and resultant effects: immediately after harvesting, different increments of new growth, and different seasons.

Other questions concern attitudes toward wetlands, perception and knowledge of wetland creation processes, and perception of water pollution control. Insights into local populations' ecological education levels as well as landscape appreciation, tastes, and values will aid in preparation of interpretive programs.

Initial activities include an inventory of visual and sensory stimuli conditions, before construction, with still and video photography and sensory notation methods from travel routes through the site and key viewpoints (overlooks and bridges). This should be done by trained professionals or through user-employed photography (scenic photographs by users[14]). This inventory gives the investigator a baseline for comparison of landscape changes.

Simulations of proposed visible landscape changes can be prepared *a priori,* and people can compare the "before" and "after" conditions. Alternately, actual photos or footage of the landscape "after alteration" can be viewed side-by-side with the "before" landscape imagery.[15,16] Local residents, samples of regional populations, resource professionals, or public policymakers can also evaluate such imagery. Responses of different groups can be compared with statistical techniques to test similarities or significant differences in perception. Retests at different intervals of landscape change may be useful.

Energy Costs and Trade-offs

Are people generally aware of costs of maintaining water quality? The project's success will be based on the ability of the average citizen to perceive the net monetary, environmental, and energy savings due to water quality improvement. A baseline understanding of local and regional ecological knowledge will be essential to successful structuring of interpretive programs.

A general approach includes assessing the local and regional populations' knowledge of basic energy and monetary costs associated with water quality treatment, then acquiring basic energy and monetary cost data from construction and operation of the project. These data can be used to construct a set of interactive simulation models and graphic portrayals to construct subsequent surveys, traveling exhibits, or an interpretive center.

Physical Site User Survey

This element addresses relations between users (recreational and other) and the site and between groups of users. One set deals with the relation of physical site conditions and recreational site usage patterns and populations. In other words, are there physical restraints to certain recreational uses (e.g., river levels too low for extensive canoeing)?

The second set of questions pertains to recreational activity on-site rather than actual usage patterns of the site. It is good to target individual recreationists' opinions of the site, the site's suitability for specific recreational activities, and whether there are existing or potential conflicts between recreational users. This information can be acquired through behavior observation and direct interviews or self-administered questionnaires that include questions about quality of recreational activities and site conditions, management preferences, and user conflicts. Other background information— socioeconomic and educational data and recreational preferences—can be acquired for more in-depth studies. Questionnaire analysis includes user/site, user/manager, and user/user conflicts or opportunities. An example is recent documentation of basic differences in user/management perceptions of high-quality aspects of the recreational experience at state beaches on eastern Lake Ontario and the St. Lawrence River.[17]

Interpretation/Communication

This element addresses transmission of information from site to local user or remote consumer. One line of inquiry involves the effect of interpretive programs on the quality of a visitor's experience at the site and the role of familiarity in the quality of this experience.[18] By exposing some site visitors to interpretive programs or information and others to none, differences in attitude, preferences, or quality of experience can be tested.

Our primary task is detailed field photography and graphic documentation of all steps and techniques used in regrading, wetland creation, vegetative community establishment, and fish and wildlife habitat enhancement. Time-lapse photography and video or movie footage of overall landscape change is also useful. Both types (and other graphic representations) become source materials (interpretive programs for site visitors and external documents and media to communicate technical material to nonlocal interested parties).

Operation and Management

The last public use element involves making decisions on design, operation, and management for a wetland site used for wastewater treatment. A basic question is whether any public use is compatible or whether only part of the site can be zoned for public use. Another question is whether public information and interpretive programs can objectively present the benefits of using wetlands for wastewater treatment and whether this translates to public acceptance. A key concept is whether user and public groups are able to perceive minimal degradation in environmental quality (odor, water quality, visual quality, vegetative patterns) with increased loadings of wastewater effluent or whether there are thresholds of perceived impact. Finally, it would be useful to survey public use of wetland areas used for wastewater treatment to document whether problems or conflicts occur and explore problem/conflict resolution through modification of operations.

SUMMARY AND CONCLUSIONS

Can we make a silk purse out of a sow's ear? The problems are formidable. Historic values and perceptions of wetlands are fraught with negative associations and images. Recent emphasis on ecological values and multifunctional aspects have "cleaned up" the image, but loading wetlands with wastewater risks resensitizing all the historical negative imagery. We know too little of human perception of environmental quality parameters of wetlands, especially odor, water quality, and health risk. We do know that perceived environmental quality hinges directly on type of wetland use, which outlines our framework for determining and enlisting public support for wetlands wastewater treatment projects. In summary, we need to:

1. understand public perception of a. specific areas to be used for wetlands wastewater treatment b. using wetlands for wastewater treatment
2. collect physical site user data and determine whether wetland site(s) restrict or enhance usage
3. design interpretation/communication packages explaining the process of wastewater treatment with wetlands, including testing and retesting to assess changes in attitudes or perception
4. incorporate information gained in design, operation, and management of wetlands wastewater treatment projects

This strategy has been successful so far for the Des Plaines River Wetlands Demonstration Project and is likely to enhance public acceptance and support for other wetland restoration programs as well as constructed wetlands used for wastewater treatment.

REFERENCES

1. Jorgensen, N. *A Guide to New England's Landscape* (Barre, MA: Barre Publishers, 1971).
2. Williams, E. "History of the Hockomock," in *Hockomock: Wonder Wetland,* E. Williams, et al., Eds. (Boston, MA: Massachusetts Audubon Society and Department of Natural Resources, 1971).
3. Smardon, R. C. "Assessing Visual-Cultural Values of Inland Wetlands in Massachusetts," in *Landscape Assessment: Value, Perceptions and Resources,* E. H. Zube, R. O. Brush, and J. G. Fabos, Eds. (Stroudsburg, PA: Dowden, Hutchinson and Ross, 1975), pp. 289–318.
4. Fritzell, P. A. "American Wetlands as Cultural Symbols: Place of Wetlands in American Culture," in *Wetland Functions and Values: The State of Our Understanding,* P. E. Greeson, J. R. Clark, and J. E. Clark, Eds. (Minneapolis, MN: American Water Resources Association, 1978), pp. 523–534.
5. Barker, M. L. "Planning for Environmental Indices: Observer Appraisals of Air Quality," in *Perceiving Environmental Quality: Research and Applications,* K. H. Craik and E. H. Zube, Eds. (New York: Plenum Press, 1976).
6. David, E. L. "Public Perceptions of Water Quality," *Water Resour. Res.* 7:453–457 (1971).
7. Willeke, G. E. "Effects of Water Pollution in San Francisco Bay," Report EEP-29, Program in Engineering-Economic Planning, Stanford University, Stanford, CA (1968).
8. Ditton, R. B., and T. L. Goodale. "Water Quality Perception and the Recreational Users of Green Bay, Lake Michigan," *Water Res.* 9:569–579 (1973).
9. Kooyoomjian, K. J., and N. L. Clesceri. "Perception of Water Quality by Select Respondent Groups in Inland Water-Based Recreation Environments," *Water Res. Bull.* 10(4) (1974).
10. Coughlin, R. E. "The Perception and Valuation of Water Quality: A Review of Research Methods and Findings," in *Perceiving Environmental Quality: Research and Applications,* K. H. Craik and E. H. Zube, Eds. (New York: Plenum Press, 1976).
11. Smardon, R. C. "Chapter 15, Public Use," in *The Des Plaines River Wetlands Demonstration Project, Vol. IV, Research Plan,* R. D. Smith and J. H. Sather, Eds. (Chicago, IL: Wetlands Research Inc., 1986).
12. Palmer, J. F. "Assessing Seasonal Aesthetic Impacts from Northern Hardwood Harvesting Practices," proposal accepted for funding by the McIntire-Stennis Cooperative Forestry Research Program, ESF/SUNY, Syracuse, NY (1983).
13. Palmer, J. F. "First Annual Report: Influence of Season and Time Since Harvest on Forest Aesthetics," CEF/SUNY, Syracuse, NY (1984).
14. Cherem, J. C., and D. E. Traweek. "Visitor Employed Photography: A Tool for Interpretation Planning on River Environments," in *Proceedings River Recreation Management Research Symposium,* USDA, Forest Service Gen. Tech. Rep. NC-28, North Central Forest Experiment Station, St. Paul, MN (1977), pp. 236–244.
15. Smardon, R. C., and M. Hunter. "Procedures and Methods for Wetland and Coastal Area Visual Impact Assessment (VIA)," in *Future of Wetlands: Assessing Visual-Cultural Values,* R. C. Smardon, Ed. (Totawa, NJ: Allenheld, Osmun and Company, 1983), pp. 171–206.
16. Smardon, R. C. "Visual-Cultural Assessment and Wetland Evaluation," in *Ecology*

and Management of Wetlands, Vol. 2. Management Use and Value of Wetlands," D. D. Hook, et al., Eds. (Beckenham, Kent, U.K.: Croom Helm Ltd., 1988), pp. 103–114.

17. Buerger, R. B. "The Perceptual Differences of Beach Users and Management Staff Towards the Recreation Attributes of the Beach," unpublished PhD dissertation, College of Environmental Science and Forestry, SUNY, Syracuse (1983).
18. Hammitt, W. E. "Assessing Visual Preferences and Familiarity for a Bog Environment," in *Future of Wetlands: Assessing Visual-Cultural Values,* R. C. Smardon, Ed. (Totawa, NJ: Allenheld, Osmun and Company, 1983), pp. 81–99.

CHAPTER 25

Preliminary Considerations Regarding Constructed Wetlands for Wastewater Treatment

R. Kelman Wieder, George Tchobanoglous, and Ronald W. Tuttle

INTRODUCTION

The unique set of hydrological, physical, biological, and chemical characteristics of wetland ecosystems may be exploited when using constructed wetlands for treating water whose quality has been degraded from anthropogenic activities. Wetland systems have been constructed to treat municipal wastewaters,[1-4] coal and metal mine drainages,[5-8] and wastewater produced from agriculture and textile and photography industries (cited in Watson et al.[9]). Use of constructed wetlands for municipal wastewater treatment has the longest history of any application and more is known about it than any other. Within the past six years, however, the number of wetlands constructed for acid coal mine drainage treatment has increased considerably, and an information base is slowly developing. Limited information on constructed wetlands for agricultural and industrial wastewater treatment precludes a general assessment of such applications at this time.

This chapter provides an overview of preliminary design factors that are important in considering constructed wetland treatment of municipal wastewaters and of coal mine drainage. In addition, the importance of maximizing aesthetics without compromising treatment effectiveness is discussed as a key component of preliminary design.

WETLANDS FOR MUNICIPAL WASTEWATER TREATMENT

This section focuses on wetland systems for secondary and/or advanced treatment of municipal wastewaters and is limited to two types of wetland systems: (1) free water surface systems with emergent plants and (2) subsurface flow systems with emergent plants.

Free water surface wetlands with emergent plants consist of basins or channels with a subsurface barrier of clay or impervious geomembrane to prevent seepage; soil or other suitable medium to support emergent vegetation; and water flowing through at a shallow depth. Cattails (*Typha* spp.), bulrushes (*Scirpus* spp.), and various sedges (*Carex* spp.) are emergent plants used commonly in free water surface wetlands.[10] The shallow water depth, low flow velocity, and presence of plant stalks and litter regulate plant flow and, especially in long narrow channels, ensure plug-flow conditions.[3]

In subsurface flow systems, unlined or lined trenches or basins are filled with a permeable packing medium, usually gravel, and are planted with emergent plants, including cattails, bulrushes, native sedges, and other local species.[3,11] The root-zone method (RZM) and rock-reed-filter (RRF) are the two principal categories of subsurface flow systems. To grow plants, a soil medium is used in the RZM system, and rock or sand is used in the RRF system. In both systems, the flow of wastewater is maintained approximately 15–30 cm below the bed surface.

Characteristics of the wastewater to be treated, as well as desired and/or mandated discharge limits, need to be taken into consideration in the early stages of designing a wetland treatment system. Key characteristics of the wastewater include carbonaceous biochemical oxygen demand ($CBOD_5$), suspended solids, nitrogen compounds, phosphorus compounds, heavy metals, refractory organics, and pathogenic bacteria and/or viruses. Although $CBOD_5$ and suspended solids are used commonly to characterize wastewater, these are nonspecific (lumped) parameters. In most cases, the nature of the constituents of the $CBOD_5$ and suspended solids is poorly known. Improved characterization of wastewater organic matter will improve understanding of $CBOD_5$ and suspended solids and support more effective wetland designs.[12] Currently, however, quality performance standards continue to be based on $CBOD_5$ and suspended solids. Well-designed wetland treatment systems can typically produce effluent water with $CBOD_5$ values of 10 mg/L and suspended solids concentrations below 10 mg/L.[13]

Sulfate concentration is a potentially important but often overlooked characteristic of wastewater. When sulfate concentrations in the wastewater are greater than 50 mg/L, sulfide can be produced by sulfate-reducing bacteria using sulfate as an electron acceptor in the anaerobic oxidation of organic matter. The sulfide produced in this manner may form hydrogen sulfide, which along with other sulfur-containing volatile organic compounds such as mercaptans can cause odor problems at constructed wetland treatment systems.

Mosquito problems may occur when wetland treatment systems are overloaded organically and anaerobic conditions develop.[14] Biological control agents such as mosquitofish *(Gambusia affinis)* die either from oxygen starvation or hydrogen sulfide toxicity, allowing mosquito larvae to mature into adults. Toxicity to fish can also result from wastewater constituents such as refractory organics and heavy metals. If toxic wastewater is suspected, bioassay tests should be performed to assess the viability of planned mosquito control agents. In some cases, a pretreatment program to eliminate or substantially reduce toxicity in the wastewater may be appropriate.[15]

For wetland systems constructed to treat municipal wastewaters, other process design considerations include operating water depths, process loading rates, process kinetics, temperature effects, and physical configuration. Because most constructed wetlands will be used for secondary or advanced wastewater treatment, some form of pretreatment is often required. Community septic tanks, recirculating sand filters, Imhoff tanks, ponds, and disk screens have been used for pretreatment. Regardless of the pretreatment scheme, effective screening of the wastewater is essential to prevent unsightly conditions from developing in the wetlands.

The area of wetland required to achieve effective treatment will depend on factors such as the operating depths, process loading rate, process kinetics, and temperature effects. Reasonably accurate approximation of the area of wetland that will be required for a particular wastewater is necessary not only for site selection but also for preliminary calculation of cost estimates.

Physical design features of wetland treatment systems are site-specific. Nevertheless, three general features should be incorporated into the design of any wetland treatment system: (1) the wetland should be divided into segments (i.e., plug-flow channels) that can be operated and drained separately; (2) provision should be made for step-feeding the influent wastewater; and (3) provision should be made for effluent recycling—a wrap-around design will minimize costs.

Two operational considerations associated with wetlands for wastewater treatment are mosquito control and plant harvesting. The objective of mosquito control is to maintain the mosquito population below threshold levels for disease transmission or nuisance. Strategies used to control mosquito populations include effective pretreatment to reduce total organic loading; step-feeding of the influent wastewater stream with effective influent distribution and effluent recycle; vegetation management; natural controls, principally mosquitofish, in conjunction with the above techniques; and application of man-made control agents. In general, natural controls are preferred. Although man-made control agents are relatively inexpensive, most local health agencies will not approve continued use because of a concern that resistant strains of mosquito might develop.

The usefulness of plant harvesting in wetland treatment systems depends on several factors, including climate, plant species, and the specific wastewater objectives. Harvesting plants to remove wastewater contaminants taken up by the plants is inefficient. However, plant harvesting can affect treatment performance of wetlands by altering the effect that plants have on the aquatic environment. Further, because harvesting reduces congestion at the water surface, control of mosquito larvae using fish is enhanced. Where a segmented wetland system is used, drying each segment separately allows harvesting with conventional equipment. Depending on location, burning the dried plant mass in place may be most economical.

Cost is often a significant factor in selecting the type of treatment system for a particular wastewater situation. Unfortunately, the availability of reliable cost data for wetland treatment systems is limited. The cost of wetland treatment systems varies depending on wastewater characteristics, the type of wetland system, and the type of bottom preparation required. Subsurface flow systems are generally more expensive than free water surface systems. Nonetheless, construction, operation, and maintenance costs suggest that constructed wetland systems are economically competitive with other wastewater treatment options.

WETLANDS FOR MINE DRAINAGE TREATMENT

Following reports of field sites where the chemistry of acid mine drainage improved upon passage through naturally occurring wetlands,[16,17] interest in constructed wetland systems as a potentially low-cost, low-maintenance alternative to chemical treatment of acid mine drainage has greatly increased. However, considerable site-to-site variability in apparent water quality improvement exists among wetlands constructed for mine drainage treatment. While some constructed wetlands appear to have been at least partially effective in treating mine drainage,[6,18-20] other man-made wetlands have been ineffective.[21,22] A U.S. Bureau of Mines (BOM) survey of mining companies, consulting firms, and federal and state agencies in Maryland, Ohio, Pennsylvania, and West Virginia also revealed both apparently effective and ineffective constructed wetlands constructed for mine drainage treatment.[5] Limited information has made it difficult to confidently predict whether a constructed wetland will effectively treat a given flow and chemical composition of mine drainage.

Nonetheless, two key topics when contemplating a constructed wetland treatment system are the area of wetland needed for a particular flow and chemistry of mine drainage, and the cost of wetland treatment relative to more conventional chemical treatment. Both topics must be evaluated within the regulatory framework requiring that discharges meet effluent water quality criteria.

A frequently used approach to sizing a constructed wetland for mine treatment was proposed in a short course at the 1986 Symposium on Surface Mining, Hydrology, Sedimentology, and Reclamation.[23] The proposed rule of thumb was that 200 ft^2 of wetland is needed for each gal/min of flow (approximately 4.9 m^2 of wetland area for each L/min of flow). Because this rule of thumb was derived empirically from the BOM survey information,[5] it will be referred to as the BOM approach. The BOM approach was derived from observations of wetlands that included a layer of organic substrate (*Sphagnum* peat, mushroom compost, hay, manure, etc.), often in conjunction with added lime or limestone. Two cautionary notes apply to this guideline. (1) It was intended for flows of 5–10 gal/min (19–38 L/min) with pH values of 4.0 or above and Fe and Mn concentrations less than 50 mg/L and 20 mg/L, respectively. (2) In synthesizing the survey data, those constructed wetland sites at which outflow metal concentrations were equal to or greater than inflow metal concentrations were excluded from the data base.[5]

The Tennessee Valley Authority (TVA) recently provided recommendations based on both the chemistry and the flow of the water to be treated.[6] Under these recommendations, to achieve an Fe concentration of 3 mg/L at the wetland discharge, if the inflow drainage has a pH <5.5, 2 m^2 of wetland is needed for each mg/min of Fe entering the wetland, whereas if the influent pH is >5.5, 0.75 m^2 of wetland is needed for each mg/min of Fe entering the wetland. Similarly, to achieve an outflow Mn concentration of 2 mg/L, if the inflow drainage has a pH <5.5, 7 m^2 of wetland is needed for each mg/min of Mn entering the wetland, whereas if influent pH is >5.5, 2 m^2 of wetland is needed for each mg/min of Mn entering the wetland. These guidelines were empirically derived from observations made of wetlands constructed by the TVA for acid drainage treatment. The TVA wetlands were constructed with in situ substrate, which typically had a relatively low organic matter content.

Data for flow and water quality for 16 wetlands constructed for mine drainage treatment are summarized in Table 1, along with the actual area of each wetland and the recommended wetland areas based on the BOM and TVA approaches. Wetland areas recommended using the TVA approach are between two and 74 times greater than the areas recommended using the BOM approach. In three sites where the constructed wetland failed to substantially improve the mine drainage water, the wetlands were sized more closely with the BOM recommendations than the TVA recommendations. One wetland, however, was only 10% larger than the BOM recommended area and was effective in lowering Fe concentrations.[18] The BOM approach and the recent TVA guidelines represent the only published quantitative design criteria for sizing constructed wetlands for acid drainage treatment.

Ultimately, the decision to adopt wetland treatment of mine drainage will depend on cost considerations. According to the BOM survey,[5] construction costs prior to 1986 averaged less than \$2.96/m^2. The TVA's wetland construction costs ranged from \$3.58/m^2 to \$32.03/m^2.[6] Both studies indicate that equipment and labor costs are major components. Subsequent maintenance costs are minimal. Thus, using either the BOM or the TVA sizing recommendations and construction cost estimates, for any given flow and chemistry of mine drainage it is possible to estimate the wetland treatment system cost and compare it with conventional chemical treatment system costs. Although this might be a fairly straightforward calculation, two cautionary notes must be emphasized. (1) A considerable expenditure of effort and funds could produce a wetland that does not adequately treat the mine drainage, necessitating additional chemical treatment. For example, the two wetlands described by Hiel and Kerins[22] cost \$27,810 and \$15,595, respectively, yet neither wetland was effective in improving mine drainage quality. (2) Data on long-term effectiveness of man-made wetlands for treating mine drainage simply do not exist. A constructed wetland has a finite capacity to retain metals, yet we have only crude estimates of that finite capacity.[24] There is no guarantee that a wetland

Table 1. Summary of Published Studies on Constructed Wetlands for Mine Drainage Treatment

Flow	Seep Quality			Wetland Area (m^2)				Is the Wetland Effective	
					Recommended by:				
L/min	pH	Fe mg/L	Mn mg/L	Actual	BOM	TVA(Fe)	TVA(Mn)	(Fe)	Reference
50	5.5	29.9	8.8	270	245	2990	3080	No	Wieder et al.[21]
38	5.5	40	50	204	186	3040	3800	Yes	Stillings et al.[18]
43	2.7	284	1.5	420	211	24424	451	No	Hiel and Kerins[22]
38	3.1	148	1.2	108	186	11248	319	No	
455	6.0	180	4.5	3000	2230	61425	4095	Yes	Stark et al.[19]
80	2.9	181	36	2597	392	28960	20160	Yes	Hedin et al.[20]
170	5.6	150	6.8	4800	833	19125	2312	Yes	Brodie et al.[6]
42	4.9	135	24.0	2000	206	11340	7056	Yes	
348	6.0	11.0	9.0	2500	1705	2871	6264	Yes	
238	5.7	45.2	13.4	7300	1166	8068	6378	Yes	
400	3.1	40.0	13.0	11000	1960	32000	36400	Yes	
87	6.3	13.0	5.0	1200	426	848	870	Yes	
492	5.6	17.9	6.9	25000	2411	6605	6790	Yes	
83	5.7	12.0	8.0	3400	407	747	1328	Yes	
53	6.3	30.0	9.1	5700	260	1192	965	Yes	
288	5.7	0.7	5.3	9200	1411	151	3053	Yes	

Note: Based on average flow and chemistry of the mine drainage at each site, estimated area requirements are given using BOM recommendations and TVA recommendations. When data in a published paper were given as ranges, the midpoint of the range is reported below. Wetland effectiveness refers to Fe only. If the constructed wetland did not result in a substantial decrease in Fe concentration in outflow water as compared to Fe concentration in inflow water, the wetland was considered to be ineffective.

effectively treating a given source of mine drainage today will be as effective in perpetuity. The issue of what happens when a constructed wetland reaches its metal retention capacity needs to be given serious consideration. Just as treatment ponds require periodic sludge removal and disposal, a wetland at its metal retention capacity may require sludge removal and disposal, incurring additional cost.

QUALITY DESIGN

Regardless of application (municipal wastewaters, mine drainage, or other), constructed wetlands have potential to enhance any landscape by introducing a water element. Unless heavily polluted, water is one of the most magnetizing and compelling of all design elements. For both utilitarian and aesthetic reasons, wildlife and people are attracted to water. Therefore, if properly designed, installed, and managed, constructed wetlands can introduce an important functional and aesthetic element into the landscape. In this context, preliminary design considerations for man-made wetland treatment systems must strive to maximize the aesthetic element as much as possible without compromising treatment effectiveness.

Aesthetic considerations include the wetland's visibility from adjacent viewpoints and compatibility with surrounding landscape and land use patterns. Man-made wetland systems should be located and designed to cause minimum disturbance and to take full advantage of the natural site features. Specific design objectives can be achieved with different options for manipulating landforms, vegetation, water, and structures at a constructed wetland site. For example, earth grading and shaping techniques can blend newly created landforms into the existing landscape. Types of vegetation planted in and around a constructed wetland can provide erosion control, screening, space definition, climate control, traffic control, and wildlife habitat. Local native plant species and varieties are good candidates for new plantings because they are adapted to site conditions and blend in with undisturbed surroundings. Instead of looking like industrial septic systems, constructed wetlands should enhance the quality of the landscape by introducing an attractive water feature.

Responsible design must also anticipate commonly voiced concerns about constructed wetlands for wastewater treatment. For example: What impact might a constructed wetland have on the productive use of the land? Will the wetland impede efficient use of the land for other purposes? Since constructed wetlands often function as a sink for contaminants, will the high concentrations of contaminants in the wetland preempt future use? The potential for surface and groundwater contamination, insect problems, and odor problems are other potential concerns that must be addressed in the design of constructed wetland treatment systems to optimize functional and aesthetic qualities.

Much remains to be learned about constructed wetlands for wastewater treatment. Constructed wetlands will not solve all of the wastewater problems facing society today. Nonetheless, with perseverance, creativity, and innovation, the bounds of this emerging technology will become increasingly well-defined. Progress will develop most rapidly through an interdisciplinary approach involving designers, engineers, and scientists.

REFERENCES

1. Kadlec, R. H., and J. A. Tilton. "The Use of Wetlands as a Tertiary Treatment Procedure," *CRC Crit. Rev. Environ. Control* 9:185–212 (1979).
2. Godfrey, P. J., E. R. Kaynor, S. Pelczarski, and J. Benforado, Eds. *Ecological Considerations in Wetlands Treatment of Municipal Wastewaters* (New York: Van Nostrand Reinhold Company, 1985).
3. Reed, S. C., E. J. Middlebrooks, and R. W. Crites. *Natural Systems for Waste Management and Treatment* (New York: McGraw-Hill Book Company, 1987).
4. Reddy, K. R., and W. H. Smith, Eds. *Aquatic Plants for Water Treatment and Resource Recovery* (Orlando, FL: Magnolia Publishing Inc., 1987).
5. Girts, M. A., and R. L. P. Kleinmann. "Constructed Wetlands for Treatment of Acid Mine Drainage: A Preliminary Review," in *National Symposium on Surface Mining, Hydrology, Sedimentology, and Reclamation* (Lexington: University of Kentucky Press, 1986), pp. 165–171.
6. Brodie, G. A., D. A. Hammer, and D. A. Tomljanovich. "Constructed Wetlands for Acid Drainage Control in the Tennessee Valley," in *Mine Drainage and Surface Mine Reclamation,* Bureau of Mines Information Circular 9183 (1988), pp. 325–331.
7. Lapakko, K., and P. Eger. "Trace Metal Removal from Stockpile Drainage by Peat," in *Mine Drainage and Surface Mine Reclamation,* Bureau of Mines Information Circular 9183 (1988), pp. 291–300.
8. Eger, P., and K. Lapakko. "Nickel and Copper Removal from Mine Drainage by a Natural Wetland," in *Mine Drainage and Surface Mine Reclamation,* Bureau of Mines Information Circular 9183 (1988), pp. 301–309.
9. Watson, J. A., S. C. Reed, R. H. Kadlec, R. L. Knight, and A. E. Whitehouse. "Performance Expectations and Loading Rates for Constructed Wetlands," Chapter 27, this volume.
10. Gearheart, R., S. Wilbur, J. Williams, D. Hull, N. Hoelper, S. Sundberg, S. Salinger, D. Hendrix, C. Holm, L. Dillon, G. Moritz, P. Greichaber, N. Lerner, and B. Finney. "Final Report, City of Arcata Marsh Pilot Project, Vol. 1, Effluent Quality Results—System Design and Management, Project No. C-06-2270," (April 1983).
11. Gearsberg, R. M., B. V. Elkins, and C. R. Goldman. "Nitrogen Removal in Artificial Wetlands," *Water Res.* 17:1009–1014 (1983).
12. Levine, A.D., G. Tchobanoglous, and T. Asano. "Characterization of the Size Distribution of Contaminants in Wastewater: Treatment and Reuse Implications," *J. Water Pollut. Control Fed.* 57:805–816 (1985).
13. Tchobanoglous, G. "Aquatic Plant Systems for Wastewater Treatment: Engineering Considerations," in *Aquatic Plants for Water Treatment and Resource Recovery,* K. R. Reddy and W. H. Smith, Eds. (Orlando, FL: Magnolia Publishing Inc., 1987).

14. Stowell, R., S. Weber, G. Tchobanoglous, B. A. Wilson, and K. R. Townzen. "Mosquito Considerations in the Design of Wetland Systems for the Treatment of Wastewater," in *Ecological Considerations in Wetlands Treatment of Municipal Wastewaters*, P. J. Godfrey, E. R. Kaynor, S. Pelczarski, and J. Benforado, Eds. (New York: Van Nostrand Reinhold Company, 1985).
15. Tchobanoglous, G., F. Maitski, K. Thompson, and T. H. Chadwick. "Evolution and Performance of City of San Diego Pilot Scale Aquatic Wastewater Treatment System Using Water Hyacinths," paper presented at the 60th Annual Conference of the Water Pollution Control Federation, Philadelphia, PA, October 1987.
16. Huntsman, B. E., J. G. Solch, and M. D. Porter. "Utilization of *Sphagnum* Species Dominated Bog for Coal Acid Mine Drainage Abatement," 91st Ann. Mtg. Geol. Soc. Amer., Toronto, Ontario, Canada (1978), p. 322.
17. Wieder, R. K., and G. E. Lang. "Modification of Acid Mine Drainage in a Freshwater Wetland," in *Proceedings of the Symposium on Wetlands of the Unglaciated Appalachian Region* (Morgantown: West Virginia University Press, 1982), pp. 43–53.
18. Stillings, L. L., J. J. Gryta, and T. A. Ronning. "Iron and Manganese Removal in a *Typha*-dominated Wetland During Ten Months Following Its Construction," in *Mine Drainage and Surface Mine Reclamation,* Bureau of Mines Information Circular 9183 (1988), pp. 317–324.
19. Stark, L. R., R. L. Kolbash, H. J. Webster, S. E. Stevens, Jr., K. A. Dionis, and E. R. Murphy. "The Simco #4 Wetland: Biological Patterns and Performance of a Wetland Receiving Mine Drainage," in *Mine Drainage and Surface Mine Reclamation,* Bureau of Mines Information Circular 9183 (1988), pp. 332–344.
20. Hedin, R. S., D. M. Hyman, and R. W. Hammack. "Implications of Sulfate-Reduction and Pyrite Formation Processes for Water Quality in a Constructed Wetland: Preliminary Observation," in *Mine Drainage and Surface Mine Reclamation,* Bureau of Mines Information Circular 9183 (1988), pp. 382–388.
21. Wieder, R. K., G. E. Lang, and A. E. Whitehouse. "Metal Removal in *Sphagnum*-dominated Wetlands: Experience with a Man-Made Wetland System," in *Wetlands and Water Management on Mined Lands* (University Park: Pennsylvania State University, 1985), pp. 353–364.
22. Hiel, M. T., and F. J. Kerins. "The Tracy Wetlands: A Case Study of Two Passive Mine Drainage Treatment Systems in Montana," in *Mine Drainage and Surface Mine Reclamation,* Bureau of Mines Information Circular 9183 (1988), pp. 352–358.
23. Kleinmann, R. L. P., R. P. Brooks, B. E. Huntsman, and B. Pesavento. "Constructed Wetlands for the Treatment of Mine Water—Course Notes," short course at the 1986 Symposium on Surface Mining, Hydrology, Sedimentology, and Reclamation, Lexington, KY.
24. Wieder, R. K. "Determining the Capacity for Metal Retention in Man-Made Wetlands Constructed for Treatment of Coal Mine Drainage," in *Mine Drainage and Surface Mine Reclamation,* Bureau of Mines Information Circular 9183 (1988), pp. 375–381.

CHAPTER 26

Selection and Evaluation of Sites for Constructed Wastewater Treatment Wetlands

Gregory A. Brodie

INTRODUCTION

Constructed wetlands are practical alternatives to conventional treatment of domestic and municipal sewage, industrial and agricultural wastes, storm water runoff, and acid mine drainage.[1–4] Siting a constructed wetland is often dictated by the location of the wastewater source, e.g., a public sewage treatment works or an acidic seep at a coal mine. The wastewater source can seldom be relocated; therefore, siting a wetland system is usually limited to the immediate area, which is often a mediocre or poor site. Nevertheless, siting can be optimized through a comprehensive site investigation process, including site selection, temporary and permanent engineered works design, environmental effects analyses, construction evaluation, remedial works design and construction, and operational and safety checks.

Site selection is based on geological, geotechnical, hydrological, and other environmental information that could affect construction, performance, and effects of a wetlands treatment system. Site selection is constrained by the availability of a suitable site and geotechnical merits, e.g., well-developed soils, good access, or low flood potential. This chapter describes ideal site selection considerations and investigative techniques and presents a methodology for selecting and evaluating sites for constructed wetlands.

SITING CONSIDERATIONS

Once a constructed wetland is proposed as an alternative for wastewater treatment, numerous siting considerations must be evaluated to ensure optimum design, construction, and operation of the facility. Depending on the project magnitude, e.g., a 900-m^2 swine waste marsh system vs a 30,000-m^2 municipal sewage treatment wetland, the degree of site evaluation will differ. Ideally, site investigation and selection includes the following:

1. preliminary office fact-finding survey
2. aerial photography interpretation
3. initial field/aerial survey
4. limited subsurface exploration, site soils classification, and collection and evaluation of environmental data
5. detailed subsurface exploration and collection of environmental data, as necessary
6. evaluation of data, potential environmental effects, and regulatory requirements

Unfortunately, due to costs and schedule delays for ideal site evaluations, the site screening process is often limited to identifying a wastewater source and designing a treatment system. At a minimum, the site screening process should (1) clearly define the wastewater management objectives and the regulatory considerations, (2) collect sufficient data to develop preliminary design of a wetland system, (3) investigate the environmental and social conditions and sensitivities to predict any adverse effects and provide mitigation, and (4) obtain legal access to the site.

Site Selection

Site selection considerations can be classified into four categories of equal importance: land use/general considerations, hydrology, geology, and environmental/regulatory considerations. These categories can be evaluated by several methodologies discussed in this chapter.

Land Use and General Considerations

Probably the most important considerations in siting a wetland are land use and access. Foremost, the wastewater to be treated must be accessible to the site. In the case of a municipal sewage treatment wetland, accessibility may be only constrained by the economics of transporting sewage to the site. For an urban runoff or mine drainage wetland treatment system, wastewater access to numerous sites may be limited if more natural, self-sustaining gravity-flow systems are desired. The site must be accessible to construction equipment, operational personnel, and chemical delivery vehicles. If public access cannot be controlled, adequate safeguards must be incorporated into the wetland design to prevent personal injury or vandalism. Also, availability of utilities, e.g., water and electricity, should be evaluated for access rights-of-way and costs.

Land availability is nearly always an issue. Immediate and surrounding landowners must be consulted so the wetland can be protected and operated for its intended life span and long-term access can be ensured. Some states may require surface control of the site by an operator prior to wetland construction. Generally, surface rights should be obtained through fee purchases, land trades, or long-term leases and easements. If land control and resulting legal costs are extensive, another site may have to be selected. Required land area is dictated by the desired wetland size, flow control structures, associated buildings or equipment storage areas, access roads and utility rights-of-way, construction material borrow areas, buffer zones, and potentially a chlorine contact basin, a flow equalization basin, or a final polishing/chemical treatment cell.

Use and values of the site and adjacent land should be evaluated with respect to public opposition to potential odors, mosquitoes, lowered property and business values, water quality degradation, aesthetic impacts, or other environmental impacts. Public involvement techniques employed to improve acceptance include local workshops, public hearings and meetings, newsletters, site visits, mass media, surveys, and personal meetings.[5] Prior to wetlands proposals and site selection, target audiences should be identified, including area residents and elected officials, media representatives, environmental and public interest groups, community and industry leaders and local service organizations, and appropriate local employees.

Site history, interpreted from sequential aerial photography,[6] maps, local interviews, and other sources of information, can often give clues to siting criteria for a wetland. Clues might include drainage modifications; historic flooding and potential for erosion/sedimentation; patterns and trends of land use; cultural resources; buried, abandoned, or reclaimed facilities such as coal mines, landfills, waste disposal sites, or road beds; and property ownership (old fence lines, cultivation, etc.).

Hydrology

Hydrologic considerations include characterizing surface and groundwater flow patterns, use, quantity, and chemistry. Hydrology, if neglected or insufficiently characterized during the siting process, can significantly impair the operation and effects of a treatment wetland.

Drainage basin characteristics for candidate sites should be evaluated. Unless the wetland is to receive only a regulated quantity of wastewater and direct precipitation, such as a lined municipal sewage treatment wetland, surface runoff can profoundly affect the constructed wetland. Flooding and scouring potential and the need for levees can be evaluated from flood hazard maps, USGS topographic maps, aerial photography interpretation, and site surveys. Minimum, maximum, and average seasonal water levels influencing the wetland hydroperiod should be determined from site data or various methodologies.[7] Wetlands should be sited away from major streams or springs because of potential flooding, scouring, sedimentation, erosion, high groundwater tables, saturated ground, and variable water quantity and quality.

The wastewater and receiving surface water bodies should be chemically characterized and evaluated for existing and potential downstream use. Downstream users should be identified, and sites that would adversely affect downstream users should be avoided unless adequate mitigation can be arranged.

Following is a list of recommended minimum water quality analyses that should be performed for baseline and wastewater characterization.

- *Acid mine or ash disposal drainage*—pH, TSS, DO, Fe, Mn, SO_4
- *Municipal and domestic sewage*—pH, TSS, DO, BOD_5, NH_3-N, NO_2 + NO_3-N, TKN, orthophosphate, alkalinity, fecal coliform
- *Urban storm water*—pH, TSS, DO, BOD_5, NH_3-N, NO_2 + NO_3-N, fecal coliforms, oil and grease, TDS, TOC, total phosphate, Pb, Fe, chlorides
- *Landfill leachate*—pH, TSS, DO, BOD_5, TKN, TDS, total phosphate, Fe, Mn, Zn, VSS, COD, heavy metals, PCBs
- *Pulp and paper mill waste*—pH, TSS, DO, TDS, Fe, Mn, TOC, chlorides, alkalinity, hardness, color, turbidity, SiO_2, free CO_2, Cl, Hg, heavy metals
- *Agricultural wastes*—pH, TSS, DO, TDS, NH_3-N, COD, BOD_5, VSS, alkalinity, fecal coliform, NO_2 + NO_3-N

In addition to these parameters, a survey of aquatic invertebrates can document stream recovery, stability, or degradation.[8] For National Pollutant Discharge Elimination System permitting, flow and relevant priority pollutants may also be required.

Any additional minor seepage or drainage that might enter the main waste stream should also be characterized. Construction operations and water ponding may alter the site hydrology and induce additional seepage downstream of the constructed wetland. Some wetland construction, e.g., acid mine drainage system substrates and dikes, if constructed of inferior material, may allow infiltration or flow-through seepage. Therefore, siting should include evaluating the sensitivity of the hydrologic environment to alterations. Sufficient area should be included to locate the permitted monitoring point downstream of the final wetland cell.

Groundwater hydrology characterization should include general flow patterns, depth, quality, high or perched water tables, existing and potential use, and nearby wells. Depending on the wastewater type, wetlands might successfully be located in either groundwater recharge or discharge areas. Sites overlying karst geology, perched water tables, arid streambeds, fissured igneous or metamorphic rock, or known groundwater recharge areas should be avoided if the wastewater has potential for recharging groundwater with deleterious contaminants. Hydrologic siting constraints for wetlands may be eased or overcome using low-permeability or impermeable liners or adequate compaction techniques in certain soils. Some wetlands systems or regulatory requirements mandate lined cells to preclude groundwater contamination. Ideally, a constructed wetland should be sited in a groundwater discharge area to minimize groundwater pollution potential. Additional advantages include base level recharge of the wetland during periods of low or no wastewater flow and dilution water to minimize ecologic stress on wetland biota or to enhance chemical, biological, and biochemical treatment mechanisms (Figure 1).

Existing and potential use of groundwater at candidate sites should be evalu-

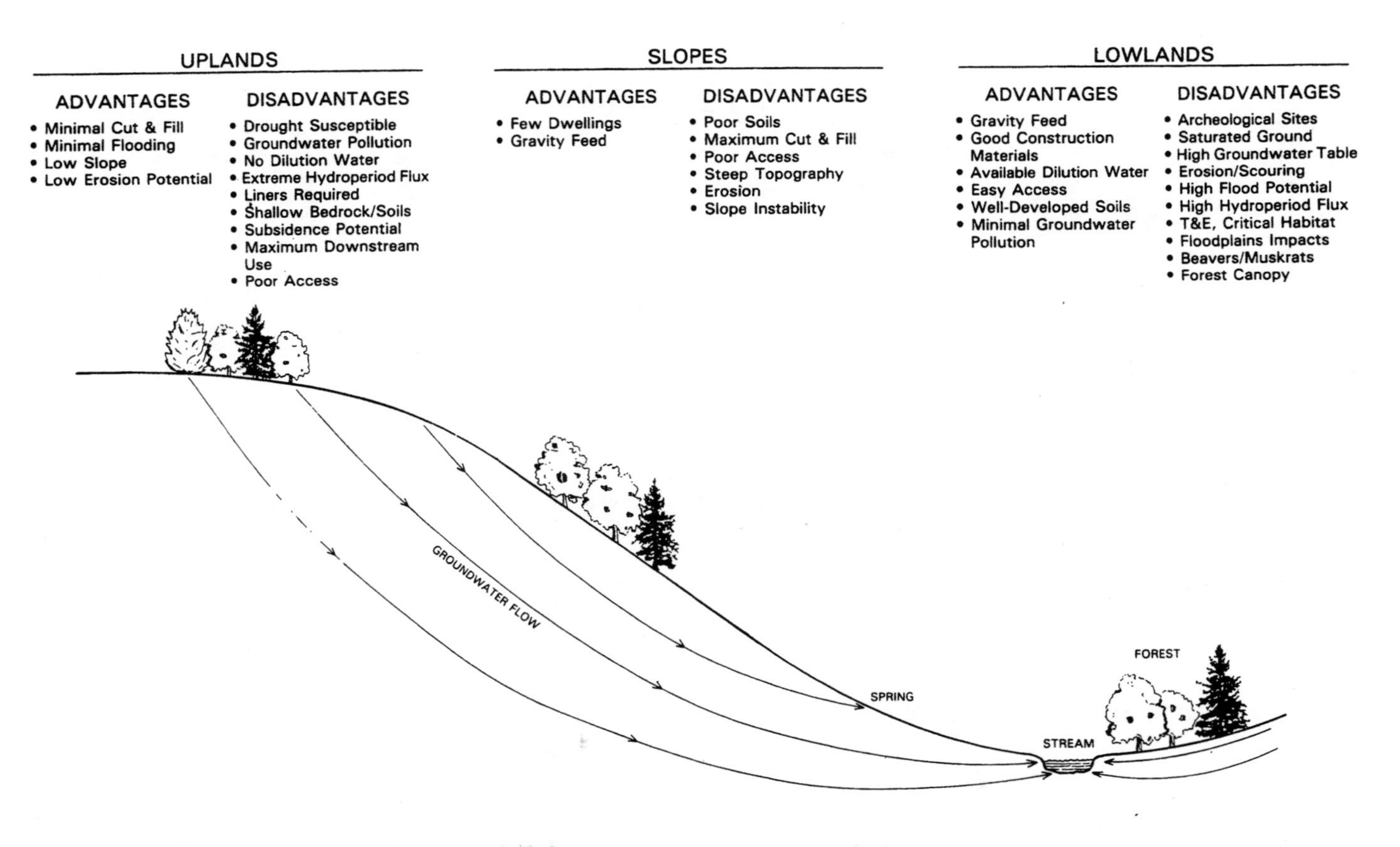

Figure 1. Advantages and disadvantages of siting constructed wastewater treatment wetlands on uplands, slopes, and lowlands.

ated. Nearby potentially affected springs and water wells should be sampled and analyzed to completely characterize baseline water quality (see p. 310). Sites that could adversely affect users should be avoided unless adequate mitigation can be ensured.

Geology

Geologic considerations include characterizing surface materials and soils, bedrock depth, topography, available construction materials, and other geological and geotechnical aspects that could affect wetlands construction and performance. Neglecting geologic considerations in the siting phase could result in increased construction and maintenance costs and/or poor performance of the wetland.

Soils and surface materials should be characterized at candidate sites for thickness and depth, classification and composition, use as construction materials, drainage characteristics, erosion potential, and variability. Investigations should include available information, e.g., soil surveys, geologic and topographic maps, aerial photography, and site investigations such as augering, test pits, percolation tests, and soils and geologic mapping.[9] Thin, poorly developed soils make poor wetland substrates, and importing of adequate substrate materials could be required. Soil composition could have important effects on wetlands performance. For example, soils with greater extractable aluminum have greater potential for phosphorus removal than do organic soils, making them better suited for a sewage treatment wetland; highly organic soils might be better suited for an acid drainage treatment wetland to enhance sulfate reduction and ionic adsorption; and soils with higher microbial activity have greater potential for nitrogen transformation.[10] Site soils and rock should be evaluated for their use as construction materials for earthdams and cores (low plasticity clay-silt loam), spillways (nonerodible soils and rock), riprap and aggregate, embankments, filters (well-graded aggregate), and pond liners (clay, silt). Such evaluations might include volume estimates, soil and rock sample analyses, permeability and compaction tests, gradation analysis, and erodibility.

Bedrock depth often eliminates a site from consideration for constructed wetlands. Shallow bedrock sites require either blasting or ripping and disposal of large quantities of rock or importing large amounts of soil. Bedrock depth should be investigated at the wetland site and at potential construction material borrow areas using existing maps and aerial photography, field surveys, or hand augers and test pits.

Topography affects cut and fill requirements, drainage and erosion potential, access, and slope stability. Ideally, sites should be flat to gently sloping, but such sites are rare. Steep slopes require maximum earth-moving activities but are amenable to terraced wetlands if a level site is not available.

In mined areas or in areas of karst geology, land subsidence or sinkholes may influence wetland siting. Soil surveys, geologic and geologic hazards maps, and aerial photography should be evaluated for potential land subsidence, mine shafts, sinkholes, faults, or other features that could affect engineered works or cause groundwater contamination from a constructed wetland.

Aerial photography interpretation[6] can often identify site lineations (e.g., buried pipelines, trenches, roads, archeological features, abandoned roadways, joint patterns, or channels) that might affect the design and construction of a treatment wetland.

Environmental and Regulatory Considerations

Environmental and regulatory considerations vary from state to state for various wastewaters to be treated and for the entity proposing to construct a wetland treatment system (e.g., private or federal). Nevertheless, general concerns should be evaluated prior to siting a wetland to minimize delays or cessations due to unanticipated environmental issues. Federal laws that should be considered in the siting phase are:

Clean Air Act
Clean Water Act
Endangered Species Act
Executive Order 11988—Floodplain Management
Executive Order 11990—Protection of Wetlands
Fish and Wildlife Coordination Act
National Environmental Policy Act
National Historic Preservation Act
Surface Mining Control and Reclamation Act

Local, regional, and state laws and requirements should be reviewed for application to siting or constructing a wetland.

Archeological and cultural resources of potential wetland sites should be generally evaluated.[11,12] Existing state and federal laws and regulations are becoming stronger, and some states require detailed surveys prior to any construction even if federal funding is not involved. For example, in Tennessee and other states, a human burial must not be disturbed until a professional archeologist has reviewed the site. Drainages are often suitable or necessary for wetlands construction sites but are also areas of potential past human habitation and burial. The prudent engineer should contact the state historic preservation officer to identify the potential for cultural sites in the area and local attitudes toward significant sites.

Critical habitat and the potential or existence of threatened or endangered species or other important wildlife should be investigated with respect to impacts on wildlife from the wetland and vice-versa. Drainages or existing wetlands at potential construction sites often harbor important species, and reviews should include contacting the appropriate U.S. Fish and Wildlife Service office or performing limited surveys. The presence of certain species, such as beavers, muskrats, and geese, may inhibit constructed wetland performance, and possible control of these species should be considered.

Low, poorly drained areas may have desirable site attributes but could entail Section 404 (dredge and fill) permitting. Early coordination with the U.S. Corp of Engineers is recommended.

Site Investigation

Generally, site selection for constructed wetlands does not entail detailed environmental resources data collection. To improve the quality of site selection and evaluation, the following should be considered:

1. Use available information from USGS topographic and geologic quadrangle maps; aerial photography; county and regional maps; aeronautical charts; previous or current permitting data; soil surveys; water, oil, or gas well drilling records; stream water surveys; hydrologic surveys; and interviews with knowledgeable people.
2. Use aerial photography at the earliest stage of the siting process. High- and low-altitude color and black-and-white, infrared, and oblique photos are perhaps the most valuable data available for initial site screening.
3. Site evaluations and field work should be directed by experienced engineers conversant with project goals and technology.
4. Site selection, investigation, and project design and construction should all incorporate flexibility, e.g., alternative sites or additional available area, to compensate for site conditions as necessary.

Site selection and evaluation should be tailored to the degree of complexity and magnitude of the project. Limited site evaluation followed by a design based on an excessive safety factor is generally wasteful and imprudent. Conversely, conducting a detailed site investigation can be very costly and is unnecessary for siting a small wetland.

A modification to Peck's observational method[13] for geotechnical investigations is appropriate to site selection and evaluation for constructed wetlands in many cases:

1. Conduct sufficient investigation and site exploration to establish the general nature, pattern, and properties of site conditions.
2. Assess the most probable conditions and most unfavorable conceivable deviations from observations based on geological, hydrological, and environmental knowledge and experience.
3. Select the site and design the wetland based on the most probable conditions expected.
4. Modify the design before, during, and after construction to suit actual site conditions.

Regardless of methodology employed in site selection, a "walkover" survey is necessary for assessment of candidate sites. This survey should include site inspection and local inquiries. The site visit should investigate any observa-

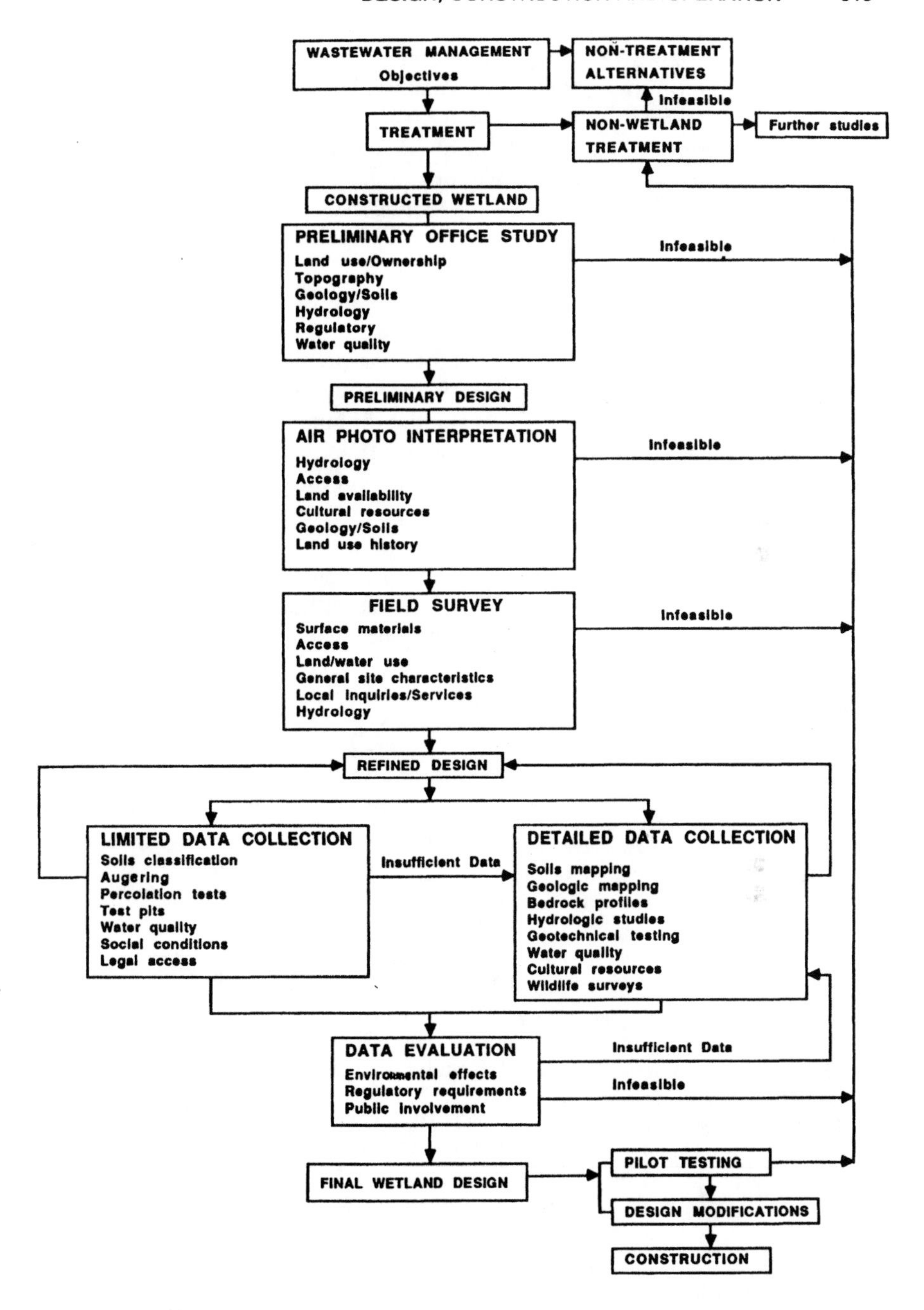

Figure 2. A generalized methodology for screening sites for constructed wetlands.

tions made from initial surveys, e.g., maps and aerial photography interpretation, presence of cultural resources, etc. Topography, surface materials, rock exposures, groundwater conditions, erosion, and access should all be confirmed. Local inquiries might include local builders and contractors, surveyors, utilities, archives and libraries, local inhabitants, clubs, societies, schools, colleges, universities, and government authorities.

SUMMARY

Proper selection and evaluation of sites for constructed wetlands for wastewater treatment are essential to optimal design, construction, and long-term operation. Site selection must include evaluation of land use, hydrology, geology, and environmental, regulatory, geotechnical, and general characteristics of candidate sites. The degree of data collection and evaluation necessary for site selection will vary with size of wetland, type of wastewater, location, and other factors.

Figure 2 illustrates a generalized methodology for site selection and evaluation to accompany detailed site-specific studies and investigations using techniques and considerations discussed herein. The methodologies and discussions are neither complete nor applicable to all cases but summarize a systematic process for screening sites for wetland construction.

REFERENCES

1. Brodie, G. A., D. A. Hammer, and D. A. Tomljanovich. "Constructed Wetlands for Acid Drainage Control in the Tennessee Valley," in *Mine Drainage and Surface Mine Reclamation,* Bureau of Mines Information Circular 9183 (1988), pp. 325–331.
2. G. R. Steiner, J. T. Watson, and D. A. Hammer. "Municipal Wastewater Treatment by Constructed Wetlands: A TVA Demonstration in Western Kentucky," in *Proceedings Wetlands—Increasing Our Resources*, J. Zelazny and J. S. Feierabend, Eds. (Washington, DC: National Wildlife Federation, 1988), p. 363.
3. Athanas, C. "Wetlands Creation for Stormwater Treatment," in *Proceedings Wetlands—Increasing Our Resources,* J. Zelazny and J. S. Feierabend, Eds. (Washington, DC: National Wildlife Federation, 1988).
4. Hammer, D. A., and J. T. Watson. "Agricultural Waste Treatment with Constructed Wetlands," in *Proceedings of National Symposium on Protection of Wetlands from Agricultural Impacts,* U.S. Fish and Wildlife Service Biol. Rep. 88(16) (1988).
5. Canter, L. W. *Environmental Impact Assessment* (New York: McGraw-Hill Book Company, 1977).
6. Avery, T. E., and G. L. Berlin. *Interpretation of Aerial Photographs* (Minneapolis, MN: Burgess Publishing Company, 1985).
7. Lyle, E. S., Jr. *Surface Mine Reclamation Manual* (New York: Elsevier Science Publishing Co., Inc., 1987).

8. Hart, C. W., Jr., and S. L. H. Fuller. *Pollution Ecology of Freshwater Invertebrates* (New York: Academic Press, Inc., 1974).
9. Acker, W. L., III. *Basic Procedures for Soil Sampling and Core Drilling* (Scranton, PA: Acker Drill Co., Inc., 1974).
10. "Freshwater Wetlands for Wastewater Management Handbook," EPA 904/9-85-135, U.S. EPA (1985).
11. *Code of Federal Regulations,* Title 36, Part 800, Advisory Council on Historic Preservation.
12. 48 *Fed. Reg.* 44716-44742, September 29, 1983, Secretary of the Interior Standards and Guidelines for Archeological and Historic Preservation.
13. Peck, R. B. "Advantages and Limitations of the Observational Method in Applied Soil Mechanics, 9th Rankine Lecture," *Geotechnique* 19(2):171-187 (1969).

CHAPTER 27

Performance Expectations and Loading Rates for Constructed Wetlands

James T. Watson, Sherwood C. Reed, Robert H. Kadlec, Robert L. Knight, and Alfred E. Whitehouse

INTRODUCTION

Constructed wetlands technology is emerging as a low-cost, easily operated efficient alternative to conventional treatment systems for a wide variety of wastewaters. The most common uses are for treatment of municipal wastewaters and acid mine drainage. Other applications include textile waste,[1] photo lab waste,[2] pulp mill effluent,[3] refinery effluent,[4] swine farrowing and feeding waste,[5] poultry rendering wastes,[6,7] landfill leachate,[8-10] and urban runoff.[11-13] Although use for a variety of wastewaters is increasing, information on design, operation, and performance is sketchy and confusing. Information is adequate for system design to reduce targeted pollutants but inadequate to optimize design and operation for consistent compliance.

This chapter summarizes information on performance expectations and loading rates and presents an overview of parameters normally regulated, loading factors for existing systems, and information on reaction kinetics. Information is segregated for the two types of systems: (1) conventional surface flow systems with an exposed free water surface and (2) the more recent subsurface flow systems with water levels below the surface of a permeable substrate.

There is a bias toward municipal wastewaters because more information is available, interest for this use is extensive, and kinetic research results are limited for other applications. The reaction kinetics presented are generic (based on first-order reactions for plug-flow, steady-state conditions) and adaptable to treatment of other wastewaters. Hydraulic concepts are applicable to all wetlands. There is a clear need for research to identify and differentiate key removal mechanisms for different wastes and flow conditions and refine numerical values.

A constructed wetland designed for surface flow consists of basins or channels, soil or another suitable medium to support emergent vegetation, and relatively shallow water flowing through the unit. If seepage needs to be prevented, a liner is incorporated into the design. Channels are typically long and narrow, ensuring approximate plug-flow conditions. Some North American municipal systems and almost all mine drainage systems are designed for surface flow.

A subsurface flow system consists of a trench or bed containing a medium that supports growth of emergent vegetation. Media used have included crushed stone, gravel, and different soils, either alone or in combination. Most beds are underlain by impermeable material to prevent seepage and assure water level control. Wastewater flows laterally and is purified during contact with media surfaces and the vegetation root zones. The subsurface zone is saturated and generally anaerobic, but excess oxygen conveyed through the plant root system supports aerobic microsites adjacent to roots and rhizomes.

Constructed wetlands with subsurface flow are widely used in Denmark, West Germany, Austria, Switzerland, and other European countries to treat screened raw sewage or primary effluent. They have been described as root-zone method, reed bed, hydrobotanical system, soil filter trench, biological-macrophytic, and marsh beds by various proponents. These systems typically use soil to support the vegetation. In North America, gravel- or rock-filled trenches or beds are used to treat primary or secondary effluents. These are described in terms of media used.

PERFORMANCE EXPECTATIONS

Wetland systems reduce many contaminants, including biochemical oxygen demand (BOD), suspended solids (SS), nitrogen, phosphorus, trace metals, trace organics, and pathogens. This reduction is accomplished by diverse treatment mechanisms: sedimentation, filtration, chemical precipitation and adsorption, microbial interactions, and uptake by vegetation. These are summarized in Table 1 and discussed in detail for major constituents.

Performance of small systems is summarized in Table 2 for municipal wastewaters and Tables 3 and 4 for acid drainage. Data for systems treating other wastewaters consist only of a few individual examples.

BOD Removal

Settleable organics are rapidly removed in wetland systems by quiescent conditions, deposition, and filtration. Attached and suspended microbial growth is responsible for removal of soluble BOD.[26] In surface flow wetlands, the major oxygen source for these reactions is reaeration at the water surface. Wind-induced water mixing and algal production will be reduced by a dense stand of vegetation or ice cover. In northern climates, development of significant ice cover is likely. If ice persists for more than a few days, reaeration is prevented, depletion of oxygen may occur, and assimilation of BOD is

Table 1. Containment Removal Mechanisms in Aquatic Systems Employing Plants and Animals

Mechanism	Contaminant Affected[a]	Description
Physical		
Sedimentation	P - Settleable solids S - Colloidal solids I - BOD, nitrogen, phosphorus, heavy metals, refractory organics, bacteria and virus	Gravity settling solids (and constituent contaminants) in pond/marsh settings.
Filtration	S - Settleable solids, colloidal solids	Particulates filtered mechanically as water passes through substrate, root masses, or fish.
Adsorption	S - Colloidal solids	Interparticle attractive force (van der Waals force).
Chemical		
Precipitation	P - Phosphorus, heavy metals	Formation of or coprecipitation with insoluble compounds.
Adsorption	P - Phosphorus, heavy metals S - Refractory organics	Adsorption on substrate and plant surface.
Decomposition	P - Refractory organics	Decomposition or alteration of less stable compounds by phenomena such as UV irradiation, oxidation, and reduction.
Biological		
Microbial metabolism[b]	P - Colloidal solids, BOD, nitrogen, refractory organics, heavy metals	Removal of colloidal solids and soluble organics by suspended, benthic, and plant-supported bacteria. Bacterial nitrification/denitrification. Microbially mediated oxidation of metals.
Plant metabolism[b]	S - Refractory organics, bacteria, and virus	Uptake and metabolism of organics by plants. Root excretions may be toxic to organisms of enteric origin.
Plant absorption	S - Nitrogen, phosphorus, heavy metals, refractory organics	Under proper conditions, significant quantities of these contaminants will be taken up by plants.
Natural dieoff	P - Bacteria and virus	Natural decay or organisms in an unfavorable environment.

Source: Stowell et al.[14]

[a]P = primary effect; S = secondary effect; I = incidental effect (effect occurring incidental to removal of another contaminant).

[b]Metabolism includes both biosynthesis and catabolic reactions.

Table 2. Performance and Hydraulic Loading Rates of Constructed Wetland Systems for Small Municipal Systems in North America and Europe

North American Systems	Ref.	Period of Record	System Type	Hydraulic Loading Rate (cm/day)	Biochemical Oxygen Demand			Suspended Solids		
					Influent Conc. (mg/L)	Effluent Conc. (mg/L)	Removal Efficacy (%)	Influent Conc. (mg/L)	Effluent Conc. (mg/L)	Removal Efficacy (%)
Listowel, Ontario	15	9/80–8/84	Surface							
			System 3	1.40	19.6	7.6	61	22.8	9.2	60
			System 4	1.40	56.3	9.6	83	111.1	8	93
Arcata, California	16	9/80–8/82	Surface							
			Cells 1–4	22.3	26	12.8	51	37	5.4	85
			Cells 5–8	11.2	26	13.3	49	37	5.7	85
			Cells 9–12	5.61	26	10.7	59	37	6.1	84
Brookhaven NL, New York	17	8/75–8/76	Surface Marsh/Pond	3.35	170	19	89	353	43	88
Santee, California	18, 19	8/83–12/84[d]	Subsurface Gravel							
			Bulrush	4.68	118.3	5.3	96	58.1	3.7	94
			Reed	4.68	118.3	22.3	81	58.1	7.9	86
			Cattail	4.68	118.3	30.4	74	58.1	5.5	91
			Control	4.68	118.3	36.4	69	58.1	5.6	90
Village of Neshaminy Falls, Pennsylvania	20	9/79–7/82	Subsurface Sand Marsh/Pond/Meadow[e]	1.26	187	8	96	193	12	94
Iselin, Pennsylvania	21	3/83–9/85	Subsurface Sand/Gravel							
			Marsh/Pond/Meadow	1.47	140	7.4	95	380	19	95
			Marsh	5.28	140	17	88	380	53	86
			Meadow	10.57	20	7.4	64	61	19	69
Benton, Kentucky	22	3/88–11/88	Surface/Subsurface							
			Surface Cattail	4.15	23	10	57	60	15	75
			Surface Woolgrass	4.27	23	11	52	60	20	67
			Subsurface Bulrush	7.97	23	8	65	60	7	88

Table 2. (contd.)

European Systems	Ref.	Period of Record	System Type	Hydraulic Loading Rate (cm/day)	Biochemical Oxygen Demand Influent Conc. (mg/L)	Biochemical Oxygen Demand Effluent Conc. (mg/L)	Biochemical Oxygen Demand Removal Efficacy (%)	Suspended Solids Influent Conc. (mg/L)	Suspended Solids Effluent Conc. (mg/L)	Suspended Solids Removal Efficacy (%)
Gravesend, England	23	4/86–1/88	Subsurface Gravel							
			Bed 1	8.16	237	84	65	131	64	51
			Bed 2	8.16	237	90	62	131	58	56
			Bed 3	8.16	237	63	73	131	38	71
Marnhull, England	23	5/87–1/88	Subsurface Soil							
			Bed 1	4.46	87	13	85	74	23	69
			Bed 2	6.90	87	17	80	74	20	73
Holtby, England	23	7/86–1/88	Subsurface Soil	4.90	223	49	79	163	24	85
Castleroe, England	23	4/87–1/88	Subsurface Gravel							
			Cell 1	4.34	157	53	66	73	30	59
			Cell 2	4.34	157	70	55	73	29	60
			Subsurface Soil							
			Cell 1	4.34	157	37	76	73	33	55
			Cell 2	4.34	157	60	62	73	37	49
Middleton, England	23	6/87–1/88	Subsurface Sand/Gravel	8.89	11.0	3.0	73	30	8	73
Bluther Burn, England	23	5/87–3/88	Subsurface							
			Fine Fly Ash	10.76	207	49	76	167	26	84
			Coarse Fly Ash	6.24	207	43	79	167	17	90
			Unclassified Gravel	9.93	207	30	86	167	26	84
			Gravel	10.09	207	51	75	167	17	90
Little Stretton, England	23	7/87–12/87	Subsurface Gravel	26.0	140.9	32.9	77	127	20	84
Ringsted, Denmark	24	9/84–10/84	Subsurface							
			Gravel	5.70	189	11	94	243	6	98
			Clay	1.71	189	15	92	243	23	91

Table 2. (contd.)

North American Systems	Ref.	Period of Record	System Type	Hydraulic Loading Rate (cm/day)	Ammonia Nitrogen Influent Conc. (mg/L)	Ammonia Nitrogen Effluent Conc. (mg/L)	Ammonia Nitrogen Removal Efficacy (%)	Total Nitrogen Influent Conc. (mg/L)	Total Nitrogen Effluent Conc. (mg/L)	Total Nitrogen Removal Efficacy (%)
Listowel, Ontario	15	9/80–8/84	Surface							
			System 3	1.40	7.2	3.8	47	12.3	6.3	49
			System 4	1.40	8.6	6.1	29	19.1	8.9	53
Arcata, California	16	9/80–8/82	Surface							
			Cells 1–4	22.3	12.8	9.8	23			
			Cells 5–8	11.2	12.8	11.6	9			
			Cells 9–12	5.61	12.8	9.6	25			
Brookhaven NL, New York	17	8/75–8/76	Surface Marsh/Pond	3.35	8.4	3.5	58	19.7	6.8	65
Santee, California	18, 19	8/83–12/84[d]	Subsurface Gravel							
			Bulrush	4.68	24.7	1.5	94			
			Reed	4.68	24.7	5.4	78			
			Cattail	4.68	24.7	17.7	28			
			Control	4.68	24.7	22.1	11			
Village of Neshaminy Falls, Pennsylvania	20	9/79–7/82	Subsurface Sand Marsh/Pond/Meadow[e]	1.26	26.1	6.4	75			
Iselin, Pennsylvania	21	3/83–9/85	Subsurface Sand/Gravel							
			Marsh/Pond/Meadow	1.47	30	3.3	83			
			Marsh	5.28	30	13	56			
			Meadow	10.57	5.2	3.3	36			
Benton, Kentucky	22	3/88–11/88	Surface/Subsurface							
			Surface Cattail	4.15	6.2	7.6	–23	14.8	10.9	26
			Surface Woolgrass	4.27	6.2	5.8	6	14.8	9.3	37
			Subsurface Bulrush	7.97	6.2	11.2	–81	14.8	13.0	12

Table 2. (contd.)

European Systems	Ref.	Period of Record	System Type	Hydraulic Loading Rate (cm/day)	Ammonia Nitrogen Influent Conc. (mg/L)	Ammonia Nitrogen Effluent Conc. (mg/L)	Ammonia Nitrogen Removal Efficacy (%)	Total Nitrogen Influent Conc. (mg/L)	Total Nitrogen Effluent Conc. (mg/L)	Total Nitrogen Removal Efficacy (%)
Gravesend, England	23	4/86–1/88	Subsurface Gravel							
			Bed 1	8.16	43	37	14			
			Bed 2	8.16	43	36	16			
			Bed 3	8.16	43	36	16			
Marnhull, England	23	5/87–1/88	Subsurface Soil							
			Bed 1	4.46	28.9	26.4	9			
			Bed 2	6.90	28.9	28.2	2			
Holtby, England	23	7/86–1/88	Subsurface Soil	4.90	34.9	32.9	6			
Castleroe, England	23	4/87–1/88	Subsurface Gravel							
			Cell 1	4.34	19.1	14.5	24			
			Cell 2	4.34	19.1	17.3	9			
			Subsurface Soil							
			Cell 1	4.34	19.1	13.9	27			
			Cell 2	4.34	19.1	18.2	5			
Middleton, England	23	6/87–1/88	Subsurface Sand/Gravel	8.89	2.8	1.5	46			
Bluther Burn, England	23	5/87–3/88	Subsurface							
			Fine Fly Ash	10.76	27	24	11			
			Coarse Fly Ash	6.24	27	20	26			
			Unclassified Gravel	9.93	27	20	26			
			Gravel	10.09	27	22	19			
Little Stretton, England	23	7/87–12/87	Subsurface Gravel	26.0	9.8	10.2	–4			
Ringsted, Denmark	24	9/84–10/84	Subsurface							
			Gravel	5.70	34	15	56	48	29.6	38
			Clay	1.71	34	15.3	55	48	18.8	61

Table 2. (contd.)

North American Systems	Ref.	Period of Record	System Type	Hydraulic Loading Rate (cm/day)	Phosphorus[a] Influent Conc. (mg/L)	Phosphorus[a] Effluent Conc. (mg/L)	Phosphorus[a] Removal Efficacy (%)	Coliforms[b] Influent Conc. (No./100 mL)	Coliforms[b] Effluent Conc. (No./100 mL)	Coliforms[b] Removal Efficacy (%)
Listowel, Ontario	15	9/80–8/84	Surface							
			System 3	1.40	1	0.5	50	1,736	53	97
			System 4	1.40	3.2[c]	0.6	81	222,990	121	100
Arcata, California	16	9/80–8/82	Surface							
			Cells 1–4	22.3				3,183	389	88
			Cells 5–8	11.2				3,183	584	82
			Cells 9–12	5.61				3,183	367	88
Brookhaven NL, New York	17	8/75–8/76	Surface Marsh/Pond	3.35	7.2	2.1	71	1,560	50	97
Santee, California	18 and 19	8/83–12/84[d]	Subsurface Gravel							
			Bulrush	4.68				67,500,000	577,000	99
			Reed	4.68						
			Cattail	4.68						
			Control	4.68				67,500,000	289,000	96
Village of Neshaminy Falls, Pennsylvania	20	9/79–7/82	Subsurface Sand Marsh/Pond/Meadow[e]	1.26				1,290,600	5,600	100
Iselin, Pennsylvania	21	3/83–9/85	Subsurface Sand/Gravel							
			Marsh/Pond/Meadow	1.47	13	2.6	80	1,400,000	150	100
			Marsh	5.28	13	4.2	69	1,400,000	3,700	100
			Meadow	10.57	3.4	2.6	23	2,100	150	93
Benton, Kentucky	22	3/88–11/88	Surface/Subsurface							
			Surface Cattail	4.15	6.0	5.3	12	3,940	515	87
			Surface Woolgrass	4.27	6.0	4.9	18	3,940	94	98
			Subsurface Bulrush	7.97	6.0	5.1	15	3,940	157	96

Table 2. (contd.)

European Systems	Ref.	Period of Record	System Type	Hydraulic Loading Rate (cm/day)	Phosphorus[a] Influent Conc. (mg/L)	Phosphorus[a] Effluent Conc. (mg/L)	Phosphorus[a] Removal Efficacy (%)	Coliforms[b] Influent Conc. (No./100 mL)	Coliforms[b] Effluent Conc. (No./100 mL)	Coliforms[b] Removal Efficacy (%)
Gravesend, England	23	4/86–1/88	Subsurface Gravel							
			Bed 1	8.16	12.6	7.2	42			
			Bed 2	8.16	12.6	7.5	40			
			Bed 3	8.16	12.6	5.0	60			
Marnhull, England	23	5/87–1/88	Subsurface Soil							
			Bed 1	4.46						
			Bed 2	6.90						
Holtby, England	23	7/86–1/88	Subsurface Soil	4.90	7.79	6.82	12			
Castleroe, England	23	4/87–1/88	Subsurface Gravel							
			Cell 1	4.34	5.0	4.0	20			
			Cell 2	4.34	5.0	4.0	4			
			Subsurface Soil							
			Cell 1	4.34	5.0	2.5	50			
			Cell 2	4.34	5.0	5.6	−12			
Middleton, England	23	6/87–1/88	Subsurface Sand/Gravel	8.89	10.8	7.6	30			
Bluther Burn, England	23	5/87–3/88	Subsurface							
			Fine Fly Ash	10.76	10.49	3.64	65			
			Coarse Fly Ash	6.24	10.49	1.71	84			
			Unclassified Gravel	9.93	10.49	0.93	91			
			Gravel	10.09	10.49	3.14	70			
Little Stretton, England	23	7/87–12/87	Subsurface Gravel	26.0						
Ringsted, Denmark	24	9/84–10/84	Subsurface							
			Gravel	5.70	15	9.6	36			
			Clay	1.71	15	6.0	60			

[a]All data are for total phosphorus except the English systems reported orthophosphate (as P).
[b]All data are for fecal coliforms except the Santee system reported total coliforms.
[c]Alum treatment provided prior to wetlands.
[d]Period of record for coliform data was January through December 1985.
[e]Influent data are for raw sewage. The sewage receives primary treatment (aeration cell) prior to the marsh. Effluent data are for the final effluent after chlorination except for the fecal coliform data, which are for the meadow effluent.

Table 3. Performance of Constructed Wetland Systems for Acid Mine Drainage

	Mean	Range[a]
Size (m^2)	1,550	93–6,070
Number of basins (ponds)	3	1–7
Size of basin (m^2)	795	19–6,070
Flow rate, L/s	1.3	0.06–12.6
Water depth, m	0.3	0–2
Water chemistry (n = 11)		
Inflow:		
pH	4.9	3.1–6.3
Acidity, mg/L	170	ND–600
Fe, mg/L	33	0.4–220
Mn, mg/L	26	8.7–54
SO_4, mg/L	950	270–1,600
Outflow:		
pH	6.0	3.5–7.7
Acidity, mg/L	40	ND–140
Fe, mg/L	1.2	0.05–7.3
Mn, mg/L	15	0.3–52
SO_4, mg/L	740	160–1,500
Construction cost	$10,000	$1,500–$65,000

Source: Kleinmann and Girts.[25]
Note: ND = Below detection limits
[a]Twenty sites.

reduced.[15,27] During summer, anoxic or anaerobic conditions may occur due to reduced reaeration and oxygen solubilities and increased oxygen use by the biological community.[15,27] Effects of these factors are more pronounced in sluggish flows caused by high evapotranspiration losses.

The major oxygen source for subsurface flow systems is via diffusion from the atmosphere through the plants to the root zone (rhizosphere).[1,28-30] Selection of plant species can be an important factor. At Santee, California, most of the root mass of cattails *(Typha latifolia)* was confined to the top 30 cm of the profile. The root zone of reeds *(Phragmites australis)* extended to 76 cm and

Table 4. TVA Acid Drainage Wetlands Treatment Summary

Wetlands System	Area m_2	Influent Concentration pH	Influent Fe (mg/L)	Influent Mn (mg/L)	Effluent Concentration pH	Effluent Fe (mg/L)	Effluent Mn (mg/L)	Removal Efficacy (%) Fe	Removal Efficacy (%) Mn	Average Flow L/min	Treatment Area m2/mg/min Fe	Treatment Area m2/mg/min Mn
King 006	9,300	6.0	170.0	4.9	2.9	17.0	3.8	96	22	1,374	0.05	1.4
WC 018	4,800	5.6	150.0	6.8	3.9	6.3	6.4	96	59	170	0.2	4.2
Imp 2	11,000	3.1	40.0	13.0	3.1	3.4	14.0	92	–8	646	0.4	1.3
Imp 4	2,000	4.9	135.0	24.0	5.0	2.3	3.5	90	85	42	0.4	2.0
Imp 3	1,200	7.1	28.7	8.8	6.8	0.7	1.3	98	85	83	0.5	1.6
RT—2	7,300	5.7	45.2	13.4	6.9	0.6	0.9	99	93	192	0.8	2.8
Imp 1	5,700	6.0	72.8	9.6	6.5	0.7	2.0	99	93	60	1.3	9.9
950 NE	2,500	6.0	11.0	9.0	7.2	0.5	0.4	96	96	75	3.0	3.7
950–1 & 2	3,400	5.7	12.0	8.0	6.5	1.1	1.6	91	80	83	3.4	5.1
Col 013	4,600	5.7	0.7	7.0	6.7	0.7	3.0	0	57	496	13.2	1.3

bulrushes *(Scirpus validus)* to more than 60 cm.[18] In the cool climates of western Europe, effective root zone depth for reeds is considered to be 60 cm.[1] The gravel bed at Santee was 76 cm deep, and water level was maintained just below the surface. Different BOD_5 removals (Table 2) may reflect the expanded aerobic zone made possible by deeper root penetration.

Removal efficacies for BOD_5 are higher (60–96%) at higher organic loads (primary or screened raw sewage, Table 2). Highest removals (>80%) occurred at hydraulic loading rates of less than 7 cm/day. Removal efficacies are generally lower for surface flow systems used for polishing treated effluents. Data for English systems and most North American systems are from immature wetlands, and performance is anticipated to improve as vegetation coverage and density increases.

Suspended Solids Removal

Both surface and subsurface wetland systems effectively remove SS (Table 2). Removal efficacies are similar to those for BOD_5. In municipal systems, most of the solids are filtered and settled within the first few meters beyond the inlet. At Arcata, California, solids removal mostly occurred in the initial 12–20% of the cell area.[16] At Listowel, Ontario, some channels received unsettled effluent from an aeration cell, and SS reached 406 mg/L.[15] Removal was still effective, but a sludge bank built up near the inlet. The remaining cell area may be important in removing solids produced by the system.

Similar patterns occur in acid mine drainage wetlands. Cell length was important in iron removal in a new *Typha*-dominated wetland.[31] The relation was attributed to oxidation rates.

Nitrogen Removal

Nitrogen is removed in surface and subsurface flow wetlands by similar mechanisms. Total nitrogen removals up to 79% are reported at nitrogen loading rates (based on elemental N) up to 44 kg/ha/day.[32] Although plant uptake of nitrogen occurs, only a minor fraction can be removed by plants. At Listowel, harvested plant material accounted for less than 10% of the nitrogen removed by the system.[15] Plant uptake at Santee, California was 12–16%.[18]

Nitrogen is most effectively removed by nitrification/denitrification.[26] Ammonia is oxidized to nitrate by nitrifying bacteria in aerobic zones, and nitrates are converted to free nitrogen in the anoxic zones by denitrifying bacteria. In subsurface flow systems, oxygen required by the nitrifiers is supplied by leakage from plant roots.[28] The depth of the bed and the type of plant used can make a significant difference. At Santee, with the same bed depth for all channels, the bulrush *(S. validus)* system removed 94% of applied nitrogen, while reeds, cattails, and unvegetated beds achieved 78%, 28%, and 11% removal, respectively.[18] The root zone of the bulrushes and reeds extended to >60 cm and 76 cm, respectively, while most of the root biomass of the cattails was confined to the top 30 cm of substrate. Thus, bed depth in subsurface flow systems should be matched to potential root penetration depth for selected vegetation. Nitrification will not occur in any flow beneath the root zone.

Nitrification was incomplete in the surface flow system at Listowel during periods of anoxia.[15] In summer, this was caused by high oxygen demand for organic decomposition. Oxygen mass transfer limits nitrification in attached growth systems.[33] This problem can be minimized by recirculation, which dilutes ammonia concentrations and increases the dissolved oxygen concentrations in the wetlands influent.

At Listowel, oxygen transfer in winter was limited by the ice cover, so low oxygen levels and low temperatures limited nitrification. In trickling filter and rotating biological contactor systems, significant levels of nitrification by attached growth organisms require a temperature of 5–7°C.[34] This may be a significant process limitation for surface flow wetlands in cold climates if year-round nitrification is required.

Nitrification cannot be completed without adequate alkalinity. Approximately 7 mg alkalinity is required for oxidation of 1 mg of ammonia nitrogen.[33] Alkalinity losses due to nitrification are partially offset by alkalinity produced by reduction of nitrate and sulfate.[33,35] If additional alkalinity is needed, it can be supplemented with a limestone substrate. Other factors important to the nitrification process are (1) minimizing carbonaceous oxygen demand so slower growing nitrifiers can compete with heterotrophic organisms; (2) maintaining pH within the optimum range of 7 to 8; (3) establishing adequate retention time (at least 5 days based on available data); and (4) limiting toxics (certain heavy metals and organic compounds inhibit nitrifiers).[33]

Fortunately, denitrification readily occurs in wetlands systems when sufficient dissolved carbon is present.[36] Nitrate removal efficiency of greater than 95% is realistic. Denitrification occurs in the reduced zones of the substrate and litter layer.

Phosphorus Removal

Phosphorus removal in wetlands systems occurs from adsorption, absorption, complexation, and precipitation. Removal efficiencies range from 0 to 90% (Table 2). Removals are highest in submerged bed designs when appropriate soils are selected as the medium.[28] A significant clay content and iron, aluminum, and calcium will enhance phosphorus removal.[1,28,37] Effectiveness is lower in surface flow wetlands because of limited contact with the soil and root zone. Burial may be important in surface flow systems.[38] Plants absorb phosphorus through their roots and transport it to growing tissues. Each autumn the aboveground portion of the plant dies, partially decomposes, and releases part of its phosphorus content. The rest is buried under plant debris and solids that settle in the following years.

Metals Removal

Only limited information is available on metals removal in wetlands. The processes include sedimentation, filtration, adsorption, complexation, precipitation, plant uptake, and microbially mediated reactions, especially oxidation. Silver[39] reviews theoretical chemical and microbial actions. Trace metal interactions with microbial biofilms are also presented by Lion et al.[40]

Data on metals removal in constructed wetlands are mostly from systems treating acid mine drainage. Early mine drainage systems were attempts to construct bogs with *Sphagnum* or other mosses as dominant plants[25] because iron and manganese removal was noted in natural bogs receiving mine drainage. Substrates were in situ materials amended with a layer of lime or limestone rock overlain by mushroom compost.[41] Presumably, compost provided an instant peat deposit with humic and other organic acids, and lime elevated the pH.

Systems with humic substrates are potentially effective because of their ion exchange capacity. A natural peatland (bog) in Minnesota removed 80% of the nickel and nearly 100% of the copper from tailings drainage.[42] Uptake by vegetation accounted for less than 1% of total metal removal. However, ion exchange capacity of humic materials or mosses will have limited functional longevity. A *Sphagnum* bed with an area of 1600 m^2 and a depth of 30 cm treating a mine drainage flow of 40 L/min and an iron concentration of 200 mg/L may be saturated in only 2.2 years.[43]

Most recent systems have attempted to construct marshes dominated by cattail *(Typha)* or other emergents[31] with in situ substrate materials[44] because of comparatively high process reactive surfaces in marshes and available knowledge on marsh construction[45] and marsh wastewater treatment.[46] Metal removal research has addressed substrate types, treatment areas, and microbial actions. Wetlands soils (natural and acid) were initially more efficient substrates than spoil, normal soil, and others, but differences were minimal at the end of the first growing season and thereafter.[47] Metallic ion encrustations occurring on stems, leaves,[48] and roots[49] and knowledge of bacterial oxidation of iron and manganese since 1888 (Winogradsky, cited in Alexander[50]) suggested an important role for microbial processes. Recently, extracellular microscopic encrustations were analyzed on filamentous algal cells,[49] and intracellular crystals were documented in suspended algal cells.[51] In both cases, algal death, precipitation, and burial immobilized these metals for long periods (comparable to processes that created "bog iron" deposits, mined to supply the early iron industry in Europe and North America).[52] Although metals removed by other processes can be released, replaced on attachment sites, or redissolved, microbially oxidized states are thermodynamically stable forms commonly found in sedimentary rock deposits.[53]

Constructed wetlands (marshes) have high removal efficacies for iron, and lower efficacies for manganese.[25,44] Iron and manganese removal efficacies ranged from 0 to 99% (median of 96%) and -8% to 96% (median of 83%), respectively, in one study at 10 constructed wetlands (Table 4).

Most systems treating acid mine drainage are surface flow systems. Subsurface flow systems can provide more contact surface for absorption, but they are likely to clog from substantial deposits of insoluble metal compounds. For example, a 0.6-ha surface flow, constructed wetland (marsh) in Alabama reduces dissolved iron of 80+ mg/L to less than 1 mg/L in 110 L/min average flow, immobilizing 4636 kg/yr of iron plus other metal oxides and hydroxides.[47]

Limited results and similar chemistries suggest possible removal of nickel, copper, lead, zinc, silver, gold, and uranium from mining or industrial drainages by constructed wetlands. Reported removal rates are high but application rates are low, and data are from pilot projects or wetlands systems that have operated for short periods.

Compared to municipal systems, constructed wetlands for metal removal suffer from (1) lack of long-term data base (few have operated three years, none more than 10); (2) most reports derive from experimental systems; (3) operating systems data is discharge sampling for permit compliance (little from influent-effluent and much less within system sampling for long periods); (4) on-site design disregarding flow, pollutant concentrations, or application rates; and (5) uninformed, poor quality construction procedures.

Pathogen Removal

Based on reductions in indicator species, pathogens are thought to be removed in both wetland types. Viruses may be adsorbed by soil and organic litter or deactivated because of time out of a suitable host. Bacteria are removed by sedimentation, ultraviolet radiation, chemical reactions, natural die-off, and predation by zooplankton.[26] Quantifying bacterial removal rates is difficult when coliforms are used as indicators because birds and mammals living in wetlands contribute coliforms.

Table 2 contains performance data for both surface and subsurface flow wetlands. Reduction in coliforms ranges from 82% to nearly 100%. Gersberg et al. suggest that the decay of total coliforms follows a first-order relationship.[19]

Changes in pH

Modification of pH is often needed in treatment of acid mine drainage and various industrial wastes. Water quality standards for receiving streams normally require wastewater discharges to maintain pH values in the neutral range of 6 to 9. The pH of acid mine drainage may be as low as 2, and industrial waste may be either acidic or alkaline. Although constructed wetlands can thrive in waters that are moderately acidic or alkaline, net change in pH between inflow and outflow is small, typically within two units. For acid mine drainage systems, pH changes in *Sphagnum* wetlands range from a decrease to no change in newly constructed systems[54] to an increase in an existing natural system.[41] Reduction of pH was postulated from exchange of hydrogen cations on moss for iron cations.[54] Cattail and mixed-species systems tend to increase pH from inlet to outlet.[44,55] Mean influent pH of the 20 systems summarized in Table 3 was 4.9 and mean effluent pH was 6.0; however, some of the systems included in these data may have used chemical treatment to raise the pH before discharge, and ineffective systems were excluded.[56] Median influent pH for the 10 systems summarized in Table 4 was 5.7, and median effluent pH was 6.5.

Change in pH within a wetland can be caused by several mechanisms. Some mechanisms lower the pH; others increase it. Both types may occur within a wetland system. One important acidity generation mechanism is production of humic substances, much of which consists of organic acids.[35] In natural wetlands with peat substrates, the humic component forms a strong buffer against pH increases. In wetlands receiving acid mine drainage, oxidation of ferrous iron to ferric iron and subsequent hydrolyzation to ferric hydroxide may be an important source of acidity.[57] Formation of carbon dioxide and its soluble forms, carbonate and bicarbonate, will increase acidity.[58] These carbonic species are generated by oxidation of organic substances. On the other hand, wetlands are largely reducing ecosystems. Reduction of nitrate and sulfate results in a net production of alkalinity.[33,35] Removal of carbon dioxide increases pH and alters forms of alkalinity but not amounts.[58] This is common in algal-dominated systems such as nutrient-rich ponds, but a reversal occurs at night.

DESIGN PROCEDURES

Summary of Loading Factors for Existing Systems

Information is generally available on only two waste types: municipal and mining wastewaters. Hydraulic loading factors are the primary design basis for most municipal constructed wetlands.[26] Guidelines for acid mine drainage are available for individual constituents.[44]

Municipal Wastewaters

Tchobanoglous and Culp[59] summarized hydraulic loading rates for several types of systems (Table 5). Rates ranged from 0.16 to 12 ha/1000 m^3/day (62 to 0.8 cm/day).

Summaries for projects in North America are presented in Table 6. Surface flow systems are typically loaded less than subsurface flow systems. An arbitrary breakline appears to be about 2.7 ha/1000 m^3/day (3.9 cm/day or 25 ac/mgd). Systems with larger acreage per unit of flow (a lower loading rate) are normally surface flow systems, while smaller acreages (a higher loading rate) are typically associated with subsurface flow systems.

Table 5. Preliminary Design Parameters for Planning Constructed Wetland Wastewater Treatment Systems

Type of System	Flow Regime	Detention Time, Days		Depth of Flow, m (ft)		Loading Rate ha/100 m^3/day (acre/mgd)	
		Range	Typical	Range	Typical	Range	Typical
Trench (with reeds or rushes)	PF	6–15	10	0.3–0.5 (1.0–1.5)	0.4 (1.3)	1.2–3.1 (11–29)	2.5 (23)
Marsh (with rushes, others)	AF	8–20	10	0.15–0.6 (0.5–2.0)	0.25 (0.75)	1.2–12 (11–112)	4.1 (38)
Marsh-pond							
1. Marsh	AF	4–12	6	0.15–0.6 (0.5–2.0)	0.25 (0.75)	0.65–8.2 (6.1–76.7)	2.5 (23)
2. Pond	AF	6–12	8	0.5–1.0 (1.5–3.0)	0.6 (2.0)	1.2–2.7 (11–25)	1.4 (13)
Lined trench	PF	4–20	6			0.16–0.49 (1.5–4.6)	1.4 (1.9)

Source: Tchobanoglous and Culp.[59]
Note: Data based on the application of primary or secondary effluent.
PF = plug flow; AF = arbitrary flow.

Selection of an appropriate design loading rate should be based on several factors:

1. *Treatment Objectives.* Compliance with permit limits is a standard consideration. Stringent limits require lower loading rates. Lower loading rates may also be needed to meet ancillary objectives, e.g., wildlife enhancement or aesthetics.
2. *System Use.* Constructed wetlands can be used to provide basic (primary and secondary) treatment following pretreatment (screening/comminution); secondary/advanced treatment following primary treatment (lagoons/septic tanks/clarifiers); or polishing treatment following any type of existing treatment facility. The relative trend in loading rates is to increase as the level of use progresses from basic to polishing treatment.
3. *System Type.* Surface flow wetlands are considered less efficient on an area basis than subsurface systems because of smaller quantities of attached microorganisms available to process the wastes. The quantity of attached microorganisms is a function of the surface area provided by the system. Subsurface flow systems provide much more attachment surface on an area basis than surface flow systems (assuming clogging does not result in surface flow), allowing higher loading rates.
4. *System Configuration.* Use of multiple cells in series, parallel, or combinations increases the potential for optimizing treatment processes, minimizing treatment area, and maximizing loading rates. Cell types can also be varied, and recirculation can further improve performance.
5. *Safety Factors.* Constructed wetlands technology is still under development. Many answers are needed before designs can be optimized and performance levels firmly established. Consequently, users may choose to design conservatively with low loading rates to reduce risks of future upgrades. Alternately, a principal advantage of the technology is the ease of expansion as needed if this

Table 6. Hydraulic Loading Rates for Constructed Wetlands in North America

Project	Ref.	Configuration	Type	Number of Cells	Flow (1000 m³/day)		Hydraulic Loading Rate (cm/day)	
					Design	Actual	Design	Actual
Anne Arundel County, MD	60	series		3				
Freshwater wetland			surface		4.27		15.1	
Peat wetland			percolation		3.46		10.2	
Offshore wetland			surface		1.89		1.09	
Arcata, CA (Pilot)	61	parallel/series	surface	12	1.07	0.119	24.0	2.66
Benton, KY	62	parallel		3	4.16	2.32	9.48	5.30
Cell 1			surface	1	1.04	0.604	7.11	4.15
Cell 2			surface	1	1.04	0.620	7.11	4.26
Cell 3			subsurface[a]	1	2.08	1.16	12.2	7.97
Brookhaven, NY	17	marsh/pond		2				
Marsh			surface			0.0379		2.26
Pond			surface			0.0757		4.53
Cannon Beach, OR	63	single	surface	1	3.44		5.71	
Cobalt, Ontario, Canada	27	serpentine	surface	1		0.0186–0.093		2.0–10.0
Collins, MS	64	serpentine	surface		1.56		3.84	
East Lansing, MI	26	marsh/ponds	surface			2.84		1.80
Emmitsburg, MD	61	single	subsurface	1		0.110		16.4
Foothills Pointe, TN	22	single	subsurface	1	0.0676		4.68	
Fort Deposit, AL	65	parallel	surface	2			1.51	
Gustine, CA	61	parallel	surface	24	3.79		3.80	
Hardin, KY	62	parallel	subsurface	2	0.379		5.92	
Harriman, NY	22	marsh/pond/meadow	surface	2		0.114		
Houghton Lake, MI	66	single	surface	1		10.0		0.143
Incline Village, NV	67	series/serpentine	surface	4	6.24		1.27	
Iron Bridge, FL	65	parallel/series	surface	2			1.56	0.615
Iselin, PA	21	marsh/pond/meadow		5	0.0454	0.0257	1.83	1.47[b]
Marsh		parallel	subsurface[a]	2	0.0454	0.0257	4.68	5.28[b]
Meadow		parallel	subsurface[a]	2	0.0454	0.0257	9.37	10.7[b]

Table 6. (contd.)

Project	Ref.	Configuration	Type	Number of Cells	Flow (1000 m³/day)		Hydraulic Loading Rate (cm/day)	
					Design	Actual	Design	Actual
Lake Buena Vista, FL	65		surface			14.0		3.99
Lakeland, FL	65	series	surface	7	53.0	29.4	0.936	0.520
Listowel, Canada	15	parallel/series	surface	5	4.54			
System 1		marsh/pond/meadow	surface	7				2.01
System 2		single	surface	1		0.040		1.89
System 3		series	surface	5		0.040		1.31
System 4		series	surface	5		0.030		1.31
System 5		single	surface	1		0.030		1.78
Mountain View Sanitary, CA	68	marsh/pond	surface	5	6.05	2.65	7.80	3.46
Neshaminy Falls, PA	20	marsh/pond/meadow			0.114	0.587	2.43	1.26
Marsh		parallel	subsurface[a]	4	0.114	0.587	4.63	2.39
Meadow		parallel	subsurface[a]	4	0.114	0.587	9.26	4.78
Orange County, FL	65		surface	3		75.7		1.53
Orlando, FL	69	series	surface	3	75.7	30.3	1.56	0.623
Paris Landing State Park, Paris, TN	22	series	subsurface	2	0.284		15.0	
Pembroke, KY	62	marsh/pond/meadow		6	0.341		2.25	
System A		series	surface	3	0.170		2.25	
System B		series	subsurface	3	0.170		2.25	
Phillips High School, Bear Creek, AL	22	single	subsurface	1	0.0757		3.74	
Santee, CA	18	parallel	subsurface	4	0.0122	0.0122	4.68	4.68
Silver Springs Shores, FL	65		surface			2.65		1.24
Vermontville, MI	26	parallel	surface	3		0.640	1.38	

[a]Plugging has resulted in substantial surface flow.
[b]Based on the use of only one of the two sets of marshes and meadows.

option is incorporated into the original planning and siting. This strategy might use higher initial loading rates, with upgrades added as necessary based on actual performance and future needs.

Initial hydraulic loading rates to be considered and modified based on the above factors are listed in Table 7. These rates are based on 4.7 cm/day (2.1 ha/1000 m^3/day or 20 ac/mgd) for subsurface flow systems and 1.9 cm/day (5.3 ha/1000 m^3/day or 50 ac/mgd) for surface flow systems to treat primary effluent to at least secondary levels. These guidelines will be refined as new information becomes available. Proper design, construction, and operation are also essential to obtain acceptable performance. For example, if the inlet area and structure, substrate depth and size, and bed slope are not properly matched for a subsurface flow system, surface flow may predominate when the system matures, resulting in performance equivalent to an undersized surface flow system.

Acid Mine Drainage

The Bureau of Mines sponsored a survey of 20 constructed wetlands for treatment of acid mine drainage.[56] The median loading rate was 928 m^2/L/s (9.31 cm/day), ranging from 61.1 to 10,700 m^2/L/s (8640 to 0.81 cm/day). Pesavento has suggested a rule-of-thumb value of 294 m^2/L/s (29.4 cm/day).[56] TVA developed preliminary guidelines from results of 11 wetlands treating acid drainage from an inactive coal preparation plant, adjacent mined areas, and three ash storage areas.[44] Because metal concentrations vary substantially, application rates were based on ion concentrations rather than hydraulic flows. Loading rates of 2.0 and 7.0 m^2/mg/min for iron and manganese, respectively, are recommended for waters with pH under 5.5. If pH is greater than 5.5, the loading rate is increased to 0.75 m^2/mg/min for iron and 2.0 m^2/mg/min for manganese.

Design Based on Reaction Kinetics

All constructed wetland systems are attached-growth biological reactors. Performance for biologically mediated reactions can be described with first-order, plug-flow kinetics if steady-state conditions are assumed. Reactions are first-order when the rate of completion is directly proportional to the first power of the reactant concentration. Organic degradation (BOD_5, COD, TOC), nitrification, disinfection, and adsorption generally follow first-order kinetics.[70] The reaction rate and constants are experimentally determined and are dependent on the treatment system characteristics. Rates for constructed wetland systems have been reported for only BOD_5.[1,46] In reality, steady-state conditions and plug-flow reactors are not typical, and modifications to basic first-order kinetic relationships should be considered. Kadlec[66] provides guidance and illustrates possible errors resulting from neglecting various phenomena. Available information is summarized below. Caution should be used

Table 7. Hydraulic Loading Rates for the Preliminary Design of Constructed Wetlands for Treating Municipal Wastewaters

	Treatment Objectives							
	Secondary Treatment				Advanced Treatment/Multiple Objectives			
	Surface Flow		Subsurface Flow		Surface Flow		Subsurface Flow	
System Use	cm/day	acres/mgd	cm/day	acres/mgd	cm/day	acres/mgd	cm/day	acres/mgd
Basic treatment	[a]	[a]	2.3–6.2	40–15	[a]	[a]	≥3.1	≥30
Secondary treatment	1.2–4.7	75–20	4.7–18.7	20–5	≥1.9	≥50	≥4.7	≥20
Polishing treatment	1.9–9.4	50–10	4.7–18.7	20–5	≥3.1	≥30	≥4.7	≥20

[a]This use has not yet been demonstrated. Surface flow systems constructed to date are preceded by at least primary treatment units (septic tanks, clarifiers, lagoons, etc.).

where specific numerical values are provided because the values are based on a limited data base. Research is needed to refine both concepts and specific values.

Constructed Wetlands with Surface Flow

The basic relationship for plug-flow reactors is given by Equation 1 for steady conditions:

$$\frac{C_e}{C_o} = \exp\,[-K_T t] \tag{1}$$

where
C_e = effluent concentration, mg/L
C_o = influent concentration, mg/L
K_T = temperature-dependent, first-order reaction rate constant, $days^{-1}$
t = hydraulic residence time, days

Reed et al.[46] suggest that experience with overland flow and trickling filter systems can be applied to constructed wetlands design. The resulting general model for BOD_5 is given in Equation 2:

$$\frac{C_e}{C_o} = A' \exp\left[\frac{-C'\,K_T(A_V)^{1.75}\,LWdn}{Q}\right] \tag{2}$$

where
A' = fraction of BOD_5 not removed as settleable solids near the head of the system (as a decimal fraction) = 0.52
C' = characteristic of the medium
= 0.7 m for a wide variety of media
K_T = $0.0057\,(1.1)^{(T-20)}$ where T is the water temperature in °C
L = length of system (parallel to flow path), m
W = width of system, m
d = design depth of system, m
n = porosity of system (as a decimal fraction)
= volume of voids/total volume
Q = average hydraulic loading on the system, m^3/day
$= \dfrac{Q_{influent} + Q_{effluent}}{2}$

The term "LWdn/Q" represents the hydraulic residence time, t. In a surface flow wetland, a portion of the available volume will be occupied by vegetation, so actual detention time will be a function of the remaining cross-sectional area available for flow. This can be defined as the porosity of the system in a manner similar to that used for soil systems. The product, dn, is equivalent to the depth of flow in a system with no vegetation present. In a vegetated system with perfect plug flow and no extraneous water losses or gains, it would also be the ratio of actual residence time to theoretical detention time. The A′ factor accounts for settleable BOD_5 removed near the head of the system.

When bed slope or hydraulic gradient is equal to 1% or greater, Reed et al.[46] use information from terrestrial overland flow systems to adjust the design model as follows:

$$\frac{C_e}{C_o} = 0.52 \exp\left[\frac{-C' K_T(A_V)^{1.75}\ LWdn}{4.63S^{1/3}Q}\right] \tag{3}$$

where S is bed slope, or hydraulic gradient (as a decimal fraction).

Kadlec proposes different constants based on data from Houghton Lake and other wetland systems.[66] The equation is based on the Ergun equation, which adds a term to Darcy's law to account for turbulent flow. In terms of hydraulic residence time, the equation is:

$$t = \frac{L}{a^{1/b}\left(\frac{Q}{W}\right)^{1-1/b} S^{c/b}} \tag{4}$$

Based on data for the Houghton Lake wetland, b = 2.5–3.0 and c = 0.7–1.0. The coefficient a is site specific, but equal to 4×10^6 m/day for Houghton Lake for b = 3 and c = 1.

Flow through the system is augmented by precipitation and (negatively) evapotranspiration. Evapotranspiration slows the water and increases contact times. Rainfall has the opposite effect. For a wetland operated at constant depth, actual residence time is given by:

$$t_a = t\left[\frac{1}{\alpha}\ln\left(\frac{1}{1-\alpha}\right)\right] \tag{5}$$

where α is fractional augmentation and t is nominal residence time, based on nominal depth and wastewater addition rate. Fractional augmentation is the net fractional water gain or loss due to precipitation, evapotranspiration, and infiltration relative to wastewater flow (see Kadlec[66] for calculation information). Evapotranspiration strongly influences residence time, whereas rain has a lesser effect. At Houghton Lake and Listowel, the range in ratios of measured residence times to nominal residence time is 40–250%. The mean for Listowel, spanning all seasons over four years, was 126%. The mean for Houghton Lake was the nominal 100% because precipitation was equal to evapotranspiration for the chosen summer period.

Substituting Equation 5 in Equation 1 gives the following expression:

$$\frac{C_e}{C_o} = \left[1 - \alpha\frac{x}{L}\right]^{(K_T t/\alpha - 1)} \tag{6}$$

where x/L is the fractional distance to outlet.

If infiltration occurs, there is water loss as with evaporation, but the concentrating effect is not present. The corresponding expression is:

$$\frac{C_e}{C_o} = \left[1 - \alpha \frac{x}{L}\right]^{(K_T t/\alpha)} \tag{7}$$

Consideration of atmospheric phenomena reveals the hydrological complexity involved in optimizing design and performance of wetland treatment systems. Additional details and examples on use of these equations are provided by Reed et al.[46] and Kadlec.[66]

Organic Loading. Organic loading has been light on most systems evaluated. The relation between organic loading and BOD_5 removal rate suggests a linear correlation, at least up to loading rates of 100 kg/ha/day, the highest value reported for surface flow wetlands. Most wetland systems have operated at loading rates ranging from 18–116 kg BOD_5/ha/day and achieve 70–95% BOD_5 removal. Organic loading has not been used as a principal design criterion for constructed wetlands. It should be checked to ensure maintenance of aerobic conditions in the system. An upper limit of 110 kg/ha/day is suggested for this purpose.

Since most settleable BOD_5 will be removed close to the inlet, organic loading may be high on this small area. At Listowel, average organic loading over the surface area of system 4 was less than 12 kg/ha/day, with occasional anoxic conditions near the head. In a constructed wetland for polishing lagoon effluent in Gustine, California, this problem was solved with wastewater distribution at the head and at the one-third point in the channels with provisions for additional points if needed.[71]

Preliminary Treatment. Preliminary treatment to at least the primary level has been used for surface flow wetlands to keep organic loading within reasonable limits and avoid localized anaerobic conditions. These problems can be minimized with inlet structures that spread waste over much of the cell.

Constructed Wetlands with Subsurface Flow

Hydraulic Considerations. The hydraulic regime in fully saturated, fine-grained soils, sands, and gravels is controlled by hydraulic conductivity of the media and hydraulic gradient of the system as defined by Darcy's law:

$$Q = k_s A S \tag{8}$$

where k_s = hydraulic conductivity of a unit area of medium perpendicular to flow direction

A = cross-sectional area

Consistent units must be used for Q, k_s, and A. Equation 8 can be rearranged to solve for the saturated cross-sectional area of the system:

$$A = \frac{Q}{k_s S} \tag{9}$$

Flow through "clean" coarse gravels and rocks is more appropriately described by Ergun's equation:[66]

$$\rho g S = 150 \frac{\mu v(1 - \epsilon)^2}{D_p^2 \epsilon^2} + 1.75 \frac{\rho v^2(1 - \epsilon)}{D_p \epsilon} \tag{10}$$

where ρ = density
g = gravitational acceleration
S = hydraulic gradient
μ = viscosity
ϵ = porosity
D_p = particle diameter
v = velocity = superficial velocity/ϵ

The term "ρgS" is the kinetic energy of the flowing water. The difference between this equation and Darcy's law is the addition of a term to account for turbulent flow, which can occur for larger particle sizes. The equation would ordinarily be used in combination with the general relationship, Q = VA, to determine slope necessary for a given flow or, for a given slope and flow rate, to determine velocity and corresponding cross-sectional area required to preclude surface flow.

Typical values for initial porosities range from 0.18 to 0.35 for coarse gravel to fine gravel;[46] however, porosity may substantially change as the system matures, due to clogging and root development. This may result in laminar or transitional flow where Darcy's equation would be appropriate if mounding does not cause surface flow. Limited information precludes specific guidelines, though work in Europe indicates a long-term hydraulic conductivity of 10^{-3} m/s or more may be appropriate for gravel beds.[23] Unit flow velocity (Q/A) through the cross section should not exceed 8.6 m/day to avoid disruption of the medium-rhizome structure and to ensure sufficient contact time for treatment.[1]

Bed depth and type of vegetation should be matched to assure all flow will be within the root zone. Aerobic microsites in the root zone appear essential for effective treatment of organics and ammonia nitrogen. Present field experience on the depth of root penetration is limited to bulrushes, reeds, and cattails. Using data from the gravel system at Santee, bed depths should not exceed 60–76 cm for bulrushes and reeds, and 30 cm for cattails. The 60-cm depth for reeds has also been recommended for subsurface flow soil systems in West Germany,[1] so climate and substrate apparently are not factors in this relationship. In Europe, design depths are reduced by 50% for industrial wastes with high concentrations of refractory organics (COD).[1]

Experience in West Germany indicates three years' growth is necessary for full 60-cm penetration of reed rhizomes.[1] Partial draining for a few months in early fall for the first two years induces downward penetration of the rootstock. It is claimed that deeper penetration may not occur if beds are continuously saturated. A fully developed reed bed with root penetration to 60 cm in soil is claimed to have produced hydraulic conductivities of 260 m^3/m^2/day when the original soil was less permeable clayey loam. Claimed hydraulic transformation by root and rhizome development is unverified, and the ability of reeds to produce hydraulic conductivity of this magnitude is doubtful.[23] When gravel or coarse sand is the medium, rhizome penetration is unnecessary to develop initial hydraulic properties of the bed. However, the root system may be critical in maintaining adequate long-term permeability (hydraulic conductivity). The root aerobic zone precludes anaerobic slimes from binding the substrate, and stabilized organics (including dead roots and rhizomes) supposedly stabilize the long-term permeability at a high level (260 m^3/m^2/day).[1] Performance of a gravel-bed system should improve during the first few years as root penetration brings oxygen to all parts of the bed.

Most soil systems are designed with a slope of 1% or slightly higher, but up to 8% may be used if local terrain permits. Bed slopes in gravel-bed systems are typically small (0–2%). Bed slopes are not equivalent to actual hydraulic gradients in most systems because reliable design values for hydraulic conductivity are not available and conductivity changes as systems mature.

Once bed depth and slope have been selected, bed width can be calculated based on Equations 9 or 10. Bed width is the cross-sectional area divided by bed depth. The cross-sectional area is independent of biochemical reactions and is controlled by hydraulic requirements. It is calculated directly by Equation 9 and indirectly by Equation 10 (velocity is calculated by Equation 10 and cross-sectional area is calculated by the general relationship $Q = VA$). For laminar flow (i.e., use of Darcy's law, Equation 9), hydraulic gradient of 1%, bed depth of 0.6 m, and long-term permeability of 260 m/day, the hydraulic loading per meter of inlet width (Q/W) is 1.56 m^3/day/m (126 gpd/ft); e.g., a bed width of 64 m is needed for a flow of 100 m^3/day. The final dimension, bed length, is determined by the hydraulic residence time required for biological reactions to remove the desired level of contaminants (next section).

Consideration of changes in flow hydraulics as the system matures and other practical system limitations indicates that the length-width ratios will be relatively small to assure subsurface flow over the long term. For gravel beds, the ratio will be less than 3:1 and perhaps less than 1:1. The ratios for soil beds will be smaller.

Biological Relationships. Removal of biodegradable constituents such as BOD_5 and organic nitrogen can be described with first-order, plug-flow kinetics:

$$\frac{C_e}{C_o} = \exp\left[-K_T t\right] \quad (1)$$

Hydraulic residence time is a function of available void spaces (V_v) and average flow rate through the system:

$$t = \frac{V_v}{Q} \quad (11)$$

where $V_v = LWdn$

To avoid design errors, average flow through the system must consider water losses due to evapotranspiration and seepage, and/or water gains due to precipitation.[66] Typical values for initial porosity range from 18% to 35% for coarse gravel to fine gravel and 37% to 44% for coarse sand to fine sand;[46] however, porosity may change from clogging and root development as the system ages. With time, much of the flow may be in the debris layer on top of the bed.

Since the product LW is equal to the surface area, A_s, the basic equation can be written as:

$$\frac{C_e}{C_o} = \exp\left[\frac{-K_T A_s dn}{Q}\right] \quad (12)$$

The temperature dependence of the rate constant for BOD_5 is described by:

$$K_T = K_{20}(1.1)^{(T-20)} \quad (13)$$

where K_{20} = rate constant at 20°C

Based on limited information from few systems,[1,18,72] the rate constant, K_{20}, for a particular system may be related to the porosity of the bed medium. This is reasonable because natural soils with high porosities have finer void spaces, providing more opportunity for surface contact compared to a more permeable gravel. The tentative relationship proposed by Reed et al.[46] is:

$$K_{20} = K_o(37.31n)^{4.172} \quad (14)$$

where K_o = the "optimum" rate constant for a medium with a fully developed root zone, $days^{-1}$
= 1.839 $days^{-1}$ for typical municipal wastewaters
= 0.198 $days^{-1}$ for industrial wastewaters with high COD

Using typical porosity values in this expression, gravel-bed systems will have a kinetic rate constant one-third to one-fourth of finer-textured sands. This difference in kinetic rate constants translates into larger surface area requirements for gravel beds.

The relationship described by Equation 14 and the K_o values is tentative and should only be used for preliminary design estimates. Further study is needed on the biological kinetics in subsurface flow wetlands.

Organic Loading. Organic loading is not used as a critical design parameter, but rather as a check to ensure sufficient oxygen will be present in the subsurface bed to sustain intended treatment reactions. Commonly used emergent plants can transmit from 5 to 45 g O_2/day/m^2 of wetland surface.[1,73]

Assuming that oxygen required is equivalent to that for partial-mix aerated ponds and a conservative oxygen loss rate for plants of 20 g/m^2/day, the oxygen balance in a subsurface flow wetland can be checked with Equations 15 and 16.

$$\text{Required } O_2 = 1.5\ L_o \tag{15}$$

$$\text{Available } O_2 = \frac{(TrO_2)(A_s)}{1000\ g/kg} \tag{16}$$

where O_2 = oxygen required or available, kg/day
L_o = organic (BOD_5) loading, kg/day
TrO_2 = oxygen transfer rate for the vegetation
= 20 g/(m^2/day)

As a further safety factor, available oxygen determined from Equation 16 should be twice the oxygen requirements. Units processing high-strength wastewaters need increased operational surface area to achieve this balance.

At the 20 g/m^2/day oxygen supply rate, organic loading could be up to 133 kg BOD_5/ha/day, similar to values for surface flow wetlands. At the theoretical maximum oxygen transfer rate (45 g/m^2/day), organic loading could be as high as 300 kg/ha/day.

Preliminary Treatment. Subsurface flow wetlands in the United States use at least primary treatment prior to constructed wetlands. Primary treatment options include septic tanks, Imhoff tanks, primary clarifiers, and lagoons. Anaerobic reactors could be used to reduce organic and solids content of high-strength industrial wastewaters. If phosphorus removal is required and gravel or coarse sand is used for bed media, a preliminary or posttreatment method for phosphorus removal must be included.

Small length-width ratios are better for high organic loading because unit loading at the inlet is reduced. Most European systems use small ratios (less than 1:1), and many apply screened and degritted wastewater to the inlet zone. Fully developed wetlands are claimed to have rapid solids removal and decomposition and no odor problem.[1]

SUMMARY AND CONCLUSIONS

Constructed wetlands are emerging as low-cost, easily operated, efficient alternatives to conventional treatment for a variety of wastewaters. Most common uses are for treatment of municipal wastewaters and acid mine drainage.

There are two major types of systems: surface flow systems with vegetation planted in a shallow pool of water, and subsurface flow systems with vegetation planted in permeable soil, gravel, or sand, and water levels below the media surface. In North America, both types are used for municipal wastewater treatment, but surface flow systems predominate for treatment of other wastes. North American municipal systems follow either primary or secondary treatment facilities. In Europe, subsurface flow systems using soil or gravel media are the most common. These systems treat screened raw sewage or primary effluent.

Wetland systems are effective on many contaminants, including BOD, SS, nitrogen, phosphorus, trace metals, trace organics, and pathogens. The effectiveness is due to the diversity of treatment mechanisms, including sedimentation, filtration, chemical precipitation and adsorption, microbial interactions with contaminants, and uptake by vegetation. Reductions in BOD_5 and SS are generally high. Highest reductions occur in systems with high organic loading and low-to-moderate hydraulic loading. Nitrogen reduction is variable: some systems have high removals and others report low removals. The critical parameter is ammonia nitrogen. To attain high ammonia reductions, the design should emphasize aspects that will support nitrifying bacteria. These include initial reduction of carbonaceous BOD_5 to low levels; maintaining adequate dissolved oxygen, alkalinity, and pH for nitrification reactions; and providing adequate retention time.

Phosphorus removal also varies from system to system. Removals are highest in subsurface flow designs where substrates contain iron, aluminum, or calcium. Metals are accumulated in many systems through adsorption, absorption, complexation, and precipitation reactions. Long-term constituent removal by adsorption, absorption, and some complexation reactions may be limited by saturation. Saturation has occurred in experimental *Sphagnum* systems treating mine drainage. Long-term immobilization of metals and trace organics by microbially mediated reactions is possible, but little empirical information is available. Pathogen reduction is believed to be high and consistent in most municipal wetlands based on indicator data. Changes in pH within most wetlands have been within two units. *Sphagnum* systems tend to decrease pH, while cattail and mixed-species systems tend to increase pH.

Current information is adequate to design systems that substantially reduce targeted contaminants but inadequate to optimize the design and operation for consistent compliance.

Hydraulic loading factors are the primary design basis for sizing municipal constructed wetlands. Loading rates vary (0.8 to 62 cm/day), indicative of various treatment and project objectives, system types and configurations, and performance levels. Surface flow systems are typically loaded less than subsurface flow systems. An arbitrary breakline is about 3.9 cm/day (2.7 ha/1000 m^3/day or 25 ac/mgd). Hydraulic loading rates of 4.7 cm/day (2.1 ha/1000 m^3/day or 20 ac/mgd) for subsurface flow systems and 1.9 cm/day (5.3 ha/1000 m^3/day or 50 ac/mgd) for surface flow systems should treat primary effluent to at least secondary levels if the systems are otherwise properly designed, constructed, and operated.

Hydraulic loading rates for acid mine drainage systems range from 61 to 11,000 m^2/L/s (8640 to 0.81 cm/day). A rule-of-thumb value of 294 m^2/L/s (29.4 cm/day) has been suggested by one source. Preliminary guidelines developed by TVA are 0.75–7.0 m^2/mg/min depending on pH and type of metal.

Many reactions responsible for pollutant reductions in wetland treatment systems follow first-order kinetics. This provides the opportunity to establish a credible, scientific design basis for many of the pollutants of interest; however, reaction rate information is currently available for only BOD_5, and values are from a limited data base. A similar situation exists with equations defining hydraulic flow regime or hydraulic residence time. The problem is compounded by changing characteristics of constructed wetland systems with time (effectiveness normally increases and stabilizes with a maturing system). Research is needed to refine the concepts and specific values before reaction kinetics and hydraulic regime equations can be used with confidence.

REFERENCES

1. Boon, A. G. "Report of a Visit by Members and Staff of WRC to Germany to Investigate the Root-Zone Method for Treatment of Wastewaters," Water Research Centre, Stevenage, England (August 1985).
2. Wolverton, B. C. "Artificial Marshes for Wastewater Treatment," paper presented at the First Annual Environmental Health Symposium, Water and Wastewater Issues in the North Central Gulf Coast, Mobile, AL, April 28–29, 1986.
3. Thut, R. N. "Utilization of Artificial Marshes for Treatment of Pulp Mill Effluents," Chapter 19, this volume.
4. Litchfield, D. K., and D. D. Schatz. "Constructed Wetlands for Wastewater Treatment at Amoco Oil Company's Mandan, North Dakota Refinery," Chapter 18, this volume.
5. Hammer, D. A., and J. T. Watson. "Agricultural Waste Treatment with Constructed Wetlands," in *Proceedings of the National Symposium on Protection of Wetlands from Agricultural Impacts,* P. J. Stuber, Coord., U.S. Fish and Wildlife Service Biol. Report 88:16 (1988).
6. Bowmer, K. H. "Nutrient Removal from Effluents by an Artificial Wetland: Influence of Rhizosphere Aeration and Preferential Flow Studied Using Bromide and Dye Tracers," *Water Res.* 21:591–599 (1987).
7. Finlayson, C. M., and A. J. Chick. "Testing the Potential of Aquatic Plants to Treat Abattoir Effluent," *Water Res.* 17:415–422 (1973).
8. Trautman, N. M., J. H. Martin, Jr., K. S. Porter, and K. C. Hawk, Jr. "Use of

Artificial Wetlands for Treatment of Municipal Solid Waste Landfill Leachate," Chapter 20, this volume.
9. Staubitz, W. W., J. M. Surface, T. S. Steenhuis, J. H. Peverly, M. J. Levine, N. C. Weeks, W. E. Sanford, and R. J. Kopka. "Potential Use of Constructed Wetlands to Treat Landfill Leachate," Chapter 41c, this volume.
10. Dornbush, J. N. "Natural Renovation of Leachate-Degraded Groundwater in Excavated Ponds at a Refuse Landfill," Chapter 41d, this volume.
11. Daukas, P., D. Lowry, and W. W. Walker, Jr. "Design of Wet Detention Basin and Constructed Wetlands for Treatment of Stormwater Runoff from a Regional Shopping Mall in Massachusetts," Chapter 40c, this volume.
12. Silverman, G. S. "Development of an Urban Runoff Treatment Wetlands in Fremont, California," Chapter 40a, this volume.
13. Livingston, E. H. "Use of Wetlands for Urban Stormwater Management," Chapter 21, this volume.
14. Stowell, R., R. Ludwig, J. Colt, and G. Tchobanoglous. "Toward the Rational Design of Aquatic Treatment Systems," Department of Civil Engineering, University of California, Davis, CA (August 1980).
15. Herskowitz, J. "Town of Listowel Artificial Marsh Project Final Report, Project No. 128RR," Ontario Ministry of the Environment, Toronto (September 1986).
16. Gearheart, R. J., S. Wilbur, J. Williams, D. Hull, B. Finney, and S. Sundberg. "Final Report City of Arcata Marsh Pilot Project Effluent Quality Results—System Design and Management, Project Report C-06-2270," City of Arcata, Department of Public Works, Arcata, CA (1983).
17. Small, M. M. "Wetlands, Wastewater Treatment Systems," paper presented at International Symposium, State of Knowledge in Land Treatment of Wastewater, Hanover, NH, August 20–25, 1978.
18. Gersberg, R. M., B. V. Elkins, S. R. Lyons, and C. R. Goldman. "Role of Aquatic Plants in Wastewater Treatment by Artificial Wetlands," *Water Res.* 20:363–367 (1985).
19. Gersberg, R. M., S. R. Brenner, S. R. Lyon, and B. V. Elkins. "Survival of Bacteria and Viruses in Municipal Wastewater Applied to Artificial Wetlands," in *Aquatic Plants for Water Treatment and Resource Recovery,* K. R. Reddy and W. H. Smith, Eds. (Orlando, FL: Magnolia Publishing Inc., 1987).
20. Shorten, G., General Manager, Village of Neshaminy Falls, 195 Stump Road, North Wales, PA. Unpublished data.
21. Watson, J. T., F. D. Diodato, and M. Lauch. "Design and Performance of the Artificial Wetlands Wastewater Treatment Plant at Iselin, Pennsylvania," in *Aquatic Plants for Water Treatment and Resource Recovery,* K. R. Reddy and W. H. Smith, Eds. (Orlando, FL: Magnolia Publishing Inc., 1987).
22. TVA, Water Quality Department, HB 2S 270C, Chattanooga, TN. Unpublished data.
23. Cooper, P. F., and J. A. Hobson. "Sewage Treatment by Reed Bed Systems: The Present Situation in the United Kingdom," Chapter 11, this volume.
24. Jacobsen, B. N. "Physical Description of the Root Zone Installation in Ringsted and Rodekro Municipalities, Experimental Plan and Preliminary Results," File No. 61.736, Water Quality Institute ATV, Horsholm, Denmark, June 11, 1985.
25. Kleinmann, R. L. P., and M. A. Girts. "Acid Mine Water Treatment in Wetlands: An Overview of an Emergent Technology," in *Aquatic Plants for Water Treatment*

and Resource Recovery, K. R. Reddy and W. H. Smith, Eds. (Orlando, FL: Magnolia Publishing Inc., 1987).
26. Hyde, H. C., and R. S. Ross. "Technology Assessment of Wetlands for Municipal Wastewater Treatment," U.S. EPA Municipal Environmental Research Laboratory, Cincinnati, OH (September 1984).
27. Miller, G. "Use of Artificial Cattail Marshes to Treat Sewage in Northern Ontario, Canada," Chapter 39i, this volume.
28. Brix, Hans. "Treatment of Wastewater in the Rhizosphere of Wetland Plants—The Root-Zone Method," *Water Sci. Technol.* 19:107–118 (1987).
29. Grosse, W. "Thermo-osmotic Air Transport in Aquatic Plants Affecting Growth Activities and Oxygen Diffusion to Wetland Soils," Chapter 37b, this volume.
30. Michaud, S. C., and C. J. Richardson. "Relative Radial Oxygen Loss in Five Wetland Plants," Chapter 38a, this volume.
31. Stillings, L. L., J. J. Gryta, and T. A. Ronning. "Iron and Manganese Removal in a *Typha*-Dominated Wetland During Ten Months Following Its Construction," in *Mine Drainage and Surface Mine Reclamation, Vol. 1, Mine Water and Mine Waste*, Bureau of Mines Information Circular 9183 (1988).
32. Zirschky, J. "Basic Design Rationale for Artificial Wetlands," Contract Report 68-01-7108, U.S. EPA, Office of Municipal Pollution Control, Washington, DC (June 1986).
33. "Process Design Manual for Nitrogen Control," U.S. EPA, Technology Transfer (October 1975).
34. Reed, S. C., C. J. Diener, and P. B. Weyrick. "Nitrogen Removal in Cold Regions Trickling Filter Systems," SR-86-2, Cold Regions Res. and Eng. Lab., Hanover, NH (February 1986).
35. Hemond, H. F., and J. Benoit. "Cumulative Impacts on Water Quality Functions of Wetlands," in *Cumulative Effects on Landscape Systems of Wetlands: Scientific Status, Prospects, and Regulatory Perspectives,* U.S. EPA, Environmental Research Laboratory, Corvallis, OR (June 1988).
36. Gersberg, R. M., B. V. Elkins, and C. R. Goldman. "Nitrogen Removal in Artificial Wetlands," *Water Res.* 17(9):1009–1014 (1983).
37. Richardson, C. J. "Mechanisms Controlling Phosphorus Retention Capacity of Freshwater Wetlands," *Science* 228:1424–1427 (1985).
38. Kadlec, J. A. "Nutrient Dynamics in Wetlands," in *Aquatic Plants for Water Treatment and Resource Recovery,* K. R. Reddy and W. H. Smith, Eds. (Orlando, FL: Magnolia Publishing Inc., 1987).
39. Silver, M. "Biology and Chemistry of Generation, Prevention and Abatement of Acid Mine Drainage," Chapter 42a, this volume.
40. Lion, L. W., M. L. Shuler, K. M. Hsieh, and W. C. Ghiorse. "Trace Metal Interactions with Microbial Biofilms in Natural and Engineered Systems," *CRC Crit. Rev. Environ. Control* 17(4) (1988).
41. Kleinmann, R. L. P., T. O. Tiernan, J. G. Solch, and R. L. Harris. "A Low-Cost, Low-Maintenance Treatment System for Acid Mine Drainage Using Sphagnum Moss and Limestone," 1983 Symposium on Surface Mining, Hydrology, Sedimentology and Reclamation, University of Kentucky, Lexington, KY, November 27–December 2, 1983.
42. Eger, P., and K. Lapakko. "Use of Wetlands to Remove Nickel and Copper from Mine Drainage," Chapter 42e, this volume.
43. Wieder, R. K. "Determining the Capacity for Metal Retention in Man-Made Wet-

lands Constructed for Treatment of Coal Mine Drainage," in *Mine Drainage and Surface Mine Reclamation,* Bureau of Mines Information Circulars 9183 and 9184 (1988).
44. Brodie, G. A., D. A. Hammer, and D. A. Tomljanovich. "Constructed Wetlands for Acid Drainage Control in the Tennessee Valley," in *Mine Drainage and Surface Mine Reclamation, Vol. 1, Mine Water and Mine Waste,* Bureau of Mines Information Circular 9183 (1988).
45. Mosby, H. S., Ed. *Wildlife Investigational Techniques* (Washington, DC: The Wildlife Society, 1963).
46. Reed, S. C., F. J. Middlebrooks, and R. W. Crites. *Natural Systems for Waste Management and Treatment* (New York: McGraw-Hill Book Company, 1988).
47. Brodie, G. A., D. A. Hammer, and D. A. Tomljanovich. "An Evaluation of Substrate Type in Constructed Wetlands Acid Drainage Treatment System," in *Mine Drainage and Surface Mine Reclamation,* Bureau of Mines Information Circular 9183 (1988).
48. Brodie, G. A., D. A. Hammer, and D. A. Tomljanovich. "Man-Made Wetlands for Acid Mine Drainage Control," in *Proceedings 8th Annual National Abandoned Mine Land Conference,* Billings, MT (1986).
49. TVA Waste Technology Center, Old City Hall, Knoxville, TN. Unpublished data.
50. Alexander, M. C. *An Introduction to Soil Microbiology* (New York: John Wiley & Sons, Inc., 1967).
51. Kepler, D. A. "An Overview of the Role of Algae in the Treatment of Acid Mine Drainage," in *Mine Drainage and Surface Mine Reclamation,* Bureau of Mines Information Circular 9183 (1988).
52. Hammer, D. A. "Constructed Wetlands for Wastewater Treatment," paper presented at the MEREC Conference, Coimbra, Portugal, October 19–24, 1987.
53. Wildeman, T. R., and L. S. Laudon. "Use of Wetlands for Treatment of Environmental Problems in Mining: Non-Coal-Mining Applications," Chapter 17, this volume.
54. Burris, J. E., D. W. Gerber, and L. E. McHerron. "Removal of Iron and Manganese from Water by Sphagnum Moss," in *Treatment of Mine Drainage by Wetlands,* contribution #264 in proceedings of a conference by Department of Biology, The Pennsylvania State University, University Park, PA, August 13, 1984.
55. Wieder, R. K., G. E. Lang, and A. E. Whitehouse. "The Use of Freshwater Wetlands to Treat Acid Mine Drainage," in *Treatment of Mine Drainage by Wetlands,* contribution #264 in proceedings of a conference by Department of Biology, The Pennsylvania State University, University Park, PA, August 13, 1984.
56. Girts, M. A., and R. L. P. Kleinmann. "Constructed Wetlands for Treatment of Acid Mine Drainage: A Preliminary Review," 1983 Symposium on Surface Mining, Hydrology, Sedimentology and Reclamation, University of Kentucky, Lexington, KY, November 27–December 2, 1983.
57. Caruccio, F. T., J. C. Ferm, J. Horne, G. Geidel, and B. Baganz. "Paleoenvironment of Coal and Its Relation to Drainage Quality," EPA-600/7–77–067, U.S. EPA Industrial Environmental Research Laboratory, Office of Research and Development, Cincinnati, OH (June 1977).
58. Sawyer, C. N., and P. L. McCarty. *Chemistry for Sanitary Engineers,* 2nd ed. (New York: McGraw-Hill Book Company, Inc., 1967).
59. Tchobanoglous, G., and G. Culp. "Aquaculture Systems for Wastewater Treat-

ment: An Engineering Assessment," EPA 430/9-80-007, U.S. EPA Office of Water Program Operations, Washington, DC (June 1980), pp. 13–42.
60. Lombardo, P., and Thomas Neel. "Wastewater Problems Solved by Natural Combination," *BioCycle* 28:48–50 (1987).
61. "Case Studies—Constructed Wetlands Systems," Draft Report No. RP0002-N, Nolte and Associates, Sacramento, CA (June 1987).
62. Watson, J. T., G. R. Steiner, and D. A. Hammer. "Constructed Wetlands for Municipal Wastewater Treatment," in *Proceedings, Mississippi Water Resources Conference, 1988,* Water Resources Research Institute, Mississippi State University, Jackson, MS.
63. Demgen, F. C. "An Overview of Four New Wastewater Wetlands Projects," Demgen Aquatic Biology, 118 Mississippi Street, Vallejo, CA.
64. Wolverton, B. C. National Aeronautics and Space Administration, National Space Technology Laboratory, NSTL Station, MS. Personal communication.
65. Knight, R. L. CH2M HILL, 7201 Northwest 11th Place, P.O. Box 1647, Gainesville, FL. Personal communication.
66. Kadlec, R. H. "Hydrologic Factors in Wetland Water Treatment," Chapter 3, this volume.
67. Williams, R. B., D. J. Reardon, and J. Shefchik. "Design and Startup of 770-Acre Wetlands Project," paper presented at the 57th Annual California Water Pollution Control Conference, Anaheim, CA, May 8, 1985.
68. Demgen, F. C. "Wetlands Creation for Habitat and Treatment at Mountain View Sanitary District, CA," paper presented at the Conference on Aquaculture Systems for Wastewater Treatment: An Engineering Assessment, U.S. EPA (1979).
69. Jackson, J. "Man-Made Wetlands for Wastewater Treatment: Two Cases," Chapter 39b, this volume.
70. Metcalf and Eddy, Inc. *Wastewater Engineering: Collection, Treatment, Disposal* (New York: McGraw-Hill Book Company, Inc., 1972).
71. "Marsh System Pilot Study City of Gustine, California," U.S. EPA Project No.C-06-2824-010, Nolte and Associates, Sacramento, CA (November 1983).
72. Jacobsen, B. N. "Physical Description of the Root Zone Installation in Ringsted and Rodekro Municipalities, Experimental Plan and Preliminary Results," Water Quality Institute (VKI), Horsholm, Denmark (1985).
73. Lawson, G. J. "Cultivating Reeds *(Phragmites australis)* for Root-Zone Treatment of Sewage, Project Report 965," Institute of Terrestrial Ecology, Cumbria, England (October 1985).

CHAPTER 28

28a Ancillary Benefits of Wetlands Constructed Primarily for Wastewater Treatment

J. Henry Sather

INTRODUCTION

A basic ecological principle is that the more complex the community structure, the more niches and the more species there will be.[1] A highly diversified wetland ecosystem may not only handle a greater variety of wastewater substances but also may attract and support wildlife for human enjoyment and provide various visual-cultural benefits.

Construction and management of wetlands to foster development of a variety of plant communities that are interspersed with respect to one another and open water and that include those species known to be important to desirable wetland-dependent animals will probably become populated by a variety of wetland-dependent animal species. What may happen to such species in the long term with continued inputs of wastewater is of great concern.[2] Consolidation of existing information and long-term intensive studies on other system components are badly needed.

Inasmuch as plant community diversity is the key to faunal diversity, primary attention should be devoted to basic factors responsible for plant community diversity. Of primary importance are nature of the substrate and hydrologic conditions—especially duration, depth, and flow patterns.[3] Much information on the hydrologic requirements of various plant species is scattered throughout ecological and taxonomic publications.[4-6] The most convenient source of information is the U.S. Fish and Wildlife Service's Annotated National Wetland Plant Species Data Base, which contains abstracted habitat requirements and references on hydrophytes and was used to develop the National Wetland Plant List.[7]

Full manifestation of ancillary benefits in constructed wetlands is contingent upon several factors:[8-10]

1. species composition and degree of interspersion of the plant communities (significant because of diverse habitat requirements of various animal species and because of several other types of benefits)

2. location with respect to human population centers (significant because of its relation to recreational, aesthetic, educational, and research benefits)
3. location with respect to other wetlands (significant because some wetland species are dependent upon wetland complexes and because nearby wetlands can serve as a source of species)

This chapter will address wildlife habitat and visual-cultural benefits as ancillary benefits of constructed wetlands for wastewater treatment.

WILDLIFE

Various species of waterfowl and other wetland-dependent vertebrates are attracted and supported by constructed wetlands,[11-13] but except for muskrats and waterfowl, there is little information associating vertebrate community dynamics with hydroperiods, water chemistry, and plant communities of wetlands.[10] However, it is quite likely that optimal habitat conditions for these animals satisfy the needs of many other wetland-dependent vertebrates.

The muskrat *(Ondatra zibethica)* is distributed widely throughout North America and has been the subject of several intensive studies.[14-27] Three major habitat factors are water depth, water quality, and emergent aquatic plants.[28] In general, muskrats require water deep enough for them to remain active under the ice in winter in more northerly latitudes and to support growth of emergent plants for food and cover. Major water level fluctuations are not desirable because they may cause flooding of bank burrows and/or destruction of muskrat houses and may increase vulnerability to predators. Water quality modifications resulting in emergent vegetation losses also adversely affect muskrat populations. Emergent vegetation suitable as forage and for house construction, such as cattails *(Typha* spp.) and bulrushes *(Scirpus* spp.), appears to be essential in supporting healthy muskrat populations. Wetland size is not important; muskrats live in wetlands varying in size from small roadside pools to marshes covering thousands of acres.

There is a large body of published and unpublished literature describing habitat requirements of waterfowl, but no attempt has been made to consolidate this information into a guide on wetland construction or management for engineers. Martin and Uhler's classic publication[4] on the food of game ducks in the United States and Canada is probably the best single source of helpful information. Bellrose[29] is probably the best current source on habitat requirements of the various waterfowl species, although he does not attempt to aggregate habitat requirements of ducks, geese, and swans.

VISUAL-CULTURAL BENEFITS

Visual-cultural benefits from wetlands construction are directly related to complexity of the plant communities and their degree of interspersion. From the perspective of ecological aesthetics, wetlands are visually and educationally unusually rich environments.[30] Ever-increasing numbers of people consider aesthetics to be one of the most valuable benefits of wetlands, but, unfortunately, these values are also the most difficult to quantify. For example, how does one assess the aesthetic value of the seasonal and daily rhythms of sounds emanating from a wetland, the flight of a family of geese or flock of pelicans across the horizon at sunset, the sight of a muskrat feeding on a cattail while sitting on a feeding platform, or the mesmeric stare of a heron watching for prey? Add to these the benefits derived from wildlife photography, bird watching, canoeing, or simply relaxing, and for many citizens no further justification is needed for constructing wetlands.

Wetlands have been described as living museums where the dynamics of ecological systems can be taught.[31] No other type of ecosystem is more suitable for demonstrating such a broad range of ecological principles within a small area. Wetlands also provide excellent outdoor research laboratories, particularly if water budgets are known (as would be true in most wastewater wetlands). This aspect is certainly one of the most valuable ancillary benefits.

RESEARCH NEEDS

During the past two decades much has been written on the creation and restoration of wetlands for wastewater treatment and mitigation.[2,32,33] Despite this increased attention, there is still a paucity of information required to design, construct, and manage a wetland that will perform the many functions attributed to natural wetlands, and considerable confusion exists about whether wetland ecosystems serve as sinks, sources, or transformers of nutrients, toxicants, and metals.

Development of the Wetland Evaluation Technique (WET), which is based upon interpretations of information found in the scientific literature, revealed significant gaps in our knowledge about functions of wetlands.[34] An exhaustive search of the wetlands literature for distinguishing features of wetlands that might be associated with various functions revealed great gaps in our knowledge of chemical mass balances. A review of recent research in the United States on the role of freshwater and saltwater wetlands as sources, sinks, and transformers of nitrogen, phosphorus, and heavy metals revealed that most research on wetland ecosystems has not generated the information needed to address the water quality functions of wetlands.[35]

The scientific knowledge base is too incomplete to say to what degree newly created wetlands will provide functions attributed to natural wetlands, particularly those functions associated with water supply, water quality, and nutrient transformation.[36] Techniques for manipulating water levels and plants to provide suitable habitats for certain birds and mammals are well documented, but there are few data describing the relation between the structure of wetland plant communities and other wetland functions.[2,37] Moreover, no firm guidelines are available to enable us with any degree of certainty to artificially create detritus and grazing food chains to replicate the full suite of functions of natural wetlands, i.e., food chain, wildlife, and fisheries habitat. We may not in the foreseeable future understand wetlands well enough to prepare definitive guidelines for their creation and management. Nevertheless, we should make careful use of all information currently available and support the establishment and operation of strategically located wetland research centers for long-term intensive studies.

SUMMARY

The magnitude of ancillary benefits realized from wetlands constructed for wastewater treatment will be primarily dependent upon the degree of complexity and the amount of interspersion of the hydrophytic plant communities. Engineering design, therefore, should be sensitive to the hydrologic regimes and soil characteristics required to produce these types of plant communities.

REFERENCES

1. Colinvaux, P. *Introduction to Ecology* (New York: John Wiley & Sons, Inc., 1973), p. 621.
2. "The Effects of Wastewater Treatment Facilities on Wetlands in the Midwest," U.S. EPA Report-905/3–84 (1984).
3. Mitsch, W. J., and J. G. Gosselink. *Wetlands* (New York: Van Nostrand Reinhold Company, 1986).
4. Martin, A. C., and F. M. Uhler. "Food of Game Ducks in the United States and Canada," USDA Tech. Bul. No. 634 (1939), p. 157.
5. Beal, E. O. "A Manual of Marsh and Aquatic Vascular Plants of North Carolina with Habitat Data," North Carolina Agric. Expt. Stat. Tech. Bul. No. 247 (1977), p. 298.
6. Tiner, R. W., Jr. *A Field Guide to Coastal Wetland Plants of the Northeastern United States* (Amherst: University of Massachusetts Press, 1987), p. 285.
7. Reed, P. G. "1986 National Wetland Plant List: Regional Indicator Compilation," U.S. Fish and Wildlife Service Publ. WELUT 86/W17.01 (1986).
8. Bevis, F. G. "Ecological Considerations in the Management of Wastewater-Engendered Volunteer Wetlands," The Michigan Wetlands Conference, July 10–12, Higgins Lake, MI (1979), p. 19.
9. Weller, M. *Freshwater Marshes: Ecology and Wildlife Management* (Minneapolis: University of Minnesota Press), 1981), p. 146.
10. Sather, J. H., and P. J. R. Stuber. "Proceedings of the National Wetland Values Assessment Workshop," U.S. Fish and Wildlife Service, WELUT FWS/OBS-84/12 (1984), p. 100.
11. Small, M. M. "Data Report, Marsh/Pond System," U.S. Energy Research and Development Adm., Brookhaven Nat. Lab. Preliminary Report No. 50600 (1976), p. 28.
12. Demgen, F. C. "Wetlands Creation for Habitat and Treatment at Mt. View Sanitary District, California," *Aquaculture Systems for Wastewater Treatment: Semi-*

nar Proceedings and Engineering Assessment, U.S. EPA, Office of Water Programs Operations, Municipal Division, Washington, DC (1979), pp. 61–73.

13. Demgen, F. C., and J. W. Nute. "Wetland Creation Using Secondarily Treated Wastewater," *American Water Works Association Research Foundation Water Reuse Symposium, Vol. 1,* (1979), pp. 727–739.
14. Johnson, C. E. "The Muskrat in New York," *Roosevelt Wildl. Bull.* 3:199–320 (1925).
15. Baumgartner, L. L. "Ecological Survey of Muskrat Populations and Habitats," Michigan Department of Natural Resources, Wildlife Division Rep. 14-R (1942), p. 45.
16. Lay, D. W., and T. O'Neil. "Muskrats on the Texas Coast," *J. Wildl. Manage.* 6:301–311 (1942).
17. Dozier, H. L., M. H. Markley, and L. M. Llewellyn. "Muskrat Investigations on the Blackwater National Wildlife Refuge, Maryland, 1941–1945," *J. Wildl. Manage.* 12:177–190 (1948).
18. Gashwiler, J. S. "Maine Muskrat Investigations," *Bul. Maine Dept. Inland Fish and Game* (1948), p. 38.
19. O'Neil, T. "The Muskrat in Louisiana Coastal Marshes," Louisiana Department of Wildlife and Fisheries (1949), p. 152.
20. Harris, V. T. "Muskrats in Tidal Marshes of Dorchester County," Chesapeake Biol. Lab., Dept. Res. and Ed. Publ. 91 (1952), p. 36.
21. Dozier, H. L. "Muskrat Production and Management," U.S. Fish and Wildlife Service Circ. 18 (1953), p. 42.
22. Sather, J. H. "Biology of the Great Plains Muskrat in Nebraska," *Wildl. Monograph* 2:35 (1958).
23. Errington, P. L. *Muskrat Populations* (Ames: Iowa State University Press, 1963), p. 665.
24. Palmisano, A. W. "The Distribution and Abundance of Muskrats *(Ondatra zibethica)* in Relation to Vegetation Types in Louisiana Coastal Marshes," in *Proceedings Southeastern Assoc. Game and Fish Commissioners* 25 (1984), pp. 160–177.
25. McCabe, T. R., and M. L. Wolfe. "Muskrat Population Dynamics and Vegetation Utilization: A Management Plan," *Worldwide Furbearer Conf. Proc.* (1981), pp. 1377–1391.
26. Proulx, G. "Relationship Between Muskrat Populations, Vegetation and Water Level Fluctuations and Management Considerations at Luther Marsh, Ontario," PhD thesis, University of Guelph, Ontario (1982).
27. Wilson, K. A. "Investigation on the Effects of Controlled Water Levels upon Muskrat Production," in *Proceedings Annu. Conf. Southeast Assoc. Game and Fish Comm.* 1–6 (1985), pp. 105–111.
28. Allen, A. W., and R. D. Hoffman. "Habitat Suitability Index Models: Muskrat," U.S. Fish and Wildlife Service Biol. Serv. Program FWS/OBS 82/10.46 (1984), p. 27.
29. Bellrose, F. C. *Ducks, Geese, and Swans of North America* (Harrisburg, PA: Stackpole Books, 1976), p. 28.
30. Smarden, R. C. "State of the Art in Assessing Visual-Cultural Values," in *Future of Wetlands: Assessing Visual-Cultural Values* (Totowa, NJ: Allanheld, Osmun & Co., 1986), pp. 5–16.
31. Niering, W. A. *Wetlands* (New York: Alfred A. Knopf, 1985), p. 638.

32. "The Ecological Impacts of Wastewater on Wetlands: An Annotated Bibliography," U.S. EPA Report-905/3-84 (1984).
33. Wolf, R. B., L. C. Lee, and R. R. Sharitz. "Wetland Creation and Restoration in the United States from 1970 to 1985: An Annotated Bibliography," *Wetlands,* Special Issue, 6(1):88 (1986).
34. Adamus, P. R., E. J. Clairain, Jr., R. D. Smith, and R. E. Young. "Wetland Evaluation Technique (WET); Volume II Methodology," U.S. Army Engineers Waterways Experiment Station, Operational Draft Technical Report Y-87 (1987).
35. Nixon, S. W., and V. Lee. "Wetlands and Water Quality: A Regional Review of Recent Research in the United States on the Role of Freshwater and Saltwater Wetlands as Sources, Sinks, and Transformers of Nitrogen, Phosphorus, and Various Heavy Metals," U.S. Army Engineers Waterways Experiment Station, Technical Report Y-86-2 (1986).
36. Larson, J. S. "Wetland Mitigation in the Glaciated Northeast: Risks and Uncertainties," in *Proceedings of a Workshop Held at the University of Massachusetts, Amherst, Sept. 29–30, 1986,* Publ. No. 87-1, pp. 4–16.
37. D'Avanzo, C. "Vegetation in Freshwater Replacement Wetlands in the Northeast," in *Proceedings of a Workshop Held at the University of Massachusetts, Amherst, Sept. 29–30, 1986,* Publ. No. 87-1, pp. 53–81.

28b Overview from Ducks Unlimited, Inc.

Robert D. Hoffman

For over 50 years, Ducks Unlimited has maintained a single purpose: "to perpetuate waterfowl and other wildlife on the North American continent principally by development, preservation, restoration, management, and maintenance of wetland areas." Incorporated in 1937 as a private not-for-profit organization, Ducks Unlimited began its enhancement program in the prairie provinces of Manitoba, Saskatchewan, and Alberta. Since that time, the growth of Ducks Unlimited has been a natural progression of planned, orderly expansion related directly to total dollar growth. In addition to enhancing wetlands in all of the Canadian Provinces and in Mexico, Ducks Unlimited is aggressively expanding waterfowl habitat work in the United States in a major thrust called Wetlands America,—as always, becoming involved through requests of state and federal wildlife management agencies.

"Wetlands America" includes three programs working directly with wetland habitat in the United States. (1) Habitat USA was established in 1984 with the opening of the Great Plains Regional Office in Bismarck, North Dakota to work with public wildlife management agencies on waterfowl production enhancement projects in Montana, North Dakota, South Dakota, and Minnesota. This program was further expanded in 1987 with the opening of the Western Regional Office in Sacramento, California. (2) In 1985, the Habitat Inventory and Evaluation System was developed in cooperation with NASA to utilize space-age technology, satellites, and the expertise of Ducks Unlimited in wetlands to develop a comprehensive inventory of our wetland resources. The goals of this program are to know where wetlands are, provide better protection for wetlands, and maximize waterfowl production from wetlands. (3) Also in 1985, the MARSH program (Matching Aid to Restore States' Habitat) was established. The MARSH program returns 7.5% of each state's fund-raising income to the state wildlife management agency to develop waterfowl habitat within state boundaries.

Ducks Unlimited would most likely become involved in utilizing wastewater for wetlands enhancement through the MARSH program, but some potential involvement may also exist through the Great Plains Regional Office in Bismarck, providing the projects occur within the four-state area. To date, the MARSH program has been involved in two sewage treatment projects. One project currently under consideration is the expansion of the Paiute Marsh located near Lancaster, California at Edwards Air Force Base. The other is Crane Marsh near Fremont, Indiana, where the quality of wastewater passed through a 9-ha marsh is monitored for comparison with pretreatment measurements.

Ducks Unlimited is interested in all forms of wetland enhancement. Utilization of sewage effluent in natural or man-made marshes is of interest specifically for the benefits it may generate for waterfowl and other wetland wildlife species. Waterfowl benefits from wastewater treatment systems have been documented in Arizona[1] and South Dakota.[2] The utilization of wastewater treatment in wetlands also demonstrates the role wetlands play in improving water quality. Many studies have shown the benefits wetlands play in improving water quality from municipal sewage plants.[3-5]

Another important benefit sewage effluent may provide is a stable water supply for wetlands in areas where the water resource is limited. This becomes extremely important in light of the fact we are losing approximately 185,500 ha of wetlands annually in the United States.[6] Caution should be taken, however, in using this justification carte blanche for pumping effluent into natural wetlands. Indiscriminant exploitation of natural marshes can lead to their degradation.[7] The addition of sewage effluent to a natural marsh may hasten eutrophication and increase exposure to toxic compounds or elements. We do not know the long-term effects on the density and diversity of the organisms in the ecosystem when it is put under these types of stresses. The design of a wastewater treatment marsh is site-specific. An evaluation of the effects of an effluent water supply on the receiving water body, and ultimately on waterfowl and other wildlife for which the system is designed to benefit, must be conducted and understood prior to constructing a facility.

SUMMARY

The utilization of wastewater treatment effluent for the creation and enhancement of wetlands has significant potential. This potential can develop into providing additional valuable habitat for waterfowl and other wetland wildlife species. Ducks Unlimited is interested in this alternative of improving wetlands through our conservation programs; however, we stress that care must be taken to ensure that wastewater utilization does not cause negative impacts on our nation's wetlands in future years.

REFERENCES

1. Piest, L. A., and L. K. Sowl. "Breeding Duck Use of a Sewage Marsh in Arizona," *J. Wildl. Manage.* 49:580–585 (1985).

2. Dornbush, J. N., and J. R. Anderson. "Ducks on the Wastewater Pond," *Water Sew. Works* 3(6):271–276 (1964).
3. Fetter, C. W., W. E. Sloey, and F. L. Spangler. "Potential Replacement of Septic Tank Drain Fields by Artificial Marsh Wastewater Treatment Systems," *Groundwater* 14:396–402 (1976).
4. Lakshman, G. "A Demonstration Project at Humbolt to Provide Tertiary Treatment to the Municipal Effluent Using Aquatic Plants," Saskatchewan Res. Counc. Tech. Rep. No. 151 (1983).
5. Sloey, W. E., F. L. Spangler, and C. W. Fetter, Jr. "Management of Freshwater Wetlands for Nutrient Assimilation," in *Management of Freshwater Wetlands for Nutrients Assimilation,* R. Good, D. Whigham, and R. Simpson, Eds. (New York: Academic Press, Inc., 1978), pp. 321–340.
6. "Draft EIS, The Sport Hunting of Migratory Birds," U.S. Fish and Wildlife Service, Washington, DC (1987).
7. Spangler, F. L., W. E. Sloey, and C. W. Fetter. "Experimental Use of Emergent Vegetation for the Biological Treatment of Municipal Wastewater in Wisconsin," in *Biological Control of Water Pollution,* J. Tourbier and R. Pierson, Eds. (Philadelphia: University of Pennsylvania Press, 1976), pp. 161–171.

CHAPTER 29

Configuration and Substrate Design Considerations for Constructed Wetlands Wastewater Treatment

Gerald R. Steiner and Robert J. Freeman, Jr.

INTRODUCTION

Components of a constructed wetlands treatment system (CWTS) include preliminary/primary treatment units and the constructed wetlands cell (or cells). The constructed wetlands (CW) includes substrate, vegetation, and biological organisms contained within a physical configuration that can be described as an attached-growth biological filter. Major pollutant removal mechanisms include sedimentation and filtration (physical), precipitation and adsorption (chemical), and bacterial metabolism (biological). A CWTS can be designed to achieve various levels of secondary and advanced level treatment for BOD_5, suspended solids, nutrients, pathogens, metals, and other substances.

Two basic types of designs for a CW are surface flow (SF) and subsurface flow (SSF). A design can incorporate one or both types. An SF system consists of a cell or cells with wastewater routed at shallow depths over a substrate supporting emergent vegetation. Flow is controlled by the shallow depth, low flow velocity, and the plant stems and litter.

An SSF system has a cell or cells with wastewater routed through and below the surface of a permeable substrate supporting emergent vegetation. SSF systems have been described as gravel marsh, root-zone, reed bed, rock/plant filter, and gravel-based emergent macrophyte systems.

Depending on specific pollutant removal needs and other factors, a variety of configurations and substrates can be used for a CW. To optimize treatment effectiveness, sound engineering decisions and practices must be used in configuration and substrate design.

CONFIGURATION

Configuration of a CW affects important hydrologic factors controlling pollutant removal processes. These factors include water velocity, water depth and fluctuation, detention time, circulation and distribution patterns, and turbulence and wave action. Configuration should enhance wastewater distribution to maximize contact between wastewater, substrate, and vegetation by minimizing short-circuiting.[1] Configuration selection should consider degree of pretreatment, required treatment area, available land shape and slope, length-to-width ratio, desired bed slope, amount of excavation and grading to obtain cell depth and slope, substrate type and cost, internal dikes and distribution/collection pipes, and operation and maintenance flexibility.

Preliminary/Primary Treatment

Current recommendation is to precede the CW with preliminary/primary treatment. Some optional flow schemes are shown in Figure 1.

"Preliminary treatment" is screening and/or comminution to remove coarse solids. "Primary treatment" removes heavier solids and reduces organic load and can include an Imhoff tank, septic tank(s), stabilization pond(s), or a primary sedimentation tank.[2]

If an existing treatment system is upgraded with a CW, cost evaluation should determine if the existing facility is to be (1) abandoned, (2) used with a CW providing polishing treatment, or (3) modified for preliminary/primary treatment with the CW providing secondary or advanced-level treatment. For an abandoned existing facility or a new system, the type of preliminary/primary treatment must be selected.

Small communities with wasteflow less than 380 m^3/day should have a simple primary treatment system preceding the CW. Septic tanks can serve individual or clusters of homes with an alternative sewer system such as septic tank effluent pumping (STEP). An Imhoff tank or septic tank(s) requires sludge removal and drying beds or disposal to a large system. Larger communities may require a more complicated sedimentation tank system with sludge digestion and sludge-drying beds.

Flow Patterns

A CW cell is designed to use one or more of three types of flow patterns: plug flow, step feed, or recirculation.[2] Plug flow (Figure 2a) is once-through flow down the cell length. Plug flow is now used for most municipal and acid drainage systems and requires minimal piping, energy use, operation, and maintenance.

Step feed (Figure 2b) may benefit pollutant removal by using more of the CW area for solids removal and by providing carbon for nitrogen removal in the lower bed area. Step feed is typically combined with recirculation. The cost-benefit of step feed needs investigation.

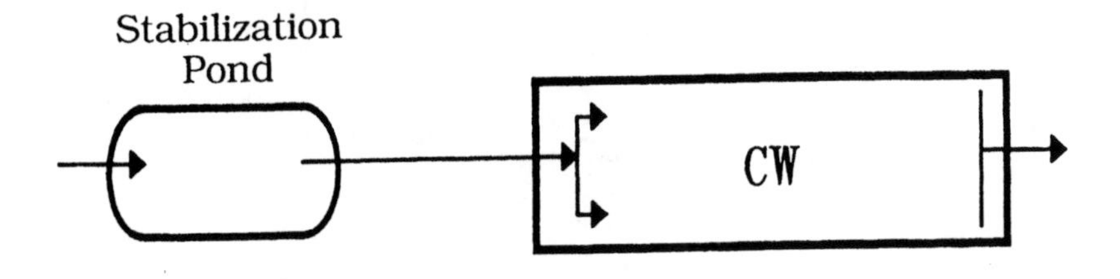

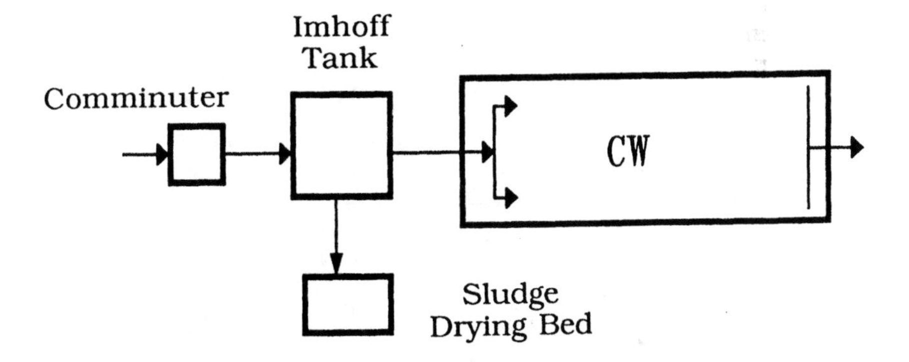

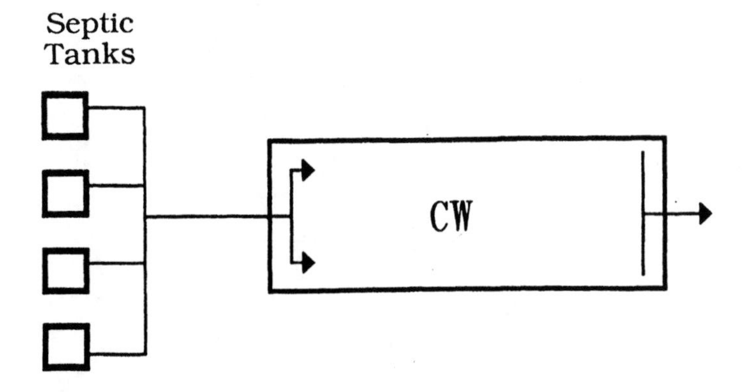

Figure 1. Optional flow schemes for preliminary/primary treatment with constructed wetlands.

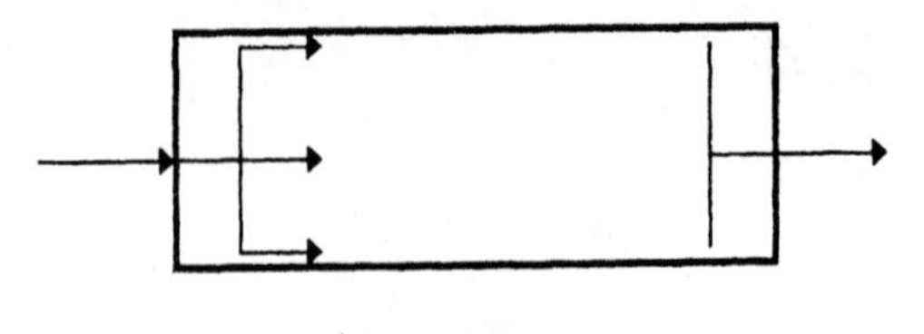

a. Plug flow

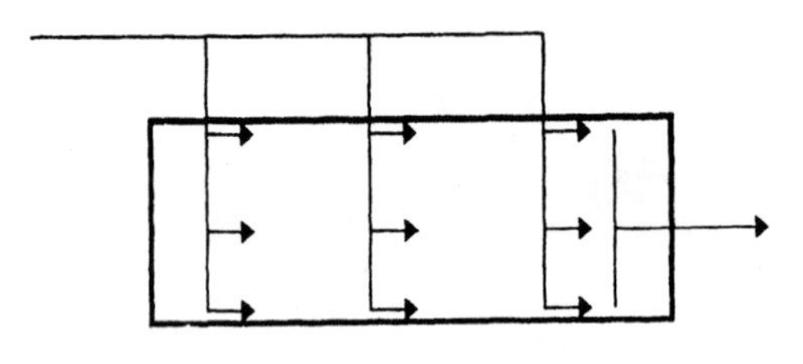

b. Step feed

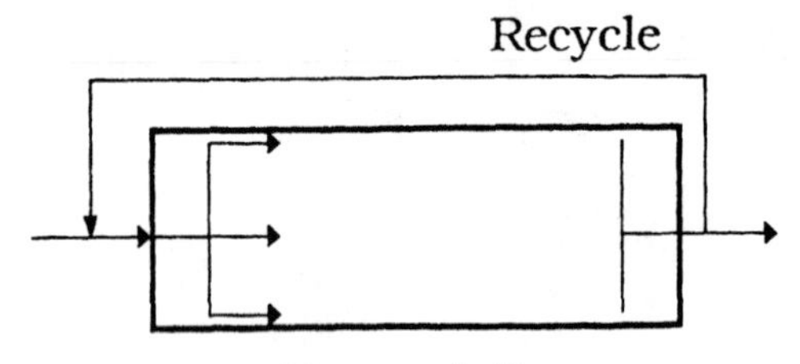

c. Recirculation

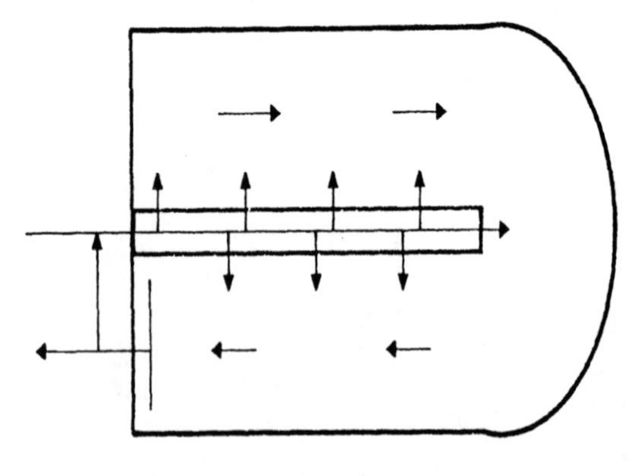

d. "Jelly roll" step feed-recirculation cell

Figure 2. Basic flow patterns for constructed wetlands.

Recirculation (Figure 2c) should be considered, and its potential needs further investigation. Recycling treated effluent will dilute influent BOD_5 and suspended solids, decreasing odor potential and increasing dissolved oxygen concentration and retention time, which will enhance nitrification and subsequent nitrogen removal. Recycled effluent can have its oxygen content increased by applying it through agriculture spray headers (using pumps that aspirate air into the flow) or cascading it into the cell.

Disadvantages to recirculation include increased construction costs and increased operation and maintenance. Although pumping is required, it is of a low solids effluent at a small head.

A possible recirculation/step feed configuration is the "jelly roll" shape (Figure 2d).[3] It reduces pumping and piping requirements by the adjacent inlet and discharge. An inlet pipe with adjustable tees will improve distribution.

Configuration Alternatives

Alternative CW configurations include a single cell, parallel cells, series cells (longitudinal or serpentine), and combinations of wetlands cell(s) and pond.

Single Cell

A single rectangular cell is the simplest design and least expensive to build (Figure 3a), but operational flexibility is limited. If a single cell needs maintenance, only the preliminary or primary treatment occurs. A single cell is only recommended for small flows (<4 m^3/day) where maintenance can be performed quickly.

Parallel Cells

At least two parallel cells are recommended for a CW because operational and maintenance flexibility is increased (Figure 3b). Flow is distributed equally or proportionally between the cells based on loading rates or other reasons. One cell may be drained for maintenance while others continue operating, though treatment effectiveness may decrease. Possible mosquito control is enhanced in an SF system by draining each cell at rotating intervals.

Parallel cells improve construction/operation cost flexibility because some cells can be SSF and some SF. Substrate for an SSF cell (gravel) costs more, but greater hydraulic loadings can be applied to it. Internal divider dikes and flow distribution boxes will increase construction costs. However, parallel cells allow CW modularization (addition of cells as system expands) and expenditure of capital funds as needed.

Series Cells

Series cells can be longitudinal or serpentine (Figures 3c and 3d). By mixing CW types in series, improved pollutant removal is obtained through a greater variety of treatment mechanisms. For example, in a three-cell SF-SSF-SSF

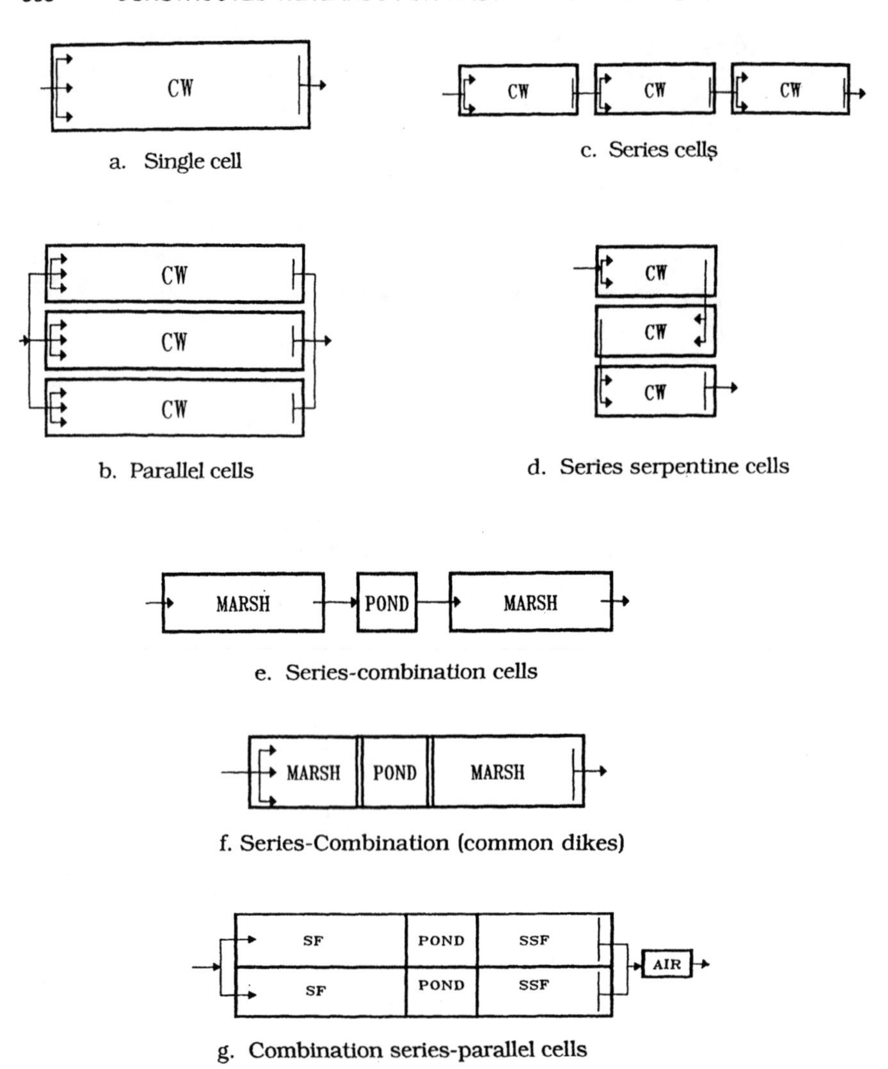

Figure 3. Alternative configurations for constructed wetlands

system most solids and BOD removal may occur in the SF cell, with a well-oxygenated wastewater entering the first SSF cell to enhance nitrification. Cascades between SF and SSF cells provide aeration to enhance nitrification or oxidize and precipitate metals in acid drainage. The final SSF cell can be designed for denitrification. Aerated effluent can be recycled to the first SSF cell, providing additional contact time for nitrification-denitrification to meet stringent ammonia or nitrogen limits. Secondary cells can have different SSF substrates to improve removal of specific pollutants such as phosphorus or trace metals.

Limited operational and maintenance flexibility, similar to single-cell systems, can be a disadvantage. All flow must bypass the system unless piping and valves are available so individual cells may be bypassed.

Combination Wetland Cells and Pond

Including a small pond or open water area with either SF or SSF cells (Figures 3e and 3f) provides pollutant removal and operational benefits. A configuration of SSF cells and open water areas performed better with secondary or primary effluents than SF or SSF only.[4,5] A pond will enhance ammonia reduction and nutrient removal through algal uptake, increased pH, and nitrification.

In an SF system, the pond can interrupt short-circuiting in an upstream cell and reestablish uniform flow distribution in a downstream cell. Pond effluent can be redistributed with piping or through a stone filter berm at the front of the downstream cell. Also, *Gambusia,* which will migrate between the pond and SF cells, can be stocked for mosquito control. *Gambusia* will survive winter and other stress periods in the pond's deep water.

A pond prior to a downstream SSF cell will provide natural reaeration to enhance nitrification or, if following an SSF cell, will increase dissolved oxygen to meet possible limits. If better oxygenation is needed to enhance nitrification either before or after the vegetation has matured, mechanical aeration can be added to the pond.

One combination, the marsh-pond-meadow design, has been successful in Pennsylvania. Treatment results for a 45 m^3/day system at Iselin, Pennsylvania have been excellent for BOD_5, ammonia nitrogen, and fecal coliform (before chlorination). The marsh and meadow have sand-gravel substrates with cattail *(Typha)* in two parallel marsh cells and reed canary grass *(Phalaris)* in two parallel meadow cells.[6] With current recommendations, a marsh-pond-meadow requires more land (5.3 ha/1000 m^3/day) than a gravel marsh (2.1 ha/1000 m^3/day), but gravel and excavation cost may be higher than land and grading cost. Parallel SF and SSF marsh-pond-meadow systems at the 341 m^3/day Pembroke, Kentucky facility will be used to compare treatment effectiveness with cost, cell size, and loading rates.[7]

Combination Series-Parallel Cells

An attractive option to series or parallel cell CW is a combination series-parallel system (Figure 3g). This will provide design flexibility to improve treatment for different pollutants and operation and maintenance flexibility.

Length-to-Width Ratio (L/W)

The configuration must minimize short-circuiting to maximize wastewater contact with the entire surface area of an SF system and the cross-sectional area of an SSF system. The L/W ratio is a key design factor to achieve this condition.

For SF systems, an L/W of at least 10 is required. At Listowel, a serpentine system with an L/W of 75 consistently outperformed a rectangular cell with an L/W of 4.5, even though the serpentine system received higher loadings.[8] Providing an L/W of at least 10 will achieve consistent internal flow distribution and reduce short-circuiting that is likely in shorter rectangular SF cells.

For an SSF system, wider beds with L/W's as low as 1 or less are recommended so that solids are distributed over a greater portion of the bed, decreasing the potential for substrate plugging. A large L/W is not as necessary because plug flow should occur in the substrate if good distribution is used and if hydraulic capacity is not exceeded. Design equations for the root-zone method using "engineered" soil substrate often result in L/W's less than 1, but poor hydraulic conductivity has caused short-circuiting across the bed surface.[9-11]

Configuration and SF or SSF Systems

Selection of SF or SSF systems must consider size, shape, and cost of available land, excavation and grading, substrate cost, climate, and pollutant removal requirements. More land area and larger L/W are required for SF than SSF systems due to hydraulic loading requirements. An SF system requires a long, narrow piece of land unless a serpentine configuration is used, whereas an SSF system requires a more compact area.

Excavation costs may be greater for an SSF system due to bed depth. Grading costs for cell floors and berms will be less for an SSF system because surface area and dike length are less than for an SF system.

Substrate is a major cost item in an SSF, depending on CW size and substrate hauling distance. An SF system typically uses in situ soil.

Climate can influence system type because an SSF cell will be less affected by cold temperatures and freezing than an SF cell. Plant cover and litter will insulate the surface of an SSF cell, reducing freezing effects. In an SF cell, ice eliminates surface reaeration and overlying snow restricts solar radiation, affecting photosynthesis and related biological processes. An SF cell requires higher berms so that water depth can be increased in winter to compensate for detention volume lost to ice cover.

Effluent quality requirements will affect CW type and configuration. An SSF system is more effective per unit area in reducing BOD and SS. Configuration combined with specific substrates can enhance specific pollutant removal. Stringent permit limits will require complex CW systems with more cells, different substrates, and recirculation.

Available Land

Configuration should take advantage of natural topography to minimize excavation and grading costs. A single SF cell may require a narrow strip of land. Parallel cells use a compact design area. For example, a CW requiring 25,000 m^2 of effective treatment area with an L/W ratio of 10 requires a single cell 500 m long and 50 m wide or four parallel cells with an effective area 250 m long by 100 m wide. However, total area of the four parallel cells will be 15% larger due to divider dikes.

Inlet Distribution

Cell inlet design must minimize short-circuiting and stagnation. With a large L/W, piping is less and inlet construction is simplified. However, to distribute solids in the influent across a greater portion of the cell, inlet design for wide beds is critical.

Series cells require redistribution at the inlet of each subsequent cell. A serpentine system with an opening between the divider dikes may have short-circuiting even in an SSF system. The discharge of each cell should be redistributed at the head of the subsequent cell. Inlet design is detailed elsewhere.[12]

Slope

A slope of 0.5% or less, as limited by construction tolerances, is recommended for an SF system. Some slope is needed to drain the cell for maintenance and possible mosquito control. Positive water flow is provided by differential water depths and inlet and outlet elevations.[9] For SSF systems, the bed slope of the bottom of the bed is critical for hydraulic flow in permeable substrates, as discussed by Watson et al.[13] SSF bed slopes should be 2% or less, based on the initial hydraulic conductivity of the substrate according to Darcy's law and Ergun's equation. A nearly flat SSF bed has the hydraulic head controlled only by the inlet and outlet elevation difference.

A level substrate surface is recommended for an SSF system to facilitate vegetation planting and so that the bed can be flooded for weed control.[9,11] In Europe, root-zone method systems have bed slopes generally of 1–4%, but up to 8%. These systems have not performed as claimed; poor hydraulic conductivity causes most of the flow to be on the surface, with channeling, treatment, and vegetation management problems.

Summary

Table 1 compares key cost, operation, and treatment factors for several configurations and can be used to guide initial configuration selection. The relative ratings are general and can differ with specific conditions. As CW technology advances, other configurations will optimize cost-effectiveness for specific treatment requirements.

Table 1. Relative Comparison of Cost, Operation, and Treatment Factors for Various Constructed Wetlands Configurations

Configuration	Land Area	Substrate Cost	Excavation/[a] Grading	Cold Climate Effect	Mosquito[b] Control	Advanced[c] Treatment	Nitrification Enhancement (NH_3 Reduction)	Denitrification Enhancement (N Removal)	Phosphorus[d] Sorption Enhancement	Effluent Dissolved Oxygen
SF	H	na	L/H	H	L	M	L	L	N	L-M
SSF	L	H	H/L	L	H	M-H	L	L	N-H	L
SF-SSF	M	M	M/M	H	L	M-H	L	L	N-H	L
SSF-SF	M	M	M/M	H	L	M-H	M	L-M	N-H	L-M
SF-P-SF	H	na	M/H	H	M	M	M-H	L	N	H
SF-P-SSF	M	M	M/M	H	M	H	H	H	N-H	L
SF-SSF-SF/P	M-H	na	H/H	H	L	M-H	M-H	M	N-H	M-H
SSF-P/C-SSF	L	H	H/M	L	H	H	H	H	N-H	L
SF-SSF-R	M	M	M/M	H	M	H	H	H	N-H	M
SSF-SSF-R	L	H	H/M	L	L	H	H	H	N-H	M
EFF-AIR	L	na	na/L	na	na	na	na	H	na	H

Note: SF = surface flow cell; SSF = subsurface flow cell; P = pond or open water area; SF/P = surface flow cell or pond, etc; R = recirculation; P/C = pond, etc., or cascade; EFF-AIR = effluent aeration; H = high; M = medium; L = low; N = none; na = not applicable.

[a]Excavation of soil without major rock.

[b]That provided by configuration only, not by operation.

[c]BOD and SS effluent concentrations 20 mg/L or less.

[d]Dependent on physical and chemical characteristics of SSF substrate.

SUBSTRATE

Whether a CW is SF or SSF is dependent on the substrate's hydraulic conductivity and the water levels in relation to the substrate surface. Substrate-water and substrate-root interfaces are critical for development of aerobic-anaerobic treatment mechanisms. The substrate supports vegetation, provides surface area for microorganism attachment, and is associated with the physical and chemical treatment mechanisms. Substrates affect treatment capability through detention time, contact surfaces for organisms with the wastewater, and oxygen availability.[9,10,14]

Vegetation

For practical purposes, CW vegetation is considered an integral part of the substrate. Vegetation provides surfaces for microbial growth on roots, rhizomes, leaves, and stems; filters solids; and transfers oxygen to provide an aerobic/oxidized environment for organic decomposition and desired microbial populations such as nitrifiers. Nutrient assimilation by plants is insignificant, and the nutrients are released when the aboveground vegetation dies seasonally.[10,15]

Plants transmit atmospheric gases to the roots, and leakage from the root hairs is an important source of oxygen for SSF systems. Since the water surface is below the substrate surface, little direct aeration occurs and submerged substrates tend to become anaerobic. Microscopic zones adjacent to the roots (the rhizosphere) are aerobic due to oxygen leakage from the roots.[9]

Vegetation helps maintain hydraulic pathways in the substrate. Plant litter accumulating on the original surface of the substrate composts and becomes additional substrate. Over time, an SF system gradually will become a partial SSF system with wastewater flowing through the litter and added rhizosphere. Similarly, an SSF system with plugged substrate may become partially SF, but treatment will continue in the permeable litter layer. Surface flow that short-circuits around clumps of litter may require maintenance.

Types

Substrate selection is based on cost, treatment requirements, and SF or SSF designs. Substrates include natural soils (clay or topsoil), soil mixtures, gravel and crushed rock (various sizes and composition), coal-fired power plant ash, and combinations.[9,16] An SSF system will more effectively remove pollutants per unit area, but costs of SSF substrates and handling (transportation and spreading) may offset the higher costs for land and earthwork of an SF system. For SF systems, local soil is used with underlying impermeable soil to protect groundwater or conserve water.

Two basic types of substrates are used for SSF systems—gravel and/or sand, and soil. Gravel and/or sand substrate is recommended for most SSF systems.

Gravel substrates allow wastewater conductance at initial operation before the vegetation has matured. Pore spaces, later filled with solids, may be opened by root and rhizome growth. Plant litter accumulating on the bed surface will increase the bed's cross-sectional area, creating additional hydraulic pathways and treatment zones.

Gravel beds are not homogeneous and will vary in hydraulic conductivity along the bed length. The inlet zone is relatively short and receives high suspended solids that accumulate. The longer downstream zone has low solids buildup and a more stable permeability.[16]

Gravel size is selected using Darcy's law to determine hydraulic conductivity after required area and desired L/W, cross-sectional area, and slope have been selected. For a 0.6-m deep bed, gravel sizes are 12–25 mm for the lower 0.45 m and pea gravel of 6–12 mm for the top 0.15 m to enhance vegetation planting and growth; however, 12–19 mm might be used throughout the total depth. Inlet and outlet zones should have 5- to 10-cm stones to distribute flow and prevent clogging. Too-small gravel will limit hydraulic conductivity, and flow may surface. Too-large gravel will decrease retention time and surface area for microbial attachment and may inhibit vegetation growth.

Common gravel substrates include washed and sized crushed limestone or river gravel. Limestone can add alkalinity needed for nitrification. Substrate type and composition can influence treatment of pollutants such as phosphorus but may have little impact on other pollutants. One study of six substrates used in treating acid mine drainage (topsoil, clay, pea gravel, natural wetland, mine spoil, and acid wetland) concluded that the substrate type was relatively unimportant to removal efficiency, and natural local substrate would minimize cost.[17]

The root-zone method, pioneered in Europe, uses soil engineered to an initial hydraulic conductivity so that wastewater travels through the soil substrate with no surface flow. Plant roots and rhizomes supposedly maintain the hydraulic conductivity in subsequent years. However, root-zone systems in Europe during the first years of performance have a low hydraulic conductivity, functioning basically as undersized SF systems. They do not have the desired treatment effectiveness, especially for nutrient removal.[10,11,18] Engineered soil substrates are not recommended for new SSF designs if advanced level treatment is required. However, if secondary level treatment is required for only BOD_5 and suspended solids, a soil substrate can be considered if the original hydraulic conductivity (not an assumed transformation value provided by root and rhizome penetration) is used to determine the required treatment area.

Substrate Depth and Type of Vegetation

Substrate depth influences retention time in an SSF system. A 0.6-m depth is common, but substrate depth and vegetation type must be compatible. Desirable vegetation should grow densely, spread rapidly, and have an extensive vertical/horizontal root system. If plant roots do not penetrate the full depth of the bed, a totally anaerobic zone will occur below the roots, decreasing nitrification, ammonia reduction, and effluent oxygen.

Common emergent vegetation includes bulrush *(Scirpus validus);* reed *(Phragmites australis);* and cattail *(Typha latifolia).* In a study with a 0.76-m-deep gravel bed, *S. validus* provided greater ammonia nitrogen reduction than *P. communis* and *T. latifolia* with only *Scirpus* penetrating the full bed depth, indicating that maximum bed depths for *Scirpus, Phragmites,* and *Typha* should be 0.76 m, 0.6 m, and 0.3 m, respectively.[15] In the United Kingdom, recommended substrate depth is 0.6 m deep at the inlet end because *Phragmites* roots weaken beyond that point. In northern locations, where a shallow bed is more susceptible to freezing, the insulating plant cover and surface litter will help prevent freezing.[11]

Pollutant Removal

Substrate type has little impact on suspended solids and organic removal in SF and SSF systems. The major removal mechanisms— sedimentation and filtration—are similar in both systems and occur in the front part of a cell due to the quiescent conditions.[9]

Substrate type is also unimportant in biological degradation of organics. However, plant species that have a large stem surface area per unit bed area will provide the greatest biological growth for wastewater contact.

Substrates influence removal of some contaminants (e.g., NH_4^+ and metals) through ion exchange and adsorption onto humic and fulvic substances and clay particles. Organic soils and clay minerals have higher exchange capacities than coarse mineral substrates such as gravel. Organic soils with high humic content readily remove metallic ions through ion exchange, but site saturation may limit longevity. Substrates containing iron, aluminum, and calcium can remove significant amounts of phosphorus through coprecipitation.[16] SSF substrate provides greater contact opportunities for these mechanisms.

After solids sedimentation and filtration, further nitrogen reduction in CW systems occurs by nitrification and denitrification. In SF and SSF systems, the design should maximize flow contacts at the water-substrate-plant interfaces. Small substrate materials in SSF systems provide greater microorganism attachment area if hydraulic conductivity is maintained. In SF systems, a dense vegetation growth may support similar amounts of attachment areas. Desired ammonia reduction may not occur until vegetation forms dense stands above the surface and extensive root/rhizome systems within the substrate. Because two to four growing seasons may be required (depending on initial planting density), a new CW system may not meet stringent ammonia or nitrogen limits until the vegetation has matured, unless oxygen is added to the wastewater by configuration or mechanical means. When efficient ammonia reduction occurs, nitrogen removal should also occur by denitrification as wastewater flows through anaerobic microscopic zones in the substrate.

Phosphorus removal in CW systems has been erratic. It depends on substrate chemical composition and obtaining good water-substrate contact to maximize chemical precipitation, bacterial action, and adsorption onto the substrate matrix. If phosphorus removal is required, clay with iron and aluminum content should be considered. However, soils with a high phosphorus removal capacity are finer textured, and sand may be added to improve hydraulic conductivity. Also, iron or aluminum added to the substrate or fed into the wastewater can improve phosphorus removal.[9,16]

Liners

If groundwater contamination or water conservation is a concern, an impermeable liner below the substrate is required. Possible materials are compacted in situ soil (permeability less than 10^{-6} cm/sec); bentonite; asphalt; synthetic butyl rubber; or plastic membranes. The liner must be strong, thick, and smooth to prevent root penetration and attachment. An option is to locate the liner below the maximum depth of root penetration; however, a zone of anaerobic wastewater may then occur. If gravel with sharp edges (e.g., crushed limestone) is used, a layer of fine-to-medium sand should be placed on synthetic membranes to prevent puncture during gravel placement. In Denmark, 0.5- to 1.0-mm polyethylene liners are successfully used.[10,11]

REFERENCES

1. Huntzsche, N. N. "Wetlands Systems for Wastewater Treatment: Engineering Applications," in *Ecological Consideration in Wetlands Treatment of Municipal Wastewaters,* P. J. Godfrey, E. R. Kaynor, S. Pelczarski, and J. Benforado, Eds. (New York: Van Nostrand Reinhold Company, 1985), p. 7.
2. Tchobanoglous, G. "Aquatic Plant Systems for Wastewater Treatment: Engineering Considerations," in *Aquatic Plants for Water Treatment and Resource Recovery,* K. R. Reddy and W. H. Smith, Eds. (Orlando, FL: Magnolia Publishing Inc., 1987), pp. 27–48.
3. Wieder, R. K., G. Tchobanoglous, and R. W. Tuttle. "Preliminary Considerations Regarding Constructed Wetlands for Wastewater Treatment," Chapter 25, this volume.
4. Bavor, H. J., D. J. Roser, and S. McKersie. "Nutrient Removal Using Shallow Lagoon-Solid Matrix Macrophyte Systems," in *Aquatic Plants for Water Treatment and Resource Recovery,* K. R. Reddy and W. H. Smith, Eds. (Orlando, FL: Magnolia Publishing Inc., 1987), pp. 227-235.
5. Bavor, H. J., D. J. Roser, and S. McKersie. "Treatment of Wastewater Using Artificial Wetlands: Large-Scale, Fixed-Film Bioreactors," *Aust. Biotechnol.* 1(4) (1987).
6. Watson, J. T., F. D. Diodato, and M. Lauch. "Design and Performance of the Artificial Wetlands Wastewater Treatment Plant at Iselin, Pennsylvania," in *Aquatic Plants for Water Treatment and Resource Recovery,* K. R. Reddy and W. H. Smith, Eds. (Orlando, FL: Magnolia Publishing Inc., 1987), pp. 263–270.
7. Steiner, G. R., J. T. Watson, and D. A. Hammer. "Constructed Wetlands for

Municipal Wastewater Treatment," paper presented at the Mississippi Water Resources Conference, Jackson, MS, March 29–30, 1988.

8. Wile, I., G. Miller, and S. Black. "Design and Use of Artificial Wetlands," in *Ecological Considerations in Wetlands Treatment of Municipal Wastewaters,* P. J. Godfrey, E. R. Kaynor, S. Pelczarski, and J. Benforado, Eds. (New York: Van Nostrand Reinhold Company, 1985), p. 26.
9. Reed, S. C., E. J. Middlebrooks, and R. W. Crites. *Natural Systems for Waste Management and Treatment* (New York: McGraw-Hill Book Company, 1988).
10. Brix, H. "Treatment of Wastewater in the Rhizosphere of Wetlands Plants—The Root-Zone Method," *Water Sci. Technol.* 19:107–118 (1987).
11. Cooper, P. F., and J. A. Hobson. "State of Knowledge on Reed Bed Treatment Systems," Report Number 581-S, Water Research Centre, United Kingdom (October 1987).
12. Watson, J. T., and J. A. Hobson. "Hydraulic Design Considerations and Control Structures for Constructed Wetlands for Wastewater Treatment," Chapter 30, this volume.
13. Watson, J. T., S. C. Reed, R. H. Kadlec, R. L. Knight, and A. E. Whitehouse. "Performance Expectations and Loading Rates for Constructed Wetlands," Chapter 27, this volume.
14. Good, B. J., and W. H. Patrick, Jr. "Root-Water-Sediment Interface Processes," in *Aquatic Plants for Water Treatment and Resource Recovery*, K. R. Reddy and W. H. Smith, Eds. (Orlando, FL: Magnolia Publishing Inc., 1987), pp. 359–371.
15. Gersberg, R. M., B. V. Elkins, S. R. Lyon, and C. R. Goldman. "Role of Aquatic Plants in Wastewater Treatment by Artificial Wetlands," *Water Res.* 20(3):363–368 (1986).
16. Venhuizen, D. "Combination Systems," briefing paper for the LCRA (Lower Colorado River Authority) Innovative/Alternative Wastewater Project (1987).
17. Brodie, G. A., D. A. Hammer, and D. A. Tomljanovich. "An Evaluation of Substrate Types in Constructed Wetlands Acid Drainage Treatment Systems," in *Mine Drainage and Surface Mine Reclamation,* Bureau of Mines Information Circular 9183 (1988).
18. Boon, A. G. "Report of a Visit by Members and Staff of WRC to Germany (GFR) to Investigate the Root-Zone Method for Treatment of Wastewaters," revised and reprinted, Water Research Centre, Stevenage, England (February 1986).

CHAPTER 30

Hydraulic Design Considerations and Control Structures for Constructed Wetlands for Wastewater Treatment

James T. Watson and John A. Hobson

INTRODUCTION

All constructed wetlands are attached-growth biological reactors. Performance is based on first-order, plug-flow kinetics (see Kadlec[1] for guidance on deviations). The basic relationship is:

$$\frac{C_e}{C_o} = \exp\,[-\,K_t t] \qquad (1)$$

where C_e = effluent concentration, mg/L
C_o = influent concentration, mg/L
K_t = temperature-dependent first-order reaction rate constant, $days^{-1}$
t = hydraulic residence time, days

The equation suggests that as hydraulic residence increases, effluent concentrations of biodegradable contaminants decrease. Consequently, hydraulic residence time becomes a key design and operational parameter for optimizing the performance of a wetland system. Hydraulic residence time is defined as:

$$t = \frac{LWnd}{Q} \qquad (2)$$

where L = length of system (parallel to flow direction), m
W = width of system (perpendicular to flow direction), m
n = porosity of the bed, as a decimal fraction
d = depth of submergence, m
Q = average flow through the system, m^3/day

The objectives of this chapter are to summarize information on these parameters, identify considerations for each parameter important to performance of wetlands systems, and identify the type and general design of structures needed to establish and control the hydraulic regime. Information is segregated, where appropriate, for the two major types of wetland systems: the conventional wetland with an exposed free water surface (surface flow system) and the submerged bed utilizing a permeable substrate (subsurface flow system). Both types have been described previously.[2]

HYDRAULIC RESIDENCE TIME PARAMETERS AND RELATED FACTORS

Length and Width

Dimensions of a constructed wetlands can be established based on biological reaction and flow kinetics or loading and dimensional guidelines derived from the literature. Details are contained elsewhere.[2,3]

Length-to-width ratios for surface flow systems are typically large, 10:1 or greater, to ensure plug-flow conditions and minimal short-circuiting. Ratios for subsurface systems are normally less than 10:1 to ensure subsurface flow. The ratio may need to be less than 3:1 for most gravel beds and much less than 1:1 for subsurface soil beds.

Once established by design, length and width of a system become constants. However, *effective* width is determined by short-circuiting and dead zones resulting from design and location of inlet and outlet structures, uneven bottom or substrate, accumulation of solids and biological slimes, and nonuniform vegetation densities. Operation and maintenance should ensure that effective width approaches actual width along the length of the system. Dyes or other tracers are typically used to determine significance and locations of channelization and plan remedial actions.

Porosity

The porosity of the system is defined as:

$$n = \frac{V_v}{V} \tag{3}$$

where V_v and V are volume of voids and total volume, respectively.

In a surface flow system, V_v is, for practical purposes, the volume not occupied by vegetation and varies with type and density of live and dead vegetation. Representative values have not been established for plants used in treatment systems. At Listowel, cattails occupied about 5% of the volume.[4] Preliminary data from TVA systems indicate the potential volumes of several common species are cattails *(Typha)*, 10%; bulrush *(Scirpus validus)*, 14%;

reeds *(Phragmites)*, 2%; woolgrass *(S. cyperinus)*, 6%; and rushes *(Juncus)*, 5%. Corresponding porosity values range between 86% and 98%.

In a subsurface flow system, void volume ranges initially from voids between the substrate (gravel, sand, or soil) to voids created by the biological reactor over time. Initial porosities range from 18% for coarse gravel to 45% for clay and silt.[5] Factors creating long-term void space include live and dead root structure of plants and live and dead microbiomass within the substrate. Microbiomass varies along the cell length due to the differential food supply (organic compounds in wastewater and corresponding microbiological community). Overall void volume for surface and subsurface flow systems may stabilize in a mature wetlands system, but typical values are not available.

Depth of Submergence, Flow, and Bed Slopes

For municipal surface flow systems, optimum detention time has been suggested to be from 7 to 14 days,[4] greater than 5 days,[6] and between 6 to 10 days.[7] The respective water depths were 10 cm in summer and 30 cm in winter,[4] 30–60 cm,[6] and 25–60 cm.[7] Evapotranspiration during the summer increases detention time and may contribute to anoxic conditions and effluent deterioration. The maximum recommended depth is 60 cm and, preferably, 30–45 cm for municipal systems.[5]

Organic loading in acid-mine drainage wetlands is light, and water depth is not critical in maintaining aerobic conditions. However, these systems must include capacity for substantial deposition of metal compounds during operational lifetimes. Iron removal efficiencies of 98% of 135 mg/L Fe in influents represent deposition of 1.5 $kg/m^2/yr$ of Fe plus associated substances.[8] Water depths should vary from 2 to 3 cm at the inlet to provide optimal habitats for emergent plants to 70–100 cm at the outlet to accommodate metal deposits.

To minimize short-circuiting, the lateral bed slope (across the width) should be zero, and a uniform longitudinal slope (from inlet to outlet) should range from 0 to 1% for surface flow systems. Lateral slopes are rarely a problem with municipal systems but may reduce efficiencies in acid drainage or agricultural systems built in existing terrain with minimal grading. Such systems develop channelized flows along paths of least resistance, typically the deepest area containing least biomass. Longitudinal gradients of up to 1% present no practical problems in surface flow systems and allow for maintenance drainage. Greater slopes inhibit balancing length-to-width ratios and water depths within desired ranges.

The objective in subsurface flow systems is to maintain the water level just below the substrate surface. This is difficult to accomplish in practice. Flow through fully saturated, fine-grained soils, sands, and gravels is defined by Darcy's law:

$$Q = k_s AS \quad (4)$$

where Q = flow per unit time, m^3/day
k_s = hydraulic conductivity of a unit area of the medium perpendicular to the flow direction, m/day
A = cross-sectional area, m^2
S = hydraulic gradient of the flow system, $\Delta h/\Delta L$, as a decimal fraction

Cross-sectional area needed for a given flow is determined by the bed hydraulic gradient and hydraulic conductivity. However, hydraulic conductivity changes as vegetation and microbiological communities mature. Even in a mature system, hydraulic conductivity may vary from inlet to outlet.

Living roots and rhizomes fill part of the void space but create aerobic conditions that preclude substrate binding from anaerobic slimes. Dead roots and rhizomes create a permeable skeletal structure important for vertical and horizontal flow on a microlevel.[9]

A transformed conductivity of 260 m/day has been suggested for all types of substrates by one European source,[10] but its validity for soils is highly suspect.[11] In Europe, most soil beds do not have hydraulic conductivities above 2.6 m/day. Conductivities this low constrain the design of a soil bed for operation without surface flow. These problems are minimized if coarse gravel is used, but confirmed values for the long-term hydraulic conductivity are not available. Values as low as 30 m/day have been measured in gravel systems in England. Appropriate values are likely between that of clean gravel (864 m/day) and the minimum measured in England (30 m/day).

Aboveground plant material may be important to the long-term hydraulic regime. During winter, this biomass falls to the bed surface and eventually decomposes. This litter is permeable, provides a large surface area for microorganisms, and may be critical to long-term performance of systems experiencing surface flow. Performance of new wetlands systems experiencing undesirable surface flow might also be improved by adding about 10 cm of straw litter to the bed surface during mid- to late winter while the vegetation is still dormant. Straw used for this purpose should not be applied until the marsh plants are established (after the second growing season) to minimize competition from weeds sprouting from seeds in the straw or inhibition of newly planted rhizomes.

The microbiological community has an indirect but important hydraulic role in subsurface flow systems. The microbes convert organic contaminants to gases and stable, permeable solids. Without the oxygen supplied by plant roots, anaerobic slimes may clog the substrate, causing surface flow.

Consequently, transformed hydraulic conductivity is critical to the long-term performance of a subsurface flow system. In soil systems, the transformed conductivity should be either similar to or higher than the initial conductivity, while in gravel or sand the initial values will decrease over time. Investigations of design values are under way in England.

In addition to conductivity, the hydraulic gradient determines the cross-sectional area needed for a given flow in a subsurface flow system. Systems in

Europe are designed to match bed slope with the hydraulic gradient, but hydraulic conductivities have not increased to the design value, resulting in hydraulic gradients greater than bed slopes and surface flow. Conductivities less than 25 m/day constrain the design by requiring excessive bed widths. Coarse gravel overcomes this problem, and bed slopes from 0 to 2% or more can be used with an outlet design providing water depth control.

The bed surface should be flat regardless of the bottom slope to provide uniform water depths from the inlet to outlet for plant management and to minimize surface flow problems. Once surface flow has developed on a downward sloping surface, the flow may not penetrate into the bed even though the true water level within the bed is well below the surface. The surface becomes a barrier to penetration more than would be expected from the bulk hydraulic conductivity of the bed. This effect is magnified if the surface has a downward slope.

Design of subsurface flow beds should allow flooding to 15 cm to foster desirable plant growth and control weeds.

Darcy's law applies to laminar flow but is also used for transitional flow conditions. Flow through coarse gravels and rocks tends to be turbulent, and Ergun's equation has been suggested as a more appropriate design equation.[1,2] However, clogging and root development as systems mature may result in laminar or transitional flow where Darcy's equation would again be appropriate if mounding is not large enough to cause surface flow. This is an area where insufficient information exists to establish specific guidelines.

Sand and gravel cells constructed in the United States have not been designed using either Darcy's law or Ergun's equation. Some are experiencing surface flow problems.

Flow rates into and within a wetlands system may vary. Typically, municipal wastewater flow varies by a factor of 2 or more between minimum hourly night flow and maximum hourly day flow. During heavy rain, infiltration/inflow may increase the daily flow by an order of magnitude. Seasonal trends are common in acid drainage or agricultural systems and may also occur in the municipal systems with severe infiltration/inflow problems. Flow within a system will also be affected by vegetation during the growing season, and evapotranspiration may be greater than base flow in conservatively designed systems.

Consideration of the above factors reveals the importance of the hydraulic regime to performance of constructed wetlands. Despite short-circuiting and other hydraulic problems, performance at most systems has been good enough to reveal the outstanding potential of the technology. To achieve this full potential, research is needed to determine appropriate design values and guidelines for hydraulic conductivity and the other hydraulic factors.

CONTROL STRUCTURES

Practical control of the hydraulic regime within desired ranges requires water level and flow control structures. Types of structures include flow splitters, inlet and outlet controls, and dikes. Many different designs are available for each type of structure, with obvious ranges in practicality, performance, and costs. Selected designs are summarized below.

Flow Splitter Structures

Municipal wetland systems are often designed with multiple parallel cells that fit length-width specifications into site constraints and provide system operational flexibility. Cells can be taken out of service for maintenance or for optimizing treatment efficiency in other cells.

A flow splitter structure is needed for parallel cells. A typical design contains parallel orifices of equal size at the same elevation in a structure. Control orifices are selected based on specific project considerations. Options include pipes, flumes, and weirs. Valves are impractical because they would require daily adjustment. Flumes and weirs need not be standardized unless flow measurements are required. Flumes minimize clogging problems in applications with high solids but are more expensive than weirs. Weirs are relatively inexpensive and can be easily replaced or modified to change flow to any of the cells. In changing the percentage of flow between cells, modifications to the control orifices must maintain a common overflow elevation to assure uniform flow distribution at all heads. A splitter structure with weirs that allows accurate flow proportioning and measurements, simple cleanout of settled solids, and simple modification of flow proportioning by changing weir plates is shown in Figure 1.

Inlet and Outlet Structures

Inlet and outlet structures differ between surface flow and subsurface flow systems. Structures for surface flow systems are typically simple: an open-end pipe or channel/spillway flowing to and from the wetlands. They are sized to handle maximum-design storm flows and are normally sited to minimize short-circuiting, but vegetation, internal dikes, water depth, solids deposition, bottom uniformity, and length-to-width ratios are the main factors affecting short-circuiting. Examples are shown in Figures 2 and 3.

Subsurface flow systems rely more on inlet and outlet structures for uniform flow distribution, particularly if the length-to-width ratio is small (less than 3:1). Inlets may be pipes or channels across the width of the cell. Uniformly spaced holes, slots, tees, or serrations have all been used. Inlets above the bed allow adjustment of flow distribution and maintenance, preclude clogging and back-pressure problems, and aerate the wastewater. Distance above the bed depends on the sludge accumulation rate (2–3 cm/yr for settled raw sewage[10]) and the available head. A minimum of 30 cm is suggested but 60 cm is preferred. Large, coarse gravel (No. 1 aggregate) is placed beneath inlets. This provides a zone of high hydraulic conductivity transversing the vertical face of the bed. Disadvantages of above-bed structures include odors if the waste-

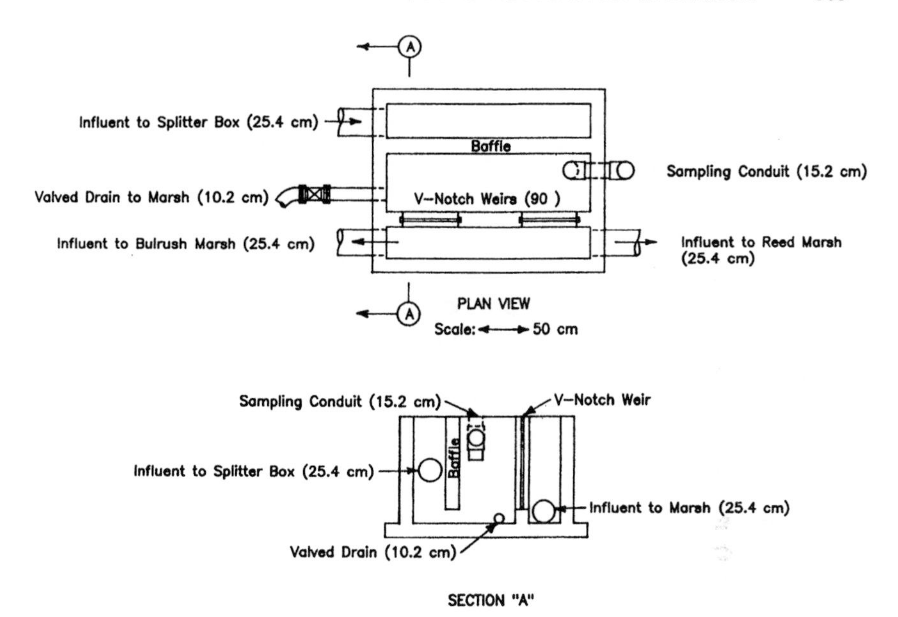

Figure 1. Influent splitter box at Hardin, Kentucky.

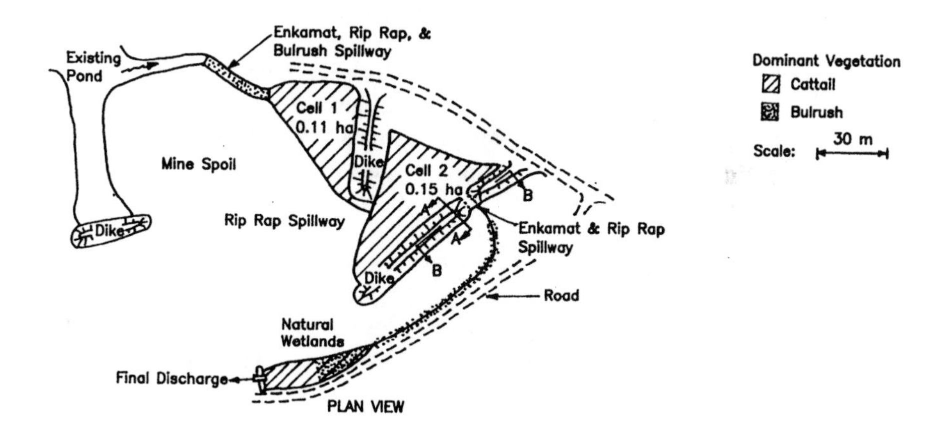

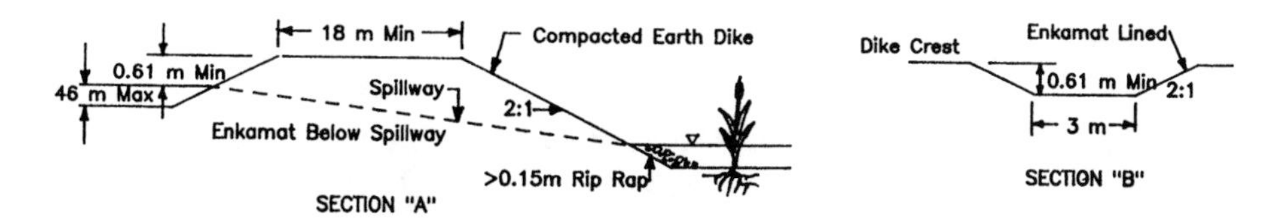

Figure 2. Fabius Coal Mines 950 NE acid drainage wetlands system.

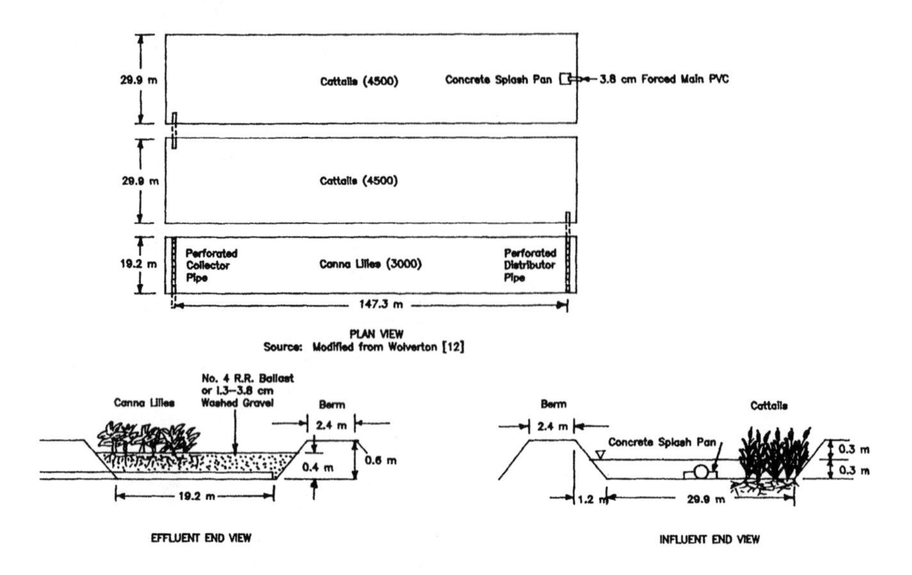

Figure 3. Constructed wetlands at Vrendenburgh, Alabama. (Modified from Wolverton.[12])

water is septic, greater freezing potential, and mosquito production if the structure is open. Advantages and disadvantages of subsurface inlets are basically opposite to those listed above. Examples of each type are shown in Figure 4. The swiveling tee concept shown in this figure is a particularly simple and effective design that (1) allows rapid adjustment of flow distribution and easy flushing of settled solids and (2) minimizes mosquito production and freezing problems. Each tee can be individually turned using a lever inserted in the mouth of the tee.

Outlet structures for subsurface flow systems typically contain perforated pipes buried in coarse gravel across the cell width and a control structure for adjusting bed water level. A flow measuring device, such as a weir or flume, may also be included. Water level control should use a direct method such as adjustable weirs or standpipes rather than valves. Standpipes are simple, inexpensive, and reliable. Examples are shown in Figures 5 and 6.

Dikes

Dikes or other barriers can be effectively used in any wetland system to control flow paths and minimize short-circuiting. Finger dikes are commonly

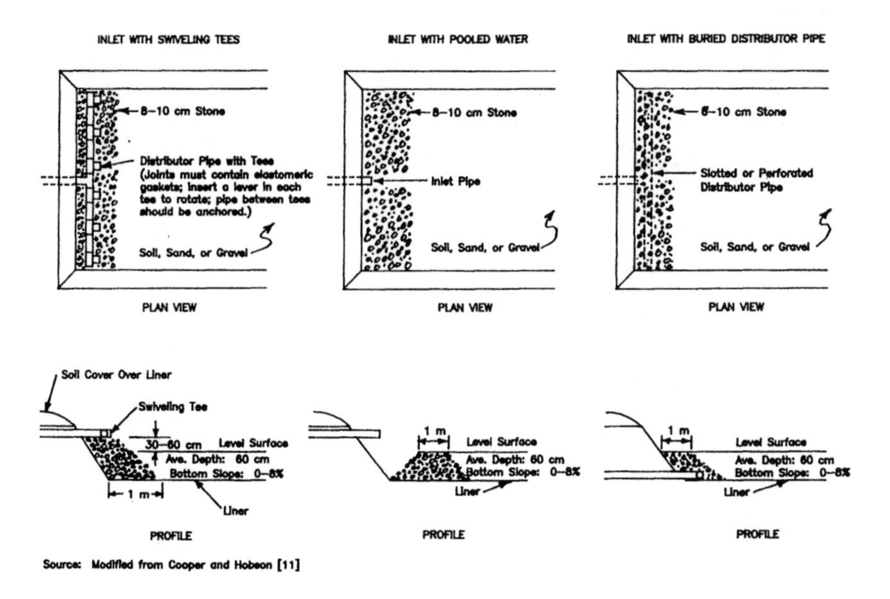

Figure 4. Inlet designs for uniform wastewater distribution.

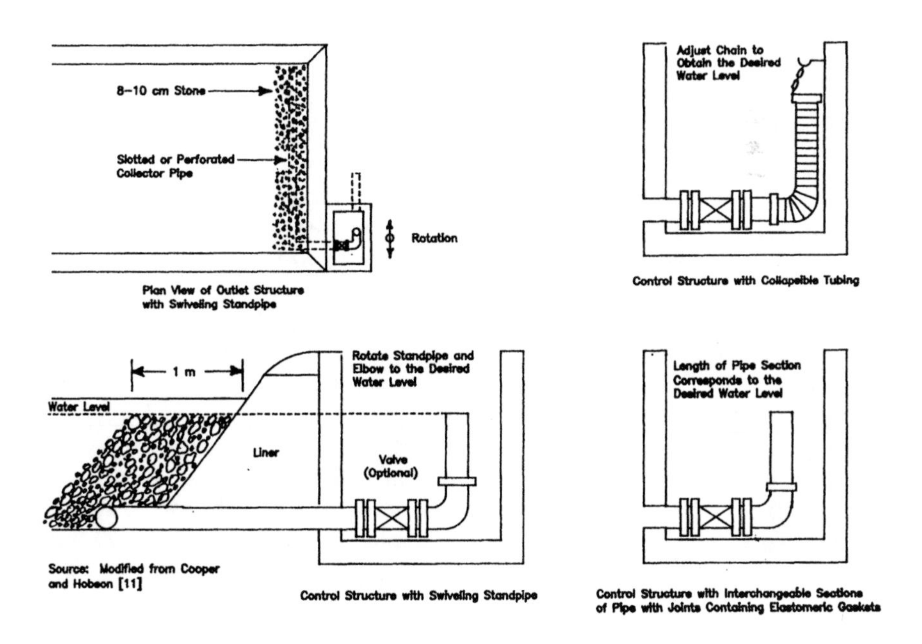

Figure 5. Outlet water level control structures.

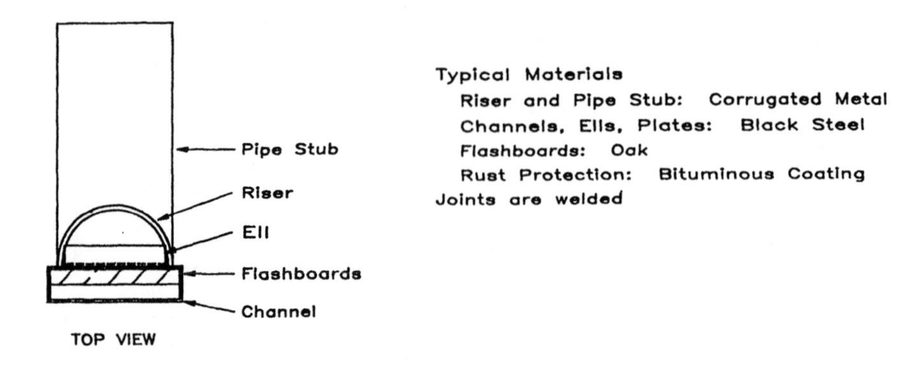

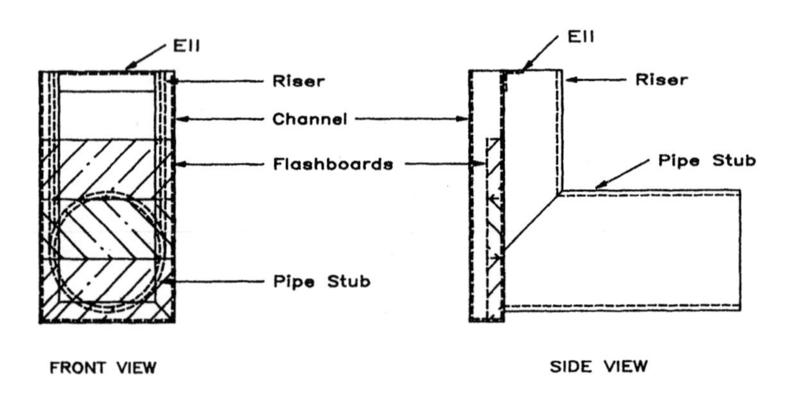

Figure 6. Flashboard riser spillway.

used in surface flow systems to create a serpentine configuration (Figure 7) and can be added in operational systems to mitigate short-circuiting. Divider dikes separate cells and attain desired length-to-width ratios within site-specific constraints (Figure 8). They are also used on steep slopes to terrace the cells (Figure 9). Details for these structures are based on site-specific needs and objectives. Most are constructed of native soils, but finger dikes are also constructed with sandbags and treated lumber. Top widths and side slopes are determined by the construction material, construction equipment, area that can be lost to dikes, and maintenance equipment. Design of external dikes is based on standard engineering practices applicable to the project area. Flood and runoff protection is a key consideration. Runoff should be routed around the wetlands, and dikes should provide protection for at least the 25-year flood.

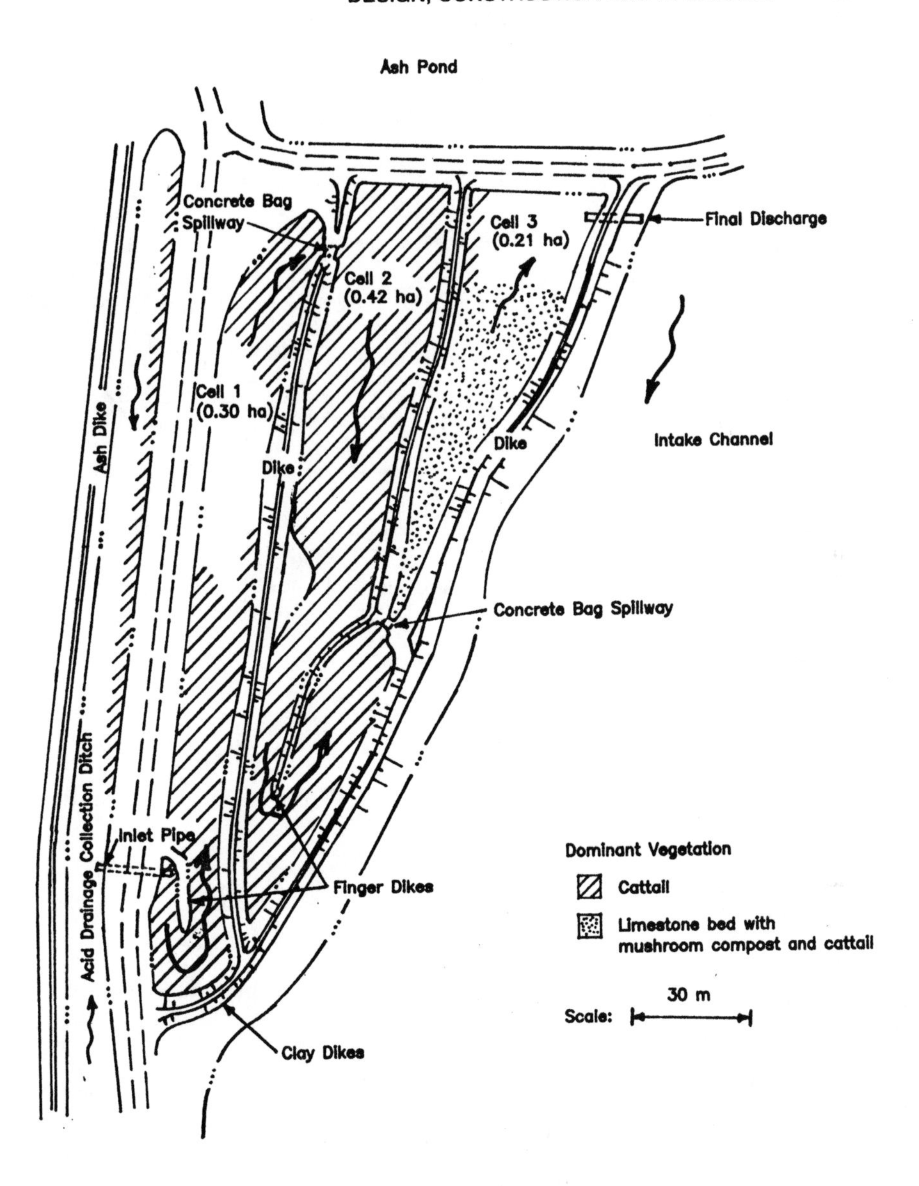

Figure 7. Acid drainage wetland (006) at TVA's Kingston Steam Plant.

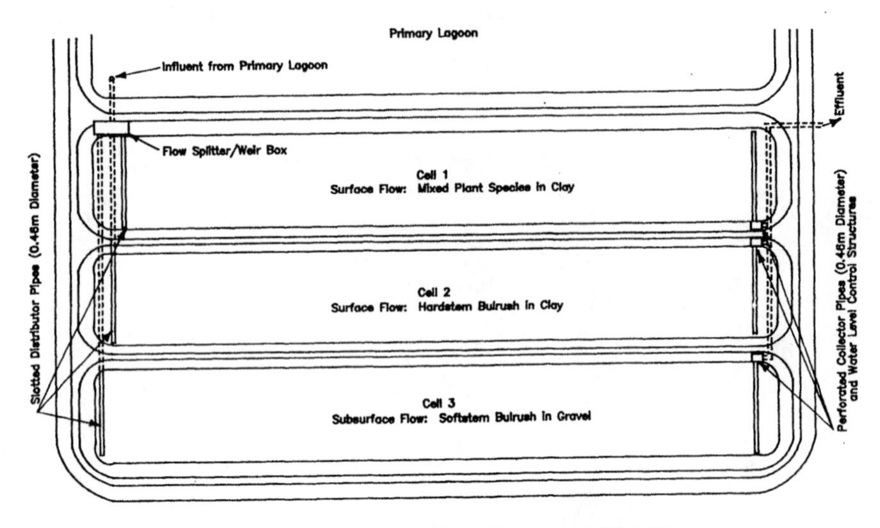

Figure 8. Constructed wetlands at Benton, Kentucky.

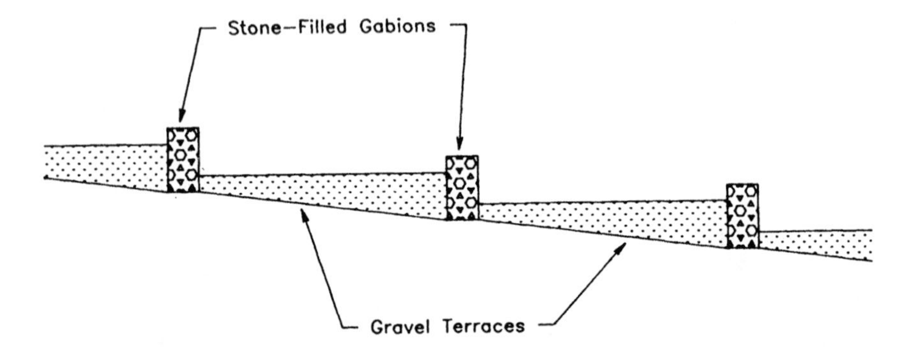

Figure 9. Terraced reed bed at Little Stretton, England.

REFERENCES

1. Kadlec, R. H. "Hydrologic Factors in Wetland Water Treatment," Chapter 3, this volume.
2. Watson, J. T., S. C. Reed, R. H. Kadlec, R. L. Knight, and A. E. Whitehouse. "Performance Expectations and Loading Rates for Constructed Wetlands," Chapter 27, this volume.
3. Steiner, G. R., and R. J. Freeman. "Configuration and Substrate Design Consider-

ations for Constructed Wetlands Wastewater Treatment," Chapter 29, this volume.
4. Herskowitz, J. "Town of Listowel Artificial Marsh Project Final Report," Project No. 128RR, Ontario Ministry of the Environment, Toronto (September 1986).
5. Reed, S. C., F. J. Middlebrooks, and R. W. Crites. *Natural Systems for Waste Management and Treatment* (McGraw-Hill Book Company, 1988).
6. Gearheart, R. A., B. A. Finney, S. Wilbur, J. Williams, and D. Hull. "The Use of Wetland Treatment Processes in Water Reuse," in *Proceedings of the Water Reuse Symposium III, Future of Water Reuse, Vol. 1* (Denver: American Water Works Association Research Foundation, 1984), pp. 617–638.
7. Tchobanoglous, G., and G. Culp. "Aquaculture Systems for Wastewater Treatment: An Engineering Assessment," U.S. EPA, Office of Water Program Operations, U.S. EPA Report-430/9–80–007 (June 1980), pp. 13–42.
8. Brodie, G. A., D. A. Hammer, and D. A. Tomljanovich. "Constructed Wetlands for Acid Drainage Control in the Tennessee Valley," *Mine Drainage and Surface Mine Reclamation, Vol. 1, Mine Water and Mine Waste,* Bureau of Mines Information Circular 9183 (1988).
9. Brix, H. "Treatment of Wastewater in the Rhizosphere of Wetland Plants—The Root-Zone Method," *Water Sci. Technol.* 19:107–118 (1987).
10. Boon, A. G. "Report of a Visit by Members and Staff of WRC to Germany to Investigate the Root-Zone Method for Treatment of Wastewaters," Water Research Centre, Stevenage, England (August 1985).
11. Cooper, P. F., and J. A. Hobson. "State of Knowledge on Reed Bed Treatment Systems," Water Research Centre, Stevenage, England (October 1987).
12. Wolverton, B. C. "Aquatic Plant Wastewater Treatment Systems," National Environmental Health Association Conference, Mobile, AL, February 7–9, 1988.

CHAPTER 31

California's Experience with Mosquitoes in Aquatic Wastewater Treatment Systems

Cecil V. Martin and Bruce F. Eldridge

INTRODUCTION

The history of wastewater treatment ponds in California is directly linked to the state and federal Clean Water Grant Program. In the early days of the grant program, concern was expressed about the possibility of mosquito production in oxidation ponds (which had been reluctantly accepted by state and local health departments) based on guidelines developed by the California State Department of Health Services to minimize mosquito production. These guidelines were as follows:

1. Ponds must be kept clear of all vegetation, and a program for regular maintenance must be developed.
2. Banks must be steep enough to discourage vegetation along the margins of the pond (minimum slope of 3:1).
3. Pond depth must exceed 4 feet to discourage emergent vegetation.

Since that time, adherence to these guidelines has successfully avoided mosquito problems in most oxidation ponds. In the late 1970s, a new technology, treatment of primary or secondary wastewater using aquatic macrophytes, appeared on the scene. One of these, water hyacinth *(Eichhornia crassipes),* was touted as a wonder plant for wastewater treatment. Merely adding them to an oxidation pond produced immediate and problem-free advanced treatment. What actually happened is that millions of mosquitoes were produced in ponds containing water hyacinths, and local public health departments threatened to outlaw use of this species of aquatic plant as a wastewater treatment technique.

Despite these problems, a number of Clean Water Grant applicants approached the California State Water Resources Control Board (state board) with requests to use grant funds to develop water hyacinth–based wastewater treatment systems that would overcome the mosquito production problem. Anticipated funding was under the Innovative/Alternative (I/A) provisions of the Clean Water Act (PL 92–500). In response, the state board established an aquaculture review section to evaluate applications and recommend additional research. Initial research done at the University of California at Davis resulted in a number of publications on use of aquatic plants in wastewater treatment.[1] Research results were applied in the design of a 1 mgd (3785 m^3/day) pilot plant for San Diego, California.

After 10 years, $1 million worth of research at the University, and $21 million in the pilot plant at San Diego, we have concluded water hyacinths can be used successfully to treat wastewater, although questions about mosquito production remain. Interestingly, design parameters we developed violate nearly all the original public health criteria for mosquito control in oxidation ponds. The solution to the mosquito problem was a combination of engineering modifications and new equipment development. In addition, the state board has funded a pilot study using a freshwater marsh for advanced wastewater treatment.[2] Arcata, California has built and is operating an award-winning treatment system/nature reserve.

PRESENTATION

Factors Leading to Growth of Created Wetlands in California

Engineering details of the San Diego project are presented elsewhere.[3] In this chapter, we discuss application of created wetlands to wastewater treatment needs in California and how the current financial situation is making these systems more attractive, especially for small communities.

Just what is the future for created wetlands in California? Here is the scenario as we see it:

1. In the past, Clean Water Grant funds paid from 87.5% to 97.5% of the cost of eligible treatment and disposal systems, and most communities paid little attention to treatment type or operation and maintenance costs. This goose will lay its last golden egg in the fall of 1988, when the grant program is scheduled to end.
2. Starting in 1989 the state board proposes to initiate the Clean Water Loan Program under the State Revolving Fund provisions of the latest Clean Water Act amendments. The interest rate, at one-half of the prevailing state bond rate of about 3.5% to 4%, is an attractive offer, but even this low-interest loan is unaffordable for many small communities.
3. State and federal clean water laws require that communities construct facilities to correct water quality problems, but there are no grant funds to pay for these facilities. As if this were not enough, these same communities are pressed to correct problems with solid wastes, toxic material, water supplies, and a myriad of infrastructural problems. Many demands are imposed upon communities that already have a high existing bonded indebtedness, placing them in a higher risk and higher interest rate category. The small communities are in a catch-22 situation.

Why the concern about small communities? For one, 98% of the treatment facilities identified in the 1984 federal "Needs Survey"[4] for the year 2000 will be for communities of less than 10,000 people. In fact, most (243 of 350) will be for communities of 1000 people or less. These small communities represent only 4.2% of the total wastewater flow in California. Results of the Needs Survey, showing the number of new plants required categorized by plant capacity in million gallons per day (mgd), are as follows:

under 0.1	243
0.1–1.0	107
1.0–10.0	3
10.0–50.0	1
over 50.0	3

This projection, along with the financial constraints already discussed, points to an increased interest in biological wastewater treatment facilities and particularly in created wetlands (including pond systems) because created wetlands:

1. can meet or exceed secondary wastewater treatment standards
2. can be constructed for one-half the cost of conventional systems
3. can be operated and maintained at one-half the cost of conventional systems
4. can remove toxic substances (including heavy metals) that conventional systems cannot handle
5. can be operated and maintained by relatively unskilled labor
6. are ideally suited to small communities where land is not expensive and wastes are primarily domestic in origin
7. are ideally suited for energy recovery and water reclamation
8. can be incorporated in conventional systems through retrofitting and conversion of some components to less energy-intensive processes, e.g., conversion of activated sludge tanks to primary clarifiers

We have the basic knowledge to design and operate created wetlands systems today. The major drawback is mosquito problems, which must be solved before created wetlands can be universally accepted by public health officials and the general public.

In our view, the following factors will influence the planning and implementation of created wetlands for small communities:

1. Water pollution problems must be solved to protect public health and the environment, in accordance with California's Porter-Cologne Act and the U.S. Clean Water Act.
2. The federal construction grants program is scheduled to end in 1988.
3. Construction costs for wastewater treatment facilities must be affordable to the people served, and conventional systems are becoming unaffordable to small communities.
4. Operation and maintenance costs must be affordable to the people served.

5. Replacement and amortization costs must be collected to pay for future repairs or replacement of facilities.
6. Treatment facilities must be considered for energy production and/or minimal energy consumption.
7. Water is a resource that must be conserved and reused to the maximum extent possible.
8. Any alternative to conventional treatment must not create more problems than it solves (specifically, such alternatives must not result in significant mosquito production or production of other pest insects such as biting or nonbiting midges).

Mosquito Problems and Their Possible Solutions

Nine pilot plants using aquatic macrophytes have been built in California since 1974.[5] Five no longer operate because of mosquito problems. Plants are operating at San Diego (San Diego County, water hyacinth); Mountain View (Contra Costa County, cattail/bulrush); Gustine (Merced County, cattail/bulrush); and Arcata (Humboldt County, cattail/bulrush).

The San Diego plant has had mosquito problems, but modifications in oxygen loading and hyacinth management appear to be abating mosquito production. The Mountain View plant has not been a source of mosquito problems, largely because of water depth management. The plant at Gustine has had mosquito problems because tall cattail *(Typha* sp.) and bulrush *(Scirpus* sp.) plants lodged late in the season, protecting mosquito larvae from mosquito predators (fish and aquatic insects). Plant replacement with lower growing forms may solve this problem. The newest plant, Arcata, has few mosquito problems to date.

Two approaches will avoid mosquito problems in wastewater treatment ponds. One is to create conditions in the pond that are not attractive to mosquitoes or are not conducive to larval development as in original guidelines for oxidation pond design. The other approach is mosquito larvae control using chemical or biological methods after they have become established. The first approach is the most effective and economical, but the least understood. Despite years of research, we know little of the relation between water quality parameters and mosquito larvae development. We can predict which mosquito species are likely to invade a particular kind of habitat—a salt marsh or catch basin in a storm drainage system—but we cannot predict whether a particular site will support mosquito breeding from measurements of pH, conductivity, certain ions, or other water quality parameters. We are learning more of the relation between aquatic vegetation and mosquito breeding. Dr. Vincent Resh and his colleagues at the University of California at Berkeley developed effective guidelines for freshwater marshes management to avoid anopheline mosquito breeding.

Mosquito control with chemicals or biological agents is difficult in wastewater treatment facilities. Materials that work well in unpolluted habitats are less effective in wastewaters because of binding to organic materials.[5] Furthermore, some biological controls (fish and certain insects) cannot tolerate low oxygen conditions in some treatment facilities. The answer to the last problem may be twofold: (1) operation of plants at levels of dissolved oxygen of at least 1 mg/L (preferably at 3 or even 5 mg/L) and (2) using fish species that can tolerate relatively low levels of oxygen yet are effective mosquito predators. Dr. Joseph Cech (University of California at Davis) has screened fish species for tolerance to low temperatures and oxygen levels. Species with good potential include guppies *(Poecilia reticulata),* Amargosa pupfish *(Cyprinodon nevadensis),* and mosquitofish *(Gambusia).*

Bacterial insecticides show considerable promise in wastewater treatment facilities, especially the unregistered species *Bacillus sphaericus.* Research by Dr. Mir Mulla (University of California at Riverside) has demonstrated its efficacy on highly polluted dairy drains and storm catch basins.

Most treatment facilities will probably need to have a tailored program for mosquito prevention and control, and each will consist of a variety of control approaches. When chemical or biological control agents do not adversely affect mosquito pathogens and predators, season-long mosquito control may be possible after only one or two treatments. The key is using highly selective mosquito larvicides at optimum dosage levels and treating only when mosquito larval density absolutely requires intervention. This requires frequent and careful monitoring of mosquito breeding activity.

Questions remain concerning the best way to prevent or avoid mosquitoes, and several factors complicate the search for answers. Some of these factors are:

1. The U.S. Environmental Protection Agency has announced a pesticide labeling plan under terms of the Endangered Species Act which may sharply restrict use of pesticides within the range of certain endangered species. Although this will probably not directly affect use of pesticides in wastewater treatment facilities, it may affect future availability of mosquito larvicides.
2. The number of new mosquito larvicides reaching the market place is shrinking to but a handful a year. Economics and pesticide laws are the main causes. Mosquito control is a very small share of the market for pesticides, and manufacturers are increasingly unwilling to finance safety and efficacy testing required for new product registration.
3. If the Federal Insecticide, Fungicide, and Rodenticide Act is overhauled by Congress within the next year or so, reregistration of existing mosquito larvicides will probably follow soon thereafter. For the above reasons, some existing larvicides will probably not be reregistered.
4. Research dollars are increasingly short for projects in the public health areas. Inflation, plus diversion of research dollars to national priorities such as AIDS research, have exacerbated the problem.
5. EPA has been slow to register new biological control agents such as the mosquito-pathogenic fungus *Lagenidium giganteum.* Registration of biological control organisms is a new endeavor for regulatory agencies, and new ground rules have yet to be completely established.

Despite these challenges, mosquitoes can be controlled with careful planning and management of wastewater facilities. Mosquito breeding management must be included in the earliest planning stages and must be included in operating procedures after plants have come on-line.

SUMMARY

Wastewater treatment facilities using aquatic macrophytes offer numerous advantages for small communities. Although problems with mosquito production have slowed development in California, careful design before construction and monitoring after construction can keep mosquito breeding within acceptable levels.

REFERENCES

1. Stephenson, M., G. Turner, P. Pope, J. Colt, A. Knight, and G. Tchobanoglous. "The Use and Potential of Aquatic Species for Wastewater Treatment, Appendix A: The Environmental Requirements of Aquatic Plants," Publication No. 65, California State Water Resources Control Board (October 1980).
2. Gearheart, R. A., J. Williams, H. Holbrook, and M. Ives. "City of Arcata Marsh Pilot Project Final Report" (January 1986).
3. Weider, R. K., G. Tchobanoglous, and R. W. Tuttle. "Preliminary Considerations Regarding Constructed Wetlands for Wastewater Treatments," Chapter 25, this volume.
4. "1986 Needs Survey Report to Congress: Assessment of Needed Publicly Owned Wastewater Treatment Facilities in the United States," U.S. EPA Report-430/9-87-001 (February 1987).
5. Eldridge, B. F., and C. V. Martin. "Mosquito Problems in Sewage Treatment Plants Using Aquatic Macrophytes in California," in *Proceedings and Papers of the 55th Annual Conference of the California Mosquito and Vector Control Association, Inc.* (1987), pp. 87–91.

CHAPTER 32

Constructing the Wastewater Treatment Wetland—Some Factors to Consider

David A. Tomljanovich and Oscar Perez

INTRODUCTION

Constructed wetlands range from simple alterations of the hydrologic characteristics of a low-lying area to creating an impoundment where no wetland existed previously. Acidic drainage treatment wetlands are typically constructed in areas where seepage has created some wetland characteristics.

Constructed wetlands for wastewater treatment may range in size from several square meters to several hectares. Design parameters vary with size, site characteristics, hydrologic soil group, pollutant type and loading rate, geographic locale, watershed characteristics, proximity to residential development, and anticipated operation and maintenance requirements. This chapter describes the construction process of wastewater treatment wetlands and discusses some important factors that influence success of the project.

CONSTRUCTION PLANS

Construction plans and specifications developed from treatment area requirements and siting investigations described in earlier chapters should be carefully reviewed. Level of detail depends on size and complexity of the wetland, site physical characteristics, and requirements established by regulatory agencies. Minimally, construction plans must have sufficient detail for accurate bid preparation and construction. The plans must include at least the following:

1. clearing and grubbing limits
2. access roads
3. utilities (overhead or underground)
4. erosion control measures
5. location and boundaries of borrow areas
6. trees and existing wetland vegetation to be left undisturbed
7. wildlife habitat enhancement structures

8. dike location, length, top width, elevation, freeboard, and upstream and downstream slopes
9. spillway location, elevation, and type (riprap, concrete, vegetation)
10. size, location, elevation, and type of water control structures
11. permeability requirements for pond bottom and dikes
12. impermeable linings (type and location of) where required
13. pond bottom contour lines and lateral slope tolerances (0.1 ft for most municipal wastewater treatment systems)
14. location of subsurface drains
15. placement of rock, gravel, soil, and limestone within the ponds by elevations and depths
16. species and spacing of wetland vegetation to be planted
17. liming and/or fertilizing requirements
18. seeding, mulching, fertilizing, and liming of dikes and adjacent disturbed land
19. inlet and outlet distribution piping (type, location, elevation)

A prebid conference with potential contractors is recommended to explain the concept, goals, and requirements of the project. This meeting can be effective in soliciting accurate bids from qualified contractors without previous experience in wetlands construction.

PRECONSTRUCTION SITE ACTIVITIES

Prior to initiating construction the following field activities will usually be required:

1. marking trees to be left standing or felled, delimbed, and left lying in the wetland to prevent channeling in mine drainage systems
2. marking the boundary of the area to be cleared if the wetland is to be built in a forested area
3. identifying need for temporarily diverting or pumping water from the construction site
4. identifying location of silt barriers (fences or straw bales) to be placed in accordance with appropriate regulations
5. conducting preliminary survey and staking the site prior to viewing by contractors at prebid conference

Construction plans, specifications, and field layout must portray to the owner and contractor the desired work. Because of wide variation in conditions and experience, plans may vary from very simple plans and a few stakes to complicated plans with detailed specifications, extensive staking, and a preconstruction conference to ensure adequate understanding. Preconstruction activities should be consistent with size and complexity of the site and adequate to ensure orderly and effective construction.

The embankment is marked by fill-slope stakes and by dike centerline stakes. Spillways are located by setting cut slope stakes or by staking the centerline. Borrow areas should be staked as necessary. Stakes should mark structural work such as water control structures and pipe installation at the proper time. Stakes should be located and marked to convey to the contractor the location and state of the work in respect to slope, elevations, and other construction plan factors.

COST ESTIMATE PREPARATION

Preparation of cost estimates vary depending on whether the wetland will be constructed by the owner/operator or contracted out. In either case, cost estimates include the following items:

1. engineering plan
2. preconstruction site preparation (items of previous section)
3. construction
   a. labor
   b. equipment
   c. bill of materials (tile, liners, control structures, pipe, valves, riprap, gravel, limestone, concrete, synthetic fabric, seed, fertilizer, mulch, and vegetation)
   d. supervision
   e. indirect and overhead charges

CONSTRUCTION

Construction includes access road building; clearing; dike building; spillway construction; piping and valving; planting; and seeding, liming, fertilizing, and mulching of dikes and disturbed areas. Valuable reference documents for constructing wetlands are "Agricultural Handbook No. 590,"[1] "Landscape Design: Ponds,"[2] *Engineering Field Manual for Conservation Practices,* Chapter 11, "Ponds and Reservoirs,"[3] "Water Quality Field Guide,"[4] and "National Food Security Act Manual,"[5] all of which are available from the Soil Conservation Service.

Close communication between owner/operator, designer, and contractor is crucial throughout the process. Changes in the plan should be clearly understood by all parties and documented.

Correct type and size of heavy equipment is crucial to ensuring cost-effective, proper construction. Under- or oversized equipment can result in cost and time overruns. A good practice is to show the equipment operator(s) the site during the planning stages to obtain his/her opinion on equipment, time to construct, and potential problems.

If the wetland is built in an existing wet area, two bulldozers (and operators) rather than one should be used if possible. Invariably, one will get mired down and require the other to free it. Valuable time is lost if a second machine is not available.

In rare instances, small sand bag dikes are a less expensive alternative to larger earthen dikes. However, in most cases, use of heavy equipment rather than manual labor will result in the most cost-effective and structurally sound dike.

Earthen dikes tend to be larger than planned. Minimum top width of dikes will be at least bulldozer-blade width with subsequent increases in the base to meet slope specifications. This is particularly true if the dikes are built in an existing wet area and should be incorporated in effective treatment area determinations.

In most municipal waste treatment wetlands, construction must precisely follow the engineering plan to achieve proper functioning of the system. For example, lateral bed slope should not vary by more than 0.1 ft from high spot to low spot. Large slope or surface variations can cause channeling, especially in systems with high width-to-length ratios. Permeability specifications must be carefully followed to prevent leakage into or out of the ponds. If native soils are adequate, dozers and sheepsfoot rollers are usually sufficient to achieve desired compaction. Where synthetic liners are required, installation should precisely follow manufacturers' instructions for bed material, sealing (liner-to-liner and liner-to-piping and control structures), and material placement on top of liner. Special care must be taken to prevent puncturing the liner.

Subsurface flow systems dependent on high hydraulic conductivities in the substrate require special provisions to avoid compacting the substrate during construction or planting. Heavy equipment must be avoided and small machine or foot traffic requirements must be clearly communicated to the contractor. Walk boards may be placed on the soil during hand planting.[6]

In acidic drainage treatment wetlands, considerable field modification may be required to achieve maximum treatment. For example, new seeps are often discovered during construction that need to be included in the treatment system. It is important that field supervisors understand tolerances for on-site modifications and that they communicate this clearly to the contractor/operator. If a bid package is prepared, anticipated degree of field modification should be clearly stated.

In most acidic drainage wetlands, some dike seepage is acceptable, and water depth in the wetland is usually less than 1 meter. Therefore, dike material is not as critical as in larger earthen dikes, where seepage is unacceptable and greater pressure is exerted by water depths. Thus, borrow material obtained on-site will usually be acceptable.

Adequate sizing and installation of spillway and discharge pipe are critical to ensuring dike integrity and safe and proper operation of the wetland. Particular attention should be given to protecting the dike against high-flow storm events.

In some areas, muskrats and beavers burrow into dikes or obstruct discharge pipes modifying the pond elevation. Muskrat damage to dikes can be minimized or prevented by installing hardware cloth (metal screen) vertically in dikes or riprapping both upstream and downstream slopes. The extra initial cost may be small compared to replacing the dike. If beavers are likely to be present, consultation with state wildlife personnel is recommended for developing a plan to preclude unwanted in-pond activity.

Acidic drainage can rapidly destroy metal tile, pipes, pumps, and valves. Corrosion-resistant metals or coatings should be used in low pH water (<5.0). The purpose of the constructed wetlands configuration is to maximize the contact between the wastewater and the substrate and vegetation by minimizing short-circuiting. Inlet distribution devices must be carefully installed to establish uniform distribution across cell width. Flow distribution at the front end of the system is particularly important in subsurface flow municipal waste treatment wetlands, where width may exceed length by up to 10 times.[7]

Emergent vegetation must be planted with care and proper guidance to ensure plant survival. Use planting materials with 20- to 30-cm stem lengths so a portion of the stem will extend above the water surface. In subsurface flow systems, roots/rhizomes must be set in water and the substrate surface should be mounded slightly around the stem to avoid creating a pool of standing water. Rhizome cuttings are usually not suitable due to the difficulty of maintaining a portion above water. A contingency or other agreement should be included in the plan and bid to cover the cost of replanting any areas where plantings failed.

INSPECTION, TESTING, AND STARTUP

Before accepting the final product, the owner/operator should require a thorough test of all components to ensure proper operation of pumps and water control structures, sealing, water level, flow distribution, and plant survival.

During initial operation of subsurface flow systems, erosion and channeling should be eliminated by substrate raking and filling by hand if necessary. Rills on the dike slopes or spillways should be filled with suitable material and thoroughly compacted. These areas should be reseeded or resodded and fertilized as needed. Should the upstream face of the earthfill wash or slough because of wave action, protective devices, such as booms or riprap, should be installed until the vegetation becomes established. If there is seepage through or under the dike, an engineer should be consulted to recommend proper corrective measures.

The vegetative cover on the dike and spillway should be maintained by mowing and fertilizing when needed. Proper mowing prevents the formation of unwanted woody growth and tends to develop a cover and root system more resistant to erosion from runoff. Fences should be kept in good repair. Trashracks, outlet structures, and valves should be kept free of trash and repaired or replaced as necessary.

Burrowing animals may cause severe damage to dikes or spillways, which if unrepaired may lead to dike failure. A thick layer of sand or gravel on the fill or wire screening inhibits burrowing. If burrowing damage continues, aggressive trapping, hunting, and poisoning may be necessary in accordance with federal, state, and local regulations.

In summary, an initial maintenance program that includes frequent and thorough inspection and immediate correction of problems is critical to ensuring successful operation of wastewater treatment wetlands.

REFERENCES

1. "Ponds—Planning, Design and Construction," Agricultural Handbook No. 590, USDA, Soil Conservation Service (July 1982).
2. "Landscape Design: Ponds," Landscape Architecture Note 2, USDA, Soil Conservation Service (September 1988).
3. "Ponds and Reservoirs," in *Engineering Field Manual for Conservation Practices,* rev. ed., USDA, Soil Conservation Service (1989), ch. 11.
4. "Water Quality Field Guide," SCS-TP-160, USDA, Soil Conservation Service (September 1983).
5. "National Food Security Act Manual," Title 180-V-NFSAM, USDA, Soil Conservation Service (March 1988).
6. Boon, A. G. "Report of a Visit by Members and Staff of WRC to Germany to Investigate the Root-Zone Method for Treatment of Wastewaters," Water Research Centre, Stevenage, England (August 1985).
7. Watson, J. T., and J. A. Hobson. "Hydraulic Design Considerations and Control Structures for Constructed Wetlands for Wastewater Treatment," Chapter 30, this volume.

CHAPTER 33

Considerations and Techniques for Vegetation Establishment in Constructed Wetlands

Hollis H. Allen, Gary J. Pierce, and Rex Van Wormer

Construction of wetlands has received increasing interest over the last 15 or 20 years, particularly for development of wildlife and fisheries habitat on dredged material,[1-3] erosion control of coastal and reservoir shorelines,[4,5] wastewater treatment,[6] and other purposes, such as mitigation of highway impacts.[7] This chapter describes considerations and techniques learned from experience in developing wetlands for wastewater treatment. It focuses on herbaceous macrophytes because most of what is known about wastewater treatment with wetlands concerns these types of plants. It also focuses primarily on in situ substrates, in contrast to substrates borrowed from other areas.

FACTORS INFLUENCING WETLAND ESTABLISHMENT

Three important factors contribute to the diversity of natural wetlands and form the basis for any wetland development protocol. These are (1) hydrologic considerations, (2) substrate, and (3) vegetation. Within the context of this chapter, "hydrology" also includes water quality. Variations in hydrology and substrate have a particularly strong influence on vegetative diversity. Through an understanding of the relation between these factors, it is possible to determine which species should be planted, and by what means, under given environmental conditions.

Hydrologic Considerations

Water depth and frequency of flooding or its periodicity are important in determining the plant species appropriate to a constructed system. Water depth causes different vegetation zones in a wetland in part because deeper water may restrict oxygen from reaching the substrate.[8] Water depth may also influence the degree of light penetration and photosynthesis.

Periodicity, duration, and seasonality of flooding are important for selecting plant species to be used for wetlands development. Wetland plants can withstand various degrees of flooding depending on when and for how long the flooding occurs. Many wetland emergent plants need a period of lower water level during the growing season, whereas in the dormant season, draw-down is not as important.

Water quality factors affecting selection of wetland plant species include such factors as water clarity, pH, salt concentration, and dissolved oxygen. Degree of water clarity is considered the most important, particularly for submerged aquatics. If water is stained or turbid, light penetration will be reduced and limit photosynthesis. Under these conditions, rooted aquatics having a floating growth form and habit should be selected for use.

Salt concentration either within the water column or the substrate has a major influence on plant species selection. In saline environments, species such as smooth cordgrass *(Spartina alterniflora)* or salt grass *(Distichlis spicata)* or some other such saltwater species may be selected. Salt tolerance levels have been determined for many plants.[2]

Substrate

Many substrates are suitable for wetland establishment. Loamy soils are especially good because they are soft and friable, allowing for easy rhizome and root penetration. But fine-textured soils such as clays may limit root and rhizome penetration. Low nutrient content may limit growth and development as may excessive nutrients (i.e., calcium levels that exceed the adaptive range of particular species). However, substrates with low nutrient concentrations may be suitable in constructed wetlands for wastewater treatment if other requirements are met, because of nutrient-enriched influent. Usually, soil amendments are not necessary in wetlands for plant growth because hydrophytes grow well in a broad range of soil types, but where a limitation exists amendments may enhance success. Specifications for wetland construction frequently call for placement of a layer of wetland or upland topsoil. Although little experimental evidence demonstrates benefits from this practice, it will probably continue because of its perception as beneficial. Evidence from a New York project suggested that drying and aerating a wetland soil is especially good for wetlands establishment because such soils have high levels of nutrients and oxidation followed by flooding enhanced macronutrient availability.[9]

Although peaty organic soils support wetland plants, they are not preferred for wetland development. They are low in nutrients because organic acids, yielding many hydrogen ions, occupy cation exchange sites, and once flooded they have a loose, soft texture that provides inadequate support for emergent aquatics. In such soils, it might be necessary to anchor each planting unit individually.

Substrates should be evaluated for calcium content. Many wetland plants such as sago pondweed (*Potamogeton pectinatus*), and muskgrass *(Chara* spp.) are known calcifiles, while others are known to be restricted to acid soils (e.g., *Sphagnum* spp.). Many species have a broad tolerance for calcium levels and can be planted in a diversity of soils; for example, softstem bulrush *(Scirpus validus)* and pickerelweed *(Pontederia cordata)* tolerate high calcium levels. In marsh construction at the Tompkins County, New York airport, wetlands with abundant calcium carbonate became dominated by muskgrass *(Chara* sp.), despite planting a variety of species.

Site excavation for wetland establishment is likely to expose a subsoil that may not be as conducive to plant growth as the topsoil. Clays and gravels frequently underlying more favorable soils may be sufficiently dense or hard to inhibit root penetration, may lack nutrients found in topsoil, or may be impermeable to water needed by roots. Emerson[10] hypothesized that establishment of a planted wetland in New York was limited by dense soils that inhibited plant growth. In a South Carolina wetland planting,[11] plants could survive in heavy clay soil, but their spread was limited to the original planting hole. However, a clay layer below the root zone is often essential to retain water.

Sandy, coarse textured, and subsoil substrates often lack nutrients and may require fertilizer.[4] However, sandy soils hold plants well and prevent them from floating out of the planting surface, in contrast to peaty or silty soils. Planting in sand media can be efficient and inexpensive because of ideal texture for hand planting; however, sand or gravel will dry out quickly and may need irrigation if water levels cannot be maintained at the root level.

Nutrient conditions in constructed wetlands are improved by natural processes and by horticultural techniques. Conversion from aerobic conditions to anaerobic conditions in a constructed wetland results in a major conversion of oxidized iron (Fe^{3+}) to reduced iron (Fe^{2+}). Subsequently, phosphorus availability changes because it is more soluble when associated with reduced iron. Potassium is also readily available in flooded soils, although the mechanism for its availability is more complex. Nitrogen, lost through nitrification in aerobic soils, is retained in wetland soils and is not usually limiting whenever organic material is present. Consequently, some researchers do not consider macronutrients limiting in aquatic systems newly constructed from upland areas.[12]

Little is known about nutrient requirements of individual hydrophytes, and consequently general fertility concepts are often applied. In coastal salt marshes, coarse-textured soils and subsoils frequently are nutrient-poor[3] and require fertilization. Water-soluble fertilizers and broadcast fertilization have obvious disadvantages in flooded wetland systems, where such applications would be highly mobile. Side dressing with time-release fertilizers, low in solubility, is frequently used to overcome this difficulty.[3] Osmocote and Magamp are granular, slow-release formulations of inorganic fertilizer appropriate to wetland plantings.[3] When water covers the plantings and cannot be lowered, fertilizer is applied through the water column, making granular formulations difficult to apply. To overcome this problem, Agriform tablets have been used successfully in a 12-ha planting in South Carolina.[11] In nutrient-enriched situations, fertilization may still be necessary until plants become well established.

Vegetation

The vegetative component is a major factor in successful wetland development. Which plants will grow given the hydrological and substrate conditions in the developed wetland? Which plants provide the appropriate attributes for wastewater treatment? How are the plants to be selected? For wastewater treatment, plants selected should (1) be active vegetative colonizers with spreading rhizome systems; (2) have considerable biomass or stem densities to achieve maximum translocation of water and assimilation of nutrients;[13] (3) have maximum surface area for microbial populations; (4) have efficient oxygen transport into the anaerobic root zone to facilitate oxidation of reduced toxic metals[14] and support a large rhizosphere; and (5) be a combination of species that will provide coverage over the broadest spread of water depths envisioned for the terrain conditions. One must couple the above attributes with the hydrologic and substrate conditions in choosing species for planting.

It is important to understand how a particular species might react to given wetland conditions before that species is nominated for use in a wetland development project. Wetland plants use a number of adaptive strategies to withstand varying edaphic and hydrologic conditions. Many strategies are unique to wetland plants or to particular wetland conditions. For example, wetland plants have air spaces (aerenchyma) in roots and stems that allow the diffusion of oxygen from the aerial portions of the plant into the roots.[15] In practice, an empirical evaluation by simply selecting species growing under conditions similar to the habitat to be developed may be the most successful.

As mentioned earlier, water depth dictates zonation of wetland plants, and species should be selected primarily on this criterion. This is more easily done by thinking in terms of planting zones. Based on the authors' experience, these zones are defined with reference to the normal water level (0.0 cm) and are divided into *deep* (–91 to –152 cm); *mid* (–15 to –91 cm); *shallow* (–15 to +15 cm); and *transitional* (+15 to +45 cm) (Figure 1). Although other wetland plants may thrive at different depths or elevations than those portrayed (Figure 1), this classification can be adopted as a general model for a broad hydrological regime for wetlands development.

SOURCES OF PLANT MATERIALS

Wetland plants can be purchased from nurseries, collected in the wild, or grown for a specific project. Each method has distinct advantages and disadvantages regarding quality of material, availability of plants, and cost for acquisition and planting. No generalization can be made which would recommend any particular source for all projects. However, the authors agree that for most projects, wild collected material is most desirable for reasons discussed below.

Only a few commercial nurseries specialize in wetland plants for wetland

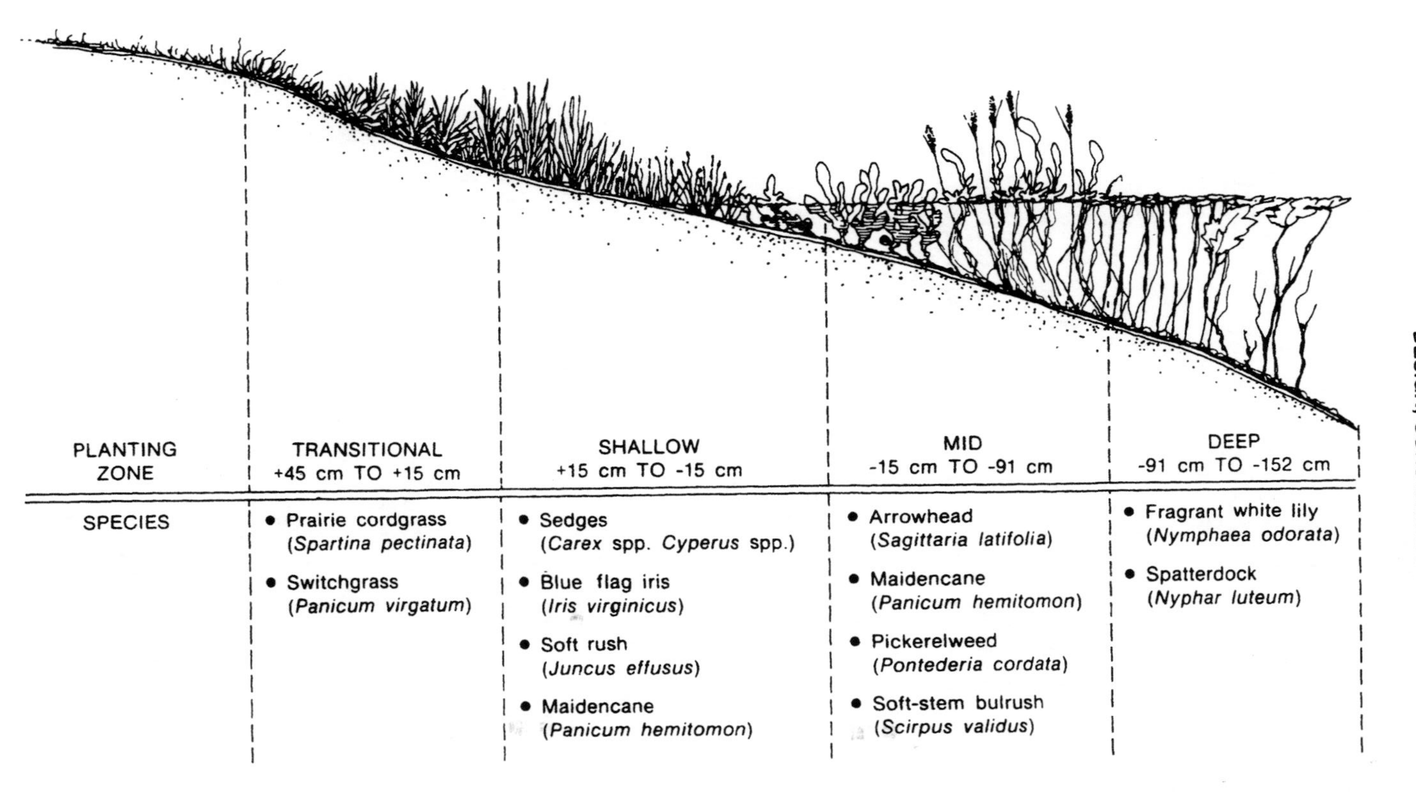

Figure 1. Typical interior United States wetland plants by planting zone, as they are related to normal water level (0.0 cm).

construction. Most can be found in Florida, Maryland, New York, and Wisconsin with almost none in the central and western United States. Other sources include state nurseries, commercial tree and shrub nurseries, seed companies, and suppliers of native flower and landscaping plants. Regional lists of dealers in plants for conservation plantings are available from the Soil Conservation Service, and these lists are invaluable for obtaining wetland adapted species.

Plants obtained from wetland nurseries are supplied in different forms. Potted materials are available from at least one nursery. These are relatively expensive, averaging $1.00 or more per plant. Other suppliers offer bare-root seedlings collected from outdoor beds and stored under refrigeration. These are typically $0.25 or less per seedling. Freshly collected growing plants from the wild are available for prices between the above costs. Commercial plants should be shipped by express package service or hand delivered by the supplier and consequently may cost several cents per plant for delivery. The extra cost is well worthwhile because a delay of one to two days can result in substantial or complete plant mortality.

The primary advantage of commercially supplied plants is quantity availability and site delivery in a suitable planting condition. This reduces logistical problems in procuring plants, but the buyer of large quantities should arrange to have material supplied in several smaller shipments coordinated with planting activities. Reducing storage time at the planting site will reduce plant mortality.

Commercial suppliers carry only a limited number of species, and there is a natural tendency to plan wetland plantings around that supply. Nursery-supplied plants are also genetically and physiologically adapted to their growing site and may be difficult to establish and maintain at locations with different edaphic and climatic characteristics. When selecting a nursery, avoid large latitudinal distances between plant source and destination; longitudinal variation is more acceptable. Finally, plants that have been packaged, shipped, and stored before planting may be stressed at time of planting.

Plants collected from the wild are more closely adapted to local environmental conditions than nursery-acquired plants. They can also be planted with limited storage; if collecting and planting are coordinated, they can normally be planted within 36 hours or less after collection. If a diverse natural ecosystem is desirable, natural populations can supply that diversity. Plants for acid mine drainage systems are often obtained from similar habitats because plants from normal habitats occasionally die or need a lengthy acclimatization period. Disadvantages of local collections are (1) undesirable weedy species may be inadvertently included or rare or endangered species populations may be decimated; (2) logistics or difficult collecting conditions may increase costs; (3) plants may not be available because of limited supply, local regulations, or difficult access to private land; and (4) plants may be unavailable early or late in the growing season. Care must be taken to avoid sites that contain noxious weeds or rare, threatened, or endangered species and to distribute collections to avoid permanent damage to the source wetland.

Plants can be grown specifically for a particular project, but this often requires a year or more of lead time, and facilities such as a greenhouse or outdoor nursery must be available. The authors have used nursery beds with soil or fertilized sand to propagate plants. Wild collected or purchased plants are placed in well-watered beds and the plants spread by vegetative growth. In cool temperate climates, a greenhouse should be used only as an oversized cold frame. This permits plants to begin growth a few weeks earlier in the season and allows hydrophyte planting when they are normally dormant and difficult to locate in the wild. The U.S. Army Engineer Waterways Experiment Station has experienced no difficulty in growing smooth cordgrass and willow *(Salix* spp.) in the warm temperate climate of Mississippi. However, in a northern climate, Southern Tier Consulting (New York) has had poor success in breaking dormancy during midwinter, and Environmental Concern (Maryland) has had similar experience.[16]

PLANTING METHODS

Establishing vegetation once a site has been prepared involves planting with a suitable propagule at the appropriate time. The planting period for herbaceous vegetation is broader than for woody plants. In temperate climates, the planting period typically begins after dormancy has begun in the fall and ends after the first third of the summer growing season has passed. Fall dormant planting is recommended for tubers and rootstock and is very successful for some species.[4] Species from the genera of burreed *(Sparganium),* dock *(Rumex),* bulrush *(Scirpus),* rush *(Juncus),* and arrowhead *(Sagittaria)* have been successfully planted in fall, whereas some manna grasses *(Glyceria* spp.), sedges *(Carex* spp.), and cattail *(Typha* spp.) have grown more successfully when transplanted after dormancy has been broken in spring. In general, early spring growing season plantings have been most successful.

Wetlands are generally planted with whole plants or dormant rhizomes and tubers. Establishment from seed typically has not been successful because of stratification requirements of wetland seed and loss of seed via water action. An exception may be when wetland turf or agricultural grasses are used (1) on upper portions of basins that are never flooded or are not flooded until after seeds are established or (2) on saturated drawdown zones of reservoirs shortly after the water has been withdrawn.[5] Also, wet prairie species have been established in the tall grass prairie province by planting wild collected seeds with a seed drill.

Dormant underground vegetative propagules are commonly planted. Tubers of duck potato *(Sagittaria latifolia),* sago pondweed *(Potamogeton pectinatus),* and softstem bulrush *(Scirpus validus)* are commonly and successfully planted. Tubers are simply placed deep enough in the substrate to prevent them from floating out of the medium.[4] Some investigators have had the best success with tubers having 20–25 cm of stem so the plants can obtain enough oxygen through the stem when flooded.

Whole plants with shoots, roots, and active buds also are frequently planted. Some species, such as water smartweeds *(Polygonum coccineum* and *P. hydropiperoides)*, planted in this manner begin to grow immediately from existing buds and root primordia on the shoots. Others, such as most grasses, begin new growth from buds at the plant base, and the existing root and shoot systems die shortly after transplanting. Loss of the shoot does not appear to inhibit establishment and early development of the plants. Use of potted plants avoids this initial phase of death and regrowth.

Cores (8- to 10-cm diameter) of wetland soil and associated propagules collected in existing marshes can be transplanted to constructed wetlands. Van Wormer uses this method because cores are sources of various seed, shoots, and roots of wetland plants that promote development of diverse marshes. The cores also produce plants that avoid the initial dieback of some species. Cores have the disadvantage of the time and cost associated with collecting, transporting intact, and planting the heavy soil mass.

We recommend allowing plantings to become well established before wastewater is introduced into the system; the plants need an opportunity to overcome planting stress before other stresses are introduced. Satisfactory establishment may take one or two full growing seasons. Gradual increase in the concentration of waste applied may also be necessary. Once the system is fully functional, vegetation should be monitored frequently; if dead or unhealthy patches of vegetation are found, they should be replanted and/or the concentrations of wastewater reduced. Simply obtain replacement plantings from the existing treatment system once it is well established.

For planting bare root plants and tubers, a tree planting bar or tile spade is a good tool. A slit is made in the substrate, the propagule is inserted, and the slit is sealed. Plant deep enough to prevent propagules from floating out of the planting hole. If a small hole is required for a wetland core or potted plant, a power auger may be needed on dry soil, or a shovel can be used if planting under water or on wet soil.

Occasionally, special anchoring methods are required when the substrate is soft, plants are buoyant, or erosion will disturb the constructed system. Those methods typically require temporary anchoring using light construction materials. Allen and Klimas[5] provide a good introduction to those techniques. Some wetland plant suppliers suggest weighted propagules can be dropped into calm water as an inexpensive means of planting, but it has been Pierce's experience this method does not provide a large proportion of successful plantings. Additional mechanical protection may be required to prevent animals from damaging newly established plants. Garbisch,[17] Environmental Laboratory,[2] Comes and McCreary,[18] and Pierce and Amerson[7] have suggested methods of preventing animal depredation. Such methods include (1) planting through chicken wire fence fastened to the surface of the substrate to prevent tuber or rhizome excavation and (2) use of unpalatable species.

WATER LEVEL MANAGEMENT

In constructed wetlands, controlling water levels will influence plant survival and desired species composition. Water level is the most critical aspect of plant survival during the first year after planting. A common mistake is to assume that because the plant is a wetland plant, it can tolerate deep water. Frequently, too much water creates more problems for wetland plants during the first growing season than too little water because the plants do not receive adequate oxygen at their roots. Wetland emergent species should be planted in a wet substrate (but not flooded) and allowed to grow enough to generate a stem with leaves that protrude above the initial flooding height. For best survival and growth during the first growing season, the substrate for small stalks (2–5 cm) should only be saturated, not flooded, and as the plants grow the water level can be raised proportionally.

For wet meadow species that grow in the shallow-to-transitional zones (Figure 1), watering during the first year should be limited to shallow sheetflow with intermittent drying periods, depending on the species. Wet meadow species such as spikerushes *(Eleocharis* spp.), fescues *(Festuca* spp.), bluegrasses *(Poa* spp.), bentgrasses *(Agrostis* spp.), and perennial foxtails *(Alopecurus* spp.) will tolerate reasonably long dry periods. Other species, such as manna grasses *(Glyceria* spp.), are not as tolerant and should not become completely dry the first growing season.

Water levels for submergent and floating leaved aquatic plants should never be lowered to the extent that the plants become exposed. The most important criterion for submergents is maintaining water level stability and keeping the plant continuously submerged the first growing season.

Water levels can be manipulated to control prolific growth and spread of weedy plants. For example, cattail may be controlled by deep flooding for several weeks during the growing season after the stems have been cut. Flooding may also inhibit establishment of undesirable opportunistic species.

SUMMARY

Wetland construction involves vegetation techniques related to species choice and plant handling. Choosing species requires knowledge of the hydrologic regime and substrate of the proposed wetland as well as the ecological characteristics of species proposed for planting. The species planted should also have attributes optimizing their desired function in the waste treatment system (e.g., efficient oxygen transmission to roots). Plants acquired from nurseries or collected in the wild must be carefully handled and planted to promote survival. Once planted, aftercare and water level manipulation will assure a well vegetated and healthy wetland.

ACKNOWLEDGMENTS

Permission was granted by the Chief of Engineers to publish this information. Thanks is extended to Dr. Charles J. Klimas and Mr. Ellis J. Clairain of the U.S. Army Engineer Waterways Experiment Station and to Dr. H. Glenn Hughes of the DuBois Campus, Pennsylvania State University, for their review of the chapter. Reference to companies and specific products in this chapter does not imply endorsement by the authors or any of their employers, such as the U.S. Army Corps of Engineers.

REFERENCES

1. Smith, H. K. "An Introduction to Habitat Development on Dredged Material," Technical Report DS-78-19, U.S. Army Engineer Waterways Experiment Station.
2. "Wetland Habitat Development with Dredged Material: Engineering and Plant Propagation," Technical Report DS-78-16, U.S. Army Engineer Waterways Experiment Station (1978).
3. Knutson, P. L., and W. W. Woodhouse, Jr. "Shore Stabilization with Salt Marsh Vegetation," Special Report No. 9, U.S. Army Corps of Engineers, Coastal Engineering Research Center (1983).
4. Kadlec, J. A., and W. A. Wentz. "State-of-the-Art Survey and Evaluation of Marsh Plant Establishment Techniques: Induced and Natural, Vol. I: Report of Research," Technical Report DS-74-9, U.S. Army Engineer Waterways Experiment Station (1974).
5. Allen, H. H., and C. V. Klimas. "Reservoir Shoreline Revegetation Guidelines," Technical Report E-86-13, U.S. Army Engineer Waterways Experiment Station (1986).
6. Brodie, G. A., D. A. Hammer, and D. A. Tomljanovich. "Constructed Wetlands for Acid Drainage Control in the Tennessee Valley," in *Mine Drainage and Surface Mine Reclamation, Vol. I, Mine Water and Mine Waste,* Bureau of Mines Information Circular 9183 (1988), pp. 325–331.
7. Pierce, G. J., and A. B. Amerson, Jr. "A Pilot Project for Wetlands Construction on the Floodplain of the Allegheny River in Cattaraugus County, New York," in *Proceedings of the Eighth Annual Conference on Wetland Restoration and Creation*, (Tampa, FL: Hillsborough Community College, 1982), pp. 140–153.
8. DuLaunie, R. D., W. H. Patrick, Jr., and J. M. Brannon. "Nutrient Transformations in Louisiana Salt Marsh Soils," Sea Grant Publication No. LSV-T-76-009, Center for Wetland Resources, Louisiana State University (1976).
9. Southern Tier Consulting, and Ecology and Environment, Inc. "Wetland Demonstration Project," Contract No. D250336-CPIN 5119.01.321 STE, Section 5P (A) for Department of Transportation, State of New York (1987).
10. Emerson, F. B., Jr., "Experimental Establishment of Food and Cover Plants in Marshes Created for Wildlife in New York State," *New York Fish Game J.* 8(2):130–144 (1961).

11. Wein, G. R., S. Kroeger, and G. J. Pierce. "Lacustrine Vegetation Establishment Within a Cooling Reservoir," The 14th Annual Conference on Wetland Restoration and Creation, Tampa, FL (1987).
12. Whitlow, T. H., and R. W. Harris. "Flood Tolerance in Plants: A State-of-the-Art Review," Technical Report E-79-2, U.S. Army Engineer Waterways Experiment Station (1979).
13. Stark, L. R., R. L. Kolbash, H. J. Webster, S. E. Stevens, Jr., K. A. Dionis, and E. R. Murphy. "The SIMCO #4 Wetland: Biological Patterns and Performance of a Wetland Receiving Mine Drainage," in *Mine Drainage and Surface Mine Reclamation, Vol. I, Mine Water and Mine Waste,* Bureau of Mines Information Circular 9183 (1988), pp. 332-344.
14. Michaud, S. C., and C. J. Richardson. "Relative Radial Oxygen Loss in Five Wetland Plants," Chapter 38a, this volume.
15. Mitsch, W. J., and J. G. Gosselink. *Wetlands* (New York: Van Nostrand Reinhold Company, 1986).
16. Garbisch, E. W., owner, Environmental Concern Inc., St. Michaels, MD. Personal communication (December 1986).
17. Garbisch, E. W. "Highways and Wetlands: Compensating Wetland Losses," Federal Highway Administration Report No. FHWA-IP-86-22, Turner-Fairbanks Highway Research Center, U.S. Department of Transportation (1986).
18. Comes, R. D., and T. McCreary. "Approaches to Revegetate Shorelines at Lake Wallula on the Columbia River, Washington-Oregon," Technical Report E-86-2, U.S. Army Engineer Waterways Experiment Station (1986).

CHAPTER 34

Operations Optimization

Michelle A. Girts and Robert L. Knight

INTRODUCTION

Constructed wetland water treatment systems integrate system design and operations management to minimize expense and maximize treatment efficiency and system longevity. In the compromise that results, design and capital investments can be minimized, but only with a concomitant increase in operating costs. Alternatively, the system design can allow for all deviations from normal conditions with little human intervention after startup to minimize operating costs.

Systems constructed to date range across the entire spectrum from those designed for "average" conditions requiring constant attention, to those designed to accommodate extreme conditions expected during the system's life span. From experience, reliance on system resiliency is often more beneficial than human intervention (i.e., management). Optimization of system design and operations is in an early developmental stage. Because of the limited number of systems and their short life spans, much information on operations changes and wetland responses is site-specific and has not been related to target treatment efficiency.

The most common approach to wetland treatment system design is to characterize influent water chemistry, environmental conditions, and target effluent water chemistry with sufficient accuracy to design for maximum treatment efficiency under "average" operating conditions. Some degree of system monitoring is required to ensure that effluent standards are met. When discharges are in danger of violating effluent standards, an evaluation of conditions and changes in operations takes place to increase treatment efficiencies and to improve effluent water chemistry.

This "best approximation" design and feedback loop has been selected to minimize costs at both design/construction and operations levels. Often, however, the design does not provide sufficient flexibility to respond to deviations from "average" operating conditions, resulting in systems that appear unmanageable and unreliable. In such a scenario, decisions are made on the basis of immediate response as opposed to long-term system viability. The overall cost of the system may thus be greater than those systems that are designed to accommodate extreme conditions and disturbances.

This chapter reviews the need for management and operating flexibility in constructed wetland water treatment systems. We examine conditions under which flexibility in operations improves treatment efficiency and longevity of a well-designed system; methods by which operation changes can help a system adapt to unanticipated demands; and associated labor requirements. We also review models of natural wetland systems and water treatment systems as the basis for developing operation models for constructed wetland water treatment systems. Finally, we present case studies of several operating systems to demonstrate the importance of planning operation alternatives that complement system design.

We assume the goals of wetland treatment system optimization are to (1) maximize treatment efficiency and capacity under all scenarios; (2) minimize system costs during design, construction, and operation; (3) maintain long-term viability of the system; and (4) achieve maximum flexibility given human, technological, and financial limitations.

PERTURBATIONS IN CONSTRUCTED WETLAND SYSTEMS

Deviations from "average" operating conditions will occur in the system lifetime with greater frequency than predicted or preferred. Perturbations generally are of two types: (1) those that can be predicted and occur periodically and (2) those that are probable but whose time of occurrence cannot be predicted. Predictable disturbances can be anticipated during initial system startup and during any subsequent period of startup following a period of maintenance or dormancy. Other predictable disturbances follow seasonal cycles associated with precipitation, temperature, and vegetation growth patterns. Seasonal changes of influent chemical or hydraulic loading caused by cycles of treatment demand also require operations changes, and therefore qualify as "disturbances." In contrast, unpredictable perturbations involve unanticipated changes in loading rate, environmental conditions, vegetation damage, or failure of a structural component of the system and are potentially catastrophic.

Predictable Disturbances

Conditions surrounding predictable disturbances, including the primary features, symptoms, and appropriate operation modifications are summarized in Table 1.

Startup Periods

Startup periods are operational phases when flora and fauna associated with treatment processes become established. The timing of startup is an important

Table 1. Predictable System Disturbances

Disturbance	Features	Symptoms	Operation Modifications
Startup	Vegetation establishment Microbial flora colonization Technical system debugging	Widely fluctuating inflow-outflow chemistries, after taking flow velocity changes into account	Control loading rates, i.e., water flow rate, chemical concentrations • water inflow rates • freshwater source Control water depths—critical for vegetation establishment and development of conditions suitable for target microbial populations Dilution/recirculation Chemical additions
Seasonal	Extreme precipitation	High loading rates	Control loading rates, i.e., water flow rates, chemical concentrations • water inflow rates • dilution • recirculation
		Decreased storage capacity	Increase storage capacity • stormwater diversion • detention pond • increase water depth
		Insufficient residence time; Channeling	Control outflow rates Installation of baffles
	Extreme low temperatures	Insufficient flow	Freshwater source, recirculation, parallel cells
		Freezing; sheet flow over ice surface	Recirculation, aeration, control water depth Preheated water
	Vegetation growth/decay	Flushing of chemicals, nutrients, and microbes from decaying vegetation, sediment	Secondary treatment pond for treatment Vegetation harvest or burning Recirculation to increase nutrient retention Drawdown and oxidation
	Population composition changes (microbial, algal)	Gradual change in treatment efficiency	Secondary treatment pond Recirculation

operations decision, as discussed in U.S. EPA.[1] Coordination of startup periods with seasonal changes in wetland hydroperiod, productivity, and wildlife use will reduce system stress and improve the likelihood of successful vegetation establishment.

During these periods, the long-term viability of the system is of primary importance. Control, delivery, and emergency systems can be debugged, assuring their performance during heavier system loading. Immediate maximization of treatment efficiency and capacity is of secondary importance given that system design guarantees some degree of treatment once system establishment is complete. Therefore, if a trade-off between water treatment and healthy vegetation becomes necessary, the latter is the optimal choice. In fact, many systems show fluctuating treatment efficiencies during this period reflected in wetland and effluent chemistry changes unrelated to loading or environmental conditions.

Operation modifications during startup periods include careful regulation of chemical and hydraulic loading rates. This may involve provision for a freshwater source or dilution of the water to be treated in order to gradually increase chemical loading. For most systems, a constant flow velocity is recommended; tidal systems are an exception. Careful control of water depths is required by many wetland plant species for germination and establishment; underwater light attenuation may also be critical, especially for open water vegetation. Recirculation of treated water is a useful option, provided that water temperature elevation during residence in the treatment system is minimal.

Chemical treatment is another operation option to be considered, especially as it aids vegetation establishment. For example, many constructed wetlands treating acid mine water in the Appalachian region incorporate lime or an alkalinity-generating substrate to ameliorate effects of acidity and metals during the startup period. Neutralization capacity decreases over time as wetland vegetation becomes established. Chemical addition may also include periodic fertilizer addition.

Seasonal Disturbance

Depending on geographic location, seasonal changes in precipitation and temperature patterns may be disruptive to the constructed wetland system and may necessitate operations adjustments.

High precipitation events result in high hydraulic loading rates, decreased storage capacity, and insufficient residence time. They also increase the likelihood of channelization within the wetland. Such events may increase contaminant concentrations in influent (by flushing of the source system) or may dilute concentrations. Appropriate operations responses will decrease inflows and increase storage capacity (e.g., by diverting to a pretreatment holding pond). Increased concentrations, particularly of suspended solids, can be partially treated in the pretreatment holding pond without long-term impact on the wetland system. Recirculation can reduce chemical concentrations by adding recirculating water to dilute the main stream flow. More often, hydraulic loading is the primary concern because influent chemical concentrations and ponded water are diluted by rainfall before entry into the wetland. Outflow as well as inflow rates can be regulated to attenuate flow-through time and to limit channelization. Installation of temporary baffles can also slow flow rates and limit channelization.

Conversely, insufficient flow or high evapotranspiration rates during the growing season will lessen plant viability and oxidize formerly anaerobic sediments, with corresponding decreases in treatment efficiency. Adding fresh water or recirculated water may alleviate the problem. Alternatively, in a parallel cell system with one cell unvegetated, all flow could be temporarily directed to the vegetated cell during periods of low flow.

High temperatures will increase evaporation plant stress, while low temperatures will result in system freezing and will increase the likelihood of sheetflow over the ice surface, short-circuiting substrate contact. High temperatures may be lowered by decreasing residence time and by dilution. Sheetflow over the ice can be avoided through proper inlet and outlet structure design. Recirculation, aeration, and fluctuating water levels will increase flow velocity and prevent freezing. Additional operations solutions offered in Reed et al.[2] apply to wetland as well as conventional water treatment systems.

During seasons of vegetation decay, nutrients and contaminants stored in plant material may be released to the water, with apparent system flushing. In systems where chemicals are not translocated to root structures at the end of the growing season and organic matter is not required as a substrate for treatment, periodic harvest of vegetative materials may limit flushing. A post-treatment detention pond enhances settling and algal growth to eliminate problems of sediment and nutrient flushing. Alternatively, effluent may be recirculated through a pretreatment detention pond and the wetland system prior to final discharge. The recirculation process will also prevent flushing of microbial communities associated with substrate loss.

In highly productive wetlands, organic matter accumulation at the end of each growing season may decrease storage capacity. If vegetation harvest is not feasible, seasonal drawdown and oxidation of organic material ("drydown") may limit cumulative losses in storage capacity.

Seasonal disturbances similar to unusual conditions during startup can be anticipated during the system design phase, and necessary operating systems can be incorporated into the primary design. Seasonal disturbances are specific to climate and vegetation of a particular site. The list of predictable disturbances and operational responses presented in Table 1 is not exhaustive but highlights the concerns of most systems in temperate regions. Additional operations responses are summarized in U.S. EPA,[1] with examples of sites and methods implemented.

Unpredictable System Perturbations

All system perturbations unforeseen in the design phase may be classified as "unpredicted," although some included occur so infrequently that incorporation into the design would entail unnecessary expense. Low probability events include record storm events; chemical spills or breakthrough at a point upstream in the system; loss of system power; events that damage system structures, such as tornadoes or earthquakes; or events that damage the system vegetation substrate, such as wind, fire, disease, insect, or wildlife damage. Similarly, pests such as mosquitoes,[3,4] while not affecting the treatment efficiency and longevity of the system, may require modification of operating procedures to discourage or eliminate the pest population. Unforeseen changes in plant operations, system malfunctions or failures, and design flaws qualify as unpredictable disturbances that may be accommodated by operation changes. In addition, with technological advances, systems may require modification or expansion and consequent revision of operation plans.

Examples of unpredictable events are presented in Table 2, along with operational features most affected, associated symptoms, and appropriate operation modifications. This table is not an exhaustive list but focuses on the operations mechanisms frequently used to increase system resiliency and maximize treatment efficiency and capacity. Many system features that provide operations mechanisms during unpredicted events are the same as those used for startup and in seasonal events (Table 1). These include the ability to interconnect and/or isolate portions of the system and provision for backup systems. The primary difference is in the required operations response time. For this reason, an operation and maintenance manual with specific contingency provisions is needed for optimal operations.

CASE HISTORIES OF OPERATING SYSTEMS

One challenge at this early stage of wetland technology is that every design is still an experiment. Either every possible system disturbance or operational mistake has not occurred, or they have not been described in the literature. This section provides some case histories of operational/optimization problems encountered in actual systems and describes corrective or preventive measures taken. Correction of actual problems has not always been successful in these systems. These case histories are from natural or seminatural wetland treatment systems; however, the concepts they convey are relevant to constructed systems as well.

Houghton Lake, Michigan

The peat-based treatment wetland at Houghton Lake, Michigan, is one of the most thoroughly described operational systems.[5] This wetland system was

Table 2. Unpredictable System Disturbances

Disturbance	Symptoms	Operation Modifications									
		Water Inflow	Water Outflow	Water Depth	Dilution	Recirculation	Pretreatment Pond	Chemical Addition	Vegetation Harvest	Replant	Predator Control
Record storm event	High hydraulic loading rates	X				X	X				
	Decreased storage capacity	X					X				
	Insufficient residence time	X	X	X	X	X	X	X			
	Channeling	X					X				
	High sediment loads			X			X				
	High chemical loads		X		X	X	X	X			
Change in chemical constituents and concentrations	High chemical loads	X			X		X	X			
	Increased toxicity (vegetation, wildlife)				X				X	X	
	Release of chemicals from sediments/vegetation		X	X		X			X	X	
	Change in chemical form			X	X			X			
Vegetation damage	Increased debris, flow hindrance	X		X					X	X	
	Elemental release from vegetation		X	X		X			X	X	
	Change in conditions for replanting	X		X			X		X	X[a]	
Pests (beavers, mosquitoes, etc.)	Complaints from neighbors			X				X			X
	Reduced flow and water level control										
Malfunctions/ Construction failures	Reduced flow and water level control	X	X	X	X	X	X	X			
	Inability to respond to need for changes in operations[b]										
Design flaw	Limited treatment capacity[b]										
	Limited lifespan[b]										

[a]New species.
[b]All operation modifications may need to be considered.

to provide nutrient removal from domestic secondary effluent, because available surface receiving waters flow to Houghton Lake, a soft water system undergoing eutrophication in the early 1970s. At that time, wetlands were considered a nutrient removal panacea with little regard for experience from other soil-based land treatment systems.

Early reports by the University of Michigan research team indicated that the Porter Ranch fen was indeed a major sink for nitrogen and phosphorus.[6] Subsequent reports after several years of continued monitoring have indicated otherwise.[5,7] Moving fronts of ammonia nitrogen and phosphorus recognized since 1979 were caused by overloading of long-term assimilation potential and saturation of available soil adsorption sites.

The Houghton Lake system has some capacity for long-term nitrogen removal through denitrification. Also, a significant adsorption capacity continues to remove nitrogen and phosphorus as long as all adsorption sites are not saturated. However, this system will have a finite life span for phosphorus removal, its primary design function. Remaining useful system life is estimated at 20 to 30 years, when phosphorus concentrations leaving the wetland to Houghton Lake will be nearly as high as levels in secondary wastewater.

The Houghton Lake example raises a difficult question for resource managers. Is it acceptable to enrich a natural wetland system of 700 ha to protect a lake for about 30 years rather than use other methods of phosphorus removal? Other area wetlands might be considered for a second phase of phosphorus removal or conventional phosphorus pretreatment might be required. On the other hand, a conventional advanced treatment plant or a no-discharge land treatment alternative might be more environmentally acceptable. Over 12 years of studies at Houghton Lake have demonstrated it cannot be effectively optimized for its primary goal of phosphorus removal.

Gainesville, Florida

Concurrent with the Houghton Lake study, University of Florida researchers led by H. T. Odum pioneered the use of another natural wetland ecosystem for wastewater discharge and treatment. Cypress domes were used during a five-year pilot study at Whitney Trailer Park near Gainesville, Florida, to receive domestic secondary wastewater at a hydraulic loading rate of 2.5 cm/week.[8,9] Phosphorus and nitrogen removal by this system was also well publicized. Unfortunately, report readers did not note the researchers' conclusion that phosphorus removal was almost completely dependent on adsorption by clay soils beneath the domes and that cypress wetlands are not a phosphorus sink if surface waters flow through them horizontally instead of vertically. Also, little attention was given to the fact that at the 2.5 cm/week annual average hydraulic loading rate, the domes overflowed seasonally during normal rainfall conditions. However, the cypress domes effectively removed all forms of nitrogen without relying on soil adsorption. This nitrogen assimilation capacity resulted from high nitrification and denitrification rates, typical of warm southern wetlands. Also, the system worked well as an infiltration system as long as the hydraulic loading rate was controlled based on an actual water balance.

At least two Florida systems began operation in the mid- to late 1970s based partially, if not entirely, on the Gainesville cypress domes research. In 1976, the Reedy Creek Improvement District near Orlando implemented a full-scale discharge of secondary wastewater to a former cypress floodplain wetland adjacent to the channelized Reedy Creek. Although this system has provided consistent removal of organic matter and nitrogen from pretreated wastewater for 10 years,[10] it has no significant assimilative capacity for phosphorus. In terms of phosphorus removal requirements, no optimization method was available within the wetland, and pretreatment for phosphorus reduction was implemented at the wastewater treatment plant.

A second cypress wetland discharge system at least partially modeled after the Gainesville prototype was constructed in 1982 at Buenaventura Lakes north of Orlando, Florida.[11] Hydraulic application rate at this system was based on an unrealistic water balance with little site-specific knowledge of actual infiltrative capacity of the cypress dome. A gradual flow increase over time to an average hydraulic application rate of 2.3 cm/week resulted in surface discharges after two years of operation. Because no alternative discharge point was available for wet season flows, unauthorized surface discharges have continued.

Problems at Reedy Creek and Buenaventura Lakes wetland systems highlight the importance of site-specific information for optimizing certain functions in treatment wetlands. Wetlands have variable phosphorus removal potential dependent on the chemical and physical nature of their soils and mode of contact between applied water and soils.[12-14] Also, infiltrative capacity is site-specific and low in wetlands, and most systems designed today rely on a surface outlet to balance water inflows and outflows. However, the Reedy Creek natural wetland as well as the wetlands at Buenaventura Lakes, Houghton Lake, and Gainesville all provide a cost-effective alternative for treating biochemical oxygen demand (BOD_5), total suspended solids (TSS), and nitrogen.

Hilton Head Island, South Carolina

The Boggy Gut wetland at Sea Pines Plantation provides examples of initially perplexing but simple operational problems and remedial actions. It was a relic, diked, overdrained rice field undergoing succession to a mixed marsh/old field plant community before 1980.[15] A local naturalist suggested that drainage impacts could be reversed and wildlife values restored if the Sea Pines Public Service District (PSD) discharged secondary effluent with a diffuser pipe in Boggy Gut. This change was implemented in 1983 following design of a three-year pilot study to assess the biological and water quality effects. The Boggy Gut wetland would provide natural filtration and assimilation of constituents remaining in the treated wastewater, groundwater recharge, and surface discharge into Lawton Canal.

During the first year of operation, eight monitoring wells were abandoned because they were placed in the area flooded by the discharge. No survey or calculations had been performed to predict the area that would be affected by the discharge. Water quality data from relocated wells indicated periodic violations for nitrate-nitrogen and fecal coliforms. Surface and groundwater monitoring revealed nitrate levels of 50 mg/L entering the wetland from the Sea Pines Wastewater Treatment Plant (WWTP), because subnate was returned from the sludge drainfield to the storage lagoon from which water was pumped to the wetland. Nitrate and total nitrogen levels in the wetland influent were reduced to less than 20 mg/L following subnate rerouting to the plant headworks.

The source of the elevated coliforms was similarly traced through additional monitoring. Treated wastewater from the Sea Pines WWTP is chlorinated and has less than 20 fecal coliforms/100 mL, but Boggy Gut surface water averages 1,600 fecal coliforms/100 mL. Monitoring determined that elevated coliform populations originated from a recreational bridal path encircling Boggy Gut wetland. Horse manure is washed from the path into the wetland during rainy conditions.

The elevated nitrate and coliform data from the monitoring wells still appeared unusual. Inadequately grouted wells probably allowed surface water leakage from the edge of the wetland along the well casings and into shallow groundwater. Improved sampling of properly installed wells over the following three-year period has not shown elevated levels of nitrate, coliforms, or other monitored parameters of concern in protection of the deeper potable aquifer.[16] Elevated chloride levels have provided a tracer of the treated wastewater plume in the shallow groundwater throughout this same period.

During 1986, at an average hydraulic rate of 13.2 cm/week, ammonia nitrogen levels increased in the Boggy Gut outflow during late summer. Because ammonia assimilation rate does not apparently diminish with time in other southern systems,[10,17] attention was turned to a possible kinetic problem at Boggy Gut. Calculated residence times based on area and water depth values estimated the system residence time ranged from 18 to 80 days under the highest loading rate tested. But flocculant organic muck had accumulated during the previous three years of discharge, and a shallow channel in the last third of the system leads directly to the outlet weir. A dye study revealed the effective hydraulic residence time in Boggy Gut at the high loading rate was less than five days during summer conditions of maximum biomass accrual. In combination with low summer dissolved oxygen values, this low residence time resulted in a temporary reduction of ammonia assimilation.

An upgrade at the Sea Pines WWTP in 1987 reduced inflow ammonia concentrations.[18] Annual average mass removal efficiencies in Boggy Gut in 1987 were BOD_5, 80%; TSS, 89%; total nitrogen (TN), 85%; ammonia nitrogen, 85%; nitrate nitrogen, 94%; and total phosphorus (TP), 50%. Assimilation rates in the wetland for TN and TP during the highest hydraulic loading period, from April through December 1986, were measured as 2.4 and 0.37 kg/ha/day, respectively.

Constructed Wetlands for Treatment of Acid Mine Drainage

Many of the wetlands constructed to date to treat high iron, manganese, and aluminum concentrations in acidic discharges from mined areas are designed to revert to natural wetlands within 20 to 50 years (the expected lifetime of the problem). Operations requirements of these systems have been minimal; however, treatment efficiency has varied widely among systems. Control of hydraulic loading rates has been implicated as a primary factor in diminished treatment efficiencies,[19] but different effects are observed over time and geographic location. For example, the Fabius Impoundment 1 acid drainage wetland constructed by TVA began successfully treating acid mine drainage (30–110 L/min; Fe, 80 mg/L; Mn, 10 mg/L) in June 1985.[20] On December 3, an additional 75 to 150 L/min of pH 3.0 slurry lake water was applied, and treatment efficiency failed by December 30 when the added influent was stopped. In May 1986, an additional 1.9 L/min of slurry lake water was applied and slowly increased to 30 L/min in September without impairing treatment efficiency. The biological system was unable to adjust to abruptly increased hydraulic loading (68–500% of previous loading) during winter but adapted to gradual increases in hydraulic loading of 2–100% of the previous loading during the growing season.

CONCLUSIONS

The importance of designing wetland wastewater treatment systems for management and operations flexibility is evident from the preceding case studies. Insufficient information about the treatment capacity and operation problems of the early designs has been available. As a result, inappropriate or inadequate monitoring programs were designed and applied to systems that could make few operations adjustments when problems were discovered.

Site-specific information on pretreatment operations, soils, hydrology, vegetation, and local land use in conjunction with literature information can now provide a basis for predicting wetland treatment system responses and problems. Designs that adjust for periods of disturbance through operations changes can extend the life and increase the efficiency of new systems. Development of a generic simulation and operations model that can be tailored to a given site will improve system optimization and reliability.

REFERENCES

1. Zedler, J. B., and M. E. Kentula. "Wetlands Research Plan," U.S. EPA Report-600/3-86/009, Environmental Research Laboratory (1985).

2. Reed, S. C., D. S. Pottle, W. B. Moeller, C. R. Oh, R. Peirent, and E. L. Niedringhaus. "Prevention of Freezing and Other Cold Weather Problems at Wastewater Treatment Facilities," Special Report 85-11, U.S. Army Corps of Engineers (1985).
3. Stowell, R., S. Weber, G. Tchobanoglous, B. A. Wilson, and K. R. Townzen. "Mosquito Considerations in the Design of Wetland Systems for the Treatment of Wastewaters," in *Ecological Considerations in Wetlands Treatment of Municipal Wastewaters,* P. J. Godfrey, E. R. Kaynor, S. Pelczarski, and J. Benforado, Eds. (New York: Van Nostrand Reinhold Company, 1985), pp. 38–47.
4. Guntenspergen, G. R., and F. Sterns. "Ecological Perspectives on Wetland Systems," in *Ecological Considerations in Wetlands Treatment of Municipal Wastewaters,* P. J. Godfrey, E. R. Kaynor, S. Pelczarski, and J. Benforado, Eds. (New York: Van Nostrand Reinhold Company, 1985), pp. 69–97.
5. Kadlec, R. H. "Northern Natural Wetland Water Systems," in *Aquatic Plants for Water Treatment and Resource Recovery,* K. R. Reddy and W. H. Smith, Eds. (Orlando, FL: Magnolia Publishing Inc., 1987), pp. 83–98.
6. Kadlec, R. H., and D. L. Tilton. "The Use of Freshwater Wetlands as a Tertiary Wastewater Treatment Alternative," *CRC Crit. Rev. Environ. Control* 9:185–212 (1979).
7. Kadlec, R. H. "Aging Phenomena in Wastewater Wetlands," in *Ecological Considerations in Wetlands Treatment of Municipal Wastewaters,* P. J. Godfrey, E. R. Kaynor, S. Pelczarski, and J. Benforado, Eds. (New York: Van Nostrand Reinhold Company, 1985), pp. 338–347.
8. Ewel, K. C., and H. T. Odum, Eds. *Cypress Swamps* (Gainesville: University Presses of Florida, 1984).
9. Dierberg, F. E., and P. L. Brezonik. "Nitrogen and Phosphorus Mass Balances in a Cypress Dome Receiving Wastewater," in *Cypress Swamps,* K. C. Ewel and H. T. Odum, Eds. (Gainesville: University Presses of Florida, 1984).
10. Knight, R. L., T. W. McKim, and H. R. Kohl. "Performance of a Natural Wetland Treatment System for Wastewater Management," *J. Water Pollut. Control Fed.* 59:746–754 (1987).
11. Sheffield Engineering & Associates. "Preliminary Report for Wetland Effluent Disposal Facilities at Buenaventura Lakes," submitted to Orange Osceola Utilities, Kissimmee, FL (1983).
12. Richardson, C. J., and D. S. Nichols. "Ecological Analysis of Wastewater Management Criteria in Wetland Ecosystems," in *Ecological Considerations in Wetlands Treatment of Municipal Wastewaters,* P. J. Godfrey, E. R. Kaynor, S. Pelczarski, and J. Benforado, Eds. (New York: Van Nostrand Reinhold Company, 1985), pp. 338–347.
13. Nichols, D. S. "Capacity for Natural Wetlands to Remove Nutrients from Wastewater," *J. Water Pollut. Control Fed.* 55(5):495–505 (1983).
14. Knight, R. L., B. H. Winchester, and J. C. Higman. "Carolina Bays: Feasibility for Effluent Advanced Treatment and Disposal," *Wetlands* 4:177–203 (1985).
15. Wilbur Smith and Associates. "Wastewater Management Study: Effluent Reuse and Disposal, Vol. 1," Technical Report to Sea Pines Public Service District (Hilton Head, SC: 1981).
16. CH2M HILL. "Boggy Gut Wetland Treated Effluent Disposal System, Hilton Head, SC," Final Report to Sea Pines Public Service District, Hilton Head, SC (1986).

17. Tuschall, J. R., P. L. Brezonik, and K. C. Ewel. "Tertiary Treatment of Wastewater Using Flow-Through Wetland Systems," in *Proceedings of Nat. Conf. Amer. Soc. Civil Eng.*, Atlanta, GA (1981).
18. Hirsekorn, R. A., and R. A. Ellison. "Sea Pines Public Service District Implements a Comprehensive Reclaimed Water System," in *Implementing Water Reuse: Proceedings of the Water Reuse Symposium IV* (Denver, CO: American Water Works Association Research Foundation, 1988), pp. 309–331.
19. Girts, M. A., and R. L. P. Kleinmann. "Constructed Wetlands for Treatment of Acid Mine Drainage: A Preliminary Review," in *Proceedings of 1986 National Symposium on Mining, Hydrology, Sedimentology, and Reclamation,* Lexington, KY (1986).
20. Brodie, G. A., D. A. Hammer, and D. A. Tomljanovich. "Man-Made Wetlands for Acid Mine Drainage Control," in *Proceedings of the 8th Annual National Abandoned Mine Lands Conference,* Billings, MT (1986).

CHAPTER 35

Pathogen Removal in Constructed Wetlands

Richard M. Gersberg, R. A. Gearheart, and Mike Ives

INTRODUCTION

An assessment of human health risks posed by municipal wastewater treatment using constructed wetlands involves delineating pathogen content of inflowing water, degree of removal/inactivation of pathogens in wetland environments, and level of exposure of receptor (human) population to treated wastewaters. Four groups of pathogens (viruses, bacteria, protozoa, and worms) may cause human waterborne diseases. Since most excreta-related diseases in North America are caused by bacteria and viruses, this chapter will focus on use of constructed wetlands to reduce human health risks associated with these two groups.

Data developed from two types of constructed wetland systems—a gravel-based constructed wetlands in Santee, California and marshes created from excavations in natural soils in Arcata, California—will be presented in this chapter. The overall objectives of these studies were to assess the degree of removal of (1) the conventional indicators of fecal contamination (total coliforms or fecal coliforms) and the specific pathogen *Salmonella* and (2) an indicator of viral pollution, MS-2 bacteriophage.

Average viral content of domestic sewage in the United States is about 7000 viruses/L, and enteric viruses can pass secondary treatment and chlorination in the infective state.[1] Over 100 different virus types are known to be excreted from human feces,[2] and the minimal infective dose (MID) may be as low as one virus particle.

The MID of most bacterial diseases is relatively high, ranging from 10^4 cells for *Shigella* sp. to 10^6–10^9 cells for *Salmonella* sp., *Vibrio* sp., and toxic strains of *Escherichia coli*.[3] Consequently, the U.S. Environmental Protection Agency (EPA) has recommended a bathing standard of 10^3 total coliform (TC) bacteria (most probable number, or MPN) per 100 mL to ensure no adverse health effects due to body contact recreation.[4] Using the most conservative estimate, assuming pathogens are equal in number to TC, the 10^3/100 mL TC standard requires minimum ingestion of 1–1000 liters to cause bacterial disease. Indeed, it appears this standard ensures a low risk level in all beneficial uses except drinking.

COLIFORM REMOVAL

Numerous studies have investigated the use of aquatic plants for the treatment of wastewater, and many studies have been encouraging.[5-11] Spangler et al.[12] determined that coliform bacteria were reduced 90–99.7% in a pilot project containing *Scirpus* sp. (bulrushes).

Coliform decline is not only due to cell die-off, but also encompasses sedimentation, filtration, adsorption, and aggregate formation. Sunlight has a lethal effect on coliform bacteria. Predators, bacteriophages, competition for limiting nutrients or trace elements, and toxins given off by other microorganisms may exert bacteriocidal effects.

In pilot constructed marsh cells dominated by *Typha latifolia* (cattail) and *Scirpus lacustris* (hardstem bulrush), average fecal coliform (FC) reduction in 12 cells was 86%.[6] Marsh cells with consistently greater than 90% reductions had reduced hydraulic loading. Additional studies showed the enteric bacterium (*Salmonella* sp.) was reduced from 94% to 96%. TC were inappropriate indicators of fecal contamination in wetlands, and the FC index was the preferred indicator of bacterial removal.

Total and fecal influent coliforms were reduced from 10% to 100%, depending upon the pattern and area of artificial marsh cells harvested of emergent vegetation.[13] Baffled and unharvested cells consistently produced superior effluents, while harvested cells demonstrated reduced pathogen removal.

FC reductions were found to be high throughout the year at the Listowel Artificial Marsh Project.[14] Three designs tested included a channelized serpentine configuration, a shallow marsh, and a complex of shallow marsh, deep pool, and channelized marsh. The channelized serpentine marsh demonstrated the highest (>99%) and most consistent FC removal.

Reduction of FC was skewed toward the influent and effluent of a wetland cell with eight baffled intercells receiving oxidation pond effluent.[15] Log_{10} reduction varied significantly from month to month, and reduction did not appear to be a function of suspended solids separation in the upper reach of the wetland cell.

Efficiency of removal/inactivation of TC bacteria in gravel-based constructed wetlands receiving primary municipal wastewaters at 5 cm/day is shown in Figure 1. Inflowing wastewaters contained a mean of 6.75×10^7 TC/100 mL. Flow through a bulrush bed reduced this level to a mean of 5.77×10^5 TC/100 mL, an overall removal efficiency of 99.1%. Pathogen removal by these constructed wetlands is superior to conventional treatment processes, where typical reductions are in the order of 1 log (90%).[16] However, even with more than 2-log (99%) removal demonstrated by the Santee wetland beds,

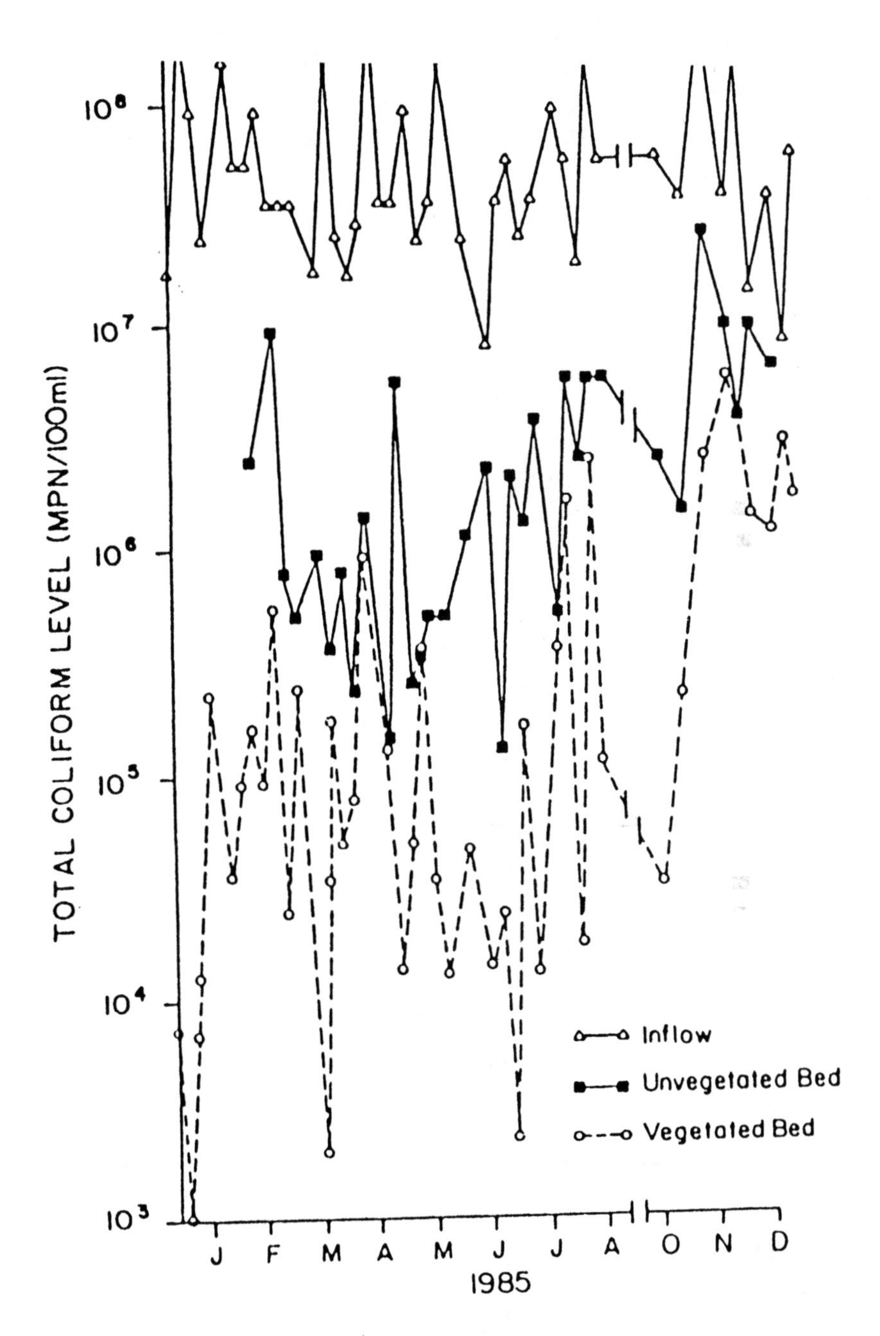

Figure 1. Concentration of total coliform in the applied primary municipal wastewater and in the effluent of a vegetated (bulrush) bed and an unvegetated bed. Hydraulic application rate was 5 cm/day for both beds.

Table 1. Survival of Total Coliform Bacteria After Chlorination of the Wetland Effluent

Date	Total Chlorine Residual (mg/L)	Total Coliform Inflow Into Wetlands (MPN per 100 mL)	Total Coliform Level After Chlorination (MPN per 100 mL)	Overall Log Removal
3/4/86	1.5	1.7×10^7	<2	>7 log
3/12	20.0	1.6×10^8	<2	>8 log
3/18	20.0	2.4×10^7	<2	>7 log
3/27	7.0	1.7×10^7	<2	>7 log
3/30	4.5	2.8×10^7	<2	>7 log
4/8	9.1	2.4×10^8	<2	>8 log
4/16	>20.0	3.5×10^7	<2	>7 log
4/23	12.0	3.5×10^7	<2	>7 log
4/30	14.0	9.2×10^7	<2	>7 log
5/28	>20.0	1.6×10^8	<2	>8 log
6/3	>20.0	2.4×10^7	<2	>7 log
6/11	19.0	2.2×10^7	<2	>7 log
6/17	>20.0	7.9×10^6	<2	>6 log
6/25	18.3	3.5×10^7	<2	>7 log
7/2	>20.0	5.4×10^7	<2	>7 log
7/23	10.0	2.4×10^7	<2	>7 log
8/6	9.0	5.4×10^7	<2	>7 log
8/15	13.0	1.8×10^7	<2	>7 log
8/18	8.0	3.5×10^7	8	>6 log
8/29	9.5	5.4×10^7	<2	>7 log
10/31	9.0	2.4×10^8	<2	>8 log
11/12	3.0	1.6×10^8	<2	>7 log
11/22	3.0	1.6×10^8	<2	>8 log
12/23	13.0	5.4×10^7	2	>7 log

Note: Chlorine contact time was 2 hours. MPN = most probable number.

further treatment is needed to reduce TC levels from 5.77×10^5/100 mL value to less than 10^3/100 mL so water reuse may be unrestricted.

Disinfection by chlorination (with combined chlorine residuals) can reduce levels of TC bacteria in wetland effluent to the drinking water standards of less than 2.2 TC (MPN)/100 mL (Table 1). At this level, risks associated with discharge or reuse of wastewater are extremely low.

SALMONELLA REMOVAL—ARCATA, CALIFORNIA

A quantitative assay technique using membrane filtration and ampicillin counterselection (ACT) for *Salmonella* in polluted water was developed at the Arcata Marsh Pilot Project.[17] ACT isolates low concentrations of *Salmonella* when high numbers of coliforms are present, can be performed quickly and easily, and was tested with *Salmonella*-spiked wastewater samples before evaluating *Salmonella* removal within the marsh (Table 2).

The main difficulty in determining *Salmonella* in aquatic systems is that when *Salmonella* are present, their concentrations can be 3–6 orders of magnitude lower than those of the coliform bacteria.[18] As a result, the lactose-positive coliform bacteria frequently mask the presence of lactose-negative

Table 2. Results of Counterselection Experiment with *Salmonella*-Spiked Marsh Water Samples

	Spiked Sample			*Salmonella* Control			Marsh Control (35°C)
Salmonella dilution	–5	–6	–7	–5	–6	–7	No *Salmonella*
Salmonella cfu/sample	197	23	5	209	19	2	0
Other cfu/sample	29	32	25	0	0	0	334,000

colonies on differential media. *Salmonella* and *Escherichia* are closely related, exhibiting 45–50% DNA homology.[19] Because they are so closely related, it is difficult to selectively remove sufficient coliform bacteria in any given sample without also removing a significant fraction of the *Salmonella* population. Most selective techniques employ compounds in the media which inhibit unwanted or competing organisms to a greater degree than the desired organisms.[20] However, almost all selective techniques, particularly when used to differentiate closely related bacteria, stress the organism one is trying to isolate. Add the fact that the enteric pathogens such as *Salmonella* are out of their environment in water and therefore are already stressed, and the problem of direct efficient quantitative recovery becomes clear.

Penicillin counterselection as an enrichment technique may allow efficient recovery and enumeration of *Salmonella* from water while largely inhibiting competition by other enteric bacteria.

The penicillins are a family of β-lactam antibiotics originally derived from the mold *Penicillium* sp. They specifically inhibit bacterial cell wall synthesis by inhibiting the transpeptidation reaction that cross-links adjacent peptidoglycan chains. As a result, growing cells are unable, after considerable metabolic expense, to complete cell wall synthesis and are subject to osmotic lysis. Penicillin counterselection depends on the fact that only growing cells (those cells actively synthesizing peptidoglycan) are killed. Since *Salmonella* are lactose-negative, i.e., they cannot use lactose as a sole source of organic carbon, they should avoid lysis in a hypotonic penicillin-containing medium where lactose is the sole carbon source. The predominate competing species, being mostly lactose-positive, should begin dividing and subsequently undergo osmotic lysis under the effects of the penicillin. If the penicillin is then removed enzymatically or by filtration and washing and the cells are placed on a solid medium that permits the growth of *Salmonella,* the resulting colonies may be counted directly. If in addition to allowing the growth of *Salmonella,* the final plating medium is differential, then probable *Salmonella* colonies can be distinguished from other lactose-negative or penicillin-resistant colonies.

PROCEDURE

Water is collected in clean 1-liter Nalgene bottles and concentrated by membrane filtration through 0.45-m Gelman GN-6 filters. The filter is placed in a flask containing 50–100 mL of LSR broth. The flask is vigorously aerated (bubbled air was used, but a shaking water bath would work as well) at 35°C for 1

Table 3. Results of *Salmonella* Removal Study in Marsh Cell 8 Using *Salmonella* #8154 and Tritium-Labeled Water as a Conservative Tracer

Time After Injection (hr)	*Salmonella* cfu/100 mL	^3H cpm	^3H dpm[a]	cfu/ dpm	% Removal
10	0	7.3	60.59		
23	10	11.6	96.28	0.10	94.44
29	14	13.6	112.88	0.12	93.33
42	6	8.3	68.89	0.09	95.00
52	4	7.2	60.09	0.07	96.11
64	0	6.0	49.80		
76	0	4.8	40.17		
100	0	3.8	31.54		

[a]12% counting efficiency.

hour to starve the *Salmonella* and allow the other bacteria present to begin growth. Ampicillin is then added to a final concentration of 1000 mg/mL from a sterile solution in Davis minimal broth prepared the same day. Incubation is continued with aeration for 15 minutes. The cells are harvested by membrane filtration of the broth. The filter is then rinsed thoroughly in sterile buffer or minimal broth to remove the ampicillin, placed face up on XLD agar, and incubated for 24–36 hours at 41.5°C.

Black colonies or red colonies with black centers are scored as positive *Salmonella* colony forming units (cfu). Representative colonies from each isolation can then be verified serologically be reaction with polyvalent antisera[21] or characterized biochemically using the API system.[22]

Salmonella #8154 was grown overnight in 2 liters of m-tetrathionate broth without aeration to a total density of 10 cfu. The broth was labeled by mixing in 1.0 microcurie of tritiated water just prior to injection into the influent manifold of marsh cell 8. Hourly samples were collected at the effluent weir and counted by LSC in Bray's solution to determine the tritium peak. Samples were counted for 50 minutes each. *Salmonella* cfu were assayed by ACT on eight samples. The colony forming unit/dose per minute (cfu/dpm) ratio was calculated based on an empirical counting efficiency for ^3H of 12%.

SALMONELLA REMOVAL FROM WASTEWATER

Table 2 shows the results of a control experiment to test the effectiveness at recovering *Salmonella* from a spiked marsh water sample. ACT effectively reduced the coliform concentration 10^4 with no loss of *Salmonella* cfu.

Survival of a *Salmonella* lab strain in one of Arcata's pilot project marsh cells was estimated with ACT (Table 3). No resident *Salmonella* were isolated from daily effluent samples of cell 8 for one week. The influent end of cell 8 was injected with a mixture of *Salmonella* and tritium-labeled water in a ratio of 1.8 cfu *Salmonella* per dpm $^3H\text{-}H_2O$. An automatic sequential sampler determined tritium dpm by liquid scintillation counting, and samples on either side of a $^3H\text{-}H_2O$ peak were examined for *Salmonella* cfu by ACT in hourly effluent samples. Reduction from dilution was computed as a decrease of ^3H dpm from the injected sample to the effluent. Further reduction in *Salmonella*

Table 4. Virus Removal Efficiencies for Various Wastewater Treatment Processes

Probably % Removal	Type of Treatment	Reported % Removal
Primary settling	0–65	50
Activated sludge	0–99.4	90
Trickling filter	0–80	50
Stabilization ponds and aerated lagoons	0–96	90

Note: Factors known to inactivate viruses include enzymatic attack; denaturation of the protein coat; loss of structural integrity; attack by oxidants; and adsorption to surfaces.[24]

cfu was detected as decrease in cfu/dpm ratio and presumed due to adsorption to fixed surfaces or die-off. The ratio of *Salmonella* cfu to ^3H dpm decreased from 1.8 injected to 0.1 at the effluent end. With marsh water mixing and dilution accounted for, a further reduction in *Salmonella* cfu of 93–96% occurred with cell residence times of 23–52 hours.

VIRUS REMOVAL FROM WASTEWATER

In a literature review on virus removals from wastewater treatment and disinfection processes, Berg[23] concluded that all conventional secondary treatment technologies (trickling filter, activated sludge, oxidation ponds) yielded erratic virus removals. Since storage time may be most effective, increasing wastewater holding time is a preferred method of virus reduction. Table 4 lists virus removal capabilities of treatment technologies as described by Gerba.[24] Important environmental stresses on viruses in a wetlands wastewater treatment process include temperature, pH, suspended solids, sunlight, redox potential, and microbial degradation.

A much different situation exists for viruses than for bacteria because the MID for many pathogenic viruses is very low—1–10 plaque forming units (pfu). Therefore, managing the risk in municipal wastewater is problematic, and there is no suitable indicator for human viruses' behavior in treatment processes. Often, the vaccine poliovirus strain is used, but its enumeration involves tissue culture methods that are expensive and technically difficult and that require three to six days for results.

BACTERIOPHAGES AS INDICATORS

Bacterial viruses (bacteriophages) were suggested as virus behavior indicators in wastewater treatment processes.[25] MS-2 bacteriophage has promise because (1) its physical size and structure closely resemble enteroviruses; (2) it is more resistant to inactivation by disinfection than enteroviruses;[26] and (3) its assay is relatively simple, fast (8–16 hr), and inexpensive.[27]

But there is no consensus on bacteriophage indicators to monitor human enteric viruses. Coliphages represent an alternative indicator, because enteroviruses were not detected unless coliphages were present in natural river water.[28] Since coliphages MS-2 and f2 are significantly more resistant to chlorine than poliovirus type 1, the former are conservative indicators of disinfection. Cramer et al.[29] used the same rationale in promoting f2 coliphage as a model of poliovirus behavior in disinfecting processes. Enteroviruses were better correlated with coliphages than with TC, FC, fecal streptococci, and standard plate count in a drinking water treatment system.

Conversely, enteric viruses were found in proportionally higher numbers than coliphages in secondary effluent from an activated sludge plant.[24] Coliphages producing plaques greater than 3 mm (the smaller morphotype including the single-stranded RNA phages) correlated with enteric viruses.[30] While Goyal and Gerba[31] argued that f2 was a poor model of adsorption to soils, the coliphage MS-2 performed better than f2 as an enterovirus behavior model in soils. None of the coliphages T2, T7, and f2 properly modeled enteric poliovirus type 1 associated with suspended solids in laboratory and field conditions.[32] Further, f2 was noninfective in the adsorbed state. These findings were confirmed,[33] and a distinct difference between poliovirus and f2 adsorption to solids was shown in an activated sludge treatment plant.[34]

Current literature does not justify use of coliphages as indicators or models of enteric viruses. However, an EPA publication[35] stated the coliphages f2 and MS-2 and members of the T-series may serve as enteric virus behavior models under restricted circumstances and may be used to assess the rate or extent of viral removal.

SITE SELECTION AND METHODS—SANTEE, CALIFORNIA

The artificial wetland site at Santee, California had beds consisting of plastic-lined excavations 18.5 m × 3.5 m × 0.76 m deep, with emergent vegetation growing in gravel.[5] Primary municipal wastewater from the Santee Water Reclamation Facility was the inflow to each vegetated bed and to an unvegetated (control) bed.

TC bacteria were enumerated by multiple-tube fermentation (presumptive and confirmed tests).[36] The host strain used for bacteriophage (MS-2) assay was *Salmonella typhimurium* strain WG 49 (phage type 3 Nal^r [F′42*lac*::Tn5]). Bacteriophages were enumerated by the double agar layer method of Melnick et al.[1] as modified by Adams[37] and Havelaar and Hogeboom.[38] Large batches of high titers of MS-2 ($5–10 \times 10^{11}$ pfu/mL) were prepared for enrichment of wetlands inflow.[39,40]

The chlorination system included a submersible pump to transfer wetland effluent to the chlorinator, a liquid metering pump that pumped the sodium hypochlorite (NaOCl) solution into the effluent stream, and a chlorine contact chamber (20.3-cm diam. PVC). Chlorine residual was measured with the DPD colorimetric method.[36] Aliquots for testing were neutralized with sodium thiosulfate after sampling to stop the chlorine reaction.

Hydraulic application rate was 5 cm/day, a hydraulic residence time of 5.5 days was regulated by valves on inflow and outflow, and beds were flooded during the study. Samples for analyses were collected from a standpipe reaching the bottom at bed effluent ends.

Mean MS-2 bacteriophage removal efficiency was nearly 99% in a bulrush wetlands, with mean (Figure 2) inflow concentration of 3.13×10^3 pfu/mL reduced to 33 pfu/mL. However, because MID for viral diseases is very low and raw wastewater may have 5×10^3 to 1×10^4 pfu/L of human enteric viruses,[41] even the 2-log (99%) removal in the wetlands beds will leave 100 pfu/L of treated wastewater. Depending on use, this level may pose an unacceptable risk to human health.

SITE SELECTION AND METHODS—ARCATA, CALIFORNIA

High concentration of MS-2 bacteriophage was continuously seeded into an experimental marsh cell at the Arcata Marsh Pilot Project (AMPP), and spatial and temporal removal followed.[15] The AMPP cells were clay-lined excavations 61 m × 6.1 m and 0.66 m deep, dominated by *T. latifolia* and *S. lacustris,* but also including *Hydrocotyle umbellata, Oenanthe saementosa, Rorripa nasturtium, Lemna* sp., and *Callitriche palustris.* Flow rate was 20 L/min with a hydraulic loading rate of 0.077 m^3/m^2/day, and nominal retention time (virus particle retention time) was experimentally determined at 3.3 days. The experimental marsh cell used for this study contained eight equally spaced baffles (sampling points), forming similarly sized subcells.

Multiple regression analysis demonstrated that suspended solids, temperature, and dissolved oxygen were predictor variables, whereas the influence of other parameters on virus removal could not be verified. Survival patterns of MS-2 virus during the six months was not linear (Figure 3). Virus removal had similar patterns in June, July, August, and September, but April and May had different removal kinetics. Overall virus removal efficiency ranged from 79% in April to 96% in August and September, with a mean removal efficiency of 91.5%.

By calculating log reductions and comparing each subcell, it was shown the subcell nearest the influent had the highest removal during all study months (Figure 4). The increase in viruses in the nonvegetated cell (subcell 5) during five months may have been due to adsorption/desorption, viral reproduction, virion conformational states, and viral aggregation. Adsorption/desorption of viruses from settled particles is most probable. Viruses removed from the water column may not be inactivated in an unvegetated cell but may be reintroduced in an infective state. This observation is important to public health.

In vitro static experiments showed that MS-2 bacteriophage was removed from water samples nearest the influent at a greater rate than samples nearest

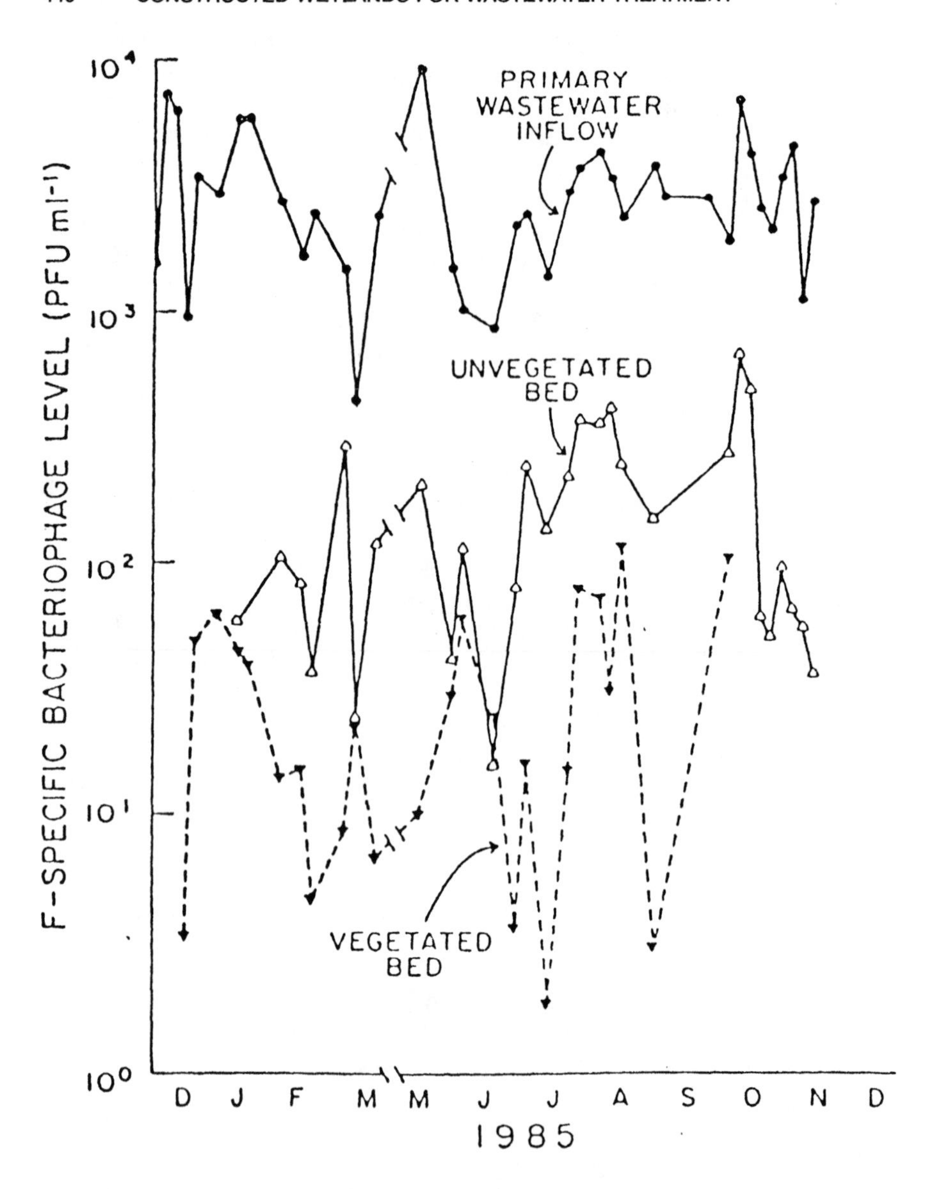

Figure 2. Concentration of indigenous F-specific bacteriophages (FRNA and FDNA phages) in the applied primary municipal wastewater and in the effluent of a vegetated (bulrush) bed and an unvegetated bed. Hydraulic application rate was 5 cm/day.

the effluent, similar to in situ results. Viruses are typically found in association with suspended solids.[32] Since there was good correlation ($r = -0.963$; $p < 0.01$) between suspended solids (algal cells) and removal rate of MS-2, suspended solids may be important in virus removal.

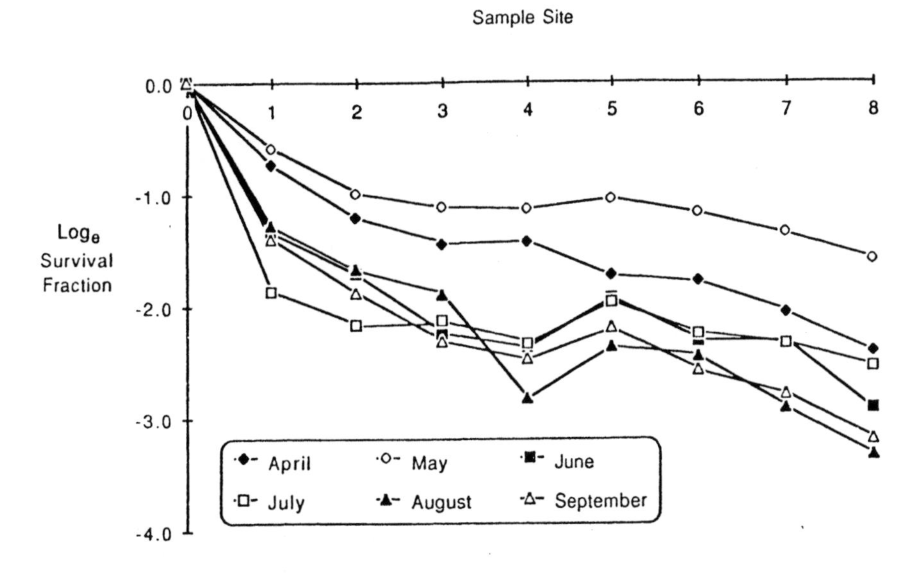

Figure 3. Survival of bacteriophage MS-2 in the Arcata Marsh Pilot Project experimental marsh (April–September 1986).

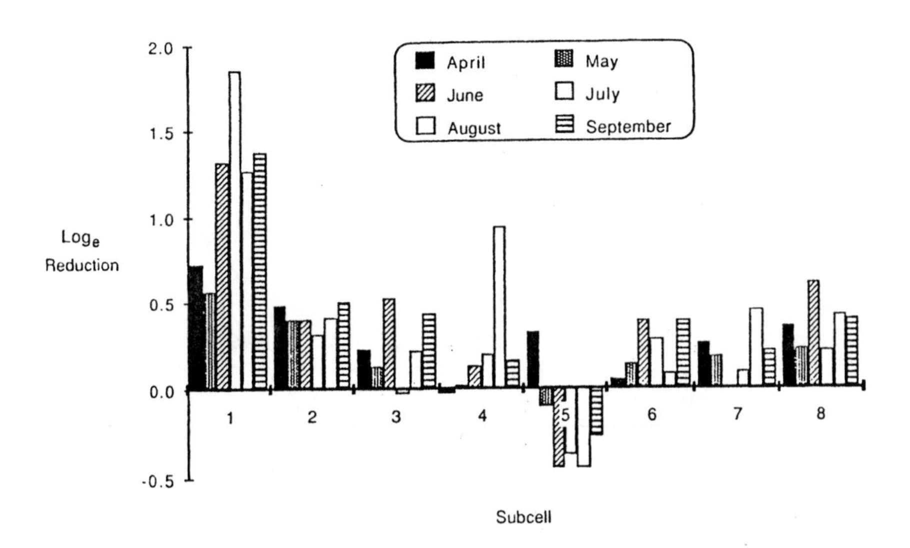

Figure 4. Reduction of bacteriophage MS-2 in the eight subcells of the experimental marsh (April–September 1986).

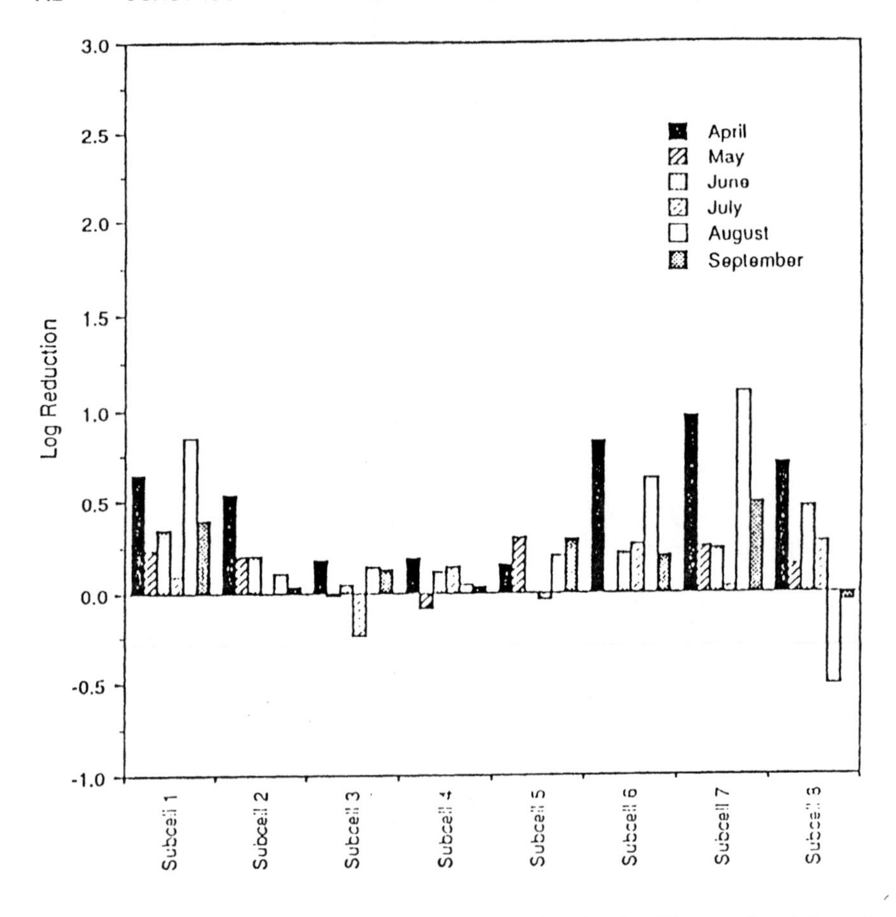

Figure 5. Average reduction of fecal coliform in the eight subcells of the experimental marsh (April–September 1986).

Reduction of FC at each subcell (Figure 5) was markedly different from reduction of the seeded MS-2 bacteriophage (Figure 3). Removal rate of FC was skewed toward the influent and effluent ends, but the marsh cell effluent end had the highest removal rate. Factors affecting virus removal may be different from FC removal factors and FC may not accurately model virus behavior in a wetland system. Bacterial indicators for predicting virus removal in wetlands is highly suspect.

Chlorination may further reduce virus levels in the wetland effluents. A total reduction of 5 logs (99.999%) was obtained when MS-2 bacteriophage was enriched in primary municipal wastewaters (10^5 to 10^6 pfu/mL) flowing into constructed wetlands followed by chlorination of the effluent.[42] If 10^4 pfu/L is the maximum number of human enteroviruses in wastewaters, then wetlands treatment plus chlorination renovate effluent to no more than 10^{-1}

pfu/L. Since human enteroviruses (poliovirus) are much more sensitive to disinfection than MS-2 (1–2 orders of magnitude),[26] postdisinfection effluents may contain levels of 10^{-2} to 10^{-3} pfu/L. At these levels, human health risks from viruses would be extremely low in reuse applications, except drinking.

SUMMARY

Both types of constructed wetlands at Arcata, California and Santee, California were shown capable of removing bacterial and viral indicators of pollution at efficiencies of 90–99%. Wastewater treatment by constructed wetlands could typically reduce total coliform levels to at or below the 10^3 total coliform per 100 mL standard when secondary wastewaters are being treated. However, for the treatment of raw or primary wastewaters, further disinfection is necessary after wetland treatment to attain this 10^3 per 100 mL total coliform standard.

Constructed wetlands can make important contributions as wastewater treatment systems not only through their ability to reduce bacteria and virus levels but also due to their ability to remove suspended solids and ammonia, both of which interfere with efficient disinfection. It appears that at hydraulic residence times of three to six days, constructed wetlands are at least equivalent and, in most cases, more effective than conventional wastewater treatment systems for the removal of disease-causing bacteria and viruses.

REFERENCES

1. Melnick, J. L., C. P. Gerba, and C. Wallis. "Viruses in Water," *Bull. World Health Org.* 56:499–508 (1978).
2. World Health Organization. "Human Viruses in Water, Wastewater and Soil," WHO, Tech. Rep. Ser. No. 639 (1979).
3. Feachem, R. G., D. J. Bradley, H. Garelick, and D. D. Mara. "Environmental Classification of Excreta-Related Infections," in *Sanitation and Disease–Health Aspects of Excreta and Wastewater Management,* World Bank Studies in Water Supply and Sanitation 3. (New York: John Wiley & Sons, 1983), pp. 23–41.
4. "Quality (Criteria for Water)," U.S. EPA, Washington, DC (1976).
5. Gersberg, R. M., B. V. Elkins, and C. R. Goldman. "Wastewater Treatment by Artificial Wetlands," *Water Sci. Technol.* 17:443–450 (1984).
6. Gearheart, R. A., S. Wilbur, J. Williams, D. Hull, B. Finney, and S. Sundberg. "City of Arcata Marsh Pilot Project: Effluent Quality Results–System Design and Management. Final Report," Project No. C-06-2270, State Water Resources Control Board, Sacramento, CA (1983).
7. Dinges, R. "Development of Hyacinth Wastewater Treatment Systems in Texas," in *Aquaculture Systems for Wastewater Treatment: Seminar Proceedings and Engineering Assessment*, R. K. Bastian and S. C. Reed, Eds., U.S. EPA Report-430/9-80-006 (1979), pp. 193–226.
8. McNabb, C. D., Jr. "The Potential of Submerged Vascular Plants for Reclamation of Wastewater in Temperate Zone Ponds," in *Biological Control of Water Pollu-*

tion, J. Tourbier and R. W. Pierson, Jr., Eds. (Philadelphia: University of Pennsylvania Press, 1976), pp. 123–132.

9. Tchobanoglous, G., R. Stowell, R. Ludwig, J. Colt, and A. Knight. "The Use of Aquatic Plants and Animals for the Treatment of Wastewater: An Overview," in *Aquaculture Systems for Wastewater Treatment: Seminar Proceedings and Engineering Assessment,* R. K. Bastian and S. C. Reed, Eds., U.S. EPA Report-430/9-80-006 (1979), pp. 35–55.
10. Wolverton, B. C. "Engineering Design Data for Small Vascular Aquatic Plant Wastewater Treatment Systems," in *Aquaculture Systems for Wastewater Treatment: Seminar Proceedings and Engineering Assessment,* R. K. Bastian and S. C. Reed, Eds., U.S. EPA Report-430/9-80-006 (1979), pp. 179–192.
11. Palmateer, G. A., W. L. Kutas, M. J. Walsh, and J. E. Keollner. Abstracts of the 85th Annual Meeting of the Am. Soc. for Microbiology, Las Vegas, NV, March 3–7, 1985.
12. Spangler, F. W. Sloey, and C. W. Fetter. "Experimental Use of Emergent Vegetation for the Biological Treatment of Municipal Wastewater in Wisconsin," in *Biological Control of Water Pollution,* J. Tourbier and R. W. Pierson, Jr., Eds. (Philadelphia: University of Pennsylvania Press, 1976), pp. 161–171.
13. Gearheart, R. A., J. Williams, H. Holbrook, and M. Ives. "City of Arcata Marsh Pilot Project Wetland Bacteria Speciation and Harvesting Effects on Effluent Quality. Final Report," Project No. 3-154-500-0, State Water Resources Control Board, Sacramento, CA (1986).
14. Herskowitz, J., S. Black, and W. Lewandowski. "Listowel Artificial Marsh Treatment Project," in *Aquatic Plants for Water Treatment and Resource Recovery,* K. R. Reddy and W. H. Smith, Eds. (Orlando, FL: Magnolia Publishing Inc., 1987), pp. 247–254.
15. Ives, M. A. "The Fate of Viruses in an Artificial Marsh Wastewater Treatment System Utilizing a Coliphage Model," master's thesis, Humboldt State University, Arcata, CA (1987).
16. Miescier, J. J., and V. J. Cabelli. "Enterococci and Other Microbial Indicators in Municipal Wastewater Effluents," *J. Wat. Pollut. Control Fed.* 54:1599–1606 (1982).
17. Wilbur, S. W. "A New Quantitative Assay Technique for Salmonella in Polluted Waters," thesis, Humboldt State University, Arcata, CA (1983).
18. Cheng, C. M., W. C. Boyle, and J. M. Goepfert. "Rapid Quantitative Method for *Salmonella* Detection in Polluted Waters," *Appl. Microbiol.* 21(4):662–667 (1971).
19. Brock, T. D. *Biology of Microorganisms,* 2nd ed. (Englewood Cliffs, NJ: Prentice-Hall, Inc., 1974).
20. Rappaport, R., N. Konforti, and B. Navon. "A New Enrichment Medium of Certain Salmonellae," *J. Clin. Pathol.* 9:261–266 (1956).
21. Ewing, W. H., and P. R. Edwards. *Identification of Enterobacteriaceae,* 3rd ed. (Edina, MN: Burgess Publishing Company, 1972).
22. Smith, P. B., et al. "API System: A Multitube Method for Identification of Enterobacteriaceae," *Appl. Microbiol.* 24:449–452 (1972).
23. Berg, G. "Removal of Viruses from Sewage, Effluents, and Waters. 1. A Review," *Bull. World Health Org.* 49:451–460 (1973).
24. Gerba, C. P. "Virus Survival in Wastewater Treatment," in *Viruses and Wastewater Treatment,* M. Goddard and M. Butler, Eds. (New York: Pergamon Press, Inc., 1980), pp. 39–48.

25. Funderburg, S. W., and C. A. Sorber. "Coliphages as Indicators of Enteric Viruses in Activated Sludge," *Water Res.* 19:547–555 (1985).
26. Grabow, W. O. K., P. Coubrough, C. Hilner, and B. W. Bateman. "Inactivation of Hepatitis A Virus, Other Enteric Viruses, and Indicator Organisms in Water by Chlorination," *Water Sci. Technol.* 17:657–664 (1985).
27. Gersberg, R. M., S. R. Lyon, R. Brenner, and B. V. Elkins. "Fate of Viruses in Artificial Wetlands," *Appl. Environ. Microbiol.* 53:731–736 (1987b).
28. Kott, Y. N., N. Rose, S. Sperber, and N. Betzer. "Bacteriophages as Viral Pollution Indicators," *Water Res.* 8:165–171 (1974).
29. Cramer, W. W., K. K. Kawata, and C W. Kruse. "Chlorination and Iodination of Poliovirus and f2," *J. Water Pollut. Control Fed.* 48:61–76 (1976).
30. Stetler, R. E. "Coliphages as Indicators of Enteroviruses," *Appl. Environ. Microbiol.* 48:668–670 (1984).
31. Goyal, S. M., and C. P. Gerba. "Comparative Adsorption of Human Enteroviruses, Simian Rotavirus and Selected Bacteriophages to Soils," *Appl. Environ. Microbiol.* 38:241–247 (1979).
32. Moore, B. E., B. P. Sagik, and J. F. Malina, Jr. "Viral Association with Suspended Solids," *Water Res.* 9:197–203 (1975).
33. Bates, J., and M. R. Goddard. "Recovery of Seeded Viruses from Activated Sludge," in *Viruses and Wastewater Treatment,* M. Goddard and M. Butler, Eds. (New York: Pergamon Press, Inc., 1981), pp. 205–209.
34. Balluz, S. A., and M. Butler. "The Influence of Operating Conditions of Activated Sludge Treatment on Behavior of f2 Coliphage," *J. Hyg.* 82:285–291 (1979).
35. Karaganis, J. V., E. P. Larkin, J. L. Melnick, P. V. Scarpino, S. A. Schaub, C. A. Sorber, R. Sullivan, and F. M. Wellings. "Research Priorities for Monitoring Viruses in the Environment," U.S. EPA Report-600/9-83-010 (1983).
36. *Standard Methods for the Examination of Water and Wastewater,* 16th ed. (Washington, DC: American Public Health Association, 1985).
37. Adams, M. H. "Assay of Phage by Agar Layer Method," in *Bacteriophages* (New York: John Wiley & Sons, Inc., 1959), pp. 450–545.
38. Havelaar, A. H., and W. M. Hogeboom. "A Method for the Enumeration of Male-Specific Bacteriophages in Sewage," *J. Appl. Bacteriol.* 56:439–447 (1984).
39. Loeb, T., and N. D. Zinder. "A Bacteriophage Containing RNA," *Proc. Nat. Acad. Sci. U.S.* 47:282–289 (1961).
40. Nathans, D. "Natural Coding of Bacterial Protein Synthesis," in *Methods in Enzymology, Vol. XII, Part B,* L. Grossman and K. Moldave, Eds. (New York: Academic Press, Inc., 1968).
41. Clark, N. A., G. Berg, P. W. Kabler, and S. L. Change. "Human Enteric Viruses in Water: Source, Survival, and Removability," in *Proceedings of 1st Int. Conf. Water Pollut. Res., London, 1962, Vol. 2* (New York: Pergamon Press, Inc., 1964).
42. Gersberg, R. M., R. Brenner, S. R. Lyon, and B. V. Elkins. "Survival of Bacteria and Viruses in Municipal Wastewaters Applied to Artificial Wetlands," in *Aquatic Plants for Water Treatment and Resource Recovery,* K. R. Reddy and W. H. Smith, Eds. (Orlando, FL: Magnolia Publishing Inc., 1987a), pp. 237–245.

CHAPTER 36

Monitoring of Constructed Wetlands for Wastewater

Delbert B. Hicks and Q. J. Stober

INTRODUCTION

Use of constructed wetlands for the disposal and treatment of wastewater is emerging as an alternative to conventional approaches for small communities and industries. Operation and maintenance of any process control system are dependent on a monitoring plan that provides information for judging the attainment of treatment objectives, performance, efficiency and the long-term viability of the system. Operation and maintenance of conventional wastewater treatment systems are linked to a systematic and proven procedure of diagnoses and adjustments to system processes. Virtually all steps in the treatment process are subject to control. Responses in biological and chemical processes to adjustments are relatively rapid and easily monitored. This is generally not the case with wetland systems.

Wetlands represent a highly diversified yet ecologically integrated system of plants and animals. No visible boundaries of compartmentalization exist in terms of the flow of energy and matter. Our knowledge of the complexity of these ecological systems is not complete; hence, early diagnoses of failing wetland functions can be difficult to recognize. Adjustments to the physical, chemical, and biological process of the wetlands are not easily accomplished nor are the effects rapidly apparent. For these reasons, the long-term monitoring of wetland treatment systems is essential to develop an information base needed to operate and maintain a healthy biological system. In short, monitoring data are essential to measure the treatment levels and to indicate the functional status and biological integrity of the wetland system.

As seen in the topics of this conference, the use of engineered wetlands for wastewater treatment has a wide range of applications including the processing of domestic waste from single housing units, small municipalities, industrial sources, urban runoff, or a combination of all of the above. In some instances, the benefits of the constructed wetlands can go beyond objectives of treatment to include fish and wildlife enhancement and recreational returns. As the complexity of wastes treated by a wetland increases so do the monitoring requirements, because more variables are introduced which may result in system failure or lead to undesirable side effects. Cost and effort of monitoring increase with chemical complexity of the influent to be treated and ecological diversity of the wetlands to be maintained. This chapter presents monitoring strategies in a hierarchical order according to increasing complexity and implementation effort.

MONITORING PLAN

Scope of a monitoring effort is linked directly to goals of the treatment project. For example, combining treatment with fish and wildlife enhancement for public benefit would require a monitoring strategy that not only provides data for judging pollution control but also public health and ecological assessments.

Basic elements of a monitoring plan include clearly and precisely stated goals of the treatment project and specific objectives of monitoring. Other elements include statements of organizational and technical responsibilities, tasks and methods, quality assurance procedures, schedules, reporting products, resource requirements, and budget. Written clarification of these elements is essential to assure that a continuum among the data sets exists through the life of the project, which could span many decades. With time, changes in project personnel, regulations, policies, results of data, and funding will foster rethinking of monitoring objectives and assessment strategies that could alter system operation. A well-conceived and clearly defined monitoring plan serves as a point of reference and source of perspective for maintaining a meaningful information base through the life of the project.

Compliance Monitoring

Monitoring for compliance with a discharge permit probably represents the minimum sampling requirements and complexity. The exception may rest with a small wetland system that features domestic wastewater treatment with no surface discharge and subsurface flow.

Where the wetland treatment system discharges to public waters, performance objectives in terms of treatment goals are established through compliance requirements of a wastewater discharge permit, i.e., the National Pollutant Discharge Elimination System. Compliance monitoring involves an array of parameters that may be both biological and chemical in nature. Parameter selection and discharge limits are a function of the level of water quality protection assigned to the receiving water body by the appropriate regulatory agency. The purpose of attaining compliance is to assure that state water quality standards of the receiving stream are maintained.

Typical parameters of interest include BOD_5, total suspended solids, pH, and fecal coliform bacteria. The suite of parameters would no doubt be expanded as the chemical complexity of the wastewater influent increases and could include toxicity assessment, nutrients, and priority pollutants. Methods for analyses are found in numerous publications.[1-4] Assessing the bacterial quality of effluent poses special considerations because of potential contributions from animals using the wetland system, such as ducks and other warm-blooded animals. Should the compliance parameters be reported in terms of mass loading rates, the flow of the discharge also will require monitoring.

Assessment of flow can be accomplished with weekly readings of a staff gauge on a simple "box" or V-notched weir.[5] The flow rate and the effluent concentration of the constituent of interest provide the basis for calculating mass loading to the receiving water body.

Wetland System Performance and Treatment Efficiency

Long-term management of a wetland treatment system requires a thorough understanding of its efficiency to remove waste constituents and hydrographic factors that affect these processes. Seasonal and possibly daily variations in hydraulic loading of the wetlands affect inundation frequency and duration, detention time, and outflow rates. Hydrographic regime of the wetland affects distribution patterns of plant and animal communities of the wetlands, their vigor, and effectiveness of the wastewater treatment system.

Determining the hydraulic loading rate is the principal means of monitoring application rate of wastewater to the wetlands. This is best accomplished by establishing a continuous flow record or a minimum record based on a minimum of weekly measurements of the wastewater stream or streams discharging to the wetland. This record, coupled with the known surface area of the wetlands, provides data to calculate daily hydraulic loading rates. For example, a total daily wastewater inflow of 3785 m^3 to a wetland of 101 ha has an application rate of 2.54 cm per week. This rate assumes the wastewater inflow is uniformly distributed across the wetland and perpendicular to the flow axis.

To monitor the efficiency of constructed wetlands to treat and remove selected chemical and biochemical constituents of the waste stream, inflow and outflow measurements of the wetland treatment system are required. Flow information, along with inflow and outflow constituent concentrations, is the basis for calculating mass loading rates and reduction efficiency across the treatment system.

Inundation frequency and duration is a matter of relating the accumulated frequency and time that the water surface elevation exceeds the land surface of the wetland. To determine this relationship, a topographic survey of the constructed wetland is required. The topographic survey best follows the final grading of the wetland site. Surface water elevations can be determined from a staff gauge or recording water level instrument in the area of the wetland with the greatest depth. The topographic survey and water surface elevations can then be used to determine the volume of the water in the wetland, the inundation frequency, and the duration of flooding. Coupling the volume determinations with the hydraulic loading rates provides estimates of detention time of wastewater in the constructed wetlands.

Circulation of wastewater through the wetland is assumed to be uniform. However, should the wetland have patchy emergence of aquatic macrophyte communities or, possibly, irregular bottom contours, uniformity of flow through the wetlands should be examined with a dye tracer study to assess possible short-circuiting flow patterns.

Monitoring Wetland Viability and Health

To optimize and sustain the long-term treatment capacity of a constructed wetland requires maintenance of a healthy and functional community of aquatic plants and animals. Monitoring these communities is the only means of judging their condition and their responses to changing hydrographic conditions, temporal effects, diseases, and varying chemical characteristics of the wastewater influent. Benthic aquatic macroinvertebrates, macrophytes, and fish are common focal points in biological monitoring.

Ideally, a biological monitoring strategy is formulated around a compare-and-contrast approach to assessing the biological condition. This approach must assume that applicable baseline or background data bases exist as a point of reference for comparison with data derived from the treatment wetland. In some cases, values from wetland literature serve this need. Another approach is to locate and monitor a nearby existing wetland of similar hydrology and wetland characteristics. An alternative is parallel construction and operation of a second wetland system to serve as a reference site until a baseline condition can be established. Following this goal, the second site can be managed in a rotational manner with the other system.

Wetland macrophytes (vegetation), a component of primary production, assimilate nutrients and produce organic matter via the photosynthetic process. The other functions include storage of chemicals in above- and below-ground tissues, transport of oxygen into the water sheet, and serving as substrate for microbial communities that treat the water. Therefore, it is necessary to monitor for changes to the vegetational community. The wetland can be viewed as an aquafarm and, as such, requires a normal amount of agricultural attention.

Water must reach all parts of the wetland surface, or there will be immediate and long-term consequences. In the short term, there will be a loss of effectiveness in proportion to the area not exposed to wastewater. In the long term, lack of nutrients and water will cause the species of vegetation to change. The depth and duration of inundation can also affect the plants; with continued deep water, some species will eventually drown, even though they are adapted to standing water during a large portion of the year.

Accumulation of dead plant material is beneficial in two respects: (1) some of this biomass will be mineralized as sediment and retain chemicals in the ecosystem, and (2) this litter becomes substrate for microbes that clean the water. Of course, most dead plant material decomposes and returns the stored chemicals to the water.

Vegetation in a man-made wetland is subject to gradual year-to-year change, just as it is in natural wetlands. There may be a tendency for some species to die out and be replaced by others. Very temporary changes, such as the appearance of algae or duckweed, can occur in response to random or seasonal climatic changes.

Knowledge of these wetland functions must be used to maintain the desired water treatment capability for the wetland. Two things easily accomplished in vegetation monitoring are determination of the current species composition and standing crop size. Species composition is determined from inspection of quadrats[6] within the wetland at selected locations, perhaps complemented by aerial photography, both color and color infrared. Because of the slow rate of vegetative changes, which may not be obvious during the tenure of a single operator, good record keeping becomes essential.

Measurements of standing crop biomass will indicate if the vegetative storage is currently increasing or decreasing. End-of-season harvest and weighing of material from clip plots provide the necessary data. Storage can be estimated from published information on chemical composition of the plants in question. Sampling strategies for the assessment of macrophytes are further detailed elsewhere.[7]

Benthic macroinvertebrates occupy nearly all levels in the trophic structure of a wetland community. They may be omnivores, carnivores, or herbivores, and in a well-balanced system all types are likely to be found (detrital and deposit feeders, scavengers, grazers, and predators). The macroinvertebrate community is sensitive to environmental stress, i.e., pollution, and can serve as a useful means of detecting subtle and gross changes in the aquatic environment. This is possible because benthic macroinvertebrates generally feature a relatively long life span, thus integrating effects of conditions during the recent past. Taxa diversity and abundance respond to environmental stresses. With increasing stress upon the community, taxa are eliminated from the community according to their tolerance to the perturbation. The remaining taxa can grow in number due to reduced competition for space and food. Methods and sampling considerations for monitoring the benthic macroinvertebrate community have been thoroughly detailed.[8] These methods include both qualitative and quantitative assessments and can involve a variety of sampling techniques ranging from simple grab samples to artificial substrates for organism colonization.

Trace metal and organic contaminants enter wetlands with all influents and partition between the water, sediment, and biota. Many contaminants bind to organic particulates either in the effluent or in the wetland and are removed from the water. Contaminant monitoring in sediments is effective for identifying the treatment occurring in any wetland; however, the sediment may represent only a temporary sink. Contaminants in sediments can be biologically concentrated and magnified through each level of the aquatic and terrestrial food chain, resulting in special monitoring requirements.

Bioaccumulation of trace amounts of chemical contaminants in aquatic organisms also serves as a monitor of constructed wetlands. Contaminants of concern should be selected on the basis of the following characteristics: high persistence in the aquatic environment, high bioaccumulation potential, high toxicity to humans and/or wildlife, known or suspected sources of the contaminant(s) entering the system, and high concentrations in previous samples of fish and invertebrates from other similar systems.

General information on persistence, bioaccumulation potential, and toxicity may be obtained elsewhere.[9,10] EPA priority-pollutant organic chemicals and selected pesticides have been ranked in descending order of bioaccumulation potential according to their octanol-water partition coefficients. Organic compounds with a log octanol-water partition coefficient greater than or equal to 2.3 are usually recommended for inclusion in monitoring programs. EPA priority-pollutant metals have been ranked in descending order of bioaccumulation potential according to bioconcentration factors.[11,12] Screening of potential contaminants of concern should be done on a site-specific basis. Methodology for the chemical analysis of 45 xenobiotic compounds in fish tissue has recently been developed[13] to allow monitoring of multiple contaminants simultaneously.

The monitoring design will vary with the type of constructed wetland to be evaluated and multiple uses that may be supported by a wetland. If the sole purpose of the wetland is wastewater treatment, the focus of sampling is to determine the health of the associated biota to ensure the proper long-term functioning of the wetland. The bioconcentration of trace contaminants in fish or invertebrate tissue may be used to indicate general health of the system, including the contaminant processing through the water and sediments to biological tissues. This monitoring provides an important feedback loop to assess the need for cleanup of the influent via source control and should be conducted over the life of the project. Whole body contaminant burdens of fish or invertebrates may have significance to the surrounding terrestrial environment, especially if top reptilian, avian, or mammalian predators become dependent on the wetland as a food source. However, if constructed wetlands are designed to support a diverse community of fish and invertebrates, the potential for human use of the aquatic resource may exist. We assume that public access can be controlled, eliminating the need to assess adverse human health effects from the wetland if significant organism contamination occurs. Depending on the wetland efficiency as a sink for trace contaminants or passage of these compounds or contaminated organisms out of the system into natural receiving waters, the potential risk to human health may require assessment.

An evaluation of the risks of consuming chemically contaminated fish generally focuses on estimation of the chance of incrementally increased risks of cancer and/or various noncarcinogenic and developmental adverse health effects in groups of people consuming various amounts of the contaminated fish over a 70-year lifetime. A draft methodology for conducting such assessments has been documented in a guidance manual[14] that provides a basis for evaluation along with guidelines for health risk assessment.[15] This methodology includes the conversion of measured values for contaminant residues in fish into daily doses to consumers based on the long-term average daily consumption.[16,17] The potential health effect resulting from estimated dose is calculated based on the USEPA carcinogenic potency factor established by toxicological testing results. The cancer risk is expressed in terms of a plausible 95% upper-bound estimate of increased lifetime incidence of cancer per unit of exposed population. The population usually at greatest risk is the local angler who may consume contaminated fish from a single source over a long period.

Sampling wetlands for contaminant accumulations in fish and invertebrate tissues should be designed to determine whether gradients exist downstream. Depending on the length of each system and water retention capacity, the monitoring design should focus on inflow and outflow areas as well as on areas where particulates tend to accumulate in the system. Many trace organics and metal contaminants are strongly adsorbed to particulates, and maximum tissue concentrations generally occur in organisms most closely associated with contaminated sediments. Benthic organisms, which inhabit and feed in contaminated sediments, and bottom-feeding fish species bioaccumulate the highest contaminant concentrations due to their close association.

If trace contaminants are carried through the wetland during flushing events or due to eventual system saturation, tissue contaminant monitoring in natural receiving waters may be necessary, including sampling of bottom fish and sport and commercial species above and below the outfall. Sampling proximity to the return flow and species selected should reflect those species spending most or all of their lives in the immediate vicinity. Both upstream and downstream samples are desirable, and more than one sampling station is usually required downstream to determine the extent of bioconcentration of specific chemical contaminants.

CONCLUSION

An appropriately designed and implemented monitoring plan is essential to the successful management, operation, and maintenance of a constructed wetlands for treatment of wastewater. The monitoring plan comprises numerous components, including clearly stated objectives, technical and management responsibilities, quality assurance procedures, resources, and schedules. Scope of the monitoring activities is a function of the treatment goals, project benefits, and diversity of plant and animal communities involved in the wetland system. Results of monitoring serve to determine compliance of the wetland discharge with permit limits established by pollution control agencies. Monitoring of the plant and animal communities provides surveillance data necessary to determine health and viability of the wetlands and identify early signs of stress to the aquatic communities of plants and animals. Early detection of a failing wetland system is essential to the operation and maintenance of the treatment system.

Engineered wetlands as treatment systems are appealing because of low costs of construction and simplicity of operation and maintenance needs. However, appropriate monitoring, though it adds costs, is a necessity and must be viewed as a priority over the life of the project.

ACKNOWLEDGMENT

We thank Robert H. Kadlec for his critical review of this chapter and constructive suggestions to its content.

REFERENCES

1. "Methods for Chemical Analysis of Water and Wastes," Environmental Monitoring and Support Laboratory, U.S. EPA Report-600/4–79–020 (Revised 1983).
2. *Standard Methods for the Examination of Water and Wastewater,* 16th ed. (Washington, DC: American Public Health Association, 1985).
3. Peltier, W., and C. I. Weber, Eds. "Methods for Measuring the Acute Toxicity of Effluents to Freshwater and Marine Organisms," 3rd ed. Environmental Monitoring and Support Laboratory, U.S. EPA Report-600/4–85–013 (1985).
4. Weber, C. I. "Short-Term Methods for Estimating the Chronic Toxicity of Effluents and Receiving Waters to Freshwater Organisms," U.S. EPA Report-600/4–89–001 (1989).
5. "Water Measurement Manual," Water Resource Technical Publication No. 2403–00086, U.S. Government Printing Office (1974), p. 327.
6. Cox, G. W. *Laboratory Manual of General Ecology* (Dubuque, IA: W. C. Brown Publishers, Inc., 1970), p. 165.
7. Dennis, W. M., and B. G. Isom, Eds. "Ecological Assessments of Macrophyton Collection, Use, and Meaning of Data," ASTM Publication 8453, Philadelphia (1983).
8. "Biological Field and Laboratory Methods for Measuring the Quality of Surface Waters and Effluents," Office of Research and Development, U.S. EPA Report-670/4–73–001 (1973).
9. Lyman, W. J., W. F. Reehl, and D. H. Rosenblatt. *Handbook of Chemical Property Estimation Methods* (New York: McGraw-Hill Book Company, 1982).
10. Callahan, M. A., M. W. Slimak, N. W. Galde, I. P. May, C. F. Fowler, J. R. Freed, P. Jennings, R. L. Durfee, F. C. Whitemore, B. Amestri, W. R. Mabey, B. R. Holt, and C. Gould. "Water-Related Environmental Fate of 129 Priority Pollutants, Vol. I and II," U.S. EPA Report-440/4–79–029a/b. (NTIS No. PB80–204373 [Vol. I] and PB80–204381 [Vol. II])
11. U.S. EPA. "Water Quality Criteria Documents; Availability," 45 *Fed. Reg.* 231, Part V, pp. 79318–79379.
12. Tetra Tech, Inc. "Bioaccumulation Monitoring Guidance: 1. Estimating the Poten-

tial for Bioaccumulation of Priority Pollutants and 301(h) Pesticides Discharged into Marine and Estuarine Waters, Final Report," prepared for Office of Marine and Estuarine Protection, U.S. EPA, Washington, DC by Tetra Tech, Inc., Bellevue, WA.
13. "National Dioxin Study, Phase II, Draft Analytical Procedures and Quality Assurance Plan for the Determination of Xenobiotic Chemical Contaminants in Fish," EPA, ERL-Duluth, MN (1988).
14. PTI, Environmental Services Inc. "Guidance Manual for Assessing Human Health Risk from Chemically Contaminated Fish and Shellfish," draft report to Battelle New England Marine Research Laboratory for USEPA Office of Marine and Estuarine Protection, Washington, DC (1987).
15. "Guidelines for Health Risk Assessment," 51 *Fed. Reg.* 79318–79379 (1986).
16. Kreiger, R. A. "Derivation of a Virtually Safe Dose (VSD) Estimate for Sport Fish Containing 2,3,7,8-Tetrachlorodibenzo-p-dioxin," Minnesota Department of Health, Health Risk Assessment Section (1985).
17. "Assessment of Dioxin Contamination of Water, Sediment and Fish in the Pigeon River System (A Synoptic Study)," U.S. EPA Region IV Environmental Services Division for U.S. EPA Region IV Water Management Division, Report #001 (1988).

SECTION IV

Recent Results from the Field and Laboratory

CHAPTER 37

Dynamics of Inorganic and Organic Materials in Wetlands Ecosystems

37a Decomposition in Wastewater Wetlands

Robert H. Kadlec

INTRODUCTION

The role of biomass accretion and decomposition in water quality improvement in constructed municipal wastewater wetlands is very important but has largely been ignored in performance discussions. Most information is presented in terms of inputs, outputs, and percent "removals," disregarding the compartment to which "removal" has occurred. Two stages in the development of a constructed wetland may be identified: (1) the startup phase, in which the vegetation and litter increase with each passing year; and (2) a stationary phase, in which vegetation and litter vary seasonally but not annually. In both stages biomass has significant effects on water quality. Harvest of wetland plants will not be considered in this discussion because it is difficult and seldom attempted.

Vegetation undergoes a sequence of steps with passage of time: accumulation of live biomass during the growing season, dieback, litterfall, litter accumulation, litter leaching, decomposition, and soil accretion. Growth and dieback occur seasonally, but the remaining processes take several years. Similarly, microflora and microfauna grow, die, decay, and accrete, but the time scale is much shorter. Further, water can more easily transport microdetritus (or sediment) than dead leaves and stems. In general, some small portion of both litter and sediment becomes a permanent component of the wetland soil substrate by becoming mineralized and by reaching an anaerobic horizon in the substrate-water column.

The purpose of this chapter is to quantify the rates of key biomass processes and the amounts of nutrients and biomass involved.

GROWTH

Growth dynamics are dependent upon climate, species, site characteristics, and water regime (operating strategy). It is not the primary purpose here to detail these effects, but some features merit attention. Many natural wetlands are nutrient-limited, and hence it is not possible to draw upon that literature. For example, peak standing crop of cattail (*Typha*) in a wastewater wetland can be three to five times that in an adjacent natural wetland.[1] Further, the nutrient content can be higher in vegetation in the wastewater environment, due to luxury uptake of selected constituents. The combined effect is use of greater quantities of nutrients by plants in the wastewater setting. A wetland receiving secondary wastewater typically has 3000 g/m^2 (dry wt seasonal max.) of aboveground biomass and a like quantity of roots. This biomass typically contains 2% nitrogen and 0.4% phosphorus.

These standing crops are not reached during year 1 of operation. If the system is planted on a spacing of 1 m during the first spring, it will take one to three years to reach full vegetation density and several years longer to reach a state of stationary nutrient dynamics. The same is true for startup of a discharge to an existing wetland, as shown in Figure 1 for the Houghton Lake system. Climatological factors will influence this time scale and cause variability from year to year, preventing a true stationary state.

There is little information on the growth rates or standing crops of the microbiomass, which includes *Lemna* (and smaller) and invertebrates (and smaller). Data from the Houghton Lake system indicate 2–400 g/m^2 seasonal generation. This amount is not found at any one time, however, due to the short life span of microorganisms.

LITTERFALL

Death of wetland plant leaves occurs over a span of several months, with the greatest amount occurring in late summer and early fall in temperate climates, leading to a similar pattern in leaf litterfall (Figure 2). In the case of the woody-stemmed species, a live standing portion remains during the nongrowing season. For reeds (*Phragmites*), sedges (*Carex*), and cattails, death of an aboveground plant part results in an increase of standing dead biomass because litterfall does not instantaneously follow death. The half-life (time for 50% to fall to the ground) of standing dead is about half a year for cattail, depending on rain, snow, and wind conditions. The result is a lag of several months between the annual peak in standing dead and that in litter (Figure 3). For woody shrub species, the length of time spent as standing dead can be considerably longer, with half-lives of a decade or longer for some species of willow. Intact dead leaves and stems are relatively immobile after reaching the ground.

In contrast, there is no standing dead compartment for microorganisms.

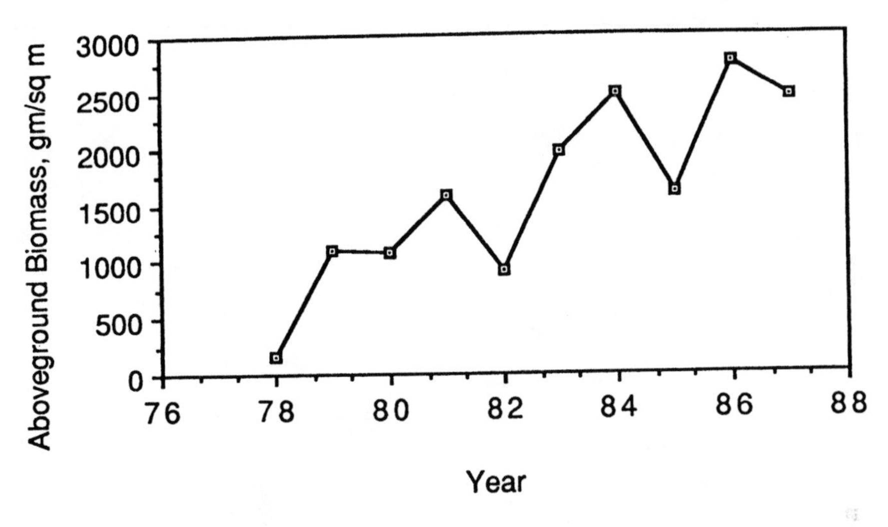

Figure 1. History of cattail biomass in the wastewater irrigation area of the Houghton Lake wetland. Data are for end-of-season standing crop, which includes a small amount of standing dead.

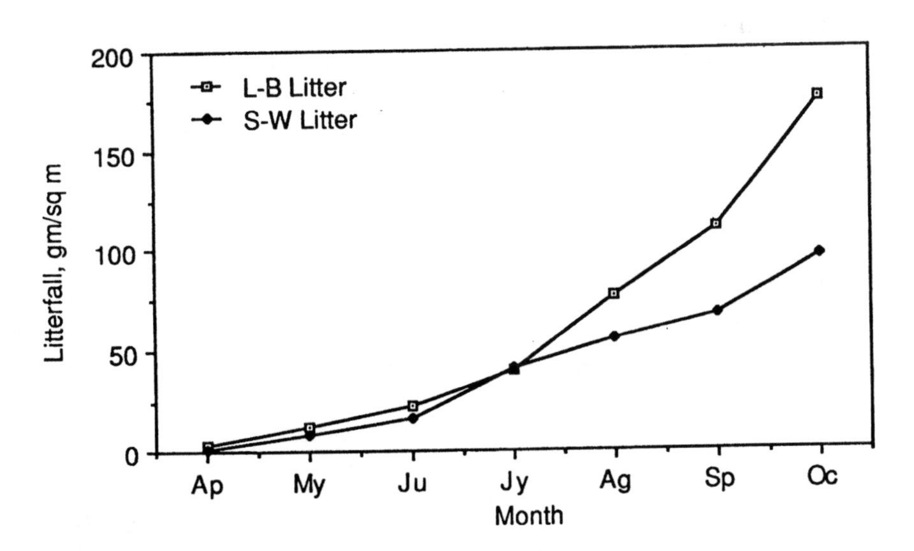

Figure 2. Litterfall versus time for the Porter Ranch natural wetland. S-W = sedge-willow community; L-B = leatherleaf-bog birch community.

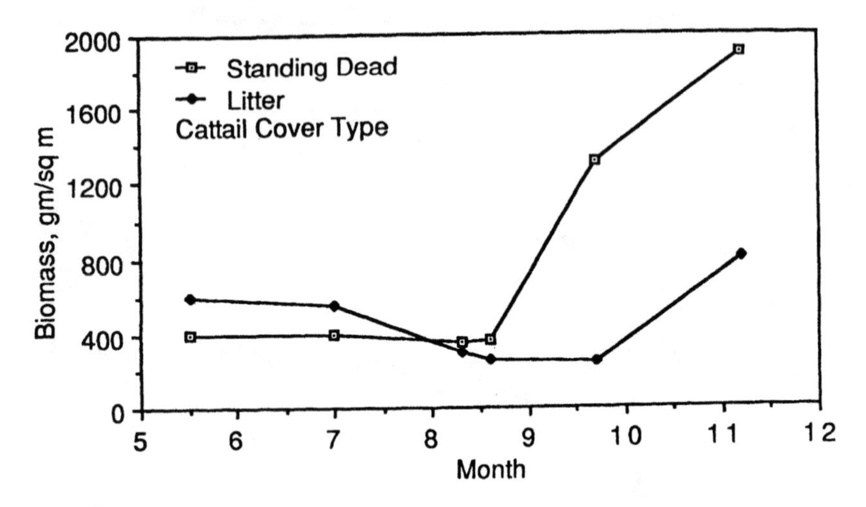

Figure 3. Contents of standing dead and litter compartments versus time for the wastewater irrigation area of the Houghton Lake wetland.

Dead bacteria, invertebrates, and algae immediately become sedimentary material. Buoyancy of such sediments varies with temperature and diurnal variables such as gas production, and sediments are transportable by flowing water. When coupled with processes of filtration, settling, and resuspension, a complicated sediment transport process results. The rate of lateral motion is much slower than the water velocity (for example, about 50 m/yr at the Houghton Lake site). These sediments are typically higher (by a factor of 2) in nutrient content than the macroscopic plant parts. The rate of deposition of sediments is shown in Figure 4. Total net deposition is less than that shown, due to resuspension, by a factor of 4. The result is that sediment deposition plays an important role in the nutrient cycle of the same order of magnitude as the plants.

LITTER DECOMPOSITION

It is relatively easy to collect aboveground plant parts, place them in mesh bags, put them in a wetland environment, and monitor weight loss as a function of time. A sizable data base is therefore available. Methods vary, including location of bags with respect to water and soil surfaces. Nonetheless, there is considerable agreement in the results from study to study.

Decomposition of leaf litter takes place in two distinct stages: a period of fairly rapid weight loss lasting about 30–60 days, followed by a period of slow, exponential decay. The first period is likely to be associated with microbial consumption of low-molecular-weight compounds (sugars and starches), while the second corresponds to utilization of celluloses. These stages are illustrated

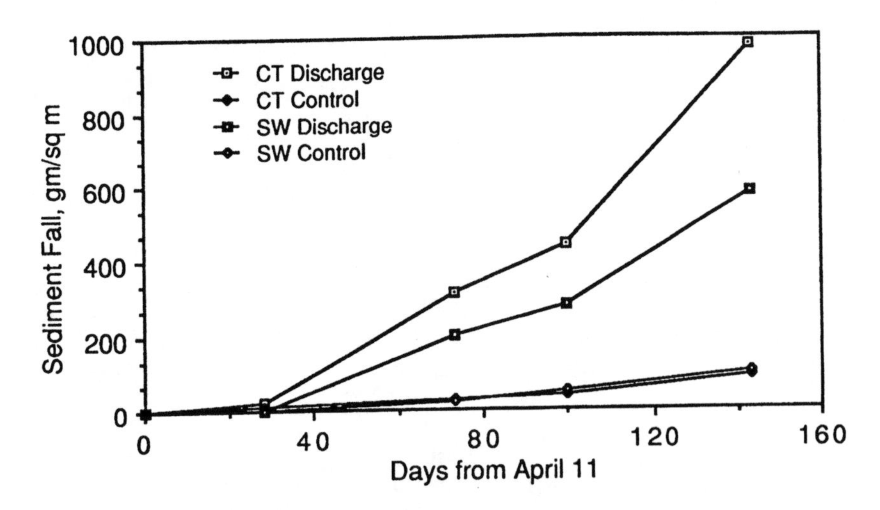

Figure 4. Sediment accumulation in wastewater discharge and control areas of the Houghton Lake wetland. CT = cattail area; SW = sedge-willow area. Results are for cup collectors and do not include leaf litterfall.

for willow leaves in Figure 5. This figure also shows no difference between exponential weight loss between a natural wetland and the same wetland receiving treated wastewater, and there was no difference between two studies spaced 12 years apart for the natural wetland. For this site, the second stage half-life of willow leaves is about two years.

Magnitude of the initial fractional weight loss depends strongly on plant

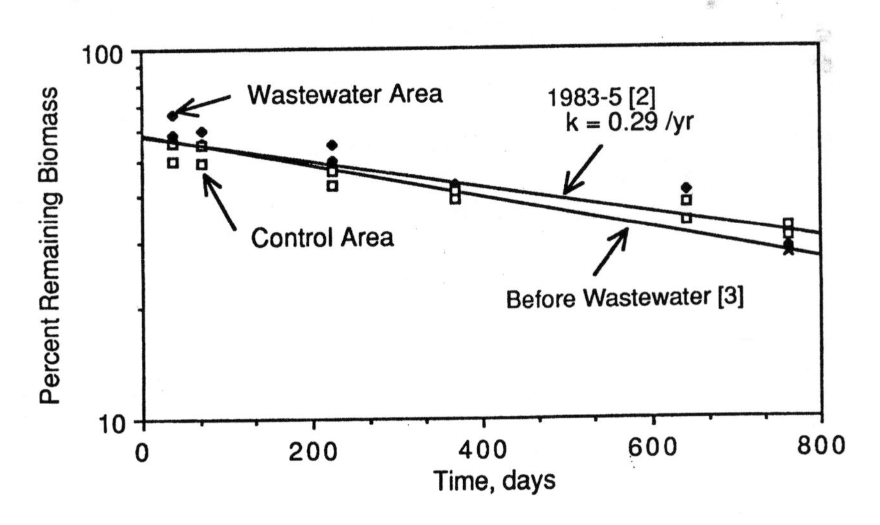

Figure 5. Litter decomposition for willow leaves in the Houghton Lake wetland under various water conditions.

Table 1. Initial Weight Loss During Litter Decomposition, First 30–60 Days.

	Wastewater			Natural		
Carex	14%	H. Lake	[2]	0–5% 0–15%	H. Lake Utrecht	[2] [4]
Typha	20%	H. Lake	[2]	10% 13%	Bradford Delta	[5] [6]
Betula leaves	38%	H. Lake	[2]	20%	H. Lake	[3]
Salix leaves	42%	H. Lake	[2]	15%	H. Lake	[3]
Sagittaria leaves	45%	Clermont	[7]	—	—	
Panicum	55%	Clermont	[7]	—	—	
Betula wood	0–6%	H. Lake	[2]	0–2%	H. Lake	[3]
Salix wood	0–10%	H. Lake	[2]	0–(–5)%	H. Lake	[3]

species and parts (Table 1). Wood displays little initial weight loss; indeed, some species have shown a slight gain over the first month or two. Grasses and leaves show initial weight loss of one-third to one-half; sedges and cattail only lose about one-fifth of their weight. It is clear from Table 1 that wastewater causes a greater initial weight loss in all cases.

Belowground biomass is much more difficult to monitor, and few studies report the death or decomposition rates for roots or peat and other organic soil components. A sampling of the data available on all compartments is given in Table 2, which lists the half-lives of several types of litter materials. These results were determined with different methods, and most include initial

Table 2. Half-Lives of Various Types of Wetland Biomass.

Type	Half-Life	Species or Subspecies	Investigators	
Plants				
Phragmites	220 ± 60 d	3	4	[8,9]
Scirpus	260 ± 80 d	3	3	[9]
Typha	400 ± 250 d	3	5	[2,6,9]
Spartina	430 ± 280 d	2	3	[9]
Juncus	460 ± 90 d	3	5	[9]
Carex	610 ± 320 d	5	6	[2,3,4]
Misc. bog plants	850 ± 600 d	10	3	[10]
Roots				
Carex		950 d	21	[4]
Caluna	14 y	1	1	[10]
Eriophorum	70 y	1	1	[10]
Wood				
Salix		5 y	11	[2]
Chamaedaphne	13 y	1	1	[2]
Betula	7 y	1	1	[2]
Peat				
Highly decomposed, oxidizing	10 y	1	1	[10]
Slightly decomposed, reducing	1700 y	1	1	[10]

weight loss in the determination of half-life. There is thus a large standard deviation among studies. However, if the sedges and bog plants are excluded, the range in half-lives is only 0.6–1.3 years, or about one year for aboveground leaf material. Similarly, the half-life of woody material is about 10 years. Lack of data prevents generalizing about roots. Peat formed from all plant parts, as the "end" product of decomposition, is itself subject to further decay. However, continued decay is contingent upon oxygen availability and under anaerobic conditions will occur only over millenia.[11] Exposure to the atmosphere can greatly accelerate peat decay and release structural constituents.

A critical gap in existing knowledge about the decay process, and thus wetland nutrient immobilization, is the rate of permanent soil formation, with buried nutrients comprising a portion of the undecomposed residual. There is some information about peat formation rates in natural wetlands, in particular northern ombrotrophic bogs. Techniques employed to ascertain the geological age of wetlands include mineral profiles, radiotracer techniques, and pollen dating. These suggest soil-building rates of several millimeters per year, with some much greater rates. For example, the Houghton Lake wetland is carbon-dated as 800 years old with a peat depth of 2 m, for an accretion rate of 2.5 mm/yr. However, this does not explain recent history or the rate which occurs under wastewater additions.

If the consequences of constant fractional (exponential) decay are explored for a litter-sediment compartment receiving a constant annual input of biomass, the following mass balance applies:

$$\frac{\partial M}{\partial t} = P - \alpha M$$

$$M = \frac{P}{\alpha}(1 - e^{\alpha t})$$

where α = decay constant, yr^{-1}
t = time, yr
P = addition rate, $g/m^2/yr$
M = biomass, g/m^2

Houghton Lake values:

$\alpha \approx 0.3$ (2 year half-life)
$t \approx 10$
$P \approx 3000$
$M \approx 10{,}000$ (about 10 cm thick)

This 10-cm zone is moving vertically upward at a rate determined by the undecomposed residual biomass. This permanent soil formation is too small to be measured by simple techniques yet may contain a significant fraction of the

Table 3. Nutrient Concentration Ratios, Litter to Original Plant, Under a Variety of Methods and Water Conditions

Source	Method	N-Ratio	P-Ratio
Houghton Lake (wastewater) [11]	clips	1.46	1.02
Houghton Lake (control) [11]	clips	1.36	0.94
Houghton Lake (control) [2]	2-yr bags	1.36	1.08
Bradford, ONT [5]	2-yr bags	1.33	0.92
Clermont, FL [7]	1-yr bags	1.25	1.25
Utrecht, NL [4]	1-yr bags	1.65	1.35
India [9]	8-mo bags	1.32	0.25

added nutrients. But in any case, this litter zone contains a large fraction of the nutrients added over the 10-year period.

LEACHING AND MINERALIZATION

Litter nutrient concentrations and those of the plant parts from which it came are not the same; old and new litter also differ. Freshly deposited litter loses some nutrients by a combination of relatively fast processes, such as desorption and lysis. As nutrient-enhanced litter materials are flushed with water, about one-fourth of the phosphorus is recovered at constant biomass. After loosely bound phosphorus is removed, further nutrient reductions are accompanied by biomass loss.

Concentrations resulting from biomass loss reflect the relative proportions of the nutrients in mineral and degradable fractions of litter. Table 3 gives results for concentration ratios under a variety of circumstances, all of which reflect the longer term, exponential biomass loss phase of decay. There is agreement that nitrogen content increases compared to the plant of origin, but no such agreement exists for phosphorus. Data for sediment nutrient content is sparse. Sediment analyses from Houghton Lake show nitrogen enrichment comparable to those in Table 3, but threefold phosphorus enhancements. These finely divided sediment materials may have important sorption potential due to their large surface area.

IMPLICATIONS FOR CONSTRUCTED WETLANDS

This summary has consequences for an unharvested constructed wetland receiving secondary municipal wastewater. Nitrogen and phosphorus will be examined here, but other assimilable constituents may be traced in a similar way if concentrations are known.

The startup phase of a constructed wetland may take several years: two to three for vegetation establishment and another two to three to establish the litter-sediment compartment. During this phase, significant nutrient quantities and other biologically active constituents are stored in new live and dead biomass pools.[12] For a wetland of 10 days' detention, a loading of 2.0 cm/day, and 10 mg/L each of nitrogen and phosphorus, establishment of the biomass pools (at 2.0% N and 0.4% P) would require five years and use 60% of the applied nitrogen and 12% of the applied phosphorus. But other processes, such as microbial denitrification and phosphorus sorption, function in parallel with biomass generation. Competition for nutrients will prolong the startup phase and cause longitudinal stratification of biomass and litter. Such phenomena have been documented and modeled for natural wetland systems.

After a stationary state is reached, only the burial component of the biomass balance contributes to removal of nutrients.[13] For the example above, deposition of 2.5 mm/yr of new soil at 2.0% N and 0.4% P would account for 1.4% of the nitrogen applied and only 0.7% of the applied phosphorus. These calculations assume that all of the applied nutrients are available to plants, which is not generally the case. Particulate removal by settling and filtration adds to the above, as do microbial processes and sorption.

It may be concluded that probable startup times for constructed wetlands are 5–10 years, which exceeds any track record yet established. Early reports are probably overly optimistic, since saturable mechanisms are still operative. It is possible, however, that microbial processes are establishing themselves at the same time that the saturable mechanisms are being depleted, because the litter compartment is the locus of attachment for the majority of the active microorganisms.

REFERENCES

1. Kadlec, R. H. "The Use of Peatlands for Wastewater Treatment," in Proceedings of Symposium '87: Wetlands/Peatlands, Edmonton, Alberta (1987), pp. 213–218.
2. Kadlec, R. H. "Wetland Utilization for Management of Community Wastewater: 1985 Operations Summary, Houghton Lake, Michigan," Report to Michigan DNR (1986), pp. 51–78.
3. Chamie, J. P. M. "The Effects of Simulated Sewage Effluent upon Decomposition, Nutrient Status and Litterfall in a Central Michigan Peatland," PhD Thesis, The University of Michigan, Ann Arbor, MI (1976).
4. Arts, H. M. H. "Decomposition of Leaf Litter and Roots in Two Mesotrophic Fens," *The Utrecht Plant Ecology News Report* 6:1–20 (1986).
5. Hershkowitz, J. "Listowel Artificial Marsh Project," Report to the Ontario Ministry of the Environment (1986), pp. 111–116.
6. Kadlec, J. A. Unpublished results (1986).
7. Zoltek, J., and S. E. Bayley. "Removal of Nutrients from Treated Municipal Wastewater by Freshwater Marshes," Report to the City of Clermont, FL (1979), pp. 146–148, 263.
8. Polunin, N. V. C. "Studies on the Ecology of Phragmites Litter in Freshwater," PhD Thesis, University of Cambridge (1979).

9. Kulshreshtha, M., and B. Gopal. "Decomposition of Freshwater Wetland Vegetation. II. Aboveground Organs of Emergent Macrophytes," in *Wetlands: Ecology and Management,* B. Gopal, R. E. Turner, R. G. Wetzel, and D. F. Whigham, Eds. (Paris: UNESCO, 1982), pp. 279–292.
10. Clymo, R. S. "Peat," in *Ecosystems of the World,* A. J. P. Gore, Ed. (Amsterdam: Elsevier Science Publishing Co., Inc., 1983), Chapter 4.
11. Kadlec, R. H. "Wetland Utilization for Management of Community Wastewater: 1982 Operations Summary, Houghton Lake, Michigan," Report to Michigan DNR (1983), pp. 62–63.
12. Kadlec, R. H., and D. E. Hammer. "Modeling Nutrient Behavior in Wetlands," *Ecological Modeling* 40:37–66 (1988).
13. Kadlec, R. H., and D. E. Hammer. "Simplified Computation of Wetland Vegetation Cycles," in *Coastal Wetlands,* H. H. Prince and F. M. D'Itri, Eds. (Chelsea, MI: Lewis Publishers, Inc., 1985), Chapter 9.

37b Thermoosmotic Air Transport in Aquatic Plants Affecting Growth Activities and Oxygen Diffusion to Wetland Soils

Wolfgang Grosse

INTRODUCTION

Many wetland species, such as yellow water lily (*Nuphar luteum*), white water lily (*Nymphaea alba*), and Amazon water lily (*Victoria amazonica*) from the Nymphaeacean family, sacred lotus (*Nelumbo nucifera*) from the Nelumbonacean family, or fringed water lily (*Nymphoides peltata*) and floating heart (*N. indica*) from the Menyanthacean family are cultivated in lakes and constructed ponds as ornamental plants. These plants supply their rhizomes and roots growing in the anoxic sediments of the lake with atmospheric oxygen to sustain growth activities in submerged organs. But distances of up to 4 m, which oxygen must travel from ambient air through the aerenchyma of the plant to buried parts, is too far to create a sufficiently high oxygen level by oxygen diffusion alone. Therefore, these plants with aerial or floating leaves generate a ventilating airflow for aeration of submerged organs, which results from the physical effect of thermoosmosis of gases, an energy-driven process requiring a special morphological feature but not physiological activities of the plant.

THERMOOSMOSIS OF GASES

The physical phenomenon of thermoosmosis of gases,[1] which is synonymous with the terms *thermodiffusion* and *thermal flow of molecules,* was reported first by the physicist Fedderson[2] more than a century ago. He observed a continuous gasflow through a porous partition made from spongy platinum, as long as a temperature difference was established between both sides of that porous partition.

Later, this phenomenon was described in more detail by the physicist Knudsen[3] as the quotient of pressures in two compartments separated by a porous partition equaling the quotient of the square roots of the temperature present in these two compartments (Figure 1, upper sketch) according to the equation: $P_a/P_b = \sqrt{T_a}/\sqrt{T_b}$.

Such an effect requires that pore diameters of the porous partition be much smaller than 1 μm to impede free gas diffusion and to give rise to the so-called Knudsen diffusion (Figure 2).

A temperature increase in one of the compartments causes a net flow of gas from the cooler side to the warmer side in an attempt to reestablish a stationary state. Resulting higher pressure in the warmer compartment can be used to initiate a flow of gas through an additional channel back to the cooler compartment (Figure 1, lower sketch). Pressure in the warmer compartment remains lower than under stationary state for the efflux of gas from the warmer compartment. Therefore, the gas enters the warmer compartment continuously through the porous partition by Knudsen diffusion and flows back through the large channel with little flow resistance. A circulating gasflow results which continues as long as a temperature difference between the two sides of the porous partition is established.

THERMOOSMOTIC GASFLOW THROUGH PLANTS

Aquatic plants with floating leaves[4–6] as well as alder *(Alnus glutinosa)*[7–9] meet all the requirements to generate such a circulating gasflow by thermoosmosis for ventilating their submerged organs.

Schematic air transport in *Nuphar luteum* along with the important physical variables is shown in Figure 3. Newly emerged leaves possess a porous partition—a monolayer of cells with intercellular gas spaces of about 0.7 μm in diameter[4]—which separates the palisade parenchyma and aerenchyma. Because of these small pores, that tissue is thermoosmotically active. When air inside the aerenchyma of a floating leaf becomes about 5 K warmer than ambient air by absorption of light energy, air enters that young leaf by Knudsen diffusion, leading to a pressurization in the aerenchyma. The rising pressure difference of about 100 pascal (1 pascal = 0.102 mm H_2O) is strong enough to establish an airflow through the connected gas space system of the whole plant and out of the most porous older leaves in which a temperature-induced pressurization never is detectable. In these old leaves the pore diameters are increased to more than 20 μm so that free airflow is possible.

Actually 40 mL air/hr in *Nymphoides peltata* to about 3.5 L air/hr in *Victoria amazonica* is transported by that thermoosmotic process through the aquatic plants when temperature differences of 2–3 K are established (Table 1).[10]

In *Nelumbo nucifera* a very elegant gasflow system exists.[6] The air enters the leaf blade and the flow causes the air to descend to the rhizome. Then the air, up to 600 mL/hr, returns to the atmosphere through two large channels of the petiole and the center plate of the same leaf.

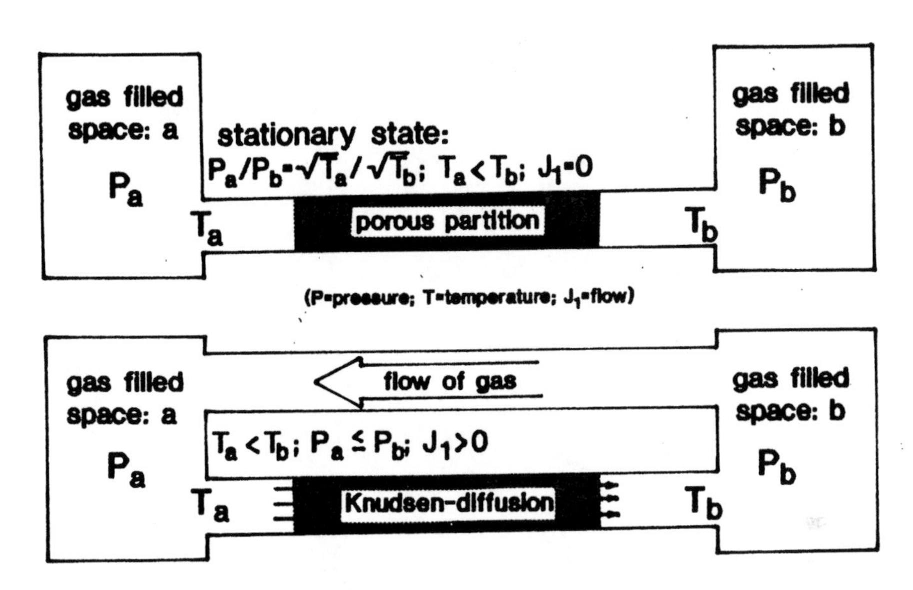

Figure 1. Thermoosmosis of gases according to Knudsen[3] at stationary state (upper sketch) and under circulating gasflow conditions (lower sketch), with pressure (P) and temperature (T) inside the two gas-filled compartments (a) and (b) and net flow of gas (J_1).

Thermoosmotic air transport has also been found in *Alnus glutinosa,* an inhabitant of the flood plains. Air transport of about 1 mL air/hr in one-year-old trees is much lower than in aquatic plants. It is directed from the stems to the roots and is driven by pressurization within the air space system of the stems. The pressure results from a temperature gradient between the stem and surrounding air caused by absorption of light energy by brownish pigments of the bark. The porous partition is proposed to be located in lenticels.[9]

In some aquatic plants the airflow results from the activities of newly developing leaves. As shown for *Nymphoides peltata* (Figure 4), efflux of air from the petioles of cut leaves increases from about 30 mL air/hr to nearly 120 mL air/hr in the youngest floating leaf when the temperature difference rises from 1 K to 8 K. This leaf functions as an influx leaf for the whole plant in conducting air transport to the rhizomes. The efflux of air from the petioles tends toward zero when the leaves reach their final expansion, and these leaves become the efflux leaves for the plant.[10]

The change from influx to efflux leaves during leaf development is shown very well with *Victoria amazonica* (Figure 5). The youngest floating leaf is too gas-tight for any air transport. As the leaf blade is expanded, the partition inside the leaf becomes porous enough to create thermoosmotic gasflow of up to 30.5 L air/hr. But in fully developed leaves with leaf blade diameters of about 1.7 m, the pores are too large and the porous partition loses its thermoosmotic activity.[10]

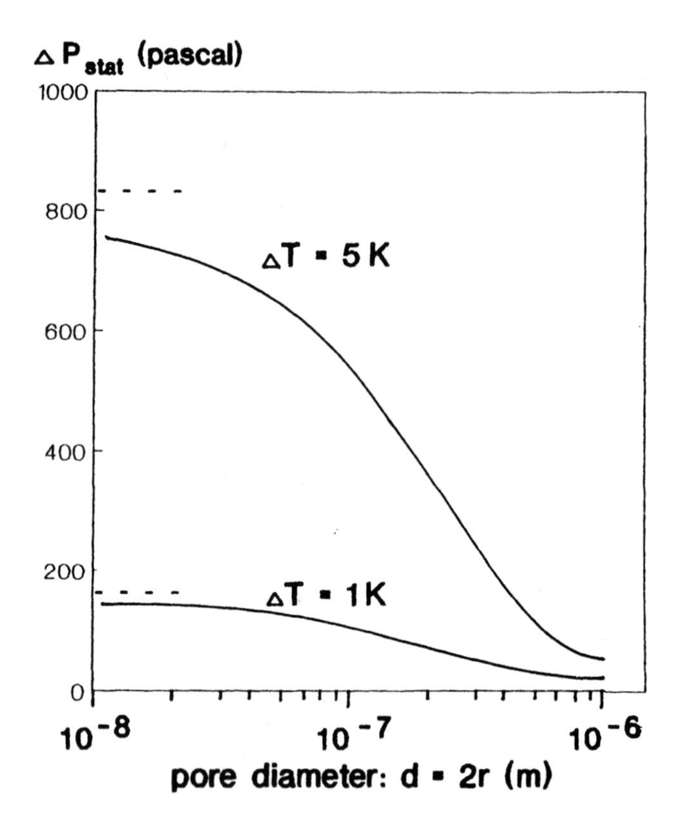

Figure 2. Thermomechanical pressure difference $(\Delta P)_{stat}$ (zero net flow) as function of pore diameter (d = 2r) at a pressure of 100,000 pascal (1 pascal = 0.102 mm H_2O) and a temperature of T = 300 K according to Schröder et al.[4] Parameter is the temperature difference between the bulk phase. The dashed lines indicating the Knudsen limits are calculated on the basis of equation: $P_a/P_b = \sqrt{T_a}/\sqrt{T_b}$.

ECOLOGICAL SIGNIFICANCE OF THERMOOSMOSIS IN PLANTS

The ecological significance of a thermoosmotic gas transport system for plants inhabiting aqueous environments lies in the increase in supply of oxygen to the root and rhizome apices. Sediments of deep water lakes and ponds are generally anoxic and are characterized by low redox potential and accumulations of Fe^{2+} hydroxide and other phytotoxic materials. Therefore, survival of aquatic plants is due partly to internal aeration from aerial parts and partly to leakage of oxygen from the submerged plant organs to the surrounding soil to protect the plant by oxidation of the phytotoxins. In the common reed, (*Phragmites australis*), the distance from the roots to the atmosphere seems to be short enough to guarantee a sufficient oxygen supply by diffusion.[11] But plants that are equipped with a thermoosmotic gas transport system are able to colonize the deep water regions.

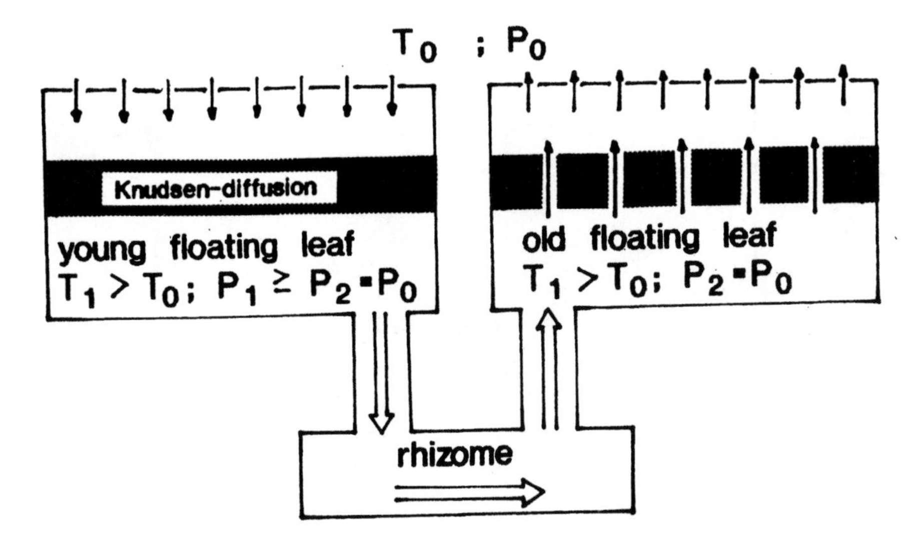

Figure 3. Circulating airflow through *Nuphar luteum* by thermoosmosis of gases. A schematic representation of the ventilation system, with pressure (P) in the young influx leaf (1), old efflux leaf (2), and atmosphere (0); temperature (T) inside the leaf (1) and in the surrounding air (0); and direction of airflow (→).

Gas analyses have shown[10] that samples from the gas spaces of submerged organs of *Nymphaea alba* contain 19.2 vol% O_2 when transport is operating. The value decreases to 3.9 vol% O_2 when transport is stopped by darkening or submerging all of the leaves. Therefore, we conclude that an adequate O_2 supply for growth activities of the submerged organs is also guaranteed by that thermoosmotic process under deep water conditions.

Table 1. Airflow Through Wetland Plants by the Physical Process of Thermoosmosis of Gases

Species	Temperature (K)	Flowrate (mL air/hr)	Ref.
Nymphaeaceae			
Nuphar luteum	2.5	260	[10]
Nymphaea alba	2.4	260	[10]
N. colorata	2.3	200	[10]
Victoria amazonica	1.6	3490	[10]
Menyanthaceae			
Nymphoides peltata	2.5	40	[10]
N. indica	3.2	490	[10]
Nelumbonaceae			
Nelumbo nucifera	2.9	600	[6]
Betulaceae			
Alnus glutinosa	1.9	<1	[9]

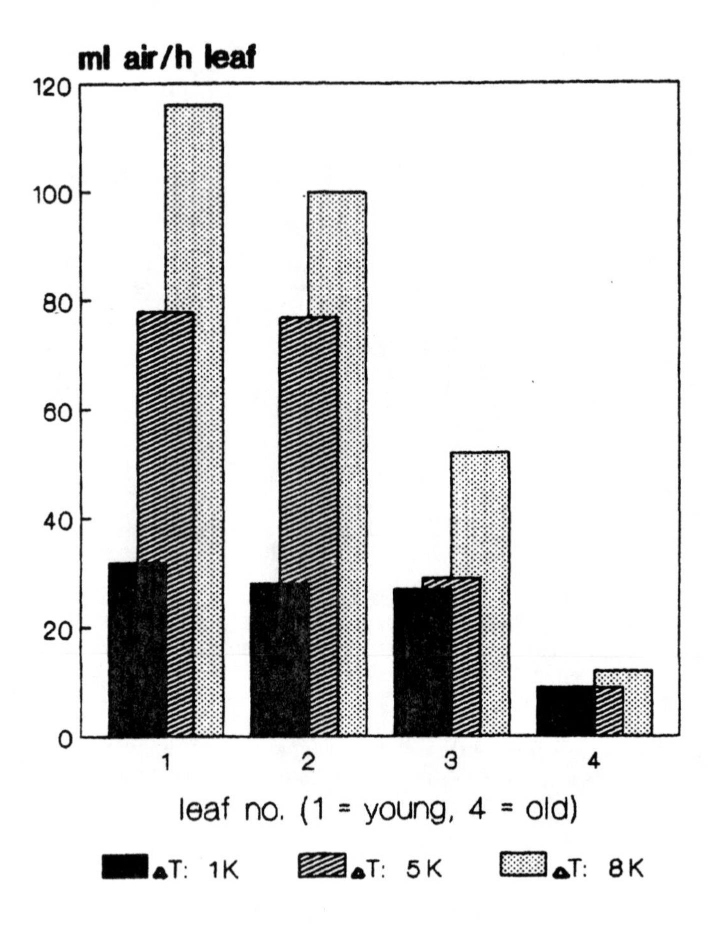

Figure 4. Efflux of air from the petioles of cut floating leaves of *Nymphoides peltata* by thermoosmosis resulting from temperature differences of 1 K, 5 K, and 8 K, respectively, between the warmer leaf and the cooler air. Nos. 1–4 represent emerged leaves of increasing age.

This gas transport may benefit the surrounding environment by enriching the rhizosphere with oxygen. Oxygen diffusion from roots into the sediment is suggested by the deposits of Fe^{3+} hydroxide often observed near the root surface and reported recently for reed.[11] Gas bubbles containing 1–6 vol% O_2 observed at the roots of alder trees under thermoosmotic conditions[12] indicate that a flow of gas out of the roots occurs. About 14–21 μmol O_2/hr exits the root system of one-year-old alder at +5°C declining to 0–3 μmol O_2/hr when gas transport is stopped.[9] In summary, thermoosmotic gas transport will enhance formation of oxygenated zones in sediments of deeper lakes or water clearing ponds which are important for purification processes.

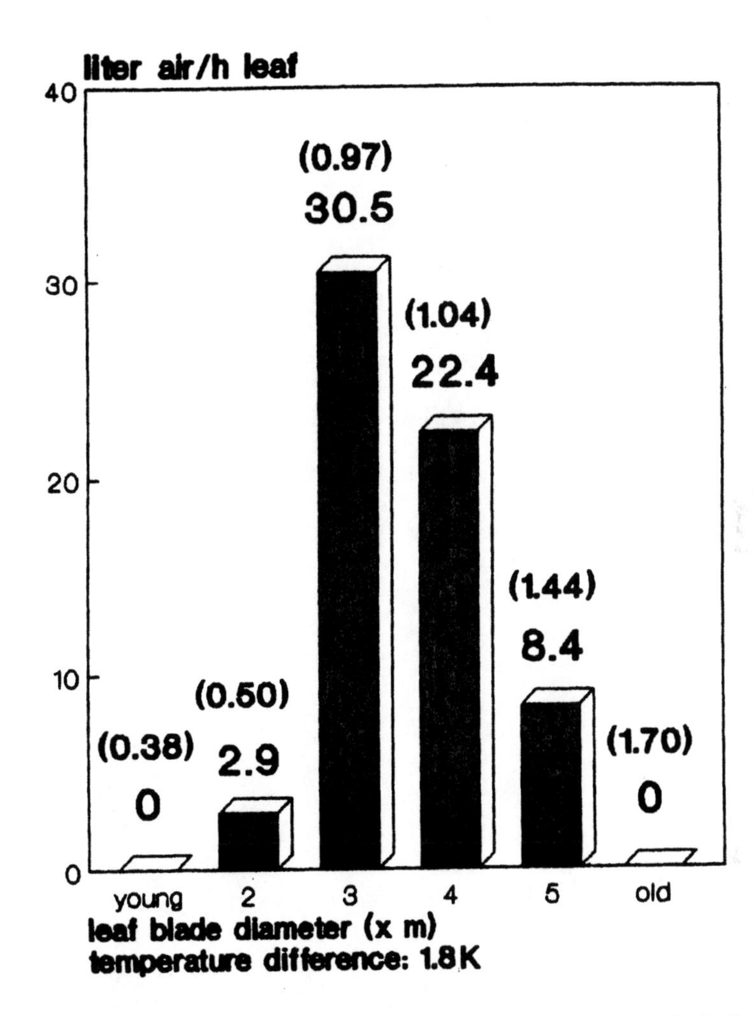

Figure 5. Efflux of air from petioles of cut floating leaves of *Victoria amazonica* by thermoosmosis resulting from a temperature difference of 1.8 K, with 26.7°C on the leaf blade surface and 28.5°C inside the leaf. Nos. 1–6 represent emerged leaves of increasing age.

REFERENCES

1. Denbign, K. G., and G. Raumann. "The Thermo-osmosis of Gases through a Membrane. I. Theoretical," *Proc. Royal Soc.* 210 A:377–387 (1951).
2. Feddersen, B. W. "Uber Thermodiffusion von Gasen," *Pogg. Ann.* 148:302–311 (1873).
3. Knudsen, M. "Eine Revision der Gleichgewichtsbedingungen der Gase. Thermische Molekularströmung," *Ann. Phys.* 31:205–229 (1910).
4. Schröder, P., W. Grosse, and D. Woermann. "Localization of Thermo-osmotically

Active Partitions in Young Leaves of *Nuphar lutea*," *J. Exp. Bot.* 37:1450–1461 (1986).

5. Grosse, W., and J. Mevi-Schutz. "A Beneficial Gas Transport System in *Nymphoides peltata*," *Am. J. Bot.* 74:947–952 (1987).
6. Mevi-Schutz, J., and W. Grosse. "A Two-Way Gas Transport System in *Nelumbo nucifera*," *Plant, Cell and Environment* 11:27–34 (1988).
7. Grosse, W., and P. Schröder. "Oxygen Supply of Roots by Gas Transport in Alder Trees," *Z. Naturforsch.* 39c:1186–1188 (1984).
8. Grosse, W., and P. Schröder. "Aeration of the Roots and Chloroplast-Free Tissues of Trees," *Ber. Deutsch. Bot. Ges.* 98:311–318 (1985).
9. Schröder, P. "Thermoosmotischer Sauerstofftransport in *Nuphar lutea* L. und *Alnus glutinosa* Gaertn. und seine Bedeutung für ein Leben in anaerober Umgegung," Ph.D. Thesis, Universität Köln, F.R.G. (1986).
10. Grosse, W., B. Büchel, and H. Tiebel. Unpublished results (1988).
11. Armstrong, J., and W. Armstrong. "*Phragmites australis*—A Preliminary Study of Soil-Oxidizing Sites and Internal Gas Transport Pathways," *New Phytologist* 108:373–382 (1988).
12. Grosse, W., and S. Sika. Unpublished results (1988).

37c Nitrification and Denitrification at the Iselin Marsh/Pond/Meadow Facility

Randal L. Davido and Thomas E. Conway

INTRODUCTION

In many areas, especially rural communities, conventional wastewater treatment systems are not feasible. Artificial wetland treatment offers an attractive alternative for treating domestic wastewater. Considerable work has been done with artificial wetlands at the Brookhaven National Laboratory. Maxwell Small found that combinations of marshes, ponds, and meadows can be effective wastewater treatment systems.[1-4]

Wetland systems are superior nutrient recycling systems. Gersberg et al. found that artificial wetlands have the capacity to remove a large percentage of the total nitrogen in wastewater.[5,6] Nitrification by chemoautotrophic nitrifying bacteria, mainly *Nitrosomonas* and *Nitrobacter,* oxidize ammonia (NH_3) to nitrite (NO_2) and nitrate (NO_3), respectively. Nitrate and NO_2 reduced by facultative bacteria to nitrous oxide (N_2O) and nitrogen gas (N_2) is the anaerobic denitrification process.

The Iselin Marsh/Pond/Meadow (MPM) is an example of an active wetlands system that has proven effective in removing nitrogen from wastewater. We intend to define major zones of nitrification and denitrification in the MPM and provide base data for future work at the site for optimizing removal capacities at Iselin or other facilities.

The Iselin MPM was designed to treat 45,360 L/d from 62 homes in Iselin, a rural mining town in Indiana County in western Pennsylvania. The project was designed by the Pennsylvania Department of Environmental Resources. Conway[7] and Breindel[8] offer a brief history and financial breakdown of the MPM system.

THE STUDY SITE

The Iselin MPM consists of six components: (1) a comminutor, (2) an aeration basin, (3) two marsh areas, (4) a pond, (5) two meadow areas, and (6) a chlorination chamber (Figure 1). The system is sealed with plastic liners and bentonite clay to prevent infiltration or leakage.

Wastewater entering the MPM is aerated for 1.5 days. Surface overflow from the aeration basin flows into one of two beds in the marsh area. Effective marsh areas are 42.7 m long, 11.3 m wide, and 0.51 m deep. The top 0.10 m of marsh strata consists of 2B stone over 0.41 m of sand. Cattails (*Typha latifolia*) were planted but growth is sparse. Grasses from overseeding adjacent banks occur in the marsh, with Meadow Fesque (*Festuca eliator*) the most prominent.

Marsh effluent flows into a 1.5 m deep, 189,000 L pond with a 15-day retention period. Duckweed (*Lemna minor*) has invaded and completely covered the pond. Water leaves the pond via two surface overflow pipes.

Pond water is discharged into one of two sides of the meadow area, which are constructed like the marshes. Size and vegetation type are the only differences. The meadows are smaller, 21.3 m × 11.3 m, and Virginia reed canary grass (*Phalaris arundinacea*) was planted. The meadow effluent is then collected, chlorinated, and discharged into Harpers Run.

SAMPLING AND ANALYSIS

Samples were collected from 10 sites throughout the MPM system (Figure 1). Samples were collected daily from August 9 to 22, 1985. Site 3 in the marsh is 10.7 m from the inlet, and sites 4 and 5 are also separated by 10.7 m. Three depths (designated *S, M,* and *B* for "surface," "middle," and "bottom,") were sampled at sites 3–5 and sites 8 and 9 by inserting stand pipes to the desired depth. Sites 8 and 9 in the meadow are spaced 7.1 m apart between influent pipes and effluent trough. In the marsh and meadow, S = 0.05 m, M = 0.25 m, and B = 0.43 m below the surface. Pond sites 6 and 7 are spaced 13.8 m apart and 13.8 m from inlet and outlet ends of the pond. S, M, and B in the pond signify depths of 0.05 m, 0.70 m, and 1.40 m, respectively.

Table 1 lists parameters affecting nitrification and denitrification processes and methods of analysis.

RESULTS AND DISCUSSION

Temperature, pH, and DO showed little variation throughout the system and are only briefly mentioned. Average pH of the influent and aeration effluent were 8.0 and 7.6, respectively, with a pH 7 average throughout the marsh and meadow. Largest variation in pH occurred in the pond, ranging from 8.0 at the surface to 6.7 at the bottom. Average temperature remained near 20°C, with a low of 16°C at the pond bottom. Average DO was less than 1.0 mg/L at all depths in the marsh. The influent had the highest DO (7.9 mg/L), and aeration effluent was next highest (5.0 mg/L). Dissolved oxygen was 2.0 mg/L at the

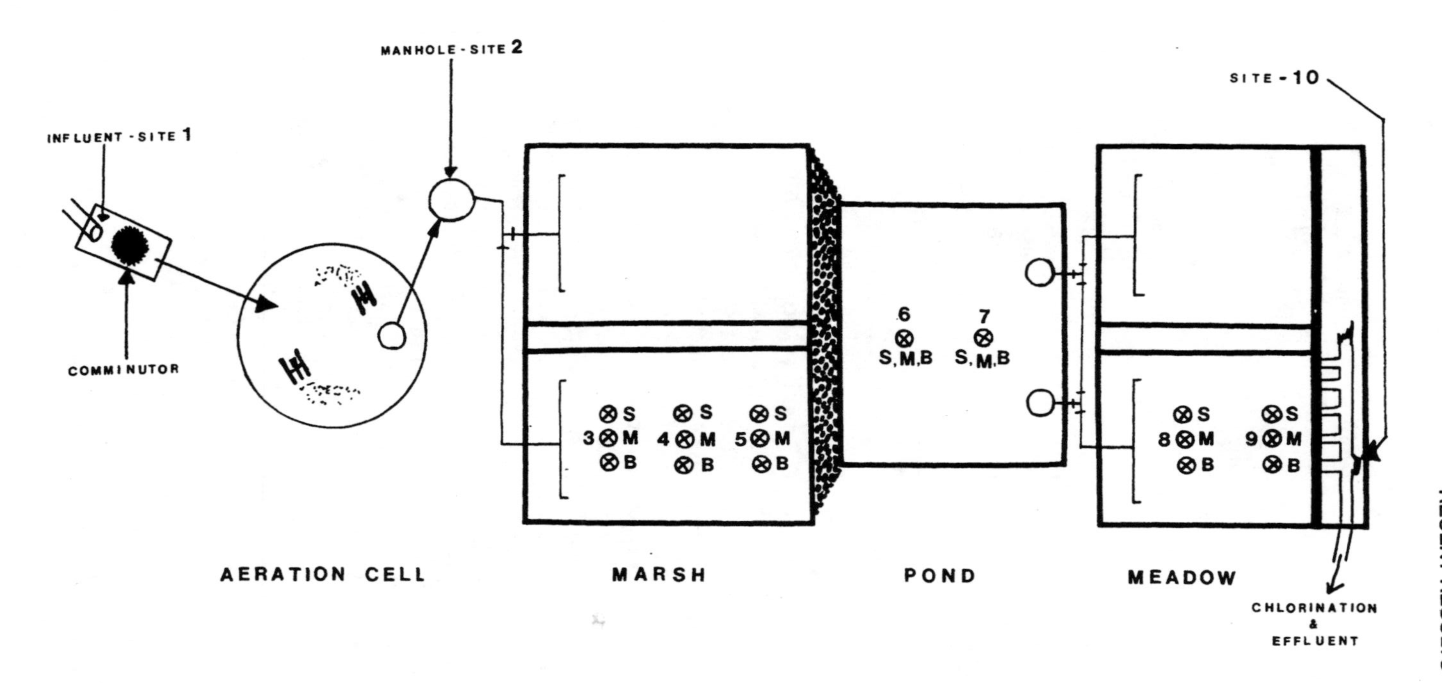

Figure 1. The Iselin MPM showing components and sample sites.

Table 1. Laboratory Analysis

Parameter	Site[a]	Method of Analysis
Ammonia nitrogen	1–10	Nessler method/specific ion electrode method [9]
Nitrate nitrogen	1–10	Cadmium reduction method using HACH Nitra Ver 5 reagent [10]
Nitrite nitrogen	1–10	Diazotization method using HACH Nitri Ver 3 reagent [10]
Total Kjeldahl nitrogen	1, 2, 5M, 7M, 10	HACH digestion and nesslerization method [11]
Chemical oxygen demand	1, 2, 5M, 7M, 10	HACH reaction/digestion method [12]
pH	1–10	Combination pH-reference electrode with an Orion 407A Ionalyzer
Dissolved oxygen	1–10	YSI 51B dissolved oxygen meter
Temperature	1–10	YSI 51B dissolved oxygen meter/temperature mode

Note: Statistical analysis of data is in progress.
[a]M = Sampling depth designated "middle."

pond surface but less than 0.5 mg/L at the bottom. Meadow DO averaged 1.0 mg/L, with slightly higher concentration at the surface.

Nitrate nitrogen (NO_3-N) decreased from influent to postaeration (Figure 2). Because carbon demand is easily met for denitrifiers in the aeration basin with chemical oxygen demand (COD) values in excess of 800 mg/L and a small anoxic zone may exist within bacterial sludge floc,[13] the decrease in NO_3-N during aeration is probably due to denitrification.

Ammonia nitrogen (NH_3-N) concentrations decreased at all depths across the marsh (Figure 3). This corresponds with an overall NO_3-N increase at all depths in the marsh (Figure 2), indicating nitrification throughout the marsh which is supported by the upward trend in nitrite nitrogen (NO_2-N) (data not shown). Nitrite data are not used as indicators because they are components of both processes. Low DO observed throughout the marsh is likely the result of activity and not its unavailability. Marsh influent DO was 5.0 mg/L, sufficient for nitrification. Denitrification may also occur in the marsh, and a decrease in total Kjeldahl nitrogen (TKN) of 51% in the marsh signifies loss from the aquatic system. Vegetative uptake as a major sink was ruled out because of sparse growth and because Gersberg et al.[6] found little difference in total N removal between vegetated and unvegetated beds. Denitrification is likely the major route for nitrogen removal even though DO is present. Denitrification occurs in anaerobic microsites common to all soils.[13]

Ammonia nitrogen levels are low at the pond surface (Figure 3), but substantial concentrations were present near the bottom. The increase in NH_3-N at the pond bottom results from ammonification of decomposing plant material. Nitrate nitrogen also increases during treatment in the pond (Figure 2). Physical mixing and oxygenation by duckweed contribute to higher DO levels near

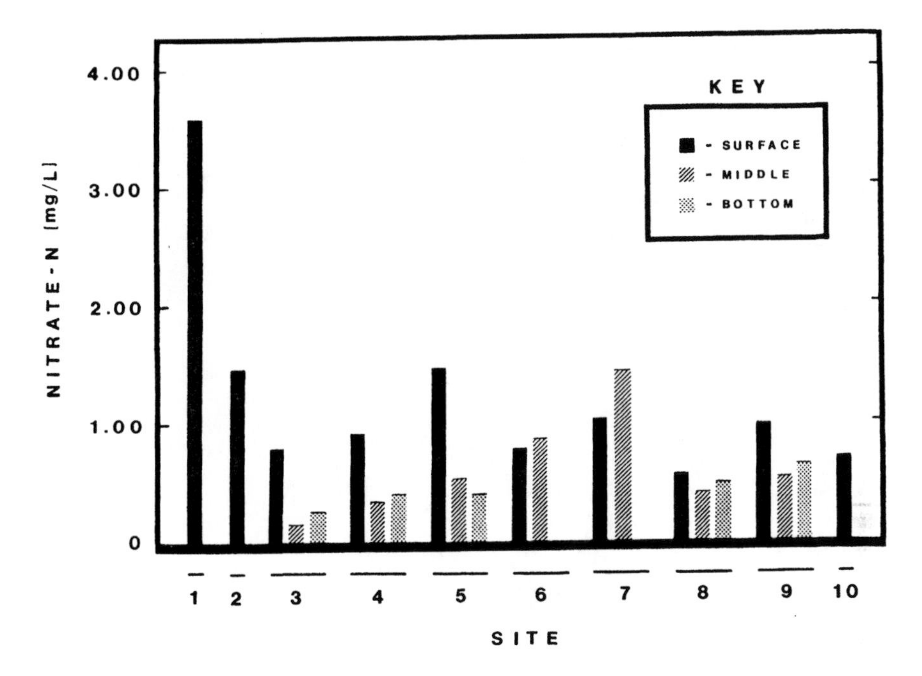

Figure 2. Average nitrate nitrogen at all sites and depths for the Iselin MPM system.

the surface. In addition to nutrient uptake, duckweed roots may provide attachment surfaces for nitrifying bacteria and increase NH_3-N removal at the pond surface. Although an analytical interference precluded NO_3-N data from the pond bottom, denitrification likely occurs there because conditions are favorable and there was a 57% TKN removal.

Ammonia nitrogen concentrations decreased to less than 0.5 mg/L at the meadow surface (Figure 3). Highest concentrations of NH_3-N at the meadow bottom near the inlet resulted from decomposing solids that had been deposited in the meadow media. The NH_3-N trend across the meadow decreases to 0.5 mg/L in the effluent, with a slight concurrent increase in NO_3-N (Figure 2). Nitrification was most effective at the meadow surface because the surface had approximately 1.0 mg/L DO compared to 0.5 mg/L at the bottom, accomplishing a 66% decrease in available TKN across the meadow. Denitrification may occur, but heavy vegetative growth, little reduction in NO_3-N, and low organic carbon levels suggest that denitrification is minimal.

Overall reduction of TKN from marsh influent to meadow effluent was 95%, producing discharge levels of less than 1 mg/L NH_3-N and NO_3-N. The MPM system was also capable of removing 93% of the average 860 mg/L COD entering the system.

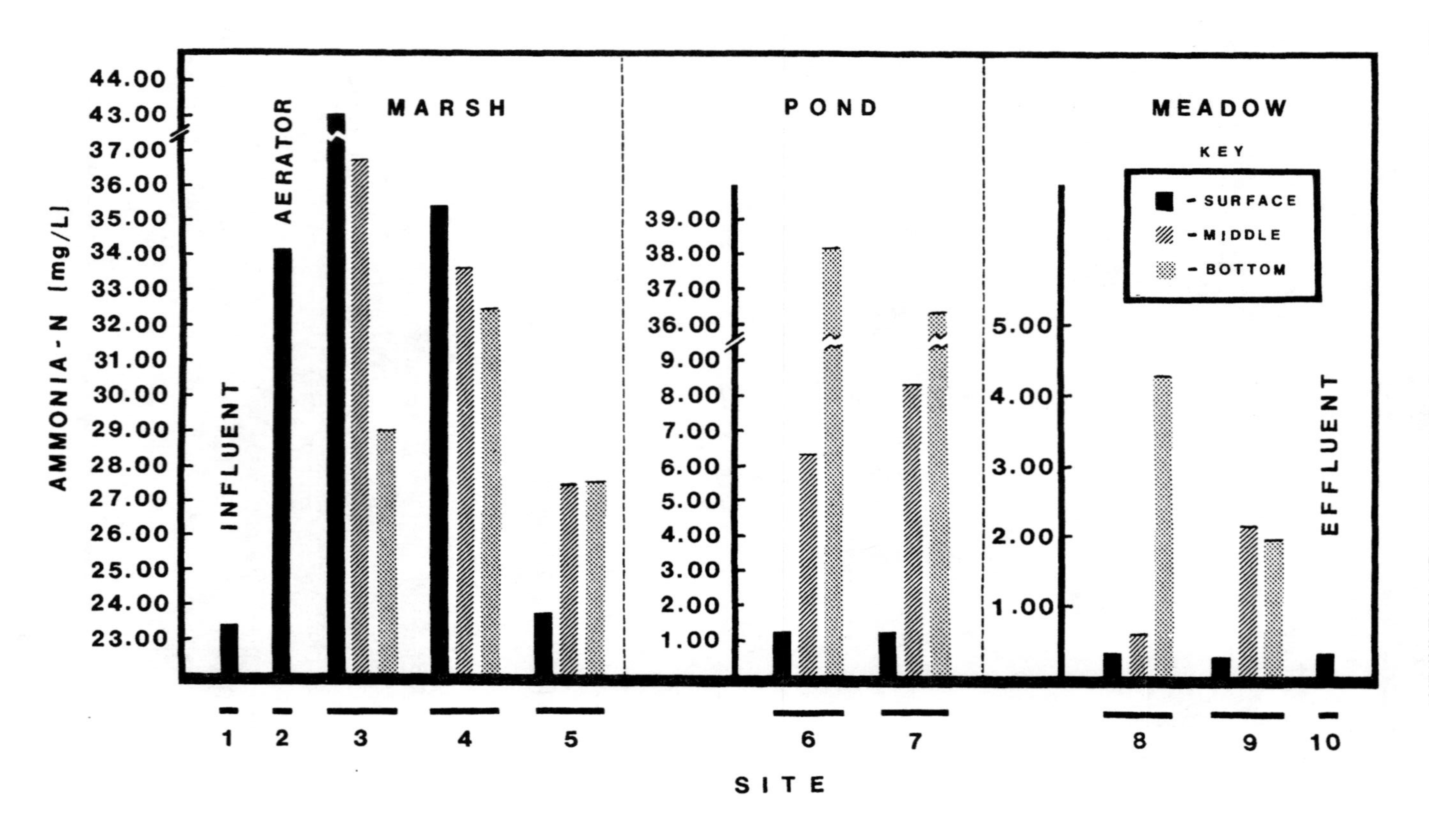

Figure 3. Average ammonia nitrogen at all sites and depths for the Iselin MPM system.

SUMMARY AND CONCLUSIONS

Nitrification and denitrification in the Iselin MPM system accomplished a removal efficiency of 95% of TKN. The most effective nitrification component is the marsh. Nitrification occurs at all depths but most effectively at the surface. Nitrification occurs in the upper zones of the pond and at all depths in the meadow.

Denitrification occurs in the aeration basin within the sludge floc and anaerobic microsites throughout the marsh and may occur in the pond sediments. Denitrification appears to be minimal in the upper pond strata and in the meadow.

REFERENCES

1. Small, M. M. "Brookhaven's Two Sewage Treatment Systems," *Compost Sci.* 16(3) (1986).
2. Small, M. M. "Marsh/Pond Sewage Treatment Plants," paper presented at the Fresh Water Wetlands and Sewage Effluent Disposal Conference, University of Michigan, Ann Arbor, Michigan, March 1976.
3. Small, M. M. "Natural Sewage Recycling Systems," Publication BNL 50630, Department of Applied Science, Brookhaven National Laboratory, Associated Universities, Inc., Upton, NY (January 1977).
4. Small, M. M. "Wetlands Wastewater Treatment Systems," presented at the International Symposium, State of Knowledge in Land Treatment of Wastewater, Hanover, NH, August 20–25, 1978.
5. Gersberg, R. M., B. V. Elkins, and C. R. Goldman. "Nitrogen Removal in Artificial Wetlands," *Water Res.* 17(9):1009–1014 (1983).
6. Gersberg, R. M., B. V. Elkins, and C. R. Goldman. "Use of Artificial Wetlands to Remove Nitrogen from Wastewater," *J. Water Poll. Control Fed.* 56(2):152–156 (1984).
7. Conway, T. E. "Iselin: Pennsylvania's Unique Research and Demonstration Project," *Water Poll. Control Assoc. PA Magazine* 19(2):8–9 (1986).
8. Breindel, R. J. "Meeting Rural Pennsylvania's Sewage Needs," *PIQ* Fall (1983).
9. *Standard Methods for the Examination of Water and Wastewater*, 15th ed. (Washington, DC: American Public Health Association, 1981).
10. *DR-EL/4 Methods Manual*, Hach Chemical Company, P.O. Box 389, Loveland, CO (1981).
11. *Digesdahl Digestion Methods Manual*, Hach Chemical Company, P.O. Box 389, Loveland, CO (1981).
12. *COD Reactor Methods Manual*, Hach Chemical Company, P.O. Box 389, Loveland, CO (1981).
13. "Biological Nitrification and Nitrogen Removal," Nutrient Control—Manual of Practice FO-7, Water Pollution Control Federation (1983), pp. 25–135.

37d Denitrification in Artificial Wetlands

Eberhard Stengel and Reinhard Schultz-Hock

INTRODUCTION

Small artificial wetlands were investigated for their potential to remove nitrates from contaminated waters to purify drinking water.[1] Since efficient nitrate elimination must be based on biological denitrification at very low oxygen concentrations, macrophytes seemed advantageous as organic carbon substrate producers but, on the other hand, probably disadvantageous because they are considered to be good oxygenators. Therefore, we started to explore the aeration function of macrophyte root horizons with horizontal flow-through.

This chapter will report on (1) denitrification in relation to oxygen concentration, organic carbon sources, and temperature; and (2) oxygen conditions in the root horizon.

MATERIALS AND METHODS

Measurements were performed on outdoor artificial wetlands with U-shaped channels 0.4 m deep, 0.6 m wide, and 16 m long and total surface area of 9.6 m^2. Channels were filled with gravel (3–8 mm) and sampling sites were located at distances between 1 and 1.5 m. Vegetated beds (dominated by *Phragmites australis* with shoot densities of 350–580/m^2 and leaf area index between 10 and 30) and an unvegetated control unit were continuously supplied with local tap water containing 30 mg/L NO_3 at 30 L/hr = 72 L/m^2/day. Retention time was about one day.

During methanol experiments in winter 1984, 48 L/m^2 day tap water was applied corresponding to 1.4 g/m^2/day NO_3, and the methanol dosage was 3.6 g/m^2/day. In 1986, the standard input, i.e., 72 L/m^2 day, was applied. The ratio of methanol to nitrate (not NO_3-N!) was 2.5. Straw percolate was generated by letting inflow water pass through an upflow reactor (total volume = 500 L, including 400 L wheat straw). Acetylene blockage experiments and gas analyses were performed afterward.[2] Low oxygen concentrations (0.7–1 mg/L)

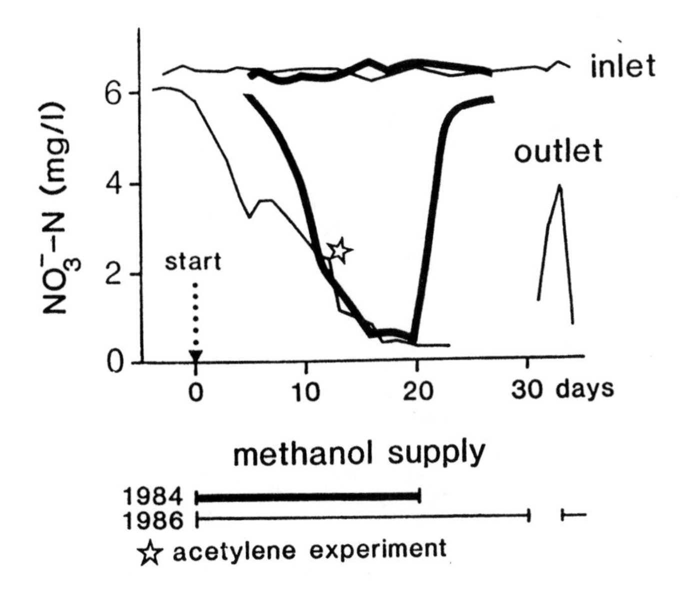

Figure 1. Enhanced NO_3 removal at average temperatures of 6–8°C by continuous addition of methanol in 1984 (29 November to 19 December 1984) and 1986/87 (5 December 1986 to March 1987) with accidental interruption for a few days. The star marks an acetylene experiment. See also Figures 4 and 5.

in inflowing water were obtained by sparging nitrogen gas in a counterstream column.

RESULTS AND DISCUSSION

Denitrification

In August 1984, indications of denitrification processes were found by comparing oxygen and nitrate profiles.[1] At average temperatures of 6–8°C, low nitrate elimination could be enhanced by supplying organic carbon (methanol, straw percolate), and denitrification—proved by application of acetylene[2]—was the primary elimination mechanism (Figures 1–3). Complete nitrate elimination also occurs at temperatures of 2–4°C when supplying methanol (unpublished results). Appreciable denitrification rates under natural and laboratory conditions are reported at similar low temperatures.[3,4] Comparison of several acetylene-blockage tests shows that denitrification occurs in the whole system independent of the particular autochthonous or allochthonous organic substrate when oxygen levels are low enough ($O_2 <$ 2–3 mg/L; Figure 3).

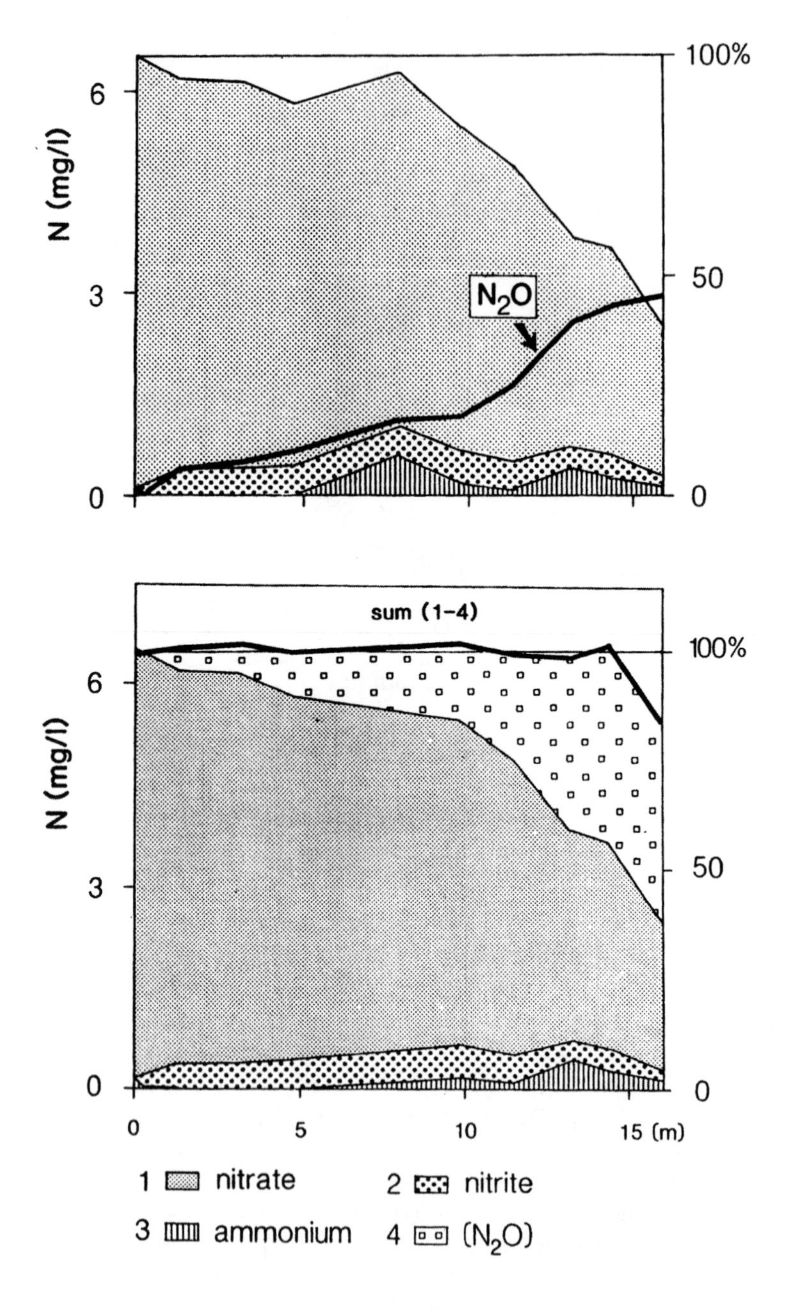

2. Profiles of the inorganic N species after 50 hours of continuous dosage of acetylene-enriched water (17 December 1986).

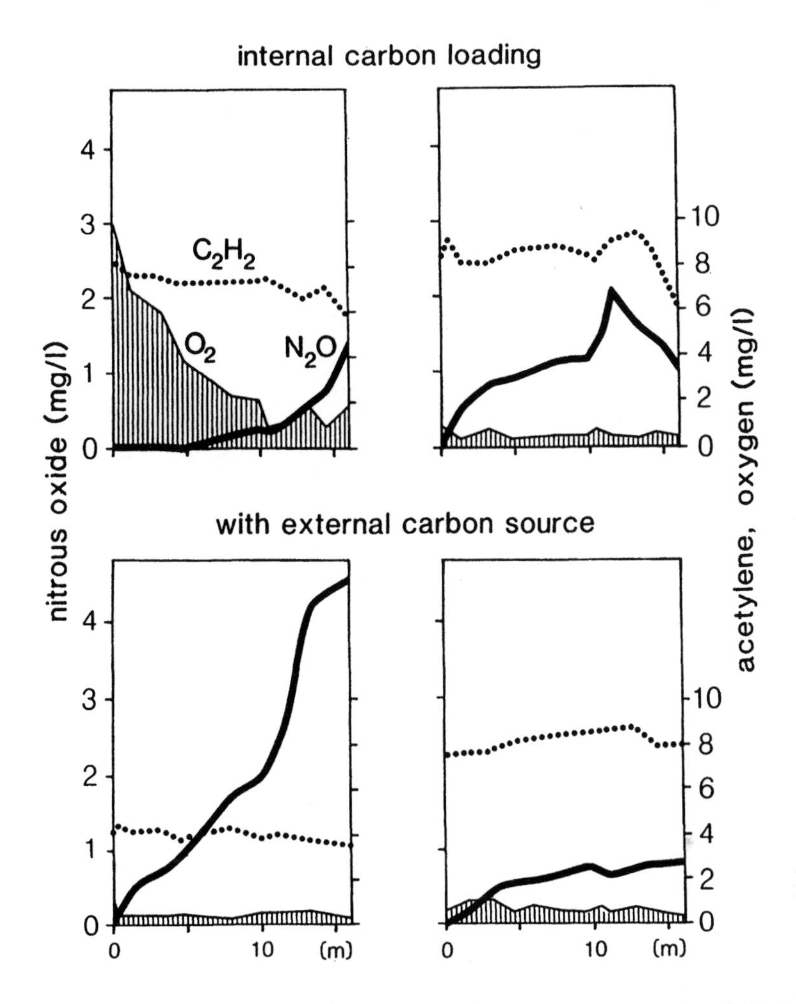

Figure 3. Denitrification in relation to oxygen conditions and carbon supply as indicated by N_2O formation after continuous acetylene dosage (cf. Figure 3) Upper left: (18 June 1986) high oxygen input. Upper right: (15 August 1987) low oxygen input. Reduced flow rate (15 L/hr) because defective pump prevented complete mixing. Lower left: (17 December 1986) continuous methanol dosage (Figures 2 and 3). Lower right: (15 July 1987) straw percolate dosage.

Oxygen Conditions

Aquatic macrophytes can act as aerators in ecosystems.[5-8] But this may not be taken for granted, because we never observed an increase in oxygen in through-flowing water in the root horizon (Figure 4). Indeed, several times gas bubbles from *Phragmites* rhizomes were released in situ (a few centimeters below water surface) and from freshly cut shoots. Gas bubble quantity was related to solar irradiation. Freshly cut young shoots produced gas pressures

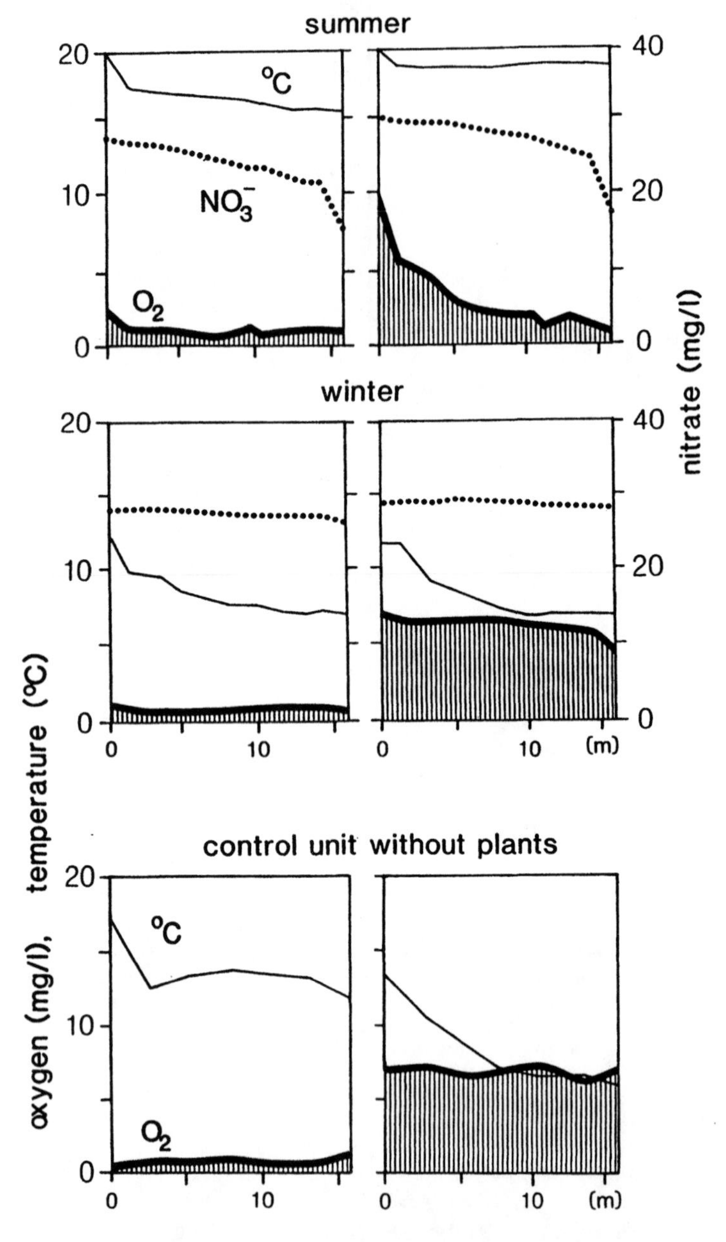

Figure 4. Oxygen conditions in vegetated and unvegetated (control) gravel beds with low and high oxygen input. Upper left: (25 August 1986); right: (7 July 1986). Middle left: (3 December 1986); right: (23 January 1986). Lower left: (24 April 1987); right: (20 December 1985).

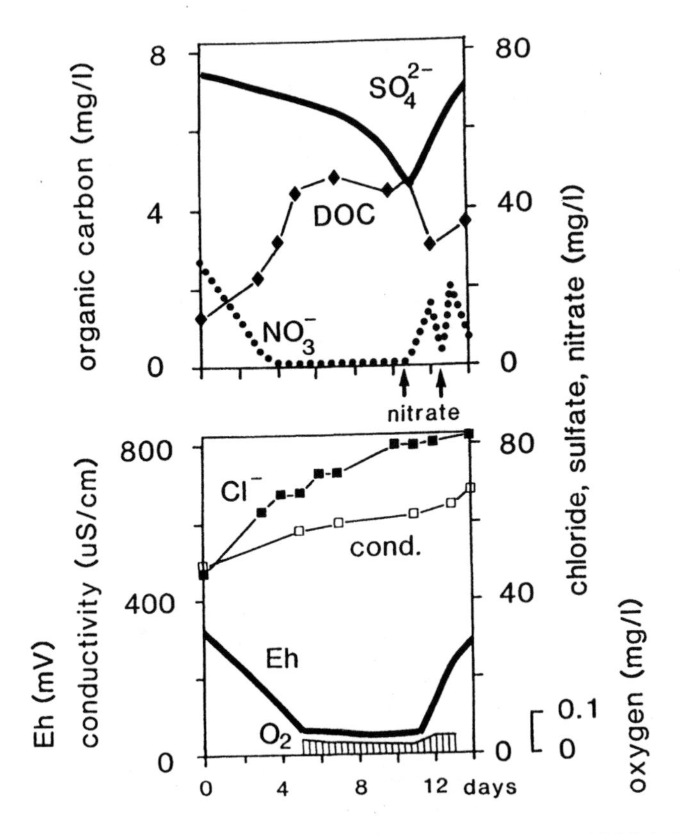

Figure 5. Closed circuit operation of an artificial wetland (pumping rate 150 L/hr); the evapotranspirated water loss is balanced by adding deionized water (14–31 August 1987). At day 11 and day 13 a pulse dosage of 40 g KNO_3 each is given.

up to 1.1 kPa. This confirms some suggestions[9,10] that gas transport in *Phragmites* is possible by mass flow. Therefore, an undefined aeration capacity was counterbalanced in our tests.

During closed circuit operation of the artificial wetland (Figure 5) we measured traces of H_2S (32 ppb on 25 August 1987 by gas chromatography; 9 ppb on 29 September 1987 by H_2S electrode). Soon after complete disappearance of nitrate at the beginning of the test, a steep decrease in sulfate concentration occurred.[11] After two pulses of KNO_3, sulfate ions reappeared to initial levels, indicating reoxidation of still-unidentified sulfur species.[12] Therefore, we need to understand the balance of oxygen supply by nitrates and internal carbon release. In Figure 6 we show six schematic situations of theoretical oxygen courses in wetland root horizons (with flow-through) understood as closed or open systems for oxygen diffusion. Assumptions include (1) resistance to gas

Theoretical oxygen profiles (root horizon)

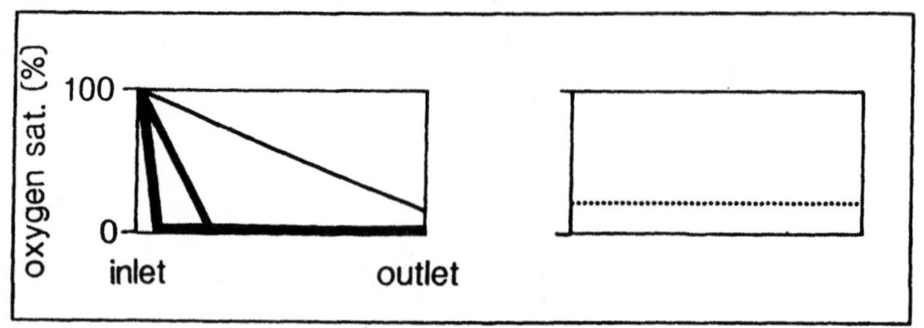

open system

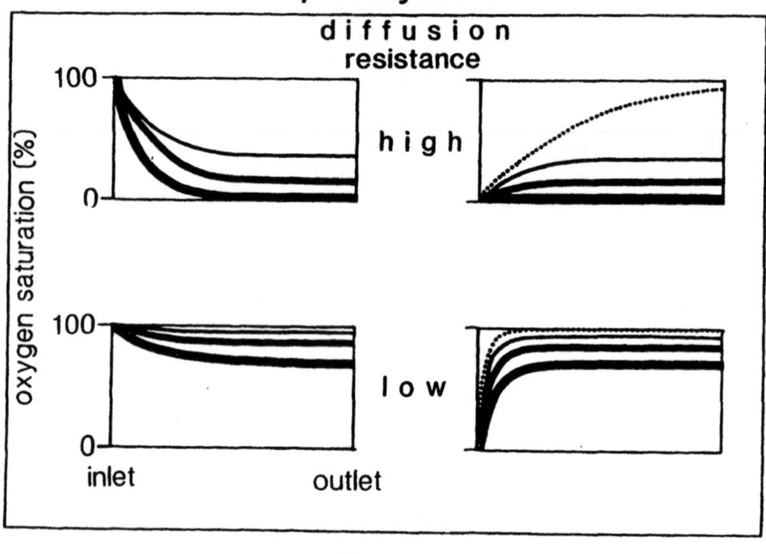

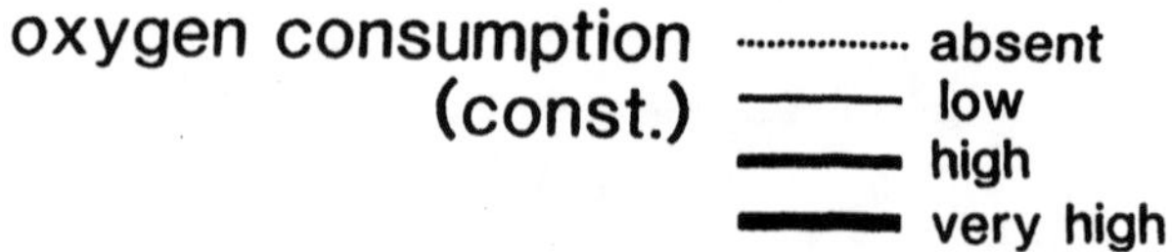

Figure 6. Theoretical oxygen profiles in the root horizon. Input: (left) oxygen saturated water (right) water without oxygen.

diffusion of vegetated beds is much lower than that of pure gravel-water systems; (2) diurnal fluctuations in oxygen downward transport are negligible (in fact, substantial diurnal internal fluctuations occur in the *Phragmites* gas conduction system[9]); (3) temperature and vegetation density are homogenous; and (4) flow-through and oxygen consumption are constant.

In comparison with theoretical situations, our vegetated units are open systems with a high diffusion-resistance in summer. During winter, they appear to be practically closed systems because water containing very low oxygen passing the root horizon never showed any substantial increase in oxygen, which could be attributed to passive "aeration."[13] In contrast, since planting *Phragmites* in 1983 we observed a more or less pronounced oxygen decrease ("deaeration"), dependent on temperature.[2] Nevertheless, our findings represent a particular case and should not be generalized. It has been well established that in natural conditions *Phragmites* can oxidize sediments.[7,14]

CONCLUSIONS

1. Nitrate elimination in artificial wetlands is possible throughout the whole year, when specific conditions are fulfilled (low oxygen concentration in the water, availability of organic carbon).
2. Even at low temperatures high denitrification rates have been verified.
3. In summer, denitrification occurs in the whole plant bed, when oxygen concentration in the inflowing water is low.
4. Our observations on oxygen conditions in the root zone did not show oxygen was added to through-flowing water.
5. The concept that particular plant species control oxygen levels in their surroundings in a predictable way has to be further examined, and if true, this concept of "controlled aeration" should assist in proper use of macrophytes for diverse ecological and technological purposes.

REFERENCES

1. Stengel, E. "Perspektiven der Nitrateliminierung in künstlichen Feuchtbiotopen," in *Grundlagen und Praxis naturnaher Klärverfahren*, Liebenburg Local Government and R. Kickuth, Eds. (Verlagsgruppe Witzenhausen, 1986), pp. 49–66.
2. Stengel, E., W. Carduck, and C. Jebsen. "Evidence for Denitrification in Artificial Wetlands," in *Aquatic Plants for Water Treatment and Resource Recovery,* K. R. Reddy and W. H. Smith, Eds. (Orlando, FL: Magnolia Publishing Inc., 1987), pp. 543–550.
3. Tiren, T. "Denitrification in Lakes 1.-A. Oxygen Consumption, Nitrate Consumption and Denitrification in Sediment-Water System," Department of Forest Ecology and Forest Soils, Research Notes 29:28–39 (1977).
4. Sutton, P. M., K. K. Murphy, and R. N. Dawson. "Low Temperature Biological Denitrification of Wastewater," *J. Water Poll. Control Fed.* 47:122–134 (1975).

5. Wium-Andersen, and J. M. Andersen. "The Influence of Vegetation on the Redox Profile of the Sediment of Grane Langso, a Danish Lobelia Lake," *Limnol. Oceanog.* 17:948-952 (1972).
6. Armstrong, W. "Aeration in Higher Plants," in *Adv. Bot. Res.* 7:226-332 (1979).
7. Anderson, F. O., and J. I. Hansen. "Nitrogen Cycling and Microbial Decomposition in Sediments with *Phragmites australis* (Poaceae)," *Hydrobiol. Bull.* 16(1):11-19 (1982).
8. Grosse, W., and P. Schröder. "Plant Life in Anaerobic Environments, the Physical Basis and Anatomical Requirements—Review," *Ber. Dtsch. Bot. Ges.* 99(3/4):367-381 (1987).
9. Brix, H. "Light-Dependent Variations in the Composition of the Internal Atmosphere of *Phragmites australis* (Cav.) Trin. Ex Steudel," *Aquat. Bot.* 30:319-329 (1988).
10. Armstrong, J., and W. Armstrong. "*Phragmites australis*—A Preliminary Study of Soil-Oxidizing Sites and Internal Gas Transport Pathways," *New Phytologist* 108:373-382 (1988).
11. Patrick, W. H., Jr. Critique of "Measurement and Prediction of Anaerobiosis in Soils," in *Nitrogen in the Environment*, D. R. Nielsen and J. G. McDonald, Eds. (England: Academic Press, Inc., 1978), pp. 449-457.
12. Gambrell, R. P., and W. J. Patrick, Jr. "Chemical and Microbiological Properties of Anaerobic Soils and Sediments," in *Plant Life in Anaerobic Environments,* D. D. Hook and R. M. M. Crawford, Eds. (Ann Arbor, MI: Ann Arbor Science Publishers, Inc., 1978), pp. 375-423.
13. Kickuth, R. "Das Wurzelraumverfahren in der Praxis," *Landschaft und Stadt* 16(3):145-153 (1984).
14. Lönnerblad, G. "Zur Kenntnis der Eisenausfällung der Pflanzen," *Botaniska Notiser. Lund* 402-412 (1933).

37e Nitrogen Removal from Freshwater Wetlands: Nitrification-Denitrification Coupling Potential

Yuch-Ping Hsieh and Charles Lynn Coultas

INTRODUCTION

Removal of nitrogen and other nutrients from secondary wastewater is costly and beyond the reach of many communities throughout the United States. Disposal of secondary wastewater into wetlands, especially palustrine wetlands, has been considered an economical and ecologically feasible alternative for many communities.[1,2] Wetlands remove nitrogen from wastewater by two pathways: (1) storage (assimilation or adsorption) in the system, or (2) removal through denitrification and ammonia volatilization. Leaching and runoff also remove nitrogen from a wetland, but they are not desirable pathways in wastewater treatments. Storage of nitrogen in a wetland is only a temporary measure, whether nitrogen is assimilated into biomass or adsorbed/fixed by soil, because sooner or later it will reach the system capacity.[3,4] Freshwater wetlands are usually acidic, and ammonia volatilization is not likely to be a significant pathway for nitrogen removal. Biological and/or chemical denitrification are key processes of long-term nitrogen removal from a freshwater wetland. Nitrification and denitrification, although two mutually exclusive processes, coexist in many ecosystems.[5] Existence of heterogeneous microsites or layers in soil systems permits coexistence of these two processes.[6] Nitrification-denitrification coupling potential of freshwater wetlands has not been studied in North Florida.

The objective of this study was to survey the nitrification-denitrification coupling potential (NDCP) of representative freshwater wetlands in North Florida and relate this potential to wetland soil properties. Four types of freshwater wetlands were chosen: savannah, cypress, black gum, and titi swamps.

Table 1. Description and Some Chemical Properties of the Wetland Soil Samples

No.	Type	Soil	OC (%)	TKN (%)	TKP (ppm)	C/N	N/P	pH
753	Savannah	Ochraquult	1.52	0.13	83	11.7	15.7	3.9
754	Savannah	Ochraquult	1.41	0.13	77	10.8	16.9	3.6
757	Savannah	Ochraquult	2.30	0.22	64	10.5	34.4	3.6
760	Savannah	Ochraquult	1.22	0.11	38	11.1	28.9	3.8
755	Cypress	Umbraquult	14.20	1.20	179	20.9	67.0	3.3
756	Cypress	Umbraquult	7.35	0.50	106	14.7	47.2	3.5
758	Cypress	Umbraquult	12.10	0.73	158	16.6	46.2	3.5
759	Cypress	Umbraquult	4.60	0.32	79	14.4	40.5	3.5
761	Gum	Medisaprist	35.10	2.09	438	16.8	47.7	3.4
762	Gum	Medisaprist	35.90	1.82	344	19.7	52.9	3.5
763	Gum	Medisaprist	35.20	2.06	350	17.1	58.9	3.7
764	Gum	Medisaprist	16.50	0.63	262	26.2	24.0	3.6
S	Titi	Haplaquod	34.90	1.80	385	19.4	46.8	3.1
N	Titi	Humaquept	34.80	1.60	296	21.8	54.1	3.6

Note: OC = organic carbon; TKN = total nitrogen; and TKP = total phosphorus.

THE SAMPLING SITES

Soils used in this study were taken from four types of freshwater wetlands in the Apalachicola National Forest of North Florida: savannah (sample Nos. 753, 754, 757, and 760), cypress (sample Nos. 755, 756, 758, and 759), black gum (sample Nos. 761, 762, 763, and 764), and titi (samples S and N). Descriptions and chemical properties of these wetland soils are given in Table 1.

Savannah soils had a dark gray sandy loam surface and a gray, highly mottled sandy loam subsoil and were classified as Ochraquults. Vegetation consisted of various grasses, sedges (*Carex* spp.), pitcher plants (*Sarracenea* spp.), St. Johnswort (*Hypericum*), and scattered pines (*Pinus* spp.). Water occurred at or near the soil surface for over six months of the year and organic matter content was low in comparison to other wetland soils. Cypress (*Taxodium distichum*) and black gum (*Nyssa biflona*) swamp sites were described previously.[7,8] Soils in the pond cypress site were classified as Umbraquults, with dark gray, sandy clay surface over a gray clay subsoil. Soils under cypress had moderate organic matter content. Soils in the black gum swamp were classified as Medisaprists.[8] Cypress and titi occasionally occurred in the black gum swamp. The titi swamp was dominated by species of *Cyrilla* and *Cliftonia* with scattered pine and cypress. Haplaquods and Humaquepts were found at this site. Haplaquods had a 10 cm accumulation of organic matter over a very dark gray, loamy sand mineral layer (the A1 horizon) followed by a gray E horizon and a thick, very dark gray spodic (Bh) horizon; Humaquepts occurred on the periphery. They consisted of 8 cm of organic matter over a dark gray, loamy sand overlying grayish brown and light grayish brown sand. Gum and titi swamps both had high soil organic matter content in the surface layer.

Savannahs had the shortest hydroperiod, cypress and black gum swamps the longest, and the titi swamp's was intermediate.

MATERIALS AND METHODS

Sample Preparation

Fresh surface soils from each site were taken to depths of 5 cm, placed in plastic bags, sealed, transported back to the laboratory, and stored in the refrigerator. Soil reaction (pH) was determined on field-moist soils, and organic carbon was determined using the Walkley-Black procedure.[9] Total nitrogen and phosphorus were determined by Technicon autoanalyzer procedures after Kjeldahl digestion.[10,11] Subsamples of each soil were taken for determination of moisture content. Some results of these analyses are listed in Table 1.

Disappearance Rates of Nitrate

Incubations Under Argon

Approximately 10 g of moist soil was placed in a 100-mL centrifuge tube. Tubes were closed with two-outlet stoppers and flushed with argon for 30 seconds, outlets closed with a clamp, and tubes were preincubated at 32°C for one week prior to treatment. After preincubation, 3 mL of 150 ppm KNO_3-N solution (prepared with oxygen-free distilled water) was added to tubes and incubated under argon at 32°C for one to seven days. Fifty mL of calcium sulfate solution (2 g/L) was used to extract nitrate from each tube. Nitrate concentration in the supernatant solution was analyzed using an ion-selective nitrate electrode (Orion model 93–07). Initial nitrate concentration of each soil before each incubation was similarly determined. Incubation periods were 1, 2, 4, and 7 days for each soil, and all treatments were duplicated.

Incubations Under Air

Procedures were the same as those for incubations under argon except the atmosphere was air instead of argon.

Disappearance Rates of Ammonium

Limed Incubation

Two soils from each type of four wetlands were selected for this experiment. To 50 g of each moist soil, 1.75 g of reagent grade calcium carbonate was added and preincubated at 32°C for one week. Before and after preincubation, pH of each soil was checked with a glass electrode. Ten g of preincubated moist soil was placed in a 100-mL centrifuge tube, and 5 mL of either distilled water or 200 ppm (v/v) N-Serve 24 (nitrification inhibitor, Dow Chemical Co.) solution was added to the tube. Tubes were incubated at 32°C for two days and 50 mL of 1 N KCl was used to extract the ammonium. Ammonium concentration of extract

Table 2. Rates of Ammonification and Nitrification of Limed and Unlimed Wetland Soils

	Ammonification		Nitrification	
	Unlimed	Limed	Unlimed	Limed
Sample No.	μg N/mL/day(pH)	μg N/mL/day(pH)	μg N/mL/day(pH)	μg N/mL/day/(pH)
Savannah				
753	5.63 (3.9)	4.40 (6.3)	4.15 (3.9)	1.73 (6.3)
754	2.66 (3.6)	4.90 (6.5)	3.14 (3.6)	2.3 (6.5)
Cypress				
755	0.61 (3.3)	0.65 (5.6)	0.50 (3.3)	1.02 (5.6)
756	1.18 (3.5)	1.10 (5.9)	0.46 (3.5)	0.00 (5.9)
Gum				
761	0.61 (3.4)	1.00 (5.2)	0.23 (3.4)	0.74 (5.2)
762	0.51 (3.5)	0.66 (5.7)	0.55 (3.5)	0.00 (5.7)
Titi				
S	1.23 (3.6)	1.82 (4.8)	0.86 (3.6)	0.00 (4.8)
N	1.29 (3.1)	1.59 (4.6)	0.30 (3.1)	0.00 (4.6)

solutions was determined by autoanalyzer.[12] Initial ammonium concentrations of soils before incubation were checked and all treatments were duplicated.

Unlimed Incubations

Procedures and soil samples were the same as limed incubation experiments except no calcium carbonate was added to the soil samples.

Due to the variable bulk densities of soils, all rates of nitrogen transformation are reported in volume of soil instead of a weight of soil basis. Bulk density of soils is calculated based on: bulk density = 2.6/(% organic carbon).[13]

Rates of Nitrification and Ammonification

Rates of nitrification were estimated by the difference in ammonium contents between the N-Serve and non-N-Serve treated samples during the two-day incubation, assuming that N-Serve totally inhibited conversion of ammonium to nitrate.[14] Rates of ammonification were estimated by the difference in ammonium contents between the initial and final N-Serve treated sample of the incubation. Denitrification rates of wetlands under aerobic incubations were estimated by disappearance rate of nitrate plus nitrification rate.

RESULTS

Rates of Nitrification and Ammonification

Estimated nitrification and ammonification rates of wetland soils are listed in Table 2. Savannahs had the highest nitrification and ammonification rates among four types of wetlands on a volume basis. On a weight basis, savannah soils had the lowest rate of nitrification due to higher bulk density. However, expression of nitrification rate on a weight basis could be misleading because we commonly consider wetlands in terms of area instead of weight of soil. Titi,

Table 3. Effect of Liming on the Ammonium Level of the Soil Samples During a Seven-Day Incubation

	Before Liming		After Liming	
Sample No.	pH	μg N/ml	pH	μg N/mL
Savannah				
753	3.9	12.7	6.3	47.5
754	3.6	8.3	6.5	43.0
Cypress				
755	3.3	1.4	5.6	10.7
756	3.5	2.4	5.9	28.9
Gum				
761	3.4	2.5	5.2	6.7
762	3.5	1.9	5.7	6.7
Titi				
S	3.1	6.1	4.6	12.3
N	3.6	10.6	4.8	7.1

cypress, and black gum swamp soils had similar rates of nitrification and ammonification despite differences in organic carbon content. Increasing soil pH by liming hardly changed the rates of ammonification after seven-day preincubations.

Liming, however, significantly reduced the rate of nitrification of some wetland soils (Table 2). The most drastic effect of liming was the increase of ammonium concentration in soils, except the Humaquept sample from titi swamp. Generally, ammonium concentration increased 2- to 10-fold after liming, which increased the soil reaction 1.2–2.3 pH units (Table 3). Ammonium concentration of the Humaquept soil sample decreased 67% after liming.

Rates of Denitrification

Denitrification potential of wetlands was estimated by the disappearance rate of nitrate during anaerobic incubations. Denitrification during anaerobic incubations followed zero-order kinetics as previously observed.[15,16] The denitrification rates varied from 2.7 to 17.4 μg/mL/day. Denitrification rates varied as much within one type of wetland as among four types of wetlands (Table 4). Cypress swamps had more uniform denitrification rates than other wetlands. Under aerobic conditions, denitrification rates were slightly lower than under anaerobic conditions (Table 5). Denitrification rate did not correlate with organic carbon content of these soils. Denitrification rate correlated poorly with C/N ratio due to the savannah data set. Excluding the savannah data set, due to high mineral content of savannah soils, denitrification rate correlated significantly ($r = 0.774$) with C/N ratio. Ammonification and nitrification rates were inversely correlated with C/N ratio (Table 5). Ammonification rate was also positively and highly correlated with nitrification rate.

Table 4. Denitrification Rates of the Soil Samples Incubated Under Aerobic and Anaerobic Incubation

Sample No.	Aerobic Incubation μg N/mL/day	Anaerobic Incubation μg N/mL/day
Savannah		
753	6.57	6.20
754	4.95	8.30
757	13.60[a]	17.40
760	10.70[a]	12.50
Cypress		
755	8.60	12.10
756	14.06	16.10
758	6.83[a]	13.40
759	12.80	15.90
Gum		
761	6.71	13.70
762	10.82	10.10
763	10.00[a]	6.90
764	5.30[a]	5.30
Titi		
S	3.20	9.40
N	2.20	2.70

[a]Not corrected for the nitrification rate.

Magnitude of nitrogen transformation rates follows the order:

denitrification > ammonification > nitrification

Ammonification and nitrification rates were in the same order, while denitrification rates were one order higher in most cases, except in savannah soils and one sample of titi soils. Nitrogen transformations in these wetlands were limited by

Table 5. Correlation Coefficient Matrix of Some Variables Investigated

		Variables							
		(1)	(2)	(3)	(4)	(5)	(6)	(7)	(8)
Org. C	(1)	1	–0.146	<u>0.621</u>	–0.299	–0.535	–0.523	<u>–0.609</u>	<u>–0.703</u>
Denitri A	(2)	–0.146	1	–0.097	<u>0.755</u>	–0.447	–0.462	–0.388	–0.172
C/N	(3)	<u>0.621</u>	–0.097	1	–0.144	<u>–0.710</u>	<u>–0.804</u>	<u>–0.803</u>	<u>–0.736</u>
Denitri B	(4)	–0.299	<u>0.755</u>	–0.144	1	–0.201	–0.382	–0.200	–0.196
Ammo.	(5)	–0.535	–0.447	<u>–0.710</u>	–0.201	1	<u>0.842</u>	<u>0.936</u>	<u>0.641</u>
Ammo. L	(6)	–0.523	–0.462	<u>–0.804</u>	–0.382	<u>0.842</u>	1	<u>0.939</u>	<u>0.814</u>
Nitri.	(7)	<u>–0.609</u>	–0.388	<u>–0.803</u>	–0.200	<u>0.936</u>	<u>0.939</u>	1	<u>0.814</u>
Nitri. L	(8)	<u>–0.703</u>	–0.172	<u>–0.736</u>	–0.196	<u>0.641</u>	<u>0.814</u>	<u>0.814</u>	1

Note 1: Org. C = organic carbon; Denitri A = anaerobic denitrification rate; Denitri B = aerobic denitrification rate; Ammo. = unlimed ammonification rate; Ammo. L = limed ammonification rate; Nitri. = unlimed nitrification rate; Nitri. L = limed nitrification rate.
Note 2: Those coefficients underlined are significant at 5% probability level.

nitrification rate, the reason why little or no nitrate was found in these wetland soils.

DISCUSSION

These results demonstrated that nitrification-denitrification coupling existed in these acidic freshwater wetlands. Nitrification rate of savannah soils, on a volume basis, is much higher than that of other wetlands, indicating that mineral constituents play an important role in nitrification. The pH values of these wetland soils were at the lower limit for nitrification.[17] Liming actually decreased the nitrification rate in most cases after seven-day preincubation, suggesting that factors other than pH might control nitrification in these soils. The drastic increase in ammonium level after liming occurred during the seven-day preincubations rather than during the two-day incubations. Liming might cause reduction of soil microbial population and subsequently increase net mineralization during the preincubation. The nitrification rate reduction by liming also contributes to the higher level of ammonium. Further investigation is needed to clarify the mechanisms of liming effects on nitrogen transformation in these wetlands.

Surprisingly, soil organic carbon content had little bearing on nitrification and denitrification rates of these soils. The C/N ratio was better correlated with nitrification and denitrification rates. Nitrification rate is, in effect, the limiting factor for nitrogen removal through denitrification in wetlands. The nitrification-denitrification coupling potential of freshwater wetlands has significant impact on the nitrogen removal from wetlands. Removal of 30–86 g NH_4-N/ha/day through nitrification-denitrification processes can be achieved by the top 1 cm soil of a titi swamp. This is equivalent to removal of 20 ppm NH_4-N from 1500–4300 L/ha/day of wastewater. Similarly, the savannah wetland can remove 20 ppm NH_4-N from 1700–20,750 L/ha/day of wastewater through nitrification-denitrification. These wetlands are capable of removing 20 ppm NO_3-N from 11,500–87,000 L/ha/day of wastewater through denitrification.

ACKNOWLEDGMENTS

We wish to thank our colleague, Dr. C. H. Yang, for assistance in autoanalyzer analysis. This study was partially supported by a USDA/CSRS research grant through the Agricultural Research Program of the Florida A&M University.

REFERENCES

1. Sloey, W. E., F. L. Spangler, and C. W. Fetler. "Management of Freshwater Wetlands for Nutrient Assimilation," in *Freshwater Wetlands,* R. E. Good, D. F. Whigham, and R. L. Simpson, Eds. (New York: Academic Press, Inc., 1978), pp. 321–340.

2. Tilton, D. L., and R. H. Kadlec. "The Utilization of a Freshwater Wetland for Nutrient Removal from Secondarily Treated Wastewater Effluent," *J. Environ. Qual.* 8:328–334 (1979).
3. Peverly, J. H. "Stream Transport of Nutrients Through a Wetland," *J. Environ. Qual.* 11:469–472 (1982).
4. Simpson, R. L., R. E. Good, R. Walker, and B. R. Frasco. "The Role of Delaware River Freshwater Tidal Wetlands in the Retention of Nutrients and Heavy Metals," *J. Environ. Qual.* 12:41–48 (1983).
5. Jenkins, M. C., and W. M. Kemp. "The Coupling of Nitrification and Denitrification in Two Estuarine Sediments," *Limnol. Oceanog.* 29:609–619 (1984).
6. Patrick, W. H., and K. R. Reddy. "Nitrification-Denitrification in Flooded Soils and Water Bottoms: Dependence on Oxygen Supply and Ammonium Diffusion," *J. Environ. Qual.* 5:469–472 (1976).
7. Coultas, C. L., and M. J. Duever. "Soils of Cypress Swamps," in *Cypress Swamps,* H. T. Odum and K. C. Ewell, Eds. (Gainesville: University of Florida Press, 1984).
8. Coultas, C. L. "Soils of the Apalachicola National Forest Wetlands, Part 2, Cypress and Gun Swamps," *Soil and Crop Sci. Soc. of Fla. Proc.* 37:154–159 (1977).
9. Jackson, M. L. *Soil Chemical Analysis* (Englewood Cliffs, NJ: Prentice Hall, Inc., 1958).
10. "Technicon AutoAnalyzer II, Method No. 696–82W. Total Kjeldahl Nitrogen," Technicon Industrial Systems, Tarrytown, NY (1979).
11. "Technicon AutoAnalyzer II, Method No. 696–82W. Total Phosphorus," Technicon Industrial Systems, Tarrytown, NY (1979).
12. "Technicon AutoAnalyzer II, Method No. 696–82W. Ammonia in Water and Wastewater," Technicon Industrial Systems, Tarrytown, NY (1979).
13. Gosselink, J. G., and R. Hatton. "Relationship of Organic Carbon and Mineral Content to Bulk Density in Louisiana Marsh Soils," *Soil Sci.* 137:177–180 (1984).
14. Henriksen, K. "Measurement of In Situ Rates of Nitrification in Sediment," *Microb. Ecology* 6:329–337 (1980).
15. Focht, D. D. "The Effect of Temperature, pH and Aeration on the Production of Nitrous Oxide and Gaseous Nitrogen—a Zero-Order Kinetic Model," *Soil Sci.* 118:173–179 (1974).
16. Focht, D. D., and W. Verstraete. "Biochemical Ecology of Nitrification and Denitrification," in *Advances in Microbial Ecology,* M. Alexander, Ed. (New York: Plenum Press, 1977), pp. 135–214.
17. Schmidt, E. L. "Nitrification in Soil," in *Nitrogen in Agricultural Soils*, F. J. Stevenson, Ed., Agronomy Monograph No. 22 (1982), pp. 253–288.

CHAPTER 38

Efficiencies of Substrates, Vegetation, Water Levels and Microbial Populations

38a Relative Radial Oxygen Loss in Five Wetland Plants

Susan Copeland Michaud and Curtis J. Richardson

INTRODUCTION

Drainage from coal mining and washing facilities frequently has a pH as low as 2.2, high concentrations of dissolved metals such as iron (Fe) and manganese (Mn), and high suspended solid concentrations.[1,2] Precipitation that falls on coal piles and slurries leaches ions from the exposed rock.[3] These pollutants contaminate streams where hydrogen and metal ions reduce water quality and inhibit plant and animal life.[4,5]

Recently, it has been demonstrated that constructed wetlands can be used to improve water quality.[2,6] The combination of chemical, hydrological, and biological processes in wetlands are able to precipitate Fe, Mn, and other metal ions, to reduce suspended solids and moderate pH.[7,8] Constructed wetlands can be built to mimic biogeochemical characteristics of natural wetlands that perform these functions.

Oxygen (O_2) diffusion to root tips is one of the physiological characteristics that permit wetland species to exist in flooded conditions.[9,10] Diffused O_2 not only supplies the roots but can oxidize the surrounding soil; this process is termed radial oxygen loss (ROL).[11,12] Increasing evidence suggests that O_2 transport provides a unique environment that can oxidize phytotoxic compounds such as ferrous iron.[12-14] This characteristic could be used to mitigate the concentration of toxic compounds in mining-related effluent. Theoretically, plants with the largest oxygenated rhizosphere and largest population of metal-oxidizing microbes would maximize the wetland's potential to remove toxic metals from the water column by oxidation. This research investigates the ability of five wetland plant species to diffuse O_2 into wetland soils.

Objectives of this study were (1) to develop a method to test for the presence of O_2 diffusion from plants' roots, (2) to determine species differences in O_2 loss to sediments and (3) to recommend plant species to use in constructed wetlands to maximize O_2 concentrations in the root zone.

MATERIALS AND METHODS

Plant Selection

Plants were obtained from a constructed wetland, TVA's Impoundment One, which receives effluent from a slurry pond. Total Fe concentration averaged 30 mg/L, Mn often exceeded 10 mg/L, dissolved oxygen (DO) was less than 2 mg/L, pH was 6.0, and suspended solids exceeded 98 mg/L.[2] Cattail (*Typha latifolia*), burreed (*Sparganium americanum*), spikerush (*Eleocharis quadrangulata*), woolgrass (*Scirpus cyperinus*), and rush (*Juncus effusus*), perennial emergent wetland plants, were selected because they were well established, healthy, and reproducing. Mature plants were collected during July and August 1987 from the third and fourth ponds of Impoundment One. *T. latifolia* and *S. americanum* grow in shallow to deep water (0.30 to 1.20 m). *E. quadrangulata* and *J. effusus* grow in shallower water (0.15 to 0.30 m), while *S. cyperinus* grows in the shallowest water (0.05 to 0.30 m).

All five species were tested with 20 replicates each. A control without plants tested the differences in DO due to O_2 diffusion down aerenchymal tissue and atmospheric diffusion. Dead plants and cut-off plants covered with paraffin to seal aerenchymal tissue were also tested to determine if O_2 presence was due to diffusion or plant enzymes.

Laboratory Procedure

To detect oxygen, we used an indigo carmine dye technique developed by Armstrong[15] for detecting ROL in some bog plant species. We deoxygenated deionized water with nitrogen gas. One-liter bottles were filled with 500 mL of deoxygenated water. Hydrogen sulfide (H_2S) saturated water (5 mL) was added to the deoxygenated solution to reduce the solution Eh below -250 mV. Five milliliters of 0.28% (weight/volume) indigo carmine dye was added to the deoxygenated solution.

The test plants' roots were carefully cleaned to remove sediment and decaying plant matter and placed in the reduced solution. Mineral oil was added to the surface to prevent O_2 from diffusing into the solution. At 6, 12, and 24 hours (h), samples were removed for spectrophotometric analysis to measure the percentage of light transmitted through the samples. Readings at 6, 12, and 24 h in conjunction with the plants' dry mass were used to determine differences in O_2 diffusion into solution. To reduce variation due to plant size, we developed a weighted index by dividing the spectrophotometer reading by the plants' total mass (% transmittance/total plant biomass = ROL index).

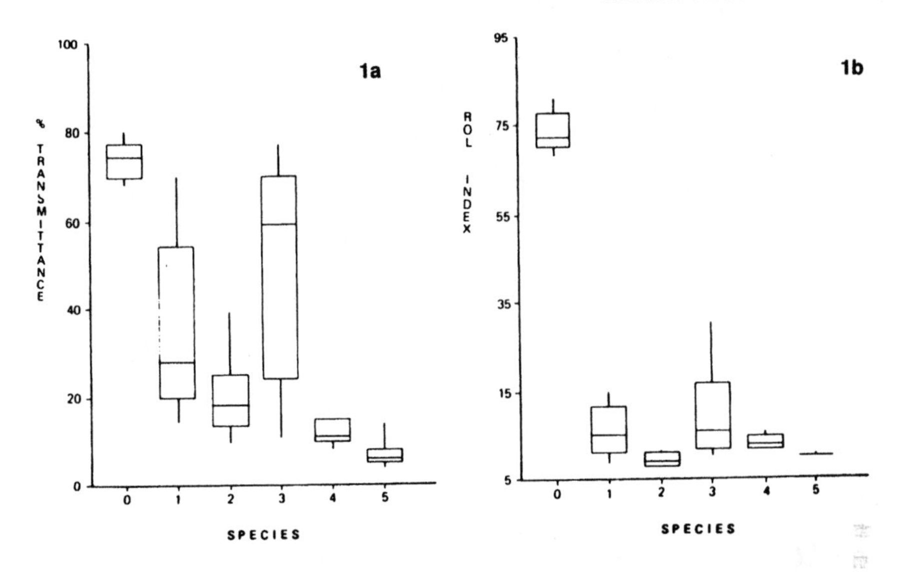

Figure 1. (1a) is a box plot of the percent light transmitted through a solution containing an oxygen-indicating dye for the control and five wetland plants tested for ROL after 12 h. Lower numbers on the y axis indicate a more oxidized solution. (1b) uses the ROL index, which divides the percent transmittance by dry weight after 12 h. 0 = control; 1 = *E. quadrangulata*; 2 = *J. effusus*; 3 = *S. cyperinus*; 4 = *S. americanum;* 5 = *T. latifolia*.

RESULTS

Transmittance taken during the experiment ranged from 80% to 7%. Controls had readings greater than 70% and only a small percentage of the dye had been oxidized, but complete dye oxidation with H_2O_2 permitted 7% light transmittance.

Figure 1a is a box plot showing relationships between species after 12 h. There is clearly a difference between controls and samples with plants. *S. cyperinus,* although statistically different from the control, did not diffuse as much O_2 into the sediments as other species tested.

Much of the variability in Figure 1a was removed by using mass as a scaling factor. Figure 1b shows relationships when transmittance readings are divided by the individual's total dry weight. Variation in *S. cyperinus* values was substantially greater than in other species.

Differences in transmittance for each plant and controls demonstrates an input of O_2 by the plants (Table 1). Experiments with cut and waxed plants had significantly higher readings than those for live, intact plants. However, waxed plant treatments did have more O_2 present than controls. This is likely due to some O_2 storage in the roots and subsequent diffusion into solution. Dead plants tested were able to transmit a significant amount of O_2 into solution but were not as efficient as live plants. For example, the mean value for dead *J. effusus* plants at six hours was 53.5; live *J. effusus* had a mean reading of 30.5

Table 1. Mean ROL from Five Wetland Plant Species Collected from a Constructed Wetland used for Coal Mine Drainage Abatement Tested with Indigo Carmine Dye as an Oxygen Indicator

		Percent Light Transmitted		
	n	6h	12h	24h
Control	**15**	**73.8 (3.3)**	**73.2 (3.8)**	**71.1 (6.7)**
Live				
E. quadrangulata	20	53.9 (15.8)	35.4 (19.2)	17.8 (2.7)
J. effusus	20	30.5 (18.9)	21.2 (10.8)	17.2 (2.2)
S. cyperinus	20	56.7 (20.3)	48.8 (23.6)	38.1 (20.8)
S. americanum	20	34.8 (20.3)	16.0 (14.6)	15.2 (4.7)
T. latifolia	20	43.2 (5.0)	27.2 (3.8)	18.1 (0.9)
Waxed				
E. quadrangulata	2	73.5	72.0	61.5
J. effusus	2	65.0	68.5	47.5
S. cyperinus	2	66.0	60.5	56.0
S. americanum	3	65.0	59.7	45.3
T. latifolia	2	61.5	58.5	45.5
Dead				
E. quadrangulata	2	67.0	67.0	30.0
J. effusus	2	53.5	31.5	23.5
S. cyperinus	—	—	—	—
S. americanum	3	36.0	31.0	21.0
T. latifolia	2	32.5	17.0	20.5
Live[a]				
E. quadrangulata	20	9.3 (4.8)	6.3 (3.9)	3.2 (1.4)
J. effusus	20	4.0 (2.1)	2.6 (4.04)	1.7 (1.4)
S. cyperinus	20	12.7 (7.9)	11.2 (12.2)	8.77 (10.1)
S. americanum	20	6.9 (6.8)	3.0 (3.5)	2.7 (1.7)
T. latifolia	20	0.04 (0.02)	0.02 (0.005)	0.01 (0.007)

Note: Lower percent transmittance corresponds to higher ROL by the plants. Live, waxed, and dead individuals from each species were tested. (X.X = S.D.)
[a]Transmittance/total plant dry weight; S.D. in parenthesis.

for the same time. Sample size for dead plants was small, and a definitive statement that dead plants have less ROL was not possible.

Since preliminary tests revealed these data are not normally distributed, we used the nonparametric Kruskal-Wallis One-Way Analysis to rank each observation.[16] The test statistic for readings at 6 h was 49.68 at a significance of 4.22^{-10} ($p = 0.05$). The high test statistic and small significance level indicate that the 6-h readings differ significantly among species. The values for 12 and 24 h were significant as well (Table 2). *T. latifolia* transferred more O_2 to solution than any other species at all times. Pairwise analysis for unequal sample sizes[17] revealed that *J. effusus* and *S. americanum* were more efficient than *S. cyperinus* at 6 h. *S. americanum* transferred more O_2 than *S. cyperinus* at 12 and 24 h.

The Kruskal-Wallis analysis for species differences corrected for differences in biomass and showed significant species differences (Table 2). *T. latifolia*

Table 2. Percent Light Transmitted for Radial Oxygen Loss in Five Wetland Plant Species Ranked by the Kruskal-Wallis Method

	Average Rankings, Unadjusted for Biomass[a]		
	6 h	12 h	24 h
E. quadrangulata	69.35	59.33	55.65
J. effusus	44.40	44.44	53.79
S. cyperinus	74.05	65.41	75.28
S. americanum	47.90	31.95	39.88
T. latifolia	16.8	12.98	11.48
Test statistic	49.68	54.22	62.10
Significance level	4.22^{-10}	4.74^{-11}	1.05^{-12}

	Average Rankings, Adjusted for Biomass[b]		
	6 h	12 h	24 h
E. quadrangulata	72.00	61.89	63.23
J. effusus	42.45	39.74	40.21
S. cyperinus	70.25	61.75	69.75
S. americanum	57.50	45.73	54.35
T. latifolia	10.50	10.50	10.50
Test statistic	61.05	56.19	59.40
Significance level	1.74^{-12}	1.82^{-11}	3.87^{-12}

[a]*S. cyperinus* ranked highest for all three periods and *T. latifolia* ranked lowest. Higher ranks indicate less oxygen was transferred to solution than lower ranks.
[b]Percent light transmittance divided by the total dry weight. *E. quadrangulata* was highest at 6 and 12 h, while *S. cyperinus* was highest at 24 h. *T. latifolia* ranked lowest at all three times.

transferred the most O_2 at all three sampling times. *J. effusus* was better at transferring O_2 than *S. cyperinus* at 6 and 24 h.

DISCUSSION

The role of dissolved oxygen in wetlands used for treating mining-related effluent is important in improving water quality. Oxygen is required for the removal of dissolved metals, to support aerobic microorganisms, and to reduce BOD_5. Results of this experiment indicate that O_2 concentration in the soil-water matrix can vary with plant species. By planting and managing for specific plants to maximize O_2 concentration, toxic metals can be oxidized more quickly and efficiently.

Our ROL experiments have shown the following trends on a per unit biomass basis:

T. latifolia > *J. effusus* > *S. americanum* > *E. quadrangulata* > *S. cyperinus*

However, individual growth characteristics, such as stand density, need to be considered when comparing ROL performance as a management tool.

J. effusus performed better in terms of ROL than *S. cyperinus* per unit biomass. Because both species grow in shallow water, planting *J. effusus* instead of *S. cyperinus* will result in greater O_2 contribution to the substrate. *T.*

latifolia and *S. americanum* both grow in deep water. *T. latifolia* consistently contributed more O_2, but *S. americanum* grows in much denser stands. Oxygen transfer per unit of substrate must be investigated before recommending which to plant. Because *E. quadrangulata* did not transfer O_2 as well as *J. effusus, S. americanum,* or *T. latifolia,* it might be advantageous to minimize the area of depth where *E. quadrangulata* grows (0.15–0.60 m).

Oxygen transport varied significantly in the emergent wetland macrophytes evaluated, suggesting that plant species can influence the amount of O_2 in the sediments. The importance of O_2 in the water column and sediments and the poor understanding of plant-oxygen transport mechanisms warrant further research into oxygen transfer per unit area and effects of oxygen transport on BOD_5 and metal oxidation. This study points toward the continued research into constructed wetlands and their potential contribution in the remediation of contaminated water.

REFERENCES

1. Mitsch, W. J., M. A. Cardamone, J. R. Taylor, and P. L. Hill, Jr. "Wetlands and Water Quality Management in the Eastern Interior Coal Basin," in *Wetlands and Water Management on Mined Lands,* R. P. Brooks, Ed. (University Park, PA: Pennsylvania State University, 1985).
2. Brodie, G. A., D. A. Hammer, and D. A. Tomljanovich. "Man-Made Wetlands for Acid Mine Drainage Control," Eighth Annual National Abandoned Mine Lands Conference, Billings, MT (1986).
3. Forstner, U., and G. Wittman. *Metal Pollution in the Aquatic Environment* (New York: Springer-Verlag, 1981).
4. Parsons, J. W. "A Biological Approach to the Study and Control of Acid Mine Pollution," *J. Tn. Acad. Sci.* 24(4):304–310.
5. Riley, C. V. "The Ecology of Water Areas Associated with Coal Strip-Mined Lands in Ohio," *Ohio J. Sci.* 60(2):196–121 (1960).
6. Richardson, C. J., and J. A. Davis. "Natural and Artificial Wetland Ecosystems: Ecological Opportunities and Limitation," in *Aquatic Plants for Water Treatment and Resource Recovery,* K. R. Reddy and W. H. Smith, Eds. (Orlando, FL: Magnolia Publishing Inc., 1987), pp. 819–854.
7. Stone, R. W. "The Presence of Iron- and Manganese-Oxidizing Bacteria in Natural and Simulated Bogs," in *Treatment of Mine Drainage by Wetlands,* J. E. Burris, Ed. (University Park, PA: Pennsylvania State University, 1984), pp. 30–36.
8. Barlett, R. J. "Iron Oxidation Proximate to Plant Roots," *Soil Sci.* 92:6 (1961).
9. Armstrong, W. "Aeration in Higher Plants," *Adv. Bot. Res.* 7:225–332 (1980).
10. Sculthorpe, C. D. *The Biology of Aquatic Vascular Plants* (London: Edward Arnold Publishers Ltd., 1967), pp. 60–65.
11. Teal, J. M., and J. W. Kanwisher. "Gas Transport in the Marsh Grass *Spartina alterniflora,*" *J. Exp. Bot.* 17:355–361 (1966).
12. Armstrong, W. "Root Aeration in the Wetland Condition," in *Plant Life in Anaerobic Environments,* D. D. Hook and R. M. M. Crawford, Eds. (Ann Arbor, MI: Ann Arbor Science Publishers, Inc., 1978), pp. 264–297.

13. Sanderson, P. L., and W. Armstrong. "Soil Waterlogging, Root Rot and Conifer Windthrow: Oxygen Deficiency or Phytotoxicity?" *Plant and Soil* 49:185–190 (1978).
14. Moorhead, K. K., and K. R. Reddy. "Oxygen Transport Through Selected Aquatic Macrophytes," *J. Environ. Qual.* 17:138–142 (1988).
15. Armstrong, W. "The Oxidizing Activity on Roots in Waterlogged Soils," *Physiol. Plant.* 20:920–926 (1967).
16. STATGRAPHICS Statistical Graphics System, Rockville, MD (1986).
17. Hollander, M., and D. A. Wolfe. *Nonparametric Statistical Methods* (New York: John Wiley & Sons, Inc., 1973), p. 125.

38b Potential Importance of Sulfate Reduction Processes in Wetlands Constructed to Treat Mine Drainage

Robert S. Hedin, Richard Hammack, and David Hyman

Dissimilatory sulfate reduction is a microbial process that commonly occurs in anoxic aquatic environments.[1] By-products of the process are hydrogen sulfide and carbonate alkalinity. Sulfate reduction in wetlands constructed to treat acid mine drainage (AMD) is desirable because hydrogen sulfide readily reacts with dissolved metals, precipitating them as sulfides, and alkalinity neutralizes drainage acidity.

The current importance of sulfate reduction in natural or constructed wetlands that receive AMD is uncertain because of contradictory findings (Table 1). Wieder and Lang[2] analyzed peat samples from Tub Run Bog, a natural *Sphagnum* wetland that had received AMD for 30 years, and found little accumulation of sulfur compounds. Wieder, Lang, and Whitehouse[3] analyzed peat samples from a wetland constructed with *Sphagnum* peat which received AMD for eight months and found minimal accumulation of sulfur. These results contrast with those for substrate samples collected from a wetland

Table 1. Sulfur Content of Organic Substrates Exposed to Acid Mine Drainage

	Total Sulfur	$S^0 + S^{2-}$ [a]
Tub Run Bog[b]		
Sphagnum peat	0.35%[c]	0.06%
(30 years of exposure to AMD)		
Constructed peat wetland[d]		
Unexposed to AMD	0.28	0.01
10 months of AMD	0.51	0.05
Constructed compost wetland		
Unexposed to AMD	1.14[e]	<0.01
6 months of AMD	2.91	>2.50

[a]Elemental and sulfide sulfur compounds.
[b]Wieder and Lang.[2]
[c]All values in percent of dry weight.
[d]Wieder et al.[3]
[e]50% organic sulfur plus 0.64% sulfate sulfur.

constructed with limestone, mushroom compost, and *Typha*. After six months of exposure to AMD, the compost had accumulated more than 2.5% (dry weight) reduced and elemental sulfur. Compared to inflow samples, substrate pore water samples had pH 3–4 units higher, 50–99% less dissolved iron, and 25–50% less sulfate.[4] Dissolved hydrogen sulfide concentrations in the substrate pore water were 1–2 mg/L.

The Bureau of Mines is currently involved in several research projects that are exploring the utilization of dissimilatory sulfate reduction and metal sulfide formation processes for the treatment of coal mine drainage. In this chapter we discuss factors which affect the importance of sulfide formation in aquatic systems and we evaluate the theoretical potential of the process in wetlands constructed to treat AMD.

DISSIMILATORY SULFATE REDUCTION AND NET SULFIDE FORMATION

The importance of dissimilatory sulfate reduction (DSR) in affecting the chemistry of flow-through water is related to the difference between rates of formation and destruction of sulfides. In many systems, sulfide formation rates are slowed by factors that limit microbial activity. In other systems, formation rates are high, but periodic destruction of the sulfides significantly reduces the net rate of sulfide accumulation. For the process to be effective in constructed wetlands receiving AMD, factors which promote DSR and sulfide formation must be maximized, and destructive factors minimized.

SULFIDE FORMATION

Dissimilatory sulfate reduction is accomplished by heterotrophic anaerobic bacteria that, in the absence of oxygen, decompose simple organic compounds using sulfate as a terminal electron acceptor. One mole of sulfate is reduced to hydrogen sulfide for every 2 moles of carbon oxidized. The reaction is:

$$SO_4^{2-} + 2CH_2O^* \rightarrow H_2S + 2HCO_3^-$$

where CH_2O^* represents a simple organic molecule such as acetate. Depending on the chemical environment, hydrogen sulfide is released as a gas, ionizes to HS^- and S^{2-}, or precipitates as a polysulfide, elemental sulfur, or iron sulfides (FeS, FeS_2).

DSR bacteria are only active under anaerobic conditions. In natural wetlands, anoxic conditions and DSR commonly occur within several cm of the water/substrate interface.[5] Similar conditions exist in wetlands constructed with compost substrates.[4]

The most common factors limiting DSR in anoxic environments are availability of suitable organic matter and dissolved sulfate. In marine systems organic matter often limits DSR, whereas in freshwater systems sulfate is the more common limiting factor.[6] Sulfate concentrations of AMD are typically greater than 500 mg/L, which is more than a magnitude greater than the 30 mg/L that limits DSR activity.[7] Most constructed wetlands contain 0.3–0.5 m of a rich organic substrate that should provide the DSR bacteria with an energy source for many years (see subsequent calculations).

DSR is inhibited by low pH. The most common sulfate reducing bacterium, *Desulfovibrio*, is not active at pH less than 5.[1] This pH limitation is seldom realized in natural wetlands because inflow water is rarely so acidic and because the generation of carbonate alkalinity by the DSR process typically results in pore water pH levels of 6–8.[1] Many mine drainages are extremely acidic (pH < 4) and might inhibit DSR. However, our preliminary work in a constructed wetland receiving AMD with pH < 3, indicates that pH of the pore water in the substrate is consistently 6–7. This alkalinity appears to result from a combination of limestone dissolution and DSR activity.[4]

Like most microbes, the activity of DSR bacteria decreases at low temperatures. Very low rates of sulfate reduction have been measured for lake sediments and natural wetland substrates in winter months, presumably because of low temperature.[5,8] The effect that slowed rates of DSR would have on constructed wetland performance is uncertain because of the ability of organic substrates to accumulate substantial amounts of elemental sulfur and polysulfides during periods of high DSR.[4] Presumably, these compounds can react with dissolved iron during periods of low DSR.

Ferric iron has been shown to inhibit sulfate reduction in laboratory experiments.[9] The inhibition is probably caused by the activity of iron-reducing bacteria, which have an energetic advantage over the sulfate reducers. This inhibition is unlikely to be significant in constructed wetlands because most systems receive AMD dominated by the reduced, ferrous form of iron.

SULFIDE DESTRUCTION

Metal sulfides can be destroyed by acidic and aerobic conditions. Acidic conditions cause the resolubilization of iron monosulfides (FeS). In most natural systems, extreme acidification of the organic substrate does not occur. However, some constructed wetlands have extremely acidic waters flowing on top of anoxic substrates with circumneutral pH. Substrate disturbance by an extreme flow event or animal movement could cause acidification and destruction of iron monosulfides. Pyrite is not destroyed by acidic conditions, so constructed systems in which FeS is rapidly transformed into FeS_2 would be more resistant to the effects of such perturbations.

The more common cause of sulfide destruction is oxidation. All forms of elemental and reduced sulfur are subject to oxidation under aerobic conditions. In wetlands with fluctuating water levels or lakes which seasonally turn over, anoxic substrates periodically are oxygenated causing cessation of DSR and destruction of sulfides. In one Cape Cod salt marsh, where very high DSR rates have been measured, almost 90% of the annual sulfide production is eventually destroyed by oxidation processes.[10] In Lake Anna, a reservoir in Virginia that receives AMD from abandoned pyrite mines, half of the annual sulfide production was oxidized during a recent year.[8] Control of sulfide oxidation in constructed wetlands requires designs that maintain several cm of water overlying the organic substrate under all weather and flow conditions. Such conditions are currently being maintained at sites where flow is never lost and the wetlands are built as a series of small ponds that drain above the level of the substrate.

In summary, constructed wetlands for AMD have conditions that should promote DSR and the accumulation of iron sulfides. They contain an excess of high quality organic matter and high concentrations of dissolved sulfate and iron. With proper maintenance, the organic substrate will always be submerged and never be exposed to aerobic conditions. Currently, however, DSR is not a dominant process in constructed wetlands, probably because wetlands are generally built to maximize aerobic biochemical processes.[11] Most water flow is on top of the organic substrate, not within it. For sulfide formation to be increased, water must be forced through the organic substrate. We suspect that if AMD could be made to enter at the base of the wetland and then percolated up through the organic substrate, DSR activity would be significantly stimulated.

SULFIDE FORMATION IN A HYPOTHETICAL WETLAND

With proper design modifications, increased rates of net sulfide formation appear possible for constructed wetlands that receive AMD. However, even if the necessary design features were accomplished, could sulfide formation significantly improve the chemistry of typical AMD in a wetland built using current sizing standards? If a sulfide-oriented system required the construction of wetlands several orders of magnitude larger than those currently being built, then the approach would clearly be impractical. This scaling question is addressed by evaluating the theoretical significance of sulfide formation in a hypothetical constructed wetland in which flow of AMD through anoxic substrates is maximized and sulfide oxidation is minimized.

HYPOTHETICAL SYSTEM CHARACTERISTICS

Characteristics of the hypothetical wetland and AMD are shown in Table 2. The hypothetical AMD flows at 50 L/min (13 gpm) and contains iron, acidity, and sulfate loadings typical of contaminated deep mine or spoil discharges in northern Appalachia. On a yearly basis, the wetland receives 2628 kg Fe, 26,280 kg SO_4^{2-}, and 13,140 kg of acidity ($CaCO_3$ equivalent).

Table 2. Hypothetical Acid Mine Drainage and Constructed Wetland

Inflow Water	Dimensions	Substrate
50 L/min flow	2500 m^2 surface area	10 cm surface water
100 mg Fe/L	1 m deep	75 cm compost
100 mg SO_4^{2-}/L		15 cm limestone
500 mg acidity/L		

The hypothetical constructed wetland is typical of systems currently being built in northern Appalachia to treat AMD. It contains a limestone base, a substrate of mushroom compost, and is planted with *Typha*. Depth of compost, 75 cm (30 in.), is twice as much as is currently used in most constructed systems. This adjustment is made to maximize the extent of the anaerobic reducing zone (which is assumed to be 50 cm). The wetland's entire organic substrate contains about 238,500 kg of carbon. This estimate is based on a density for fresh mushroom compost of 500 kg/m^3 (personal communication, Moonlight Mushroom Farm spokesperson, Worthington, PA), 50% moisture content (unpublished BuMines data), and a dry weight fraction that is 40% carbon.

The surface area/flow ratio for the hypothetical wetland is 50 m^2/L/min. This is two to three times larger than that commonly observed for recently constructed systems in western Pennsylvania. This increase in size is necessary where high loads of metals are being treated. The inflow iron concentration, 100 mg/L, is twice as high as was considered feasible only two years ago,[11] but is consistent with observations made at recently constructed wetlands in northern Appalachia.

Construction costs, estimated by mining company and state reclamation officials with experience in building wetlands, generally range from $10–$20 per square meter of wetland. Thus, the cost of the hypothetical wetland would be $25,000–$50,000.

HYPOTHETICAL SYSTEM OPERATION

Complete removal of all the dissolved iron by pyrite formation requires a minimum reduction of 9010 kg of sulfate and oxidation of 2253 kg of carbon.

$$Fe^{2+} + 4C + 2SO_4^{2-} \rightarrow FeS_2$$

$$2628\ \text{kg} + 2253\ \text{kg} + 9010\ \text{kg} \rightarrow 4828\ \text{kg}$$

Both the sulfate and carbon needs of the DSR process are substantially exceeded in the constructed system. Annual inflow sulfate loads, 26,280 kg, are three times higher than necessary. The carbon of the original substrate (238,500 kg) is 100 times the annual needs. Inputs of new organic carbon via the net primary production (NPP) of the wetland should almost meet the DSR needs. Natural temperate wetlands and swamps have an average NPP of

1000 g carbon/m²/yr.[12] Using this figure as a basis, the hypothetical wetland should produce 2500 kg carbon per year. If 75% of the NPP is used by sulfate reducers, as occurs in a Cape Cod salt marsh,[10] then 1975 kg of carbon will cycle through the wetland's DSR system annually. This is 83% of the carbon needed for basic sulfide formation.

The last calculation involves rates of DSR. In order to remove all the dissolved iron as pyrite, the system must produce 93,857 moles of H_2S, or 37.5 moles/m²/yr. Measurements of DSR are routinely expressed as nanomoles (nmol = 10^{-9} moles) of H_2S produced per cm³ of substrate per day. Assuming that the wetland has 50 cm of DSR activity, then the DSR bacteria must reduce sulfate at a rate of 206 nmole H_2S/cm³/day on a 365-day basis, or 626 nmole H_2S/cm³/day on a 120-day basis. The latter incorporates the expectation that a majority of the DSR will occur in the summer months when water temperatures are highest.

Rates of DSR for several natural systems are shown in Table 3. Much variability exists between systems, between seasons, and between sampling points within systems. Salt marshes represent the best comparison to constructed wetlands because they are highly productive organic systems that contain high sulfate concentrations (2300–2500 mg/L). DSR rates in both salt marshes cited meet or exceed the needs of the hypothetical constructed system. The Cape Cod salt marsh rates exceed the 120-day figure by 1.6–11 times. Thus, there is enough excess DSR potential to account for the oxidation of some sulfides that would undoubtedly occur in even the best designed and constructed systems.

In summary, dissimilatory sulfate reduction and iron sulfide formation are biogeochemical processes that could significantly affect mine drainage chemistry. The sizing of wetlands being constructed today is within an order of magnitude of satisfying the theoretical organic needs of such a system. These processes are not significant in most existing constructed wetlands probably because of designs that maximize surface flow and oxidizing conditions. If the potential of sulfide formation processes is to be realized, wetland designs must be modified to increase the movement of drainage through anoxic sediments. The Bureau is currently experimenting with a design that inputs water through perforated PVC pipe into limestone gravel lying at the base of the constructed

Table 3. Measured Rates of Sulfate Reduction in Natural Systems

	nmole/cm³/day[1]	
	Winter	Summer
New Jersey Cedar Swamp[a]	1–11	3–123
Lake Anna Sediments[b]	7–21	366–4430
Georgia Salt Marsh[c]	NA	500–2000
Cape Cod Salt Marsh[c]	NA	1000–7000

[a]Spratt et al.[7]
[b]Herlihy et al.[8]
[c]Howarth and Merkel.[5]

wetland. The plan calls for the acid inflow to be partially neutralized by the limestone, subjected to DSR processes as it diffuses upward through the organic substrate, and subjected to oxidizing, "polishing" processes at the wetland surface.

REFERENCES

1. Postgate, J. R. *The Sulfate-Reducing Bacteria,* 2nd ed. (New York: Cambridge University Press, 1984).
2. Wieder, R. K., and G. E. Lang. "Fe, Al, Mn, and S Chemistry of *Sphagnum* in Four Peatlands with Different Metal and Sulfur Input," *Water Air Soil Poll.* 29:309–320 (1986).
3. Wieder, R. K., G. E. Lang, and A. E. Whitehouse. "Metal Removal in *Sphagnum*-Dominated Wetlands: Experience with a Man-Made Wetland System," in *Wetlands and Water Management on Mined Lands*, R. P. Brooks, Ed. (University Park, PA: Pennsylvania State University, 1985), p. 353.
4. Hedin, R. S., D. M. Hyman, and R. W. Hammack. "Implications of Sulfate-Reduction and Pyrite Formation Processes for Water Quality in a Constructed Wetland: Preliminary Observations," in *Mine Drainage and Surface Mine Reclamation, Vol. I: Mine Water and Mine Waste*, Bureau of Mines Information Circular 9183 (1988), p. 382.
5. Howarth, R. W., and S. Merkel. "Pyrite Formation and the Measurement of Sulfate Reduction in Salt Marsh Sediments," *Limnol. Oceanog.* 29:598–608 (1984).
6. Berner, R. A. "Sedimentary Pyrite Formation: An Update," *Geochim. Cosmochim. Acta* 48:605–615 (1984).
7. Spratt, H. G., M. D. Morgan, and R. E. Good. "Sulfate Reduction in Peat from a New Jersey Pinelands Cedar Swamp," *Appl. Environ. Microbiol.* 53:1406–1411 (1987).
8. Herlihy, A. T., A. L. Mills, G. M. Hornberger, and A. E. Bruckner. "The Importance of Sediment Sulfate Reduction to the Sulfate Budget of an Impoundment Receiving Acid Mine Drainage," *Water Resour. Res.* 23:287–292 (1987).
9. Loveley, D. R., and E. Phillips. "Competitive Mechanisms for Inhibition of Sulfate Reduction and Methane Production in the Zone of Ferric Iron Reduction in Sediments," *Appl. Environ. Microbiol.* 53:2636–2641 (1987).
10. Howarth, R. W., A. Giblin, J. Gale, B. J. Peterson, and G. W. Luther III. "Reduced Sulfur Compounds in the Pore Waters of a New England Salt Marsh," *Environ. Biogeochem.* 35:135–152 (1983).
11. Kleinmann, R. L. P., R. P. Brooks, B. Huntsman, and B. G. Pesavento. "Constructing Wetlands for the Treatment of Mine Water," course notes, Lexington, Kentucky Workshops (1986).
12. Whittaker, R. H., and G. E. Likens. "Carbon in the Biota," in *Carbon and the Biosphere*, G. M. Woodwell and E. V. Pecan, Eds. (Springfield, VA: U.S. Atomic Energy Commission, 1973), p. 281.

38c Evaluation of Specific Microbiological Assays for Constructed Wetlands Wastewater Treatment Management

Ralph J. Portier

INTRODUCTION

A major difficulty in the assessment of toxic chemical impact in aquatic/marine environments is obtaining reproducible, valid field test information. This chapter presents cost-effective approaches for in situ analyses of related soil/sediment microenvironments to evaluate wastewater impact and effect in constructed wetlands.

Employing a modification of the extraction procedure outlined by Van de Werf and Verstraete,[1] microbial ATP was analyzed along a salinity gradient reflecting microenvironments of considerable diversity and biomass. Comparisons were made with specific enzyme assays and standard microbial diversity tests. Also evaluated were the preliminary toxicological implications of a hazardous waste material spill in a freshwater aquatic microenvironment. Application of the assays in a spill situation provided indications of the sensitivity of this assessment protocol in examining stressed aquatic microenvironments and/or managing constructed wetlands for wastewater treatment.

MATERIALS AND METHODS

Sample Collection

Soil and sediment collections were made in the Terrebonne-Timbalier Bay and Barataria Bay drainage basins in Louisiana. These watershed regions comprise distinct vegetation zones and areas of contrasting salinity, all affected by gulfward movement of water. Brackish and freshwater regions located adjacent to intensively cultivated agricultural fields and industrial sites are most directly affected by runoff. Aluminum cylinder cores 60 cm tall × 7.6 cm diameter were inserted into the sediment to the water level, capped and sealed, and stored on ice for transportation to the laboratory for microbial analysis.

Enumeration of Microorganisms

Four general groups of microorganisms (bacteria, actinomycetes, filamentous fungi, and yeasts) were enumerated using colony forming units (CFU) criteria.[2,3] For tabulation of bacteria and actinomycetes, Jensen's[4] agar medium was inoculated with replicate 1.0-mL aliquots, and cycloheximide (Sigma) was added at 40 μg/mL to inhibit overgrowth by molds. For enumeration of filamentous fungi and yeast, Martin's[5] agar medium was inoculated with streptomycin (Sigma), and chlortetracycline (Sigma) at 30 μg/mL was used to retard bacterial growth.

Adenosine Triphosphate Assay

A modification[6] of the adenosine 5′-triphosphate (ATP) assay as advanced by Holm-Hansen and Booth,[7] and further presented by Van de Werf and Verstraete[1] and Karl,[8] was used to determine microbial biomass. The following procedure was used: Soil/sediment aliquots (30 g wet weight) were transferred to salinity-adjusted *tris*-EDTA ($MgSO_4$) buffer solutions (270 mL), homogenized for 1 min using a Sorval™ mixer, and allowed to equilibrate under field incubation conditions for 15 min. Following dilution depending upon soil/sediment texture, 100 μL of each equilibrated suspension was transferred to 3-mL plastic vials, inserted into the luminescence apparatus (Biocounter M2010, LUMAC Systems, USA), and allowed to incubate at 30°C for 1 min. Luminescence background measurements were performed during this time. Following injection of 100 μL of a luciferin-luciferase solution (LUMIT™, LUMAC Systems, USA) into the vial, light output was determined over a 10-sec integration period. Light output was expressed as relative light units (RLU, or μg ATP/g dry weight using standardized 10-μL aliquots of a known ATP standard).[6]

Enzyme Assays

Phosphatase activity was determined using the procedure outlined by Tabatabai and Bremner[9] as modified by Atlas et al.[10] Replicate 2-g aliquots of control and pesticide-treated sediments were placed in 150-mL Erlenmeyer flasks with 0.1 mL of toluene, 4 mL of sterile, distilled deionized water, and 0.5 mg/mL of *p*-nitrophenol phosphate. The resulting yellow solution was assayed by spectrophotometer (PYE-UNICAM 1700) at 410 nm for units of *p*-nitrophenol produced.[11]

Dehydrogenase determination was based on formation of 2,3,5-triphenyltetrazolium formation (TPF). Replicate 10-g aliquots of control and toxicant-treated soils/sediments were placed in 25-mL Erlenmeyer flasks brought to 100% water-holding capacity with addition of a 0.5% solution of

2,3,5-triphenyltetrazolium chloride (Sigma), as described by Bartha et al.[11] The TTC was converted by dehydrogenase enzymes to 2,3,5-triphenyltetrazolium formazan, forming a characteristic red color. Absorption was measured at 485 nm, and dehydrogenase activity units expressed as units of TPF mg/mL.[11]

In Situ Investigations

Related in situ investigations were conducted within the Barataria Bay drainage basin and the adjacent Terrebonne-Timbalier system. To identify and define important system design parameters for subsequent laboratory microcosm investigations, microbial diversity and related enzymatic activity of selected aquatic microenvironments along a salinity gradient were surveyed. (A similar approach is valid for preliminary design of a constructed wetland.) Replicate cores were analyzed at 0–5 cm, 10–15 cm, and 25–30 cm for variations in bacteria, actinomycetes, filamentous fungi, and yeasts. Phosphatase, dehydrogenase, and microbial ATP levels for these depths were also determined, and salinity, pH, Eh, and temperature were measured at each site. The following designations were used to describe specific soil/sediment sampling locations:

A—agricultural, sugarcane
P—pasture
SW—swamp floor, cypress, tupelo gum, intermittent flooding
IMP—impounded swamp forest floor, intermittent flooding, controlled drainage
NS—natural swamp forest, streamside, 0 o/00
10 o/00—brackish marsh
20 o/00—brackish-saline marsh, *Spartina patens*
26 o/00—saline marsh, *Spartina alterniflora*

Stressed Microenvironment Study

A second series of in situ measurements were taken within the Lac des Allemands site in the Barataria Bay basin—sites previously studied for salinity gradient characterization— following the spillage of commercial oil exploration drilling fluids within the impounded swamp area. This provided an opportunity to discern under in situ conditions the responsiveness of these field assay protocols. Replicate cores in the impounded swamp site were collected for analysis approximately 30 days following the spill. Cores were removed at distances of approximately 1 m, 10 m, 100 m, and 200 m from the rupture point in the earthen dike system constructed to retain these fluids. Analyses included microbial diversity, phosphatase and dehydrogenase activity, and microbial ATP. The spilled drilling fluid was primarily crude oils, paraffins, and barium-based drilling lubricants or "drilling muds" used in maintaining deep well flow control. Of the petroleum hydrocarbons and bulk chemicals present, phenanthrene and acenaphthalene concentrations ranged from 500 to 1200 ppm in disturbed sediments.

RESULTS AND DISCUSSION

Epiphytic Microenvironment Assessment

Variations along a salinity gradient of four major morphological groups (bacteria, filamentous fungi, yeasts, and actinomycetes) are shown in Figure 1 for the 0–5 cm depth. Also shown are corresponding enzymatic activity variations, i.e., phosphatase, dehydrogenase, and adenosine triphosphate (ATP) levels. In general, microbial populations declined with depth but varied considerably between soil/sediment/salinity microenvironments. Bacteria and actinomycetes constituted a more significant portion of the microbial community in upland soils, whereas bacteria and filamentous fungi were found at higher levels in freshwater swamp forest and impounded swamp forest sites. Actinomycetes were also significant, indicative of the high organic material accumulation on the forest floor as a result of litter fall.

Further along the salinity gradient at more brackish and saline sites, yeasts and filamentous fungi propagules were statistically more responsive to salinity and anaerobiosis than bacteria or actinomycetes. This again may have been attributable to organic content, for at the 10–15 cm and 25–30 cm depths (not shown) these groups were much lower in number at this mid-salinity range. There appeared to be less stratification in these two morphological groups with depth at the 26 o/00 salinity level. Investigations by Stevenson et al.[12] alluded to the possible importance of monitoring levels of filamentous fungi propagules in characterizing the flushing phenomena of detrital-based ecosystems. Actinomycetes and bacteria did not significantly change in total CFU/g dry wt for both salinity and depth. However, it is certain that species variation and replacement was occurring, as shown by the increasing number of gram-negative bacteria; a more pronounced stratification in actinomycetous groups was apparent at the higher salinities.

Specific levels of phosphatase and dehydrogenase were significantly higher in microenvironments, reflecting higher organic content. These levels were less stratified with depth. Again, increases were noted for sites with higher organic content but were more homogeneous with salinity fluctuations. It is difficult to reach a definitive conclusion about this phenomenon. Perhaps, these assays are in actuality reflecting relative enzyme activity by microbial populations under more anoxic conditions. Another consideration is a loss in sensitivity by these two assays for anaerobic conditions.

Adenosine triphosphate (ATP) levels, reflecting estimates of relative microbial biomass along the salinity gradient, are also shown. Higher microbial biomass levels were noted for upland soil microenvironments and for the agriculturally affected freshwater habitats. Composite statistical analysis of total fungal and bacterial populations showed high correlations with microbial ATP. These and other specific correlations are shown in Table 1. Relative ATP levels were lower at mid-salinity ranges (where pronounced flushing processes predominated) and tended to increase at the higher salinities (where streamside

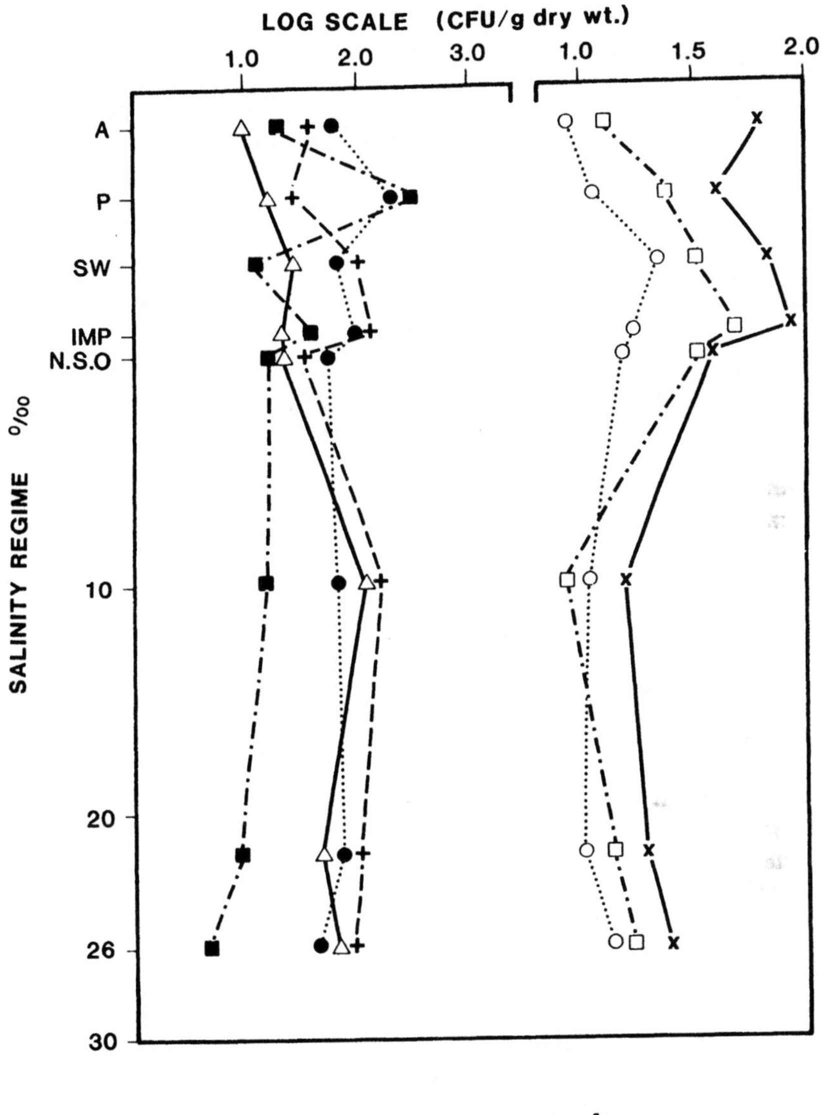

Figure 1. Microbial diversity, enzymatic activity, and microbial ATP variations along a salinity gradient, depth 0–5 cm.

Table 1. Specific Data Correlations for In Situ Test Protocols

	Bacteria	Fungi	Yeast	ACT	pH	Depth	Salinity
Microbial ATP	0.9241**	0.8394**	0.7710**	0.6240*	0.6478*	–0.7840*	–0.6700*
Phosphatase	0.8371*	0.7810*	0.7440*	0.5240	0.4148	–0.4140	–0.4140
Dehydrogenase	0.8633*	0.8202*	0.8410	0.6110	0.5580	–0.6256*	–0.5150

Note: ACT = actinomycetes.
* = significance to the 5% probability level.
** = significance to the 1% probability level.

organic matter occurred). Yeast and filamentous fungi populations appeared to peak at the mid-salinity ranges in contrast with microbial ATP levels. However, statistical analysis indicated a strong correlation with bacterial numbers. Thus, while microbial ATP estimates reflected and confirmed specific enzymatic trends for salinity and depth and for composite microbial activity, this assay more generally reflected bacterial population fluctuations. Nevertheless, a fairly concise analysis of microenvironment characteristics could be ascertained from this assay when complemented with other specific assays.

Application of Field Approaches in Ascertaining Toxicant Effects

Figures 2 and 3 present data on the relative effects of several industrial source pollutants, drilling fluids, and crude oil on microenvironments at distances of 1 m, 10 m, 100 m, and 200 m from the point source in the impounded freshwater swamp forest site.

Microbial diversity variations for populations of bacteria, filamentous fungi, actinomycetes, and yeast are shown in Figure 2. At a distance of 1 m from the ruptured dike wall, minimal numbers of bacteria and filamentous fungi (log CFU/g dry wt) were noted for all depths. Actinomycetous populations attained sufficient concentration at a distance of 10 m. Beyond the general work site for drilling operations at 100 m, all four morphological groups were isolated and exhibited a characteristic depth profile. However, these numbers were lower than those at 200 m, the site of earlier sampling efforts mentioned previously. Bacterial populations predominated at the more severely impacted site, and filamentous fungi and yeast populations experienced a significant shift in total numbers along the relative impact transect shown.

Phosphatase, dehydrogenase, and microbial ATP profiles for each sample site are shown in Figure 3. At 1 m, minimal levels of microbial ATP with depth were noted, and phosphatase and dehydrogenase were not detectable. At 10 m, higher levels of microbial ATP and minimal levels of phosphatase and dehydrogenase were noted. At 100 m, a more pronounced enzyme profile with depth was apparent. At 200 m, levels of dehydrogenase and microbial ATP had increased substantially at the 0–5 cm depth. Microbial ATP demonstrated a greater response for toxicant effect than did phosphatase, and dehydrogenase a more moderate response.

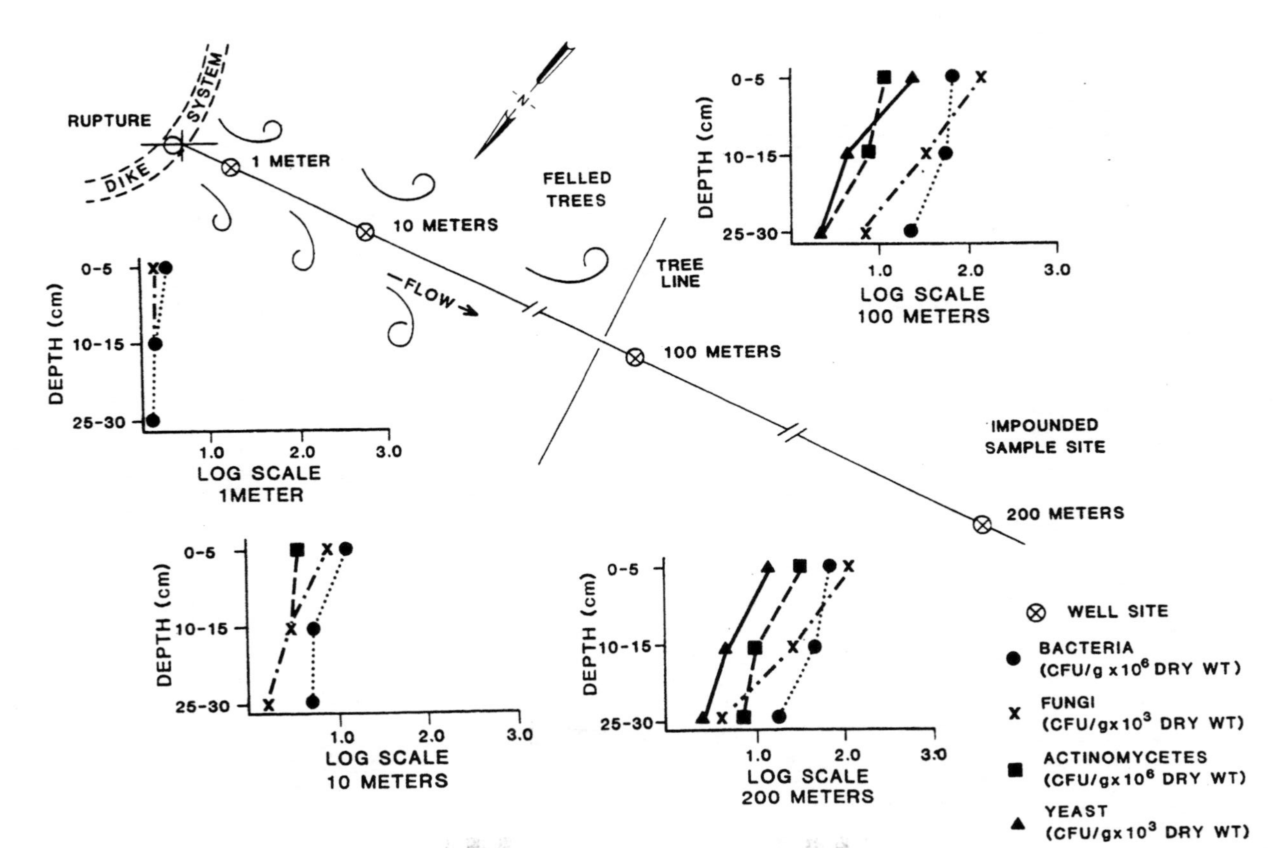

Figure 2. Microbial diversity profiles for depth and distance in impounded swamp forest site spill.

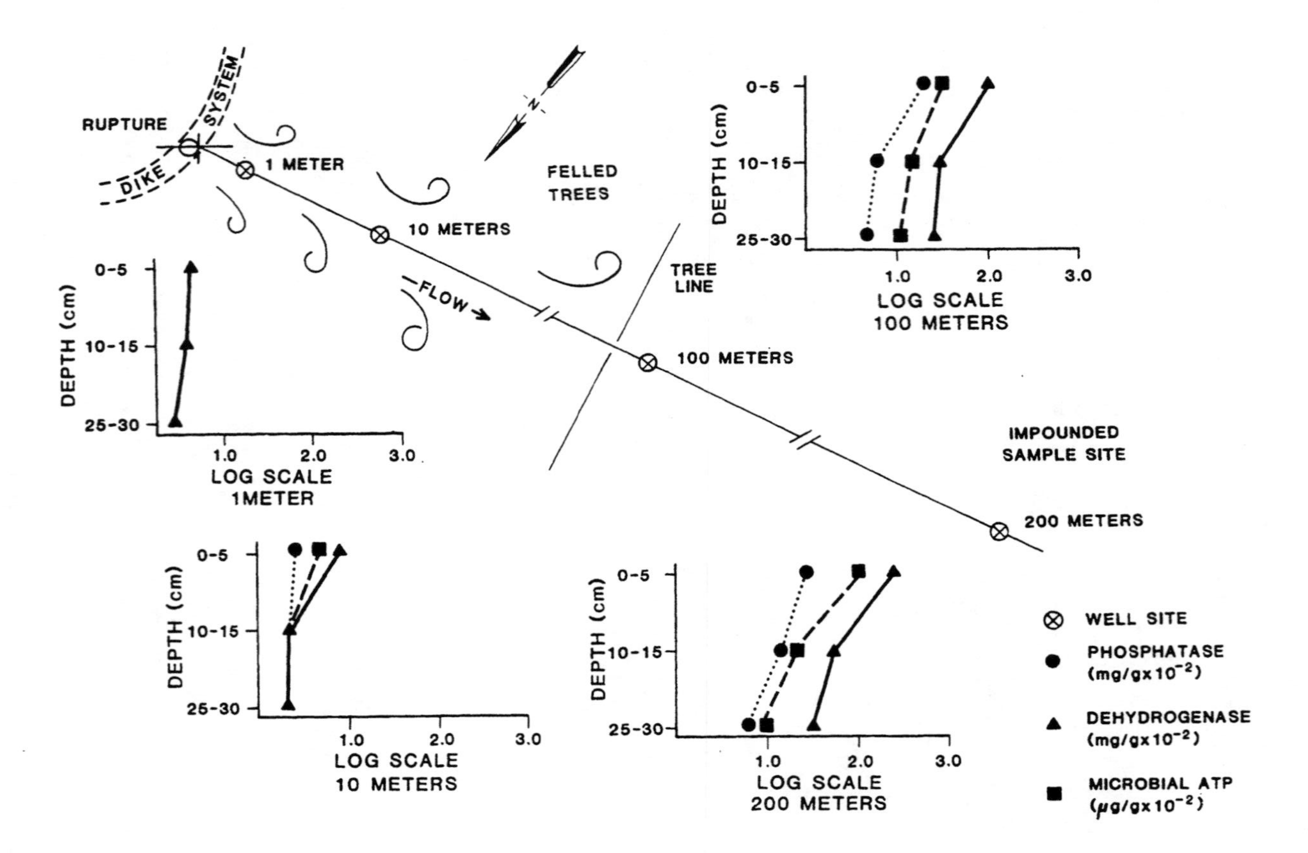

Figure 3. Enzymatic activity and microbial ATP profiles for depth and distance in impounded swamp forest site spill.

CONCLUSIONS

In situ analyses of interrelated soil/sediment microenvironments are relevant to understanding determinants of productivity in aquatic regions or biotreatment efficiency in a constructed wetland. Total microbial biomass, as determined by the luciferin-luciferase method outlined, is useful in understanding in situ physiological states of naturally occurring microbial assemblages. Furthermore, total microbial biomass and species diversity may serve as valid indices of biodegradable substrate turnover and productivity. This should provide information on microhabitat features as affected by addition of a variety of toxic substances in a wastewater treatment scenario.

Karl,[8] in concluding comments on cellular nucleotide measurements, suggested the need for acquisition of corroborative data to obtain "nucleotide fingerprinting" of diverse soil/sediment environments. The microbial ATP method developed here was only oriented toward fingerprinting an in situ toxic response by microbial populations. However, both relative levels of microbial ATP and measurement responses in highly organic soils appeared to be in agreement. These responses were more discernable in subsequent aquatic microcosm studies using a variety of toxic substances.[13,14]

Variations in microbial diversity, enzyme activity, microbial ATP, and substrate uptake in ecologically diverse but related soil/sediment microenvironments in coastal wetland systems provide a vehicle for aquatic microenvironment parameter identification and assessment.[14] These variations reflect autochthonous parameters of microbial viability for toxicological studies and characterize land management practices between hydrologically linked localities. Such observations provide a baseline for monitoring possible impact of toxicant runoff and introduce methods for establishing criteria for productivity assessment in constructed wetlands management.

ACKNOWLEDGMENTS

Portions of the work reported in this chapter were funded from grant NA 81-AA-D-0046, Office of Sea Grant. Additional support was obtained from a research and development program funded by the Louisiana Board of Regents (contract B20–21).

REFERENCES

1. Van de Werf, R., and W. Verstraete. "Direct Measurement of Microbial ATP in Soils" Proceedings International Symposium Analytical Applications of Bioluminescence and Chemiluminescence, Schram et al., Eds. (1979), pp. 333–338.
2. Portier, R. J., and S. P. Meyers. "Analysis of Chitin Substrate Transformation and Pesticide Interactions in a Simulated Aquatic Microenvironmental System," *Devel. Ind. Microbiol.* 21:543–555 (1981).
3. Portier, R. J., and S. P. Meyers. "Use of Microcosms for Analyses of Stress-

Related Factors in Estuarine Ecosystems," International Wetlands Conference, New Delhi, India (1981), pp. 375–387.

4. Jensen, H. Z. "Actinomycetes in Danish Soils," *Soil Sci.* 30:59–77 (1930).
5. Martin, J. P. "The Use of Acid, Rose Bengal, Streptomycin in the Plate Count Method for Estimating Soil Fungi," *Soil Sci.* 69:215–233 (1950).
6. Portier, R. J. "Correlative Field and Laboratory Microcosm Approaches in Ascertaining Xenobiotic Effect and Fate in Diverse Aquatic Microenvironments," Ph.D. dissertation, Louisiana State University, Baton Rouge, LA (1982).
7. Holm-Hansen, O., and C. R. Booth. "The Measurement of Adenosine Triphosphate in the Ocean and Its Ecological Significance," *Limnol. Oceanog.* 11:510–519 (1966).
8. Karl, D. M. "Cellular Nucleotide Measurements and Applications in Microbial Ecology," *Microbiol. Rev.* 44:739–796 (1980).
9. Tabatabai, M. A., and J. M. Bremner. "Use of p-Nitrophenol Phosphate for Assay of Soil Phosphatase Activity," *Soil Biol. Biochem.* 1:301–307 (1969).
10. Atlas, R. M., D. Pramer, and R. Bartha. "Assessment of Pesticide Effects on Nontarget Soil Microorganisms," *Soil Biol. Biochem.* 10:231–239 (1977).
11. Bartha, R., P. Lanzilotta, and D. Parmer. "Stability and Effects of Some Pesticides in Soil," *Appl. Microbiol.* 15:67–75 (1967).
12. Stevenson, L. H., T. H. Crzanowski, and C. W. Erkenbrecher. "The Adenosine Triphosphate Assay: Conceptions and Misconceptions," American Society of Testing and Materials Special Technical Publication 695 (1979), pp. 99–116.
13. Portier, R. J., H. M. Chen, and S. P. Meyers. "Environmental Effect and Fate of Selected Phenols in Aquatic Ecosystems Using Microcosm Approaches," *Devel. Ind. Microbiol.* 24:409–424 (1982).
14. Portier, R. J., and S. P. Meyers. "Coupling of In Situ and Laboratory Microcosm Protocols for Ascertaining Fate and Effect of Xenobiotics," *Toxicity Testing Using Bacteria*, B. J. Dutka and D. Liu, Eds. (New York: Marcel Decker, 1983), pp. 345–379.

38d Secondary Treatment of Domestic Wastewater Using Floating and Emergent Macrophytes

T. A. DeBusk, P. S. Burgoon, and K. R. Reddy

INTRODUCTION

Recent studies in Florida demonstrated that shallow ponds containing large-leaved floating macrophytes, such as pennywort *(Hydrocotyle umbellata)* and water hyacinth (*Eichhornia crassipes*), can remove biochemical oxygen demand (BOD_5) from domestic wastewaters at high rates (100–200 kg/ha/day).[1,2] However, because neither pennywort nor water hyacinth are tolerant of prolonged subfreezing temperatures, treatment systems that utilize these species are limited in year-round operation to the southernmost regions of the United States. For temperate climates, some success has been reported in using gravel-bed systems for treating wastewaters.[3] Gravel-bed systems (GBS) consist of gravel-filled trenches planted with one or more species of emergent macrophytes. Wastewater is usually passed through GBS in a subsurface flow to eliminate problems associated with surface ponding (mosquito production, odors).

Comparative data are not available on the utility of GBS and floating aquatic macrophyte-based treatment systems (FAMS) for providing secondary treatment of domestic wastewater. Therefore, we examined BOD_5 and suspended solids (SS) removal rates from primary effluent using floating and emergent macrophytes cultured in pond and gravel-bed systems.

METHODS

This study was conducted in central Florida using 3000-L outdoor raceways fed primary domestic effluent at a hydraulic loading of 10 cm/day. During November 1987, four treatments were established in duplicate tanks: a gravel (1–3 cm river rock) filled bed with no macrophytes; a gravel bed planted with swordgrass (*Scirpus pungens* = *S. americanus*); a gravel bed planted with

Table 1. Wastewater Biochemical Oxygen Demand (BOD_5) and Suspended Solids (SS) Concentrations in Raceways Containing Floating and Emergent Macrophytes

Sample Location	Plant/ Substrate[a]	BOD_5 (mg/L)	SS (mg/L)
Influent	—	169	44
Effluent	Pennywort/FAMS	27	4
Effluent	Arrowhead/GBS	43	15
Effluent	Swordgrass/GBS	56	16
Effluent	None/GBS	62	19

Note: Values represent means of weekly measurements (duplicate tanks) collected over a six-month period. Average raceway influent contaminant loadings were 172 kg BOD_5 and 45 kg SS/ha/day.
[a]FAMS = floating macrophyte system; GBS = gravel-bed system.

arrowhead (*Sagittaria latifolia*); and an open water system stocked with the floating macrophyte pennywort (*Hydrocotyle umbellata*).

BOD_5 and SS concentrations were measured in grab samples collected weekly from tank effluents.[4] Influent wastewater BOD_5 and SS concentrations were measured twice weekly during the six-month study (November 1987 to April 1988). Each month, plant tissues harvested from within three randomly placed 0.25 m^2 quadrats were dried (72 hr at 60°C) and weighed to provide a measure of shoot standing crop.

RESULTS AND DISCUSSION

Greatest reductions in wastewater contaminant concentrations were attained in raceways containing the floating macrophyte pennywort (Table 1). BOD_5 and SS removal rates in pennywort raceways averaged 145 and 41 kg/ha/day, respectively, during the six-month study. Contaminant removal rates in gravel-bed systems were highest in raceways containing arrowhead, followed by those containing swordgrass, and those without vegetation (Table 1). The BOD_5 removal rates observed in arrowhead and nonvegetated raceways (115 kg and 98 kg/ha/day, respectively) are higher than the mean annual removal rates of 53 kg BOD_5 and 39 kg BOD_5/ha/day reported for bulrush *(Scirpus validus)* and nonvegetated gravel beds in California.[5]

While mean treatment performance of the GBS was poorer than that of pennywort tanks, effluent quality in the raceways containing gravel gradually improved during the study. Indeed, following six months of operation, effluent wastewater BOD_5 and SS concentrations from the arrowhead raceways were similar to those obtained in the pennywort systems (Figure 1).

Shoot standing crop in the GBS systems increased sharply from the period December 1987 to April 1988 (from 0.18 to 1.31 kg/m^2 for arrowhead and 0.03 to 1.27 kg/m^2 for swordgrass). A concomitant increase in root biomass also was observed in these systems during this period. Although this suggests a positive correlation between plant standing crop (e.g., root penetration into the gravel bed) and treatment efficiency, it should be noted that effluent BOD_5

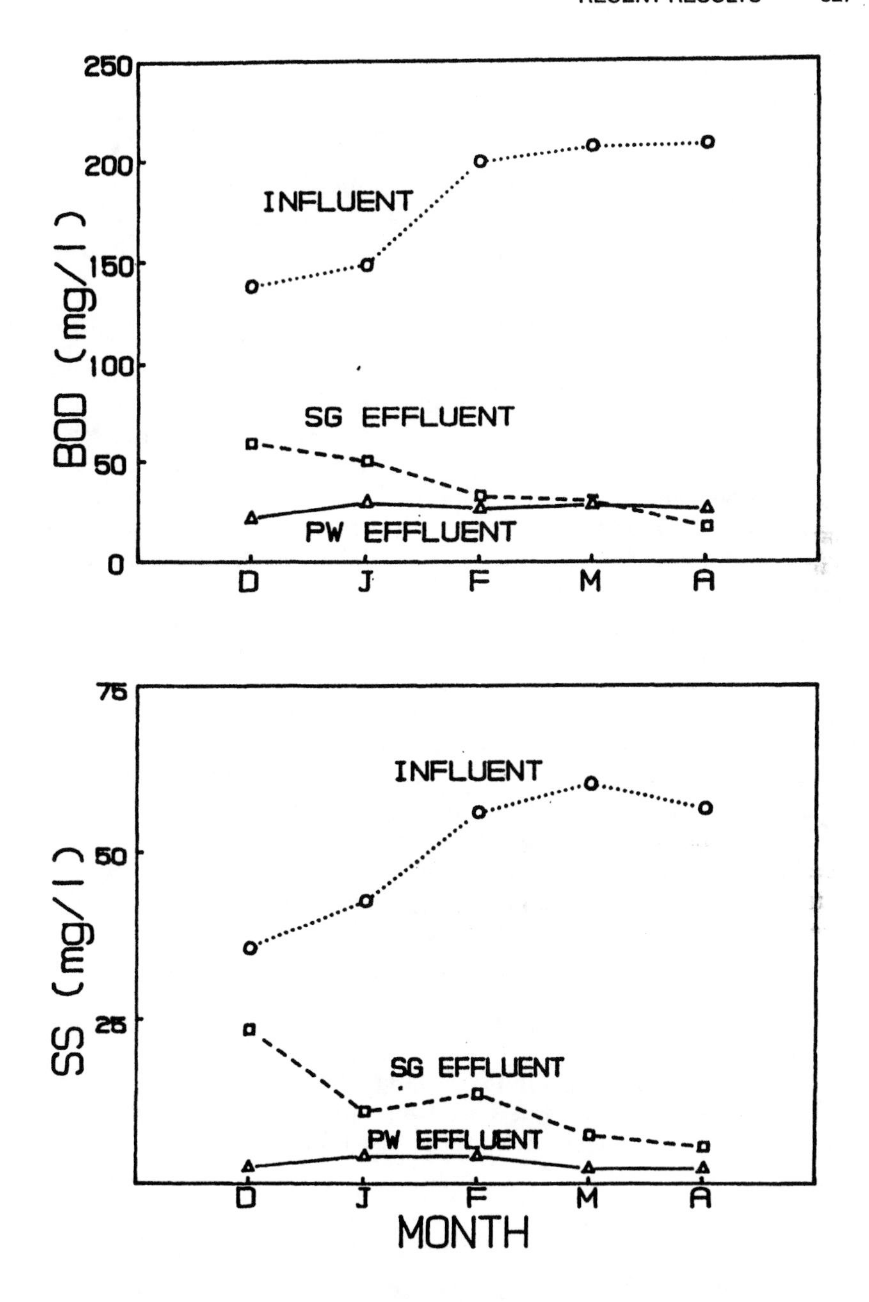

Figure 1. Monthly mean influent and effluent wastewater BOD_5 (top) and SS (bottom) concentrations in raceways containing the macrophytes arrowhead (SG) and pennywort (PW).

and SS concentrations in the unvegetated gravel bed also improved during the six-month study (from 86 mg BOD_5/L in November 1987 to 26 mg/L in April 1988). Apparently, several months are required for development of full treatment capacity in both vegetated and unvegetated beds. This time lag may represent a colonization period for bacteria throughout the gravel system.

The efficiency of floating macrophytes in removing BOD_5 from primary effluent is related to their capability of transporting O_2 from the foliage to the rhizosphere.[1] Oxygen transport rates may also dictate BOD_5 removal rates of emergent species cultured in gravel. In laboratory studies, we observed oxygen transport rates of 112, 19, and 4 mg O_2/g(root)/day for pennywort, arrowhead, and swordgrass, respectively. It is difficult to extrapolate such measurements of O_2 transport by aquatic macrophytes to field conditions because of wide interspecific and intraspecific variability in plant (root) morphology and standing crop. Moreover, little is known of the effects of physiological (e.g., plant age) and environmental parameters on O_2 transport in aquatic plants.

The results of this study demonstrate a gradual improvement in treatment efficiencies of vegetated and nonvegetated GBS during the initial six months of operation. This increased efficiency appears to be related to bacterial colonization in the gravel bed. In contrast, FAMS appear to operate at peak efficiency immediately following startup. Although FAMS have been shown to provide secondary treatment of domestic effluent for several years without operational problems, questions remain regarding the long-term performance of GBS. For example, it is unknown whether treatment efficiency becomes impaired by the accumulation of wastewater-borne solids and plant detritus, which can clog the bed and cause short-circuiting. Moreover, the relation between plant harvest and contaminant removal in GBS is unknown. To facilitate design and operation of GBS, effects of wastewater loading rate, plant species, and harvest regime on the removal of various wastewater contaminants should be investigated.

ACKNOWLEDGMENTS

This study was supported in part by a grant entitled "Methane from Community Wastes" from the Gas Research Institute, Chicago, Illinois to Walt Disney Imagineering (contract no. 5082–223–0762).

REFERENCES

1. DeBusk, T. A., and K. R. Reddy. "BOD Removal in Floating Aquatic Macrophyte-Based Treatment Systems," *Water Sci. Technol.* 19:273–279 (1987).
2. DeBusk, T. A., and K. R. Reddy. "Wastewater Treatment Using Floating Aquatic Macrophytes: Contaminant Removal Processes and Management Strategies," in *Aquatic Plants for Water Treatment and Resource Recovery,* K. R. Reddy and W. H. Smith, Eds. (Orlando, FL: Magnolia Publishing Inc., 1987), pp. 643–656.

3. Brix, H. "Treatment of Wastewater in the Rhizosphere of Wetland Plants—The Root Zone Method," *Water Sci. Technol.* 19:107–118 (1987).
4. *Standard Methods for the Examination of Water and Wastewater,* 16th ed. (Washington, DC: American Public Health Association, 1985).
5. Gersberg, R. M., B. V. Elkins, S. R. Lyon, and C. R. Goldman. "Role of Aquatic Plants in Wastewater Treatment by Artificial Wetlands," *Water Res.* 20:363–368 (1986).

38e Amplification of Total Dry Matter, Nitrogen and Phosphorus Removal from Stands of *Phragmites australis* by Harvesting and Reharvesting Regenerated Shoots

Takao Suzuki, W. G. Ariyawathie Nissanka, and Yasushi Kurihara

INTRODUCTION

Earlier investigations showed that sand filtration pot systems with *Phragmites australis* were more effective in removing nitrogen (N) and phosphorus (P) from wastewater.[1,2] Obviously, part of N and P was incorporated in biomass of *Phragmites*. A larger fraction was associated with easily harvestable aboveground plant parts suggesting that harvesting shoots alone could remove a large quantity of N and P from the system.

Studies on natural stands of *Phragmites* have shown higher concentrations of N and P in younger tissues and decreasing concentrations with maturity.[3,4] This loss of N and P in mature plants suggests that N and P removal could be increased by repeated harvesting to maintain young vegetational stages. Harvesting should be planned to obtain maximum removal through a combination of higher concentrations and biomass.

This experiment was designed to establish the best timing for harvest and reharvest of regenerated shoots of *Phragmites* to amplify removal of total dry matter, N, and P from a *Phragmites* stand growing under field conditions.

MATERIALS AND METHODS

Harvesting at a freshwater stand of *Phragmites australis* (Botanical Garden, Tohoku University, Sendai, Japan) was conducted from mid-May 1983 to early November, when the leaves were almost dried and some had fallen. Shoots were harvested at short intervals during the growing season and once a month after flowering in August. At each sampling, an area of more than 4 m^2 was cleared at ground level to prevent interference from unharvested plants, and shoots in a few 50-cm^2 quadrats were collected. Leaves were separated

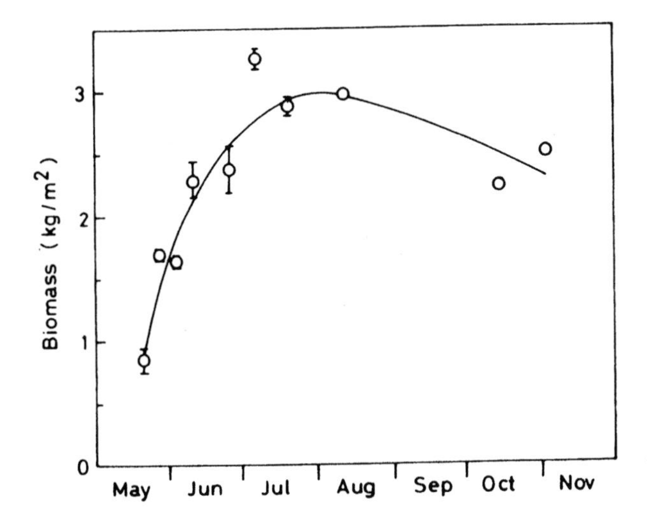

Figure 1. Biomass changes through the growing season. Dry weight averages are plotted. Ranges are shown by vertical bars.

from shoots, and samples were taken for determination of biomass, carbon (C), N, and P. When regenerated shoots appeared, portions were harvested in mid growing season and processed similarly. At final sampling in November, regenerated shoots in all plots were harvested, and 1-m^2 samples were processed similarly.

Plant samples were dried at 65°C for 72 hr and powdered using a grinder (NRK R-8 Type, Nippon Rikagaku Kikai Co. Ltd.). Carbon and N concentrations were determined with a CN Corder (Yanako Model MT-500, Yanagimoto MFG Co. Ltd.), and P was estimated by ignition.[5]

RESULTS

Biomass

Phragmites shoot biomass increased rapidly early in the growing season, peaked in early August, and then decreased (Figure 1). Shoot regeneration was better in the early-harvested plots than in the late-harvested plots, but continued into September. Total plot production was calculated combining the original plot biomass and the biomass of regenerated shoots from the same plot (Figure 2). Even though plots harvested early in the season produced higher regenerated-biomass than those harvested later, maximum total biomass was obtained from the plot that was harvested at the peak of original growth curve. Total biomass obtained by double harvesting was 3.4 kg/m^2, 0.5 kg/m^2 greater than the maximum obtained if shoots were harvested only once in the growing season.

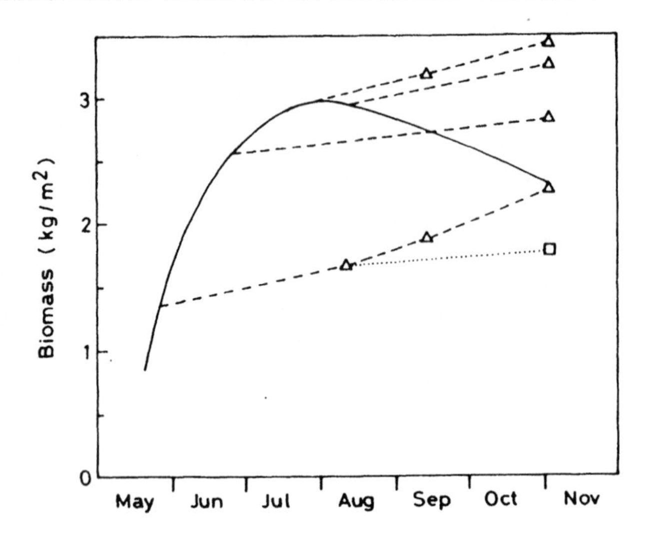

Figure 2. Seasonal changes in the dry weight biomass (solid line) and increased amounts from regeneration at different points (broken and dotted lines). Δ = growth after first harvest; □ = growth after second harvest.

Nitrogen and Phosphorus

Seasonal changes of C, N, and P concentrations in *Phragmites* stem and leaf tissues are shown in Figure 3. Nitrogen and P in leaf tissue were always higher than those in the stem. Younger tissues had higher concentrations that reduced

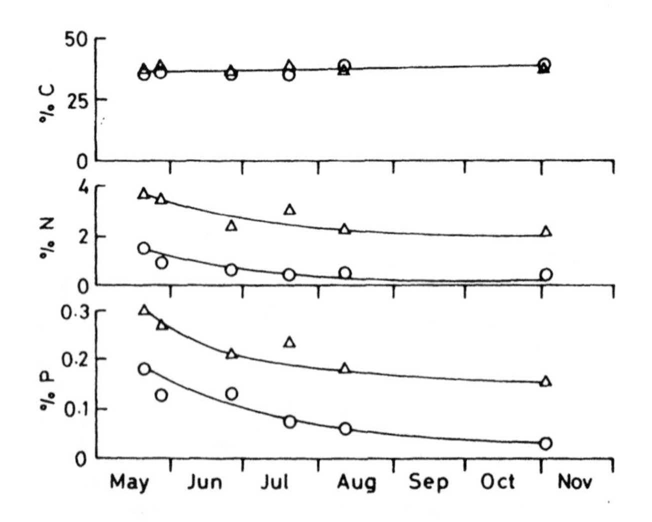

Figure 3. Seasonal changes in concentrations of carbon, nitrogen, and phosphorus in stem (O) and leaf (Δ) tissues.

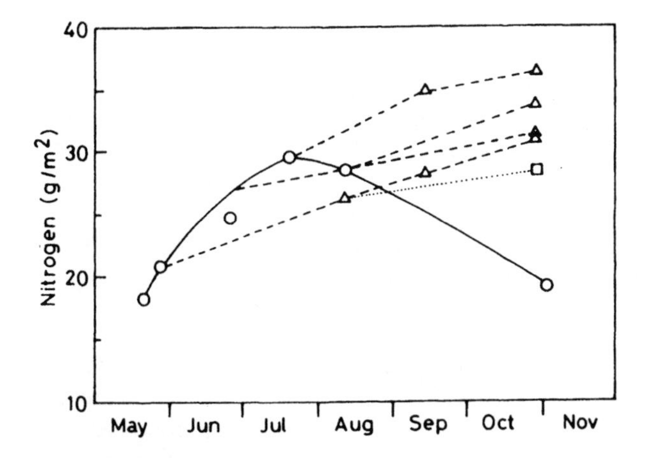

Figure 4. Seasonal changes in nitrogen amounts (solid line) and increases from regeneration at different points (broken and dotted lines). Δ = growth after first harvest; □ = growth after second harvest.

with maturity, and P declined more rapidly than N. Carbon concentrations in stem and leaf tissues were similar and changed very little with maturity.

From these data, total amounts of N and P that could be removed from each plot were calculated (Figures 4 and 5). Because of higher concentration of N in younger tissues and rapid biomass increase in early growing season, N increased very rapidly to peak in late July, slightly before the biomass peak. The subsequent decline was due to decreasing N concentrations and biomass

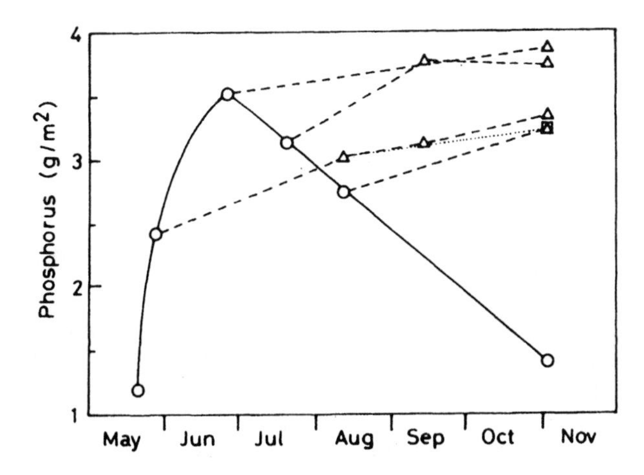

Figure 5. Seasonal changes in phosphorus amounts (solid line) and increases from regeneration (broken and dotted lines). Δ = growth after first harvest; □ = growth after second harvest.

with maturity. When the N harvested with the regenerated shoots was added to the original amount obtained from each plot, similar to biomass, the highest removal of N was obtained when shoots were harvested at the peak N content and by reharvesting the regenerated shoots. This combination could remove 36.5 g/m^2 total N, 6.8 g/m^2 more than the maximum obtained if shoots were harvested only once during the growing season.

Similarly, higher concentrations of P in younger tissues and the rapid biomass increase of *Phragmites* shoots in early growing season caused P content to peak in late June—much earlier than biomass or N peaks. Because of the biomass decrease and rapid decrease in P concentration in mature tissues, P amounts dropped very rapidly. Adding P quantities obtained from first harvest to quantities from regenerated shoots in respective plots showed that shoots should be harvested when P content peaks, followed by reharvest of regenerated shoots at the end of the growing season to enhance P removal from a *Phragmites* stand. This harvesting strategy could remove 3.9 g/m^2 P, 0.4 g/m^2 more than the highest amount removed if shoots were harvested only once during the growing season.

DISCUSSION

Harvesting *Phragmites* shoots twice during the growing season increased the total amount of biomass, N, and P removal from a plot. To amplify removal, the first harvest should occur when maximum nutrient content is reached, with a second harvest when the plant growth ceases in late autumn.

Total harvestable biomass with this strategy was 34,000 kg/ha dry wt, 14% higher than the highest amount that could be obtained by a single harvest, or 34% more than a single end-of-growing-season harvest.

The maximum total amount of N that could be removed from a stand was 365 kg/ha, 22% greater than the highest amount that could be removed by a single harvest, or 90% more than a single end-of-growing-season harvest.

For P, the maximum total removal was 38.8 kg/ha. This was 10% higher than the highest amount removed by a single harvest and 175% higher than the amount removed by a single end-of-growing-season harvest.

Higher concentrations of N and P in younger tissues and decreases with maturity have been observed by others.[3,4] Studies on P removal from mixed natural stands of aquatic plants by harvesting at different intervals revealed that total biomass removed from quadrats harvested biweekly and monthly was less than that from control quadrats. However, because of higher concentrations in younger tissues, more than three times as much P could be removed by biweekly harvest than a single end-of-season harvest.[6]

When harvesting *Phragmites* shoots, the level of cutting and harvest season affect the shoot regeneration. If reeds are flooded after harvest, air supply to the roots is interrupted and the stand may be killed.[7] In a *Phragmites* wastewater treatment system, root mortality may adversely affect nutrient removal efficiency. Shoot harvest during the growing season may also affect the quality of effluent. Consequently, harvesting methods in such systems should consider effluent quality maintenance along with amplifying N and P removal from the system by harvesting shoots of *Phragmites*.

REFERENCES

1. Suzuki, T., A. G. Wathugala, and Y. Kurihara. "Preliminary Studies on Making Use of *Phragmites australis* for the Removal of Nitrogen, Phosphorus, and COD from the Waste Water," research related to the UNESCO'S Man and the Biosphere Programme in Japan, Coordinating Committee on MAB Programme (1985), pp. 95–99.
2. Wathugala, A. G., T. Suzuki, and Y. Kurihara. "Removal of Nitrogen, Phosphorus and COD from Waste Water Using Sand Filtration Systems with *Phragmites australis*," *Water Res.* 21(10):1217–1224 (1987).
3. Dykyjova, D. "Content of Mineral Macronutrients in Emergent Macrophytes During Their Seasonal Growth and Decomposition," *Ecosystem Study on Wetland Biome in Czechoslovakia*, Czech. IBP/PT-PP Report No. 3, Trebon (1973), pp. 163–172.
4. Kvet, J. "Mineral Nutrients in Shoots of Reed (*Phragmites communis* Trin.)," *Pol. Arch. Hydrobiol.* 20:137–147 (1973).
5. Anderson, J. M. "An Ignition Method for the Determination of Total Phosphorus in Lake Sediments," *Water Res.* 10:329–331 (1976).
6. Spangler, F., W. Sloey, and C. W. Fetter. "Experimental Use of Emergent Vegetation for Biological Treatment of Municipal Wastewater in Wisconsin," in *Biological Control of Water Pollution*, J. Tourbier and R. W. Pierson, Eds. (Philadelphia: University of Pennsylvania Press, 1976), pp. 161–171.
7. Haslam, S. M. "The Performance of *Phragmites communis* Trin. in Relation to Water-Supply," *Ann. Bot.* 34:867–877 (1970).

38f Domestic Wastewater Treatment Using Emergent Plants Cultured in Gravel and Plastic Substrates

P. S. Burgoon, K. R. Reddy, and T. A. DeBusk

INTRODUCTION

Constructed wetlands used for the treatment of wastewater have three basic components: emergent aquatic plants, a substrate (generally gravel or sand), and heterotrophic and autotrophic bacteria. The bacteria colonize the plant root and substrate environment. Wastewater passed through a constructed wetland is "cleaned" by the bacteria and plants.

The substrate (except in the case of a soil matrix) is generally thought of as inert material that provides surface area for colonization of bacteria. Specific surface area (m^2/m^3) required for optimal wastewater treatment in an artificial marsh is unknown. Use of high-specific-surface-area substrates in the trickling filter process has improved BOD_5 removal and nitrification when compared to the traditional gravel substrates.[1] Use of plastic substrates was recommended by Wolverton[2] because, depending on size of the gravel, the plastic may have two to four times the specific surface area of gravel substrate. Moreover, the high porosity of plastic substrates may reduce clogging of the planting media by wastewater-borne solids. If treatment is improved using the high-specific-surface-area substrates, the volume of substrate and the area needed for the constructed wetland could be reduced significantly (assuming plants will grow as well in plastic as in gravel). Decrease in land costs may compensate for high costs of plastic media.

This study compared plant growth and wastewater treatment in two plastic substrates and in 1-cm-diameter gravel, each of which had different specific surface areas.

Table 1. Physical Properties of Substrates

Substrate	Specific Surface Area[a] (m^2/m^3)	Porosity (%)	Surface area/ Bucket (m^2/bucket)
River gravel	394.0[b]	48	6.8
Tri-Paks			
2.5 cm[c]	278.7[d]	66	4.8
5.0 cm[c]	137.8[d]	83	2.4

[a]Square meters of surface area per cubic meter bulk volume.
[b]Based on volume displacement and estimate of pebble surface area.
[c]Diameter of the plastic spheres.
[d]As reported by manufacturer (Jaeger Products, Inc., Spring, TX).

METHOD AND MATERIALS

Three substrates were used: gravel (high specific surface area, low porosity) and two sizes of plastic media (low and medium specific surface area, both with high porosity) (Table 1). Microcosms (22-L) were filled with substrates and then planted with an emergent species. Four emergent plant species were evaluated: *Typha latifolia, Phragmites australis, Scirpus pungens,* and *Sagittaria latifolia.* All species of plants were grown in the gravel substrate. *T. latifolia* was not grown in either plastic substrate. Three replicates were set up for each plant/substrate combination. Substrate replicates with no plants were designated as control treatments. Plants were grown for six months in primary sewage effluent prior to this experiment.

Treatments were batch loaded with primary sewage pumped directly from the effluent weir of the primary sedimentation tank at the University of Florida Wastewater Treatment Plant. Effluent was applied at BOD_5 and TKN mass loading rates of 210 and 45 kg/ha/day, respectively. Because pore volume differed among the three substrates, hydraulic retention times (HRT) were varied for each substrate treatment so that the hydraulic loading rate (HLR)* would be the same for all microcosms (Table 2). Six weeks after this loading test was started, influent and effluent samples were collected before

Table 2. Hydraulic Retention Times for Plant/Substrate Treatments During the High Loading Rate

Substrate	HLR[a] (cm/day)	HRT[b] (days)
Gravel	14.1	0.5
Plastic (1 in.)	14.1	0.9
Plastic (2 in.)	14.1	1.1

[a]Hydraulic Loading Rate. The flow rate (cubic centimeters per day) of primary effluent applied per unit surface area of microcosm (640 cm^2).
[b]Hydraulic Retention Time. The pore volume of each microcosm divided by the flow rate of primary effluent. This represents the actual time the wastewater remained in the buckets.

*Hydraulic Loading Rate is defined as the flow rate of wastewater (m^3/day) divided by the surface area (m^2) of the gravel bed.

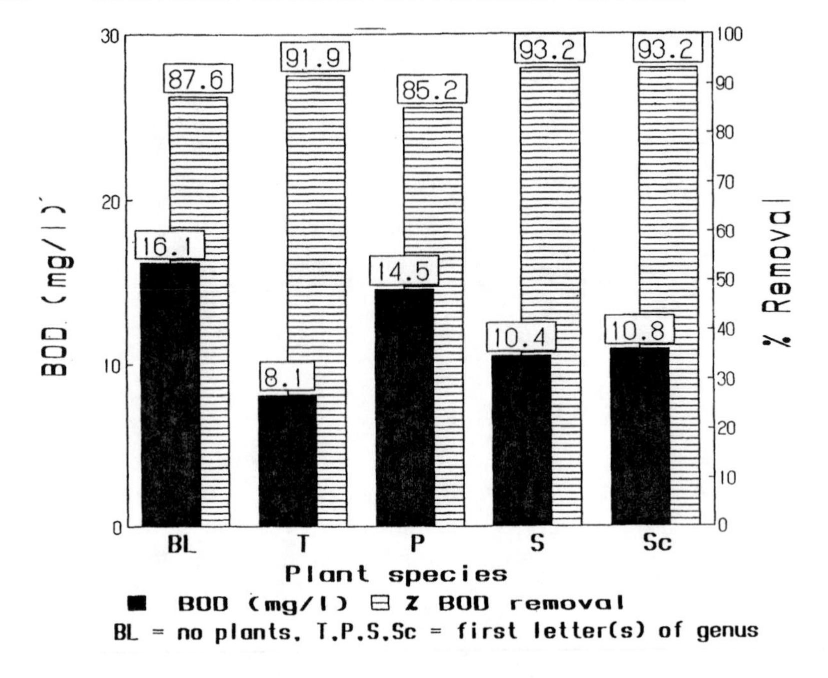

Figure 1. Effluent BOD_5 concentration and removal—gravel.

and after three separate HRTs for each treatment. The samples were analyzed for BOD_5, TKN, TP, $N\text{-}NH_4$, $N\text{-}NO_3$, and $P\text{-}PO_4$.[3] After the study (eight weeks), aboveground biomass of all species was harvested.

RESULTS

Average influent concentrations of BOD_5, TKN, and TP were 120, 25.4, and 5.1 mg/L, respectively. Effluent BOD_5 was below 30 mg/L for all treatments. *Sagittaria, Scirpus,* and *Typha* removed a significantly larger percentage of BOD_5 than the control and *Phragmites.* Effluent concentrations of *Sagittaria* and *Typha* was significantly lower than those of the other treatments (Figure 1). Percent BOD_5 removal was higher and effluent concentrations lower in the gravel compared with the plastic substrates (Figure 2).

Removal of TKN and TP was highest in the gravel treatments. Percent removal of TKN (based on a mass balance accounting for evapotranspiration losses in each plant/substrate treatment) for the *S. latifolia* and *T. latifolia* planted in the gravel was significantly ($p = 0.05$) higher than the *P. australis* and control. The control and *Scirpus* treatments in gravel removed higher percentages of TKN than the same plastic treatments (Table 3). Based on treatment comparisons of TKN mass removal, *Sagittaria, Scirpus,* and *Typha* removed similar amounts of TKN (2.63, 2.52, 2.99 $g/m^2/day$, respectively).

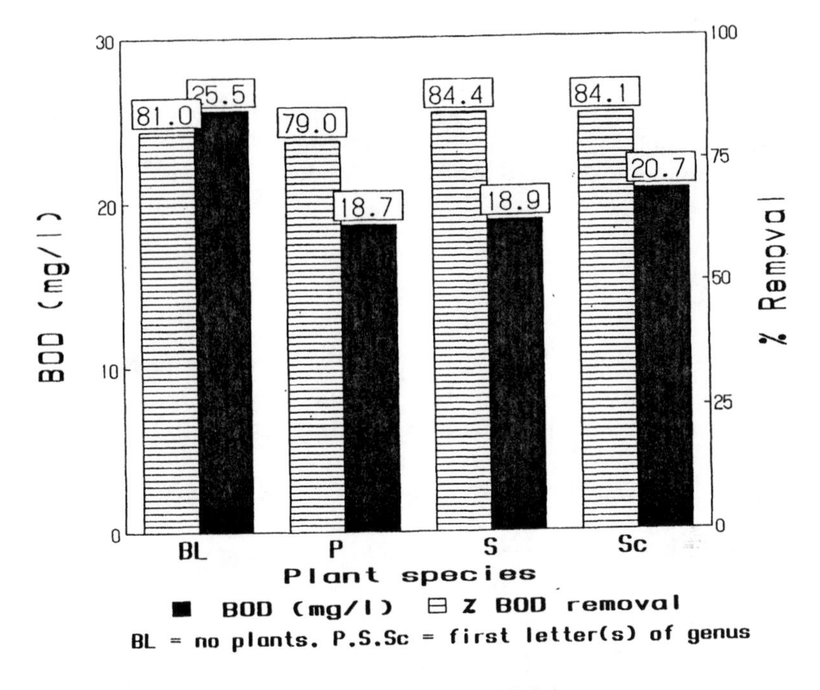

Figure 2. BOD_5 concentration and removal—2.5 cm Tri-Pak.

Mass removal of TKN by *Typha* (2.99 g/m²/day) was significantly (p = 0.05) higher than that by *Phragmites* (1.95 g/m²/day) and the control (1.97 g/m²/day). Total phosphorus removal averaged 45% in the gravel control. *S. latifolia* planted in gravel removed a higher percentage of TP than the other treatments. Mass removal of TP was highest in the gravel treatments: *Sagittaria* (0.52 g/m²/day), *Typha* (0.54 g/m²/day), and the control (0.51 g/m²/day) removed more TP than *Phragmites* (0.27 g/m²/day) and *Scirpus* (0.31 g/m²/day). Effluent concentrations of TKN and TP for *S. latifolia* and *T. latifolia* in gravel were both less than 5 mg/L TKN and 3 mg/L TP.

S. latifolia and *S. pungens* grew poorly in the plastic substrates, whereas *P. australis* grew well in all three substrates. *S. latifolia* was very susceptible to a fungus, which was controlled with biweekly applications of the fungicide

Table 3. Total Kjeldahl Nitrogen (TKN) Removal as a Function of Plant Species and Substrate Type

	TKN Removal (%)		
Treatments	**Gravel**	**Plastic (2.5 cm)**	**Plastic (5.0 cm)**
No plants	55.2 ± 13.5	29.0 ± 2.4	33.4 ± 3.3
S. pungens	75.5 ± 7.3	35.1 ± 10.7	34.7 ± 7.1
S. latifolia	91.8 ± 2.2	52.7 ± 6.9	75.8 ± 15.5
P. australis	67.5 ± 9.7	58.6 ± 19.5	85.7 ± 3.3
T. latifolia	85.6 ± 2.4	NA	NA

Benomyl®. Biomass production for *S. latifolia*, *P. australis*, and *S. pungens* in gravel averaged 344.2, 223.2, and 10.8 kg/ha/day, respectively. Of the nitrogen and phosphorus removed from the wastewater, 22% and 18%, respectively, were stored in aboveground biomass of the *S. latifolia*.

DISCUSSION

Wastewater contaminant removal generally was better in the high-specific-surface-area gravel substrate than in the plastic substrates. This was consistent in the three substrate controls, where the gravel effluent was of highest quality. *P. australis* grew well in all three substrates; however, there was no significant ($p = 0.05$) difference in water quality improvement between the three substrates containing this species. Several replicates of *Sagittaria* in the plastic substrates were killed by fungus, but the 5.0-cm plastic replicate treatments with *Sagittaria* grew back. These plastic treatments removed amounts of TKN and BOD_5 similar to those removed by *Sagittaria* in gravel. It is not possible to conclude that wastewater treatment in the microcosms was limited by specific surface area. Surface area of the roots may have been large enough to negate any treatment effects due to an increase in specific surface area of the substrate. The fact that plant growth in general was best in the gravel substrate implies that plastic substrates are not good substitutes for gravel in submerged vegetated beds.

The effect of increased surface area needs to be evaluated in a continuous-feed system instead of a batch-loaded system to confirm these results. Different sizes of gravel with different specific surface areas should be evaluated. A larger gravel (i.e., 5- to 7-cm) is less expensive than the 1.25-cm gravel but has one-fourth the specific surface area (420 vs 105 m^2/m^3, respectively)[4] and may clog quickly. Work with continuous-feed systems is ongoing at the Reedy Creek Utilities Energy Services Community and Waste Research Facility. These long-term studies will allow us to better evaluate plastic substrate for plant growth, wastewater treatment, and potential for reduced plugging due to high porosity of the substrate.

Common arrowhead (*S. latifolia*) is clearly an excellent species for evaluation in constructed wetlands and submerged vegetated beds. *S. latifolia, S. pungens,* and *T. latifolia* removed similar amounts of BOD_5, TKN, and TP. Comparisons of *Scirpus validus, S. pungens,* and *S. latifolia* in continuous-flow wastewater treatment systems are underway. Gersberg[5] determined that *S. validus* is better at removing nitrogen than *P. australis* and *T. latifolia*. Common arrowhead is damaged in freezing weather, so its use may be limited to the subtropical and tropical areas of the world.

REFERENCES

1. "A Comparison of Trickling Filter Media," CH2M HILL, Engineering Consultants, P.O. Box 22508, Denver, CO (1984).
2. Wolverton, B. C. "Hybrid Wastewater Treatment Using Anaerobic Microorganisms and Reed (Phragmites communis)," *Econ. Bot.* 36(4):373–380 (1982).
3. *Standard Methods for the Examination of Water and Wastewater*, 16th ed. (Washington, DC: American Public Health Association, 1986).
4. Muir, J. F. "Recirculated Water Systems in Aquaculture," in *Recent Advances in Aquaculture*, J. F. Muir and R. J. Roberts, Eds. (Boulder, CO: Westview Press, 1983), pp. 359–446.
5. Gersberg, R. M., B. V. Elkins, S. R. Lyons, and C. R. Goldman. "Role of Aquatic Plants in Wastewater Treatment by Artificial Wetlands," *Water Res.* 20(3):363–368 (1986).

38g Aquatic Plant Culture for Waste Treatment and Resource Recovery

J. B. Kingsley, J. J. Maddox, and P. M. Giordano

INTRODUCTION

Innovative and cost-effective waste treatment and resource recovery methods are needed to reduce the environmental degradation associated with liquid waste streams from municipalities, industries, and agricultural enterprises. Artificial wetland systems that use aquatic macrophytes have been developed to treat liquid wastes with a minimum of capital and energy input.[1-3] However, there have been limited efforts to recover useful products from these aquatic systems.

Aquatic macrophytes with potential uses in industry and agriculture are Chinese water chestnuts, cattails, and common reeds.[4-6] Chinese water chestnuts (*Eleocharis dulcis*) produce a root crop called a corm that has been used for food in Asia for hundreds of years and is imported into the United States.[7] Research at TVA also indicates that water chestnut corms have as much available starch for fuel alcohol production as corn (TVA unpublished data). Research supported by TVA indicates that Chinese water chestnut hay is a good forage for ruminant animals.[8] Due to high starch content in rhizomes and potential high yields, cattail (*Typha* spp.) has potential as a source of starch for fuel alcohol production.[9] Common reeds (*Phragmites australis*) are used for fiber and thatching in Europe and may offer a source of cellulose for industrial purposes. Incorporated into the wetland waste treatment concept, commercial uses of these crops could increase the cost-effectiveness of waste treatment and recover valuable plant nutrients.

This project demonstrated the potential of three aquatic plants to remove pollutants from wastewater and produce useful crops.

METHODS AND MATERIALS

Demonstrations were conducted at Cypress Creek Sewage Treatment Plant, Florence, Alabama. Two types of earthen facilities were built: (1) five units to demonstrate sludge as plant fertilizer in soil flooded with secondary effluent and (2) four units to conduct flow-through demonstrations (Figure 1). Flow-through units were constructed of native soil with earthen weirs to maintain water depth. Flow meters, ball valves, and float valves maintained depth in each sludge unit. *E. dulcis* sets were planted May 1, 1980, on 76-cm centers in the demonstration units and immediately flooded with secondary effluent. Flow-through units were continuously irrigated with secondary effluent (Figure 1).

Secondary sewage sludge was applied to dry soil in sludge units at one-time rates of 0, 4.5, 11.2, and 22.4 mt/ha dry wt. One comparison unit received only commercial granular fertilizer at rates of 280 kg/ha N as urea, 112 kg/ha P as P_2O_5, and 168 kg/ha K as K_2O.

In 1981, flow-through units were reduced in size due to drainage problems encountered in 1980. In addition to water chestnuts, common reeds (*P. australis*) and cattails (*T. latifolia*) were also planted in sequential order (Figure 2).

Cattails and reeds were planted on 61-cm centers from May 25 to May 29 and irrigation began May 29, 1981. Due to poor survival of initial transplanting, common reeds were replanted on June 15. Sludge was not applied in 1981 in order to measure residual effect of sludge application. The inorganically fertilized unit was fertilized at the same rate as in 1980.

RESULTS AND DISCUSSION

Sludge Demonstration

1980

Chestnut yields in the sludge demonstration ranged from 4.3 mt/ha in the low sludge unit to 18.5 mt/ha in the 11.2 mt/ha sludge unit (Table 1). The suppressed yield at the high sludge rate may be due to the high organic loading in the soil, causing anaerobic conditions in the root zone and inhibiting root growth early in the season. Sludge treatments averaged 38% higher yield than the inorganic fertilizer treatment, and the high sludge was almost three times higher than the fertilizer treatment.

Water chestnut hay yields for sludge treatments in 1980 showed little variation between units (Table 1). Variation in hay yield was small when compared to the variation in chestnut yields. Hay yields may have reached a maximum for water chestnuts under these conditions. After about 100 days of growth, water chestnut foliage can become so dense that it begins to fall over (lodge), inhibiting further foliage growth and limiting water chestnut hay production.

1981

Chinese water chestnut yields. In 1981, chestnut yields ranged from 8.6 mt/ha in the no-sludge unit to 1.4 mt/ha in the 4.5 mt/ha unit (Table 2). Corm yields were less than 1980 levels in all units. Average yields in static sludge

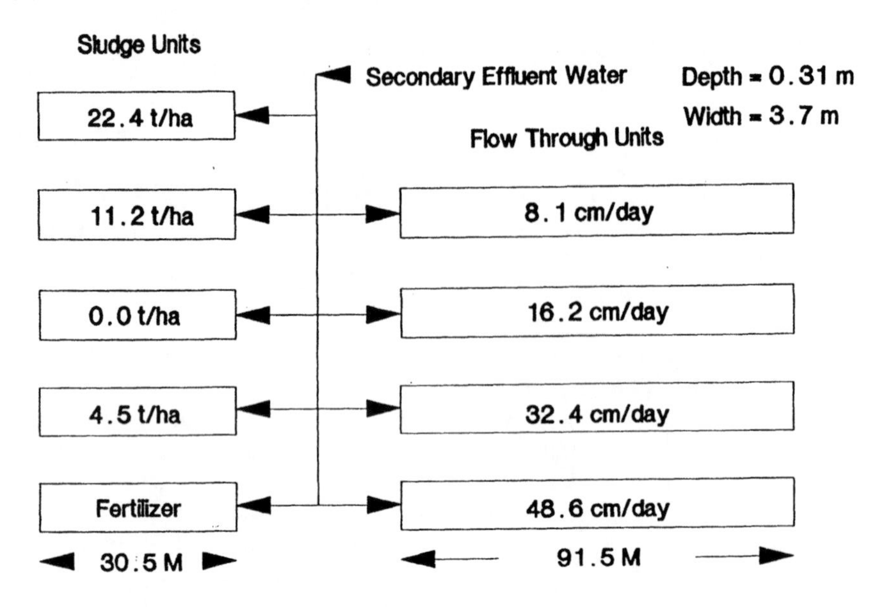

Figure 1. Design for 1980 demonstrations.

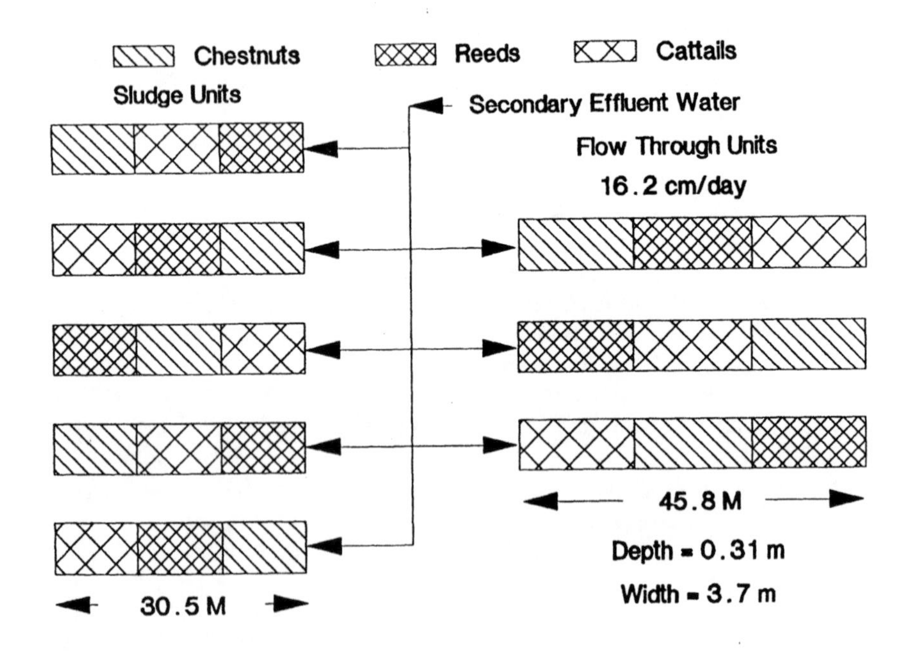

Figure 2. Design for 1981 demonstrations.

Table 1. Yield Data for Aquatic Plants in Wetlands Demonstration, 1980

Treatment	Yield (mt/ha)	
	Water Chestnut Corms	Water Chestnut Hay
Sludge (mt/ha)		
0.0	12.1	7.1
4.5	4.3	8.2
11.2	18.5	5.7
22.0	10.1	6.6
Inorganic Fertilizer (kg/ha)	6.3	6.9
Flow-through secondary effluent (cm/day)		
8.1	4.1	7.2
16.2	9.7	7.7
32.4	7.2	8.1
48.6	3.8	6.0

mt = metric tons.

treatments were 60% less in 1981 than in 1980. Lack of nutrients may have limited chestnut yield; however, inorganic fertilizer was applied in 1981 at the same rate as in 1980, and 1981 yields were 42% lower than in 1980. A more plausible explanation for reduced chestnut yields in 1981 was severe weed infestation in all culture areas. Competition from weeds established in 1980 effectively reduced nutrients, increased shading, and lowered yield.

Hay yields for 1981 (Table 2) were slightly lower than 1980. However, these yield figures are misleading because of weed contamination. Commercial hay harvesting equipment was used to harvest hay as in 1980, and visual inspection of the hay revealed a high percentage of weed contamination.

Table 2. Yield Data for Aquatic Plants in Wetlands Demonstration, 1981

	Yield (mt/ha)			
	Water Chestnut Corms	Water Chestnut Hay	Cattail Foliage	Cattail Rhizomes
Treatment (mt/ha)				
Sludge applied				
0.0	8.6	6.7	4.6	1.03
4.5	1.4	9.8	6.7	1.12
11.2	2.2	6.6	2.0	0.74
22.4	3.9	7.4	4.2	0.81
Inorganic fertilizer (kg/ha)	4.0	6.0	2.3	0.64
Flow-through secondary effluent (16.2 cm/day)				
Unit 1	12.5	5.3	9.9	1.5
Unit 2	6.0	8.2	0.90	no sample
Unit 3	6.3	7.8	0.30	0.04
Average	8.3	7.1	3.7	0.51

Table 3. Water Quality of Effluent from Flow-Through Units, 1980

	BOD_5 (mg/L)	Suspended Solids (mg/L)	Dissolved Oxygen (mg/L)	pH	Fecal[a] Coliform Bacteria
Influent	8.9	9.3	6.7	7.1	>20,000
Effluent					
(8.1)	3.1 (65)	4.0 (57)	6.3 (6)	6.8 (5)	500 (>97)
(16.2)	3.7 (59)	3.3 (64)	6.1 (10)	6.8 (4)	3,900 (>81)
(32.4)	3.9 (58)	5.3 (43)	5.0 (25)	7.2 (−2)	16,600 (>17)
(48.6)	3.8 (57)	5.0 (46)	5.3 (21)	7.2 (−1)	17,900 (11)
Average	3.6 (60)	3.5 (63)	5.7 (16)	7.0 (2)	9,725 (51)

Note: Percent reduction is given in parentheses.
[a]Colonies/100 mL.

Cattail yields. Cattail foliage yields in sludge-applied units averaged 3.7 mt/ha and ranged from 6.7 to 2.0 mt/ha (Table 2). No relationship between cattail yield and sludge application rate was apparent. Andrews and Pratt[10] report cultured cattail shoot yields in fertilized plots as high as 25.7 mt/ha dry wt. However, these results were obtained when initial planting density was 24 rhizomes/m^2 and weeds were controlled. In this demonstration, planting density was 2.4 shoot + rhizomes/m^2, and the average rhizome yield was 0.87 mt/ha. Andrews and Pratt[10] and Moss et al.[11] (as cited in Andrews and Pratt[10]) reported root-to-shoot ratios of 1.0 and 1.7, respectively. The low root/shoot ratio in this instance (0.21) is probably due to plant sparsity, causing rhizomes to be widely dispersed.

Flow-Through Demonstration

1980

Chinese water chestnut yields. Higher yields occurred in the lower flow rates, indicating a relationship between yield and flow rate (Table 1). The low yield in the lowest flow rate was probably due to limited nutrient availability at the low flow rate. When the flow rate was doubled, the chestnut yield more than doubled. Further increases in the flow rate reduced the yield because channelization reduced availability of nutrients to plants.

Hay yields varied little between flow rates but increased as flow increased until the highest rate was reached. Low variation among hay yields can be attributed to lodging. Faster growing units were first inhibited by lodging, and low-yielding units continued growing until a similar degree of lodging inhibited their growth (Table 1).

Water quality. Water quality parameters monitored by the Florence, Alabama sewage treatment plant are 5-day biochemical oxygen demand (BOD_5); suspended solids (SS); dissolved oxygen (DO); and pH (Table 3). BOD_5 concentration in secondary effluent water was reduced by an average of

60% after passage through the wetland. The greatest reduction in BOD_5 occurred at the lowest flow rate, and the smallest reduction at the highest flow rate. SS reduction ranged from 43% to 63%, increasing with increasing flow rate. DO concentration in wetland effluents decreased by an average of 1.0 mg/L. Reductions in DO were inversely related to increasing flow, with the smallest decrease occurring in the lowest flow rate and the largest decrease in the next-to-highest flow rate. The pH was slightly decreased, from 7.1 in the secondary effluent to an average of 7.0 in the wetland effluent.

The Florence, Alabama sewage treatment plant does not normally monitor fecal coliform bacteria (FCB) because it is not required to chlorinate the secondary effluent. The FCB count regularly exceeds 20,000 colonies/100 mL. FCB was checked in mid-July of 1980 (Table 3). Change in FCB count in wetland effluent compared to secondary effluent on this date was negatively correlated with flow rate. FCB removal rate decreased as flow rate increased. Longer hydraulic retention time in lower flow rate units allowed sufficient time for FCB to be removed. As the hydraulic retention time decreased, more FCB were flushed out.

While there were some differences between individual flow rates, average concentration of five water quality parameters was either greatly improved (BOD_5, SS, FCB) or not adversely affected (DO, pH) by passing through artificial wetland systems in 1980.

1981

Chinese water chestnut yields. In contrast to decreased yields observed in the sludge demonstration in 1981, average chestnut yields in flow-through units increased by 8% (Table 2). Chestnut yields in units 2 and 3 were almost identical (6.0 and 6.3 mt/ha, respectively). However, the yield in unit 1 was twice as high (12.5 mt/ha). The higher yield in unit 1 might have resulted from water chestnut plants being located near the water inlet, allowing plants with greater access to available plant nutrients to produce higher yields (Figure 2).

Water chestnut hay yields averaged 7.1 mt/ha for the three flow-through units, ranging from 8.2 to 5.3 mt/ha (Table 2).

Cattail and reed yields. Cattail foliage yield in flow-through units averaged 3.7 mt/ha and ranged from 0.30 to 9.9 mt/ha (Table 2). Stand establishment in unit 1 was much better than in units 2 and 3. A high degree of yield variability in managed and natural cattail stands is reported in the literature (Moss et al.[11] as reported in Andrews and Pratt[10]; Fox[12]). Rhizome yield in flow-through units averaged 0.51 mt/ha and root/shoot ratio was low (0.14), as in sludge-applied units. Reasons for this result are similar to those discussed for the sludge-applied units.

Reed yield data could not be reported due to poor stand establishment in 1981.

Table 4. Water Quality of Effluent from Flow-Through Units, 1981

	BOD_5 (mg/L)	Suspended Solids (mg/L)	Dissolved Oxygen (mg/L)	pH	Fecal[a] Coliform Bacteria
Influent (16.2 cm/day)	9.1	16.8	8.0	7.2	>20,000
Effluent (16.2 cm/day)					
Unit 1	3.0 (67)	1.7 (90)	7.2 (11)	6.9 (5)	1,615 (92)
Unit 2	5.3 (42)	6.0 (64)	7.4 (8)	6.9 (5)	10,000 (>50)
Unit 3	3.8 (58)	0.7 (96)	6.6 (17)	6.9 (5)	2,965 (>85)
Average	4.05 (55)	2.78 (83)	7.0 (12)	6.9 (5)	4,860 (76)

Note: Percent reduction is given in parentheses.
[a]Colonies per 100 mL.

Water quality. Because flow rate in all units was constant at 16.2 cm/day in 1981, only averages will be discussed. Water quality improvements in 1981 showed trends similar to those observed in 1980. BOD_5, SS, and FCB showed dramatic reductions, and DO and pH were not adversely affected. Secondary effluent BOD_5 was reduced by 55% from 9.06 to 4.06 mg/L, SS was reduced 83% from 16.75 to 2.78 mg/L, and FCB was reduced more than 75% from >20,000 to <5000 colonies per 100 mL. DO and pH were reduced by 12% and 5%, respectively (Table 4).

CONCLUSIONS

Artificial wetland water treatment systems using aquatic plants to absorb nutrients from water can effectively remove pollutants from wastewater and produce an economically valuable crop. Plants recover valuable nutrients lost in conventional waste treatment systems, and crops produced can help offset waste treatment costs. Optimum flow rate of secondary effluent for waste treatment was 8.1 to 16.2 cm/day, which reduced BOD_5 60%, SS 72%, and FCB >80%. Aquatic plants can recover nutrients from wet sludge (3.5% solids), and sludge is at least as effective as inorganic fertilizer for aquatic plant production.

REFERENCES

1. Steiner, G. R., J. T. Watson, and D. A. Hammer. "Municipal Wastewater Treatment by Constructed Wetlands—A TVA Demonstration in Western Kentucky," paper presented at the National Wildlife Federation Conference on Increasing Our Wetland Resources, Washington, DC, October 4–7, 1987.
2. Watson, J. T., F. D. Diodato, and M. Lauch. "Design and Performance of the Artificial Wetlands Wastewater Treatment Plant at Iselin, Pennsylvania," in *Aquatic Plants for Water Treatment and Resource Recovery,* K. R. Reddy and W. H. Smith, Eds. (Orlando, FL: Magnolia Publishing Inc., 1987).
3. Wolverton, B. C. "Aquatic Plant Wastewater Treatment Systems," paper presented

at National Environmental Health Association Conference, Mobile, AL, February 7–9, 1988.

4. Maddox, J. J., L. L. Behrends, D. W. Burch, J. B. Kingsley, and E. L. Waddell, Jr. "Optimization of Biological Recycling of Plant Nutrients in Livestock Waste by Utilizing Waste Heat from Cooling Water," Report EPA-600/7-82-041, prepared for the Office of Research and Development, U.S. Environmental Protection Agency, Washington, DC (1982).
5. Brodie, G. A., D. A. Hammer, and D. A. Tomljanovich. "Constructed Wetlands for Acid Drainage Control in the Tennessee Valley," paper presented at the National Wildlife Federation Conference on Increasing Our Wetland Resources, Washington, DC, October 4–7, 1987.
6. Wolverton, B. C., C. C. Myrick, and K. M. Johnson. "Upgrading Septic Tanks Using Microbial/Plant Filters," *J. Mississippi Acad. Sci.* 29:19–25 (1984).
7. Hodge, W. H. "Chinese Water Chestnut or Maitai—A Paddy Crop of China," *Econ. Bot.* 10:65 (1956).
8. Tayer, S. R. "Evaluation of Aquatic Plant Materials Using Lambs and Laboratory Techniques," MS Thesis, Mississippi State University, Mississippi State, MS (1981).
9. Ladenburg, K. "Ethanol Production from Cattails," paper no. 82-3600, presented at winter meetings of the American Society of Agricultural Engineers, Chicago, IL, December 14–17, 1982.
10. Andrews, N. J., and D. C. Pratt. "Energy Potential of Cattails (*Typha* spp.) and Productivity in Managed Stands," *J. Minnesota Acad. Sci.* 44(2):5–8 (1978).
11. Moss, D. N., C. A. Fox, and S. Hsi. "Biomass Yield of Managed Cattails," unpublished data (1977).
12. Fox, C. A. "Capture of Radiant Energy by Plants," M.S. Thesis, University of Minnesota, St. Paul, MN (1975).

38h Bacteriological Tests from the Constructed Wetland of the Big Five Tunnel, Idaho Springs, Colorado

Wafa Batal, Leslie S. Laudon, Thomas R. Wildeman, and Noorhanita Mohdnoordin

INTRODUCTION

Acid mine drainage is one of the most persistent industrial pollution problems in the United States. Streams and rivers have been adversely affected by underground mines that have been abandoned for decades. Acid mine drainage originates from metabolic activity of iron-oxidizing bacteria (*Thiobacillus ferrooxidans*), which catalyzes the oxidation of pyrite.[1] Control methods include chemical treatments to neutralize toxic constituents and make them insoluble and physical storage to create anoxic environments to inhibit growth of iron-oxidizing bacteria. These methods are costly and have many limitations. Creative treatment solutions need to be studied for application in remote mountain environments, where harsh winters and difficult access make conventional methods infeasible.

Wetlands are a potential treatment for small flows of acid mine waters. Metal and nutrient contents are modified by adsorption and exchange in peat soils, plant uptake, microbial activity, and other geochemical processes. For metal removal, intrinsic bacterial microflora in a wetland system are important. In the aerobic zone, two groups of heterotrophic metal-utilizing bacteria are present: the iron and manganese oxidizers and one group of obligate autotrophic iron oxidizers (*T. ferrooxidans*).[2] In the anaerobic zone, sulfate reducers are known to be present.

This chapter reports on the occurrence, depth, and position of these bacteria in the wetland pilot system at the Big Five Tunnel at Idaho Springs, Colorado.

BACTERIA IN WETLANDS IMPORTANT TO METAL REMOVAL

T. ferrooxidans is a motile acidophilic, mesophilic chemoautotroph. It is a gram-negative, non-spore-forming rod-shaped organism that lives autotrophically on thiosulfate.[3] It is incapable of heterotrophic growth when inoculated onto an organic nutrient medium. Some strains have adapted to growing heterotrophically on glucose, but this occurrence is variable even with different cultures of a single strain.[4] Energy requirements for carbon dioxide fixation and other metabolic functions are derived from oxidation of ferrous iron and reduced sulfur compounds.

Fine-grained pyrite is oxidized by *T. ferrooxidans* and air according to the reaction:

$$4FeS_2 + 15O_2 + 2H_2O = 2Fe_2(SO_4)_3 + 2H_2SO_4 \tag{1}$$

Ferrous iron is released into solution and oxidized to the ferric state according to the reaction:

$$4Fe^{2+} + O_2 + 10H_2O = 4Fe(OH)_3 + 8H^+ \tag{2}$$

Ferric hydroxide is precipitated as "yellow boy," and pH of the water decreases. Abiotic oxidation of Fe^{2+} is slow, and *T. ferrooxidans* catalyzes the oxidation of Fe^{2+} producing ferric hydroxide at pH less than 3.5.[5] Another bacterium, *Metallogenium,* was isolated from acidic mine drainage at pH of 3.5 to 5.0.[6] *Gallionella* is also capable of iron oxidation at neutral pH values.[3]

Sulfate-reducing bacteria are strict anaerobes that respire by reducing sulfate to sulfide.[7] The two important genera are spore-forming types (*Desulfotomaculum*) and heterotrophic non-sporulating types (*Desulfovibrio*). They reduce sulfate to sulfide through dissimilatory sulfate reduction, with the sulfide ion as an intermediate.[8] The reduction involves a sequence of biochemical steps and many enzymes.[3] The sulfide produced reacts with metal ions, precipitating amorphous metal sulfides.[9,10]

Mixed cultures of sulfate-reducing bacteria have reduced sulfate ions and increased the pH of acid mine waters when supplied with a source of carbon and energy.[11,12] Bacterial growth is favored by neutral pH and retarded by pH less than 5.0. Sulfate was reduced by bacteria in the laboratory at pH 2.8 when sawdust was in the substrate.[11]

Manganese removal, promoted by manganese-oxidizing bacteria, is divided into two major classes: enzymatic and nonenzymatic.[3] The enzymatic oxidation can proceed by a number of paths. *Arthrobacter* strain B isolated by Bromfield,[13] *Pseudomonas* spp., and *Citrobacter freundii* can oxidize manganese as follows:

$$Mn^{2+} + {}^1/_2\,O_2 + H_2O = MnO_2 + 2H^+ \tag{3}$$

Leptothrix pseudoochracea, Arthrobacter siderocapsulatus, and *Metallogenium* spp. can also catalyze the oxidation using prebound Mn(IV):[14]

$$Mn_2O_3 + 1/2\ O_2 + 2H_2O = 2H_2MnO_3 \quad (4)$$

Nonenzymatic manganese oxidation is promoted by a large number of bacteria.[3] In addition, some microorganisms such as *Metallogenium* spp. cause precipitation of oxidized manganese.

MATERIALS AND METHODS

The pilot wetland ecosystem constructed at Big Five Tunnel has three 19-m^2 cells.[15] Soils in the 1-m-deep cells are mushroom compost in cell A; equal amounts of aged manure, decomposed wood products, and Colorado peat in cell B; and the same soil as cell B underlain with 10–15 cm of 5- to 8-cm limestone rock in cell C. Portions of the soil components and each cell's soil mixtures were sampled during design and construction.[15]

Microbial Tests

For both autotrophic iron oxidizers and sulfate reducers, 1 g of soil components and actual soil samples were taken to determine initial bacterial content. Two months after mine drainage flowed in the system, samples were collected from the top 15 cm and 0.9 m below the surface at two sites in each cell (Figure 1 and Table 1).

Soils were collected in sterilized screw-capped Nalgene jars filled to the top. Samples were mixed for 20 min, and bacterial analyses were performed within 24 hr on the well-mixed, wet soil. Gentle grinding with a sterilized mortar and pestle was sometimes necessary to provide a slurry. Moisture content was determined by placing a portion of the sample in a tared jar, air drying, and weighing. For nonfluid samples, 200 mL of sterile water was added, creating a slurry, and a slurry aliquot was taken. Serial dilutions of the sample with 0.2% NaCl solution (autoclaved) were prepared to 10^{-7}. For autotrophic iron-oxidizing bacteria, inoculations were made by adding 1 mL of the diluted solution to sterile 14 mL test tubes containing 4 mL ferrous sulfate (9 K) solution of Silverman and Lundgren.[16] Test tubes were incubated at 30°C for three weeks. Formation of orange ferric hydroxide and microscopic verification (presence of rod-shaped greyish cells) indicated bacterial growth, and the most-probable-number method (MPN)[1,17] was used to enumerate bacteria. Autotrophic iron oxidizers were also tested in acid mine water flowing in and out of cell A, using ferrous sulfate (9 K) solution and the MPN method. For sulfate reducers, inoculations were made by adding 1 mL of diluted solution to 8 mL of Postgate-modified medium B containing resazurin.[8] Screw-capped, autoclaved 9-mL test tubes were inoculated and tubes filled to the top with Postgate medium. Tubes were incubated at 30°C for two to three weeks and checked every three to four days for positive tests, indicated by a black precipitate (FeS) in the tubes and the odor of H_2S. The MPN method was used for the bacterial enumeration.

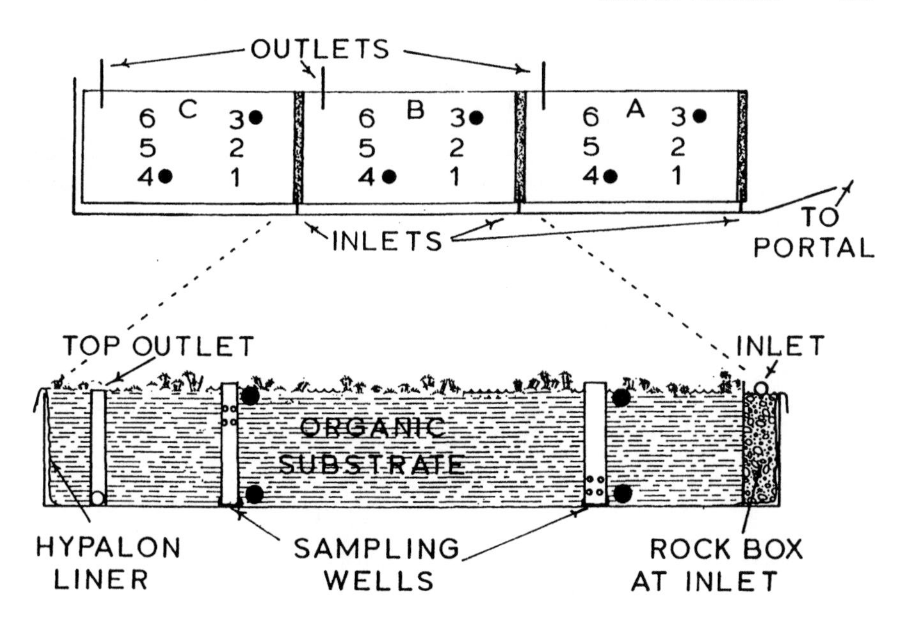

Figure 1. Plan view of the wetland pilot system at the Big Five Tunnel and cross section of cell B. The dots show the sampling locations for the substrate for the iron-oxidizing autotrophs and the sulfate reducers.

Wet soils were collected in association with filamentous algae (*Ulothrix*), matted algae (*Euglena*),[18] and in the absence of algae (Figure 2). Samples a and d contained primarily algae, while b, c, e, and f contained primarily substrate from the top 10 cm. Sampling was performed by allowing soil slurries to flow into sterilized Nalgene jars held just below the substrate surface. In the laboratory, jars were placed on a shaker for 20 min, and 1 mL of each sample was diluted with sterilized water to 10^{-4}. The dilution was then plated out aseptically in duplicates on agar media.[19] Agar media (5–7 mL) were poured at 45°C into sterilized petri dishes; one-half of the plates contained 1.0 mL $FeSO_4$ solution (1.0 g/L) each, and the other half contained 1.0 mL of $MnSO_4$ (1.0 g/L) solution each. Plate inoculation was performed with a bent glass rod (flamed) to spread 1.0 mL of the dilution sample on the agar; plates were then incubated at 30°C for five to seven days. Total heterotrophic count was obtained with a colony counter. The agar surface was then flooded with 0.2% tetramethylbenzidine in 2% acetic acid, and colonies that turned a blue color were enumerated to determine the number of heterotrophic iron and manganese oxidizers per gram of substrate.[18]

Table 1. Bacteria per Gram in Initial Substrate Components; Substrate Samples in the Top 15 cm and 0.9 m from the Surface in Wetland Cells After Two Months of Mine Drainage Flow at the Big Five Tunnel, Idaho Springs, CO

	Iron oxidizers		Sulfate Reducers	
Initial Component	**MPN/g $\times 10^3$**	**Factor of Confidence**	**MPN/g $\times 10^4$**	**Factor of Confidence**
Aged manure	0		9	3.3
Wood products	0		0.3	3.3
Mushroom compost	0		50	3.3
Peat	0.2	3.3	0.3	3.3
Peat/manure/wood[b]	0		2	3.3
Two-month samples from 15 cm				
Cell A Well 3	20	3.8	1000	3.3
Cell A Well 5	900	3.3	>3000	3.3
Cell B Well 3	5	3.3	>1000	3.3
Cell B Well 5	8	3.3	>1000	3.3
Cell C Well 3	40	3.3	>1000	3.3
Cell C Well 4	5	3.3	>1000	3.3
Two-month samples from 0.9 m				
Cell A Well 3	30	3.3	1000	3.3
Cell A Well 4	600	3.3	1000	3.3
Cell B Well 3	10	3.3	500	3.3
Cell B Well 5	2	3.3	200	3.3
Cell C Well 3	4	3.3	>800	3.3
Cell C Well 4	18	3.3	900	3.3

[a]Average of three samples.
[b]Average of two samples.

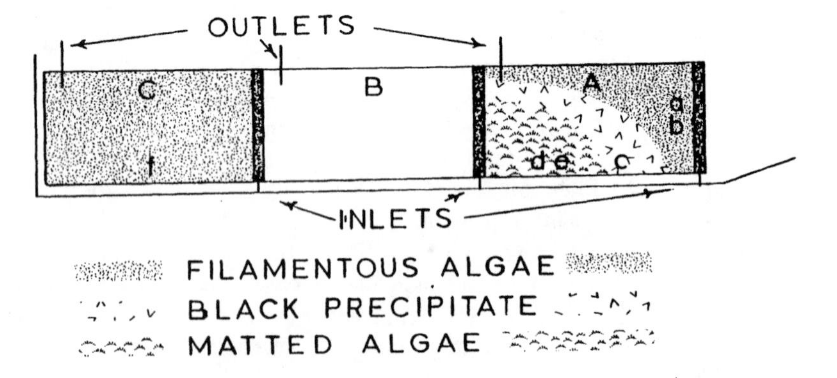

Figure 2. Plan view of the wetland pilot system showing the distribution of surface features in cells A and C. The letters mark the locations where samples were taken from iron- and manganese-oxidizing heterotrophs.

Table 2. Colony Count for Heterotrophic Bacteria (10^4 Colonies/g)

Cell	Sample	Total Heterotrophic Count: Plates with $FeSO_4$	Total Heterotrophic Count: Plates with $MnSO_4$	Heterotrophic Count for: Manganese Oxidizers	Heterotrophic Count for: Iron Oxidizers
A	(a)	>2000	10	0	200
A	Duplicate	>2000	30	50	7
A	(b)[a]	> 300	50	3	100
A	(c)[a]	> 800	>800	>800	>800
A	(d)[a]	> 700	30	20	200
A	(e)[a]	> 700	60	10	300
C	(f)[a]	> 400	>400	20	200

[a]Average of two samples.

RESULTS AND DISCUSSION

Different soil components and soils were good sources of sulfate-reducing bacteria. Table 1 shows that sulfate reducers appeared in all components in relatively high numbers, and summarizes bacterial populations found in the cell substrates after exposure to 4.4 L/min of mine drainage for two months. Cells B and C exhibited variations in sulfate-reducer populations; however, growth at dilutions of 10^{-7} was common. Mushroom compost in cell A contained the highest populations of sulfate reducers in the original soil and after two months of flow.

Autotrophic iron oxidizers were initially present only in peat, and the population was only 200 bacteria/g (Table 1). After two months of mine drainage flow, iron-oxidizing autotrophs were present on the surface and at depth in all the substrates (Table 1). To explain the presence of iron oxidizers in the anaerobic zone, samples of mine drainage input and output water from cell A were tested for iron-oxidizing autotrophs. In mine drainage, the bacterial population was 50/mL, and the number in cell A output was not significantly different. Because this population is orders of magnitude lower than in anaerobic zones of cells, the high population of iron oxidizers cannot simply be explained by mine drainage flowing through the cell depths. How high populations of iron-oxidizing autotrophs can exist in anaerobic zones in this constructed wetland is unknown but warrants further investigation.

Preliminary results indicate that iron- and manganese-oxidizing heterotrophs exist on the surface environment in cells A and C (Table 2). Samples a and b in Table 2 were taken near the *Ulothrix* algae. Sample a contains primarily algae and sample b primarily substrate. *Ulothrix* do not promote growth of iron-oxidizing heterotrophs and appear to severely depress growth of the manganese oxidizers. Samples d and e were taken near *Euglena*. Sample d contains primarily algae and e primarily substrate. Here also, growth of heterotrophs is not promoted by the algae. On the other hand, sample c is black precipitate sometimes found in wetlands receiving mine drainage. This material promotes the growth of both iron- and manganese-oxidizing heterotrophs.

On the inoculated petri dishes, fungal growth appeared in association with manganese oxidizers. The mycelium was dark, and brown crystals were observed under a microscope. Similar crystals observed by Vail et al.[20] in their experiment for the isolation and culture of the manganese-oxidizing bacterium were assumed to be manganese dioxide. Further research will determine growth behavior of the iron- and manganese-oxidizing heterotrophs.

CONCLUSION

Results reported herein are preliminary. Natural wetlands contain bacterial populations with the ability to remove metals. In this constructed wetland, presence of iron and manganese oxidizers and sulfate reducers was determined in various substrates. In previous studies, numbers of these bacteria and the removal of metals by wetlands was correlated;[2] therefore, this man-made wetland should be a promising treatment facility.

Further investigations should include (1) occurrence of iron oxidizers in the anoxic zone, (2) presence of sulfate reducers in the aerobic zone, (3) effects of the two algal groups on bacterial populations, and (4) effects of fungus growth in association with manganese-oxidizing bacteria.

REFERENCES

1. Silver, M. "Distribution of Iron-Oxidizing Bacteria in the Nordic Uranium Tailings Deposit, Elliot Lake, Ontario, Canada," *Appl. Environ. Microbiol.* 54(4):846–852 (1987).
2. Stone, W. R. "The Presence of Iron- and Manganese-Oxidizing Bacteria in Natural and Simulated Bogs," in *Treatment of Mine Drainage by Wetlands,* J. E. Burris, Ed. (University Park: Pennsylvania State University Press, 1986), pp. 30–36.
3. Ehrlich, H. L. *Geomicrobiology* (New York: Marcel Dekker, Inc., 1981), p.393.
4. Kelly, L. P., and O. H. Tuovinen. "Recommendation that the Names *Ferrobacillus ferrooxidans* Leathen and Braley and *Ferrobacillus sulfooxidans* Kinsel Be Recognized as Synonyms of *Thiobacillus ferrooxidans* Temple and Colmer," *Int. J. Systematic Bacteriol.* 22(3) (1972).
5. Schnaitman, C. A., M. S. Korazynski, and D. G. Lundgren. "Kinetics Studies of Iron Oxidation by Whole Cells of *Ferrobacillus ferrooxidans,*" *J. Bacteriol.* 99:552–557 (1969).
6. Walsh, F., and R. Mitchell. "A pH Dependent Succession of Iron Bacteria," *Environ. Sci. Technol.* 6:809–812 (1972).
7. Postgate, J. R. "Media for Sulphur Bacteria," LABP 5–74 Article (1985).
8. Postgate, J. R. *The Sulphur-Reducing Bacteria* (New York: Cambridge University Press, 1979), p. 151.
9. Brierley, J. A., and C. L. Brierley. "Biological Methods to Remove Selected Inorganic Pollutants from Uranium Mine Waste Water," in *Biogeochemistry of Ancient and Modern Environments,* P. A. Trudinger, M. R. Walter, and B. J. Ralph, Eds. (Canberra: Australian Academy of Science, 1980), pp. 661–667.
10. Brierley, J. A., C. L. Brierley, and K. T. Dreher. "Removal of Selected Inorganic

Pollutants from Uranium Minewaste Water by Biological Methods," in *Uranium Mine Waste Disposal,* C. O. Brawner, Ed. (New York: American Institute of Mining, Metallurgical and Petroleum Engineering, 1980), pp. 365–376.
11. Tuttle, J. H., P. R. Dugan, and C. I. Randles. "Microbial Sulfate Reduction and Its Potential Utility as an Acid Mine Water Pollution Abatement Procedure," *Appl. Microbiol.* 17(2):297–302 (1969).
12. Wakao, N., T. Takahashi, H. Shiota, and Y. Sakurai. "A Treatment of Acid Mine Water Using Sulfate-Reducing Bacteria," *J. Ferment. Technol.* 57(5):445–452 (1979).
13. Bromfield, S. M. "Oxidation of Manganese by Soil Microorganisms," *Austr. J. Biol. Sci.* 9:238–252 (1956).
14. Silver, M., H. L. Ehrlich, and K. C. Ivarson. "Soil Mineral Transformation Mediated by Soil Microbes," in *Interactions of Soil Minerals with Natural Organics and Microbes* (Madison, WI: Soil Society of America, 1986), pp. 497–519.
15. Howard, E. A., J. C. Emerick, and T. R. Wildeman. "Design and Construction of a Research Site for Passive Mine Drainage Treatment in Idaho Springs, Colorado," Chapter 42b, this volume.
16. Silverman, M. P., and D. G. Lundgren. "Studies on the Chemoautotrophic Iron Bacterium *Ferrobacillus ferrooxidans.* I. An Improved Medium and a Harvesting Procedure for Securing High Cell Yields," *J. Bacteriol.* 77:642–647 (1959).
17. Page, A., R. Miller, and D. Keeney. *Methods for Soil Analysis Part II,* 2nd ed. (Madison, WI: American Society of Agronomy, Inc., Soil Society of America, Inc., 1982), Chapter 39.
18. Brierley, J. Personal communication (1988).
19. Silver, M. Personal communication (1988).
20. Vail, W. J., S. Wilson, and R. K. Reiley. "Isolation and Culture of a Manganese-Oxidizing Bacterium from a Man-Made Cattail Wetland," in *Mine Drainage and Surface Mine Reclamation, Vol. I, Mine Water and Mine Waste,* Bureau of Mines Information Circular 9183 (1988), p. 399.

38i Use of Periphyton for Nutrient Removal from Waters

Jan Vymazal

INTRODUCTION

Many lakes and streams are showing signs of excessive fertilization due to the input of aquatic plant nutrients from anthropogenic sources. Although nitrogen and phosphorus are not the only nutrients required for algal growth, it is generally agreed they are the two main nutrients involved. Phosphorus is often found to be the key element. Eutrophication is emerging as one of the most significant causes of water quality deterioration. Eutrophication of lakes, and especially of drinking water reservoirs, leads to profound changes and has considerable detrimental effects on the quality of water.

Drinking water treatment difficulties presented by algal development in reservoirs can be prevented, first of all, by lowering nutrient input into drinking water reservoirs. One possible intervention is to use naturally growing periphyton communities on artificial substrata to lower nutrient loads in reservoir tributaries. This study evaluated the efficiency of periphyton communities in nutrient elimination in a continuous-flow trough in the field.

MATERIALS AND METHODS

The continuous-flow trough was made of wood and was 5 m long, 0.5 m wide, and 0.7 m deep. The inside walls and the bottom of the trough were laminated. Fine-mesh silon (a kind of nylon) screens (50 × 50 cm, hole size ca. 1 mm^2) were fixed in plastic frames and placed perpendicular to the direction of flow. Screens were held in place 20 cm apart by grooves on the side walls of the trough.

Vltava River water was used for all experiments in the field trough. Retention time in the trough was 4 hr, based on the results of preliminary laboratory experiments. Flow was 312.5 L/hr (total volume of water in the trough was 1250 L).

The first experiment was aimed at examining nutrient elimination with time, periphyton community growth, and composition. Efficiency of nutrient elimination was monitored with NH_4^+, NO_2^-, NO_3^-, and PO_4^{3-} analyses. Triplicate influent and effluent samples were analyzed to determine the amount of nitrogen and phosphorus taken up in the trough, and results were expressed as elimination (E), which was determined from the relation:

$$E = \frac{C_0 - C_1}{C_0} \times 100\ (\%) \tag{1}$$

where C_0 = influent concentration of observed chemical
C_1 = effluent concentration

Elimination was observed at two- to three-day intervals, and average values were plotted.

The algal growth potential (AGP) was determined to measure the extent of total decrease in water fertility. As the test alga, *Scenedesmus quadricauda* (Turp.) Breb., strain Greifswald/15 was used, and triplicate samples were taken for each datum plotted.

The saprobic index (S) of the periphyton community was computed using the following formula:

$$S = \frac{\Sigma S_i h_i}{\Sigma h_i} \tag{2}$$

where S = saprobic index of the community
S_i = saprobic value of individual species
h_i = abundance of a species in the sample

with relative abundance on a scale of 1 to 5 (1 = rare, 2 = occasional, 3 = frequent, 4 = abundant, and 5 = very abundant). Samples were scraped from six sites, both in inflow and outflow parts of the trough. For calculation of the saprobic index (S), average values were used. When calculating S, only periphytic organisms were taken into account, i.e., planktonic organisms were not used for calculations in spite of their presence, which is primarily due to mechanical holding. However, planktonic organisms significantly added to the overall metabolism of the periphyton community.

Species composition of inflow and outflow communities was compared by computing Sørensen's similarity coefficients (C_S):

$$C_S = \frac{2n_{jk}}{n_j + n_k} \tag{3}$$

where n_{jk} is the number of species occurring in both sample j and in sample k, and n_j and n_k are the numbers of species occurring in sample j and k, respectively. If there is no correspondence between two samples, then C_S = 0. If

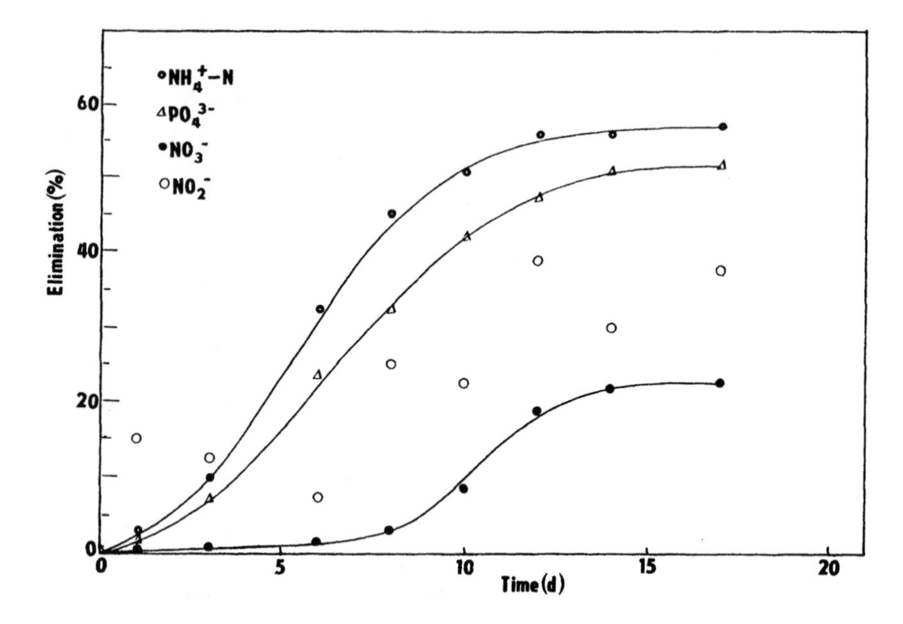

Figure 1. Elimination of nutrients during first field experiment.

there is maximal correspondence, then C_S = 1. Operation of the first field experiment was May 6–23, 1984.

The second experiment was aimed at the course of nutrient elimination and composition of periphyton communities in the trough. Nutrient elimination was controlled by NO_3^-, NH_4^+, and PO_4^{3-} analyses. Methods for calculation of S and C_S were similar to those used during the first experiment. The operation period was July 22 to August 15, 1984.

RESULTS

The course of nutrient elimination is given in Figures 1 and 2. The maximum values of elimination during the first field experiment were NO_3^-, 24%; NH_4^+, 59%; and PO_4^{3-}, 54%. The course of nitrate removal fluctuated greatly and was difficult to evaluate. Average influent concentrations were NO_3^-, 14.1 mg/L; NO_2^-, 0.17 mg/L; NH_4^+, 2.65 mg/L; and PO_4^{3-}, 1.35 mg/L. Water temperature ranged between 12.0°C and 14.2°C.

The course of nutrient removal during the second field experiment was similar to the first, but the maximum levels of NH_4^+ and PO_4^{3-} elimination were much higher (almost 80% and 70%, respectively). Elimination of nitrates was at approximately the same level (22%). The average influent concentrations were similar to those in the first experiment. The temperature of water during this period ranged between 16.1°C and 18.2°C.

Elimination of nutrients stabilized in spite of the increase in periphyton

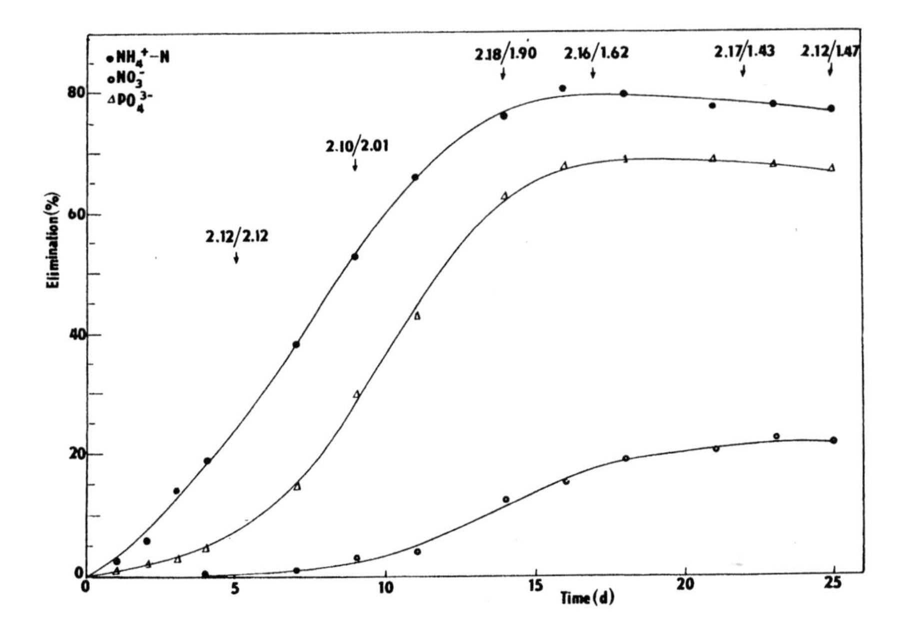

Figure 2. Elimination of nutrients during second field experiment. Numbers indicate values of the saprobic index of periphyton communities—inflow/outflow.

biomass. For practical applications, these results indicate it is best to replace highly colonized screens promptly, allowing biofilm formation for only a limited period.

In both experiments, NH_4^+ and PO_4^{3-} elimination appeared almost immediately (after a very short lag phase for ammonium and a bit longer for orthophosphate). The lag phase for nitrate elimination was much longer, and the increase in elimination appeared after filamentous greens were observed in the end of the trough. This timing reinforces the suggestion that elimination of nitrate is connected with activity of filamentous algae species.

Trophic state of the water was determined during the first experiment on the 15th day, after a steady state of nutrient elimination was reached. The value of AGP decreased by 44% in the outflow versus the inflow water (Figure 3). In the second experiment, AGP was also determined when nutrient elimination was at a maximum (20th day). Decrease in the AGP value in the effluent was about 61% in comparison with the influent. The greater decrease in trophic state in the course of the second field experiment corresponds with the higher level of nutrient elimination.

Composition of the influent community was quite steady and identical to the most common naturally occurring periphyton of the Vltava River. The influent community indicated lower β-mesosaprobity and was dominated by the filamentous diatom *Melosira varians*. Other more abundant species were

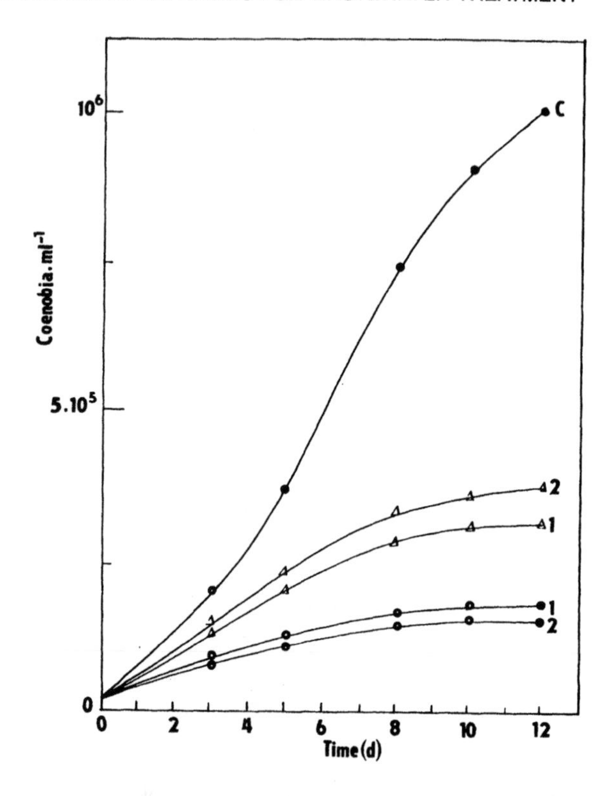

Figure 3. Decrease in trophic state of water in the trough. C = control grown in the artificial medium; 1 = first experiment; 2 = second experiment; Δ = inflow; O = outflow; coenobium = assemblage of four cells.

Oscillatoria sp., *Diatoma vulgare, Gomphonema olivaceum, Navicula cryptocephala* var. *veneta, N. cuspidata, Nitzschia acicularis,* and *Synedra ulna.*

Effluent community structure changed during the course of the experiment (Table 1) in response to a decrease in inorganic nutrients and organic content in the trough. Certain species observed in the inflow community gradually decreased in abundance in the outflow community during the course of the experiment and, in some cases, were not present at the end of the experiment. Such species were *Oscillatoria* sp., *Cymatopleura librilis, Cymbella tumida, Navicula cryptocephala* var. *veneta, N. cuspidata, N. gregaria, Nitzschia acicularis,* and *Nitzschia sigmoidea,* as well as the colorless flagellate *Antophysa vegetans* and stalked, ciliated protozoan *Carchesium polypinum.* All mentioned species prefer high levels of nutrients and organics and belong to the worst β-mesosaprobic or α-mesosaprobic communities.

Some species not present in the inflow community during the course of the experiment gradually appeared and increased in abundance in the outflow community. This was again caused by water quality changes in the trough. Species belonging to this group were: *Achnanthes minutissima, Caloneis bacil-*

Table 1. Periphyton Community Structure Changes During Field Experiments

	Inflow						Outflow					
I. Experiment												
Day of Operation	6	10	12	17			6	10	12	17		
S	2.20	2.23	2.23	2.21			2.17	2.01	1.82	1.71		
C_S	1.00	0.789	0.583	0.489								
II. Experiment												
Day of Operation	5	9	14	17	22	25	5	9	14	17	22	25
S	2.12	2.10	2.18	2.16	2.17	2.12	2.12	2.01	1.90	1.62	1.43	1.4
C_S	0.970	0.732	0.465	0.326	0.254	0.254						

Note: S = saprobic index; C_S = similarity coefficient.

lum, Cymbella helvetica, Diatoma tenue, Fragilaria construens, Melosira italica, Navicula exigua, N. gracilis, N. minima, N. oblonga, N. paleacea, Nitzschia fonticola, Pinnularia borealis, P. subcapitata var. *subcapitata, Synedra acus,* the filamentous greens *Hormidium flaccidum, H. tribonematoideum, Spirogyra* cf. *porticalis,* and *Ulothrix tenuissima* as well as the stalked, ciliated protozoans *Epistylis digitalis* and *Vorticella margaritata,* testacean *Difflugia limnetica,* and rotifer *Cephalodella gibba.*

The periphyton community growing on artificial substrata has proved to be a useful means of nutrient removal from polluted waters. Periphyton growth could be used either in waterworks pretreatment, especially in small eutrophic tributaries to drinking water reservoirs, or in the tertiary treatment process. Before use, however, it would be necessary to check the response under different environmental conditions such as current velocity, influent concentrations of nutrients, size and composition of artificial substrata, or position of the substrata in the trough. At present, these results are successfully being tested for the treatment of agricultural drainage waters with high content of phosphorus and nitrogen.

Note

Detailed results including literature appeared in *Hydrobiologia* 166:225–237 (1988).

CHAPTER 39

Management of Domestic and Municipal Wastewaters

39a Danish Experience with Sewage Treatment in Constructed Wetlands

Hans Brix and Hans-Henrik Schierup

INTRODUCTION

The concept of treating wastewaters with emergent aquatic macrophytes (the root-zone process) was introduced in Denmark in 1983. Basically, this process depends on a horizontal subsurface flow through the common reed (*Phragmites australis* Cav. Trin. ex Steud.) rhizosphere. During passage of wastewater through the rhizosphere, organic matter content of wastewater should theoretically be decomposed by aerobic and anaerobic microorganisms; microbial nitrification and subsequent denitrification should release nitrogen (N) to the atmosphere; and phosphorus (P) should be removed by chemical coprecipitation with iron, aluminum, and calcium compounds in the soil. The most important functions of macrophytes in the reeds beds are (1) to supply oxygen to the aerobic microorganisms in the rhizosphere and (2) to increase/stabilize the hydraulic permeability of the soil. Direct assimilation of nutrients by vegetation is considered to be of no significance for the purification ability of the systems because the maximum amount of nutrients which can be removed by harvesting the aboveground biomass is less than 5% of the load on a yearly basis. Theoretical purification processes in reed beds with subsurface flow have been described.[1]

The root-zone process was introduced as a low-cost, low-technology decentralized solution capable of producing an effluent quality equivalent to, or even exceeding, conventional tertiary treatment technology. Performance for BOD_5 as well as nitrogen and phosphorus was claimed to be better than 90%. The process was a very attractive solution for local municipalities, and the first full-scale treatment facilities in Denmark were constructed during winter 1983–84.

DESIGN OF THE DANISH SYSTEMS

All Danish systems are constructed as subsurface flow systems designed to percolate sewage horizontally through the macrophyte rhizosphere. Most of the systems are constructed to treat domestic sewage and have surface areas of less than 4000 m^2. Some beds have been constructed to act as a tertiary system until mature, when they eventually will be switched to treating mechanically pretreated sewage.[2]

In principle, treatment plants consist of a plastic-lined excavation filled with approximately 1 m of soil and with a surface slope of 1–5%. Original soil at the plant site is usually used as the rooting medium, but a few facilities are established with gravel or sand. Beds are planted with reeds or a combination of reeds and cattails (*Typha latifolia* L.). Sewage is mechanically pretreated in septic tanks, sedimentation tanks, or stabilization ponds before it is led into the beds. For even distribution across the bed and throughout its depth, pretreated sewage is led into an inlet trench filled with stones. Cleaned sewage is collected in a similar trench at the outlet end and released via a pipe at the trench bottom so that the discharge level of water can be varied to regulate the water table in the bed.

Bed dimensions vary because different construction companies use different design criteria with little or no input from researchers or regulatory review by federal agencies. In general, systems receiving sewage from combined drainage systems or dilute sewage, with relatively high hydraulic loading rates, are built as relatively short and wide beds to prevent overland flow.

PERFORMANCE OF THE BEDS

Local authorities that own and operate reed beds are obliged to perform 8–12 inlet and effluent annual quality control analyses. Samples are generally taken proportional to volume over a 24-hour period, and analyses are performed according to standard methods by independent laboratories. Results are stored in a central data base at the Botanical Institute, University of Aarhus. This evaluation concentrates on performance with respect to BOD_5, total phosphorus (TP), and total nitrogen (TN) of 25 facilities. Basic design data for the facilities are presented in Table 1. The performance data presented for most of the facilities are for the actual reed bed *excluding* the mechanical pretreatment. Thus, the performance of the entire system, i.e., septic tank *and* the reed bed itself, is better than the data presented.

Biochemical Oxygen Demand (BOD_5)

Mean inlet and outlet concentrations of BOD_5 in the 25 reed beds are shown in Figure 1a. Facilities at sites 1–4 receive rather concentrated sewage from separate drainage systems and some additional farm waste. Facilities at sites

Table 1. Basic Design Data for 25 Danish Constructed Reed Beds

Site	Construction period	Area (m^2)	Vegetation[a]	Feed	Loading rate (cm/day)
1. Ingstrup	Spring 1984	100	Phr	Septic tank effluent	1.0
2. Hobjerg	Autumn 1987	4100	Typ. Phr.	Septic tank effluent	4.3
3. Rugballegard	Spring 1984	120	Phr.	Septic tank effluent	1.3
4. Brondum	Summer 1986	420	Phr.	Septic tank effluent	3.8
5. Ferring	Summer 1984	2000	Phr.	Screened settled sewage	1.7
6. Focusing	Autumn 1984	1175	Phr.	Septic tank effluent	0.9
7. Karstoft	Spring 1985	2400	Phr.	Septic tank effluent	0.5
8. Rudbol	Summer 1986	1800	Phr.	Septic tank effluent	2.1
9. Sdr. Thise	Spring 1985	560	Phr.	Septic tank effluent	5.4
10. Hjordkaer	Spring 1984	1125	Phr.	Settled sewage	15.4
11. Borum	Autumn 1987	2500	Typ. Phr.	Septic tank effluent	10.6
12. Moesgard	Spring 1984	500	Typ.Car.Phr.	Septic tank effluent	2.0
13. Knudby	Autumn 1984	361	Phr.	Septic tank effluent	6.4
14. Kalo	Autumn 1984	940	Phr.	Septic tank effluent	3.8
15. Lyngby	Summer 1987	4000	Typ. Phr.	Septic tank effluent	10.0
16. Uggerhalne	Autumn 1985	2640	Phr.	Septic tank effluent	5.2
17. Borup	Autumn 1984	1500	Phr.	Septic tank effluent	4.4
18. Jaungyde	Autumn 1985	800	Phr.	Septic tank effluent	7.0
19. Sabro	Autumn 1986	2650	Typ. Phr.	Settled sewage	10.7
20. Lunderskov	Autumn 1984	1800	Gly.Car.Phr.	Secondary effluent	3.6
21. Ormslev	Spring 1986	1850	Typ. Phr.	Septic tank effluent	7.5
22. Thise	Autumn 1985	800	Phr.	Septic tank effluent	4.4
23. Egebaek	Autumn 1986	5000	Phr.	Secondary effluent	8.0
24. Homa	Summer 1987	1700	Typ. Phr.	Septic tank effluent	8.8
25. Egeskov	Autumn 1984	3600	Phr.	Secondary effluent	7.7

[a]Phr. = *Phragmites australis* (Cav.) Trin. ex Steud; Typ. = *Typha latifolia* L; Car. = *Carex acutiformis* Ehrh; Gly. = *Glyceria maxima* (Hartm.) Holmb.

5–14 receive predominantly municipal sewage from separate drainage systems, and the facilities at sites 15–25 receive sewage from combined systems or secondary effluent. The BOD_5 effluent concentrations are generally less than 20 mg/L and often less than 10 mg/L irrespective of inlet concentrations. Effluent limits for BOD_5 in these systems are 15 or 20 mg/L, depending on site, and BOD_5 discharge criteria generally can be fulfilled by these reed beds.

Total Phosphorus

The general effluent limit for TP in Danish sewage treatment plants serving 5000 population equivalents or more is 1.5 mg/L. Similar standards apply for many smaller facilities that discharge into sensitive receiving waters. This effluent standard cannot be fulfilled by reed beds because the concentration of TP in effluents typically is in the range of 4–8 mg/L (Figure 1b). Removal efficiency is generally 20–40%. Facilities with higher removal efficiencies have very low hydraulic loading rates.

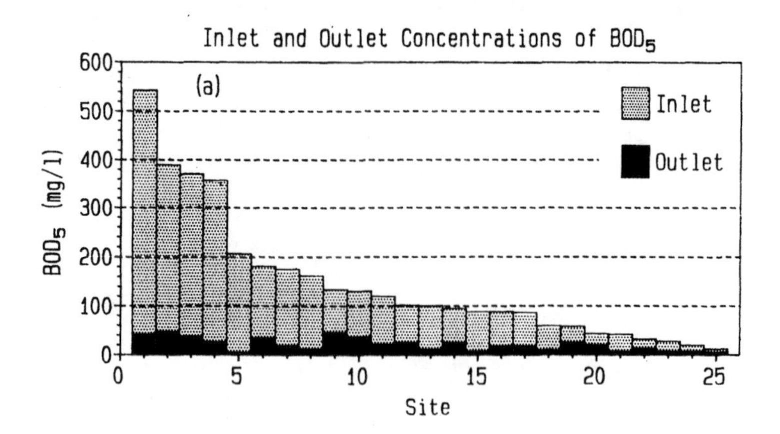

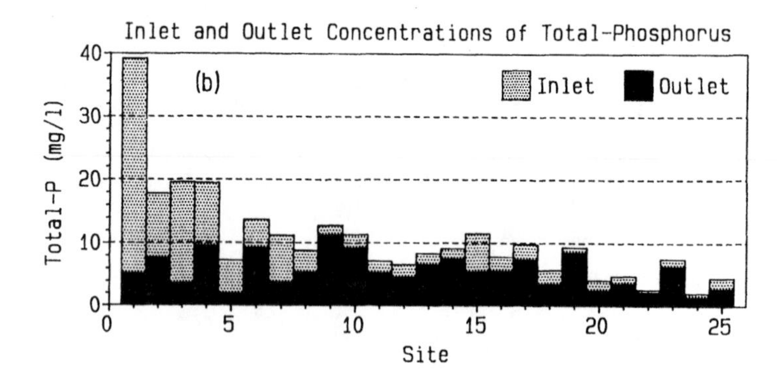

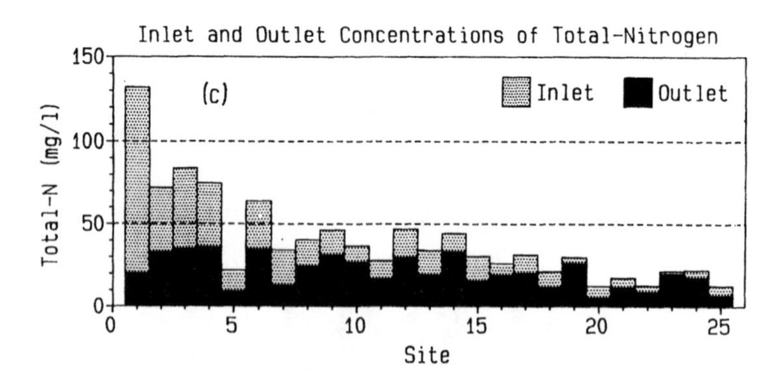

Figure 1. Mean inlet and outlet concentrations (mg/L) of (a) BOD_5, (b) total phosphorus, and (c) total nitrogen of 25 Danish constructed reed beds.

Total Nitrogen

The Danish effluent standard for TN in large sewage plants is 8 mg/L. This criterion cannot be fulfilled by reed beds because effluent concentrations for TN are typically 20–30 mg/L (Figure 1c). Nitrification seems to be the critical process because the principal form of N in the effluent is ammonia.

DEVELOPMENT AND SEASONAL VARIATION IN PERFORMANCE

The macrophyte root system in the beds gradually develops during the initial years, and performance is supposed to improve accordingly with time. Purification of sewage in reed beds is mediated partly by microbiological processes. Therefore, one would expect performance with respect to BOD_5 and N in particular to decrease during winter due to low ambient temperatures and reduced microbiological activity. Figure 2 shows development and seasonal variation in performance for the reed bed at Kalo (site 14), which was constructed in autumn 1984. Performance with respect to BOD_5 was poor during the first growing season but improved during the second and third years (Figure 2a). Poor effluent quality during winter 1985–86 coincides with a period of severe frost. Performance for TP and TN was poor during the initial year but seemed to improve thereafter (Figures 2b and 2c). However, there was no improvement from the second year to the third year. The principal form of N in the influent was ammonia (NO_2 + NO_3 $<$ 0.05 mg/L), and the same was true for the effluent (Figure 2d). During the second and third years, some NO_3 occurred in the effluent indicating that nitrification occurred in the bed, but the rate was too low to be of quantitative significance for removal of N in the bed.

DETERMINING FACTORS FOR THE PERFORMANCE

Bed design had no obvious relation to removal efficiency, and type of substrate (i.e., texture; organic matter content; and Ca, Fe, and Al content) showed no correlation with performance, except for P removal in gravel beds. On the other hand, removal efficiencies did correlate significantly with hydraulic loading rates of the systems (Table 2). Hydraulic loading rate is particularly important regarding performance for N and P. Only systems with loading rates of less than 2 cm/day have removal efficiencies for N and P better than 50% (Figure 3). These removal efficiencies can be ascribed to the generally low hydraulic conductivities of reed bed soils. If the loading rate is increased beyond 2 cm/day, a major proportion of the sewage runs off on the surface as overland flow, and theoretical purifying processes for N and P work only to a limited extent. No correlation between organic loading rate and removal efficiency was observed (Table 2).

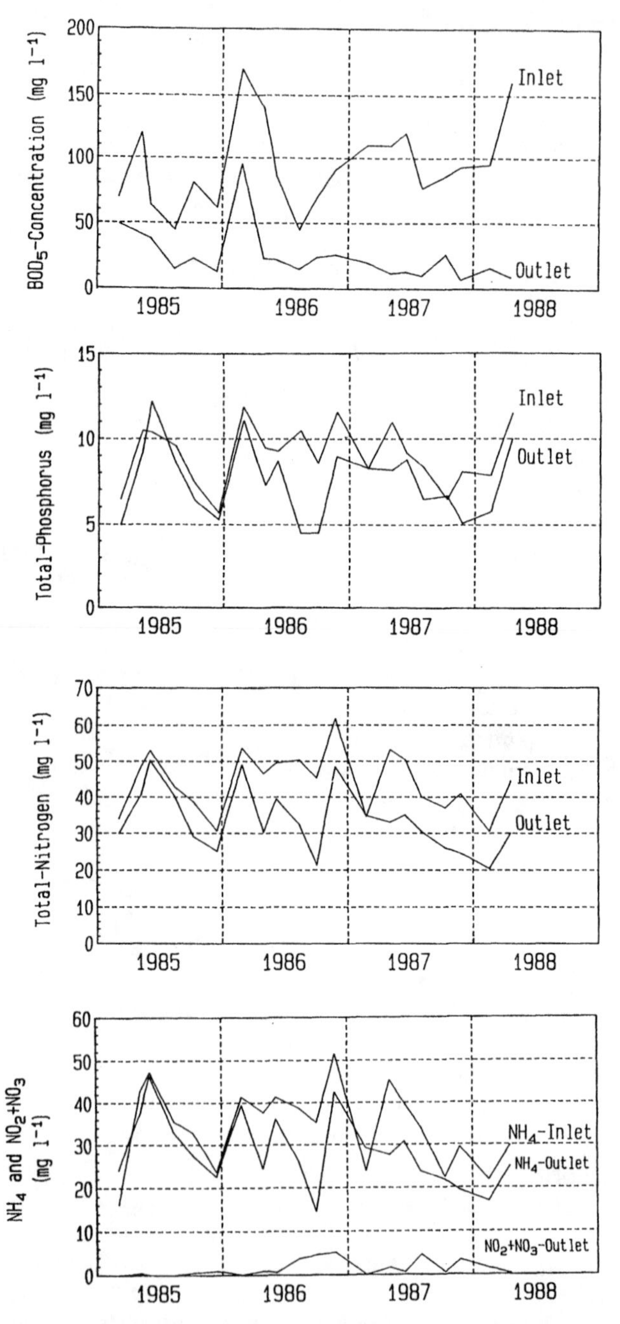

Figure 2. Seasonal variations in inlet and outlet concentrations (mg/L) of (a) BOD_5, (b) total phosphorus, (c) total nitrogen, and (d) NH_4 and $NO_2 + NO_3$ in the constructed reed bed at Kalo (site 14). (Inlet concentrations of $NO_2 + NO_3$ were <0.5 mg/L).

Table 2. Correlation Matrix of Performance of 25 Danish Constructed Reed Beds

	BOD_5 (% removal)	Total N (% removal)	Total P (% removal)
Hydraulic loading rate (cm/day)	0.32^{NS}	0.55^{**}	0.59^{**}
Area usage (day/cm)	0.42^{*}	0.61^{**}	0.65^{***}
BOD_5 load (g $BOD_5/m^2 \cdot$ day)	0.19^{NS}	0.01^{NS}	0.08^{NS}
N load (g $TN/m^2 \cdot$ day)	0.02^{NS}	0.32^{NS}	0.37^{NS}
P load (g $TP/m^2 \cdot$ day)	0.01^{NS}	0.24^{NS}	0.30^{NS}

* = $p < 0.05$
** = $p < 0.02$
*** = $p < 0.001$
NS = Not significant.

SUMMARY AND CONCLUSIONS

The initial four years of experience from 25 Danish constructed reed beds can be summarized as follows:

- Performance with respect to BOD_5 is reasonably good (typically 70–90%), producing a consistent effluent concentration of less than 20 mg/L after one growing season.
- Typical reduction of TN and TP is 25–50% and 20–40%, respectively. Poor performance with respect to N and P can be explained by (1) surface runoff (low soil permeability preventing sewage from reaching the rhizosphere) and (2) insufficient release of oxygen from reed root systems to support quantitatively significant nitrification.
- N and P removal is dependent on hydraulic loading rate and consequent retention time within the reed beds. Only facilities with loading rates of less than 2 cm/day show removals of N and P of over 50%.
- Hydraulic permeability of the soil develops slowly, if ever. Even after four growing seasons, overland flow predominates, forming a pattern covering only part of the reed bed area.

However, as most Danish reed beds are constructed on sites with no requirement for nutrient removal, we conclude that generally reed bed function is satisfactory for fulfilling quality criteria set by authorities for the specific sites. Designs used in Denmark, i.e., one-basin constructed reed beds with horizontal subsurface flow, may not be the most cost-effective low-technology model. The general opinion among specialists working in this area is that future construction of macrophyte-based treatment systems should combine advantages of different process types. Such integrated systems may include several small units containing different kinds of macrophytes, different substrates, different flow patterns (horizontal vs vertical flow), and some degree of water

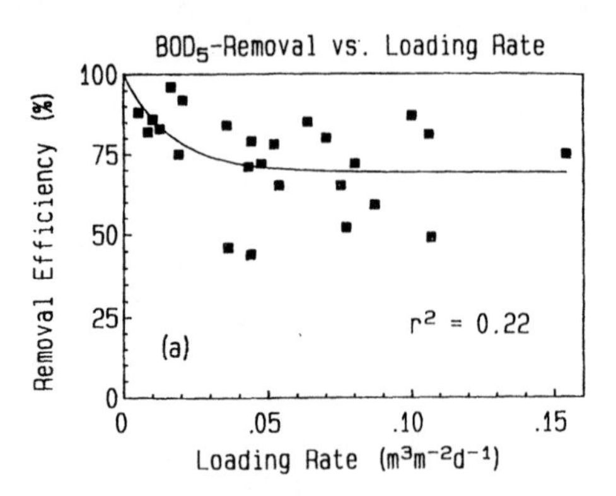

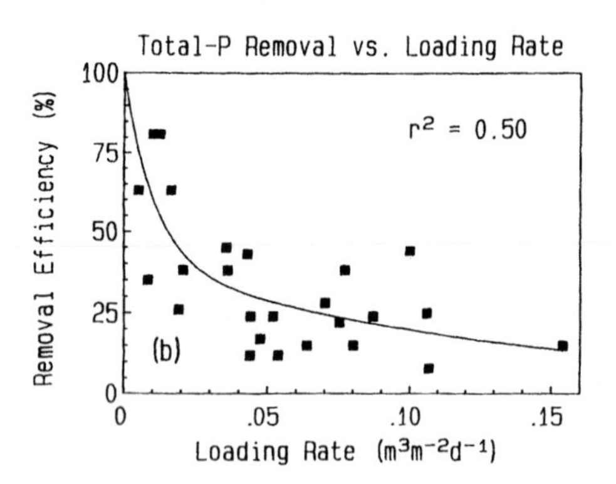

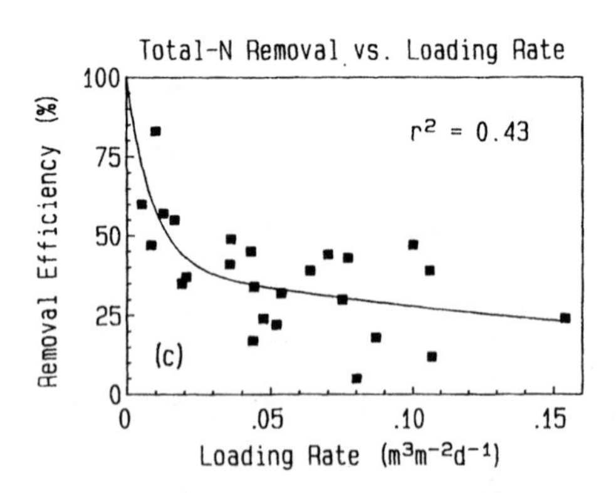

Figure 3. Mean removal efficiencies (%) for (a) BOD_5, (b) total phosphorus, and (c) total nitrogen plotted against mean hydraulic loading rates of 25 Danish constructed reed beds. r^2 indicates the goodness-of-fit of the curves.

recirculation. However, we need to monitor the large number of Danish full-scale reed beds for some years until biological steady-state conditions develop with subsequent improvements in hydraulic conductivity.

ACKNOWLEDGMENTS

We wish to thank the owners of the reed beds for supplying the design and water quality data.

REFERENCES

1. Brix, H. "Treatment of Wastewater in the Rhizosphere of Wetland Plants—The Root-Zone Method," *Water Sci. Technol.* 19:107–118 (1987).
2. Brix, H., and Schierup, H.-H. "Root-Zone Systems—Operational Experience of 14 Danish Systems in the Initial Phase," Report to the Danish Environmental Protection Agency (1986).

39b Man-Made Wetlands for Wastewater Treatment: Two Case Studies

JoAnn Jackson

INTRODUCTION

Many of Florida's lakes, rivers, bays, and estuaries are threatened by development, so state and federal regulatory agencies are discouraging the traditional practice of discharging treated wastewater directly to surface waters. Land application is the most permittable disposal option, but land most suited to conventional land application methods is becoming scarce and costly. Wastewater treatment plant owners and engineers must develop more creative effluent disposal methods using lower-lying or otherwise less desirable lands while adequately protecting surface water resources. Two innovative designs that meet both these needs have been constructed in Lakeland and Orlando, Florida. They represent the largest known created wetlands used for wastewater treatment/disposal and the first created systems in Florida.

CITY OF LAKELAND SYSTEM DESCRIPTION

Background

In 1983, the City of Lakeland's Glendale Street Wastewater Treatment Plant was ordered by the United States Environmental Protection Agency (EPA) to cease discharge to Banana Lake, one of the most eutrophic lakes in Florida. After an extensive evaluation of disposal options for up to 0.61 m^3/s (14 mgd) of treated wastewater, an abandoned phosphate mine clay settling area was selected to be used as a wetlands to provide additional treatment prior to discharge to the Alafia River.

Prior to city purchase, 566 ha (1400 ac) of the 650-ha site was used as clay settling ponds by the phosphate mining industry. As water passed through a series of seven constructed ponds, clays settled out forming an impermeable liner on the pond bottoms. In shallower areas, cattails (*Typha* spp.) and willow (*Salix* spp.) and other wetlands vegetation took root. After abandonment by

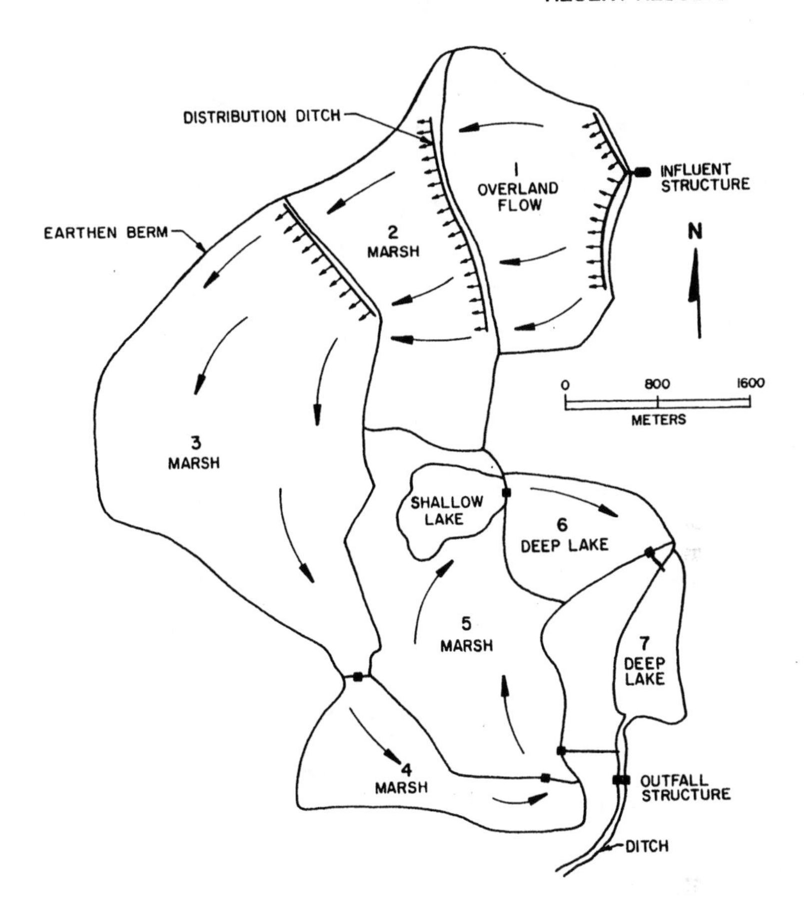

Figure 1. City of Lakeland man-made wetlands system.

the phosphate industry, portions of shallower ponds dried up, and wetlands vegetation became stressed or died.

System Design

To prepare the site for domestic wastewater treatment use, the city regraded, improved berms, constructed a distribution system, and planted herbaceous and hardwood species. The first five wetland areas consist of a series of shallow, clay-lined ponds planted with cattails, other herbaceous species, and (to a lesser degree) hardwood swamp tree seedlings (Figure 1). From the shallow ponds, water flows into a shallow lake and then to a series of deep lakes prior to discharge to a ditch and the Alafia River.

The constructed wetlands were designed for a flow of 0.61 m^3/s, equivalent to a hydraulic application rate of 1 cm/day (2.6 in./wk). Design criteria included reducing influent concentrations of 20 mg/L biochemical oxygen

Table 1. Monthly Operating Performance: City of Lakeland Man-Made Wetlands System

	Influent Conditions					Effluent Conditons				
Month	Flow	BOD_5	TSS	NH_3-N	TN	Flow	BOD_5	TSS	NH_3-N	TN
April 1987	0.04	18.0	22.5	14.9	20.5	0.04	2.0	4.0	0.3	1.8
May	0.39	15.6	16.6	12.6	15.0	0.01	2.0	3.3	0.2	1.9
June	0.28	20.2	21.0	13.7	21.3	0.01	3.8	18.6	0.7	5.4
July	0.31	22.3	31.5	13.8	21.7	0.20	2.0	1.5	0.3	2.2
August	0.29	14.0	33.4	13.6	13.8	0.41	1.0	8.0	0.1	1.7
September	0.27	3.2	8.8	4.6	19.8	0.30	2.0	3.4	0.1	2.4
October	0.26	0.8	3.8	0.5	13.9	0.21	2.0	2.8	0.3	0.7
November	0.32	1.5	2.8	0.5	13.2	0.47	2.0	7.3	0.2	1.8
December	0.28	3.2	3.4	0.5	16.0	0.29	1.0	1.0	0.2	1.2
January 1988	0.32	4.5	5.0	0.6	13.3	0.18	2.0	1.5	0.2	1.1
February	0.33	5.0	4.0	1.4	5.4	0.26	2.0	2.5	0.2	2.5
March	0.35	7.8	4.8	1.2	4.6	0.28	3.4	3.0	0.4	1.6
Average	0.29	9.7	13.1	6.5	14.9	0.22	2.1	4.7	0.3	2.0
Permit requirements	0.44	20.0	20.0	8.0	—[a]	0.44	5.0	10.0	1.0	4.0

Note: All parameters expressed in mg/L, except flow in m^3/s; improvements in wetlands influent quality in September 1987 correspond to startup of improvements at the wastewater treatment plant.
[a]No permit requirements for this parameter.

demand (BOD_5), 20 mg/L total suspended solids (TSS), and 8 mg/L total ammonia as nitrogen (NH_3-N) to 5 mg/L BOD_5, 10 mg/L TSS, and 1.0 mg/L NH_3-N. Phosphorus (TP) removals of approximately 50% were also expected.

Project Performance

The Lakeland wetland system began operating in March 1987. The system has consistently complied with the effluent requirements even during initial startup, except for one occurrence in June 1987 when the TSS concentration exceeded the 10 mg/L limit (Table 1). The high TSS concentration was attributed to flushing of decayed plant material as the final cells were first flooded. Influent phosphorus concentrations of 8.5 mg/L are reduced to 3.5 mg/L, a 60% removal.

Although wastewater was first applied in April 1987, the first significant discharge did not occur until July 1987. Additionally, in September 1987, improvements to the treatment plant resulted in improved wetlands influent quality. Despite these conditions, wetlands effluent quality has remained fairly consistent.

Although influent flow has been approximately 65% of the permitted capacity (0.44 m^3/s or 10 mgd) and 50% of the design capacity (0.61 m^3/s), effluent quality is expected to remain consistent as influent flows increase. Periodic sampling between cells has found final effluent quality after cell 2, about one-third of the entire area.

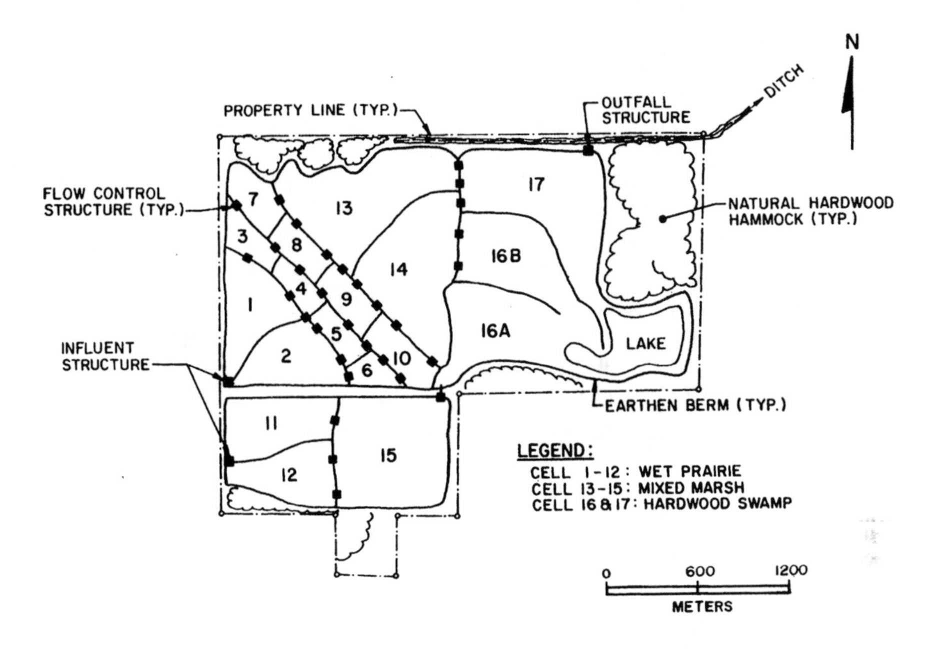

Figure 2. City of Orlando man-made wetlands system.

CITY OF ORLANDO SYSTEM DESCRIPTION

Background

In 1984, the City of Orlando's Iron Bridge Water Pollution Control Facility (Iron Bridge plant) was quickly approaching design capacity and required expansion to accommodate projected growth. Under an EPA wasteload allocation, the Iron Bridge plant discharges effluent to the Little Econlockhatchee River. Since this allocation was fully utilized by loading from the existing plant, additional disposal means were investigated to accommodate up to 0.88 m^3/s (20 mgd) of flow from an expanded facility. The selected disposal option was creation of a 494-ha (1220-ac) man-made wetlands on an improved pasture to further treat water prior to discharge to the St. Johns River.

System Design

The pasture was segmented by berms into small cells, with weir structures to control flow distribution, depth, and detention time. The segmented design, as illustrated in Figure 2, allows individual cells to be taken out of service for maintenance with little effect on the rest of the system. Three major vegetative communities were established on the site: wet prairie, mixed marsh, and hardwood swamp.

The wet prairie is 170 ha (420 ac) and was planted primarily with cattail

(*Typha latifolia*) and bulrush (*Scirpus californicus*). This area receives the greatest nutrient load and needed species that were hardy and had high growth rates and nutrient removal capabilities. The wet prairie is divided into 12 cells, any of which can be isolated from flow for maintenance purposes. Maintenance schedules are not developed, but studies are proposed to evaluate whether harvesting will be required and at what frequency and by what method (burning, mechanical harvesting, or biological harvesting). Water depths in the wet prairie range from 0 to 1 m.

From the wet prairie, water flows into a 154-ha (380-ac) mixed marsh. Although only 10 different species were planted, approximately 60 have established themselves. Planted species, which are indigenous and have a high wildlife value, include maidencane (*Panicum hemitomen*), soft rush (*Juncus effusus*), arrowhead (*Sagittaria lancifolia*), and bulrush (*Scirpus validus* and *S. americanus*). The mixed marsh is less subdivided than the wet prairie because nutrient loadings are lower and the potential need for maintenance is reduced. Water depths in the mixed marsh range from 0 to 0.6 m.

The final community is a 162-ha (400-ac) hardwood swamp that includes a 49-ha (120-ac) lake. The swamp was planted with seedlings of bald cypress (*Taxodium distichum*), red maple (*Acer rubrum*), loblolly bay (*Gordonia lasianthus*), water oak (*Quercus nigra*), dahoon holly (*Ilex cassine*), and other tree species. An herbaceous understory will provide additional nutrient removal and wildlife habitat while trees are developing. Once the hardwood swamp is established, it will maintain a diverse wildlife community. The 9-m-deep lake created from an area excavated for fill within the swamp provides additional final treatment through sedimentation. The hardwood swamp/lake is essentially one cell with low interior berms to direct most of the flow into the lake. No harvesting or extensive maintenance will occur in this area.

Water leaves the hardwood swamp through an effluent discharge structure with a control weir into a ditch on the northern boundary of the site. The ditch, historically discharging stormwater runoff directly to the St. Johns River, altered the hydrology of adjacent natural wetlands and caused high nutrient loadings from agricultural runoff. As part of the city's wetlands project, this ditch was blocked and water forced to sheet-flow across approximately 240 ha (600 ac) of natural wetlands prior to discharge to the St. Johns River.

The constructed wetlands site was designed to receive 0.88 m^3/s of water from the Iron Bridge plant with 6 mg/L TN and 0.75 mg/L TP. For the 494-ha site, this is equivalent to nutrient loadings of 0.92 kg/ha/day (0.81 lb/ac/day) and 0.11 kg/ha/day (0.10 lb/ac/day) for TN and TP, respectively. Based on a literature review conducted by the University of Florida Center for Wetlands Research for plant uptake and litter stabilization rates,[1] combined with denitrification rates,[2] the system is capable of removing up to 1.5 kg/ha/day (1.3 lb/ac/day) TN and 0.15 kg/ha/day (0.13 lb/ac/day) TP. The TP removal rate does not include soil sorption of phosphorus. Although TP removal can be a major removal mechanism in some systems, because of difficulty in predicting

Table 2. Monthly Operating Performance: City of Orlando Man-Made Wetlands System

	Influent Conditions					Effluent Conditions					
Month	Flow	BOD_5	TSS	TN	TP	Flow	DO	BOD_5	TSS	TN	TP
October 1987	0.34	4.0	5.8	2.4	0.35	—[a]	5.7	3.6	4.2	1.13	0.15
November	0.35	3.9	8.4	3.3	0.42	—[a]	5.9	2.3	4.1	0.96	0.13
December	0.35	4.7	10.1	4.4	0.48	0.25	6.7	2.0	3.7	0.92	0.11
January 1988	0.35	4.9	12.9	4.9	0.63	0.47	8.1	1.2	3.1	0.91	0.11
February	0.35	5.9	13.3	5.9	0.60	0.36	8.0	1.5	3.2	0.88	0.09
March	0.35	7.5	12.2	6.1	0.61	0.44	7.7	2.0	2.8	0.90	0.11
April	0.35	6.3	12.2	5.8	0.57	0.18	6.6	3.5	5.1	1.22	0.14
Average	0.35	5.3	10.7	4.7	0.52	0.34	7.0	2.3	3.7	0.99	0.12
Permit requirements	0.35[b]	10.0	15.0	6.0	0.75	—[c]	3.5	—[c]	15.0	2.3	0.20

Note: All parameters expressed in mg/L, except flow in m^3/s.
[a]Effluent flowmeter was not functioning.
[b]System startup is limited to 0.35 m^3/s. Based on proven system performance, flows will be gradually increased to 0.88 m^3/s.
[c]No permit requirements for this parameter.

the actual rate and low permeability soils at the Orlando site, only plant uptake/litter stabilization rates were estimated.[3] Concentration limits for water leaving the man-made wetlands (established in state and EPA permits based on background concentrations in adjacent natural wetlands) are 2.3 mg/L TN and 0.2 mg/L TP.

Project Performance

In June 1987, treated wastewater from the Iron Bridge plant was introduced to the wetlands. The first discharge of wastewater from the system did not occur until late September 1987. Although designed for 0.88 m^3/s, the system was started with 0.35 m^3/s (8 mgd) of flow under state permit conditions, and the flow will be gradually increased to design flow based on proven system performance.

A summary of the wetlands performance for October 1987 through April 1988 is presented in Table 2. A more detailed analysis of wetlands water quality indicates treatment for TN and TP is occurring in the first cells of the wetlands. Water leaving cell 1, representing the first cells of the wet prairie system, has average TN and TP concentrations of 1.04 mg/L and 0.10 mg/L, respectively, from September 1987 through April 1988. Effluent from cell 1 is similar to final effluent quality, and little treatment is occurring in the remaining portions of the wetlands. Although flows are only 40% of design flow, sufficient treatment is occurring in only 10% of the system. Better than expected phosphorus removal performance can be attributed to soil sorption, which was not included in design estimates.

SUMMARY

Both Lakeland and Orlando wetland systems have achieved permitted effluent requirements, although loaded at less than design conditions. Based on present treatment occurring in the first cells, as flows are increased, more of the system will be utilized for treatment, but continued compliance with permitted effluent limits is expected. Through continued operation and monitoring, effects of increased loading will be studied. Other proposed studies include long-term phosphorus removal capabilities, management needs, nutrient removal pathways, and comparative vegetational performance studies. Results of these studies and continued operation of the Lakeland and Orlando projects will provide valuable information for design and operation of future projects.

REFERENCES

1. Best, G. R., and M. T. Brown. "Establishment of Wetland Communities for Wastewater Renovation," in *Engineering Report: Orlando Easterly Artificial Wetlands Wildlife Management and Conservation Area*, Post, Buckley, Schuh & Jernigan, Inc., Orlando, FL (1985).
2. Graetz, D. A., P. A. Krottje, N. L. Erickson, J. G. A. Fiskell, and D. F. Rothwell. "Denitrification in Wetlands as a Means of Water Quality Improvement," Publication No. 48, Water Resources Research Center, University of Florida, Gainesville, FL (1980).
3. Zoltek, J., and R. J. Mestan. "The Fate of Phosphorus in Secondarily Treated Wastewater Applied to a Freshwater Marsh," Technical Completion Report No. B-050-Fla, Water Resources Research Center, University of Florida, Gainesville, FL (1985).

39c Research to Develop Engineering Guidelines for Implementation of Constructed Wetlands for Wastewater Treatment in Southern Africa

A. Wood and L. C. Hensman

INTRODUCTION

In recent years, increasing production and disposal of wastewaters have caused accelerated eutrophication of many of South Africa's impoundments,[1-3] necessitating imposition of stringent effluent nutrient discharge standards.

African conditions, however, seriously constrain the ability of rural communities to reliably achieve required discharge standards. If wastewater treatment is practiced at all, it is likely to be pit latrines, septic tanks, oxidation ponds, or biological trickling filters—systems designed to meet basic discharge standards rather than high-level nutrient polishing and solids removal. Unfortunately, water scarcity for effluent dilution or posttreatment means these effluents can significantly contribute to the eutrophication of downstream impoundments. Natural wetlands form an important barrier against nonpoint source discharges of most heavily industrialized areas of South Africa.[4] Research is now underway at a number of locations to evaluate the potential for constructed wetlands in wastewater treatment as a simple, low-cost, low-technology approach to the problem.[5,6]

This chapter reports current research designed to provide engineering data on the biological and physiogeochemical constraints of the constructed wetland concept.

EXPERIMENTAL INVESTIGATIONS

Division of Water Technology, CSIR

The Division of Water Technology (DWT) of the Council for Scientific and Industrial Research initiated a constructed wetland research program in late 1985. Recognizing the inherent cautious approach of design engineers to unproved technologies, a consulting engineering company, Stewart Sviridov & Oliver, is working with DWT to establish guidelines for constructed wetland design implementation. Research is underway in a series of laboratory, bench, and pilot-scale facilities under the guidance of a coordinating committee comprising scientific, industrial, and consulting personnel interested or active in the field.

Pilot Studies, First Stage

Studies investigating the controlling factors responsible for efficient operation and design of wetland systems include (1) the substrata responsible for most nutrient removal and hydraulic capacity of the system; (2) macrophyte species responsible for aeration and enhancing permeability; and (3) effluent characteristics, loading rates, and retention times in the operational regime. Substrate types are evaluated for permeability (K_f) and nutrient removal capacity under controlled conditions with desired effluent types. For example, seven soil types demonstrated K_f values of 0.71 to 45.4 cm/day for stabilization pond effluent compared with 23.2 to 239.9 cm/day for stabilization pond effluent that had passed through a rough ash filter. Settled sewage caused surface blockages of hydraulic cells, leading to impermeability problems. In the field, microbiological activity and rhizosphere structure should degrade this layer and promote permeability. For effluents with higher organic or solids concentrations, coarser substrata, higher hydraulic heads, or lower loading rates are required to achieve adequate passage through the bed.

Phosphate uptake capacity of the soils studied ranged from 4.0×10^{-3} $\mu g/g$ to 896.5 $\mu g/g$, with power-station waste ash at 54.9 $\mu g/g$. This range was (1) related to soil clay content as well as the metals Fe, Al, Ca, and Mg and organic matter and (2) related inversely to hydraulic permeability. Evaluating a local soil or media type will require a balance between permeability and desired nutrient removal. Power-station waste ash is a promising substrata because it combines high permeability with a degree of phosphate removal associated with high salt concentrations and alkalinity.

Potential plant species have been established in simple pot reactors containing waste ash as substrata and receiving oxidation pond effluent or settled sewage over a 12-month period. Table 1 indicates the effluent treatment efficiency of each system. Each of the plant species utilized demonstrated rapid development of both above- and belowground structure and occupation of the available space in the pots. At a stabilization pond effluent loading rate of 84

Table 1. Nutrient Levels in the Effluent and Removal Efficiencies for Macrophyte Species Receiving Stabilization Pond Effluent or Settled Sewage at 84 and 28 L/m²/day, Respectively

			COD		PO_4-P		NH_4-N		NO_3-N	
Pot	Species	Influent	mg/L	% Rem	mg/L	% Rem	mg/L	% Rem	mg/L	% Rem
1	T	SPE	16.3	54	0.77	51	0.54	61	1.3	70
2	S	SPE	17.1	51	0.77	54	0.36	74	0.7	84
3	C	SPE	17.1	51	0.71	51	0.35	74	0.3	93
4	K	SPE	13.36	62	1.06	32	0.38	72	0.25	94
5	S	SPE	23.1	34	0.4	74	0.35	74	0.5	89
6	P	SPE	16.19	54	0.84	46	0.33	76	0.77	82
7	A	SPE	23	35	0.3	81	0.45	67	0.37	92
8	T	SS_1	19.96	94	0.63	92	1.24	94	0.46	40
9	S	SS_1	21.07	93	0.99	88	0.73	97	0.35	54
10	C	SS_1	16.8	95	0.96	88	0.74	97	0.98	–27
11	K	SS_1	15.48	95	1.36	83	0.63	97	0.36	53
12	S	SS_1	16.16	95	1.48	82	0.89	96	0.43	44
13	P	SS_1	16.3	95	1.1	86	1.45	93	0.27	65
14	A	SS_1	24.9	92	0.47	94	1.25	94	0.45	41

Note: T = *Typha capensis;* S = *Scirpus lacustris;* C = *Cyperinus platycaulis;* K = *Pennisetum clandestinum;* P = *Phragmites australis;* A = *Arundo donax;* SPE = stabilization pond effluent (COD = 35.4 mg/L; PO_4-P = 1.6 mg/L; NH_4-N = 1.4 mg/L; NO_3-N = 4.6 mg/L); SS_1 = settled sewage (COD = 332 mg/L; PO_4-P = 8 mg/L; NH_4-N = 20.7 mg/L; NO_3-N = 0.9 mg/L).

L/m²/day and settled sewage rate of 28 L/m²/day, effluent discharge consents were within general standards and close to special discharge consents of 1 mg/L PO_4-P.

The low effluent concentrations mask any significant difference between species, suggesting that any of the species could be appropriate. However, other studies have shown that the grass *Pennisetum clandestinum*, cultivated as a lawn material, and *Arundo donax* do not readily produce subsurface rhizome proliferation and are restricted in their capacity to oxygenate the wetland system. The occurrence of nitrification and denitrification is evidence of the simultaneous presence of aerobic and anoxic zones within the bed, while significant removal of phosphate is a response to salt and alkalinity associated with waste ash, resulting in pH levels as high as pH 11 initially. Macrophytes did not appear to be detrimentally affected by high pH and benefit from the presence of micronutrients.

Twelve pilot-scale reactors, each with 4 m² of surface area, have also been constructed to investigate the systems' operational regimes and effects on wastewater treatment potential. These studies compare two macrophytes (*Phragmites australis* and *Scirpus*); three effluents (stabilization pond, ash-filtered stabilization pond, and septic tank); and two substrata (soil or ash). As with the pot studies, the low effluent pollutant concentrations of each system mask significant differences between species or conditions, although low ammonia and COD levels of *Scirpus* tend to suggest a superior oxygenation

Table 2. Effluent Results for the Pilot Reactors Receiving 20 cm/day of Influent During 12-Month Period

Species	Strata	Influent	COD (mg/L)	PO_4-P (mg/L)	NH_4-N (mg/L)	NO_3-N (mg/L)	Suspended Solids (mg/L)
	Soil	FSPE	42	0.3	3.4	0.9	16.0
S	Soil	FSPE	35	0.4	1.9	0.9	15.2
P	Soil	FSPE	29	0.2	2.3	0.7	18.4
S	Soil	SPE	34	0.3	1.5	1.2	28.8
P	Soil	SPE	37	0.6	2.1	1.7	21.6
P	Soil	SPE	30	0.2	3.9	1.2	31.6
P	Ash	SPE	27	0.4	4.6	2.1	4.8
S	Ash	SS_2	21	0.6	9.4	0.7	14.4
	Ash	SS_2	31	0.7	27.5	0.4	4.8
P**	Ash	SS_2	29	1.7	32.5	0.4	11.2
P	Soil	SS_2	39	0.3	9.3	0.5	27.2
P*	Soil	SS_2	47	0.2	11.4	0.4	25.6

Note: S = *Scirpus lacustris;* P = *Phragmites australis;* FSPE = ash-filtered stabilization pond effluent (COD = 75 mg/L; PO_4-P = 4.4 mg/L; NH_4-N = 11.2 mg/L; NO_3-N = 3.1 mg/L); SPE = stabilization pond effluent (COD = 116 mg/L; PO_4-P = 4.5 mg/L; NH_4-N = 12.0 mg/L; NO_3-N = 3.4 mg/L); SS_2 = septic sewage (COD = 166 mg/L; PO_4-P = 7.6 mg/L; NH_4-N = 42.4 mg/L; NO_3-N = 0.9 mg/L).
* = reactor receiving effluent from reactor marked "**".

capacity of this plant. This may be a result of a more rapid establishment of this species than *Phragmites* and will require long-term operation to evaluate true efficiency. Again, nitrification and denitrification have been accomplished, while phosphate levels with soil and ash substrate generally remain within special discharge consent. Relatively high suspended solids levels are a result of carryover of soil and ash particles into the effluent and are not thought to pose a serious pollutant threat to the receiving waters.

Reactors were planted in May 1987 and have established a viable macrophyte stand. Early after planting, *Phragmites* was almost destroyed by aphids, requiring insecticide (Malasol) treatment. *Scirpus* was ignored by aphids and maintained a dense, actively growing stand. However, heavy rainfall and wind have blown over *Scirpus,* necessitating stem cutting to allow new growth. *Scirpus* died back in early winter (April), while *Phragmites* remained green and upright, indicating a greater temperature tolerance. Table 2 provides effluent levels from the systems after a year of operation.

Pilot Studies, Second Stage

A former fish dam 5 m × 5 m × 1 m was lined with hyperplastic membrane (Carbofol 500 μm) protected by a 50-mm sand layer prior to filling with coarse gravel (19 mm) and planted with local *Arundo donax* receiving septic sewage at 10 cm/m^2/day. *Arundo* fully occupied the bed within four months, and the system was operated for one full season. Although sus-

Table 3. Pollutant Level Removal Through the Combined Systems

Parameter	Influent (mg/L)	After Artificial Wetland		After Algal Pond		Cumulative % Removal
		mg/L	% Removal	mg/L	% Removal	
NH_4-N	40	26	34	6	47	85
PO_4-P	8	5	31	4	32	50
NO_3-N	0	2	—	13	—	—
COD	223	91	59	46	49	79
TSS	186	41	77	105	—	—

Combining NH_4-N and NO_3-N, total % removal of influent NH_4-N = 49%.

pended solids (90%) and COD (65%) were significantly reduced to 34 and 95 mg/L respectively, these did not meet the effluent discharge standards of 20 and 75 mg/L, respectively, while only 18% ammonia and 7% phosphate removal was achieved. Local *Scirpus* and *Phragmites* were planted, and the influent changed to raw sewage. Close to total occupancy of bed space was achieved within six months, and effluent quality has improved, with 73% COD removal (to 68 mg/L) and 38% ammonia reduction (to 22.7 mg/L). Suspended solids and P were unaffected. The influent is now supplemented with molasses to simulate high COD loadings associated with rural communities, and effluent quality is remaining close to or below COD discharge consent as the plant community fully develops.

Two other lined dams have been converted into wetlands in upgrading studies of 12 small-scale units. One system with power-station waste ash is planted with *Phragmites* and loaded with raw sewage. Effluent flows over into a *Phragmites* bed with a local soil covered in a layer of coarse ash. The first bed is designed for primary treatment of solids and carbonaceous removal, and the second bed is designed for removal of P and N from 10 cm/m²/day of raw domestic sewage. The first six months of operation have achieved consent, meeting effluent qualities except for ammonia, which remains higher due to low initial oxygenation capacity of the plants.

Combined Artificial Wetland, High-Rate Algal Pond Studies

The system consists of two units, each 22 m × 11 m × 40 cm. Septic sewage is loaded to the wetland at a rate of 135 L/m²/day. Effluent from the bed is pumped from the sump into the algal pond. The coarse gravel bed is planted with *Arundo donax* and two 2-m bands of *Typha*. *A. donax* resembles *P. australis* yet has a more vigorous growth, which makes it potentially more efficient at oxygen transfer to the rhizosphere. Two bands of *Typha* were included as denitrification zones because they are reported to be less efficient at oxygenating the rhizosphere and anoxic conditions would prevail.[5]

Over a year's period, effluent COD levels were reduced 59%, NH_4-N 34%, PO_4-P 31%, and suspended solids 77% (Table 3). Inability to achieve total organics removal and nitrification results from the high loading rate, short-circuiting, and inability of macrophytes to meet oxygen demands. *Arundo*

Table 4. Comparison of Reed Bed Influent, Effluent, and General Standard Regulations at Mpophomeni

Deteminand	General Standard	Reed Bed Influent Mean	Max	Min	Reed Bed Effluent Mean	Max	Min	% Removal
pH (pH units)	5.5–9.5		8.5	6.1		7.9	5.3	
Alkalinity as $CaCO_3$		43.2	92.7	4.7	86.1	160.7	25.3	−99.3
Conductivity as mS/m	75 above intake	59.8	92.6	30.2	46.8	63.2	17.3	20.1
NO_3 as N		46.0	109.6	3.9	15.1	54.2	0.34	67.1
NH_3 as N	10	2.8	18	0.01	1.2	3.62	0.01	57.1
PO_4 as P	1	7.0	13.5	—	0.87	6.5	0.3	87.7
TSS	25	39.1	220	4	13.7	54.6	4	65.0
TOC		11.6	27.6	1.9	5.4	9.8	2.2	53.4
COD	75	64.1	168	8	22.5	54.5	8	54.9
PV	13	10.5	—	—	5.0	—	—	75.9
Coliforms/100 mL		5.6×10^5	18×10^5	1×10^4	5.5×10^4	10×10^5	Nil	90.1
E. coli/100 mL	Nil/100	1.6×10^5	6×10^5	1×10^3	3.4×10^4	5×10^4	Nil	97.9
F. strep/100 mL		3.7×10^4	4×10^5	Nil	1.3×10^3	2.7×10^4	Nil	96.5

Note: All concentrations are in mg/L unless otherwise indicated. mS/m = millisemens/meter; PV = Permanganate Value (oxygen absorbance).

roots extended down to 30 cm, while the root structure was black, indicating reducing conditions around the rhizomes. Subsequently, *Typha, Phragmites,* and *Scirpus* were planted and the influent introduced along the sides. *Typha* has occupied much of the bed, while *Scirpus* and *Phragmites* have taken longer to become established, with unoccupied space present in the *Phragmites* zone after nine months. Present discharge quality from the wetland is 78 mg/L COD, 28 mg/L SS, 16 mg/L NH_4-N, 0.2 mg/L NO_3-N, and 5.6 mg/L PO_4-P.

Mpophomeni Sewage Works

Mpophomeni, a Kwazulu town on the eastern shore of a sensitive P-limited impoundment, has a biofilter sewage works with an artificial reed bed for polishing final effluent. Surface area of the reed bed is 2500 m^2 and 1.5 m deep with a maximum design loading of 20 cm/m^2/day. Lined with a hyperplastic membrane with a bottom drainage level of crushed gravel and a geotextile filter layer, the bed consists of acidic soil (Doveton series) of high P-fixing capacity and *Phragmites.*

Inconsistent loading rates (5–23.5 cm/m^2/day) have led to some inability to assess performance, although the reed bed has a marked effect on the quality of effluent (Table 4). Effluent concentrations, except *Escherichia coli,* comply with general standards. Reduction in physicochemical constituents is between 20% (for conductivity) and 88% (for orthophosphate). Good nitrification and denitrification are performed with associated alkalinity increase.

The system was taken off line twice due to poor bed permeability. Core sample analyses have shown that 85% of total root development occurred in the 0–60 cm region, probably due to the high water table. Lack of subsurface root development has seriously reduced expected permeability improvement, while decay of plant biomass and effluent suspended solid appears to be sealing the surface in this vertical-flow application. On-site infiltrometer studies are presently evaluating the benefits of increasing the calcium content of the soil with gypsum to enhance permeability.[7]

Grootvlei Power Station, Transvaal

Trial systems at Grootvlei are also designed to remove nutrients and pathogens from a biofilter plant effluent. A horizontal-flow stone-media stage is divided into a zone planted with *Scirpus* (for nitrification), a zone with mulched plant material (for denitrification), and a zone with *Typha*. The second stage is a vertical-flow soil bed planted with *Scirpus*, with a gravel underdrain for nutrient polishing. At a loading rate of 7.5 $cm/m^2/day$, mean effluent parameters were 1.9 mg/L total N, 0.8 mg/L PO_4-P, 19 mg/L COD, 10.8 mg/L suspended solids, and 2×10^4 fecal coliforms/100 mL. Influent total N levels were reduced 79%, PO_4-P 86%, COD 67%, and fecal coliforms 99.9%.

Olifantsvlei Sewage Works, Johannesburg

A system at Olifantsvlei sewage works near Johannesburg is intermediate between natural wetlands and wetlands constructed with gravel or soil. Three units are geotextile-lined, 120 m × 20 m × 2 m. Two beds were filled with peat from an adjacent natural system, while the third has a woven rope net supporting the peat mass above the channel bottom. Each bed is dominated by *Phragmites* and receives approximately 50 $cm/m^2/day$ secondary effluent. The units have run intermittently for three years, and available data indicates significant nitrification, denitrification, and removal of suspended solids and pathogens. Phosphate removal is limited due to the lack of suitable binding to the peat and problems with surface rather than subsurface flow through the peat.

Additional Studies

Reed beds have been established in two rural areas of Gazankulu. At Giyani, population 90,000, outflow from biological seeping beds is channeled to reed beds, where retention time is four days. Seven beds are alternately isolated and drained down 10 cm for one day to destroy mosquito eggs. Harvested reeds are used in basket weaving, a traditional industry of the area. A similar system has been constructed at Nknowankowa.

A horizontal-flow pilot-scale unit 100 m × 20 m filled with waste and coarse ash and planted with *Typha* at the inlet and *Phragmites* at the outlet (effluent end) is treating petrochemical effluents at a design loading of 50 L/m^2/day. Coarse ash neutralizes acidic effluent, filters oil and petrochemical residues, and provides high permeability.

A vertical-flow pilot-scale unit to be constructed in the near future for secondary effluent treatment is likely to be a combination of waste mining slime (for its high phosphate uptake) and local soils of medium phosphate uptake but improved permeability characteristics. A gravel or coarse stone system will be designed principally for suspended solids removal and nitrification prior to final discharge to a nature reserve ecosystem.

Construction of several full-scale wetlands are also planned as total treatment processes or integrated into pond or conventional systems, and in Zimbabwe water hyacinth (*Eichhornia crassipes*) will be investigated.

CONCLUSION

Constructed wetlands have considerable potential in southern Africa for treatment of raw wastewaters emanating from rural communities, for upgrading oxidation pond and secondary effluents to general and special discharge standards, and for treatment of industrial effluents. This chapter has been a brief synopsis of experiments underway in southern Africa to evaluate various applications.

ACKNOWLEDGMENTS

The authors are indebted to the Water Research Commission for provision of funds for preparation of engineering design guidelines, to Mr. D. A. Kerdachi of the Umgeni Water Board for his assistance and data from the Mpophomeni reed beds, Messrs. P. Rosser and H. Smit of the Electricity Supply Commission for their work and data on the Grootvlei reed beds, and Dr. K. Rogers of the University of Witwatersrand.

REFERENCES

1. Alexander, W. V. "The Potential of Artificial Reedbeds for the Treatment of Wastewaters in Southern Africa," paper presented at the Seminar Technol Transfer Water Supply & Sanitation in Developing Areas, February 12–15, 1985.
2. Wood, A., and M. Rowley. "Artificial Wetlands for Wastewater Treatment," paper presented at the Symposium Ecology and Conservation of Wetlands in South Africa, Council for Scientific and Industrial Research, South Africa, October 15–16, 1987.
3. Zohary, T. "Hyperscums of the Cyanobacterium *Microcystis aeruginosa* in a Hypertrophic Lake," *J. Plankton Res.* 7:399–409 (1985).
4. Viljoen, F. C. "An Input Output Study of Various Physical and Chemical Constituents in Water After Passage Through a Section of the Natalspruit Wetland," paper

presented at the IWPC Biennial Conference, Port Elizabeth, South Africa, May 12–15, 1987.

5. Alexander, W. V., and A. Wood. "Experimental Investigations into the Use of Emergent Plants to Treat Sewage in South Africa," paper presented at the Symposium Macrophytes in Wastewater Treatment, IWPC, Piracicaba, Brazil, August 1986.
6. Wood, A., J. Scheepers, M. Hillis, and M. Rowley. "Combined Artificial Wetland and High Rate Pond for Wastewater Treatment and Protein Production" (1987).
7. Furness, H. D., K. J. Healey, D. A. Kerdachi, and W. N. Richards. "The Performance of an Artificial Wetland for the Treatment of Biological Filter Effluent," paper presented at the Symposium Ecology and Conservation of Wetlands in South Africa, Council for Scientific and Industrial Research, South Africa, October 15–16, 1987.

39d Constructed Wetlands: Design, Construction, and Costs

Kelly J. Whalen, Pio S. Lombardo, D. Bruce Wile, and Thomas H. Neel

INTRODUCTION

Constructed wetlands have successfully treated settled domestic wastewater;[1-4] however, most wetlands have been pilot studies or small-scale treatment works. Information in the literature focuses primarily on performance. Practical design, construction, and costing guidance is not readily available to engineers involved in development of constructed wetlands from planning to operation phases. Design and construction issues discussed in this chapter include process design, basin design, process control features, storm impact, construction, and construction costs. Experience gained in the design and construction of a 1.1-ha bulrush wetland forms the basis.

The constructed bulrush wetland, owned by the Anne Arundel County Department of Utilities (AACDU) and designed by Lombardo & Associates, Inc., Boston, Massachusetts, is a principal component of the 1770 m^3/day Mayo Water Reclamation Subdistrict Large Communal Water Reclamation Facility (LCWRF). The LCWRF will treat septic tank effluent collected from 2000 homes and discharge the treated wastewater to the Chesapeake Bay. The LCWRF will consist of recirculating sand filters, bulrush wetlands, ultraviolet disinfection, peat wetlands, a postaeration aspirator, and an offshore wetland. The purpose of the bulrush wetlands is to denitrify the wastewater.

DESIGN

A summary of the design parameters for the LCWRF bulrush wetlands and their values or descriptions follows:

- Process/conceptual
  Purpose—denitrification
  Flow—1770 m^3/day
  Target removal—70% (total nitrogen)

Carbon source—septic tank effluent
Carbon/nitrogen ratio—1.4:1 (BOD_5 to total nitrogen)
Areal hydraulic loading rate—16 cm/day
Detention time—1.5 days

- Basin
  Description—lined and baffled earthen basin wetland divided into 3.5 clusters consisting of six modules each
  Liner—30-mil PVC with geotextile lining
  Baffle—30-mil PVC with plywood/lumber support
  Length/width ratio—9:1
  Media—top layer = 1.3-cm peastone gravel; bottom layer = 3.8-cm coarse gravel
  Vegetation—bulrush (*Scirpus olneyi*)
  Harvesting—vehicular access on all sides
  Storm impact—site drainage away from wetland; outlet pipe sized to restrict discharge; freeboard provided for temporary storage; downstream process units designed for equalized storm flow
- Process control
  Carbon/nitrogen ratio—adjustable at upstream distribution box
  Loading—independently operated diffusers in each module for step loading; adjustable hydraulic loading for each module
  Drain—independently drainable modules
  Odor control—loading can be distributed along module; ratio of sand filter filtrate and septic tank effluent adjustable

Process/Conceptual Design

The bulrush wetlands will be the second process unit in the treatment chain (Figure 1). Approximately 75% of the collected septic tank effluent will flow to sand filters, and 25% will flow to the bulrush wetlands. Septic tank effluent flowing through sand filters will be nitrified. Septic tank effluent flowing to the bulrush wetlands will supply carbon for denitrifying the sand filter filtrate as it flows through the wetland. Carbon/nitrogen ratio will be 1.4:1 (BOD_5 to total nitrogen).

The bulrush wetlands will consist of three and a half separate and independently operated clusters of six modules each with a distribution box to apportion flow among the clusters. An inlet box at each cluster will apportion flow between the six modules of each cluster, and flow can be directed to any of four separate inlets within modules. Areal hydraulic loading rate will be 16 cm/day and detention time will be 1.5 days, based on an assumed media pore volume of 40% of total volume.[1,2]

Basin Design

Each cluster will consist of a lined and baffled earthen basin forming a parallelogram to fit onto the site. The basin will be filled with gravel and

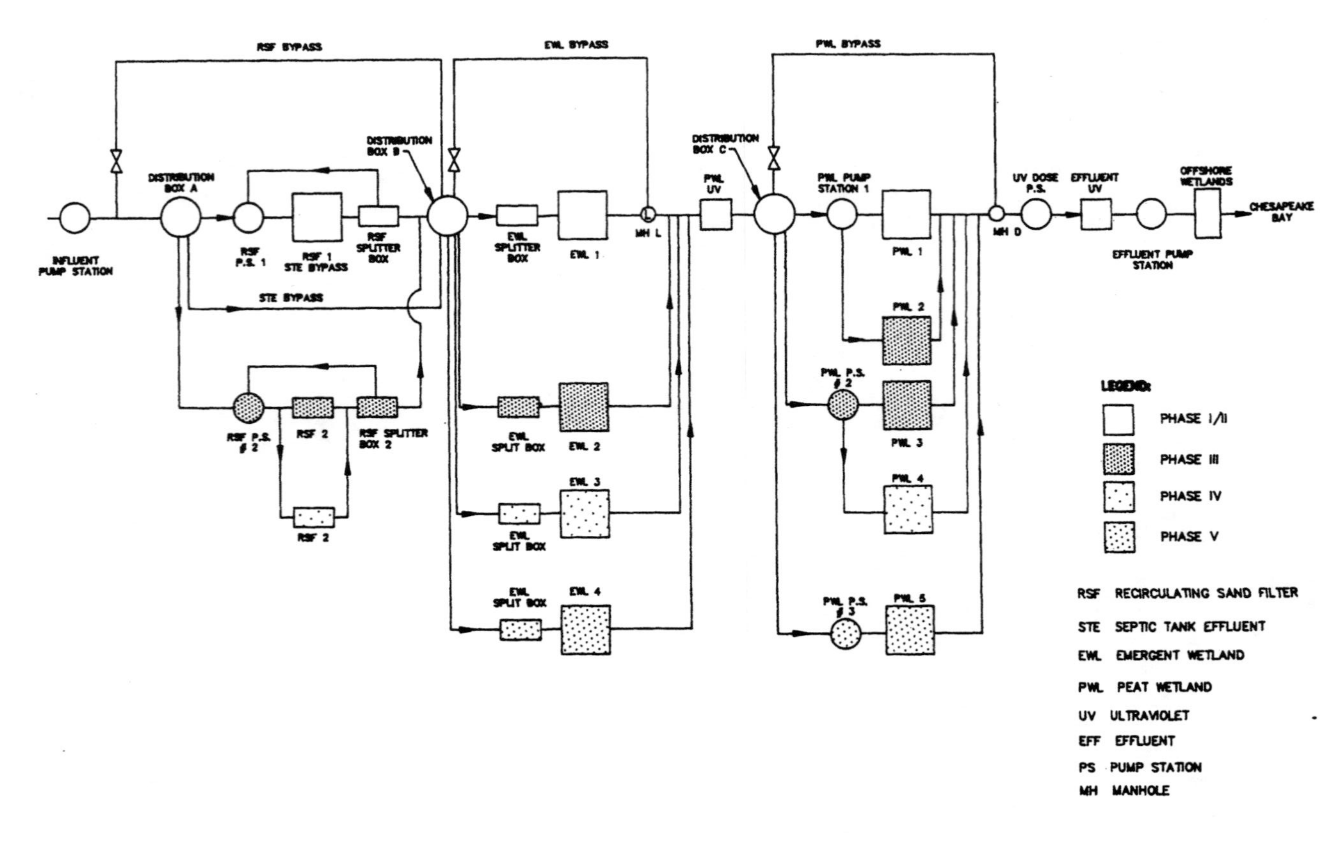

Figure 1. Process flow design.

planted with bulrush (*Scirpus olneyi*), and adequate freeboard will be provided for capacity during severe storms.

Liner

The basin liner will be industrial-grade 30-mil PVC liner with an 80-mil geotextile liner on top to protect the PVC liner from puncture by gravel. The PVC liner in the freeboard will be covered with stone to protect the PVC from degradation from sunlight.

Baffles

Baffles installed to separate modules and to attain a 9:1 length-to-width ratio for each module will consist of sections of PVC liner seamed to the bottom liner. The baffle liner will be supported by a plywood/lumber support to facilitate gravel installation, and it will be allowed to rot in place.

Media

Wetland media will consist of a bottom layer of 3.8-cm stone 24 cm thick and a top layer of 1.3-cm pea stone 15.2 cm thick. The bottom layer functions as a process medium, and the top layer contributes to aesthetics and mosquito control. Water level will be maintained 7.6 cm below the top layer.

Vegetation

Bulrushes (*Scirpus olneyi*) will be single plants 30–60 cm in height with leaves and roots attached and planted 90 cm apart. Vegetation harvesting will be by corn harvester from a service road on the perimeter of each cluster. Harvested vegetation will be composted on-site, and compost will be used by the county or offered to the public.

Storm Impact

Severe storms are expected to contribute significant amounts of flow via direct rainfall on the 0.65-ha sand filter and 1.1-ha bulrush wetland. The wetland freeboard will accommodate an 18.3 cm/day or 100-yr 24-hr storm in the Annapolis, Maryland area.

Freeboard requirements were based on hydraulic analysis of the facility pipe network and assumed rain intensities for 0.1- to 12-hr time increments. The wetland discharge pipe was sized to mitigate excessive flows due to direct rainfall. Excessive flows can be stored in downstream peat filters and pumped out at design rates through the final UV disinfection units to ensure fecal coliform permit compliance.

PROCESS CONTROL

Process control will include carbon/nitrogen ratio control, variable loadings between clusters and modules, step loading, and odor control. Carbon/nitrogen ratio will be controlled by adjusting weir lengths in the distribution box that apportions septic tank effluent between the sand filters and bulrush wetlands. Hydraulic loadings can be varied between clusters at the distribution box apportioning flow among the wetland clusters by adjusting weir lengths. Hydraulic loadings can be varied among modules of a cluster at the inlet box. Each module will be equipped with four inlet pipes located at the head end and at quarter points along the length of the module. Each inlet will be equipped with shut-off valves. We expect that step loading will make possible higher loadings in the wetland. Odor control will be achieved by increasing the fraction of sand filter filtrate or by increasing the number of operable diffusers to reduce local overloading.

CONSTRUCTION

Construction issues include project phasing of clusters and construction sequencing of individual clusters. Special construction problems unique to wetland construction include baffle installation, pipe installation, media placement, and startup.

The wetlands were designed to be modular to facilitate matching treatment capacity with phased collection system construction and to promote developer financing of future wetland additions.[5] The sequence of wetland construction is survey control; site preparation; rough grading; installation of structures and pipes in berm; fine grading; installation of liner, baffles, gravel, and interior pipes; filling the wetland; and planting vegetation. Site and basin preparation is similar to that for a lined pond. Baffle and media installation should follow liner installation as quickly as possible to minimize potential for liner damage. The wetland should be filled with water before vegetation is planted.

Wetland startup will simply involve allowing a mixture of septic tank effluent and sand filter filtrate to flow through the wetland. It is expected that denitrification will occur within a month of startup. The wetland will not be seeded with denitrifiers from another wastewater treatment facility.

A hydraulic budget including evapotranspiration, rainfall, wastewater flow, and outflow was calculated for the wetland to ensure that during the low flow period in which sewer hookups are made, a critical water level will be maintained. Startup of the wetlands will occur in late fall; rainfall and wastewater flow is expected to exceed evapotranspiration.

COSTS

Due to the phased construction of the water reclamation facility, construction of the wetlands was divided into three separate contracts, two of which have been awarded. Bids were solicited for Contract No. 1 in February 1987 and for Contract No. 2 in March 1988. Design capacity of the Contract No. 1 system was 511 m^3/day and included one wetland cluster. Design capacity of the Contract No. 2 system was 1012 m^3/day and included two wetland clusters. Notice to bidders was posted in local and regional newspapers, trade journals, and contractor reports systems. Contractor associations and trade associations were also notified. Contractor response was low: four bids were received for Contract No. 1, and one for Contract No. 2. Prices bid were averaged to calculate media unit costs.

Average construction cost for wastewater treatment capacity is \$530/m^3/day. Distribution of cost items on a percentage basis is as follows:

piping and valves	13.7
sitework	16.4
gravel (all sizes)	27.2
structures	23.0
miscellaneous	6.1
mobilization	13.6

Unit costs for major items (including furnishing and installation, i.e., backfill, cover, and all appurtenances) are as follows:

peastone gravel (1.3-cm diameter)	\$34.50/m^3
coarse gravel (3.8-cm diameter)	37.50/m^3
liner (30-mil PVC with 80-mil geotextile)	6.00/m^2
vegetation (bulrush)	7.50/m^2
clearing and grubbing (site)	2.50/m^2
embankment (30-cm lifts)	3.50/m^3
excavation (favorable conditions)	3.50/m^3

Prices are assumed to be inflated due to the small number of bids submitted and construction cost uncertainties. However, unit prices are expected to drop 10% to 20% as experience is gained by contractors. Crushed stone was used for media because no natural deposits of suitable gravel occur in the vicinity. Media costs may be substantially less in areas where gravel is available. Site work costs (excavation and embankment) are representative of generally favorable conditions for construction at the facility site.

SUMMARY/RECOMMENDATIONS

We recommend that design of community scale wetland treatment facilities for nitrogen removal include consideration of the following issues:

- *Process/conceptual*—target removal, carbon source, carbon/nitrogen ratio
- *Basin design*—liner, baffles, length/width ratio, media, vegetation, storm impact
- *Process control*—adjustment of carbon/nitrogen ratios, variable loadings, adjustable hydraulic loadings, odor control

Construction of wetlands is very similar to construction of lined ponds. Construction of wetlands should be modular to match treatment capacity to flows and to facilitate recovery of capital cost of expanding the treatment facilities to serve future users. Wetlands startup should consider hydraulic balance to ensure that a critical water level is maintained in the wetland to maintain vegetation. Construction cost was equivalent to $530/m^3/day of treated wastewater capacity and was somewhat inflated due to uncompetitive contractor environment and the use of crushed rock media rather than natural gravel.

REFERENCES

1. Bavor, H. J., D. J. Roser, and S. McKersie. "Nutrient Removal Using Shallow Lagoon-Solid Matrix Macrophyte Systems," in *Aquatic Plants for Water Treatment and Resource Recovery*, K. R. Reddy and W. H. Smith, Eds. (Orlando, FL: Magnolia Publishing Inc., 1987), pp. 227–236.
2. Gersberg, R. M., B. V. Elkins, and C. R. Goldman. "Use of Artificial Wetlands to Remove Nitrogen from Wastewater," *J. Water Poll. Control Fed.* 56(2):152–156 (1984).
3. Herskowitz, J., S. Black, and W. Lewandowski. "Listowel Artificial Marsh Treatment Project," in *Aquatic Plants for Water Treatment and Resource Recovery*, K. R. Reddy and W. H. Smith, Eds. (Orlando, FL: Magnolia Publishing Inc., 1987), pp. 247–254.
4. Wolverton, B. C. "Hybrid Wastewater Treatment System Using Anaerobic Microorganisms and Reed (Phragmites communis)," *Econ. Bot.* 36(4):373–380 (1982).
5. Hurley, J. M. "Developer Financing of Public Wastewater Service Infrastructure," *J. Water Poll. Control Fed.* 60(5):608–613 (1988).

39e Wastewater Treatment/Disposal in a Combined Marsh and Forest System Provides for Wildlife Habitat and Recreational Use

Beverly B. James and Richard Bogaert

The Mt. View Sanitary District in Martinez, California established 3.6 ha of freshwater wetlands in 1973 to process secondary effluent. It was expanded to 8.5 ha in 1977. In 1979, a pilot project combining a forest area and marsh pond was added. In 1984 and 1987 respectively, 9 ha and 16 ha of seasonal wetlands were added, bringing the total size to approximately 37 ha. This chapter summarizes 15 years of operating experience on two wetland areas ("wetland") constructed in the 1970's and the marsh/forest pilot project ("marsh/forest"), receiving secondary effluent from the district's wastewater treatment plant as the sole water source.

The Mt. View Sanitary District provides sewerage service to approximately 16,000 people. The service area is mostly residential with some commercial and light industrial discharges. A strictly enforced source control policy prohibits discharge of industrial waste to the sewer system. The district's treatment plant treats sewage flows averaging 5300 m^3/day with comminution, primary sedimentation, two-stage high-rate biofiltration, secondary sedimentation, chlorination, and dechlorination.

OBJECTIVES

The primary purpose of the wetland operation (8.5 ha) is to provide an environmentally beneficial use of treated wastewater. To meet this objective, the system has been designed to closely approximate a natural system; it is not designed as a wastewater treatment system. Water quality parameters are monitored along with flora and fauna to assure no adverse effect on the plant and animal community.

The marsh/forest pilot program (4.74 m^2) has a dual objective of simultaneously providing an environmentally beneficial use along with wastewater treatment and disposal. This dual objective makes it suitable for applications where the downstream water body might be adversely affected by discharge of nutrient-rich waters typical of the other wetland.

PROJECT DESCRIPTION

Wetland

The wetland area is divided by dikes into six separate ponds. Water flow through the ponds is controlled by weirs that allow the flow pattern through the ponds to be varied if necessary. Roughly half the total area is in ponds approximately 1 m deep. Emergent vegetation—cattails (*Typha*) dominate—grows along the pond edges, extending well out into the center in some places. The remaining pond area is divided into a large area operated at 0.3 m deep in winter months and allowed to dry during the summer to provide wading bird habitat and suitable conditions for growing bulrushes (*Scirpus*). The second area is a series of channels 1.5 m deep dominated by cattails, and the third pond provides 2-m-deep open water with nesting islands in the center.

Marsh/Forest

The pilot project consists of a 74 m^2, 1-m-deep marsh pond and 400 m^2 forest area (Figure 1). Artificial floating habitats (Ecofloats™) are provided on the pond surface to encourage growth of aquatic invertebrates. Water is circulated from the pond through underground irrigation units in the forest to benefit the marsh by decreasing sunlight and nutrients to the algae in the water, thus reducing algal blooms and increasing water quality in the marsh. A secondary benefit is that the trees use the water, creating a reduced or zero discharge.

The forest area has 26 irrigation chambers (K-6 units™) spaced at 3-m intervals along parallel pipelines with a distribution box at each end. Openings in the end of each chamber allow tree roots to penetrate. The crucial soil-air interface at the surface is left undisturbed and water contacting tree roots remains aerobic. Water is pumped from the pond to the forest by a low-head solar-powered pump, and excess water returns to the pond. Redwood trees (*Sequoia sempervirens)* were planted next to each of the irrigation units in October 1978. A small eucalyptus (*Eucalyptus globulus*) was already on site, and two alders (*Alnus rhombifolia)* and three Monterey pines (*Pinus radifa*) were added in 1981.

PLANT AND ANIMAL COMMUNITIES

A measure of the success of the wetland project is its ability to sustain a diverse community of plants and animals. Over 70 species of plants and 123 species of waterfowl have been identified on the site.[1] The area also supports a

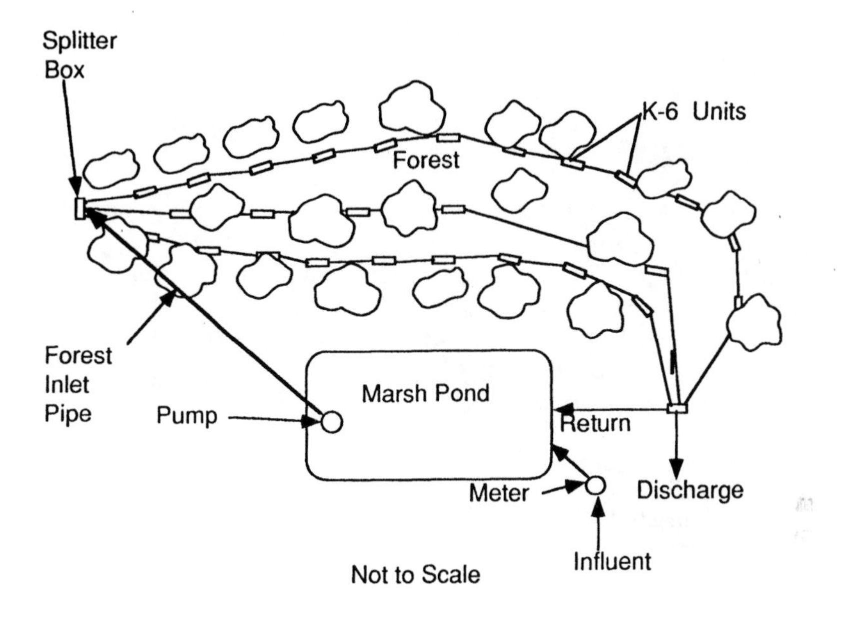

Figure 1. Marsh forest layout.

diverse population of invertebrates, fish, amphibians, reptiles, and small mammals. River otters (*Lutra canadensis*) have been observed on-site.

PUBLIC USE

The district encourages public use of the wetlands, particularly for educational purposes. The wetlands are open to the public from 8:00 a.m. to 4:30 p.m. The district provides a printed guide to the wetlands area that identifies stations where particular features may be observed. Visitors sign in and out, allowing the district to keep records of the public use of the area. As can be seen from Table 1, the biggest attraction of the wetlands is its abundant birdlife. The wetlands area is not extensively advertised, but the district does

Table 1. Number of Visitors/Year to the Mt. View Marshes

	Visitors/Year				
Year	**Birdwatchers**	**Educational**	**Officials**	**R&D**	**Total**
1977					242
1983	75	215		1	291
1984	192	90		4	286
1985	251	63		6	320
1987	194	79	13	15	301

hold periodic open houses, and local conservation and birding groups list it as an excellent birding site.

WATER QUALITY

The wetland and marsh/forest influents consist entirely of secondary wastewater from the district's treatment plant. The influent to the two projects is monitored regularly for pH, BOD_5, suspended solids (SS), nitrogen compounds, phosphorus, heavy metals, pesticides, phenols, toxicity, and coliforms. Heavy metals, phenols, and pesticides in the influent, as shown in the following list in mg/L, are consistently below requirements (effluent is not monitored for these constituents):

arsenic	0.001
cadmium	0.0001
chromium	0.0015
copper	0.082
lead	0.001
mercury	0.0003
nickel	0.006
silver	0.0044
zinc	0.08
cyanides	0.02
phenols	0.028
total identifiable chlorinated hydrocarbons	<detectable

Monitoring data have been consistent from year to year after the first year of startup conditions. Figures 2 and 3 present data from 1981 because that is the only year for which samples were run simultaneously on the marsh/forest and the wetlands.

The influent to the marsh generally just meets the secondary treatment standard of 30 mg/L for SS (Figure 2). Discharge from the wetlands has a higher SS concentration, ranging from 45 to 178 mg/L due primarily to growth of algae in the ponds. The marsh/forest, by controlling algal growth, is able to maintain discharge SS in the range of 2 to 25 mg/L. Influent BOD_5 to the marsh/forest and wetlands ranges from 23 to 59 mg/L. BOD_5 levels are somewhat lower, leaving the wetlands with a range of 17 to 48 mg/L. In the marsh/forest, the BOD_5 of the discharge ranges from 6 to 25 mg/L. The lowest SS and BOD_5 values for the marsh/forest are recorded during the summer months, when algal blooms are causing high readings in the wetlands.

Nitrogen enters the wetlands and marsh/forest in five forms: ammonia, nitrite, nitrate, organic nitrogen, and N_2 fixation by algae. In the wetlands environment, some ammonia is oxidized to nitrite and nitrate, some organic nitrogen is broken down into ammonia, and some nitrogen is converted to

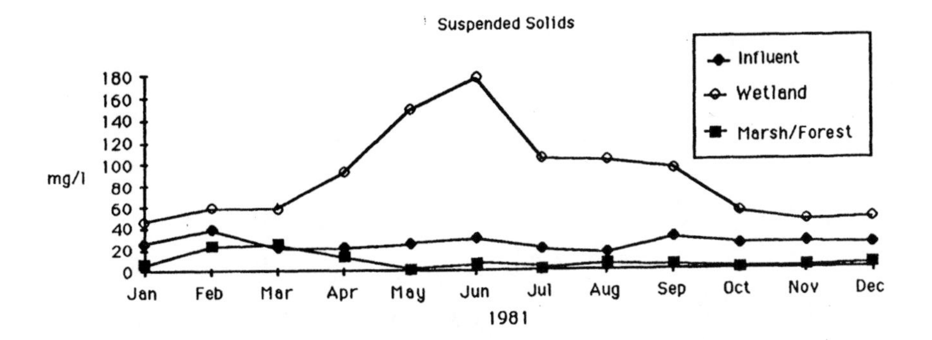

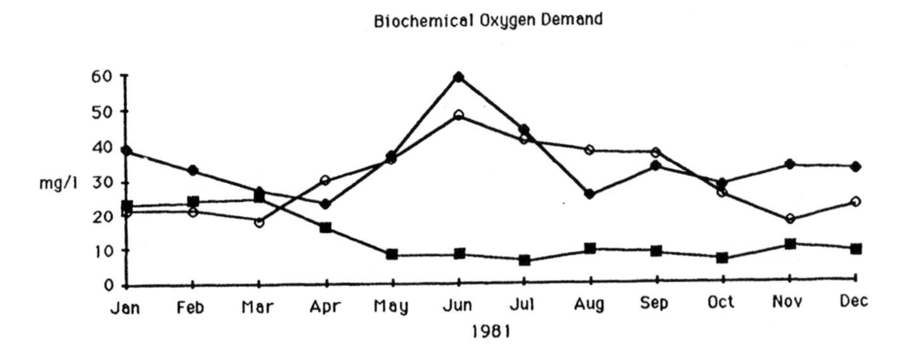

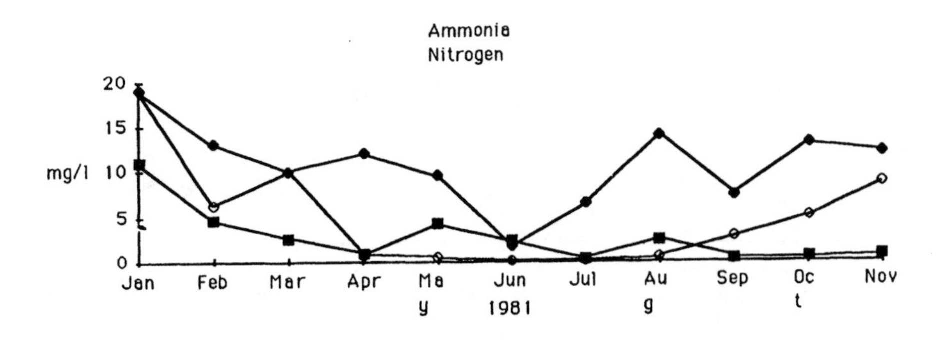

Figure 2. Suspended solids, BOD_5, and ammonia nitrogen in the wetlands and marsh/forest influent and discharges.

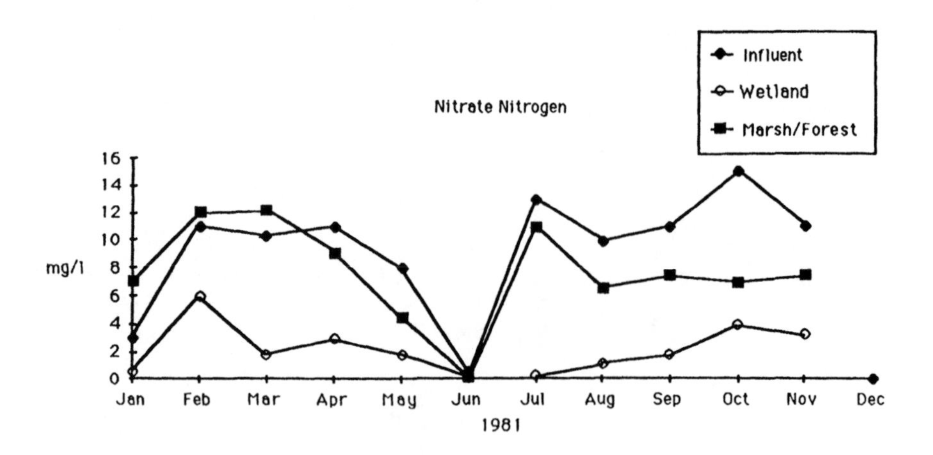

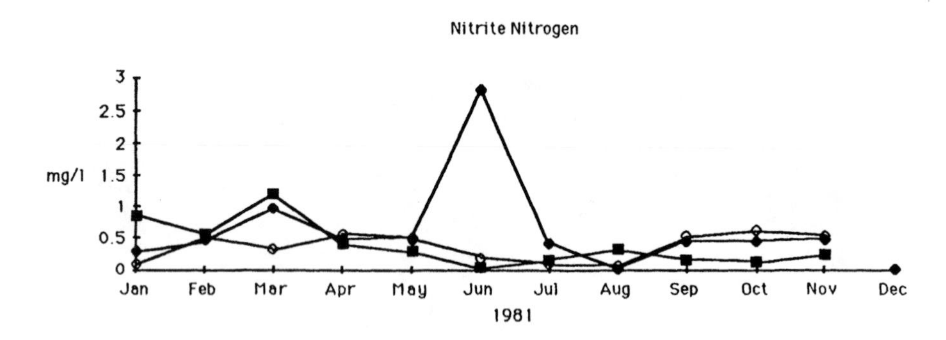

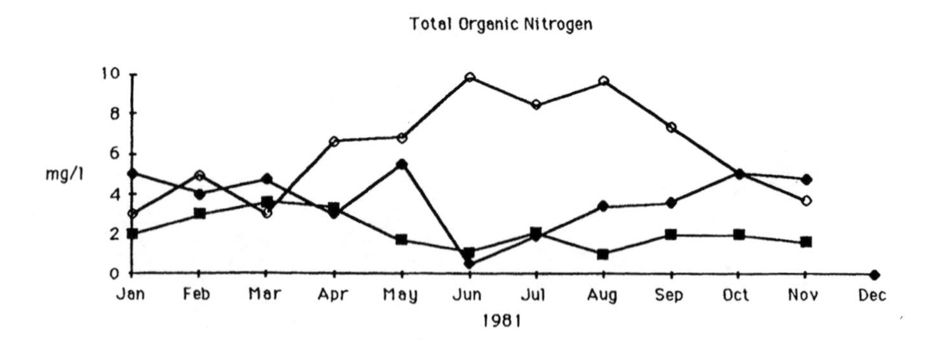

Figure 3. Nitrate, nitrite, and total organic nitrogen in the wetlands and marsh/forest influent and discharges.

nitrogen gas and plant matter. Relative importance of these mechanisms in the two systems is apparent from Figure 3. In the wetlands, the dominant impact is from uptake of nitrogen by algae, especially during summer months when ammonia levels approach zero in the discharge and organic nitrogen levels are higher in the discharge than in the influent. The marsh/forest does not reduce the ammonia nitrogen levels as much in the summer months and, in fact, shows the greatest reduction in nitrogen compounds in the fall when the trees are more active.

WATER CONSUMPTION

Water use data is available only for the marsh/forest. Water is lost by the following processes: (1) transpiration—use of water by the vegetation; (2) evaporation—loss of water from pond, soil, and plant surfaces; (3) percolation—loss of water from the upper soil layer to groundwater; and (4) runoff—water that flows off-site. Transpiration and evaporation from the soil and plant surfaces are often combined in one term, *evapotranspiration,* because it is difficult to measure them separately. Evapotranspiration is difficult to measure directly, but it can be estimated by tracking all the measurable water flows into and out of the system and developing a water balance for the whole system. Water enters the system from the metered wastewater flow and rain falling on the pond and forest area. Water leaves the system by evaporation from the surface of the pond, percolation down into the ground, evapotranspiration from the forest, and discharge. Runoff from the forest area returns to the marsh pond.

Studies by others[2] have shown that very high rates of annual water use can be expected from certain trees when it is possible to supply them with abundant water year-round. In mild climate areas, evergreen trees actively transpire even during the winter months.[3] These findings were confirmed by the data at the marsh/forest, where redwood, alder, eucalyptus, and pine trees were irrigated via the underground irrigation units.

In 1980, evapotranspiration ranged from a low of 11 m^3/day/ha in February to a high of 41 m^3/day/ha in April (Figure 4). After April 1980, the supply of wastewater to the marsh/forest was restricted, so water demand generally exceeded the available supply. It is probable that the water use rates in the summer would have exceeded the April values if sufficient water had been available. In 1980, the system had been in operation for just over a year. The trees had recovered from transplant shock and were just starting to grow. Measurements of water use taken over a two-week period in August 1985 averaged 52 m^3/day/ha (Figure 4). In 1985, height of the trees averaged 4.5 m and they were growing vigorously.

The irrigation demand of the forested area compares favorably with typical pasture irrigation rates of 8–11 m^3/day/ha in the summer season under similar climate and soil conditions. More important, the system continues to consume

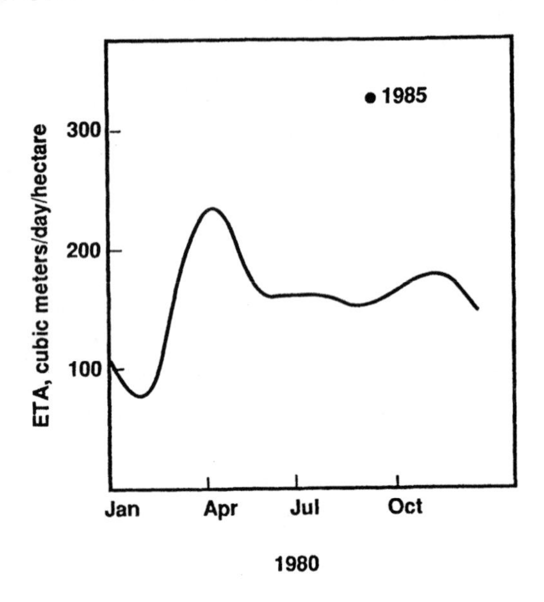

Figure 4. Marsh forest average evapotranspiration.

substantial volumes of water even in winter, when other irrigation systems are inactive.

SUMMARY

Fifteen years of experience operating wetlands and marsh/forest reclamation systems has demonstrated that it is possible to provide a richly varied aquatic habitat through careful management of the reclaimed wastewater. The area now provides a valued wildlife habitat that is a popular recreation spot for hundreds of birdwatchers.

Nine years of experience with a prototype marsh/forest evapotranspiration system has shown that a marsh system can be designed and managed to provide very high quality water as well as wildlife habitat. It also demonstrates that it is possible to take advantage of the high water demand of selected trees to reduce or eliminate reclaimed water discharge to sensitive receiving waters.

REFERENCES

1. Bogaert, R. "Wetlands Enhancement Program: Status Report: August, 1986," Mt. View Sanitary District, Martinez, CA (1986).
2. Stewart, H. T. "A Review of Irrigated Forestry with Australian Tree Species," in *Proceedings of Papers Contributed and/or Presented and Histories of Australian*

Forestry and Forest Products Institutions and Associations, Vol. II, The Australian Forest Development Institute, Albury-Wondonga, Australia, May 1988.

3. Emmingham, W., and R. Waring. "An Index of Photosynthesis for Comparing Forest Sites in Western Oregon," *Can. J. Forest Res.* 7(1) (1977).

39f Root-Zone System: Mannersdorf—New Results

Raimund Haberl and Reinhard Perfler

INTRODUCTION

The Mannersdorf experimental sewage treatment system is one of the first research-oriented wetland plants. Uncertainties regarding wetland systems and the population structure in Lower Austria (predominantly small communities) led to the decision of the Lower Austrian provincial government to construct a full-size experimental treatment plant and to manage it while allowing concurrent scientific studies.

Scientific supervision of this investigation was entrusted to the Institute for Water Resources of the Universität für Bodenkultur Wien, Department for the Protection of Water Quality and for Public Water Supplies. To allow a multifaceted study of the many problems regarding botany and soil properties as well as their relation to sewage technologies, two further University institutes are involved in this project: the Institute for Soil Research and Construction Geology Soil Science Department, Universität für Bodenkultur Wien; and the Institute for Plant Physiology, Universität Wien.

EXPERIMENTAL SYSTEM

The plant has been described in detail.[1,2] A few necessary observations are included for proper understanding of the results subsequently reported.

The site had to fulfill several conditions: (1) proximity to an existing municipal biological sewage treatment plant to have opportunity to treat different kinds of sewage in the experimental plant (raw sewage, settled sewage, and biologically treated sewage effluent); and (2) soil conditions which did not require an impermeable membrane below ground, because a fairly impermeable layer exists at certain depths beneath the plant. These conditions were fulfilled at the Mannersdorf/Leithagebirge site.

In the available area, four experimental plots of equal size (10 × 15 m) were laid out and sealed with vertically placed membrane barriers (Figure 1). Indi-

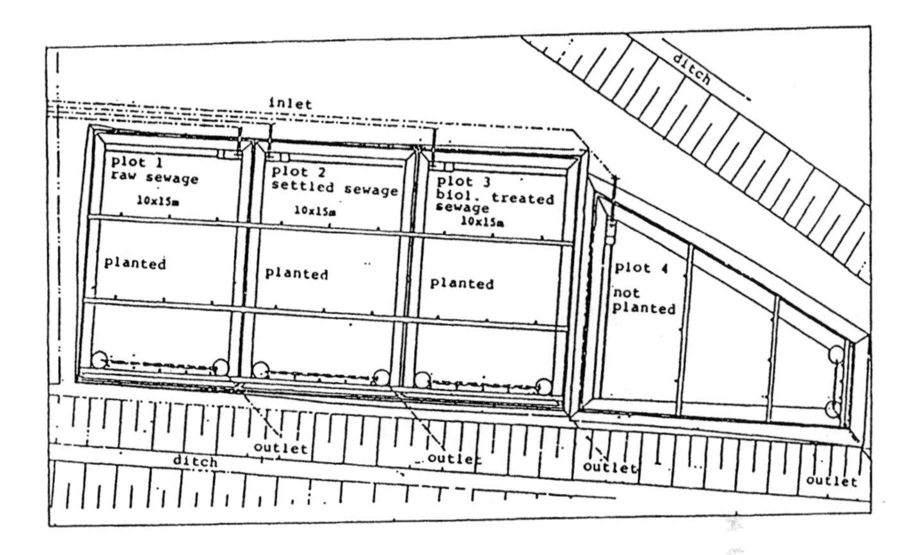

Figure 1. Plan of site.

vidual plots were supplied from a receptacle with a measuring weir having aboveground distribution by PVC pipeline with branching manifolds. A filter and drainage system was employed for the effluent. Outlet sump level controlled depth of liquid in each plot. Eleven gauging boreholes were located in each plot for monitoring water levels and liquid sampling (Figure 1).

Humus deposits interlaced with reed rhizome shoots from the bankside area of the neighboring receiving stream were applied to a depth of 0.3 m. The surface of plot 4 was allowed to remain in its original state. Reeds (*Phragmites australis*) were selected because they are the most successfully competitive species among the higher aquatic plants, have vigorous propagative activity and growth rates, seem to develop the deepest rhizome system and encompass the largest treatment volume, and are indigenous to the site, for cost and convenience.

Since June 1984, plot 1 has been fed with raw sewage, plot 2 with settled sewage, and plots 3 and 4 with biologically treated sewage effluent. Feed rate per plot was initially 20–30 mm/day (1.0 mm/day equals 1 PE for a per capita volume of 150 L/day and a plot size of 150 m^2). The feed rate has been increased in line with treatment performance of the system to 50–60 mm/day.

Since 1983, the annual growth of vegetative tissue has remained within the system.

EXPERIMENTAL RESULTS

The following presents a summary of previous results included in annual reports[2–5] and several other publications.[1,6,7]

Table 1. Sewage Characteristics

		Raw Sewage	Settled Sewage	Biologically Treated Sewage
COD	(mg/L)	100–700	100–500	40–100
BOD_5	(mg/L)	50–300	50–200	10–50
N_{TOT}	(mg/L)	15–70	15–50	10–30
P_{TOT}	(mg/L)	5–25	5–20	5–15

Results of Sewage Technology Studies

Inlet concentrations of communal sewage to the plots vary widely (Table 1). But daily variation causes no immediate influence on effluent quality. This is also true of COD and BOD_5 and N_{TOT} and P_{TOT}. This buffering capacity of the plant can be observed in both warm and cold seasons (Figure 2).

Behavior of the reed plant during the experimental period from 1984 to 1988 is shown in Figures 3–5 as average values of removal efficiencies of diurnal profiles of COD, BOD_5, N_{TOT}, and P_{TOT}. Feed rate varied from 20 to 60 mm/day. Removal efficiencies for COD and BOD_5 amounted to 80–98%, and elimination of nutrients (N_{TOT} and P_{TOT}) reached 20–70%. Treatment performance of the plant has been almost stable since its start in 1984, although in some cases feed rate has been increased significantly (Figures 3–5). The slight exacerbation of P_{TOT} was probably caused by increase of hydraulic charge. Conspicuously, reduction of N_{TOT} and P_{TOT} had considerable variation and decreased significantly during the cold season (Figure 6). Treatment performance for COD and BOD_5 was considerably less temperature dependent.

Varying feed rates (20–60 mm/day) hardly affected the plant removal efficiency. Increased influent caused enhanced superficial runoff and required a longer travel distance for a given reduction in contamination (Figure 7). The same effect appeared in winter because of lower rates of infiltration and a pronounced temporal shift of discharge volumes induced by ice cover of plot surfaces. In warm weather, a noticeable portion of the input (approximately 10 mm/day)[3,4] was lost through evapotranspiration. However, values of removal efficiency were calculated only from concentration of input and output (without regard to the volumes). Accordingly, the loss by evapotranspiration between May and October provided a sizeable safety cushion for the plant effectiveness. In October 1987, the hydraulic charge was increased from 30 to 90 mm/day for three days (to simulate high storm water) in the combined sewer system with no significant deterioration of the effluent quality (Table 2).

Hydraulic Investigations

Hydraulic measurements performed to date[8-11] permit preliminary conclusions on soil permeability, infiltration rate, and retention times in the plots. But soil mass inhomogeneity complicates interpretation of the results. Pro-

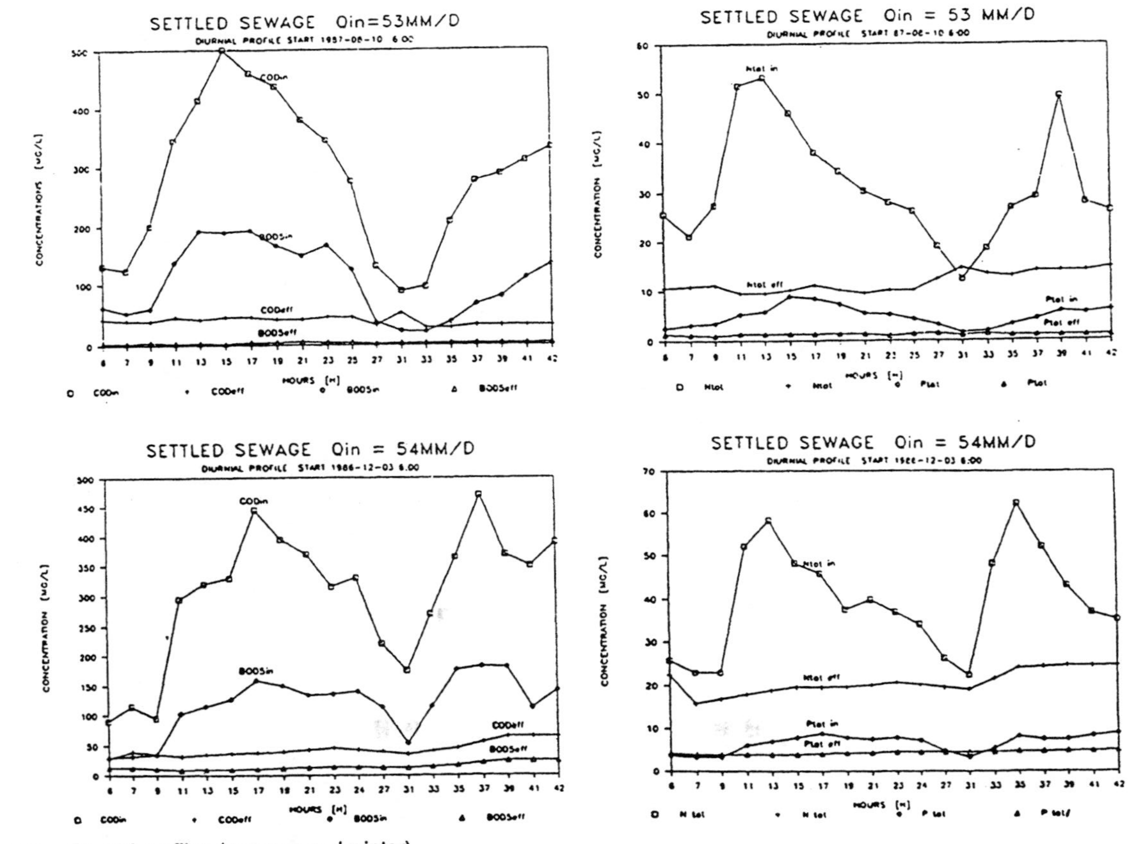

Figure 2. Diurnal profiles (summer and winter).

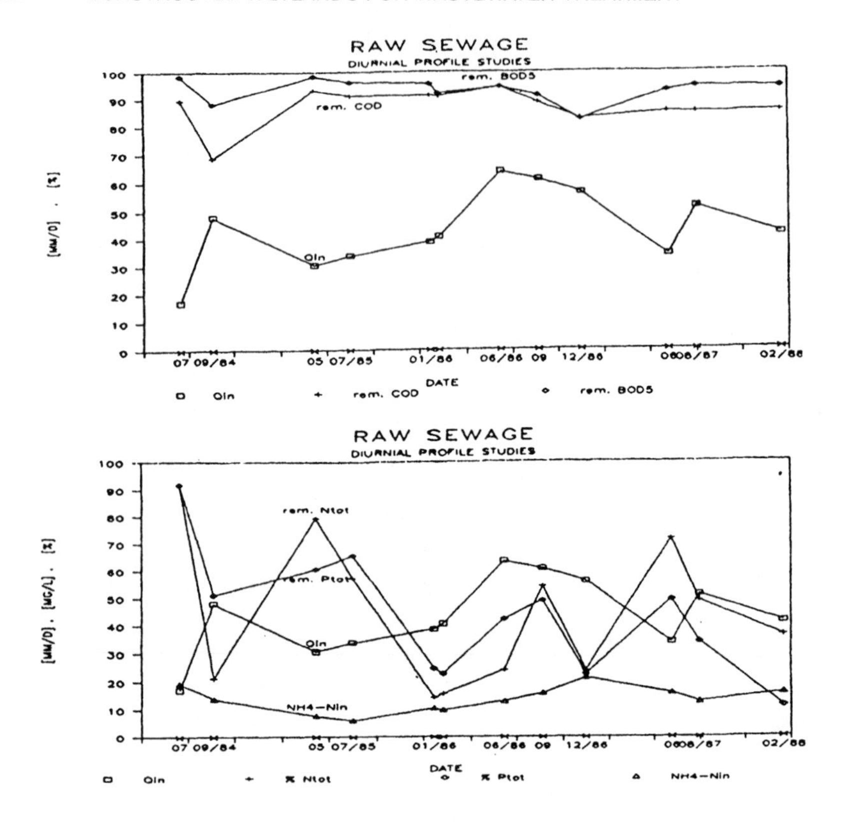

Figure 3. Average values of the removal efficiencies for plot 1.

nounced effects of boundary conditions—feed rate, height of water table, sewage temperature, and instantaneous evapotranspiration—render repetition of measurements under identical conditions as difficult as they are necessary. Deposition of sewage constituents—especially obvious in the inlet zone of the raw sewage plot (10–15 cm)—reduces the vertical infiltration rate and causes greater areal requirement for infiltration. Thus, rates of infiltration/day were lowest for the raw sewage plot, although this has not exerted any adverse effect on the treatment performance till now.

Soil permeability varies so widely that no general statement can be made of trends from 1984 to 1987. The k_f values determined for the three plots ranged from 10^{-5} to 10^{-7} m/s—comparable to k_f values of the original soil profile (3×10^{-5} to 7×10^{-6} m/s).

Retention time, determined from tracer experiments with uranine and NaCl, showed no consistent pattern, varying from 80 hr at a feed rate of 30 mm/day to 13 hr at 50 mm/day (Table 3). Maximal values may be severely affected by water losses from evapotranspiration.

From the retention time for a feed rate of 50 mm/day (1987), theoretical

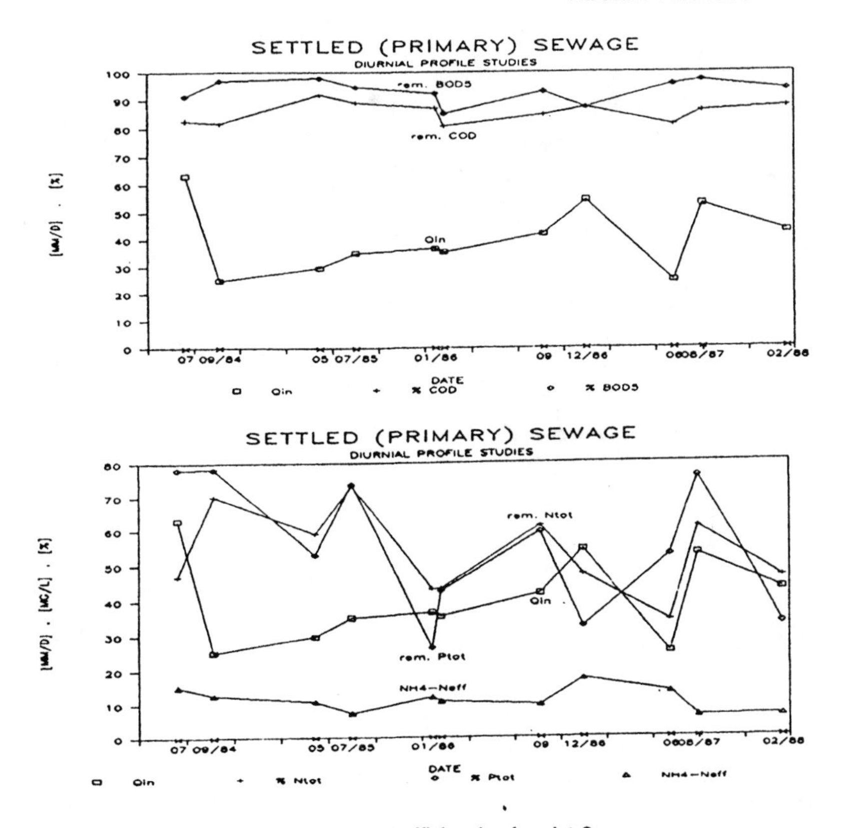

Figure 4. Average values of the removal efficiencies for plot 2.

calculation of retention time for actual BOD_5 reduction in plots 1 and 2—following the kinetic degradation model[12]—produced the actual value for the reaction coefficient of k = 0.11.

$$C_t = C_o e^{-kt}$$

where t = retention time
C_o = inlet concentration, mg/L
C_t = outlet concentration, mg/L
k = reaction coefficient, 0.085 [12], 0.023 [13]

Tracer experiment results in plots 1 and 2 indicated that most of the flow (60–80%)—depending on rate of feed—occurs in the near-surface region or in the litter layer. Only plot 3 has predominately deeper flow layers (80%). The existing soil mass is not fully utilized in either the vertical or transverse directions (Figure 8). Adverse flow distribution in plots 1 and 2 may have been caused by irregular settlement or sewage constituents deposition, which caused a surface incline toward the outlet.

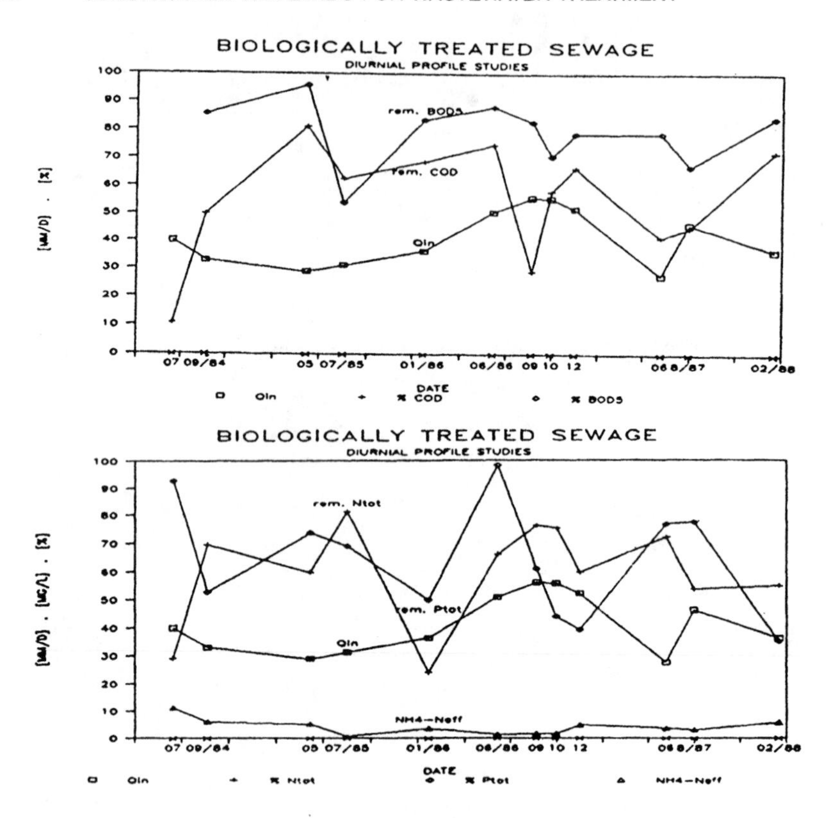

Figure 5. Average values of the removal efficiencies for plot 3.

Microbiological Investigations

Sewage samples are examined for the presence of saprophytes, coliforms (spot samples and daily and weekly series) and, beginning in 1987, salmonellae (monthly). Rates of elimination for saprophytes and coliforms were:[14] (1) raw sewage plot—96-98% and 92-96%; (2) settled sewage plot—95-99% and 98%; and (3) biologically treated sewage plot—89-95% and 93-97%. In comparison with results of conventional treatment plants, with elimination rates of 90-98% the experimental plant at Mannersdorf does very well. Of a total of 72 samples, 24 were contaminated with *Salmonella* (*S. agona*, *S. hadar*, and *S. heidelberg*). For experimental reasons, a rate of elimination is not specified.

Plant Physiology Studies

Continuous records of reed coverage in the plots showed the following progression: September 1984—35%; May 1985—69%; September 1985—86%; August 1986—89%; and August 1987—83%.

Density at present is about 70 stems/m^2—about 80 in 1985, 100 in 1986, and

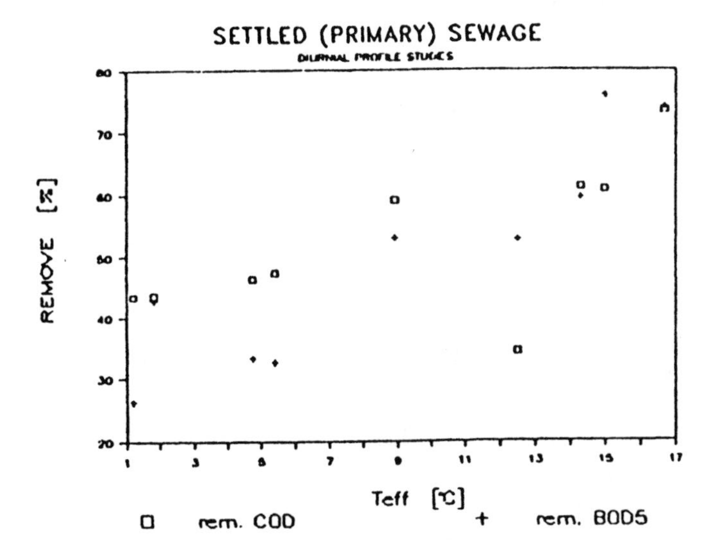

Figure 6. Removal efficiency as a function of effluent temperature.

80 in 1987—with highest density near inlets and field centers and lowest near the outlets.

Development of stem height is similar to plant density—longer stems at the inflow, and middle and shorter at the outflow. Maximum stem length (without panicle) was 3.3–3.5 m in 1986 and 1987. In general, plots 1 and 2 behave similarly; plot 3 remains somewhat inferior in all respects (in 1986 there was also a difference between plots 1 and 2). Average stem length of all the plots was 1.92 m in 1986 and 1.71 m in 1987.

No significant difference between 1986 and 1987 was found for stem morphology (Table 4). Leaf area averaged 75.8 cm^2/leaf in August 1986 and 96.6 a year later. Leaf biomass contributes to a third of the total biomass above ground. Total biomass above and below ground on a dry matter basis (DM) is given in Table 5. Leaf area index (LAI = m^2 leaf area per m^2 soil surface) ranges from 5 to 8.

To characterize mechanical qualities of reed, various parameters were surveyed, e.g., ideal Young's modulus, ideal breaking stress, bending stiffness, and ultimate load. There are only slight differences in ideal breaking stress between the three plots and the unloaded sites (Figure 9).

Measurements with cuvettes of the oxygen transferred into the soil by *Phragmites australis* gave an average value of 0.5 g O_2/m^2/day (maximum 3 g O_2/m^2/day) relating to m^2 surface area. Roots, turions, leaf casings, and damaged spots were identified as oxygen transition zones of the reed to the environs. In the setup of these experiments, it is unlikely that thermodiffusion occurred, which numerous authors describe for aquatic plants.[15]

Transpiration coefficients for July to September 1987 obtained by weigh-

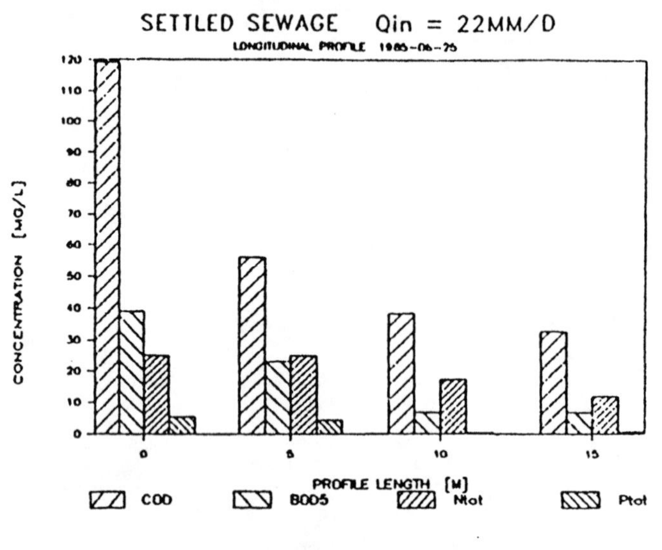

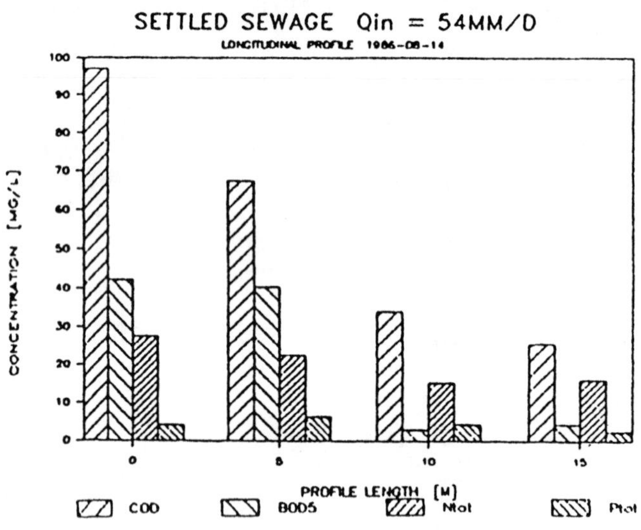

Figure 7. Longitudinal profile of concentrations.

ing[15] were from 8.4 to 10.4 mg/dm²/min in relation to leaf area. Extrapolated to the entire planted area, this represents a water loss of 6 to 8 mm/day (L/m²/day).

Nutrients and heavy metal analysis results—differentiated in surface and underground parts of the reed—are included in Figures 10 and 11.[16] Highest concentrations of P, Zn, Cd, and Pb occurred in flowers and root shoots.

Table 2. Hydraulic Charge Experiment

	Q		COD			BOD_5			O_2	N_{TOT}			P_{TOT}		
Date	inf mm/day	eff mm/day	inf (mg/L)	eff (mg/L)	rem. %	inf (mg/L)	eff (mg/L)	rem. %	eff (mg/L)	inf (mg/L)	eff (mg/L)	rem. %	inf (mg/L)	eff (mg/L)	rem. %
RFW sewage															
10/20/87	41	27	231	49	79	96	6	94	0.3	35.3	15.6	56	5.8	1.9	67
10/21/87	60	41	379	43	89	82	4	95	0.1	23.6	18.3	22	5.0	0.7	86
10/22/87	55	39	254	42	83	77	4	95	0.1	25.0	14.7	41	5.3	1.4	74
10/23/87	89	42	437	44	90	99	4	96	0.2	32.8	20.3	38	6.4	1.6	75
10/27/87	28	22		48			7		1.3		20.6			3.1	
10/28/87	89	78	299	42	86	102	6	94	1.7	30.0	19.9	34	5.2	1.9	63
10/29/87	52	37	265	39	85	101	3	97	0.8	28.7	16.9	41	5.2	1.6	69
10/30/87	77	40	345	44	87	97	5	95	0.3	26.7	18.0	33	4.9	2.5	49
Primary treated sewage															
10/20/87	28	17	414	48	88	156	4	97	0.7	29.2	7.7	74	10.1	0.7	93
10/21/87	59	40	500	46	91	127	4	97		40.2	9.1	77	5.5	0.7	87
10/22/87	59	36	322	35	89	125	3	98	0.5	39.9	8.6	78	7.0	0.4	95
10/23/87	90	62	460	55	88	121	4	97	0.1	38.3	12.8	67	7.3	1.0	86
10/27/87	28	10		30			2		0.2		11.0			1.9	
10/28/87	60	38	505	41	92	152	4	97	0.6	41.2	15.4	63	8.2	1.0	88
10/29/87	62	40	609	39	94	172	3	98	0.3	34.6	11.5	67	5.5	1.3	76
10/30/87	87	50	437	44	90	125	8	94	0.1	36.9	15.6	58	7.3	2.7	63
Biologically treated sewage															
10/20/87	32	15	77	24	69	22	4	82	0.3	20.9	4.2	80	7.0	1.3	81
10/21/87	62	40	80	41	49	20	3	85	0.2	20.3	6.1	70	3.4	0.4	89
10/22/87	58	29	76	37	51	16	2	88	0.2	21.4	4.3	80	4.6	0.4	92
10/23/87	90	62	83	46	45	22	4	82	0.3	24.3	9.4	61	4.3	1.3	70
10/27/87	27	12		41			2		0.1		4.2			0.4	
10/28/87	59	30	76	36	53	29	1	97	0.7	24.4	5.9	76	5.5	0.7	87
10/29/87	57	29	75	36	52	22	1	95	1.4	21.2	4.8	77	4.9	2.2	55
10/30/87	85	58	77	37	52	19	4	79	0.4	21.7	9.3	57	4.9	2.1	57

Table 3. Comparison of Retention Times at Various Feed Rates

	Plot 1		Plot 2		Plot 3	
	1986	1987	1986	1987	1986	1987
Q_{in} = 30 mm/day	40 h		80 h	56h		
Q_{in} = 40 mm/day	50 h				96 h	
Q_{in} = 50 mm/day	13 h	30 h		32h		16 h

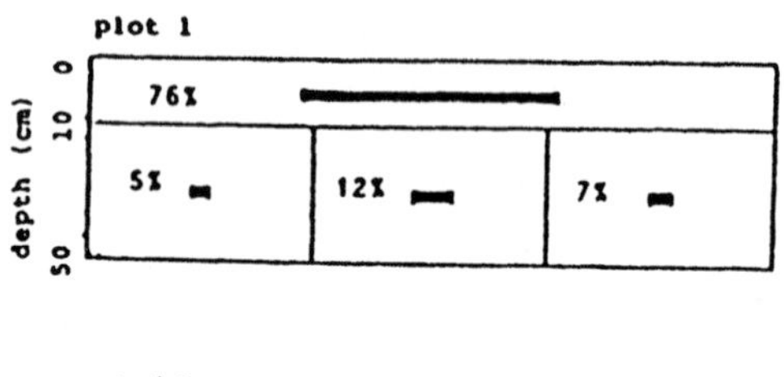

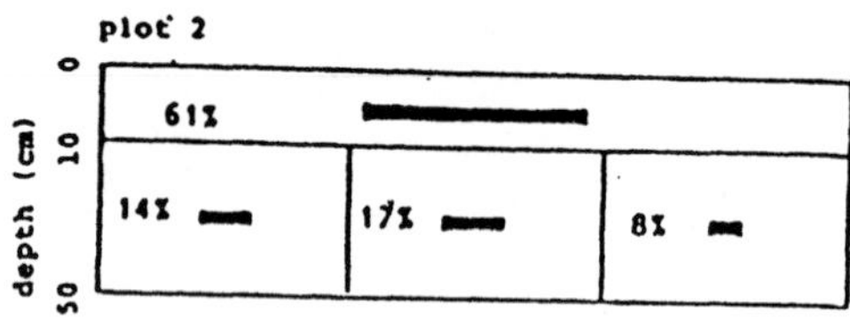

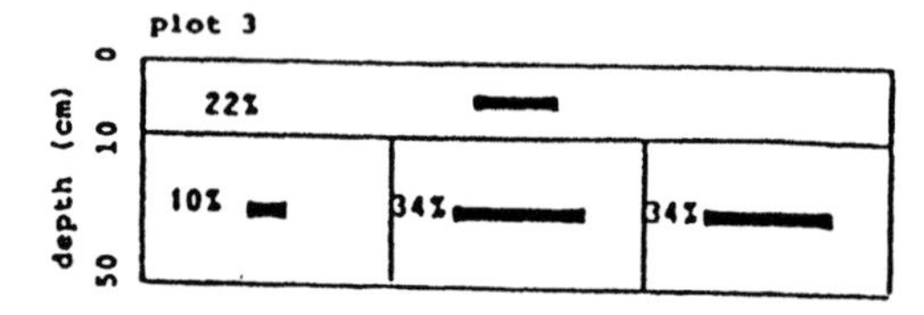

Figure 8. Flow distributions in the three plots.

Table 4. Stem Morphology

	Diameter at Base		Average Diameter of the Stems		Internodes			
					Number		Length	
	1986	1987	1986	1987	1986	1987	1986	1987
Plot 1	7.3		6.5		17		11.6	
Plot 2	7.5	7.0	6.6	6.1	19	21	12.2	11.0
Plot 3	6.1		5.4		18		10.5	

Table 5. Total Biomass

	Above Ground (kg DM/m²)		Below Ground (kg DM/m²)
	1986	1987	1987
Plot 1	1.3	1.2	1.8
Plot 2	1.6	1.2	2.1
Plot 3	1.3	0.6	2.1

Note: DM = dry matter.

Flowers and leaves had the most N, and root shoots and leaf casings had the highest Cu concentrations.

Nutrients (analyzed as N_{TOT}, P_{TOT}, NH_4^+, NO_3^-, and PO_4^{3-}) during the progress of the vegetation period from July to November showed the expected trends: maximal values in early summer and minimal concentrations at the end of the vegetation period.

Concentration of Zn ranged from 10 mg/kg DM to 100 mg/kg DM and is about double or triple the concentration of Cu. The heavy metals Pb and Cd are ubiquitous toxic substances in communal sewage. Their resorption and accumulation in various parts of the reed depends on the mobility of the substances.

No consistent trend was detectable in distribution of nutrients and heavy metal content of *Phragmites* in the three plots or in different zones of each plot (inlet, middle, and outlet).

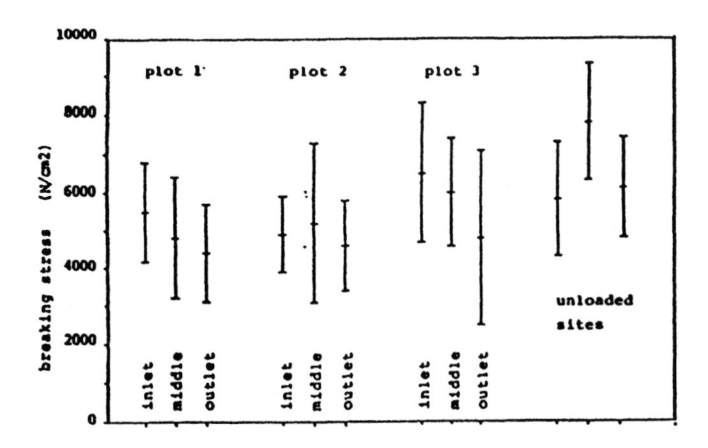

Figure 9. Breaking stress in the three plots and in unloaded sites.

Figure 10. Values for N_{TOT} and P_{TOT}.

Soil Science Investigations

For soils analysis, individual plots were subdivided into inlet, middle, and outlet zones, and within these zones soil samples were taken at numerous points from different depths (0–30, 30–60, and 60–90 cm), and average values were later determined for each zone in each plot and for the overall system at three depths (Table 6).

At the initial conditions in 1983,[17] the soil could be described as ranging between moderately heavy and heavy, with a relatively high content of humus (0.5% and 5.3%). Therefore, a high level of microbial activity and a high adsorption capacity for heavy metals is present. There was an increase in the N, humus, and P contents during 1984–1987, which was particularly marked during the first year. The pH value has slightly increased during the same period. No clear trend in heavy metals has been observed in either the aqua regia extract or the readily soluble metals in EDTA extracts. Heavy metal contents in "litter" and in "sludge" sediment also show no clear trend. A

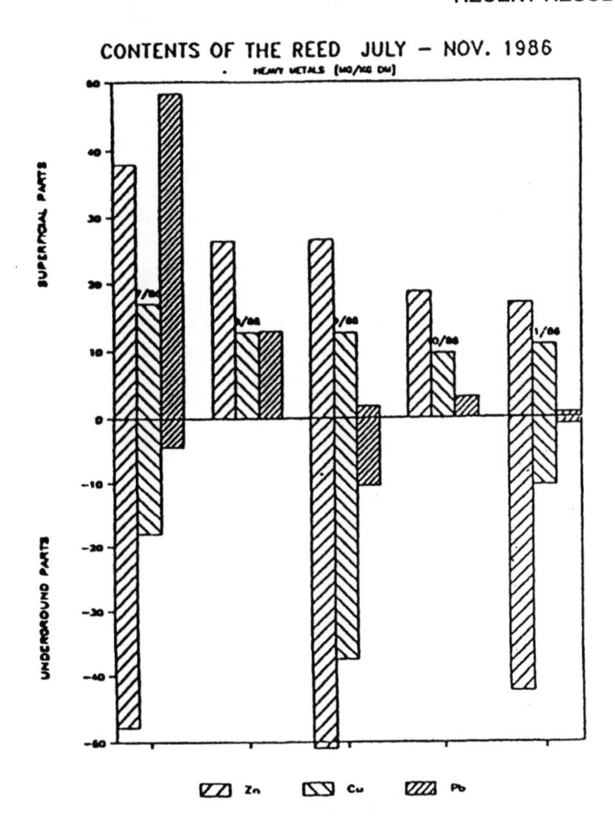

Figure 11. Values for Zn, Cu, and Pb.

significant decreasing tendency toward the soil can be discerned for Zn, Cu, and Cr, whereas concentrations of Fe, Pb, and Ni decline in the opposite direction. Cd shows no homogeneous behavior. In general, concentrations of heavy metals in litter and sludge sediment diminish in proportion to their distance from the inlet zone.

SUMMARY

Results of the research plant in Mannersdorf of Lower Austria demonstrate that root-zone systems may contribute to the solution of wastewater treatment in areas with 100 to 500 PE. It seems necessary to have a mechanical pretreatment before the wetland plant. The treatment area for 1 PE should be 5–10 m^2. Nevertheless further research should be done on the yet unsolved problems with root-zone systems, particularly the hydraulic aspects and removal of nutrients.

Table 6. Soil Science Investigations

Year	Depth (cm)	General Characteristics (mg/100 g dry soil)				Elements in Aqua Regia Extract (mg/100 g dry soil)							Elements in EDTA Extract (mg/100 g dry soil)								
		pH (H_2O)	N_{TOT}	Humus (%)	P_2O_5	Zn	Pb	Cd	Cr	Cu	Ni	Fe	Zn	Pb	Cd	Cr	Cu	Ni	Fe	Al
1983	0–50	7.7	216	3.2	13.6	8.2	7.9	0.07	1.8	7.3	1.5	1754	1.9	2.5	0.04	0.14	4.1	0.4	13.6	0.5
	50–100	7.8	220	3.1	5.1	8.1	4.2	0.09	3.0	3.8	3.1	2450	0.8	1.7	0.02	0.14	1.8	0.7	21.0	0.8
	0–30	6.9	363	14.7	33.6	18.8	15.2	0.23	4.1	16.6	3.8	2029	4.3	2.6	0.03	0.15	4.6	0.32	23.7	0.9
1984	30–60	7.1	351	13.7	15.8	13.1	11.1	0.23	4.6	8.1	5.5	2176	1.5	1.7	0.02	0.15	1.7	0.52	17.5	0.8
	60–90	7.3	143	8.4	12.4	8.5	7.3	0.20	4.6	3.1	4.7	2271	0.6	1.1	0.01	0.16	0.6	0.45	19.9	1.3
	0–30	7.2	354	7.4	34.4	16.2	10.0	0.17	3.4	12.2	3.7	2054	4.5	3.3	0.01	0.13	5.9	0.72	57.5	1.3
1985	30–60	7.2	330	6.2	11.9	10.7	6.9	0.16	3.7	5.6	5.3	2187	1.3	2.1	0.05	0.13	2.3	1.27	42.1	1.2
	60–90	7.4	158	2.8	9.9	8.3	5.4	0.14	3.9	3.3	4.5	2231	0.6	1.7	0.05	0.13	1.0	0.89	32.5	1.5
	0–30	7.4	384	7.9	33.0	17.9	8.7	0.15	2.7	11.7	2.9	1766	5.5	3.6	0.04	0.14	5.7	0.54	62.6	1.6
1986	30–60	7.6	328	6.8	20.7	10.9	5.8	0.12	2.6	6.2	3.4	1598	2.2	2.3	0.04	0.14	2.7	0.71	41.0	1.2
	60–90	7.8	229	4.1	16.8	0.8	4.2	0.12	3.0	3.3	3.7	1702	1.0	1.6	0.03	0.14	1.2	0.76	26.0	1.0
	sludge	7.1	514	10.5	48.7	22.0	7.7	0.13	2.4	11.5	8.6	1782	8.9	3.4	0.02	0.13	4.7	0.45	42.2	1.0
1987	0–30	7.4	415	12.3	38.0	18.0	9.2	0.14	2.6	14.1	3.1	1892	5.1	3.3	0.04	0.14	5.5	0.58	44.8	1.4
	30–60	7.6	365	11.7	19.2	13.2	7.6	0.14	2.8	7.8	3.9	2228	2.3	2.6	0.03	0.14	3.1	0.78	43.6	1.2
	60–90	7.8	205	8.3	18.6	8.8	4.5	0.12	2.9	3.3	4.0	1914	0.8	1.5	0.03	0.14	1.2	0.68	32.7	1.3
1987	litter					elements in HNO_3 /$HClO_4$ extract														
						77.4	4.0	0.15	2.6	19.5	1.4	913								

REFERENCES

1. Haberl, R. “Erfahrungen mit dem Bau und dem Betrieb der Pflanzenanlage Mannersdorf,” *WAR* 26, Darmstadt (1986).
2. “Pflanzenklaeranlage Mannersdorf, Jahresbericht 1984,” Institut f. Wasserwirtschaft der Univ. f. Bodenkultur Wien (1985).
3. “Pflanzenklaeranlage Mannersdorf, Jahresbericht 1985,” Institut f. Wasserwirtschaft der Univ. f. Bodenkultur Wien (1986).
4. “Pflanzenklaeranlage Mannersdorf, Jahresbericht 1986,” Institut f. Wasserwirtschaft der Univ. f. Bodenkultur Wien (1987).
5. “Pflanzenklaeranlage Mannersdorf, Jahresbericht 1987,” Institut f. Wasserwirtschaft der Univ. f. Bodenkultur Wien (1988).
6. Perfler, R., und R. Haberl. “Erfahrungsbericht über die Versuchspflanzenanlage Mannersdorf,” *Wiener Mitteilungen* 71 (1987).
7. Perfler, R., und R. Haberl. “Pflanzenanlage Mannersdorf—Aktuelle Ergebnisse,” *Wiener Mitteilungen* 77 (1988).
8. Diebold, W. “Voruntersuchung zu Planung und Betrieb einer Versuchsanlage zur Abwasserreinigung mit Makrophyten,” Diplomarbeit am Institut für Wasserwirtschaft, Univ. f. Bodenkultur Wien (1984).
9. Moosbrugger, R. “Bodenwasserbewegung in Pflanzenklaeranlagen,” Diplomarbeit am Institut für Wasserwirtschaft, Univ. f. Bodenkultur Wien (1986).
10. Dihanich, R. “Untersuchungen an der Pflanzenklaeranlage Mannersdorf unter besonderer Beruecksichtigung der Art der Probenahme,” Diplomarbeit am Institut für Wasserwirtschaft, Univ. f. Bodenkultur Wien (1988).
11. Raschke, T. “Untersuchungen an der Pflanzenklaeranlage Mannersdorf unter besonderer Beruecksichtigung der hydraulischen Verhaeltnisse,” Diplomarbeit am Institut für Wasserwirtschaft, Univ. f. Bodenkultur Wien (1988).
12. Kickuth, R. “Abwasserreinigung in Mosaikmatrizen aus anaeroben und aeroben Teilbezirken,” *Grundlagen der Abwasserreinigung*, GWF, Schriftenreihe Wasser-Abwasser 19, Oldenbourg Verlag (1981).
13. Haider, R. “Ermittlung der Aufenthaltszeit im bewachsenen Bodenfilter der Klaeranlage Saldenburg, Abwasser-Nachreinigung und Naehrstoffelimination durch einen bewachsenen Bodenfilter,” Bayrische Landesanstalt für Wasserforschung, München, Bericht Nr. 2 (1986).
14. Theil, M. “Mikrobiologische Untersuchungen an einer Pflanzenanlage,” Diplomarbeit am Institut für Wasserwirtschaft, Univ. f. Bodenkultur Wien (1987).
15. Janauer, G. “The Mannersdorf root-zone plant,” Kursus plantebaserede anlaeg til rensning of spildevand, Aarhus Universitet (1987).
16. Grafl, E. “Naehrstoffe und Schwermetalle im Schilfbestand der Pflanzenanlage Mannersdorf/Leithagebirge,” Diplomarbeit am Institut für Pflanzenphysiologie, Universität Wien (1988).
17. Rampazzo, N. “Bodenkundliche Untersuchungen für die Abwasserentsorgung mittels Schilfbeeten,” Diplomarbeit am Institut für Bodenforschung und Baugeologie, Univ. f. Bodenkultur Wien (1984).

39g Constructed Wetlands for Secondary Treatment

Thomas J. Mingee and Ronald W. Crites

INTRODUCTION

The use of constructed wetlands for wastewater treatment has become more prevalent in recent years as a result of a wider understanding of the workings and advantages of these natural systems. Constructed wetlands can provide a low-cost wastewater treatment alternative to achieve secondary treatment for small- to medium-size communities. This chapter presents a case study of a constructed wetlands system at Gustine, California utilizing emergent aquatic vegetation. History, pilot-study effort, construction problems, construction cost, and initial performance data are included.

The City of Gustine, California is a small agricultural town on the west side of the San Joaquin Valley with a population of about 4000. The city is surrounded by agricultural land used for orchards, field crops, and dairy pasture. The land is nearly flat with a gentle eastward slope and is about 25 m above sea level. Climate in the area is characterized by hot, dry summers and cool winters in which low temperatures occasionally drop below freezing. Average annual rainfall in Gustine is about 0.29 m, of which 90% falls in the months of November through March.

The city treats approximately 4542 m^3/day of wastewater, of which one-third originates from domestic and commercial sources and the remainder from three dairy products industries. The high-strength wastewater reflects the industrial component of the waste, and averages over 1200 mg/L biochemical oxygen demand (BOD_5) and 450 mg/L suspended solids (SS).

Until recently, the city's wastewater treatment plant consisted of 14 oxidation ponds operated in series on 21.8 ha that provided 54 days' detention time. Treated effluent was discharged without disinfection to a small stream leading to the San Joaquin River. As with many oxidation pond systems, mandatory secondary treatment levels were not achieved consistently. The discharge regularly exceeded the 30 mg/L SS standard and periodically violated the 30 mg/L BOD_5 standard, both, at times, by large margins.

The city applied for and received funding under the Clean Water Act to analyze alternatives and develop a facilities plan. Alternatives included: (1) treatment followed by land application (slow rate irrigation); (2) treatment followed by reuse by seasonal flooding of local duck clubs to attract migrating waterfowl; (3) oxidation pond treatment followed by effluent polishing using sand filters, microscreens, or submerged rock filters; (4) conventional mechanical treatment (aeration); and (5) oxidation pond pretreatment followed by effluent polishing in a constructed wetland. The last three alternatives were required to meet secondary treatment standards for river disposal.

Analysis of alternatives showed that the oxidation pond/constructed wetland system was the most cost-effective alternative. Advantages included availability of suitable land, compatibility of the treatment method with the surrounding area (a low land area with naturally occurring aquatic vegetation and no development), lowest life-cycle cost, and consumption of very little energy.

PILOT STUDY

A pilot study was conducted from December 1982 to October 1983 to verify the efficacy of constructed wetlands and develop design criteria. An existing stand of cattail near the treatment plant was modified by constructing enclosure berms, piping, and pumps to simulate a portion of the constructed wetlands envisioned for the full-scale project. The pilot wetland cell measured 12 m by 275 m.

Effluent was applied to the pilot cell using different ponds as the influent source to regulate mass loading of BOD_5 and SS and to avoid applying high concentrations of algae. Changing the source pond resulted in detention time in the oxidation ponds ranging from 28 to 54 days, shorter times generally occurring in summer and longer times in winter. Water depth in the wetland cell was about 0.15 m and detention time, determined through dye tracer studies, was 60% of theoretical (1.3–3.8 days).

Influent to the wetland cell averaged 180 mg/L BOD_5 (range 18–400 mg/L) and 118 mg/L SS (range 30–300 mg/L). Mass loading in the cell was 11–220 kg/ha/day BOD_5 and 20–145 kg/ha/day SS. Generally, loading of both BOD_5 and SS was $\leq$ 112 kg/ha/day. Values for BOD_5 and SS in the effluent of the wetland cell each stabilized at or near 30 mg/L after the startup period. Effluent quality varied nearly directly with mass loading. Relationships between effluent quality and influent concentration, detention time, or temperature were not discernible from the data. An inverse relationship between effluent quality and detention time would be expected but was not apparent, probably due to the small range of detention times used.

The pilot-test results demonstrated the efficacy of the constructed wetlands system and were used to develop criteria for the full-scale design (listed below):

Effluent requirements	
BOD_5	<30 mg/L
SS	<30 mg/L
Design Flow	3785 m^3/day
Number of Cells	24
Area per Cell	0.4 ha
Aspect Ratio	20:1 minimum
Hydraulic Loading Rate	380 m^3/ha/day
Depth	0.1–9.45 m
Detention Time	4–11 days
BOD_5 Mass Loading Rate	<112 kg/ha/day
SS Mass Loading Rate	<112 kg/ha/day
Inlets	head end of channels and 1/3 point
Outlets	adjustable weirs

DESCRIPTION OF TREATMENT SYSTEM

The wastewater treatment system for Gustine consists of 11 oxidation ponds operated in series followed by a constructed wetlands system and chlorination/dechlorination facilities (Figure 1).

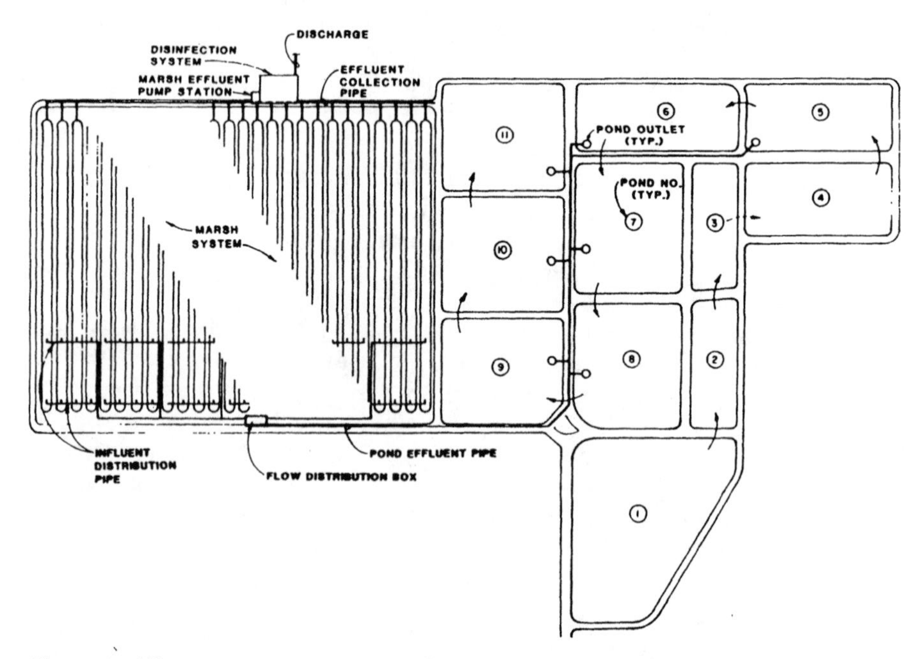

Figure 1. Wastewater treatment plant at Gustine, California.

The oxidation ponds have a net area of 17 ha and provide pretreatment of Gustine's strong waste prior to wetland application. Operators can draw pond effluent from any one of ponds 5 to 11, adjusting detention time in the ponds (24–54 days), regulating waste loading on the wetland, and avoiding application of heavy algae concentrations during summer.

The wetlands system consists of 24 cells operated in parallel, each measuring 11.6 m by 337 m (0.4 ha). Depth is adjustable at cell outlet weirs over a range of 0.1 to 0.45 m. Levees, 3 m wide and 0.6 m high, separate the cells. Influent flow to the wetland passes through a distribution box, where V-notch weirs divide the flow into six portions for equal distribution to six groups of four cells.

Gated aluminum irrigation pipe distributes flow equally across the width of each cell both at its head end and at one-third of the length. Gates in the pipe provide flexibility to apply equal amounts of flow at both distribution points or different flows at either point. This piping arrangement enhances the wastewater distribution over the wetland area and avoids an overloaded inlet zone. Effluent is collected from the wetland cells in a common sewer and pumped to the disinfection process.

The number of cells and variable water depths provide flexibility to attain detention times in the marsh from four days in the summer to 11 days in the winter. Cells may be sequentially taken out of service each summer for vegetation management and other maintenance requirements.

WETLAND CONSTRUCTION AND STARTUP

The wetlands system and other treatment plant improvements were constructed in 1986 and 1987. Project specifications required 18 cells to be planted with cattail (*Typha*) rhizomes on 0.9-m centers and six cells with bulrush (*Scirpus fluviatilis*) rhizomes on 0.5 m centers. Although cattails were the preferred plant because they were used in the pilot study, bulrush was added for side-by-side evaluation.

Few problems occurred in construction, except for planting and establishment of wetland vegetation. Planting was complicated by lack of experience and fresh water for initial irrigation. The construction contractor's first attempt involved mechanical harvesting of cattail and bulrush rhizomes from natural stands in a nearby wildlife refuge. Rhizomes were spread and disced into wetland cells in September 1986. Few plants emerged the next spring, undoubtedly due to rough treatment and lack of irrigation after planting.

The second planting attempt in June 1987 used a tomato planter with cattail seedlings from a nursery. Nursery stock was easily handled by the machine, and planting went smoothly; however, nearly all plants failed within a short time. Shock due to high air temperatures during planting (38°C) and irrigation with poor-quality pond effluent may have been responsible. The strong efflu-

Table 1. Performance During Startup of Constructed Wetland

	BOD_5			SS			
Date	In (mg/L)	Out (mg/L)	Mass Load (kg/ha/day)	In (mg/L)	Out (mg/L)	Mass Load (kg/ha/day)	Number of Cells in Service
April 1988	351	177	174	305	409	151	16
May 1988	248	84	123	275	127	137	8

Note: Data for April from an average of five grab samples; for May, three grab samples. Data provided by City of Gustine.

ent (>300 mg/L BOD_5) and flooding by irrigation probably prevented oxygen from reaching the seedling rhizomes through the soil.

The contractor also seeded the wetland area by broadcasting seed. By late summer of 1987, random stands of thriving sedge (*Carex*) existed over half the site. Some cells were almost completely covered.

Further success was obtained by hand-transplanting live cattail plants from drainage areas adjacent to the treatment plant. Mature plants have acclimated to the wetland treatment system.

In spring 1988, 12 of 24 cells had reasonably dense growth, covering 75% or more of the cell area with patches of bulrush, and isolated clumps of cattail. Other cells had some growth but lacked coverage and plant density needed for effective treatment. In all cells, cattail was expected to slowly spread and drive out the bulrush.

WETLAND PERFORMANCE

Without full plant coverage, the wetland portion of the treatment plant is still in the startup phase of operation. Performance of the system as a whole has not reached design expectations, but positive results have been achieved. Performance has been hampered by lack of dense growth in more than a minimum number of cells. Additionally, flow and BOD_5 entering the treatment plant are significantly higher than design values; flow has increased by 20% (from 3785 m^3/day to 4542 m^3/day) and BOD_5 concentration has doubled (from 600 mg/L to 1200 mg/L).

Recent operating data for the constructed wetland are presented in Table 1. While significant reduction of BOD_5 and SS is achieved (66% and 54%, respectively), the effect of high-strength influent wastewater and incomplete plant coverage is evident. The mass loading of BOD_5 and SS is significantly higher (>50%) than maximum design loading of 112 kg/ha/day.

CONSTRUCTION COST

Costs for the wetland system portion of the project are summarized in Table 2. City-owned land for the Gustine project was available, so no land

Table 2. Capital Costs for Gustine Constructed Wetlands Project

Item	Cost ($) (August 1985)
Pond Effluent Piping[a]	192,000
Earthwork[b]	200,000
Flow Distribution Structure[c]	16,000
Flow Distribution Piping in Marsh[d]	205,000
Marsh Cell Water Level Control Structures[e]	27,000
Marsh Effluent Collection Piping[f]	83,000
Planting[g]	69,000
Paving[h]	90,000
Land[i]	-0-
TOTAL	882,000

[a]Included 793 m of 0.53-m PVC gravity piping, five manholes, and seven pond outlet control pipes with wooden access platforms.
[b]Total earthwork volume, approximately 34,400 m^3. Cost includes clearing and grubbing, extra effort to work in area of very shallow groundwater and to construct a 2-m-high outer levee to enclose the wetland area and protect it from the 100-yr flood.
[c]A concrete structure with V-notch weirs, grating, access stairs, and handrail.
[d]Approximately 854 m of 0.20-m PVC gravity sewer pipe, 763 m of 0.20-m gated aluminum pipe, and wooden support structures with concrete base slabs for the gated pipe installed at the one-third-of-length point.
[e]Small concrete structures in each cell with weir board guides and 0.006-m mesh stainless steel screen.
[f]Approximately 458 m of 0.1–0.4-m PVC gravity sewer pipe plus three manholes.
[g]Based on mechanical planting of bulrush and cattail rhizomes on 0.45 m and 0.90 m grid, respectively. Total bulrush area of about 2.4 ha, 7.2 ha for cattails; due to lack of experience with wetland planting, contractor's bid (shown in table) was twice his actual estimated cost.
[h]Aggregate base paving of the outer levee and selected inner levees of the marsh area.
[i]Land already owned by city was used; gross land area required was 14.5 ha.

costs were involved. Land requirements for this system were 9.7 ha net for the area actually planted and about 14.5 ha gross for the whole marsh system, including all interior cell divider levees and the outer flood protection levee. Project costs of $882,000 to treat design loading of 3785 m^3/day amount to $233/$m^3$/day ($0.88/gallon/day).

39h Hydraulic Considerations and the Design of Reed Bed Treatment Systems

John A. Hobson

INTRODUCTION

The U.K. program on reed bed treatment systems is concentrating almost exclusively on beds planted with *Phragmites australis,* the common reed. All of the information in this chapter refers specifically to such systems.

The ultimate aim of any research program into reed bed treatment systems (RBTS) is to demonstrate the feasibility of such systems and to produce successful designs for their construction. In practice the two must run hand in hand. A reed bed, even if only experimental, must be designed before it can be tested. Ideally it should be possible to design experimental systems to test particular aspects of RBTS, but because of the interactions between the different features of such treatment systems, this is not easy. In particular, if RBTS do not behave hydraulically as they are designed, there is very little chance that they will perform as they are designed. Although this chapter concentrates on the hydraulics of RBTS, it is necessary to discuss the possible mechanisms of treatment that take place, because they define the objective of any design.

MECHANISMS OF TREATMENT IN RBTS

Oxidation of organic matter and ammonia in the rhizosphere or root zone of *P. australis* relies on the ability of *P. australis* along with other aquatics to supply their roots with atmospheric oxygen. Oxygen then diffuses out of the roots into the surrounding rhizosphere, where it becomes available to bacteria involved in the purification processes found in more conventional wastewater purification systems. Reeds probably do not transfer oxygen into standing water nor into the layer of composting debris on the reed bed surface (see below). Physicochemical interaction between wastewater and soil or other growing medium has been suggested as a major route for the removal of phosphorus if the correct growing medium is used. Plant nutrients may be taken up by *P. australis.* In this way, nitrogen and phosphorus species might be removed, but the green upper parts of *Phragmites* must be removed at the end of each growing season or the stored nutrients will be released into the wastewater being treated as the aerial parts of the plants die back. As reeds die, they form a layer of composting stems and leaves on the surface of the reed bed (this has been termed the F-horizon). There is evidence that wastewater flowing through this layer is subject to a considerable degree of purification. A reed bed can be operated with permanent standing water. In this case, all the purification mechanisms of a lagoon are likely to operate. Reeds enhance these purification mechanisms by providing surface area for attachment of bacterial slimes and by providing surfaces for intercepting settling particles. Rhizosphere oxidation, soil absorption, and plant uptake can only act if the flow of wastewater to be treated is through the medium in which the reeds are growing. Rhizosphere oxidation and plant uptake require the flow to be through the rhizosphere.

Lagoons have never been popular for sewage treatment in the United Kingdom. While there is no intention to dismiss the use of reed-enhanced lagoons, they will not be further discussed. For similar reasons, reed beds were not designed for most of the treatment to occur in the F-horizon. However, it now appears that most reed beds built on soil in the United Kingdom are operating in this manner. Only beds built with gravel (or sand) have wastewater flowing through (within) the medium in which the reeds are growing, excepting certain pilot- or small-scale soil beds. Only coarse-medium beds have any chance of rhizosphere treatment as the dominant mechanism.

HYDRAULIC CONDUCTIVITY

Darcy's law governs flow through a packed or porous bed.

$$Q_s = A_c ks \left(\frac{dh}{ds}\right)$$

where Q_s = flow, m^3/s
A_c = saturated cross-sectional area perpendicular to flow, m^2
ks = saturated hydraulic conductivity, m/s
$\frac{dh}{ds}$ = hydraulic gradient (slope of the water table)

If there is to be any flow in a RBTS, the hydraulic gradient must be nonzero. However, if the hydraulic gradient is nonzero, A_c is rarely constant. If A_c is not constant, the hydraulic gradient cannot be constant. Therefore, a proper solution of Darcy's law for a reed bed is not straightforward, though it is usually possible to estimate a satisfactory solution.

For example, Figure 1 represents an inlet zone and an outlet zone (both of infinite hydraulic conductivity) and a zone in between. Curve 1 represents zero

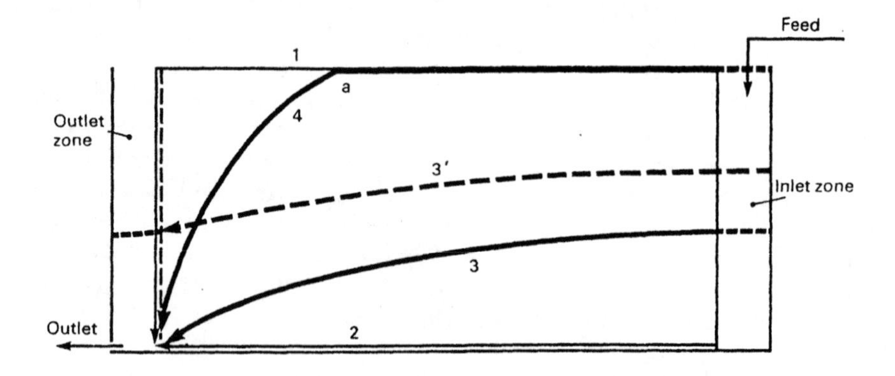

Figure 1. Water level curves for reed-bed treatment systems with different hydraulic conductivities. Curve 1 = zero conductivity; curve 2 = infinite conductivity; curve 3 = high conductivity; curve 3′ = same hydraulic conductivity as curve 3 but with the outlet level artificially raised; and curve 4 = low conductivity.

conductivity in this zone; the zone is impervious to water. The inlet zone fills up and water flows over the surface and cascades into the outlet zone. Curve 2 represents the case of infinite conductivity; the zone could be considered to be air, and water simply flows along the bottom. In both cases, there is no flow restriction from the bottom of the outlet zone. All other cases must be intermediate between these. Curve 3 shows a case of high conductivity, where the inlet zone partially fills. Curve 4 represents a much lower conductivity. The inlet zone completely fills. The curve represents decreasing surface flow up to point "a" at which point the water table falls to the outlet point. This case, with surface flow in the first portion and a dry surface toward the outlet, has been observed in operating RBTS. Curve 3′ represents the same hydraulic conductivity as curve 3 but where the outlet level is artificially raised.

If there is no inlet zone, Figure 2 applies. Curve 1 for zero conductivity is the same as before, but curve 2 for infinite conductivity is now different. Curves 3, 4, and 5 represent intermediate conductivities. Curve 3 is particularly interesting. If a concave curve of this type is found in practice, it suggests that an inlet zone is either blocked or being bypassed and that the feed is being applied onto the surface of the reed bed.

So far, these examples have all been for beds with horizontal surfaces and horizontal bases. The situation is not very different for sloping beds. Figure 3 shows a sloping bed. The different curves represent water table surfaces for the case where the outlet level is artificially controlled for a bed of given hydraulic conductivity with a given flow rate. In this case, bed conductivity is not quite high enough to accept the flow at a hydraulic gradient equal to the bed gradient of 2%. Only the three curves marked 1 to 3 are legitimate water level curves. They correspond to setting a variable head outlet to the three levels shown. Even with the outlet set at its lowest level, there will still be some

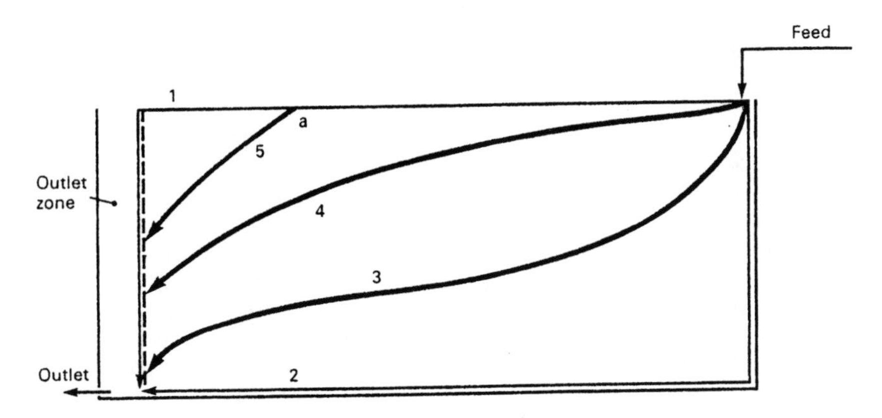

Figure 2. Water level curves for reed-beed treatment systems with different hydraulic conductivities: no inlet zone. Curve 1 = zero conductivity; curve 2 = infinite conductivity; and curves 3, 4, and 5 = intermediate conductivities.

surface flow on the front portion of the bed. In this example, the flow rate was set numerically equal to the hydraulic conductivity. Flow will be contained within the bed when A_c × hydraulic gradient = 1. For a depth of 1 m and a width of 30 m, the required hydraulic gradient is 0.033 (about 3%) as shown. As the saturated depth diminishes, the velocity of flow must increase and the required hydraulic gradient must increase.

Due to an oversimplified application of Darcy's law, many early U.K. reed beds were built with a slope. It seems likely that all future UK reed beds will

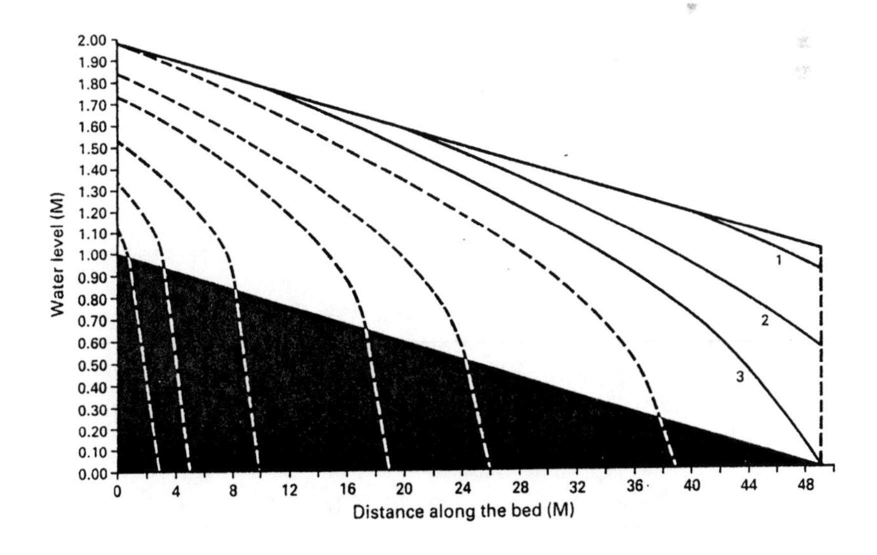

Figure 3. Water levels in a sloping bed (2%) assuming a variety of inlet heights.

have level surfaces and level or very gently sloping bases. There are several reasons for this.

1. The range of possible hydraulic conductivities for reed bed media is very high, from 10^{-7} m/s for clay soils to 10^{-1} m/s for coarse gravels. Generally, it will be possible to operate a level bed with a hydraulic gradient in the region of 1% (up to 5% for short beds) because inlet and outlet levels can be different. By sloping a bed, it will only be possible to increase the hydraulic gradient by 5 at the most. It is only necessary to build a sloping bed if the hydraulic conductivity of the medium is too low for a 1% hydraulic gradient. However, the possible shortfall on hydraulic conductivity could be several orders of magnitude, whereas at most only a factor of 5 can be gained. Many sloping beds have been constructed both in the United Kingdom and Europe which still have surface flow.
2. Many beds constructed from soil do have surface flow but still give useful rates of treatment. However, if a reed bed was designed with the expectation of treatment within an F-horizon, it would be best to use a zero or even very slight upwardly sloping surface. Once surface flow develops on a downward sloping surface, it tends to erode the surface, forming channels and damaging reed growth. Retention times of wastewater can be very low once channeling occurs. At three large gravel beds at Gravesend (U.K.), feed bounces off stones in the inlet zone directly onto the 2% sloping surface of the bed. The water levels within the bed are similar to curve 3 in Figure 2, but there is surface water flowing over substantial areas of these beds. It seems that once surface flow has developed on a bed with a downward sloping surface it can have difficulty penetrating the bed even though the true water level within the bed is significantly below the surface. The surface of a porous bed over which wastewater has been flowing can become a barrier to penetration greater than would be expected from the bulk hydraulic conductivity of the bed. This effect is magnified on a downward sloping surface.
3. Possibly the greatest problem of having a sloping surface to a reed bed is that water level control becomes very difficult. Water level control is needed for control of weeds and may be useful to encourage rhizome penetration. Even more important is providing proper conditions for reeds to flourish throughout the bed. Reed beds with a surface slope generally have an area where reeds thrive and an area of poor growth due to excessive standing water at the outlet end or dry conditions at the inlet.

Figure 4 uses Darcy's law to calculate what length of inlet is required per person for reed beds of different hydraulic conductivities and with different hydraulic gradients in order to maintain full subsurface flow. Experience has shown that most soil beds will not have hydraulic conductivities much above 3×10^{-5} m/s. Even with hydraulic gradients of several percent, it will be difficult to reduce the inlet requirement to less than 1 m/person. If the requirement for total plan area for treatment is 3–6 m^2/person, there is no point in having the bed any longer than 3–6 m. In any case, if the bed is 0.6 m thick, it is only possible to maintain a hydraulic gradient of 5% in a flat bed for a distance of 5 or 6 m. Constraints on soil bed design without surface flow are considerable,

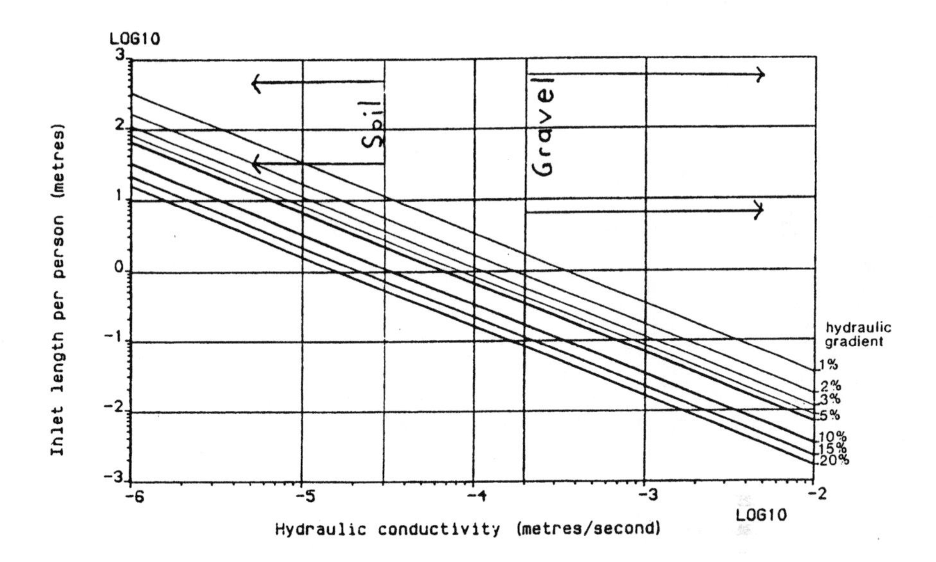

Figure 4. RBTS inlet length per person based on Darcy's law and an assumed flow of 180 L/person-day and a depth of 0.6 m.

but many of these problems disappear if coarse gravel is used. At a hydraulic conductivity of 10^{-2} m/s, requirements for inlet length are less than 10 mm/person. This relation means beds can be very long (300 m to give an area of 3 m^2/person), or alternatively hydraulic gradients can be very low. In most gravel beds, inlet and outlet levels will be about the same and 10 cm below the gravel surface to maintain conditions wet enough for reed growth. For soil beds, to maximize the hydraulic gradient, the outlet level should be kept as low as possible (capillary action will wet soil enough to support reed growth at the outlet end). Reeds are sensitive to drying out in gravel beds, where the effect of capillary action is much less than in soil beds.

From the above, gravel beds seem favored over soil beds and are much easier to design. However, in side-by-side comparisons of soil and gravel beds, soil beds give superior treatment if surface flow is avoided. Figures 5–7 show this in early results from 6-m-long beds at Audlem, and similar differences have been noted in 1-m-long beds at Stevenage in the United Kingdom. The soil beds are producing low BOD_5 values but high figures for suspended solids. This may imply a washout of soil particles for considerable periods after a soil bed is commissioned. In coarse-medium beds, the position is reversed. Effluents are often clear with low suspended solids but with comparatively high BOD_5 values, suggesting that in the early days of a gravel bed, treatment is largely by physical filtration.

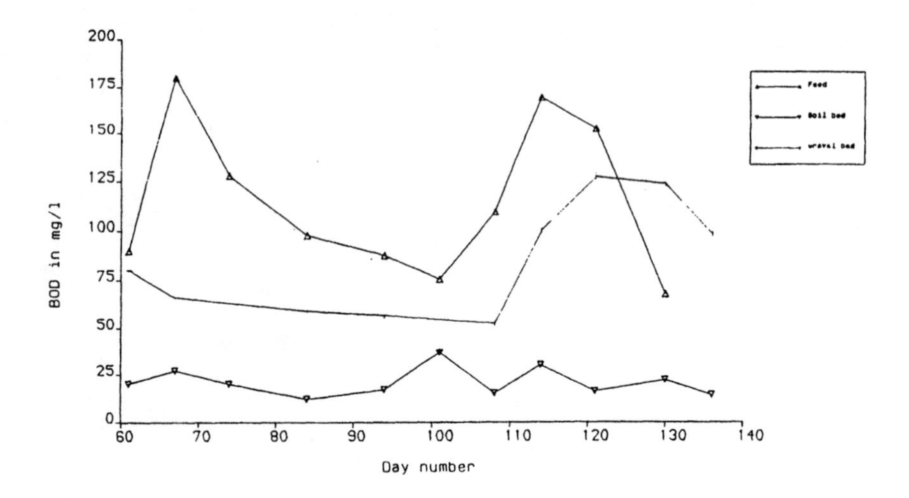

Figure 5. Audlem reed beds, BOD_5 values.

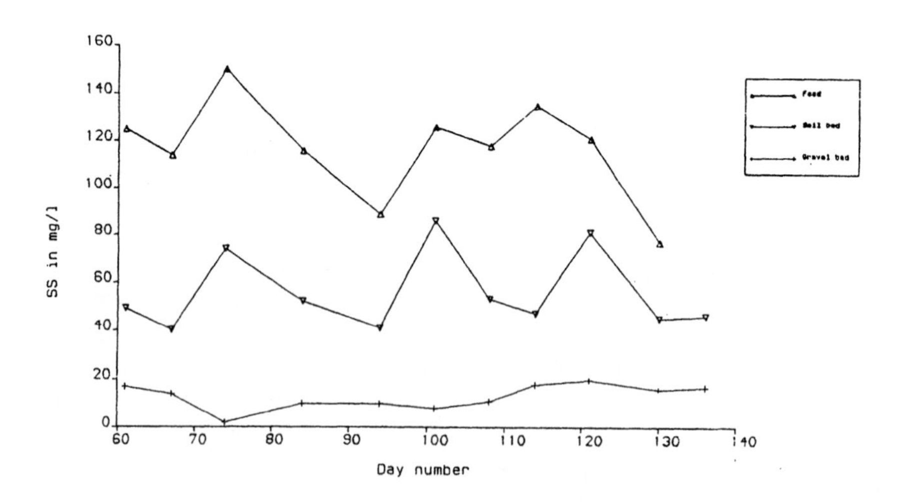

Figure 6. Audlem reed beds, suspended solids values.

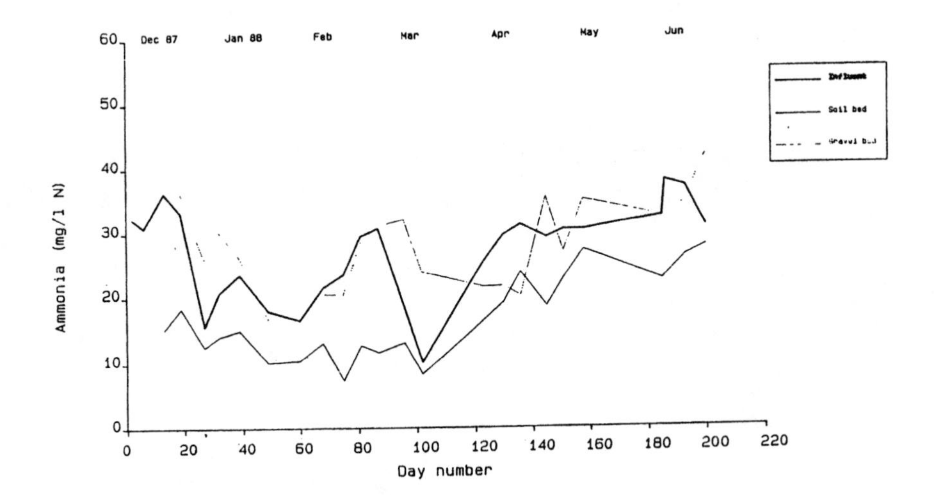

Figure 7. Audlem reed beds, ammonia concentrations.

39i Use of Artificial Cattail Marshes to Treat Sewage in Northern Ontario, Canada

Gordon Miller

INTRODUCTION

In July 1980, the Ontario Ministry of the Environment (MOE) started the Listowel Marsh project in southwestern Ontario (Figure 1). The project established that properly configured cattail (*Typha*) marshes have the capacity to significantly improve the quality of sewage wastewaters.[1]

In 1981, inquiries were made to MOE from the Ontario Ministry of Northern Affairs on the feasibility of marsh treatment technology for communities in northern Ontario. Many small northern communities cannot afford operating costs of conventional secondary sewage treatment facilities. In addition, many had poorly maintained combined storm and sanitary systems with dilute effluents. Thus, capital costs of conventional treatment were prohibitive. As a result, some northern communities had inadequate sewage treatment or none at all.

The inexpensive marsh treatment technology seemed ideally suited to these communities if the technology could be transferred to the cold northern climate. Adequate treatment had persisted through the winter months at Listowel, but oxygen stress had been evident.[1] The northern community of Cobalt was selected to test applicability of a marsh wetlands for wastewater treatment in Northern Ontario.

METHOD

Marsh design was based on information from the Listowel facility. A cell of in situ materials lined with off-site clay was constructed 310 m by 3 m (0.1 ha) with a serpentine configuration. At Listowel, it had been concluded that a length-to-width ratio of greater than 100:1 was desirable.[1]

Plot loading approximated 0.20 L/s (170 m^3/ha/day, 19.0 mm/day) of raw sewage. Depth was maintained at 15 cm to keep retention time near the seven-

Figure 1. Location map, Listowel and Cobalt artificial cattail marshes. Isolines show growing season length in days.

day period used at Listowel[1] and increased in winter to provide 15 cm of water under an ice layer.

The bulk of the sewage bypassed the plot by overflowing a large rectangular weir which dampened severe fluctuations in head in the sewer system. Flow of sewage to the plot was controlled by a 22.5-degree V-notch in the same structure. The head of water behind the weir was measured by a mechanical stage recorder and stage records were calculated by timed bucket discharges during regular visits to the site. An identical weir and stage recorder were installed at the plot discharge flow control structure.

Vegetation was established in 1982 by planting individual cattail shoots with an attached piece of rhizome (>5 cm) at 1-m intervals throughout the plot. Cattails were transplanted from adjacent natural stands.

Raw sewage influent and effluent from the marsh system (if present) were sampled for one year starting on July 11, 1983. Samples taken weekly during the summer months and alternate weeks through the winter were analyzed for BOD_5, suspended solids, and total phosphorus.

From July 11 to August 8, 1983, only the plot effluent was analyzed for fecal streptococci and fecal coliforms. Subsequently, samples of both influent and effluent were monitored. In November 1983, the lab changed its dilution technique and method of reporting, increasing the maximum level of detection for bacterial counts an order of magnitude.

Due to variation in ambient temperature and handling, samples taken on different dates may not be directly comparable. Interpretation must be limited to a paired comparison basis for samples on the same date. Marsh performance was evaluated on design criteria (15 mg/L of BOD_5, 15 mg/L of suspended solids [annual average], and 1 mg/L of total phosphorus [monthly average]).

RESULTS

Cattail growth and development was assessed on July 5, 1983, one year after planting. The marsh was filled in completely and seedling establishment and continued vegetative reproduction were evident. The cattails were approximately 1.3 m high, and while small voids existed, average density was estimated at 20 stems/m^2. The marsh plot had been invaded by well-established bulrush (*Scirpus*) and smartweed (*Polygonum*).

During the winter of 1982–83, muskrats tunneled through berms of the last channel; the channel dried out in the early spring and the cattails died. The holes were sealed in May and the area reflooded. Some replanting was done, but it proved superfluous because seedlings had vigorously recolonized the area by July 5, 1983.

Hydrology of the Marsh

Fluctuations in water in the splitter box, solids accumulation in the splitter box which interfered with both stage recorder floats and the V-notch weir, and stage recorder mechanism failures prevented measurements of influent and effluent flow. Consequently, the numerous bucket discharge measurements taken were used to characterize flows. Initially, the marsh received three times the design volume, but this gradually increased to five times the design loading after the first four months. Marsh outflow was always substantially less than inflow, especially during fall periods when the water level was deliberately raised prior to ice cover. The marsh was covered with ice and snow from the first week in November until the second week in April.

Table 1. Results of Water Chemistry Analysis for BOD_5, Suspended Solids and Total Phosphorus Entering and Leaving the Marsh

Month	Number of Samples	BOD_5		Suspended Solids		Total Phosphorus	
		Influent	Effluent	Influent	Effluent	Influent	Effluent
July 1983	3	18.9	1.7	27.2	2.7	1.5	0.7
August	5	22.7	5.3	33.4	3.9	1.9	0.4
September	4	25.5	2.8	37.0	28.7	1.2	0.3
October	5	19.4	2.1	23.1	14.3	1.5	0.6
November	4	40.9	6.5	45.4	52.2	3.0	1.2
December	1	27.0	11.3	16.5	270.0	1.4	1.3
January 1984	2	16.6	7.8	16.6	11.7	1.1	0.8
February	1	23.4	3.1	281.0	7.8	0.8	0.7
March	2	27.0	10.8	28.9	20.1	0.9	0.9
April	2	8.8	6.9	9.8	63.3	0.2	1.0
May	3	16.8	2.3	18.4	6.6	1.6	0.3
June	4	13.3	2.4	17.0	4.5	1.2	0.5
July	4	18.8	2.8	12.0	5.5	1.5	0.7

Note: All data in mg/L.

Water Quality of Influent and Effluent

The BOD_5 in raw sewage averaged 21.7 mg/L (Table 1). Passage through the marsh reduced the BOD_5 by 80%, producing an average effluent of 4.1 mg/L with maximum levels below 15 mg/L BOD_5 (Figure 2). Suspended solids in raw sewage averaged 31.9 mg/L. Excluding three anomalies, effluent suspended solids averaged 10.9 mg/L, well below the secondary effluent objective of 15 mg/L. Total phosphorus in untreated sewage averaged 1.53 mg/L. Marsh outflow averaged 0.66 mg/L TP, 57% less than raw sewage and within the design objective of 1 mg/L.

Bacterial counts were consistently high going into the plots, but output counts varied (Table 2). Out of 33 paired input and output samples for fecal streptococci, 29 showed lower bacterial populations in the outflow. Three could not be distinguished because input and output levels were above maximum detectable limits and one sample showed a slight increase. Similarly, fecal coliforms were lower in the outflows in 32 out of 33 paired comparisons, with one undistinguishable pair. Because there is a reasonable possibility that the samples were accidentally switched on this date, data for September 14, 1983, are not considered in the analysis.

Numbers of fecal streptococci and fecal coliform discharged were usually low during cool periods, and in the best of conditions the marsh is capable of discharging an effluent that meets swimming and bathing water quality objectives.[2]

DISCUSSION

Marsh growth and development paralleled marsh development at Listowel. Cattail height and density after one year were not as great as in Listowel, but

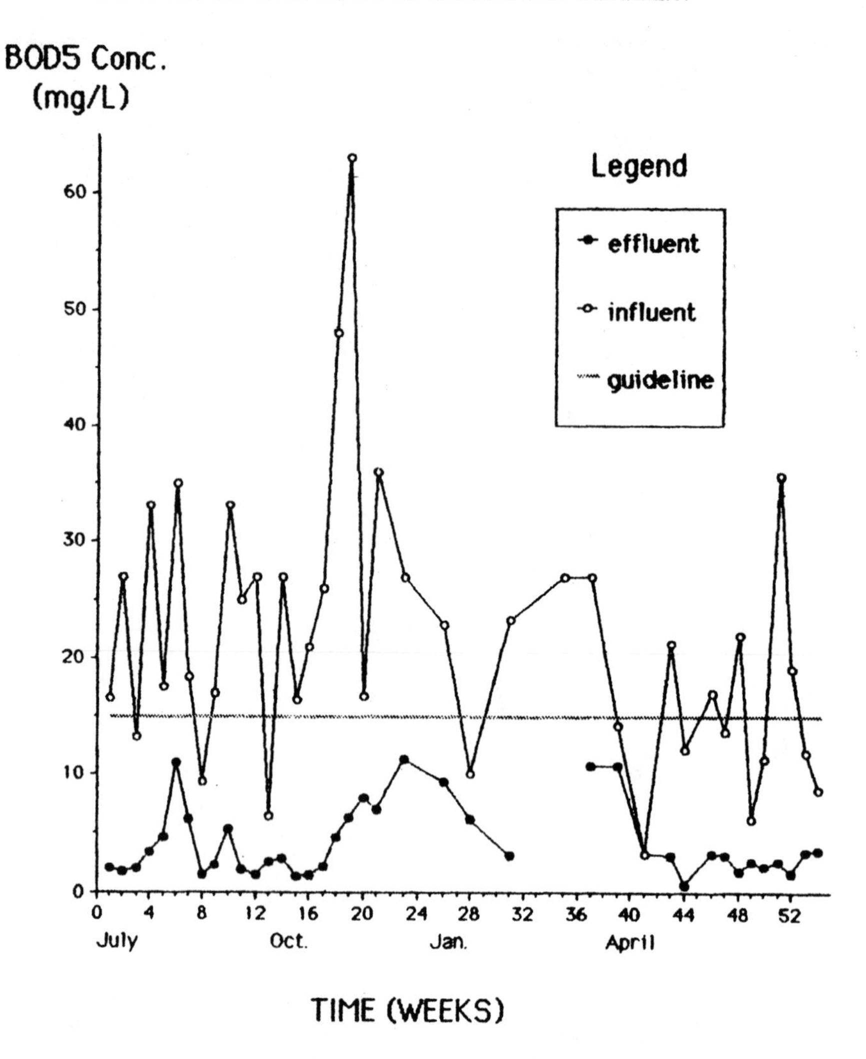

Figure 2. BOD_5 concentrations in the influent and effluent of the artificial marsh over a one-year period.

the vigor with which the marsh established itself was similar. The technology is not limited in this climate by cattail biology.

Unlike mechanical systems, marshes may interact hydraulically with the atmosphere and groundwater in a positive and negative manner. Bucket discharge data revealed a large water loss even in winter months. This must have been due to exfiltration despite the clay lining. Fortunately, this loss does not interfere with the critical criteria for the assessment of marsh performance because these are expressed as concentrations, not loadings. Exfiltration can-

Table 2. Results of Bacteriological Sampling of the Water Entering and Leaving the Marsh

Date	Fecal Streptococci		Fecal Coliforms	
	Influent	Effluent	Influent	Effluent
Jul 11, 1983	#NA	9.00	#NA	12.00
Jul 18, 1983	#NA	8.00	#NA	0.28
Jul 25, 1983	#NA	9.00	#NA	12.00[a]
Aug 2, 1983	#NA	3.84	#NA	1.20
Aug 8, 1983	#NA	<0.04	#NA	0.04
Aug 15, 1983	20.00[a]	<0.10	12.00[a]	12.00[a]
Aug 22, 1983	#NA	#NA	12.00[a]	0.04
Aug 29, 1983	20.00[a]	4.00	12.00[a]	0.96
Sep 6, 1983	20.00[a]	20.00[a]	12.00[a]	0.08
Sep 14, 1983	20.00[a]	20.00[a]	1.80	12.00[a]
Sep 20, 1983	20.00[a]	3.08	12.00[a]	0.04
Sep 26, 1983	20.00[a]	1.16	12.00[a]	0.04
Oct 4, 1983	3.60	<0.04	12.00[a]	0.60
Oct 11, 1983	20.00[a]	0.48	12.00[a]	0.04
Oct 18, 1983	20.00[a]	0.40	12.00[a]	0.04
Oct 24, 1983	20.00[a]	<0.04	12.00[a]	0.04
Oct 31, 1983	20.00[a]	0.20	12.00[a]	0.32
Nov 7, 1983	2.70	1.72	12.00[a]	2.56
Nov 14, 1983	20.00[a]	2.00	12.00[a]	3.40
Nov 21, 1983	20.00[a]	20.00[a]	12.00[a]	12.00[a]
Nov 28, 1983	200.00[a]	6.60	600.00[a]	1.00
Dec 12, 1983	60.00	60.00	60.00	14.00
Jan 3, 1984	200.00[a]	62.00	290.00	10.00
Jan 16, 1984	8.00	2.80	86.00	16.00
Jan 30, 1984	16.00	5.40	44.00	20.00
Feb 13, 1984	31.00	2.00	27.00	2.20
Mar 13, 1984	22.00	#NA	220.00	#NA
Mar 27, 1984	1.80	3.00	53.00	20.00
Apr 9, 1984	18.00	4.00	150.00	<10.00
Apr 23, 1984	15.00	0.80	200.00[a]	3.00
May 7, 1984	200.00[a]	24.00	600.00[a]	<10.00
May 15, 1984	34.00	<0.20	250.00	<10.00
May 28, 1984	130.00	1.00	450.00	10.00
Jun 5, 1984	200.00	0.80	600.00[a]	<10.00
Jun 12, 1984	130.00	3.20	600.00[a]	<10.00
Jun 19, 1984	70.00	24.00	360.00	<10.00
Jun 25, 1984	40.00	0.20	11.00	<10.00
Jul 3, 1984	0.20	110.00	340.00	<10.00
Jul 9, 1984	30.00	10.00	410.00	<10.00
Jul 16, 1984	50.00	0.80	520.00	<10.00
Jul 24, 1984	110.00	1.60	600.00	<10.00

Note: All values are reported as (counts <times> 10,000/L.
[a]Denotes count greater than maximum detectable number.

not decrease the effluent concentration, but it may increase suspended solids or bacteria.

The design criteria for a secondary sewage treatment plant may not be the most appropriate factors to consider when evaluating successful marshlands treatment. BOD_5 and suspended solids discharging from a marsh are often, in reality, zooplankton and phytoplankton. These organisms are not normally considered ecologically damaging to the receiving waters, and it is misleading to consider the production of life forms as "incomplete" sewage treatment. However, when it is necessary, as in this study, to make a decision about comparative performance, adherence to the traditional parameters is most expedient.

The marsh maintained an effluent BOD_5 concentration well within the 15 mg/L standard expected of a secondary sewage treatment system during the entire study period. As at Listowel, effluent quality of the marsh seems more closely related to seasonal influences rather than influent quality. During periods of oxygen stress in August and midwinter, BOD_5 removal was diminished.

Marsh performance for suspended solids was variable. At Listowel, marshes were excellent settling and filtering mechanisms, yet in the Cobalt data high levels of SS were reported in the effluent on some occasions. Caution is advised in interpreting this because marshes have autochthonous production of detritus, and many of the high SS discharges were coincident with water level adjustments. Disturbance-caused suspension of sediment and detrital material as a possible explanation is reinforced by the lack of coincidence of high suspended solids with evaluated BOD_5 concentrations.

Phosphorus concentrations leaving the marsh during the growing season were very low, although during the oxygen-stressed periods of winter under ice cover, much of this treatment was lost. Although the marsh effluent varies widely through different seasons and conditions, bacteriological quality of the effluent in the better operating conditions far exceeds the quality expected of a traditional secondary treatment plant.

It seems reasonable to conclude the marsh sewage treatment technology can be transferred successfully to northern climates. Results from this study compare favorably to secondary sewage treatment standards despite the marsh's loading at three to five times the intended rate.

Original design loadings were based on conservative estimates of what the marsh could handle, but higher loadings of this particular effluent were satisfactorily treated. The maximum treatment capability cannot be strictly defined due to the problems with flow regulation and exfiltration. However, it appears that areal loadings of 100 mm/day are within the capability of this technology in this climate.

REFERENCES

1. Wile, I., G. Miller, and S. Black. "Design and Use of Artificial Wetlands," in *Ecological Considerations in Wetland Treatment of Municipal Wastewaters*, P. J. Godfrey, E. R. Kaynor, S. Pelczarski, and J. Benforado, Eds. (New York: Van Nostrand Reinhold Company, 1985), pp. 26–37.
2. "Water Management—Goals, Policies, Objectives and Implementation Procedures of the Ministry of the Environment," Ontario Ministry of the Environment, Toronto, Ontario, Canada (1978).

39j Some Ancillary Benefits of a Natural Land Treatment System

A. Larry Schwartz and Robert L. Knight

The Grand Strand Water and Sewer Authority (GSW&SA) is a countywide water and sewer provider to South Carolina's most rapidly growing county, including the area called the Grand Strand. Horry County has several major rivers with large swamp floodplains and extensive inland wetland areas drained by swampy streams. The rivers have limited wastewater assimilative capacities due to swamp influences contributing to naturally low dissolved oxygen (DO) conditions. Assimilative capacities are so low that it was cost-effective for the City of Myrtle Beach to pump secondary wastewater 26 km rather than treat to advanced levels for nearby discharge.

Existing discharges to surface waters are at permitted assimilative capacity for secondary treatment, while flows over the next 20 years are projected to increase from 75.6 million L/day to over 283.5 million L/day. These increased flows will be required to maintain the present discharge loadings due to the finite assimilative capacity of the area's receiving waters.

As an alternative to costly advanced conventional treatment, alternative upland application analyses were conducted by GSW&SA and their consultants, CH2M HILL. Suitable land application sites are scarce in the Grand Strand area because of high groundwater levels, large natural wetland areas (over 30%), and fragmented ownership patterns. Because of extensive coverage, wetlands and other poorly drained areas became an element of the alternative land application analysis. In addition to riverine wetlands, a large number of unique oval-shaped, natural land forms called Carolina Bays are located here. These bay areas are poorly drained and have well-defined drainage areas confined by raised sand rims around the perimeter, apparently ideal for wastewater treatment. After four years of 201 Facilities Plan analysis, a cluster of four Carolina Bays was selected for one treatment plant service area and a riverine swamp with well-defined sloughs for another facility service area. Over 336 ha (830 acres) of Carolina Bay areas are contained in the cluster and over 142 ha (350 acres) in the riverine swamp. The riverine swamp and three of the four Carolina Bays were classified as wetlands. One bay and the riverine slough receiving discharges of secondary effluent are permitted as pilot projects and have performed as expected after at least 15 months of discharge. The only measurable vegetative changes were slightly increased tree growth rates.

Only four of 2700 Carolina Bays over 1 ha in size in South Carolina are publicly owned and only one publicly owned facility exists in Horry County. Several large coastal reserves in the region and almost all reserves and other publicly owned areas are operated as "protected" areas with limited access. Less than 0.1% of the wetlands in South Carolina are protected through any state, federal, or local programs. Unfortunately, over 90% of the Carolina Bays in South Carolina have been significantly disturbed by agriculture or silviculture.

The GSW&SA through its Natural Land Treatment Programs owns 336 ha of Carolina Bays and 142 ha of riverine wetlands. Opportunities these areas provide for public access and education are significant, and the GSW&SA in conjunction with CH2M HILL has outlined a conceptual plan for a "Carolina Bay Nature Park—A Native Plant and Wildlife Interpretive Center." The center will serve as a focus for natural education programs by providing natural areas access for the public and interested researchers. While these areas are not pristine preserved areas, they represent a balancing of development needs and environmental conservation of natural areas. They will have boardwalks (pipe support structures) that provide access into the heart of thickly vegetated Carolina Bays and deep swamp riverine wetlands. Individuals, school groups, environmental groups, and civic groups being allowed easy access to previously inaccessible areas along with native plant interpretive trails is unique in our region and should provide varied opportunities for all levels of interest.

Another ancillary benefit of the GSW&SA Natural Land Treatment Program is the development of a data base and communication among agencies on objectives of natural land treatment projects. These projects are among the first in South Carolina to raise questions and issues about wetlands discharges. Lack of an adequate data base has been a major factor in the level of concern expressed by regulatory agencies and other interested groups. Predicting long-range impacts of similar projects and deciding what level of impact is acceptable has been a key element to negotiations with the various agencies. Region IV of the United States Environmental Protection Agency (EPA) has sufficient interest in the direction of the Carolina Bay Project and its implications that they have funded four additional years of the monitoring program to collect intensive water quality and vegetative data. The information generated will create a data base from which to evaluate future projects. At the present time, in the absence of water quality criteria and standards for wetlands, each project is evaluated on a case-by-case approach with a minimum of background performance data from similar systems.

Permitting issues associated with wetlands discharges range from water quality impacts and standards for naturally low DO wetland to "permissible" levels of vegetative changes. South Carolina has no wetlands policy or coordinated program to establish criteria for use of wetlands. The GSW&SA projects have provided a stimulus for developing goals and objectives for such projects. The pilot study nature provides flexibility for agencies to establish specific criteria for these projects yet avoids prematurely defining policy that could guide future projects. Negotiations leading to permitting of the Carolina Bay project provided a stimulus for discussions on the value and uses of wetlands and, specifically, Carolina Bays in South Carolina. A key discussion element was the criteria for "success" of the projects. The criterion initially proposed was "no significant change" in the plant community as well as in populations of any threatened or endangered plant or animal species.

The "no change" approach seemed overly restrictive because expectant changes will make these Carolina Bays more like wetlands. Also, this "no change" criterion was considered unrealistic by GSW&SA because all of the selected Carolina Bays had previously been affected by human activities to various extents. A compromise will allow greater or lesser change in each bay depending upon its previous history of human alteration. The most altered bays will be permitted to higher wastewater loading rates than less affected bays that have more value for preservation.

Specific criteria selected for each bay are directly measurable and are linked to plant density, dominance, and diversity and to bird diversity. Specific criteria are expected to prevent future disagreements over meanings of ambiguous general wording in permits. Success criteria protect GSW&SA from changing regulatory priorities and protect the public interest that the agencies are entrusted to maintain.

In summary, the GSW&SA is conducting pilot studies in riverine wetlands and Carolina Bays that will confirm feasibility of long-term use of natural systems while conserving their integrity and function as measured by a set of biological criteria. In addition, these projects provided EPA, S.C. Department of Health and Environmental Control, S.C. Wildlife and Marine Resources Department, U.S. Fish and Wildlife Service, and the S.C. Coastal Council an opportunity to define goals and criteria for protection of specific wetland areas throughout the state. Primary benefits will accrue to citizens of Horry County and South Carolina who will have the opportunity to venture into the heart of Carolina Bays or into a swamp. They will be able to view plant and animal communities in their natural habitat without significant alteration or destruction of form or function and observe the positive beneficial uses that these areas can provide.

39k Performance of Solid-Matrix Wetland Systems Viewed as Fixed-Film Bioreactors

H. J. Bavor, D. J. Roser, P. J. Fisher, and I. C. Smalls

INTRODUCTION

Wastewater treatment performance of solid-matrix, constructed wetland systems has been investigated during the design, operation, and maintenance of seven large-scale units at Richmond, Australia over a 3.5-year period. Performance models were developed by treating the systems as fixed-film bioreactors. Systems consisted of lined trenches planted with *Typha orientalis* or *Scirpus validus* in gravel and a floating macrophyte, *Myriophyllum aquaticum,* growing in open water with no solid matrix. Domestic sewage and primary (settled) and secondary effluent from a conventional trickling filter/maturation pond were used at hydraulic loadings from 0.2 to 1.4 mL/ha/day. Effective removal of BOD_5 (annual mean of 95%), suspended solids (94%), and total nitrogen (67%) was achieved by the systems when operating at hydraulic detention times of two to 10 days. Indicator bacteria were reduced by up to 5 orders of magnitude. Phosphorus removal, however, was low (15%). The *Myriophyllum* system was found to be the least effective design format tested. Initial studies and detailed performance results have been reported in Bavor et al.[1] and Roser et al.[2]

MATERIALS AND METHODS

The systems consisted of a series of seven rectangular lined trenches, each 100 m long, 4 m wide, and 0.5 m deep (Figure 1). They comprised:

- two systems filled with gravel and planted with emergent macrophytes *T. orientalis* (cumbungi, trench 5) and *S. validus* (bulrush, trench 3), respectively
- one system filled with water and planted with the floating macrophyte *M. aquaticum* (parrot feather, trench 4)

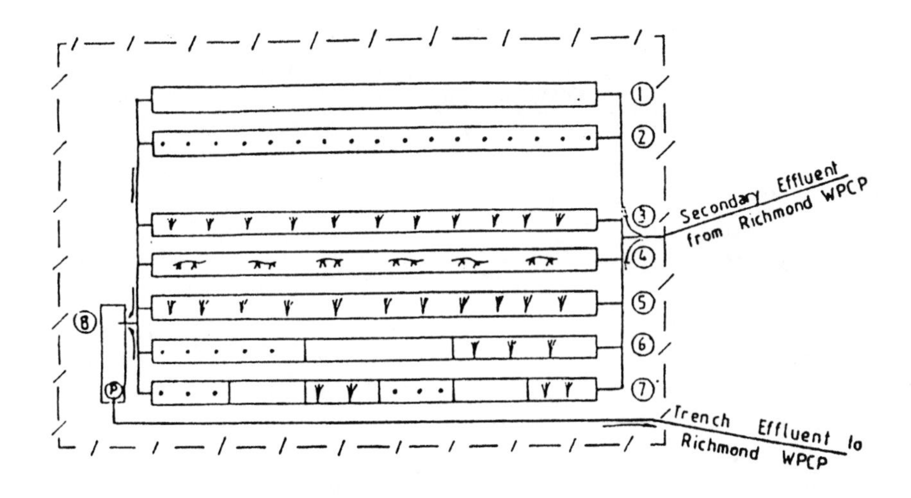

① Open Water Control Trench
② Gravel Control Trench
③ Scirpus Trench
④ *Myriophyllum* Trench
⑤ *Typha* Trench
⑥ Mixed = Zone "A" (with *Typha*)
⑦ Mixed = Zone "B" (with *Typha*)
⑧ Effluent Collection Sump
Ⓟ Pump Station

Figure 1. Major features of artificial wetland-trench systems at Richmond N.S.W., Australia. All trenches are 100 m × 4 m × 0.5 m.

- two "mixed-zone" systems (A—trench 6 and B—trench 7) comprising sections of gravel planted with *Typha*, unplanted gravel, and open water
- two control systems, open water (trench 1) and gravel (trench 2), without plants

Between January 1984 and July 1986, the systems continuously received secondary clarified effluent from a trickling filter at rates of 5–85 m^3/day/system (0.5–8.0 ha/1000 m^3/day), corresponding to detention times of two to 20 days. During the major experimental periods, however, rates were set to nominal retention times of three to four or six to eight days, corresponding to loading rates of 200–1400 m^3/ha/day. In September 1986, trenches 2, 5, 6, and 7 were switched to settled, primary effluent input at rates varying from 300 to 400 m^3/ha/day.

Composition of the input effluent varied diurnally; consequently, "composite" influent pollutant levels were estimated for modeling purposes from rou-

Table 1. Composition of Input Effluent as Estimated for Modeling Purposes

	Type of Effluent	
	Secondary (mg/L)	Primary (mg/L)
BOD_5	40	171
SS	50	157
TKN	40	50
NH_4-N	35	35
Tot-P	9.6	9.8

Note: Complete nutrient analyses were carried out as described in Roser et al.[2] Only primary pollutants are noted above.

tine (weekly) and continuous (24–48 hr) sampling runs as described in Bavor et al.[1] Values used are shown in Table 1.

Design modifications, because of difficulties in controlling water levels and hydraulic gradient formation, included replacement of a simple ball cock arrangement, using a float control on the inlet and a gate-valve outlet, with continuously operating gate valves and electronic level sensor/pump activators at the outlets. Dye studies and water table monitoring were used to characterize effluent flow.[2]

Fouling of gravel-water interfaces (from input SS and decaying algal material) below open water sections of mixed-zone systems, was remedied by replacement of the first 5 m of 5- to 10-mm gravel with 30- to 40-mm gravel and mounding gravel directly over inlet sparger pipes, adding 5- to 10-m lengths of 50- and 100-mm agricultural drainage pipe in subgravel areas downstream of open water sections, shading open water/gravel interfaces to discourage surface algal film growth, and inserting under- and overflow baffles to reduce movement of large algal aggregates into the interface gravel.

To determine potential nitrification rates, nitrifier populations in 100-g surface cores were incubated at 25°C, with agitation, in an artificial effluent medium[3] with a final NH_4-N concentration of 45 mg/L, and levels of NH_4-N, NO_2-N, and NO_3-N were monitored. Potential rates of denitrification were estimated using the acetylene blockage technique.[3]

RESULTS

System performance varied in response to variations in influent application rate and makeup, local temperature variation, system size and design, and macrophyte type and cover. Examination of effluent data showed that the extent of removal of SS, BOD_5, TKN, NH_4, TP, and fecal coliforms could be mathematically modeled and that output concentrations could be predicted with a high degree of certainty. Derived relationships were based on first-order kinetics, the Arrhenius temperature/rate constant, and hydraulic mixing/flow equations.[1]

Effect of Loading Rate on Pollutant Removal

Simple, first-order kinetics were used to model removal of solids, carbon, phosphorus, nitrogen, and microorganisms. Relationships involved in the removal of pollutants are complex and first-order kinetics may not adequately describe many situations, but acceptable model verification was obtained for nitrogen species, coliforms, and other pollutants. Removal performance formulae were derived from regression analyses performed on the logarithm of final concentration values for SS, BOD_5, TOC, TKN, NH_4-N, TP, and fecal coliforms with the following equation:

$$\ln\left(\frac{C}{C_0}\right) = -K \cdot RT \tag{1}$$

$$\text{or } \ln(C) = \ln(C_0) - K \cdot RT$$

$$\text{or } C = e^{(\ln(C_0) - K \cdot RT)}$$

where RT = hydraulic retention time within macrophyte system
C_0 = initial pollutant concentration
C = effluent concentration
K = reaction constant
ln = natural log

Estimates of K were obtained by regression of ln C (or $\log_{10}$ with appropriate correction) against retention time of water sampled or ln (internal system concentration) (or $\log_{10}$ with appropriate correction) against distance traveled by the effluent to each sampling point. In the latter case, the retention-time-based K value was estimated by assuming that time taken to travel 100 m was equal to mean system retention time measured over the preceding week.

In all the gravel-based systems, there was a negative correlation between retention time and effluent solids levels, but the values were consistently low. BOD_5 and TOC behaved similarly, with BOD_5 displaying a slightly higher correlation to retention time.

Sampling results suggest that solids and BOD_5 levels decrease with increasing retention time (possibly via first-order kinetics). However, most removal took place in inlet zones and there was insufficient data to estimate removal rates in these very active regions.

The best correlations between pollutants versus retention time were obtained for reductions in fecal coliforms, TKN, and NH_4-N levels (Table 2). Regression analyses of P removal did not show significant trends of increasing removal with retention time. Predicted and actual effluent levels were virtually identical to influent levels, but a wide scatter of effluent P concentration may have reflected cycles of P accumulation on gravel or in sediment, followed by desorption.

Table 2. Correlation Between Removal Rate and Retention Time

System	Date	K[a] (day^{-1})	Correlation Coefficient (R)[b]
Open water	11/5/85	0.164	–0.69
Gravel	3/26/86	0.089	–0.92
	7/16/86	0.076	–0.77
Scirpus	2/4/86	0.118	–0.89
Myriophyllum	1/21/86	0.072	–0.49
	4/22/86	0.130	–0.70
Typha	12/3/85	0.103	–0.93
	3/4/86	0.160	–0.95
	7/10/86	0.105	–0.73
Mixed-zone A	10/22/85	0.153	–0.009
	6/3/86	0.094	–0.94
Mixed-zone B	11/19/85	0.189	–0.96
	4/8/86	0.088	–0.49

Note: Removal rates are presented for TKN Nitrogen from secondary effluent inputs determined by lengthwise intensive sampling on selected dates.

[a]K (reaction constant) values were determined by plotting $\log_{10}$ of the effluent TKN-N concentration against the distance from the inlet point and assuming that the distance traveled was directly proportional to the detention time.

[b]For trenches 2, 3, 5, 6, and 7, the mean R value was –0.86 (SD = 0.09). The mean Y intercept (C_0 value on the assumption of first-order kinetics) for all regressions corresponded to an influent [N] of 35.0 mg/L (SD = 13.3). Poor correlations obtained on two occasions in the mixed-zone systems corresponded to high TKN levels in the open water sections presumably due to algal blooms.

Effect of Seasonality on Pollutant Removal

Examination of solids and BOD_5 removal data showed that most removal occurred in the first 10–20 m of the gravel systems (Figure 2), and little change was observed thereafter. Expected seasonal change in removal rate (in inlet zones) was not followed by marked change in pollutant levels at system outlets. As effluents consistently contained solids and BOD_5 levels 10 mg/L throughout the year, a temperature variation factor was negligible and could be omitted from the first-order rate equation for prediction of removal of SS, BOD_5, and TOC.

Seasonal variation was apparent in comparisons of removal of nitrogen and longitudinal sampling indicated a continual decrease in N and fecal coliforms along the system length. Therefore, temperature/seasonal influence was considered in removal performance analysis for a number of sewage constituents, as described below.

When theoretical rate constants $[\ln (C_0/C) \cdot (1/RT)]$ were plotted against time of sampling, a seasonal periodicity was observed, particularly with nitrogen removal. The variation in rate constant could be modeled according to the van't Hoff–Arrhenius equation:

$$K = A \cdot e^{(-E/R \cdot t)} \tag{2}$$

where K = the reaction constant
A = van't Hoff–Arrhenius coefficient

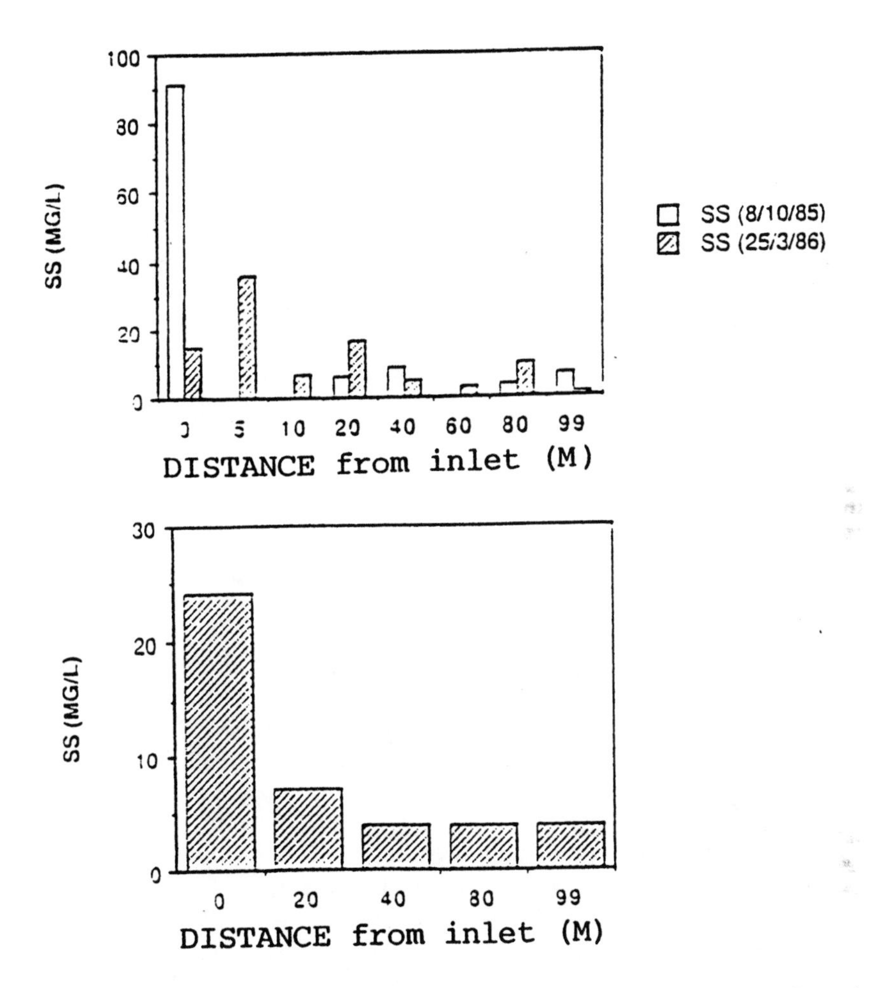

Figure 2. Suspended solids reduction along the length of selected gravel-filled trenches. Median input was 50 mg/L. (a) gravel control, trench 2; (b) *Scirpus,* April 2, 1986, trench 3.

E = activation energy
R = gas constant
t = temperature in K

Since the observed removal rates appeared to increase with temperature, seasonal variations in TKN, NH_4-N, and fecal coliform removal rate constants could be modeled using an exponential regression of ln $(C_0/C)/RT$ versus 1/t. From this, coefficients a and b (equivalent to A and E/R) could be obtained. Due to the presence of some negative values, the data set needed to be edited prior to conducting the regression routine and only coefficient b was determined from this step. Coefficient a was determined by linear regression of $RT \cdot e^{(b/t)}$ versus ln (C).

Finally, the removal rates of TKN, NH_4-N, and fecal coliforms were modeled as equations of the form:

$$\ln C = \ln C_0 - RT \cdot a \cdot e^{(b/t)} \quad (3)$$

Model verification, indicated by the closeness of fit between the model predicted and actual pollutant values, is shown in Figure 3 and coefficients a and b for the different systems in Tables 3 and 4. Predictive models were developed having correlation coefficients up to 0.85.

Nitrification Potential of the Macrophyte System Microbial Populations

Enumeration of microbial populations present in macrophyte systems during application of secondary effluent demonstrated that most of the total microflora and nitrifier populations were associated with gravel particles and root mass regions. Similar association of nitrifier populations with particulates and surfaces was noted by Bavor et al.[3]

Surface core samples (oxidized gravel zone, 0–5 cm) taken at the commencement of primary effluent application (1986) showed the presence of low numbers of nitrifiers, around 10^1/100 g of gravel. Samplings during 1987, however, indicated a much higher population, 10^2–10^7/100 g of gravel in all sites and on all occasions where samples were taken.

Populations were distributed throughout both unplanted and planted surface gravel zones. Although limited in number, results obtained were consistent with the hypothesis that larger populations (and, hence, higher "potential" nitrification rates) were concentrated in more aerobic zones closer to trench surfaces further downstream from inlets and near macrophyte roots.

Following an acclimation period of 5–15 days, nitrogen transformation occurred rapidly in all samples, with complete NH_4-N transformation occurring at 10–20 mg N/L/day (or 2–4 g N/m^2/day), i.e., rapidly enough to completely nitrify typical input NH_4-N levels within a six-day retention time (assuming ideal conditions). A 20-cm deep oxidized gravel zone of potential nitrifying activity was assumed in rate calculations. Considerably lower ammonium oxidation rates would be expected under in situ conditions at much lower oxygen transfer efficiencies, mixing activity, temperature, and so on.

Denitrification

Rate measurements on gravel cores showed that potential rates of 0.5–1.5 g N/m^2/day (5–15 kg N/ha/day) were common. A gravel depth of approximately 20 cm had suitable reducing conditions for denitrification and was assumed in estimates for rate calculations.

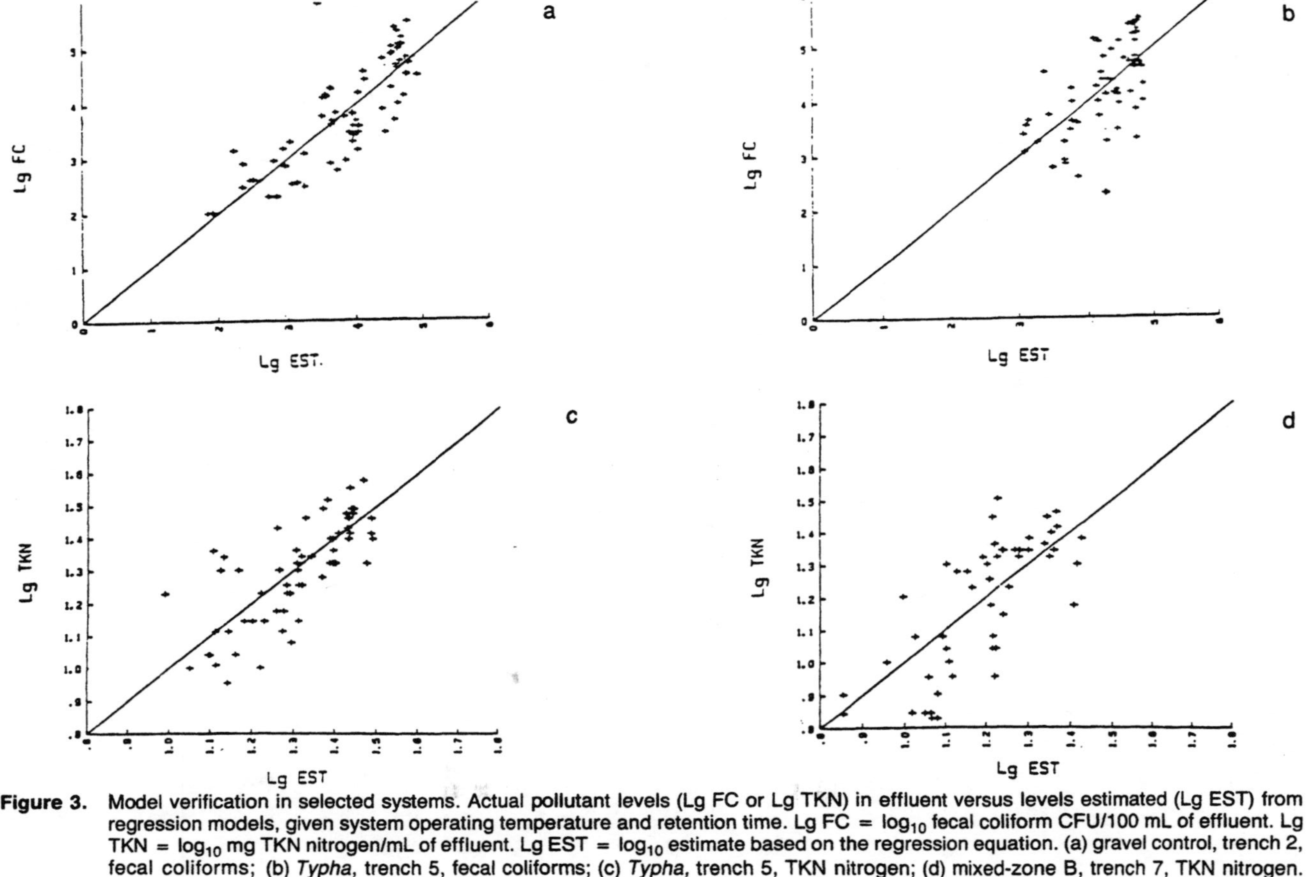

Figure 3. Model verification in selected systems. Actual pollutant levels (Lg FC or Lg TKN) in effluent versus levels estimated (Lg EST) from regression models, given system operating temperature and retention time. Lg FC = $\log_{10}$ fecal coliform CFU/100 mL of effluent. Lg TKN = $\log_{10}$ mg TKN nitrogen/mL of effluent. Lg EST = $\log_{10}$ estimate based on the regression equation. (a) gravel control, trench 2, fecal coliforms; (b) *Typha*, trench 5, fecal coliforms; (c) *Typha*, trench 5, TKN nitrogen; (d) mixed-zone B, trench 7, TKN nitrogen.

Table 3. Regression Equations Describing the Removal of Fecal Coliforms by Macrophyte Systems as a Function of Temperature, Retention Time, and Influent Concentration

	Equation Coefficients[a]			Correl. Coeff.	Removal Rate[b]	Effluent Conc.[c]
Sample Site	$\ln(C_0)$	a	b	R	$K_{20°C}$ (d^{-1})	$\log_{10}$ CFU /100 mL
Open water	12.90	1207	–2300	0.46	0.47	4.37
Gravel	13.25	5.49×10^5	–3840	0.79	1.12	2.84
Scirpus	12.72	6.88×10^5	–4070	0.63	0.64	3.86
Myriophyllum	15.62	1.32×10^4	2585	0.71	0.90	4.45
Typha	12.15	1.22×10^5	–3600	0.60	0.56	3.81
Mixed-zone A	13.85	66	–1150	0.81	1.30	2.61
Mixed-zone B	13.15	571	–1780	0.68	1.31	2.29

[a]Coefficients are for equations of the form:

$\ln(C) = \ln(C_0) - RT \cdot a \cdot e^{(b/t)}$

where $\ln(C)$ = natural log of the final coliform concentration in CFU/100 mL

$\ln(C_0)$ = natural log of the initial coliform concentration as determined by regression analysis

RT = system retention time in days

t = system operating temperature in K

"a" and "b" are coefficients describing the temperature dependency of the rate constant. The general equation described above was obtained by combining the Arrhenius expression for the temperature dependency of rate constants with the equation for first-order kinetics. Estimates of "a" were obtained by the regression of $RT \cdot e^{(b/t)}$ against ln (C). Estimates of "b" were derived from exponential regression of day-to-day estimates of $[\ln(C_0/C) \cdot (1/RT)]$ against 1/t. The $\ln(C_0)$ values correspond to a mean initial fecal coliform level of 6.5×10^5 CFU/100 mL.

[b]$K_{20°C}$ is the rate constant estimate for 20°C.

[c]Effluent concentrations ($\log_{10}$ values) given were derived from each equation given the coefficients listed, a retention time of six days, and an operating temperature of 20°C.

High denitrification potential was shown for sites where high available carbon (SS and BOD_5) was present. In all systems, it appeared that denitrification activity was maximum near the inlet zones and in the sludge of open water sections and rapidly declined away from these regions. Denitrification might be improved in design formats having a more even influent distribution with available C in reduced regions downstream of nitrifying zones.

Hydraulic Behavior

Investigations of hydraulic behavior of macrophyte systems indicated that the gravel beds have a narrow inlet zone, characterized by high solids levels and reduced hydraulic conductivity, followed by a downstream zone with low solids buildup and stable permeability of approximately 0.3 m/s. Dye studies demonstrated preferred flow paths in the gravel around and beneath the root zone of emergent macrophytes.

Table 4. Regression Equations Describing Removal of Nitrogen by Macrophyte Systems as a Function of Temperature, Retention Time, and Influent Concentration

	Equation Coefficients[a]			Correl. Coeff.	Removal Rate[b]	Effluent Conc.[c]
Sample Site	$\ln(C_0)$	a	b	R	$K_{20°C}$ (d^{-1})	[N] mg/L
Open water	3.64	1.9×10^8	–6300	0.60	0.087	22.6
Gravel	3.75	168	–2000	0.72	0.182	14.2
Scirpus	3.52	1445	–2760	0.79	0.117	16.7
Myriophyllum	3.75	9.59×10^{-2}	–13.7	0.65	0.092	24.6
Typha	3.58	137	–2060	0.75	0.121	17.3
Mixed-zone A	3.55	7.26×10^4	–3870	0.78	0.133	15.6
Mixed-zone B	3.53	4.61×10^4	–3710	0.75	0.146	14.2

[a]Coefficients are for equations of the form:

$\ln (C) = \ln (C_0) - RT \cdot a \cdot e^{(b/t)}$

where $\ln (C)$ = natural log of the final coliform concentration in CFU/100 mL

$\ln (C_0)$ = natural log of the initial coliform concentration as determined by regression analysis

RT = system retention time in days

t = system operating temperature in K

"a" and "b" are coefficients describing the temperature dependency of the rate constant. The general equation described above was obtained by combining the Arrhenius expression for the temperature dependency of rate constants with the equation for first-order kinetics. Estimates of "a" were obtained by the regression of $RT \cdot e^{(b/t)}$ against ln (C). Estimates of "b" were derived from exponential regression of day-to-day estimates of $[\ln(C_0/C) \cdot (1/RT)]$ against 1/t. The $\ln (C_0)$ values correspond to a mean initial fecal coliform level of 6.5×10^5 CFU/100 mL.

[b]$K_{20°C}$ is the rate constant estimate for 20°C.

[c]Effluent concentrations ($\log_{10}$ values) given were derived from each equation given the coefficients listed, a retention time of six days, and an operating temperature of 20°C.

CONCLUSIONS

Removal of suspended solids, BOD_5, nitrogen, phosphorus, and fecal coliforms was investigated with respect to loading, detention time, and temperature parameters to allow predictive modeling of system performance, assuming first-order removal kinetics and treating the units as fixed-film bioreactors. For a number of effluent constituents, first-order removal kinetics may not adequately describe removal performance. Improved models are being developed through examination of removal data from discrete compartments in the systems and/or analysis using removal kinetics of increased complexity. Intensive examination of a number of removal mechanisms and removal rates have been undertaken.

ACKNOWLEDGMENT

Sole funding by the Water Board, Sydney, is gratefully acknowledged.

REFERENCES

1. Bavor, H. J., D. J. Roser, and S. A. McKersie. "Nutrient Removal Using Shallow Lagoon-Solid Matrix Macrophyte Systems," in *Aquatic Plants for Water Treatment and Resource Recovery*, K. R. Reddy and W. H. Smith, Eds. (Orlando, FL: Magnolia Publishing Inc., 1987), pp. 227–236.
2. Roser, D. J., S. A. McKersie, P. J. Fisher, P. F. Breen, and H. J. Bavor. "Sewage Treatment Using Aquatic Plants and Artificial Wetlands," *Water* 14(3):20–24 (1987).
3. Bavor, H. J., N. F. Millis, and A. J. Hay. "Assimilative Capacity of Wetlands for Sewage Effluent," Ministry for Conservation, Victoria, Environmental Studies Program ESP 363 (1981).

39I Fate of Microbial Indicators and Viruses in a Forested Wetland

Phillip R. Scheuerman, Gabriel Bitton, and Samuel R. Farrah

INTRODUCTION

Land application of sewage effluents and sludges offers many benefits, including water conservation by aquifer recharge, protection of surface waters from pollution, and supply of valuable nutrients to crops.[1] Wetlands have been suggested as an inexpensive means for tertiary treatment of sewage effluents.[2] However, concern regarding potential contamination of ground and surface waters with heavy metals, trace organics, nitrates, and microbial pathogens must be considered.

Although little is known regarding the fate of microorganisms in wetland systems, more is known about the fate of bacteria than viruses, and improvement in bacteriological water quality of sewage effluents has been observed. High initial levels of fecal coliforms (10^6 MPN/100 mL) were reduced to background levels (10^1 MPN/100 mL) in a mixed hardwood swamp near Wildwood, Florida.[3] Similar reductions occurred in a cypress strand near Gainesville, Florida.[4] Wellings et al.[5] found virus in a 10-m well, 7 m from the discharge point at the same site studied by Boyt.[3] Butner[6] studying two wetland systems—one the system described in this study—found a high removal of total and fecal coliforms with flow through the system.

In contrast, studies in a peat marsh in central Michigan observed poor removal of indicator bacteria.[7] Viruses were present in two aeration ponds but were below detection limits in a pilot wetland irrigation site and a natural wetland. Vaughn and Landry[8] also found poor removal of bacteria and viruses in two artificial wetland systems.

MATERIALS AND METHODS

Site Description

In 1934, the town of Waldo, Florida started treatment of its wastewater in a large concrete septic tank. Septic tank effluent flows through a ditch and into a 2.6-ha cypress wetland and subsequently flows downstream into a 607-ha cypress strand.

To obtain an unimpacted cypress strand for studies, septic tank effluent was pumped 1.6 km and discharged into two parallel corridors. These corridors were constructed with fiberglass panels. The three-sided corridors (exit end open for outflow) were 10 m wide and 40 m long. Fiberglass panels were 1.5 m in height above the peat surface and were inserted 0.5 m into the peat (below peat surface) to restrict subsurface outflow. Septic tank effluent was applied continuously at 49 L/min from May 1 to November 18, 1983.

Monitoring of Bacterial Indicators

Total coliforms, fecal coliforms, and fecal streptococci in water were enumerated using membrane filter procedures.[9] Sediment samples (either 1 or 11 g) were mixed with 99 mL of phosphate-buffered saline (PBS). Dilutions of the mixtures in PBS were added to tubes in a most probable number (MPN) procedure, using five tubes per dilution.[9]

Monitoring of Bacteriophages

Bacteriophages in the water were isolated using the magnetic-organic flocculation (MOF) procedure[10] and assayed using the double layer technique[11] with *Escherichia coli* C-3000 as host bacteria. Bacteriophages were recovered from sediment by mixing 10 g of wet sediment with 100 mL of 1% purified casein, pH 9.0. Tween-80 was added to a final concentration of 0.1%. The pH of the mixture was adjusted to pH 9.0 with 1 M glycine, pH 11.5, and stirred for 15 min. Following centrifugation at 4080 $\times$ g for 5 min, the supernatant fraction was adjusted to 150 mg/L magnetite, and the pH was lowered to 4.5 with 1 M glycine, pH 2.0. An additional 50 mg/L magnetite was added, and the sample was settled for 30 min on a magnet. Flocs were collected and processed as described for bacteriophages in water.

Monitoring of Enteroviruses

Enteroviruses were monitored using the method described previously.[12] Enteroviruses in 37–300 L of water were recovered by adsorption onto 25.4-cm Filterite (Filterite Corp., Timonium, MD) filters (0.45 μm or 0.25 μm). The filter was transferred to the laboratory on ice, and enteroviruses were eluted from the filter with 800 mL of 0.6 M sodium trichloroacetate + 0.1 M lysine,

Table 1. Levels of Total and Fecal Coliforms (CFU/100 mL) in Experimental Corridors Before, During, and After Addition of Sewage

Prepumping[a]		Pumping[b]		Postpumping[c]	
Total Coliform	Fecal Coliform	Total Coliform	Fecal Coliform	Total Coliform	Fecal Coliform
Corridor A					
Station 1					
8.7×10^5	15	2.2×10^7	7.5×10^6	3.0×10^2	3.0×10^2
Station 3					
2.3×10^2	25	1.4×10^6	3.6×10^5	1.3×10^3	2.0×10^2
Corridor B					
Station 1					
1.5×10^2	21	9.2×10^6	7.9×10^6	1.6×10^3	4.0×10^2
Station 3					
4.6×10^2	25	6.0×10^5	1.8×10^5	7.0×10^2	3.0×10^2

[a]Prepumping analysis conducted April 1980.[6]
[b]Pumping analysis mean values May to November 1983.
[c]Postpumping analysis five months after pumping ended (April 1984).

pH 9.0. The sample was then concentrated using inorganic flocculation with 1 M aluminum chloride, followed by organic flocculation with 5% beef extract. The pellet was suspended in a small volume of fetal calf serum and assayed for viruses using tube cultures of Buffalo Green Monkey (BGM) kidney cells. Tubes showing cytopathic effects were frozen and passed on BGM cells to confirm the presence of viruses. The 50% tissue culture infective dose ($TCID_{50}$) was calculated according to the procedure of Reed and Muench.[13]

RESULTS

Fate of Bacterial Indicators and Phages After Pumping Ended

Survival of total coliforms, fecal coliforms, and bacteriophage was determined by sampling in the experimental corridors five months after sewage addition had ended (Tables 1 and 2). The concentration of total coliforms was similar to concentrations reported by Butner[6] before sewage addition. Bacte-

Table 2. Levels of Bacteriophage in Experimental Corridors During and After Addition of Sewage

	Water (PFU/L)		Sediment (PFU/g dry sediment)	
	Pumping	Postpumping	Pumping	Postpumping
Corridor A				
1	4.2×10^5	9.6×10^4	3.8×10^2	1.4×10^2
3	6.1×10^4	2.6×10^4	1.9×10^2	ND[a]
Corridor B				
1	3×10^5	5.0×10^3	3.4×10^2	2.3×10^2
3	1.1×10^5	3.4×10^3	4.1×10^1	ND

[a]ND = none detected.

Table 3. Mean Levels of Indicator Bacteria in Water and Sediments Downstream from the Sewage Outfall

Station[a]	Total Coliform	Fecal Coliform	Fecal Streptococci
Water (CFU/100 mL)			
Outfall	1.8×10^7	5.0×10^6	6.0×10^4
S_0	5.7×10^6	2.6×10^6	5.9×10^4
Underpass	3.3×10^6	7.7×10^5	3.6×10^4
Midcanal	3.5×10^3	1.0×10^3	4.2×10^2
Culvert	1.9×10^2	3.5×10^2	4.0×10^2
Sediment (MPN/g dry sediment)			
Outfall	7.2×10^6	7.0×10^6	5.2×10^7
S_0	2.9×10^4	2.9×10^4	1.8×10^5
Underpass	1.4×10^5	1.6×10^4	3.2×10^4
Midcanal	ND[b]	ND	9.5×10^2
Culvert	2.7×10^2	1.1×10^2	6.0×10^4

[a]Each station reflects some distance further downstream from the outfall.
[b]ND = none detected.

riophage concentrations in water did not decrease after sewage addition was stopped. Two samples were taken from a strand water control site during the pumping period. One sample contained no detectable bacteriophage, and the second contained 4.2×10^2 PFU/L. Sediment samples taken from the control sight contained no detectable bacteriophage. Bacteriophage in sediment samples taken after sewage addition was stopped remained high at station 1 in the corridors and were not detectable at the outlet of the corridors.

Removal of Bacterial Indicators, Enteroviruses, and Phages Downstream from the Outfall

Water samples taken at various locations (0–1.6 km) downstream from the outfall contained high concentrations of bacterial indicators at stations near the outfall and declined rapidly with distance from the outfall (Table 3). Total and fecal coliforms were high in sediment samples near the outfall and declined similarly with distance. Fecal streptococci in sediments remained high at all stations.

Bacteriophages and enteroviruses declined in the water with distance from the outfall (Table 4). Bacteriophage concentrations also declined in the sediments with distance from the outfall. Enteroviruses were found only in sediments at the outfall.

Decay Rates of Laboratory Strains in Survival Chambers and Microbial Indicators in Experimental Corridors

Decay rates were calculated for laboratory strains contained in McFeters[14] type survival chambers suspended in the experimental corridors. Decay rates were also calculated for bacterial indicators monitored in the experimental corridors.[12] Laboratory strains were removed faster than the microbial indicators (Figure 1). The trend for each group was the same. Fecal streptococci and

Table 4. Levels of Bacteriophages and Enteroviruses in Sediments Downstream from Sewage Outfall

Station[a]	Bacteriophage	Enterovirus
	PFU/L	$TCID_{50}/L$[b]
Water		
Outfall	5.7×10^4	1.1×10^4 (100)
S_0	4.6×10^4	0.26
Underpass	2.8×10^6	3.7×10^2 (100)
Midcanal	2×10^2	ND[c]
Culvert	1.2×10^2	ND
	PFU/g dry sed.	$TCID_{50}$/g dry sed.
Sediment		
Outfall	4.8×10^2	2.3 (50)
S_0	ND	—[d]
Underpass	4.1×10^2	ND
Midcanal	7.1	ND
Culvert	ND	ND

[a]Each station reflects some distance further downstream from the outfall.
[b]Numbers in parenthesis are percent samples positive when multiple samples analyzed.
[c]ND = none detected
[d]Not done

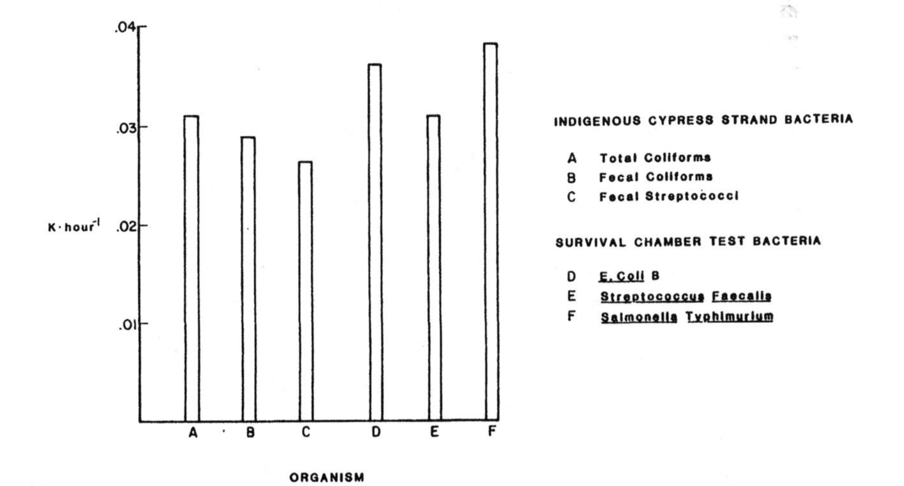

Figure 1. Decay rates for indigenous and laboratory bacteria in experimental corridors.

Streptococcus faecalis were removed the slowest. *Salmonella typhimurium* was removed the fastest.

DISCUSSION

We have previously shown that microbial indicators are removed with sewage flow through the experimental corridors and strand B.[12] The results reported herein, similar to results of other researchers,[15] suggest that initial rapid removal is followed by a slower decline and that part of the microbial population is able to undergo long-term survival.

Previous results[12] that fecal streptococci accumulated in sediments of the experimental corridors were confirmed in this study. Decay rates showed that fecal streptococci and *S. faecalis* are removed the slowest. Our earlier conclusion that fecal streptococci may be a more conservative indicator for fecal pollution is reinforced.

Bacteriophage were removed at a slower rate than bacterial indicators and enteroviruses. Enteroviruses appeared to decline at the greatest rate and did not accumulate in the sediments.

In summary, with sufficient time and distance, microbial quality of sewage effluents is improved after discharge into wetlands. Since at least part of the microbial population can undergo long-term survival, a more conservative indicator may be prudent.

ACKNOWLEDGMENT

Although the research described in this chapter has been funded by the U.S. Environmental Protection Agency under grant number R809402, it has not been subjected to the agency's review and therefore does not necessarily reflect the views of the agency, and no official endorsement should be inferred. The authors would like to thank R. J. Dutton, John Bossart, John Daron, and Jeff Kosik for excellent technical assistance.

REFERENCES

1. Thabaraj, G. J. "Land Spreading of Secondary Effluents," *Florida Scientist* 38:222–227 (1975).
2. Odum, H. T., and K. C. Ewel. "Cypress Wetlands for Water Management, Recycling and Conservation," 3rd Annual Report to NSF and Rockefeller Foundation by the Center for Wetlands, University of Florida, Gainesville, FL (1976).
3. Boyt, F. L. "A Mixed Hardwood Swamp as an Alternative to Tertiary Wastewater Treatment," Master's Thesis, University of Florida, Gainesville, FL (1976).
4. Fox, J. L., and J. Allison. "Coliform Monitoring Associated with the Cypress Dome Project," 3rd Annual Report to NSF and Rockefeller Foundation by the Center for Wetlands, University of Florida, Gainesville, FL (1976).

5. Wellings, F. M., A. L. Lewis, W. W. Mountain, and L. V. Pierce. "Demonstration of Virus in Groundwater After Effluent Discharge onto Soil," *Appl. Microbiol.* 29:751–757 (1975).
6. Butner, J. "Public Health Aspects of Wastewater Recycling Through Wetlands," Master's Thesis, University of Florida, Gainesville, FL (1983).
7. Kadlec, R. H., D. L. Tilton, and B. R. Schwegler. "Wetlands for Tertiary Treatment: A Three-Year Summary of Pilot Scale Operations at Houghton Lake," NSF-RANN AEN 75–08855, The University of Michigan Wetlands Ecosystem Research Group, School of Natural Resources, College of Engineering, Ann Arbor, MI (1979).
8. Vaughn, J. M., and E. F. Landry. "Data Report: The Fate of Human Enteric Viruses in a Natural Sewage Recycling System," Report BNL 51281, Department of Energy and Environment, Brookhaven National Laboratory, Upton, NY (1980).
9. *Standard Methods for the Examination of Waters and Wastes,* 15th ed. (Washington, DC: American Public Health Association, 1982).
10. Bitton, G., L. T. Change, S. R. Farrah, and K. Clifford. "Recovery of Coliphages from Wastewater Effluents and Polluted Lake Water by the Magnetite-Organic Flocculating Method," *Appl. Environ. Microbiol.* 41:93–96 (1981).
11. Adams, M. H. *Bacteriophages* (New York: John Wiley & Sons, 1959).
12. Scheuerman, P. R., S. R. Farrah, and G. Bitton. "Reduction of Microbial Indicators and Viruses in a Cypress Strand," *Water Sci. Technol.* 19:539–546 (1987).
13. Reed, L. J., and H. Muench. "A Simple Method for Estimating Fifty Percent Endpoints," *Am. J. Hyg.* 27:493–497 (1939).
14. McFeters, G. A., and D. G. Stuart. "Survival of Coliform Bacteria in Natural Waters: Field and Laboratory Studies with Membrane-Filter Chambers," *Appl. Microbiol.* 24:805–811 (1972).
15. Liang, L. N., J. L. Sinclair, L. M. Mallory, and M. Alexander. "Fate in Model Ecosystems of Microbial Species of Potential Use in Genetic Engineering," *Appl. Environ. Microbiol.* 44:708–714 (1982).

39m Wastewater Wetlands: User Friendly Mosquito Habitats

Charles H. Dill

I have borrowed a term from computer science to emphasize the problem inherent in treated effluent wetlands. We are all familiar with the meaning of *user friendly* regarding computers but not with regard to mosquito control. For us, it means "immediately attractive as an oviposition site for female mosquitoes." When in need of large numbers of mosquito eggs or larvae for test purposes, we commonly prepare a container of dechlorinated water that is combined with cut grass or hay and allowed to "season" for a few days in a warm laboratory. In a few days of outdoor exposure, thousands of eggs or larvae can be collected.

In California, as in most areas, toxic chemicals and water pollution are major concerns. Additionally, habitat destruction in general and loss of wetlands in particular are of increased importance to Californians. Seasonal zero discharge of treated effluent and use of that effluent to construct wetlands is a marriage that seemingly solves two of today's major concerns. Zero discharge of effluent to freshwater streams during periods of low stream flow protects the riverine biotic community as well as residential water sources for many communities. Preservation of biotic communities and protection of sport and commercial fisheries are the major goals of a zero discharge policy around bay and coastal wetlands. Sanitary districts are, by necessity, adjacent to the areas they serve, and thus constructed wetlands are close to both plant and residential communities. Early effluent disposal projects were designed to accommodate the anticipated dry weather flow by buying or contracting for the necessary acreage needed for irrigation, containment, or both.

Today wildlife habitat needs are at the top of the list of project goals. We agree with this priority as long as mosquito prevention design (Figure 1) and water management considerations are part of the overall project plans. Where this has been done, mosquito problems, while present, are minimal and manageable. Along with design and water management, vegetation selection is also important. Many of the vegetation selection principles delineated by U.S. Public Health Service and TVA scientists for malaria control in the late '40s are

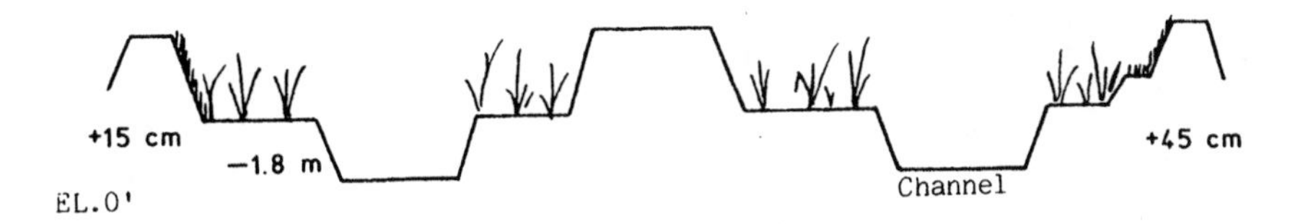

Figure 1. Wildlife marsh profile. From "O and M Manual," Las Gallinas, S.D. Effluent Disposal Project.

still valid today.[1] With the ability to raise, lower, and if necessary eliminate the water, vegetation selection becomes less critical. Most types of emergent vegetation are tolerable if water levels can be drawn down below the vegetation and a good population of mosquito fish is present. If projects are designed with diversity in mind, the combination of predators present can keep mosquito populations at tolerable levels or at levels that require only small amounts of pesticides to achieve control.

Communities in the San Francisco Bay area are surrounded by critical habitat (Figure 2) and must serve large human populations. Many and varied approaches to wastewater treatments are underway, and each presents a

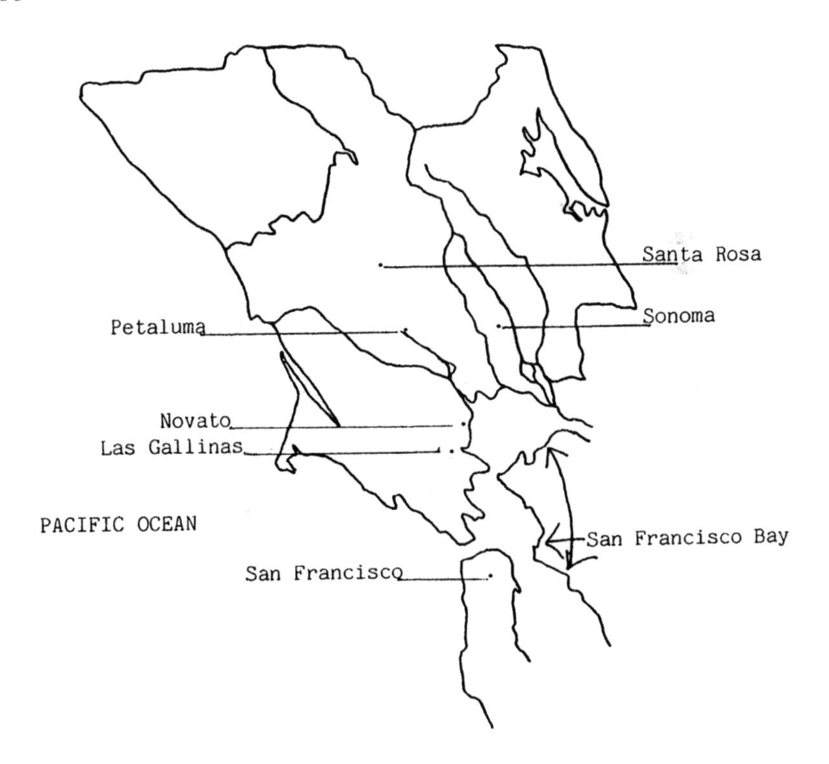

Figure 2. San Francisco Bay area.

Table 1. Marin and Sonoma Sanitary Districts

District	Pop. (thous.)	Flow (1000 m^3d)	Irrigation (ha)	Wildlife (ha)	Storage (1000 m^3)
Marin County					
Las Gallinas	27.0	7.6	89	8	45
Novato	54.0	15.9	332	4	1110
Sonoma County					
Santa Rosa	175.0	64.3	1768	2	5605
Sonoma	27.0	11.4	607	61	537
Petaluma	40.0	10.6	223	113	2072

unique set of problems. In Marin and Sonoma counties, within which lies my 2486-km^2 mosquito control district, there are five (and soon six) major treated-effluent irrigation projects. All of the projects have a major impact on mosquito control due to their proximity to large residential populations (Table 1). The problem created by treated-effluent wetlands in the San Francisco Bay area is that they provide breeding habitat for the encephalitis mosquito (*Culex tarsalis*) and extend the breeding season into the warmer summer months when the possibility of encephalitis virus propagation and transmission is enhanced. The diseases of concern are western equine encephalomyelitis (WEE) and St. Louis encephalomyelitis (SLE). These diseases occur naturally in wild birds and are transmitted from bird to bird by mosquitoes. Humans and, in the case of WEE, horses are dead-end hosts, i.e., the disease cannot be picked up and transmitted by a mosquito feeding on an infected host. Humans and horses can and do, however, fall victim to the disease. Without preventive design criteria and water management, projects could pose public health risks by the end of the first irrigation cycle. Table 2 illustrates the impact these projects have on mosquito populations. The Las Gallinas project had considerable input from my staff prior to implementation. The relatively minor increase in the adult mosquito population in June and July is tolerable because it occurs during a period of low viral activity, creating only a localized pest problem. Population numbers for the other months do not differ appreciably. The pond design in Figure 1 was utilized, and no significant mosquito production has

Table 2. Average Daily Adult *Culex* Caught per Trap Night

	Santa Rosa Irrigation			Las Gallinas Irrigation		
	Pre	Post		Pre	Post	
	1982	1986	1987	1982	1986	1987
June	0.0	—[a]	26.6	0.0	5.1	4.4
July	0.0	32.2	42.0	0.7	7.5	4.0
August	0.3	34.5	51.9	14.4	7.4	13.4
September	4.1	15.8	37.1	5.1	5.6	8.1
October	0.9	7.4	4.3	2.1	1.3	12.7

Note: Standard New Jersey light trap.

[a]Trap relocation.

occurred. Problems have only been encountered in areas where vegetation along channels has prevented larval predation by mosquito fish and where ditch design prevented good drainage. As problems arose they were reported to sanitary district staff, and problems have been corrected.

The Santa Rosa project was completely planned and constructed before we became involved. The basic need met by the project was enough acres to irrigate in order to dispose of the effluent generated. Numbers of mosquitoes involved are tremendous. Additional live trapping using CO_2 baited traps for viral surveillance commonly produced catches of from 500 to 600 female *Culex tarsalis* per night in 1986. In 1987, cooperative planning sessions with the agencies involved and source reduction efforts by my staff saw those numbers drop to approximately 50 per night. Still high, but significantly better. In 1987, we cleaned or constructed 10,700 m of drainage ditches to move runoff water to a return pond, where it can be used again for irrigation. Return ponds are stocked with mosquito fish and have not created any problems to date as long as the fish survive and vegetation is managed. While the cooperative efforts of the agencies involved has greatly reduced the existing mosquito problems, much work is still needed. In 1987, another aspect of viral surveillance, a state health department sentinel chicken flock, showed signs of WEE virus activity in the irrigation project. Our efforts this year will see approximately 15,000 m of new ditches emplaced for water control. To date, the Santa Rosa project has only 2 ha of wildlife wetlands apart from the crop irrigation. However, we are working on plans for an 81-ha parcel to be used for wildlife involving irrigation, a collection-and-return system, and fish-rearing facilities.

In conclusion, it is evident to us that early input by mosquito-control professionals can keep an environmental asset from becoming a public health risk. Good preventive design coupled with water management and vegetation control will normally be enough to minimize mosquito problems. Continued surveillance is necessary, though, because problems at treatment plants or unusually hot weather can easily kill off resident mosquito fish populations, resulting in emergence of very large numbers of mosquitoes in a short period of time. Mosquito control must be a basic element of the preproject planning as well as in operation and management documents for all treated-effluent systems.

REFERENCE

1. *Malaria Control on Impounded Water* (Washington, DC: U.S. Public Health Service and TVA Health and Safety Department, 1947).

CHAPTER 40

Treatment of Nonpoint Source Pollutants—Urban Runoff and Agricultural Wastes

40a Development of an Urban Runoff Treatment Wetlands in Fremont, California

Gary S. Silverman

INTRODUCTION

A 20-ha fallow agricultural field in Fremont (Alameda County), California was converted into a seasonal freshwater wetlands. Impetus for constructing this system was multiobjective, but driven by recognition that urban stormwater runoff is a significant pollution source to San Francisco Bay. Developing wetlands to treat urban stormwater runoff presents a different set of problems than developing a system to treat wastewater. This chapter will address these differences and creation of a particular wetlands.

There is a growing body of literature describing techniques for constructing wetlands to treat wastewater. However, little has been done to develop wetlands to treat stormwater runoff. From a technical perspective, both the chemical and hydrological nature of input differs dramatically between urban runoff and municipal wastewater treatment wetlands systems. Municipal wastewater (from an area with separate storm and septic systems) tends to have a consistent flow with characteristic water quality. Changes in flow or quality normally follow well-defined limits after precipitation events or discharge of industrial waste into municipal sewers. Biological degradation processes can be sized and maintained to optimize removal efficiency. Urban storm water, however, is quite variable, generating large flows during and after precipitation events, with rapid changes in quality not exclusively dependent on flow. Biota characteristic of a stormwater treatment wetlands may be present more as a function of the biota's ability to tolerate extremely variable conditions than its ability to make optimal use of energy sources supplied in the pollutant stream.

A nontechnical but no less important obstacle to development of urban stormwater treatment wetlands is the lack of responsibility for controlling pollution from urban nonpoint sources. There is no one to select urban stormwater wetlands as the preferred treatment technology, fund and develop an appropriate system, and be responsible for operations and maintenance. However, this may be changing. The Water Quality Act of 1987, while postponing promulgation of U.S. Environmental Protection Agency (EPA) stormwater regulations, established deadlines for development of a comprehensive program to control stormwater point source discharges. Implementation may create a demand for control measures to improve urban stormwater runoff quality and provide advocates for treatment wetlands. Data from the system described in this chapter should prove useful in assessing the potential of urban runoff treatment wetlands for water quality benefits, cost effectiveness, and other community benefits.

BACKGROUND

Interest in developing this urban stormwater treatment wetlands grew out of work started in the mid-1970s by the Association of Bay Area Governments (ABAG). ABAG is the designated agency in the San Francisco Bay area for regional water quality planning as specified by Section 208 of the Federal Water Pollution Control Act of 1972 (FWPCA). A significant threat to the health of the Bay results from nonpoint source runoff from urban areas, and existing programs do little to solve the problem. This project was initiated to explore the potential of wetlands to mitigate effects of urban stormwater runoff.

While local stormwater runoff is responsible for a small percentage of total flow to the Bay, it contributes substantially to pollutant loading (Table 1).[1] Local, primarily urban, nonpoint sources contribute over one-third of the heavy metal load reaching the Bay. Substantial quantities of oil and grease from urban drainages discharge into the Bay, with estimates of 2400 to 4500 metric tons from this source during the year 1985, increasing to between 2800 and 5100 metric tons by the year 2000.[2] In comparison, about 5200 metric tons of oil and grease entered the Bay in 1982 from all point sources.[3]

While not contributing much of the total pollutant load to the Bay, other pollutants from urban runoff adversely affect localized areas. For example, several shellfish beds in San Francisco Bay have elevated bacterial levels due to urban runoff.[4] Control of urban stormwater runoff appears vital to protect vulnerable offshore areas and to limit the total mass of pollutant discharge.

Table 1. Pollutants to San Francisco Bay (Percent Contribution for Year 1800, 1978, 2000)

Source	Flow			SS			BOD_5			Total N			Total P			Metals		
Surface runoff	2	4	6	5	6	40	3	8	16	5	6	7	9	4	3	13	42	37
Aerial fallout	2	2	2	2	2	3	6	2	4	4	3	3	<1	1	1	<1	6	3
Delta outfall	96	91	83	94	72	56	91	32	35	95	35	22	91	19	9	77	49	26
Point sources	<1	3	7	<1	10	1	<1	58	44	<1	56	68	<1	77	87	<1	2	34

TREATMENT MECHANISMS

Wetlands enhance water quality through a variety of physical, chemical, and biological processes that trap and degrade pollutants. The physical processes of sedimentation, adsorption, and filtration are key in capturing pollutants. Once in the system, pollutants may be degraded, stored indefinitely, or removed periodically (usually through dredging).

Sedimentation within most wetlands is responsible for removal of suspended material. However, urban runoff typically contains very fine particles not effectively removed by sedimentation within reasonable wetlands system size constraints. Filtration and adsorption as water passes through vegetation and infiltrates through the soil provide alternative means for particle removal. However, if soils are very fine clays, as near San Francisco Bay, infiltration will be minimal. This situation contrasts with many wetland wastewater treatment systems: influents contain larger sized suspended particles, options exist for adding coagulants to assist flocculation of fine suspended material, and the site may be developed on permeable soils (with impermeable deposits dredged periodically to facilitate infiltration).

Traditional biological treatment systems rely on degradation of organic pollutants by well-conditioned flora. Wastewater treatment wetlands, receiving a consistent influent, provide an analogous situation. In a well-operated system, degradation rates are equivalent to pollutant input rates. In a wetlands receiving urban runoff, input will occur episodically. Full treatment of pollutants entrained in the water column ("plugs" passing through the system) cannot be expected. Pollutants must be captured, with most degradation occurring later during dry periods with little or no flow. Thus, more design emphasis needs to be on retaining pollutants for subsequent degradation in an urban runoff treatment wetlands than in a wetlands receiving wastewater.

CASE STUDY

A wetlands was developed in Fremont, California as part of Coyote Hills Regional Park along the southwestern border of San Francisco Bay. The site is about 20 ha, partly owned by the East Bay Regional Park District and partly by the Alameda County Flood Control District, with lease-back to the Park. While the system concept grew out of the need to control urban stormwater runoff, identification of other wetlands values was necessary to obtain project support. Institutional arrangements made to provide resources and permits for the project and key design features are described below.

Institutional Considerations

The wetlands was conceived as a prototype system and research facility to study wetlands creation for stormwater treatment in the Bay Area. The site for

Table 2. Funding Sources for Wetlands Development

California License Plate Fund	$ 10,000	Planning & Design
U.S. EPA	42,500	Planning & Design
Association of Bay Area Governments	7,500	Planning & Design
East Bay Regional Park District	10,000	Planning & Design
California License Plate Fund	200,000	Construction
East Bay Regional Park District		Use of parkland
Alameda County Flood Control District		Use of flood basin

the wetlands, part of Coyote Hills Regional Park, facilitated obtaining resources from sources interested in other wetlands values.

Funding sources for the system are shown in Table 2. The largest source of revenue, a California "Environmental License Plate Fund" grant administered by the Department of Parks and Recreation, was obtained in response to the potential of providing additional wetlands—not to the potential of the wetlands to mitigate water quality problems. Similarly, the East Bay Regional Park District's contribution was due more to its interest in developing parks than in dealing with long-term nonpoint source runoff problems. Only the U.S. EPA's and ABAG's interest derived from a water quality perspective.

Permits to develop the system are shown below.

Permit type:	*Permit activity:*
California Environmental Quality Act: Determination of Environmental Significance	Conditional negative impact declaration after modifying plan to ensure transitional area not lost.
Army Corps of Engineers Section 404 Permit	Concept plan submitted with negative declaration for public review.
California Department of Fish & Game Stream Alteration Permit	Concept plan submitted with negative declaration. Construction period limited to protect fish and wildlife.
Alameda County Flood Control District Encroachment Permit	Concept plan submitted with negative declaration. Reduction in storage capacity or flow prohibited.
City of Fremont Grading Permit	Exempt: no fill input or export from site.

The State Water Resources Control Board (SWRCB), the state agency responsible for setting state policy on water quality control and for water appropriations, is not listed. SWRCB lacks an applicable policy, reflecting not only the project's innovative nature but also the uncertainties facing a project proponent regarding state response to proposals. The San Francisco Bay Basin Plan, in describing policies to protect regional water quality, allows wastewater use to create or enhance wetlands with specific restrictions. No reference exists to wetlands created from stormwater runoff or to policies for diverting stormwater runoff for this purpose. A letter from the executive director of the San Francisco Regional Water Quality Control Board (the agency responsible for implementing and enforcing SWRCB policy in the region), claims that newly created marshland used to treat urban runoff may not be considered water of the state (thus minimizing SWRCB responsibility), although discharges must meet requirements consistent with Basin Plan objectives.[5]

As stormwater discharges become subject to National Pollutant Discharge Elimination System (NPDES) requirements under developing U.S. EPA regulations, performance and discharge requirements may be clarified. Paradoxically, stringent discharge requirements could discourage collection of nonpoint source discharges to create a treatment wetlands.

Three major concerns were raised during permitting of the wetlands. The local mosquito abatement district expressed concern that the salt marsh mosquito (*Aedes squamiger*) might become a nuisance. This concern was mitigated by planning for mosquito fish (*Gambusia affinis*) introduction, creating deep ponds to sustain fish habitat during the dry season, and providing vehicle access should other treatment be necessary. The Alameda County Flood Control District (ACFCD) expressed concern that no flood control capacity be lost from the site. Design included a net removal of material from the flood basin and providing adequate flow capacity to decrease risk of upstream flooding. The California Department of Fish and Game (CDF&G) was concerned that habitat transitional between aquatic and terrestrial systems would be lost. This concern was mitigated with "before" and "after" maps showing that transitional and open water areas would be increased at the expense of uplands.

The key to successful permitting of the project was good communications between interested parties rather than strict adherence to formal policies. For example, CDF&G did not have guidelines specifying preference of transitional areas over uplands, so this concern was not identified until meetings were held. Lack of formal policies led to some delays and potentially could lead to inconsistencies among permitting decisions.

Design

Before development into the urban runoff treatment wetlands, the site contained an abandoned agricultural field, a dense willow grove, an area of pickleweed (*Salicornia virginica*), and a meandering slough with no surface outlet, which drained a small agricultural area. Water was diverted onto the site from Crandall Creek, draining a 12-km^2 area characterized by 75% suburban/residential development and 25% agricultural and open space.

Three distinct systems were incorporated into the wetlands to test performance of different designs. Influent is diverted fairly equally into two initial systems. One is a long, narrow pond containing a long island. Considerable area was devoted to shallow edges to encourage growth of rooted aquatic vegetation (mainly cattails, *Typha latifolia*). The other system is more complex, using a spreading pond draining into an overland flow system (inundated only during storms), followed by a pond with berms supporting rooted aquatic vegetation. This system allows testing of water quality effects of overland flow characterized by different vegetation and flow patterns than those of the pond and effects of "combing" water through cattail stands.

These systems drain into a common third system, which provides an area of shallow, meandering channels, maximizing contact with various types of wetlands vegetation. The discharge is into another section of Coyote Hills Regional Park and flows back into the channel that Crandall Creek discharged into before diversion. Hydraulic considerations included sizing the diversion structure and channels to accommodate the 10-yr, 6-hr storm, with greater flows causing diversion structure failure with most of the flow remaining in Crandall Creek.

POTENTIAL FOR URBAN STORMWATER TREATMENT WETLANDS

Development of this wetlands development has a number of benefits. An attractive wetlands has been created in an urbanized region badly needing additional "natural" areas, and a facility to research the potential and future designs for urban runoff treatment systems has been provided. Another important benefit is the practical demonstration for implementation of other wetlands development projects.

Federal and state agencies have limited ability to actively promote or develop urban stormwater treatment wetlands and instead serve in a regulatory or review capacity. Local governments and special interest groups, by developing systems that satisfy objectives of several groups, have a much greater potential for creating urban stormwater treatment wetlands.

Local governments typically have policies that favor park development and give high priorities to maintaining aesthetic environments and open spaces and conserving natural resources. For example, in the Alameda County General Plan, 10 open space objectives have direct applicability to wetlands. Added incentive for local government action should develop as stormwater discharges become subject to NPDES regulations. While wetlands creation may have been viewed by local government as a luxury, new regulatory obligations might foster active efforts to develop systems that meet water quality needs coincident with other objectives.

A variety of environmental and other special interest groups may be active participants in wetlands creation and restoration projects. For example, in the San Francisco Bay area a "wetlands coalition" has been formed, including 24 private groups and two government agencies. These groups rarely have substantial financial resources, but they serve as project advocates, provide technical assistance, and may be instrumental in maintaining a facility.

Development of this wetlands required active efforts from the regional council of governments and park district, financial support from the state and federal governments, and cooperation from a host of local agencies and special districts. It demonstrated the need and potential for obtaining diverse support of wetlands projects. One group may not be successful in developing a wetlands to meet a single goal without promoting multiple values and integrating efforts with other groups. Anticipated regulatory requirements for stormwater discharges will encourage local governments to develop treatment systems, and local governments may provide significant project funding. Determining construction and operating criteria, including design controlled trade-offs between different wetlands values, remains challenging. Monitoring the Fremont wetlands systems should provide some answers. Clearly, urban stormwater wetlands systems provide an outstanding opportunity to meet water quality objectives and provide a multitude of other benefits to society.

REFERENCES

1. Russell, P. P., T. A. Bursztynsky, L. Jackson, and E. Leong. "Water and Waste Inputs to San Francisco Estuary—A Historical Perspective," in *San Francisco Bay: Use and Protection,* W. Kockelman, T. Conomos, and A. Leviton, Eds. (American Association for the Advancement of Science, Pacific Division, 1984).
2. Silverman, G. S., M. K. Stenstrom, and S. Fam. "Evaluation of Hydrocarbons in Runoff to San Francisco Bay," Association of Bay Area Governments completion report to U.S. EPA under grant C06000-21, Oakland, CA (December 1985).
3. "Toxics in the Bay. An Assessment of the Discharge of Toxic Pollutants to San Francisco Bay by Municipal and Industrial Point Sources," Citizens for a Better Environment, San Francisco, CA (1983).
4. Jarvis, F. E., A. W. Olivieri, T. G. Rumjahn, M. J. Ammann, and S. R. Ritchie. "Preliminary Sanitary Survey Report for the East Bay Study Area," San Francisco Bay Shellfish Program, Regional Water Quality Control Board, San Francisco Bay Region (1981).
5. Dierker, F. H., Executive Director, California Regional Water Quality Control Board, San Francisco Bay Region to Terry Bursztynsky, Water Quality Program Manager, Association of Bay Area Governments, January 11, 1982.

40b Urban Runoff Treatment in a Fresh/Brackish Water Marsh in Fremont, California

Emy Chan Meiorin

WETLAND DESIGN

Temporary detention of stormwater runoff can improve water quality.[1] The detention system study was constructed for the "Demonstration of Urban Stormwater Treatment" (DUST) Marsh in Fremont, California. The wetland system and control structures were built in 1983 to receive runoff from approximately 1200 ha with land uses as follows: low-density residential, 66%; agricultural/open space, 28%; high-density residential, 5%; and urban road and commercial, 1%. The marsh was monitored during the wet seasons of 1984–85 and 1985–86 to document marsh development and treatment effectiveness.[2]

The wetland system covers approximately 22 ha and is divided into three separate subsystems (Figure 1). Major features are presented in Table 1 and described by Silverman.[3] Stormwater runoff enters the DUST Marsh at the 0.15-ha "debris basin," which retains gross debris and serves as a distribution structure for flows into the parallel A and B systems. Systems A and B flow into System C, which discharges to a parkland and wildlife habitat area that eventually drains into San Francisco Bay. Each system was designed to allow testing of soil/vegetation/flow configurations for comparative effectiveness at removing water pollutants. System A simulates pretreatment at a wastewater treatment plant by reducing suspended particles. System B provides a combination overland flow (soil-water interface) and pond system divided into four cells. Secondary treatment is accomplished by System C, which provides storage, detention between storms, and a place for biological activity to reduce suspended and dissolved constituents.

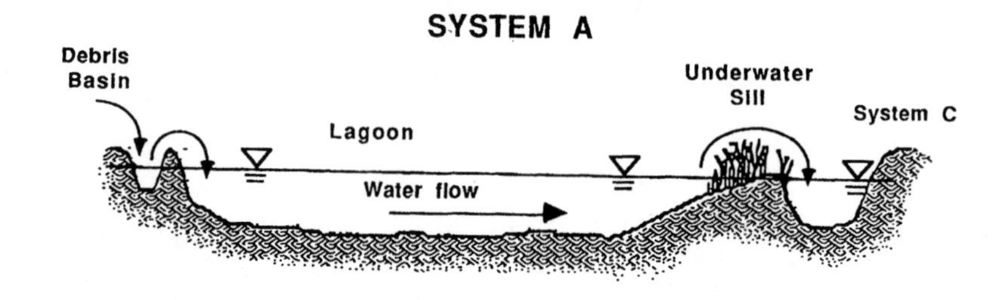

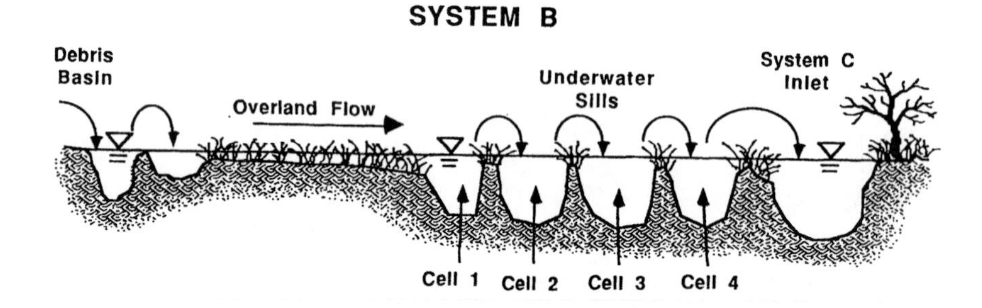

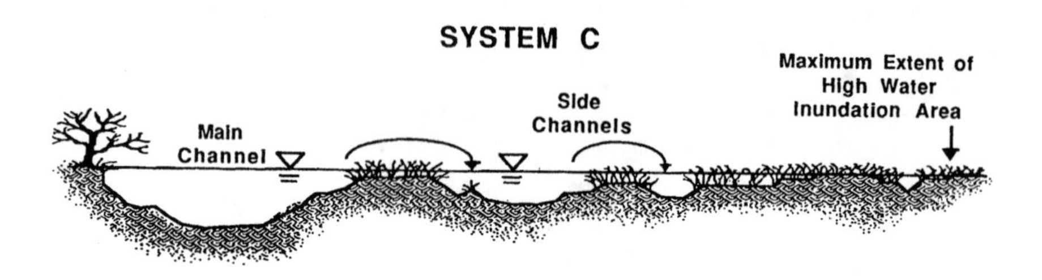

Figure 1. Schematic of DUST Marsh system types.

WETLANDS MONITORING PROCEDURES

Automatic instruments collected water stage data and discrete hourly water samples at various points in the marsh. Velocity and flow were calculated from water stage/flow rating curves developed during the project, and water samples were composited to reflect changes in flow conditions. Constituent mass loads passing through (1) the DUST Marsh inlet and (2) the outlets to Systems A, B, and C were computed, and changes in loads and efficiency of Systems A, B, and C and the DUST Marsh overall were determined.

Eleven storms were monitored between November 1984 and March 1986. Rainfall ranged from 0.88 cm to 4.4 cm per storm with storm durations of 2.7

Table 1. DUST Marsh System Types

System	Type	Size (ha)	System Description
A	Lagoon	2	Flow-through system with 1:4 slope on lagoon margins; 1.8 m maximum water depth (dry weather); 2.4 m maximum depth (wet weather).
B	Overland flow	1.6	Overland flow with 0.5% slope; 30 cm maximum depth (wet weather); dry between storms; ground with vegetated cover, 20%; open water, 25%; and bare ground, 55%.
	Pond with underwater sills	0.68	Linear pond transected by underwater sills to form 4 cells; average pond depth = 1.0–1.2 m; sill depth = 0.3–0.6 m; sills are vegetated with cattails that intercept flow through system.
C	Braided channels	8.5	Linear system with open water, 4.7 ha and vegetated area, 4.2 ha. Between storm water depth = 2.0 m; maximum wet weather depth = 2.3 m, with wet area = 6.5 ha; heavy cattail and alkali bulrush vegetation.

to 38.0 hr. Measured stormwater runoff volumes entering the marsh ranged from 13.4×10^3 to 69.2×10^3 m^3, with calculated peak flow rates of 0.22 to 1.2 m^3/s. Static storage capacity of the DUST Marsh is 71.7×10^3 m^3, which will hold the total flow from a large storm or several small, successive storms—assuming that previous storage is pushed out by recent stormwater inflows. Based on dye tracer studies, stormwater detention times were 5–48 hr in Systems A and B and 1–12 days in System C, which has slow-moving flows and abundant emergent vegetation. These detention times indicated some short-circuiting of stormwater inflows through Systems A and B but confirmed long-term detention within the marsh for one or more storm cycles.

WETLANDS TREATMENT PERFORMANCE

Trap efficiency was defined as the estimated long-term decrease in loads through the system expressed as a percentage of input load. Because the DUST Marsh could accumulate stormwater volumes through one or more storm cycles with up to several weeks' detention time, seasonal mass loadings were calculated as a total of all monitored storms during each wet season.

Suspended Solids and Heavy Metals

The large lagoon in System A and the broad wetland area in System C generally reduced suspended constituent loads. Trap efficiency for suspended constituents was 42–45% for solids, 30–83% for Pb, 40–53% for Cr, 12–34% for Ni, 6–51% for Zn, and 5–32% for Cu (Table 2). These reductions are probably due to the settling of heavier suspended particles that are affected by residence time and turbulence. System C exhibited the greatest efficiency, due to the long residence time and vegetative uptake of metals in the littoral zone

Table 2. DUST Marsh Trap Efficiencies (percent)

Constituents	System A	System B	System C[a]	Overall[b]
TDS	-9	-20	-50	-49
TSS	42	24	45	64
BOD_5	-26	-22	-8	-35
NH_3-N	-22	27	12	10
NO_3-N	9	5	8	15
TKN	7	-32	-17	-28
Orthophosphate	53	19	28	56
Total phosphate	17	-44	51	48
Chromium (Cr)	40	20	53	68
Copper (Cu)	5	-10	32	31
Lead (Pb)	30	27	83	88
Manganese (Mn)	-22	-1	-86	-111
Nickel (Ni)	34	-30	12	20
Zinc (Zn)	6	-22	51	33

[a]System C inflow = composite of System A and B outflows.
[b]Overall trap efficiencies may be greater than cumulative reductions by individual systems because System C provides secondary treatment for System A and B discharges.

or by absorption in the sediments. Lead, in particular, is less soluble compared to Zn and exhibits little remobilization once deposited. The greatest Pb reduction (83%) was observed in System C, where soil concentrations of Pb were 1.5–2.5 times higher than those in System A or B.

System B, the combination overland flow and pond system, frequently had larger loads of metals leaving the system than entering—particularly for Ni, Zn, and Cu. These metals probably exist in particulate form in the watershed and are washed into the drainage channels, where they are eventually deposited in the marsh. The more soluble metals equilibrate into the dissolved form throughout the water column and the next storm flushes the water out of the marsh, resulting in more dissolved Ni, Zn, and Cu leaving System B than entering. Turbulent flows between cells of System B may also scour pond bottoms and resuspend bottom sediments, reducing pond efficiency. Manganese is also relatively soluble, contributing to higher levels of dissolved solids leaving than entering the marsh. However, stormwater inputs and sediment remobilization are probably not sufficient to cause significant changes in Mn levels. Concurrent with Mn level increases from System A to System C is a rise in electrical conductivity (2000–6000 μmhos/cm) in System C. Electrical conductivity and chlorides in the groundwater zone directly below the marsh indicate that the groundwater is saline (15,000–25,000 μmhos/cm) and is probably contiguous with the marsh at System C. Thus, there is a gradation from freshwater conditions at the System A and B inlets to brackish-water conditions at the System C outlet.

Nutrients

Nitrogen is transported into the Marsh primarily as ammonia and organic nitrogen. Trap efficiency for ammonia in Systems B and C was moderate

(12–27%) due to plant uptake and absorption of ammonia into the sediments. Concentrations of nitrate were relatively low, and corresponding efficiencies were less than 10%. Relatively high efficiencies were observed for orthophosphate (53%), which may be used by vegetation or may react with clay in pond sediments. On the other hand, turbulent flows in the System A lagoon and between pond cells in System B probably lead to flushing of suspended phosphorus and organic N such that organic N concentrations were often greater than influent levels.

Uptake of nutrients in the project area was low because most storms occurred in winter, when plant growth was reduced by ambient temperatures, short day lengths, and low light levels. Conversely, senescence of annual and some perennial plants during the winter increases the littoral zone and organic N, P, and other materials. Organic material, as measured by BOD_5, showed a net increase through all systems (8–26%), indicating a resuspension of plant material and detritus.

ACCUMULATIONS OF HEAVY METALS IN MARSH SYSTEM

Plant species, cattail (*Typha latifolia* and *T. angustifolia*) and alkali bulrush (*Scirpus robustus*), were sampled and analyzed for uptake and accumulation of metals Cd, Cr, Pb, Mn, Ni, and Zn. Plant samples were segregated into root/rhizome, leaf/stem, and fruit/seed parts for analyses. Soil samples from the upper 5 cm layer near the sampled plants were also analyzed. Filter feeders—Sacramento blackfish (*Orthodon microlepidotus*)—and bottom detritus feeders—carp (*Cyprinus carpio*)—were collected and analyzed for metals in flesh and liver tissues. Three other species occurring in the marsh—mosquito fish (*Gambusia affinis*), threespine stickleback (*Gasterosteus acculeatus*), and sculpin (*Cottus* sp.)—were analyzed when available. A comparison of bioaccumulation of heavy metals among various sample media is shown in Figure 2. Cadmium levels were generally at or below the detection limit (<0.002 mg/L) and are not included. The bioaccumulation index is based on the soil concentrations that are defined as the background level. Indices for plant and fish tissues were derived by dividing specimen concentration by background soil concentration. An index of 1.0 in plant or fish tissue indicates the same metal concentration as the soil's.

Wetlands Vegetation

Heavy metal concentrations in vegetation had greatest uptake in plant roots, with decreasing levels in leaf and seed parts (Figure 3). Generally, subsurface soil concentrations were less than half of surface soil amounts, and leaf and seed levels were one-half to one-fourth of root amounts. Manganese uptake was consistently highest within leaf tissue for both species, with statistically significant uptake associated with *Typha*. Leaf tissue in *Typha* took up

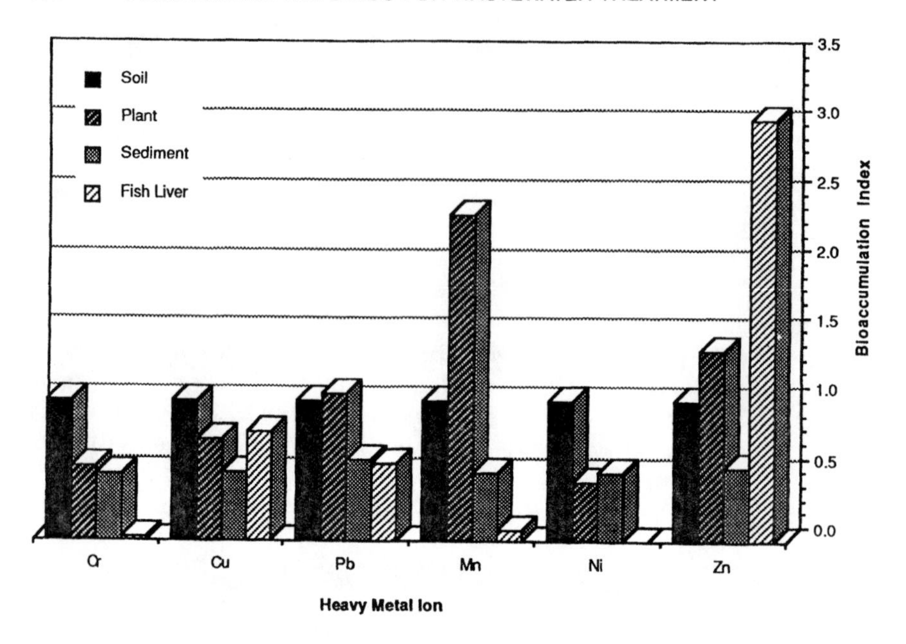

Figure 2. Bioaccumulation of heavy metals in the DUST Marsh.

447–1220 mg Mn/kg dry wt (34–185% of the background soil level). Manganese is an important plant micronutrient and appears to be heavily used by *Typha*.[4-6] Zinc uptake in *Scirpus* and *Typha* was greatest among root parts, with the highest observed levels above the debris basin (81.3 mg Zn/kg dry wt), reaching 105% of the background soil concentration.

Overall, *Typha* exhibits a greater ability to accumulate heavy metals than *Scirpus*. Metal accumulations appear to be highest within Systems A and B, areas with lower pH (5.7–6.5) and probably higher Mn and Zn solubility. *Typha* m^2 plots typically contain two to four times higher metal loadings than *Scirpus* due to higher unit biomass and differential physiology that confers a high tolerance to heavy metals (possibly through a cell wall metal precipitation mechanism.)[7,8] Short-term or new-growth bioaccumulation of Cr, Cu, Pb, Ni, and Zn was significant in *Scirpus* root tissue and various *Typha* parts. Metal concentrations in mixed-age or mature stands were generally lower, indicating either lower uptake rates or rerelease of metals from aging plant material. Manganese uptake appeared to be continuous through all age stands, and long-term bioaccumulation in mature stands, particularly *Scirpus*, was significant.

Fish Tissues

Heavy metal concentrations in fish tissues were generally below the ambient soil and sediment concentrations for Cr, Pb, Mn, and Ni (Figure 2). Carp liver

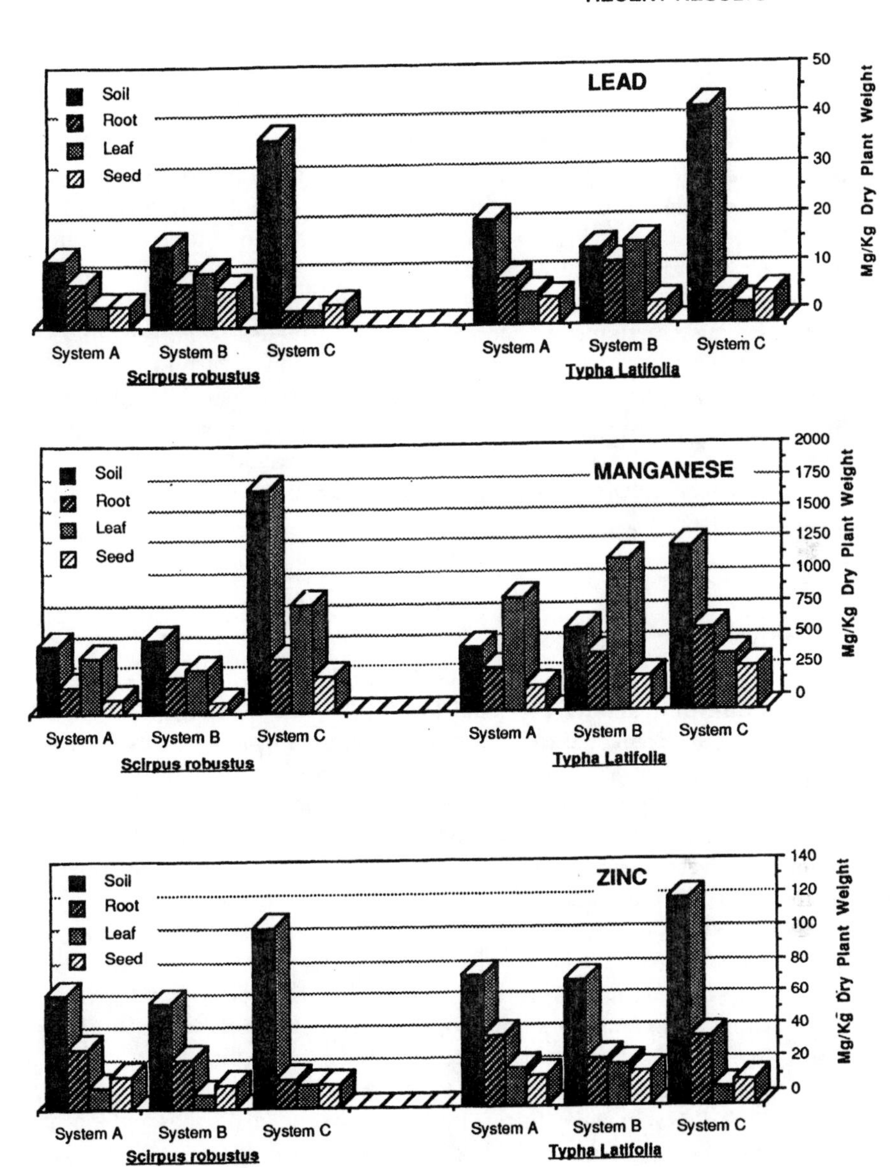

Figure 3. Heavy metal accumulations in plant parts.

samples from Systems A and B and *Gambusia* samples from System B had relatively high Zn concentrations. Carp liver concentrations were 28 times greater than the flesh and about five times higher than the sediments. Zinc levels in *Gambusia* and *Cottus* were lower than values reported in other studies and comparable to or less than levels found in adjacent areas without stormwater inputs. Blackfish and carp livers exhibited long-term bioaccumulation of Cr, Cu, and Pb. Carp also accumulated Pb and Zn in mature and larger fish specimens. The source of metal contamination appears to be the sediments. Because these metal concentrations are found in *Gambusia* and *Gasterosteus* from areas with and without stormwater input, factors such as saline soil conditions and/or groundwater intrusion may be involved. Additionally, there is a high probability that blackfish and carp specimens (3–10 years old) had migrated from other parts of the adjacent parkland and water channels, and metals accumulations may not be solely due to stormwater-affected sediment exposure.

CONCLUSIONS

At the time of this study, the marsh system was two to three years old. As the DUST Marsh system matures, the following conditions are anticipated: (1) development of more complete and dense vegetative cover on the overland flow area of System B and the perimeter of System A; (2) stabilization of marsh sediments; (3) establishment of an equilibrium for dissolved/suspended material between the soil-water interface; and (4) improved water treatment performance from a combination of the first three conditions.

Positive treatment performance in a mature marsh system is shown by loading data on System C, which performed most consistently of three systems studied. Systems A and B were newly constructed from upland areas, and plant colonization and marsh development were incomplete. System C, with heavy vegetation and extensive inundated areas during high water, efficiently reduced incremental BOD_5, organic P, and all heavy metals except Mn. Overall, the DUST Marsh was effective in reducing suspended solids and inorganic N, P, and Pb regardless of system. As the marsh matures, differences in treatment levels between systems due to design variations will become more apparent.

Based on the results of this study, the use of wetlands to treat urban stormwater runoff should be limited to constructed wetlands. Because the degree and significance of bioaccumulations of pollutants in the food chain is as yet unclear, such risks should not be imposed upon natural wetlands. These risks are more appropriately taken in constructed wetlands, where conditions may be better controlled and periodic maintenance such as dredging or vegetation harvesting would be acceptable.

REFERENCES

1. Chan, E., T. A. Bursztynsky, N. Hantzsche, and Y. J. Litwin. "The Use of Wetlands for Water Pollution Control," EPA-600/S2-82-086, U.S. Environmental Protection Agency, Municipal Environmental Research Laboratory, Cincinnati, OH (1982).
2. Meiorin, E. C. "Urban Stormwater Treatment at Coyote Hills Marsh," Association of Bay Area Governments, P.O. Box 2050, Oakland, CA (1986).
3. Silverman, G. S. "Development of an Urban Runoff Treatment Wetlands in Fremont, California," Chapter 40a, this volume.
4. Wells, J. R., P. B. Kaufman, and J. D. Jones. "Heavy Metal Contents in Some Macrophytes from Saginaw Bay (Lake Huron, USA)," *Aquat. Bot.* 9:185-193 (1980).
5. Estabrook, G. F., D. W. Burk, D. R. Inman, P. B. Kaufman, J. R. Wells, J. D. Jones, and N. Ghosheh. "Comparison of Heavy Metals in Aquatic Plants on Charity Island, Saginaw Bay, Lake Huron, USA with Plants Along the Shoreline of Saginaw Bay," *Am. J. Bot.* 72(2):209–216 (1985).
6. Taylor, G. J., and A. A. Crowder. "Uptake and Accumulation of Heavy Metals by *Typha latifolia* in Wetlands of the Sudbury, Ontario Region," *Can. J. Bot.* 61(1):63–73 (1983).
7. Mudroch, A., and J. A. Capobianco. "Study of Selected Metals in Marshes on Lake St. Clair, Ontario," *Arch. Hydrobiol.* 84(1):87–108 (1978).
8. Mudroch, A., and J. A. Capobianco. "Effects of Mine Effluent on Uptake of Co, Ni, As, Zn, Cd, Cr and Pb by Aquatic Macrophytes," *Hydrobiologia* 64(3):223–231 (1979).

40c Design of Wet Detention Basins and Constructed Wetlands for Treatment of Stormwater Runoff from a Regional Shopping Mall in Massachusetts

Paula Daukas, Dennis Lowry, and William W. Walker, Jr.

INTRODUCTION

Controlling stormwater runoff quality is an increasing environmental concern significantly influencing design and permitting of large development projects. Developers must clearly demonstrate that parking lot runoff will not degrade the quality of receiving water bodies. Runoff from parking lots and roadways contains high concentrations of suspended solids, nutrients, trace metals, oil and grease, and deicing salts. This case study presents the water quality mitigation measures for the 83,600-m^2 Emerald Square Mall constructed by New England Development, Newton, Massachusetts. The project site encompasses 23 ha in the Town of North Attleborough, 80 km south of Boston. The project has come under close scrutiny from local, state, and federal agencies because the shopping mall is located within the watershed of the Sevenmile River, which contributes to the drinking water supply for the City of Attleboro.

The Sevenmile River from its source to the outlet of Orr's Pond is targeted as a Class A water body by Massachusetts regulations; primary use is designated as public drinking water supply. The subdrainage area above Orr's Pond is approximately 1620 ha (Figure 1). The water supply system begins with water flowing from the Hoppin Hill Reservoir downstream past the project site into Luther Reservoir. Water either spills from Luther or is released as needed to maintain adequate storage volume in Orr's Pond. During periods of high flow and adequate volume in Orr's Pond, water is pumped from Luther Reservoir to Manchester Reservoir for off-stream storage. As water is needed in Luther, water is gravity-fed back to the reservoir. Water leaving Luther passes through three gravel pit ponds and a filter pond prior to entering Orr's

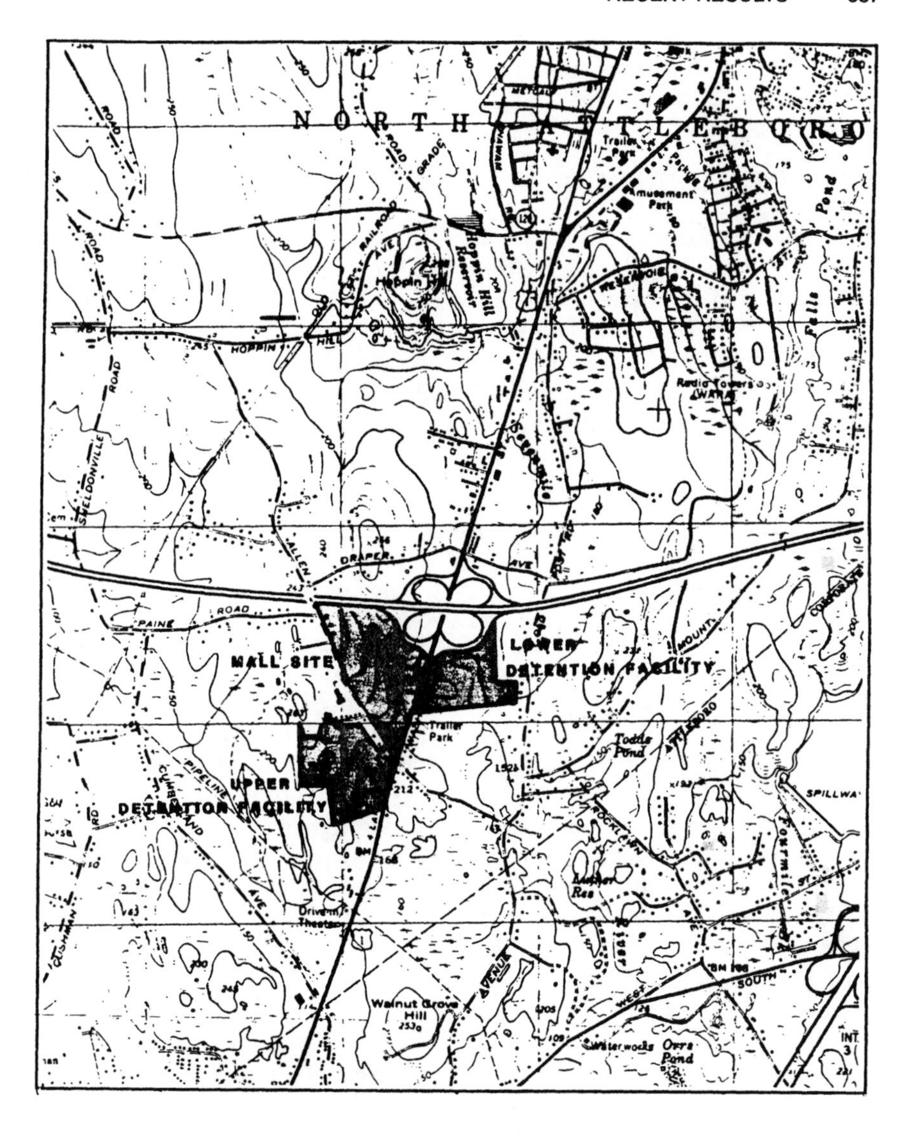

Figure 1. Site locus, Emerald Square Mall, North Attleborough, Massachusetts, constructed by New England Development, Newton, Massachusetts. Scale: 1 in. = 2083 ft. From USGS Attleboro, Massachusetts quadrangle.

Pond. Nine wells surrounding Orr's Pond provide water for a clear well for treatment and distribution to the City of Attleboro.

DESIGN OF THE STORMWATER MANAGEMENT SYSTEM

The sensitive use of the Sevenmile River necessitated that the drainage design objectives meet three basic criteria: (1) to protect the water quality of the river and its use as a public water supply, (2) to maintain the existing quantity of flow within the watershed, and (3) to control the peak rate of runoff and protect against downstream flooding.

For controlling stormwater runoff quality, the design and operation of Emerald Square Mall includes the following mitigation measures:

- wet detention ponds
- constructed wetland basins
- catch basins equipped with oil and grease traps
- parking lot sweeping
- sodium-free deicing salts
- restricted use of herbicides, pesticides, and fertilizers

Principal control mechanisms for attenuating peak flows and reducing contaminant loadings are detention ponds and created wetland basins. Stormwater runoff from the shopping mall will be directed through the wetland/detention systems prior to discharging into tributaries of the Sevenmile River. The watershed of the mall site is functionally divided into two separate basins as shown in Figure 2. Total drainage area of the system is 36 ha.

Due to space constraints on the mall site, runoff from the upper watershed will be directed through a box culvert to a detention facility located on an adjacent parcel of the mall (Figure 3). Runoff will collect in a wet detention pond and then flow through a series of three created wetland basins totaling 0.3 ha. Treated runoff is then routed back under the mall site and discharges into an existing small pond located east of the site to maintain adequate flow through the pond. Discharge from this pond flows through an extensive wooded swamp (1.0 ha) before reaching the Sevenmile River.

Runoff from the lower watershed and existing drainage from one quarter of the Route 295 interchange will be directed into two detention ponds located on the mall site. Discharge will be routed under the Route 1 highway through an existing natural wetland channel, diverted into three constructed wetland basins totaling 0.4 ha, and then dispersed into a large wooded swamp tributary to the Sevenmile River.

CREATION OF THE WETLAND BASINS

Wetland basins were established during the first construction phase to provide at least one full growing season prior to receiving parking lot runoff. The

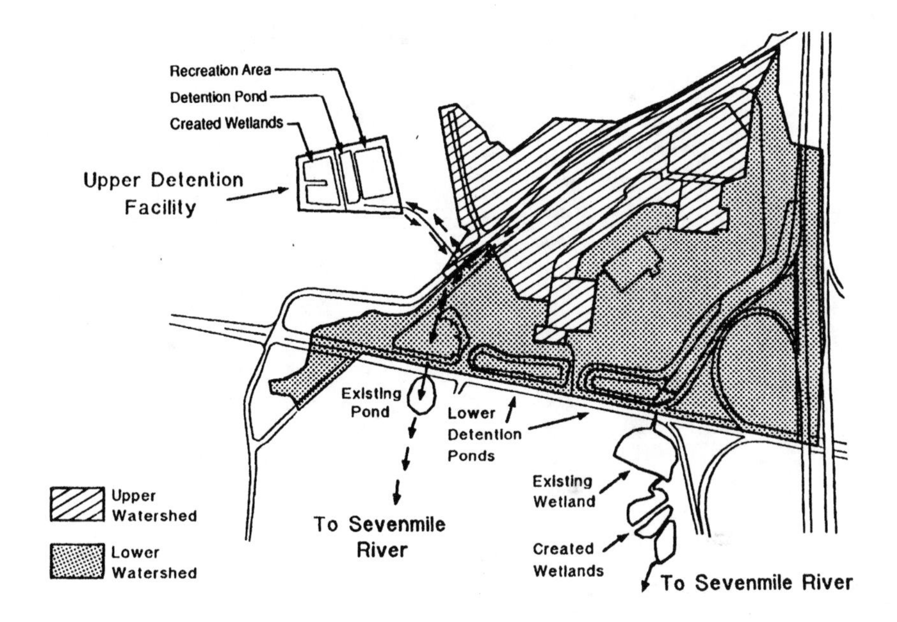

Figure 2. Drainage areas, Emerald Square Mall, North Attleborough, Massachusetts, constructed by New England Development, Newton, Massachusetts. Not to scale.

basins are designed as shallow marsh communities on organic soil. In addition, a selection of wetland shrub species will be planted on side slopes of some basins to enhance edge diversity for wildlife habitat. Configuration of the wetland basins (Figure 3) is designed to promote a long flow path and dispersion through the wetlands to maximize exposure to organic soils and vegetation. Basin outlet structures are equipped with stop logs for regulating water levels and wetness in the wetlands.

Wetland basins contain 46–60 cm of organic soil (>12% organic carbon), hand-raked to smooth top surfaces. Sumps were located in the basins to draw down the water table as necessary during construction. Approximately 35,000 tubers of indigenous marsh emergents were planted at the following percentages: 30% cattail (*Typha latifolia*), 25% arrowhead (*Sagittaria latifolia*); 25% bulrush (*Scirpus validus*), and 20% sweet flag (*Acorus calumus*). In addition, millet (*Echinochloa* sp.), reed canary grass (*Phalaris arundinacea*), and soft rush (*Juncus effusus*) were hand-sown over organic soils at 17–22 kg/ha. Side slopes of the basins with 10–15 cm of loam were hydroseeded with millet for rapid stabilization and reed canary grass for permanent establishment of vegetative cover. Wetland shrub species planted on side slopes will include red-osier dogwood (*Cornus stolonifera*), arrowwood (*Viburnum recognitum*), winterberry (*Ilex verticillata*), highbush blueberry (*Vaccinium corymbosum*), sweet pepperbush (*Clethra alnifolia*), and willow (*Salix* sp.).

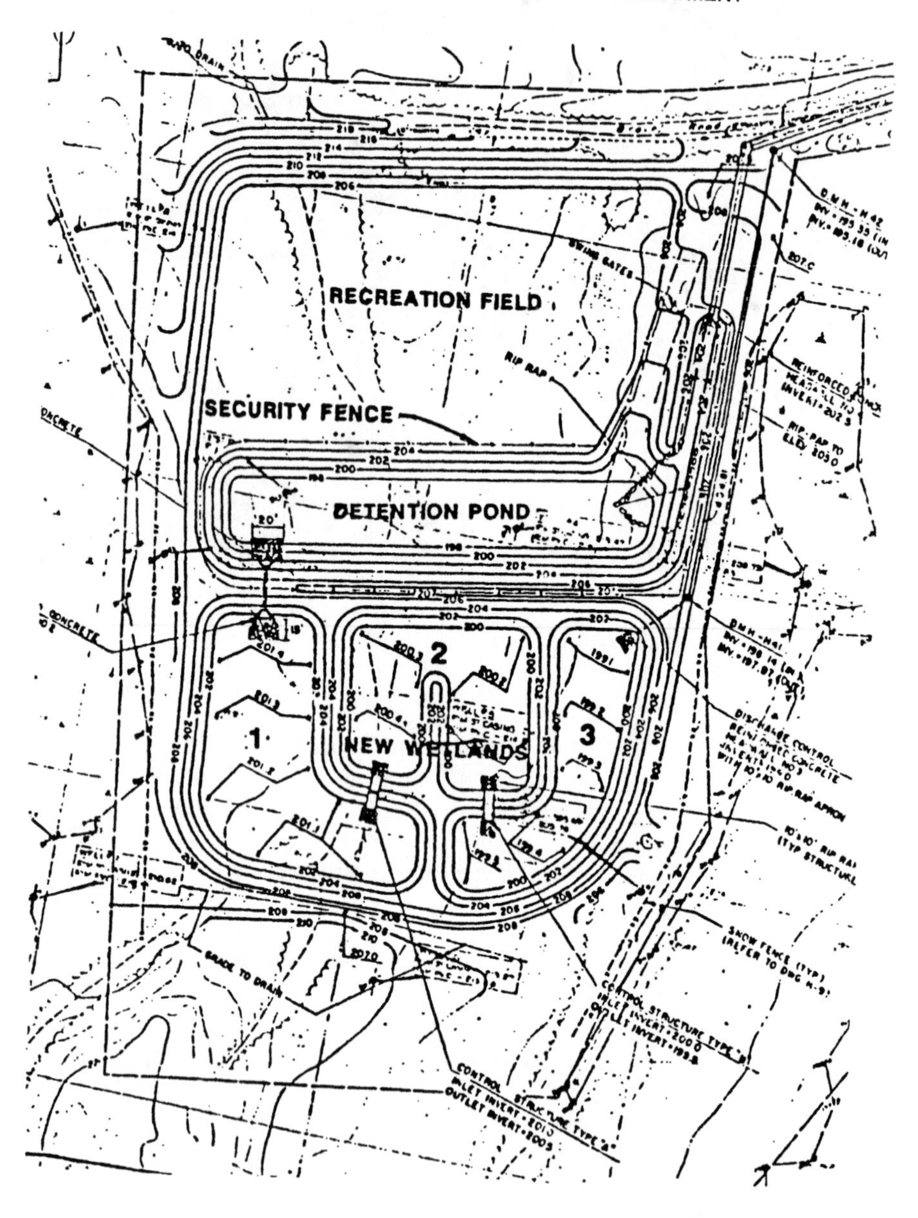

Figure 3. Upper detention pond and created wetlands, Emerald Square Mall, North Attleborough, Massachusetts, constructed by New England Development, Newton, Massachusetts. Not to scale.

EFFECTIVENESS OF THE WET DETENTION/WETLAND SYSTEM

Wet detention ponds and constructed wetlands are incorporated into the drainage network for reducing contaminant levels. The U.S. EPA's Nationwide Urban Runoff Program (NURP) concluded that detention basins with a permanent pool are among the most effective methods for controlling loadings of sediment, organics, nutrients, and heavy metals from urban watersheds.[1] Capabilities of wetlands for removing pollutants from runoff and domestic sewage have been widely demonstrated. Primary pollutant removal mechanisms include sedimentation, filtration, precipitation, adsorption, and biological uptake.

A significant portion of pollutant loadings in urban runoff results from the washoff of accumulated solids during storm events. The bulk of suspended solids will be removed by the detention ponds. Performance of these ponds is dependent on hydraulic characteristics, including pond area/drainage area ratio, mean depth, storm intensity and duration, and antecedent dry-weather period. Based on NURP guidelines, detention ponds in the upper and lower watersheds have been sized to represent 1% of their respective catchment areas, with a permanent pool mean depth exceeding 0.9 m. The ponds are also designed to attenuate peak flood flows for storms up to the 100-year, 24-hr event. During large storm events, flood waters in the upper facility can overflow onto an adjacent field that serves as a recreation field during dry periods. Effective residence time in the detention ponds for an average range of storms is 9–20 days. Runoff from an average storm (0.4 in.) will displace only one-fifth of the permanent pool volume. Therefore, during periods between storms, runoff will continue to have benefit from treatment in the ponds. The effective settling velocity of suspended solids and urban contaminants tends to decrease over time because remaining contaminants are associated with the smaller particulate fractions or are dissolved. Following the detention basins by vegetated wetlands provides additional treatment by other physical, chemical, and biological processes of the system.

Detention ponds provide several important functions:

1. dissipation of kinetic energy associated with runoff
2. improved hydraulic control and even distribution of flows over wetland surfaces
3. removal of coarse particulates which reduces the sediment load reaching the wetlands

The created wetlands provide additional functions:

1. physical filtering and sedimentation
2. biochemical processes such as reduction of nitrates to gaseous nitrogen in anaerobic soils
3. direct uptake by wetland vegetation
4. adsorption and chelation processes in the organic matter

EVALUATION OF THE POLLUTANT REMOVAL EFFICIENCY

Based on the hydraulic loading characteristics of the wet detention basins and values reported in the literature for wetland treatment, a range of pollutant removal efficiencies was estimated for the Emerald Square Mall (listed below, average percent reduction):

suspended solids	80–95%	cadmium	50–90%
total phosphorus	60–85%	chromium	50–90%
total nitrogen	40–70%	copper	50–90%
BOD_5	50–80%	lead	80–95%
sodium	0	mercury	50–90%
		zinc	50–90%

Effectiveness of runoff control measures to attenuate contaminants was evaluated by mass balance modeling derived from models and data bases developed by the NURP study.[2] The assessment provides a comparison of state and federal water quality criteria for human consumption and protection of aquatic life with projected contaminant concentrations in the (1) site runoff, (2) Sevenmile River upstream of the site, and (3) Sevenmile River downstream at the inlet to Luther Reservoir. Mass balance calculations evaluate concentrations under existing and future watershed conditions and for various hydrologic regimes. Existing conditions in the river were established through an extensive water quality monitoring program conducted at over 20 sampling stations throughout the watershed. Model results for monthly and storm-event mean concentrations in the Sevenmile River at the inlet to Luther Reservoir under existing and future conditions are presented in Table 1. Violation probabilities for water quality standards and criteria under the maximum storm-event conditions are included.

In all cases, predicted future concentrations at Luther Reservoir inlet are similar to or slightly below the existing concentrations. Slight improvements in water quality reflect future treatment of runoff from the existing urban watershed, in particular the highway interchange, and from mixing treated runoff with untreated runoff from existing nonpoint sources in the watershed. These results demonstrate that event mean concentrations of the effluent will meet the state and federal drinking water standards. Violation probabilities for aquatic life criteria tend to be higher under both existing and future conditions because the criteria tend to be more stringent than drinking water standards. Although these violations are typical of urban streams, especially for copper, it has been difficult to demonstrate specific biological impacts associated with heavy metals in urban runoff.[1] Mass balance calculations assume metals are conservative in the receiving stream, but further losses can be expected through adsorption to the river and lake sediments.

Development of Emerald Square Mall with these runoff control measures should not adversely affect critical use of the river as a water supply. Integrity of the water supply is further protected by natural water quality processes that

Table 1. Results of Mass Balance Calculations for the Sevenmile River

Variable		Storm-Event Mean Conc[b]	Monthly Mean Conc[b]	Criterion/Standard[a] Human Health	Violation Prob (%)[c]	Aquatic Life[d]	Violation Prob (%)
Suspended solids (ppm)	Existing	45.2	9.5	—	—	—	—
	Future	37.2	9.0	—	—	—	—
Phosphorus (ppm)	Existing	0.2	0.04	—	—	—	—
	Future	0.1	0.04	—	—	—	—
Nitrogen (ppm)	Existing	1.8	1.0	10	<1	—	—
	Future	1.6	1.0	10	<1	—	—
BOD_5 (ppm)	Existing	6.0	2.0	—	—	—	—
	Future	5.4	2.0	—	—	—	—
Sodium (ppm)	Existing	19.1	33.1	20	47	—	—
	Future	17.1	32.9	20	38	—	—
Cadmium (ppb)	Existing	2.8	0.6	10	14	1.5	70
	Future	2.4	0.6	10	10	1.5	66
Chromium (ppb)	Existing	6.4	1.3	50	2	16.0	16
	Future	5.5	1.3	50	<1	16.0	12
Copper (ppb)	Existing	31.2	6.3	1000	<1	7.1	93
	Future	27.0	6.2	1000	<1	7.1	92
Lead (ppb)	Existing	44.2	8.9	50	45	24.0	74
	Future	36.3	8.4	50	37	24.0	67
Mercury (ppb)	Existing	0.7	0.1	2	14	2.4	10
	Future	0.6	0.1	2	10	2.4	7
Zinc (ppb)	Existing	108.1	21.7	5000	<1	140.0	40
	Future	93.5	21.6	5000	<1	140.0	34

[a]Maximum contaminant levels based on Massachusetts Regulations (310 CMR 22); USEPA Primary and Secondary Drinking Water Regulations (40 CFR 141 & 143). Nitrogen standard refers to nitrate-nitrogen only.
[b]Median (50%) estimates of monthly mean and storm-event mean concentrations in the Sevenmile River at the inflow to Luther Reservoir.
[c]Violation probabilities determined for maximum storm-event concentrations which are not shown.
[d]Freshwater aquatic life criteria (49 FR 79318) based upon average hardness of 38 ppm.

occur in stream channels and several intervening impoundments between the point of mall discharge and water supply wells. In addition, routine monitoring of site discharges will be conducted under the guidance of U.S. EPA's National Pollutant Discharge Elimination System permit program. Effluent limitations are proposed which meet, at a minimum, drinking water standards for suspended solids, nutrients, petroleum hydrocarbons, and heavy metals. Long-term monitoring of treated runoff will detect unanticipated water quality problems that may result from the shopping mall operation.

REFERENCES

1. Athayde, D. N., P. E. Shelly, E. D. Driscoll, D. Gadboury, and G. Boyd. "Results of the Nationwide Urban Runoff Program: Volume I—Final Report," U.S. EPA, NTIS PB84-185552 (December 1983).
2. "Final Environmental Impact Report—Emerald Square Mall, North Attleborough, MA," New England Development, Newton, MA (April 1986).

40d Creation of Wetlands for the Improvement of Water Quality: A Proposal for the Joint Use of Highway Right-of-Way

Lewis C. Linker

INTRODUCTION

This chapter decribes a proposal to incorporate public lands through joint use of highway right-of-way. The proposal identifies a potential highway site in Maryland for joint use as an engineered wetland to control urban nonpoint source (NPS) pollution from highly developed, established urban areas and provides preliminary analysis of the site's control effectiveness and design life costs.

BACKGROUND

The 1987 Chesapeake Bay Agreement, signed by the states of Maryland, Pennsylvania, and Virginia, the District of Columbia, and the federal government (EPA), has a goal of a 40% reduction by the year 2000 of nitrogen and phosphorus entering the main stem of the bay. Urban NPS pollutants, a substantial portion of the harmful nutrient problem in the bay, are increasing in importance as the basin becomes more heavily urbanized. Numerous NPS management programs are directed toward agricultural lands and significant control programs have been recently implemented by the states for developing urban areas, but few control measures have been successfully applied to established, highly developed urban area nonpoint sources.

Urban NPS pollutants include sediment, trace metals, toxicants, hydrocarbons, nitrogen, and phosphorus that originate from atmospheric deposition, metal corrosion, material from worn brake linings and tires, organic matter, litter, and debris. These materials accumulate rapidly on impervious surfaces and are easily washed off by rain and other means into streams and water bodies.

Concentration of many pollutants in urban storm water is at least an order of magnitude above background levels.[1] Below the fall line, urban NPS nitrogen is 6.5% and phosphorus is 7.7% of the total point source and NPS load (1985 loads, Chesapeake Bay Program data base). In the Patuxent basin, for example, urban land contributes three times more nitrogen and 13 times more phosphorus per acre than forest land in an average rainfall year. Relative to agricultural land, urban land contributes 1.2 times more nitrogen and 2.8 times more phosphorus per hectare in an average rainfall year (Watershed Model, HSPF 8.0).

Achieving the nutrient reduction goal of the 1987 Bay Agreement requires an effective program of urban NPS control. Difficulties associated with control in established urban areas include the high cost of control technology, lack of available public land, and inappropriate topography for control structures.

Creation of large engineered wetlands, strategically located, offers an effective means of reducing a substantial portion of the urban NPS pollutants entering the bay. Locating and constructing these wetlands through joint use of highway right-of-way removes many difficulties normally associated with control in established urban areas. Economies of scale indicate that this management practice is the most cost-effective urban nonpoint control technology available.

ENGINEERED WETLANDS

An engineered wetland (wet pond, retention pond) is a sink for pollutants (Figure 1). Engineered wetlands are a proven technology, and there is a considerable body of literature on their design and effectiveness.[1-3] Wetland pollutant removal processes include physical mechanisms such as settling or adsorption and biological processes such as uptake or degradation. Properly sized and maintained, engineered wetlands achieve high removal efficiencies of nutrients, sediment, BOD_5, metals, and hydrocarbons. Removal efficiencies depend on the design characteristics of the wetland. A primary design determinant is the volume of the wetland pond relative to the volume of runoff. The reported ranges of removal efficiencies of appropriately designed wetlands are:[1]

sediment	75–90%
total phosphorus	55–65%
total nitrogen	~40%
BOD_5	~40%
metals	0–80%

Ancillary benefits of wetland creation include flood control, stream bank erosion control, wildlife habitat creation, and aesthetics. Recreation and educational opportunities may also be improved.

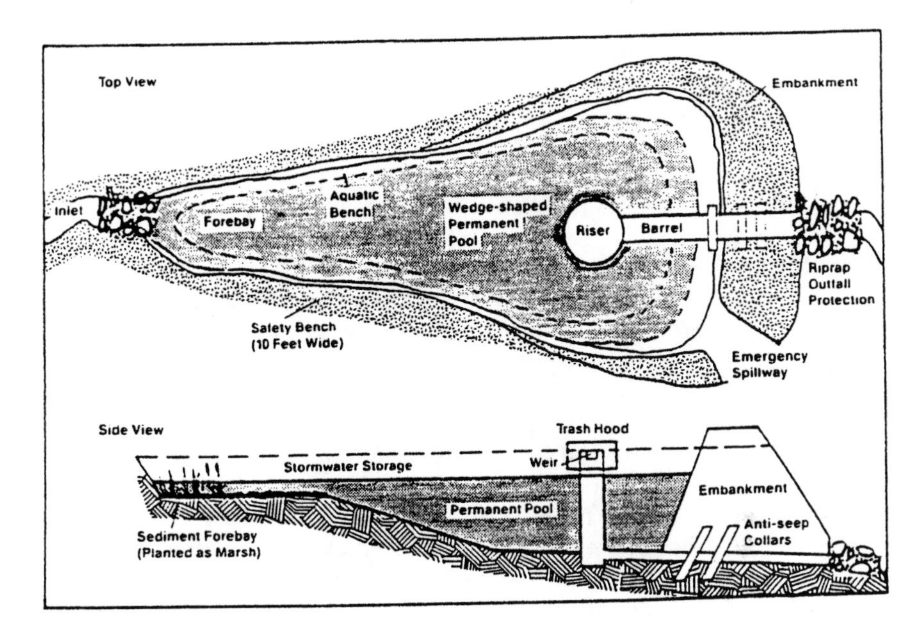

Figure 1. Schematic of an engineered wetland. From Scheuler.[1]

HIGHWAY JOINT USE ISSUES

Interstate and primary highway systems are owned and administered by the states under both federal and state laws and regulations. Highway right-of-way often includes extensive areas to protect and service the highway facility, mitigate environmental impacts, and provide an aesthetic setting. State highway agencies also acquire land for borrow and disposal areas during construction, as uneconomic remnants of partial takings from landowners, and as a result of design changes. These excess parcels range from minuscule to very large.

State highway agencies have supported appropriate joint use projects on rights-of-way and have sought public uses for the excess parcels or have sold them off. In recognition of the immense public investment in the nation's rights-of-way, the Secretary of Transportation has directed the Federal Highway Administration to encourage the states to hasten disposal of excess lands, to convert them to public uses, or to encourage joint use projects, with emphasis on public uses.

Where large tracts of highway lands coincide with urban streams, there are excellent opportunities for joint use projects to improve water quality. Development of engineered wetlands within highway right-of-way and on excess lands is compatible with transportation uses. The vertical and horizontal alignment of modern highways produces a very appropriate topography for wetland creation. These incidentally created basins reduce wetland creation costs and help it blend with the environment. Wetlands can be engineered to assist the hydrology functions of the highway facility and reduce highway mowing and other right-of-way management costs.

EFFICIENCY AND COST ANALYSIS OF A PROPOSED SITE

Several right-of-way sites within the Chesapeake Bay basin have been identified which (1) have large area and (2) intersect an urban stream. A portion of the I-695 intersection with the Southeast Throughway and the Windlass Freeway in Middle River northeast of Baltimore was chosen as representative. This site is located within Stemmers Run, a developed urban basin of 23.2 km^2 in Baltimore County. For 1978, the basin was 70% urban land, with the balance evenly split between forest and agriculture. The basin outlet is tidal Chesapeake at the head of Back River. Just upstream of the basin/bay boundary, I-695 makes a sweeping curve from the northeast to the northwest. Enclosed in the radius of the curve, between the north and south lanes, is an area of about 0.25 km^2.

The following wetland design information is based on design criteria, hydrology equations, estimated loads, and cost from Schueler.[1] Overall imperviousness of the watershed is 55%. The calculated storm runoff for the design storm (mean storm of 1 cm, 0.2 cm/hr, six hours' duration, Ft = 82 hr) was checked against two USGS gaging stations in the watershed and found to be consistent with their data.

size of basin	23.2 km^2
size of pond	14 ha
average depth	2 m
range of depth	0.3–2.4 m
volume of permanent pool (V_p)	2.8×10^5 m^3
volume of runoff from mean storm (V_s)	1.1×10^5 m^3
V_p/V_s	2.5

As the downstream area of the example wetland is tidal, freeboard for floodwater management is unnecessary in the design. The primary objective of water quality management is satisfied by maximizing the area of permanent pool with the critical design factor: the ratio of the permanent pool volume to the mean storm runoff volume. One can estimate pollutant quantities removed annually by using this ratio and Nationwide Urban Runoff Program (NURP) average pollutant loads (U.S. EPA, 1983) and Stemmers Run land use information (Table 1).

Based on the quantities of pollutants removed, a 50-year design life, and estimates of routine and nonroutine maintenance, the estimated total cost—or net present cost (NPC)—is $1,420,000 (Figure 2). Economies of scale, construction economies, and land acquisition economies make this the most cost-effective urban best management practice (BMP). Alternative technologies

Table 1. Pollutant Removal Efficiencies and Pollutant Removal Cost

Constituent	Storm Flow Loads (kg)[a]	Removal Efficiencies (%)	Load Removed (kg)	Cost per kg of Pollutant Removed[b]
TP	5,400	53	2,860	$ 9.90
TN	38,700	41	15,900	$ 1.80
TP + TN	44,100	—	18,760	$ 1.50
BOD_5	139,000	40	55,700	$ 0.50
Pb	2,100	72	1,500	—
Zn	2,100	40	860	—
Cu	540	40	230	—
Total metals	4,740	—	2,590	$11.00
Sediment	—	77	—	—

Note: TP = total phosphorus; TN = total nitrogen; BOD_5 = biological oxygen demand; Pb, Zn, Cu = lead, zinc, and copper.
[a]Annual loads for average year assumed. Baseflow loads are assumed negligible.
[b]Based on total pollutants removed over the design life of the structure.

that remove an equivalent amount of nutrients are several times more costly. Only agricultural BMPs are less expensive per pound of nutrients removed. Tables 1 and 2 give planning level unit costs for nutrient removal. A design life of 50 years is based on local data on similar concrete structures[1] and considers the high level of planned maintenance included in the test site cost analysis. The lower design life of other BMPs is in part a reflection of their lower level of planned maintenance and high failure rate.[5]

Large urban wetlands in highway right-of-way would address a major problem of water quality control and create an aesthetic area of wildlife habitat close to high-density urban areas. Vandalism and liability problems may be decreased because access is controlled on interstate right-of-way. Problems associated with the joint use approach of large engineered wetlands include high cost of construction and a high level of planning and design necessary for

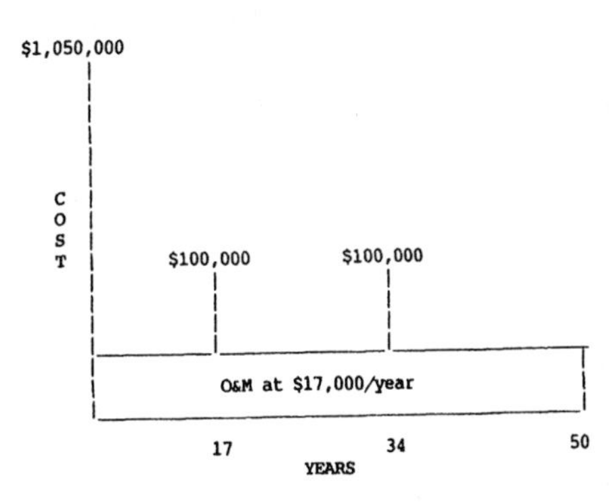

Figure 2. Cost diagram of site 695 discounted for 50 years. Assume routine maintenance = $1200/ha or $17,000/yr, sediment clean-out every 17 years at $100,000/clean-out, and 5% discount rate. Based on NURP data for construction of large-scale engineered wetlands. Construction cost (C): C = $34(V_p)0.64$ = $1,050,000. Total cost (NPC) = construction cost + O&M + sediment clean-out = $1,420,000.

Table 2. Unit Cost for Nutrient Control

Control Method	Design Life (y)	Cost of Total N and P Removed($/kg)
Agricultural BMP		
Animal waste	25	0.97
Cropland	5	0.29
Urban stormwater BMP		
Commercial	20	2.62
Residential	20	6.94
Regional, joint-use[a]	50	1.52
Point source treatment		
Chemical P removal		
1 mgd	20	5.58
10 mgd	20	4.83
BNR(A_2O)		
1 mgd	20	5.67
10 mgd	20	2.16

Source: Modified from EPA report.[4]
Note: BMP = best management practice; BNR (A_2O) = biological nutrient removal based on the A_2O system.
[a]Based on Stemmers Run test site.

regional BMPs. Maintenance and liability agreements and implementation of an air space agreement would require considerable cooperation and effort between federal and state transportation and environmental administrations.

CONCLUSION

The large areas of land required for highway rights-of-way have been assembled at great public cost. Wetlands engineered to improve urban water quality on a basinwide scale also require large tracts of land. Objectives of transportation and water quality improvement may be compatible in many areas of highway right-of-way. A quality program of joint use for these two objectives maximizes the public benefit from this land resource. Public benefits accrue from three aspects of a joint use program:

1. Basinwide control provides considerable economies of scale from both construction and maintenance phases.
2. Vertical and horizontal alignment of major interstate interchanges ("cut and fill" construction) provides an appropriate topography for engineered wetlands. Basins created incidentally in the construction of highways can be utilized in the construction of engineered wetlands at considerable public savings.
3. Joint use of public land avoids the social and economic expense of condemning private land to assemble a separate tract of land for NPS control.

It is in the public interest to maximize the quality use of highway right-of-way. A program of joint right-of-way use for the improvement of urban water quality should be implemented if plans can be developed to the satisfaction of the Maryland State Highway Administration and the Federal Highway Administration.

REFERENCES

1. Schueler, T. R. "Controlling Urban Runoff: A Practical Manual for Planning and Designing Urban BMPs," Metropolitan Washington Council of Governments, Washington, DC (1987).
2. "BMP Handbook for the Occoquan Watershed," Occoquan Basin Nonpoint Pollution Management Program, Northern Virginia Planning District Commission, Annandale, VA (1987).
3. Harrington, B. W. "Feasibility and Design of Wet Ponds to Achieve Water Quality Control," Maryland Water Resources Administration, Annapolis, MD (1986).
4. "A Commitment Renewed: Restoration Progress and the Course Ahead Under the 1987 Bay Agreement," U.S. EPA report no. CBP/TRS 21/88 (June 1988).
5. Lindsey, G. "A Planning Guide to Stormwater Management Utilities," Sediment and Stormwater Administration, Maryland Department of the Environment, Annapolis, MD (1987).

40e Wetlands Treatment of Dairy Animal Wastes in Irish Drumlin Landscape

C. J. Costello

INTRODUCTION

Lough Gara Farms Limited established an intensive dairy farm to produce milk for direct retail sale in 1961. The existing treatment system uses a natural wetland formed as a result of successive drainage schemes carried out in the lake and its feed and discharge rivers between 1816 and 1953. The system has remained in use since then but was working at a reduced volume between 1979 and 1983. The Sligo County Council is the regulatory authority in the area under the Local Government (Water Pollution) Act.[1] This act does not establish performance requirements for pollution control, and the opinion of the regulatory authority regarding existence or potential threat of pollution prevails. In forming its opinion, the regulatory authority relies on the Irish Department of Agriculture's "Guidelines and Recommendations on Control of Pollution from Farmyard Wastes."[2]

In February 1985, the company was charged with "on the grounds that, contrary to Section 171 (b) of Fisheries Act, they emptied, permitted or caused to fall into the waters of a watercourse flowing into Lough Gara deleterious matter." The circuit court judge, having heard the evidence, said, "I am satisfied that the system of effluent control used by Lough Gara Farms Limited is a perfectly good system."

On July 1, 1985, Sligo County Council issued a notice under the Local Government Water Pollution Act.[1] This notice was withdrawn and subsequently reissued on February 7, 1986, because the system was not one described in the Department of Agriculture's guidelines.[2]

Consequently, Lough Gara Farms decided to (1) commission an independent environmental study on Lough Gara in relation to the physical, chemical, microbiological, and ecological characteristics of the lake to determine the impact, if any, of the existing treatment system on the lake; (2) study the design, functioning, and operation of the existing system with a view to its improvement; (3) prepare, and have costed, plans to carry out the works requested by Sligo County Council; and (4) reinvestigate the availability and suitability of land for the spreading of slurry as requested by Sligo County Council. This chapter gives the results of the wetland treatment system portion of these studies.

LOCATION

Lough Gara Farms are located on the southern shores of Lough Gara, a shallow limestone lake situated between County Sligo and County Roscommon in the northeast section of the Great Carboniferous limestone plain drained by the River Shannon. Lough Gara is a 1700-ha expanse of shallow-bog-fringed limestone lake fed by canalized and fast-flowing Breedoge and Lung rivers. The maximum lake depth is 13 m. The Breedoge drains an area of sparsely populated dairy and cattle farmland with large tracts of raised bog. It enters the lake at upper Lough Gara. The villages of Tulsk and Frenchpark are located south of the lake.

The Lung drains similar terrain, including the town of Ballaghaderreen, the Regional Dairy Cooperative's main milk processing plant, and animal feed mills and meat plants near the town. The town's treated sewerage is discharged into the river. The river enters the lake through an extensive raised bog into middle Lough Gara. The combined waters of the Lung and Breedoge flow into lower Lough Gara at Clooncunny Bridge through a short, canalized river.

Landscape on the southern shore of lower Lough Gara has been molded by glacial ice sheets into typical drumlin whaleback hills of boulder clay and poorly drained peaty hollows. Drumlins also create islands in the lake and many peninsulas of land jutting out into the lake, upon one of which Rathtermon Farm is situated.[3,4] Soil on the tops of drumlins is well-drained loam earth of sandy clay/loam texture and at the bottoms is poorly drained gley of clay/loam texture. The bottoms are drained with mole drains and longitudinal main drains. The farm is in an area of high mean annual rainfall (100 cm) and of low-to-very-low winter rain acceptance potential and a low estimated soil moisture deficit for the region.

Lough Gara is classified by An Foras Forbortha[5] as an area of regional scientific interest in respect to its ornithology and botany. It was once an important commercial eel fishing lake, but drainage works and the consequent silting of the lake have destroyed this resource. The lake is not frequented by fishermen and is off the tourist track.

Drainage schemes for agricultural reclamation between 1816 and 1953 have deteriorated the quality and appearance of the lake. The now-shallow lake waters have been colonized by *Nuphar luteum* (yellow water lily), *Iris pseudocorus* (yellow flag), and notably *Typha latifolia* (cattail), whose roots and stems trap peat sediment now being washed into the lake by the rapidly flowing Breedoge and Lung rivers. Buildup of vegetation is most noticeable in the upper lake and in the Rathtermon inlet. Large amounts of peat debris carried

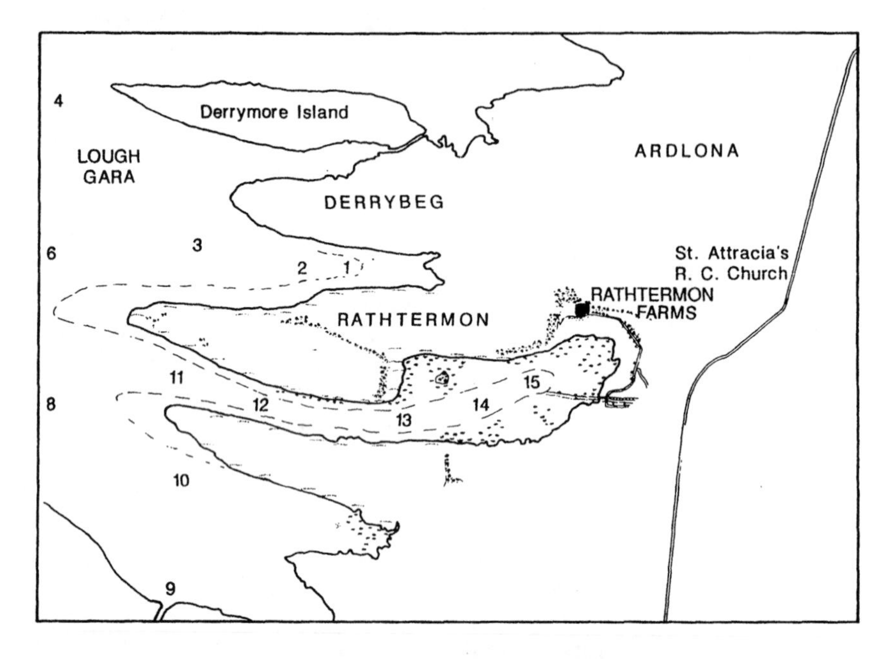

Figure 1. Location of Rathtermon Farms on Lower Lough Gara, County Roscommon, Ireland showing environmental study lake sampling locations.

into the lake by the Lung and Breedoge rivers and rapid inflow of these rivers results in flash floods on the lake.

Until 1979, the farm carried up to 165 cows and replacements. Due to a brucellosis outbreak, milk production ceased, and the farm changed to a suckling/cattle rearing system with the help of the Dairy Herd Conversion Scheme. A clause in the agreement under this scheme prevented the sale of milk until the end of July 1983, when milk production recommenced. Total farm area is about 161 ha in two separate portions. Rathtermon Farm, of about 100 ha adjusted to 71 ha, is used for the dairy enterprise (Figure 1). Most of the adjusted area can be cut for silage, and the soil quality is good. Lurgan Lodge Farm—5 km from Rathtermon, containing 61 ha, 50 adjusted—is used for summer grazing. Silage is cut from 20 ha, 25 ha is reclaimed peat, and 10 ha is reclaimed from the lake.

Farm buildings at Rathtermon consist of 96 rubber-matted cow cubicles, one 50-cow kennel, four covered silage silos with a combined capacity of 1150 metric tons, an eight-unit 16-stall milking parlor, two bull boxes, and 20 loose boxes for cows and calves.

Approximately 450,000 L/year of milk is produced on the farm. Culled cows and heifers, in-calf heifers, and dropped calves are sold off the farm. Cows, calves, and other livestock are housed for up to 23 weeks during the winter.

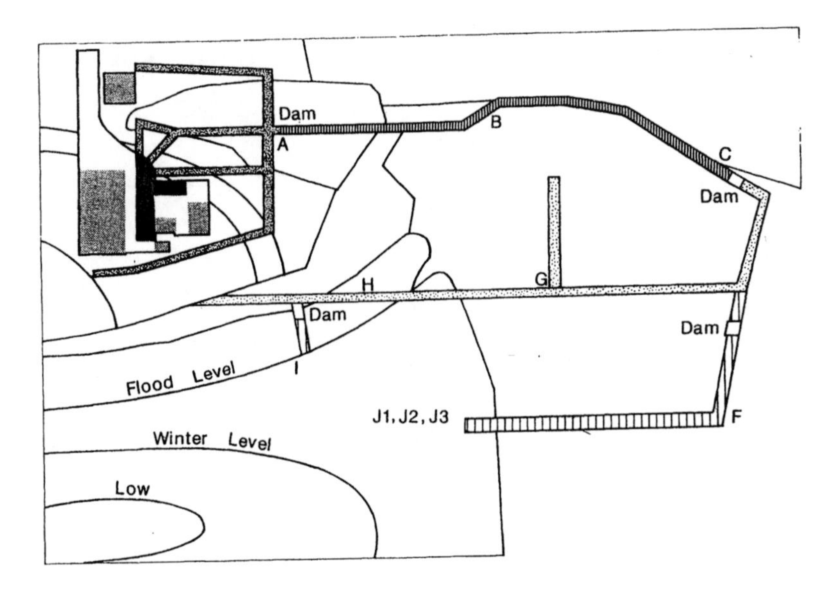

Figure 2. The Sligo Regional Technical College environmental study sampling locations on Rathtermon Farm.

Waste Treatment System

Waste from the cow cubicle house and cow kennels is removed with a mounted scraper to an uncovered external concrete yard along the full length of the buildings. The yard drains into the primary collection concrete apron and is discharged in turn by a tractor-mounted scraper down chutes leading to the first stage of the treatment system (Figure 2).

Anaerobic decomposition of the slurry occurs in interlinked 2-m-deep-by-2.5-m-wide trenches formed in the poorly drained clay soil at the lower slope of the drumlin. These soils are underlain by marls, and the whole forms an impervious series of collection chambers containing 800 m^3.

Daily volume of material fed into this system consists of neat excreta (feces plus urine), 7500 L; milking parlor washings, 75,000 L; rain and washings in unroofed yard areas (800 m^2), 5000 L; and rainfall in slurry collection channel areas (1000 m^2), 4000 L—totaling 24,000 L/day. Slurry in the anaerobic collection chambers separates into a floating hard crust of solids, a middle liquid fraction, and a second fraction of solid that settles to the bottom.

Interconnecting anaerobic collection chambers feed into one channel with concrete dam (A) containing an adjustable sluice gate. This gate is manually adjusted to allow the liquid fraction to flow into a second anaerobic fermentation chamber of approximately 200 m^3 with a limestone filter at the outlet. Filtered liquid (3600 m^3/yr) flows into the main channel (B,C), feeding the wetlands treatment area.

Table 1. Water Quality Analysis from Various Locations on Lough Gara Farms on February 5, 1985

Sample Location	Water Quality Parameters				
	BOD (mg/L)	SS (mg/L)	DO (mg/L)	Temp (°C)	pH (s.u.)
1	640.0	168.0	4.0	4.0	7.4
2	1.5	7.0	9.3	4.0	6.9
3	56.0	764.0	1.4	5.0	6.8
4	140.0	78.0	0.6	4.0	7.2
5	192.0	64.0	0.4	3.5	7.2
6	112.0	496.0	0.5	3.0	6.9
7	1.2	1.0	3.3	2.5	6.2
8	84.0	40.0	1.0	4.0	6.9
9	19.6	29.0	3.0	5.0	7.0
L. Gara	1.9	5.0	11.2	5.0	7.1

Wetlands Treatment

The wetlands wastewater treatment system is divided into two zones. The northern zone lies in the former lake bed and has a total area of approximately 6.4 ha. Peat (60–75 cm) in this zone overlies a 50-cm layer of marl. This area of peat is fairly level and 30–40 cm higher than the second zone. The system was planned so that the treatment would take place in this area of peat and vigorous vegetation. The southern and part of the eastern boundary drain (H,G,C) receives outlet waters from the first treatment zone to avoid short-circuiting and to serve as a distribution channel to the second zone.

The second zone has an area of approximately 5.5 ha and a fall of 1 m to the lake level at the southwest. Perimeter drains (F,J) collect the final outfall from the system to facilitate performance monitoring.

RESULTS

Results from the sample collections by the water pollution offices of the Regional Fisheries Board on February 5, 1985, are shown in Table 1. Although precise locations could not be determined from the officer's hand-drawn maps (Figure 3), the drop in BOD_5 and SS concentrations from sites 8 to 9 indicated significant treatment by the wetlands.

The objective of the environmental study conducted by Sligo Regional Technical College between June 15 and July 24, 1987, was to determine the impact, if any, on the southeast shore of Lough Gara of the farm effluent arising from the intensive dairy operation present on Rathtermon Estate. The portion on wastewater treatment system performance is included herein.

The effluent treatment system at Rathtermon produced a 99.1–99.95% reduction in BOD_5 concentrations, although at sites I and J3 BOD_5 varied greatly, occasionally exceeding 200 mg/L (Figure 3 and Table 2). The stagnant nature of site I and decomposition of plant material from the extensive vegeta-

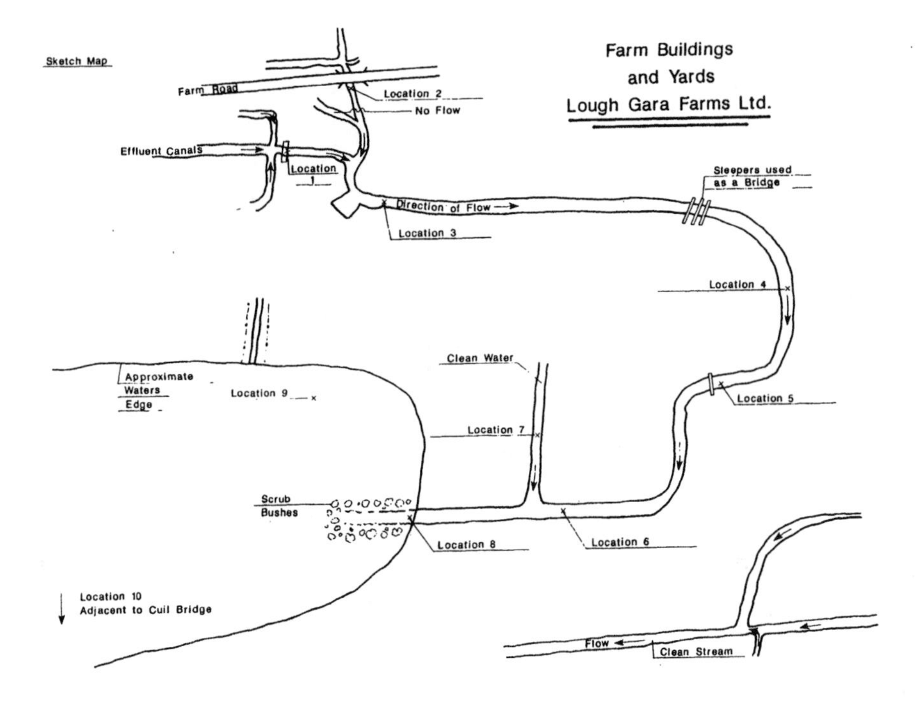

Figure 3. Water pollution officer's map of water sampling locations on Rathtermon Farm on February 5, 1985.

tion in the region contributed to the high BOD_5 values. Sites F, G, and H were within the Royal Commission standards of 20 mg/L BOD_5.

Chloride and potassium concentrations at sites H and I were high, and NH_3 (1.36–3.49 mg/L), NO_3 (>1.5 mg/L), and oPO_4 (1.75–2.78 mg/L) concentrations at sites H, I, and J3 were markedly higher than the other drain sites. Phosphates or nitrates were not higher in this angle than in the remainder of the lake. Ammonia at site 15 is influenced by ammonia concentrations from I and J3, which are in turn influenced by the extensive vegetation in the area. Total mesophilic bacteria in effluent samples showed a wide variation, but all sites had an FC-FS ratio <4.0, indicating animal contamination as expected. During times of high water levels, part of the second-phase treatment area can be covered by lake water.

Sampling of the effluent treatment system at Rathtermon by Sligo RTC found it to be working effectively with regard to (1) BOD_5—the system produced a 99.95% reduction in BOD_5 concentrations at best and 99.1% reduction at worst during the period of the investigation; (2) ammonia—concentrations were reduced by up to 95.4%; (3) nitrate—concentrations were reduced by up to 95.4%; and (4) orthophosphate—concentrations were reduced by up to 91.3%.

Table 2. Characteristics of Effluent and Water Samples at Rathtermon (15 June to 24 July 1987)

Parameter	Input			Control		Phase I		Output				
	A	B	C	D	E	F	G	H	I	J3	J2	J1
pH	7.5	7.5	6.9	7.3	7.3	7.0	7.0	7.0	7.3	7.3		
Conductivity	1,848.0	1,731.0	989.0	542.0	542.0	611.0	622.0	812.0	906.0	581.0		
Temp (°C)	18.0	18.5	16.0	15.0	15.5	17.0	14.0	15.0	20.0	20.0		
Turbidity (FTU)	250.0	140.0	60.0	10.0	5.0	5.0	2.0	25.0	65.0	20.0		
Color (Hazen)	225.0	150.0	50.0	10.0	10.0	10.0	10.0	15.0	15.0	15.0		
Suspended solids (mg/L)	0.0	0.0	0.0	0.0	0.0	0.2	0.0	0.0	0.0	0.1		
Nitrate (mg/L NO_3)	40.0	12.6	9.0	2.3	2.5	2.0	1.6	1.5	1.8	1.6		
Orthophosphate (mg/L PO_4)	20.0	18.0	3.5	0.5	0.9	0.8	1.0	2.8	1.9	1.8		
Ammonia (mg/L NH_3)	19.7	—	10.7	—	—	1.0	0.1	1.3	3.5	0.9	2.2	
Chloride (mg/L Cl)	73.0	11.5	30.0	28.0	22.0	22.0	31.0	49.0	63.0	19.0		
Potassium (mg/L K)	200.0	195.0	110.0	4.0	2.2	32.0	5.0	35.0	62.0	5.6		
Sodium (mg/L Na)	57.0	50.0	38.0	19.0	16.0	18.0	20.0	32.0	36.0	16.0		
K/Na Ratio	3.5	3.9	3.0	0.2	0.1	0.2	0.2	1.1	1.7	0.3		
Hardness (mg/L $CaCO^3$)	50.0	40.0	42.0	36.0	40.0	32.0	54.0	32.0	32.0	46.0		
Chlorophyll(a) (mg/L m^3)	834.0	523.0	208.0	—	1-	182.0	77.0	46.0	44.0	—	29.0	18.0
BOD_5 (mg/L) June 25, 1987	200.0	7,400.0	13,800.0	10.1	2.8	12.0	5.6	9.0	200.0	5.4	—	—
July 6, 1987	24,000.0	24,000.0	800.0	—	—	12.0	6.6	20.6	10.0	205.0	—	—
Total mesophilic bacteria (CFU/100 mL)	40,000	173,000	470,000	—	1,330	902	1,383	63,300	24,500	17,000	358	465
Fecal coliforms (CFU/100 mL)	905	7,930	41,200	—	295	5	0	0	70	340	0	0
Fecal strep (CFU/100 mL)	1,100	6,220	23,700	—	147	15	0	40	515	465	0	0
FC, FS (CFU/100 mL)	0.9	1.2	1.7	—	2.0	0.3	—	—	0.2	0.7		

The Sligo RTC report conclusion that "Lough Gara Farms Limited at Rathtermon appear to be having a slight eutrophic impact on the waters of Lough Gara" is based on the BOD_5 measurement and the increase in phytoplankton, but both these measurements are a reflection of the aquatic and emergent plants in this region.

REFERENCES

1. Local Government (Water Pollution) Act (1977).
2. "The Department of Agriculture Guidelines and Recommendations on Control of Pollution from Farmyard Wastes."
3. Whitlow, J. B. *Geology and Scenery in Ireland* (Pelican Books).
4. Praezer, R. L. *The Way That I Went* (Dublin: Allen Figgis, 1937).
5. Flanagan, P. J., and P. F. Tomer. "A Preliminary Survey of Irish Lakes," An Foras Forbartha (1975).

40f Potential Role of Marsh Creation in Restoration of Hypertrophic Lakes

Edgar F. Lowe, David L. Stites, and Lawrence E. Battoe

INTRODUCTION

The use of wetlands to reduce concentrations of dissolved nitrogen (N) and phosphorus (P) in secondary sewage effluent is well documented.[1] We propose a novel application: the use of wetlands for the restoration of hypertrophic lakes. In a typical sewage treatment wetland, water high in inorganic nutrients passes through the system once, and the management goal is to maximize the nutrient removal efficiency. In our application, water to be treated has high concentrations of nutrients associated with particles and dissolved organic molecules. Water is passed through the system many times rather than once. Goals of this approach are to maximize *power* (nutrient quantity removed per unit of time) and *capacity* (nutrient permanently stored) rather than efficiency (nutrient fraction removed in a single pass).

LAKE APOPKA—AN EXAMPLE

The Lake

Lake Apopka (Figure 1), located in central Florida, is large (A = 124 km^2), shallow (Z = 1.7 m), and hypertrophic (TSI = 86).

Lake Morphometry and Hydrology[2]

area	124 km^2
avg. depth	1.7 m
maximum depth	11.0 m
volume	2.108 ee + 8 m^3
hydraulic loading (1977)	190 ee + 6 m^3
hydraulic retention time (1977)	4–6 years

Chemical Characteristics[3]

parameter:	*mean (std. dev.):*
alkalinity (mg/L)	122 (13)
hardness (mg/L)	163 (7)
spec. cond. (μmhos/cm)	391 (32)
pH	9.1 (0.3)
chloride (mg/L)	42.2 (3.4)
total iron (μg/L)	<50 (not available)
total nitrogen (mg/L)	4.5 (0.6)
total phosphorus (mg/L)	0.22 (0.05)
inorg. nitrogen (mg/L)	0.031 (0.019)
ortho-phosphorus (mg/L)	0.059 (0.024)
Chlorophyll a (μg/L)	56.6 (11.3)
turbidity (NTU)	31 (12)
Secchi disk (m)	0.32 (0.10)

In a survey of 573 Florida lakes, it had the highest trophic state index value of any large lake.[4] Because Lake Apopka is (1) close to Orlando, a population center; (2) the fourth largest lake in Florida; and (3) the headwater of the Oklawaha Chain of Lakes, its pollution is especially problematic. It may be the primary cause for eutrophication of five downstream lakes.[2]

N- and P-laden agricultural wastewater (76 billion L/yr) is the primary external nutrient source (Table 1). The water is discharged from farms that have occupied the lake's floodplain (formerly a 73-km^2 sawgrass marsh) since 1942, when that area was diked and drained for vegetable production. Extensive continuous drainage is required because the land surface lies below lake level. This drainage and other farming practices have resulted in soil subsidence and release of nutrients stored in the soil.[6]

The Proposal

The St. Johns River Water Management District has proposed state acquisition of up to 32 km^2 of muck farmland and creation of the wetland on the acquired land. The project has three salient goals: (1) establishment and maintenance of a hydrologic regime and vegetation that maximizes the wetland's power and capacity for nutrient storage; (2) creation of a broad and persistent biological connection between the lake and the created marsh; and (3) establishment and maintenance of a diverse wetland habitat to support aquatic and wetland wildlife. To achieve the first goal, direct discharge of lake water through the Apopka-Beauclair Canal (the sole surface outlet) will be eliminated. All water leaving the lake will enter the created wetland by gravity and will move through the wetland as sheetflow. The wetland will be shallow (mean depth ca. 75 cm) and almost continuously inundated. It will be divided into several operational units that can be isolated and drained for soil compaction and vegetation maintenance. Because the marsh soil surface has subsided from its original level, we can encourage soil creation (with consequent nutri-

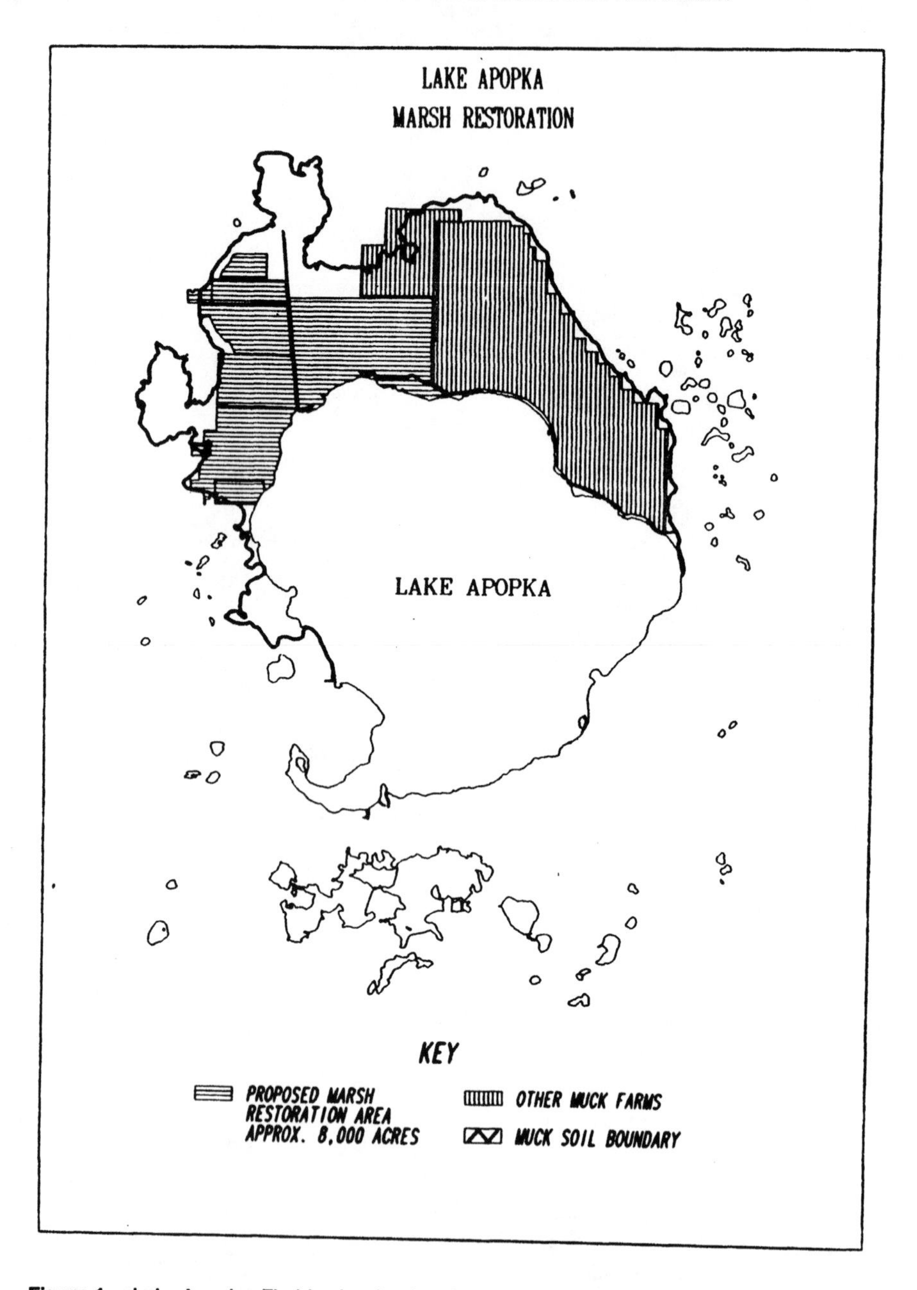

Figure 1. Lake Apopka, Florida showing farming areas and proposed marsh restoration site.

Table 1. Estimated Volumes of Water, Mean Concentrations of Total Nitrogen and Total Phosphorus, and External Loads of Nitrogen and Phosphorus for Nutrient Sources of Lake Apopka

Component	Water Volume (millions m^3)	Concentration (mg/L)		Load (metric tons)	
		N	P	N	P
Rainfall	157	1.21	0.048	190	8
Springflow	22	2.55	0.045	55	2
Lateral flow	4	6.13	0.440	24	2
Outflow	29	4.49	0.220	−130	−7
Farm discharges	76	5.18	2.000	392	151

Source: Brezonik et al.,[2] St. Johns River Water Management District,[3] and Central Florida Agricultural Institute.[5]

ent storage) for many years. A small portion of the flow (ca. 6%) will exit by gravity downstream; most (ca. 94%) will be pumped back into the lake via the canal (Figure 2). At 14.4 m^3/s, about twice the lake volume could be passed through the marsh annually.

Wetland vegetation will enhance the power and capacity of the marsh for nutrient removal. Near the inflows, where nutrient loading will be highest, species with high production rates (e.g., *Typha latifolia* and *Eichhornia crassipes*) will be grown and periodically harvested to increase the rate of nutrient removal. Distant from the inflows, species with capacity to store large amounts of nutrients in woody biomass (e.g., *Taxodium distichum*) will be planted.

The second goal will be achieved by careful design of the interface between the wetland and the lake. Water control structures will be baffled to decrease the velocity of flow in order to permit bidirectional movement of aquatic fauna. Each structure will be the focus of a network of shallow channels reaching into the lake and the wetland (Figure 2). This "biological manifold" is expected to promote the movement of aquatic organisms through the structures and increase the effective biological breadth of the connection between wetland and lake without increasing the size of the hydrologic connection.

Habitat diversity, the third goal, will stem from the existing and created topographic relief and concomitant spatial variation in the hydrologic regime. Planting may be required to augment species diversity, but there is evidence that a diverse seed bank resides in the peat. To discourage early dominance by opportunistic species (e.g., *Typha latifolia*), a dense crop of grass will be grown on the site prior to inundation.

Expected Benefits

The first benefit will be the elimination of nearly 45% of the agricultural discharge, approximately 41% (66 metric tons) of the total annual loading of P (Table 2). A second benefit will be increased nutrient export from the lake. With an inflow of 14.4 m^3/s, the site would receive about 100 metric tons P/yr. Assuming a treatment efficiency of 30%, annual P removal from the lake

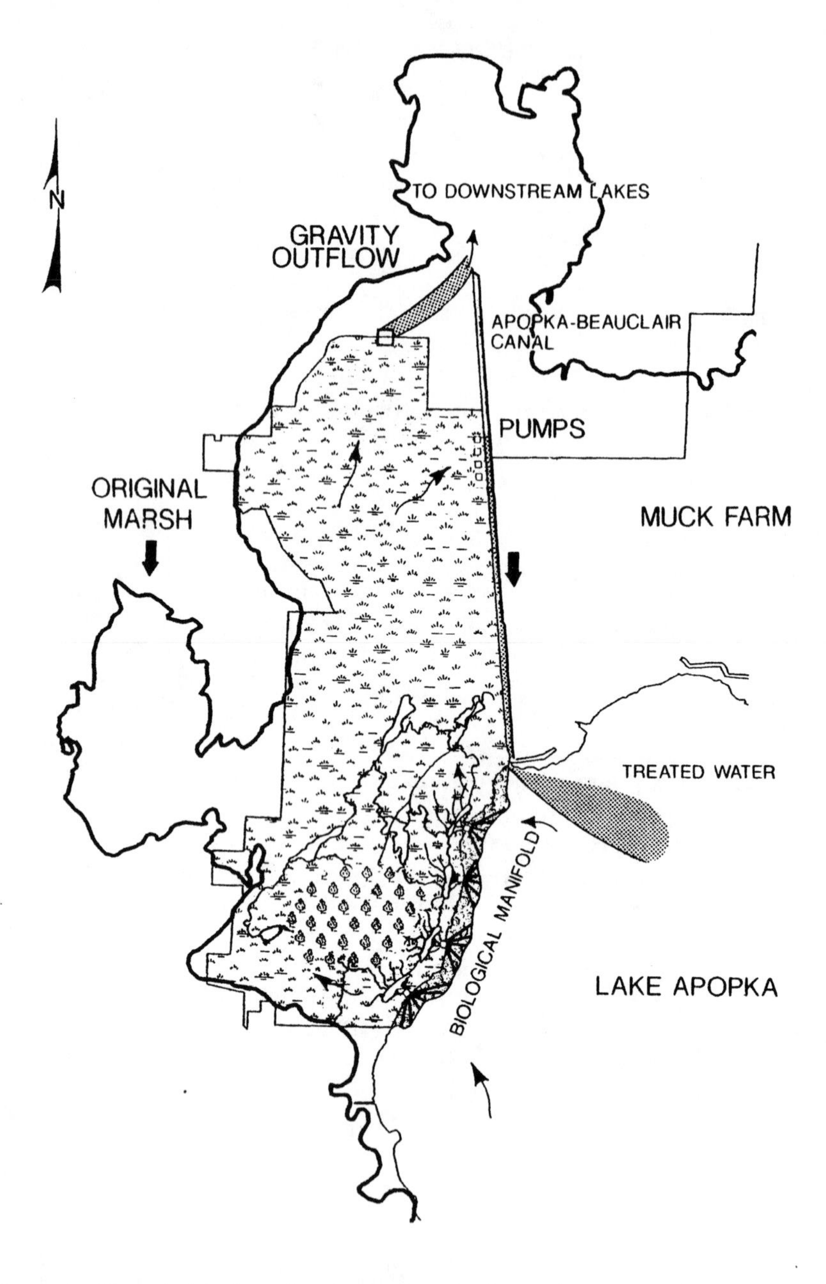

Figure 2. Phase I of the Lake Apopka marsh restoration project. Arrows depict direction of water flow.

Table 2. Estimated Phosphorus Budget of Lake Apopka Under Three Conditions

	Condition		
Component	1	2	3
Rainfall	8	8	8
Springflow	2	2	2
Lateral flow	2	2	2
Outflow	–7	–5	–5
Farm discharges	151	85	4
Wetland storage	0	–30	–30
Gross load	163	97	16
Net load	156	62	–19

Note: Condition 1 is without modification. Condition 2 is with creation and operation of the proposed 32.4 km^2 wetland treatment system. Condition 3 is condition 2 with a 95% reduction in the phosphorus load from the remaining farms. All values are metric tons of phosphorus.

would be 30 metric tons. An additional 5 metric tons would be exported downstream. Thus, annual export of P would increase from 7 to 35 metric tons (Table 2). The combined effect would be a reduction in the net P budget from 156 to 62 metric tons. Although substantial, this reduction will not be sufficient to restore the lake; we believe that restoration depends upon creation of a negative P budget. This will require a substantial reduction in P loading from the remaining farms. The farms are under court action by state regulatory agencies seeking to dramatically reduce their nutrient loading. Assuming the remaining agricultural load will be reduced 95%, the annual P loading would be approximately 16 metric tons. At this point, operation of the proposed wetland system would produce a negative budget for P (–19 metric tons, Table 2), and lake restoration would begin.

Some 63 km^2 of wetland habitat, once contiguous with the lake, have been lost to agricultural development. Creation of 32 km^2 of wetland would restore 52% of the lost habitat. If adequately connected to the lake, the marsh would markedly improve sport fish populations by providing spawning and growth habitat. The restored marsh also would provide a regionally significant habitat for wading birds and waterfowl.

DISCUSSION

The efficiency assumed above (30%) is conservative. At least 50% of the P in Lake Apopka's water is associated with living algae;[5] soluble reactive P accounts for another 1–20%. The balance is probably associated with suspended sediments. Thus, 80% or more of the P is particulate and would be filtered[7] by the wetland. Natural wetlands can remove 60–90% of suspended solids in wastewaters;[8,9] 30% removal is below the minimum expected efficiency (60% efficiency × 80% particulate P = 48% removal). In addition, some soluble P would be expected to be transformed and stored in soil and in plant biomass.[10] South Florida wetlands receiving agricultural wastewater high

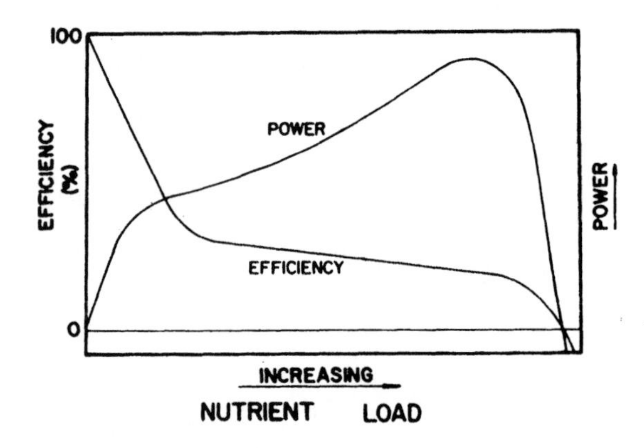

Figure 3. Theoretical relationship of areal loading (g nutrient/m^2) to removal efficiency (g nutrient removed per g nutrient received × 100) and power (g nutrient removed per unit time).

in soluble reactive P (SRP = 58–67% of total P)[11] removed a minimum of 26% of the P and averaged 55% removal.[12]

Although treatment efficiency must be considered, the primary concern in the proposed system is the amount of nutrient that could be removed annually rather than removal efficiency. As P loading increases, P removal efficiency decreases asymptotically,[13] but the power of nutrient removal increases (Figure 3). For example, at 0.85 m^3/s the areal loading rate would be 0.18 $g/m^2/yr$, the removal efficiency would be about 90%, and the power of nutrient removal would be 5 metric tons. At 14.4 m^3/s, the areal loading rate would be 3.1 g P/m^2/yr and the removal efficiency would fall to 50%, but 50 metric tons would be removed (a 10-fold increase in power).

Maximum power would be attained at a relatively low efficiency. We believe, however, that the complete efficiency curve is sigmoidal (Figure 3). We expect rapidly declining efficiency at high hydraulic loading due to low retention time and high current velocity. Power would also drop precipitously at these levels.

Wetland creation could be appropriate for restoration of many other hypertrophic lakes. In any application, water must be moved against a hydraulic gradient; thus, economic constraints will always apply. The realized power of nutrient removal will most likely be determined by the cost of moving water. In large-scale applications, funding may limit power to a level well below the maximum. Lake Apopka should provide a useful example of the efficacy of wetland creation in hypertrophic lake restoration.

REFERENCES

1. "The Ecological Impacts of Wastewater on Wetlands—An Annotated Bibliography," EPA-905/3-84-002, EPA Region V, Chicago, IL and Fish and Wildlife Service, Eastern Energy and Land Use Team, Kearneysville, WV (1984).
2. Brezonik, P.L, C. D. Pollman, T. L. Crisman, J. N. Allinson, and J. L. Fox. "Limnological Studies on Lake Apopka and the Oklawaha Chain of Lakes. I. Water Quality in 1977," Report No. ENV-07-78-01, Department of Environmental Engineering, University of Florida, Gainesville (1978).
3. St. Johns River Water Management District. Unpublished data (1988).
4. Huber, W. C., P. L. Brezonik, J. P. Heaney, R. E. Dickinson, S. D. Preston, D. S. Dwornik, and M. A. DeMaio. "A Classification of Florida Lakes," Report No. ENV-05-82-1, Department of Environmental Engineering Sciences, University of Florida, Gainesville (December 1982).
5. "Central Florida Agricultural Institute Water Quality Monitoring Program," Document No. 07-025.00, prepared for the Central Florida Agricultural Institute by Post, Buckley, Schuh & Jernigan, Inc. (1986).
6. Reddy, K. R. "Soluble Phosphorus Release from Organic Soils," *Agric. Ecosys. Environ.* 9:373-382 (1983).
7. Richardson, C. J. "Freshwater Wetlands: Transformers, Filters, or Sinks?" *FOREM Magazine* 11(2):3-9 (1988).
8. Day, J. W., Jr., and G. P. Kemp. "Long-Term Impacts of Agricultural Runoff in a Louisiana Swamp Forest," in *Ecological Considerations in Wetlands Treatment of Municipal Wastewaters*, P. J. Godfrey, E. R. Kaynor, S. Pelczarski, and J. Benforado, Eds. (New York: Van Nostrand Reinhold Company, 1985), p. 317.
9. Reed, S. C., R. K. Bastian, and W. J. Jewell. "Engineering Assessment of Aquacultural Systems for Wastewater Treatment: An Overview," in *Aquaculture Systems for Wastewater Treatment: An Engineering Assessment,* S. C. Reed and R. K. Bastian, Eds. U.S. EPA Report-430/9-80-007 (1981).
10. Dolan, T. J., S. E. Bayley, J. Zoltek, Jr., and A. J. Hermann. "Phosphorus Dynamics of a Florida Freshwater Marsh Receiving Treated Wastewater," *J. Appl. Ecol.* 18:205-219 (1981).
11. Lutz, J. R. "Water Quality and Nutrient Loadings of the Major Inflows from the Everglades Agricultural Area to the Conservation Areas, Southeast Florida," Technical Publication No. 77-6, Resource Planning Department, South Florida Water Management District, West Palm Beach, FL (November 1977).
12. Davis F. E., A. C. Federico, A. L. Goldstein, and S. M. Davis. "Use of Wetlands for Water Quality Improvements," in *Stormwater Management—An Update, Proceedings of Symposium,* M. P. Wanielesta and Y. A. Yousef, Eds. University of Central Florida Environmental and Systems Engineering Institute, Publication No. 85-1 (July 1985).
13. Nichols, D. S. "Capacity of Natural Wetlands to Remove Nutrients from Wastewater," *J. Water Pollut. Control Fed.* 55(5):495-505 (1983).

CHAPTER 41

Applications to Industrial and Landfill Wastewaters

41a Utilization and Treatment of Thermal Discharge by Establishment of a Wetlands Plant Nursery

M. Stephen Ailstock

INTRODUCTION

The Nevamar Corporation, Odenton, Maryland manufactures decorative building surfaces for counter tops, furniture, and other interior applications. The product is made by curing phenolic/melamine resins in presses at 6900 kPa for 60 min. Heat generated by this exothermic reaction is extracted by heat exchange with noncontact groundwater, pH 5.0–5.2. Water temperature at the presses reaches 51.6°C. Temperature of the cooling water is lowered to 43.3°C by dilution with an additional 32,000 L/hr groundwater. The pH of all process water is adjusted over a range of 6.6–6.8, and water is diverted to a holding pond for further temperature reduction. Treated water is discharged into Picture Spring Branch, a small tributary of the Severn River. Presses are activated sequentially at 15 min intervals and operate 7 day/wk, 24 hr/day. Flow from the presses into and subsequently out of the pond is continuous at 284,000 L/hr.

Discharge of cooling water is regulated by the Maryland Department of Environmental Protection through issuance of a National Pollution Discharge Elimination System (NPDES) permit. The original permit (1976) set maximum temperature at the pond outfall at 36.7°C. In 1986, a second permit required discharge temperature not to exceed 32.2°C and pH to be 6.0–8.5. As a part of the permit process, Nevamar studies to improve holding pond thermal efficiency identified increasing pond depth, surface area, and equalization and retention time as suitable modifications. These improvements were compatible with and would be optimized with a wetlands plant nursery.

The ecological significance of wetlands is widely recognized.[1-4] Despite the importance of wetlands for maintaining environmental quality, total acreage of this resource has been greatly reduced.[5,6] To encourage wetlands preservation, legislation has been enacted by federal, state, and local governments to protect existing wetlands and encourage efforts to create wetland areas.[7]

The creation and restoration of wetlands requires native plant materials sources, propagation and production systems, evaluation of plant materials and site characteristics to achieve objectives, and a procedure for installing plants at the preselected locations.[8] In 1985, the Nevamar Corporation funded Anne Arundel Community College to design and manage a wetlands plant nursery to use the thermal discharge from their manufacturing facility. Harvested seeds and vegetative structures from nursery-grown plants treated and stored at the college were grown in a greenhouse to produce transplant materials for wetlands projects.

Initial stock acquisition, propagation systems development, and laboratory manipulation provided employment/learning opportunities for college students; the nursery effort was then expanded to include the mentally handicapped to help with the repetitive tasks involved in growing plants. The Providence Center is a nonprofit service organization providing training and employment to developmentally disabled adults. State and federal agencies provide 80% of their budget, and 20% is self-generated through product assembly and a recycling service. Adding aquatic plant production to Earthtones, an existing program for producing and marketing horticultural crops, would increase employment opportunities and generate positive cash flow. In 1986, Providence Center clients planted propagation materials—seeds, tubers, rhizome cuttings, and divisions—and grew plants in containers to sizes appropriate for field planting. Plants were of high quality and performed well in field installations. Costs were funded by the Nevamar Corporation, and many were donated to community groups for wetlands projects.

This chapter reports thermal treatment pond/aquatic nursery design, efficiency of modifications for improving wastewater treatment, nursery productivity during the first year, and a summary of potential applications.

POND DESIGN

Prior to renovation, the pond had approximately 1600 m^2 of surface area and average depth of 0.25 m. Discharge into the pond was direct via two 0.91-m-diameter concrete pipes. Distribution of thermal flow within the pond was minimal. Short-circuiting resulted from severe channelization of the bottom from source to outfall. The outfall consisted of a shored break in a sandbag retaining wall adjacent to the receiving stream.

The pond as it currently exists is shown in Figure 1. Thermally affected water enters a treated wood distribution box that diverts one-third of the flow through a 1-m wooden trough (primary channel) serving the planting rows

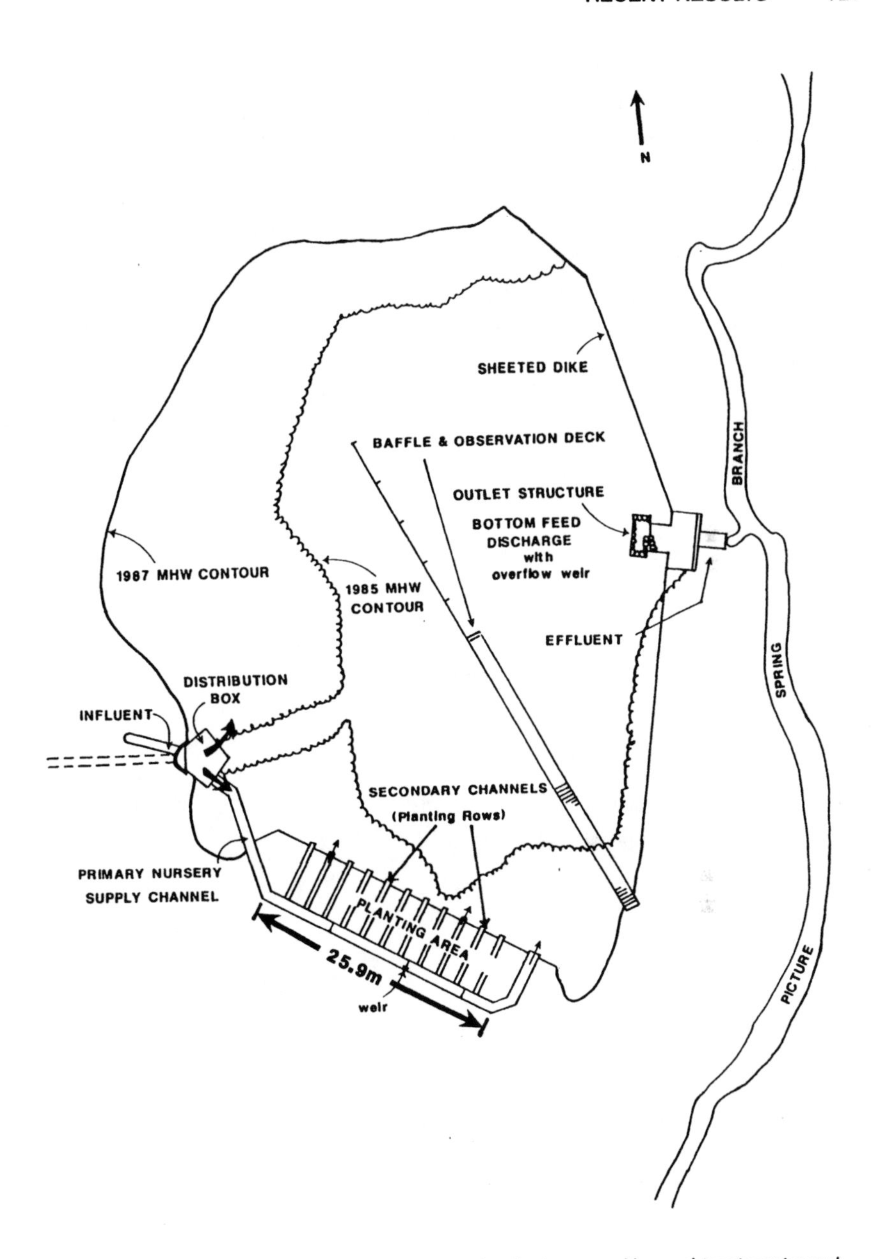

Figure 1. Design features of the Nevamar aquatic plant nursery/thermal treatment pond.

(secondary channels). Water entering the pond and primary channel passes beneath oil absorbent booms installed to remove petrochemicals from storm drains in parking areas. Diffusion from primary and secondary channels and buffering the sheeted wood baffle facilitate mixing by directing flow to all parts of the pond. The sheeted dike increased average depth to 0.50 m and pond surface area to 5300 m^2. The outlet structure incorporates bottom feed discharge with an overflow weir to retain warmer water. Weir effluent passes through a 1.5-m-by-5.0-m gabion-lined channel to increase oxygenation.

Pond discharge temperatures are recorded daily and weekly, and Nevamar continuously monitors influent for temperature, pH, and turbidity. If necessary, sealing pipes with compressed air containment bags stop pond influent and the water discharge interrupted by weir adjustment at the outfall.

TREATMENT EFFICIENCY

Prior to pond renovation, temperatures during the summer months approached the permit limit of 32.2°C (Figure 2). Modifications for short-circuiting problems decreased effluent temperatures by an average of 5°C from April 1987 to April 1988, even though construction was not completed until after August. Temperature variations following construction range from 0.0°C to 1.2°C compared with 10.6°C to 14.7°C in previous years. Lower and more uniform temperatures of the thermal discharge into Picture Spring Branch should minimize impacts to resident biota.

NURSERY OPERATION

Production methodologies for many species of aquatic plants used for wetlands creation and restoration are well established.[8-11] Procedures used for growing aquatic plants in containers prior to field establishment are simple, paralleling those used for more conventional types of horticulture crops.[12] Plants are grown in 4- to 12-L containers and placed in the secondary channels filled with gravel (Figure 3). Channel water heights are adjusted by modified weirs according to requirements of the particular species. Flow provides continuous irrigation and, with the exception of an occasional supplementary application of slow-release fertilizer, plants are maintenance free. Availability of warm water greatly extends the growing season and significantly enhances belowground growth. Species and number produced during the first year of operation are given in Table 1. Efforts are to maximize production efficiency and quantify differences in reproductive performance between nursery plants and those in the wild.

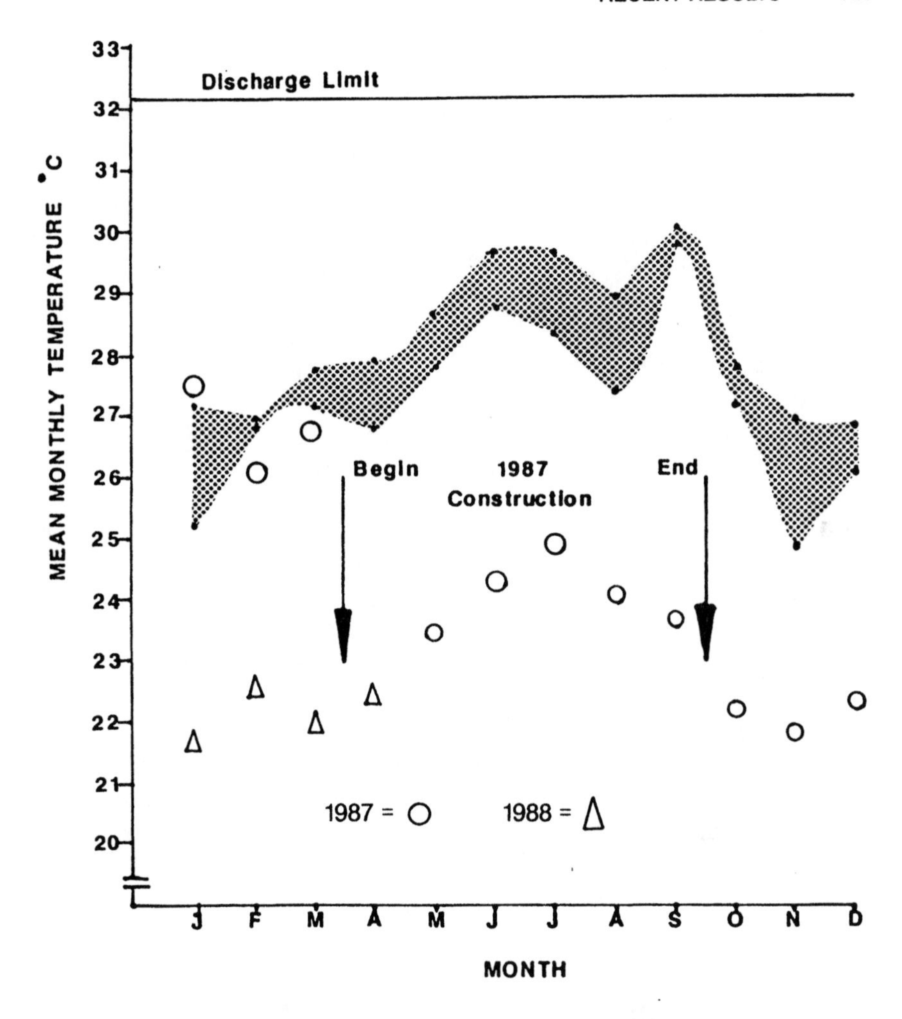

Figure 2. Mean monthly temperatures of effluent from the aquatic plant nursery/thermal treatment pond from 1985 to present. High-low monthly means for 1985 and 1986 shaded.

WETLANDS CREATION

Two wetlands projects have been installed with plants produced during spring 1988. A small *Spartina* marsh was created for Annapolis, Maryland to stabilize a parcel of shoreline, provide stormwater management, and minimize human activity in the area. *S. alterniflora* and *S. patens* (500 each) planted in the project area of 111 m^2 by community volunteers were donated by the Nevamar Corporation.

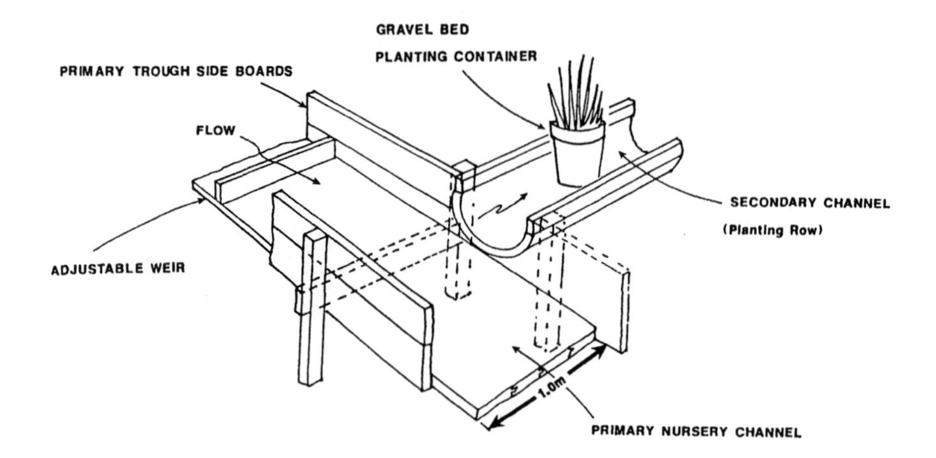

Figure 3. Cross section through the primary channel and secondary channel. Arrows indicate direction of flow toward the outlet structure.

The second project, also a tidal wetland, was funded by the U.S. Navy. In the first phase, 15,000 *S. alterniflora* and 10,000 *S. patens* were planted to stabilize 610 m of shoreline along the Severn River on the David Taylor United States Naval Station property. During the second phase, 500 each of *Hibiscus palustris*, *Panicum virgatum*, *Scirpus robustus*, and *Myrica pensylvanica* will be planted to maximize the habitat value of the site. This project has publicized the value of nonstructural shoreline stabilization, and proceeds have employed college students and clients of the Providence Center.

Table 1. Plants Produced During the 1987–1988 Growing Season from the Nevamar Aquatic Plant Nursery

	Number Produced by Propagation Structure		
Species	**Seeds**	**Vegetative**	**Total Produced**
Spartina alterniflora	16,500	2,500	19,000
Spartina patens	10,000	6,050	16,050
Spartina cynosuroides	1,000	—	1,000
Pontedaria cordata	—	250	250
Hibiscus palustris	1,000	—	1,000
Panicum virgatum	—	500	500
Scirpus robustus	1,000	—	1,000
Misc. species	5,000	2,000	7,000
Total			45,800

Note: All plants are perennial, reproducing by both seeds and vegetative structures.

SUMMARY

Inclusion of an aquatic plant nursery with treatment of thermally affected industrial wastewater has (1) reduced Nevamar Corporation's effluent temperatures to permit limitations; (2) provided plants for constructing wetlands to meet local environmental objectives; and (3) developed a self-supporting employment program for the Providence Center. Applications of this model system extend to other types of wastewater management. Residential wastewater systems that include aquatic plants for water polishing benefit from periodic plant harvesting.[13-15] Aquatic plants may be used by industries for transpirational dewatering of fine-textured ore residues and to reduce spoil disposal costs, and harvested plants may contribute to environmental restoration projects. Given availability of competent developmentally disabled workers, high demand for aquatic plants, and responsibility of industry to maximize beneficial use of by-products, establishment of multiuse treatment systems should be encouraged.

REFERENCES

1. Odum, W. E., J. T. Smith III, J. K. Hoover, and C. C. McIvor. "The Ecology of Tidal Freshwater Marshes of the United States East Coast: A Community Profile," U.S. Department of the Interior, FWS/OBS-83/17 (1984).
2. Martin, A. C., H. S. Zim, and A. L. Nelson. *American Wildlife and Plants: A Guide to Wildlife Food Habits* (New York: Dover Publications, Inc., 1961).
3. Eleuterius, L. N. "Submergent Vegetation for Bottom Stabilization," in *Estuarine Research, Vol. II Geology and Engineering* (New York: Academic Press, Inc., 1975), pp. 439–456.
4. Ward, L., W. M. Kemp, and W. R. Boynton. "The Influence of Waves and Seagrass Communities on Suspended Sediment Dynamics in an Estuarine Embayment," *Mar. Geol.* 59:85–103 (1984).
5. Cowardin, L. M., V. Carter, F. C. Golet, and E. T. LaRoe. "Classification of Wetlands and Deepwater Habitats of the United States," U.S. Department of the Interior, FWS/OBS-79/31 (1979).
6. Tiner, R. W., Jr. "Wetlands of the United States: Current Status and Recent Trends," USFWS National Wetlands Inventory, U.S. Government Printing Office (March 1984).
7. Dillman, B. A. "Wetlands Regulation: An Effective Approach," in *Proceedings of the 12th Annual Conference on Wetlands Restoration and Creation*, F. J. Webb, Ed. (Tampa, FL: Hillsborough Community College, 1985).
8. Lewis, R. R., III, Ed. *Creation and Restoration of Coastal Plant Communities* (Boca Raton, FL: CRC Press, Inc., 1982).
9. Ailstock, M. S. "Summary Waterfowl Habitat Improvement Using Propagated Native Plants," Maryland Department of Natural Resources, Forest, Park and Wildlife Service, Annapolis, MD (1987).
10. deMond, J. D., D. R. Clark, and B. E. Spicer. "A Review of Wetland Restoration, Enhancement, and Creation Practices in the Louisiana Coastal Zone," in *Proceed-*

ings of the 13th Annual Conference on Wetlands Restoration and Creation (Tampa, FL: Hillsborough Community College, 1986), pp. 64–74.

11. Madsen, C., R. Dornfeld, and J. Piehl. "A Pilot Program for Private Lands Wetland Restoration in Minnesota," in *Proceedings of the 13th Annual Conference on Wetlands Restoration and Creation* (Tampa, FL: Hillsborough Community College, 1986), pp. 130–133.
12. Boadley, J. W. *The Commercial Greenhouse* (Albany, NY: Delmar Publishers, 1981).
13. Cerra, F. J., and J. W. Maisel, Ed. *Wastewater Engineering: Treatment, Disposal, Reuse,* 2nd ed. (New York: McGraw-Hill Book Company, 1979).
14. Johnson, R., and C. W. Sheffield. "Wetlands for Effluent Disposal," in *Proceedings of the 12th Annual Conference on Wetlands Restoration and Creation* (Tampa, FL: Hillsborough Community College, 1985), pp. 159–168.
15. Johnson, R., E. I. Sheffield, and C. W. Sheffield. "Innovative Flow-Through Wetland for Effluent Disposal," in *Proceedings of the 13th Annual Conference on Wetlands Restoration and Creation* (Tampa, FL: Hillsborough Community College, 1986), pp. 180–188.

41b Experiments in Wastewater Polishing in Constructed Tidal Marshes: Does It Work? Are the Results Predictable?

Vincent G. Guida and Irwin J. Kugelman

INTRODUCTION

Natural tidal salt marshes may have limited use in wastewater treatment applications. High-strength, low-toxicity, saline wastewater from seafood processing are good candidates for such treatment.[1] Varying strength and corrosive salinity disrupt combined industrial and domestic treatment systems.[2] Exclusive treatment of clam processing wastewater for direct discharge is common, but new state standards have made this difficult. Extended aeration activated sludge treatment cannot meet direct discharge standards for BOD_5 or TSS in New Jersey,[3] and no better process is known. Furthermore, this process is inefficient at removing nitrogen and phosphorus.

We investigated salt marsh "polishing" of effluent from activated sludge treatment of clam processing waste to meet state regulations for BOD_5 and TSS and anticipated standards for N and P. Constructed salt marshes might be used, but abundant natural marshes are a more obvious choice. Natural freshwater marshes are effective in reducing BOD_5, TSS, N, and P in a number of cases.[4-6] With salt-tolerant microbiota, plants, and animals, salt marshes seem ideal for treating saline clam processing water.

Salt marshes exposed to semisolid sewage sludge[7,8] and fertilizers[9] retained substantial N over the short term, but further investigations with sludge were discouraged by large heavy metal fluxes.[10] Complex tidal hydrology makes quantitative assessment of treatment for aqueous wastes efficacy difficult. Simple comparisons of upstream and downstream concentrations are meaningless. Tidal flooding is discontinuous and of brief duration, and depth is variable. Flow patterns are multidirectional and rapidly changing, and flow boundaries are poorly defined.

U.S. EPA investigated BOD_5 and TSS in a salt marsh treating shrimp processing wastewater.[2] However, conclusions were site-specific because material fluxes were not determined and effects of environmental factors on efficacy were not assessed. Factors influencing material fluxes dictate impacts of polishing on estuarine waters, of major regulatory concern. We used constructed marsh microcosms with simple hydrology to assess wetlands' ability to polish effluents from activated sludge units treating clam processing waste. This allowed simultaneous replicate treatments, increasing statistical confidence, and aqueous mass balance (flux) calculations.

In this chapter, three environmental factors are addressed using experimental results: (1) Does tidal flooding frequency, which shortens wastewater residence to a few hours, prevent effective treatment? (2) Natural salt marshes can demonstrate either net import or export of organic material and nutrients to surrounding waters.[11,12] Is the behavior of pristine marshes a good indicator of their ability to polish wastewater? (3) Is the outcome of polishing readily predictable in a variable natural marsh? Data from experiments are presented to answer these questions.

MATERIALS AND METHODS

Our experiments were performed at the Wetlands Institute near Stone Harbor, New Jersey from August 1985 to January 1987. Clam processing wastewater was treated by extended aeration activated sludge in two 440-L fill-and-draw units (units A and B). Feeding and sludge wasting was done daily. Unit temperature was maintained >14°C at all times. Solids retention times (SRT) for units A and B were seven and 15 days, respectively, in winter and three and seven days from April to October. Unit effluent was settled for 20 min, decanted, and applied to marsh microcosms by sprinkling measured volumes five days per week.

Polishing experiments were performed in outdoor marsh microcosms. These semiisolated plots were subject to controlled wastewater application and tidal flooding while exposed to ambient climate. Microcosms were planted in marine plywood chambers 75 × 59 × 60 cm. Chambers held intact cordgrass sods (30 cm deep) from a natural salt marsh. Drains at the sod surface level allowed drainage of standing water from enclosed marshes but not soil water.

Tidal floodwater was pumped from a creek into two 570-L fiberglass tanks and distributed to each box via PVC pipe. Tidal flooding simulated the schedule of natural marsh flooding. All results presented here are from microcosm marshes in steady-state conditions.[3] We randomly assigned five of 15 low marsh microcosms as controls without effluent. Treatments were randomized among the 10 treated microcosms. On January 22, all 10 treated microcosms received the same treatment. In the May 8 and August 23 trials, five received effluent from unit A (shorter SRT), and five received effluent from unit B (longer SRT) at 2.3 L/m². During other experiments included in regression analyses, effluent rates up to 11 L/m² and raw processing wastewater at 2.3 L/

Table 1. Concentrations of Water Quality Parameters in Activated Sludge Effluent, Tidewater, and Runoff Water from Control and Treated Marsh Microcosms

Date (Temp.)	Water Quality Parameter	Activated Sludge Effluents A	Activated Sludge Effluents B	Tidewater	Marsh Runoff Means Control	Marsh Runoff Means Treatment A	Marsh Runoff Means Treatment B
January 22 (0.5°C)	BOD_5	380	—	1.5	0.8	2.1	—
	TSS	1540	—	13	16	13	—
	$NO_2 + NO_3$	6.51	—	0.08	0.08	0.28	—
	NH_3	124.81	—	0.06	0.04	1.50	—
	org. N	158.15	—	0.03	0.27	0.88	—
	oPO_4	13.35	—	0.01	0.01	0.45	—
	org. P	47.40	—	0.08	0.09	0.11	—
May 8 (23°C)	BOD_5	89	60	0.4	0.9	2.0	1.7
	TSS	191	120	9	13	15	13
	$NO_2 + NO_3$	3.80	24.18	0.03	0.02	0.19	0.32
	NH_3	118.96	59.55	<0.01	0.02	0.67	0.24
	org. N	42.13	18.55	0.72	0.55	0.85	0.76
	oPO_4	26.53	24.32	0.02	0.03	0.26	0.27
	org. P	4.13	1.40	0.02	0.04	0.05	0.03
August 23 (24°C)	BOD_5	60	70	1.7	3.0	3.0	4.0
	TSS	164	243	7	7	8	8
	$NO_2 + NO_3$	5.64	0.97	0.01	0.01	0.03	0.02
	NH_3	55.20	45.21	<0.01	0.17	0.20	0.20
	org. N	<0.01	<0.01	0.13	0.47	0.48	0.41
	oPO_4	23.85	31.35	0.04	0.07	0.21	0.16
	org. P	5.63	1.20	<0.01	0.07	0.10	0.21

Note: Values are in mg/L, mg N/L, or mg P/L.

m^2 were used. Runoff following tidal flooding was the only source of aqueous output from the marsh boxes.

Samples of activated sludge effluent applied to microcosms, tidewater, and tidal runoff were drawn from mixed batches and analyzed for BOD_5, TSS, total Kjeldahl N (TKN), nitrate plus nitrite N ($NO_2 + NO_3$), ammonia N (NH_3), orthophosphate P (oPO_4), and total P (TP).[13,14] Organic nitrogen (org. N) is TKN minus NH_3, and organic phosphorus (org. P) is TP minus oPO_4. Subsamples of runoff from all microcosms within each treatment group, including controls, were pooled for the purposes of BOD_5 and TSS analysis, and statistical analysis was not possible. N and P analyses were performed for runoff from each microcosm and analyzed with Student's t-test.

RESULTS AND CONCLUSIONS

Water Quality

Activated sludge effluents have substantial BOD_5, TSS, NH_3, and oPO_4 concentrations (Table 1), and tidewater has low concentrations. Polishing degrades the quality of flooding tidewater. In the January 22 and May 8 results, BOD_5, $NO_2 + NO_3$, NH_3, org. N, and oPO_4 concentrations were

Table 2. Daily Values and Statistical Significance of Mass Changes

Date	Water Quality Parameter	Treatment A Total Loading	Treatment A Mean Mass Change ± SD	Treatment B Total Loading	Treatment B Mean Mass Change ± SD
January 22	BOD_5	994	–804 —	—	— —
	TSS	4655	–3480 —	—	— —
	$NO_2 + NO_3$	22.3	+3.4 ± 0.8	—	— —
	NH_3	287.3	–151.7 ± 4.5**	—	— —
	org. N	359.7	–280.2 ± 0.7**	—	— —
	oPO_4	31.4	+9.5 ± 0.0	—	— —
	org. P	39.0	–28.9 ± 2.4**	—	— —
May 8	BOD_5	237	–56 —	172	–18 —
	TSS	1246	+110 —	1085	+90 —
	$NO_2 + NO_3$	10.9	+7.6 ± 0.9	56.9	–27.9 ± 5.6**
	NH_3	268.9	–208.2 ± 8.4**	134.7	–113.3 ± 6.0**
	org. N	160.1	–94.1 ± 8.0**	106.8	–38.2 ± 12.9**
	oPO_4	61.3	–36.7 ± 2.9**	56.3	–31.7 ± 2.9**
	org. P	11.3	–7.0 ± 0.8**	5.1	–2.2 ± 1.3**
August 23	BOD_5	328	+11 —	350	+102 —
	TSS	1456	–609 —	1426	–522 —
	$NO_2 + NO_3$	13.4	–9.3 ± 1.3**	2.8	–1.3 ± 1.6
	NH_3	125.1	–106.7 ± 7.9**	102.8	–85.1 ± 9.7**
	org. N	11.5	+32.2 ± 9.7	11.5	+24.8 ± 19.2
	oPO_4	58.4	–47.4 ± 5.1**	75.4	–52.7 ± 4.6**
	org. P	12.7	–3.3 ± 3.4	2.7	+6.4 ± 3.2

Note: (–) = removal; (+) = generation (corrected for control values in treated microcosms). Values are in mg/m^2, mg N/m^2, or mg P/m^2. ** = mass removals significantly greater than 0.0 mg/m^2 ($p_{V=0.0} < 1.0\%$).

higher in treated runoff than in control marsh runoff. On August 23, only BOD_5 and org. P in the treatment B group and oPO_4 in both treatment groups were higher than control levels. Less dramatic effects on water quality on August 23 may have been due to the weakness of the applied effluent on that date.

Mass Balances and the Residence Time Hypothesis

Total loading values include loading from applied effluent and from tidewater. On January 22, large proportions of BOD_5, TSS, NH_3, org. N, and org. P loadings were removed by the marsh despite the 0.5°C water temperature (Table 2). Removal of NH_3-N, org. N, and org. P was statistically significant. Small amounts of $NO_2 + NO_3$ and oPO_4 were generated by marshes on that date (positive mean mass changes). On May 8, both treatments had small net removals of BOD_5 and small net additions of TSS. Removal of all forms of N and P in both groups, excepting $NO_2 + NO_3$ in the treatment A group, was significant. On August 23, there was net generation of BOD_5 and removal of TSS by both groups, significant removal of NH_3 and oPO_4 in both groups, and significant removal of $NO_2 + NO_3$ in the treatment A group. These data suggest that net removal occurs at high loading rates, whereas net generation occurs at low loading rates.

Table 3. Daily Values and Statistical Significance of Mass Changes

Date	Water Quality Parameter	Tidewater Loading	Mean Mass Change ± SD	Implications for H_0 #2	
				Treatment A	Treatment B
January 22	BOD_5	136	−63 —		
	TSS	1175	+271 —		
	$NO_2 + NO_3$	7.6	−0.6 ± 0.4		
	NH_3	5.3	−2.0 ± 0.9		
	org. N	2.3	+16.1 ± 0.5**	X	
	oPO_4	1.2	+0.2 ± 0.1		
	org. P	9.1	−2.9 ± 2.5		
May 8	BOD_5	36	+45 —		
	TSS	814	+361 —		
	$NO_2 + NO_3$	2.3	−0.1 ± 0.9		
	NH_3	0.1	+2.0 ± 0.7**	X	X
	org. N	64.9	−15.0 ± 12.3		
	oPO_4	1.4	+1.5 ± 0.3**	X	X
	org. P	2.0	+1.3 ± 1.3		
August 23	BOD_5	192	+147 —		
	TSS	814	−80 —		
	$NO_2 + NO_3$	0.6	+0.5 ± 1.0		
	NH_3	0.3	+15.4 ± 5.2**	X	X
	org. N	11.5	+30.5 ± 9.5**		
	oPO_4	3.6	+2.8 ± 1.3**	X	X
	org. P	<0.1	+6.7 ± 1.3**		

Note: (−) = removal; (+) = generation of water quality in control microcosms and implications for hypothesis #2. Values are in mg/m^2, $mg\ N/m^2$, or $mg\ P/m^2$. ** = significant mass generation of the parameter by the marsh. X = significant mass generation in control microcosms plus significant mass removal in the treatment group (Table 2), thus disproving H_0 #2.

Thus, short residence time (6 hr) of wastewater in a salt marsh due to tidal hydrology did not preclude effective treatment in tidal marshes, even at near-freezing temperatures.

Control Marsh Mass Balances and the Pristine Behavior Hypothesis

Runoff concentrations from control microcosms were often higher than concentrations in influent tidewater (Table 1). These differences show statistically significant net generation (positive mean mass changes) of materials (Table 3). Organic nitrogen was generated by control marshes on January 22, NH_3 and oPO_4 were generated on May 8, and NH_3, org. N, oPO_4, and org. P all had significant net generation on August 23. Generation of TSS occurred during the first two experiments, and generation of BOD_5 occurred during the last. Several instances where control marshes showed net generation coincide with simultaneous net removal of the same parameters in treated marshes (Table 3). In fact, when largest removals occurred in treated marshes, org. N on January 22 and NH_3 on May 8 and August 23 (Table 2), control marshes consistently demonstrated net generation (Table 3). Treated marshes did not show net generation concurrent with control marshes; thus, marsh behavior under pristine conditions is not a good indicator of the marsh's ability to polish wastewater.

Table 4. Least Squares Linear Regressions of Removal (R) vs Input (I) for Seven Warm Season Marsh Polishing Experiments

Water Quality Parameter	Regression Equation	x-Intercept Value (break-even loading)	Correlation Coefficient (r)
BOD_5	R = 0.9751 * I – 130.88	134.22	0.98643
TSS	R = 0.8727 * I – 972.65	1114.53	0.58100
$NO_2 + NO_3$	R = 0.7646 * I – 7.15	9.35	0.98142
NH_3	R = 0.8659 * I – 4.80	5.54	0.99625
org. N	R = 0.9214 * I – 53.85	58.44	0.98954
oPO_4	R = 0.7129 * I – 4.51	6.33	0.96241[a]
org. P	R = 1.0066 * I – 4.20	4.17	0.99601

Note: All values are in mg/m^2, mg N/m^2, or mg P/m^2.
[a] Outlier at high loading (235 mg P/m^2) excluded from regression.

Predictability

Linear regressions of removal vs input for results of seven experiments from April to August yielded correlation coefficients ranging from 0.981 to 0.996 (Table 4). Only TSS had a low correlation ($r < 0.600$) due to data scatter, not curvilinearity. Despite the good fit to BOD_5 and org. N data, much scatter occurred at lower loading rates. This scatter results from the heterogeneous nature of BOD_5, TSS, and org. N lumped together from two distinct sources: wastewater and tidewater. Other N and P parameters lacked data scatter. For homogeneous parameters $NO_2 + NO_3$, NH_3, and oPO_4, source is of no consequence, and removal data are close to a regression line. Only in the case of oPO_4 is there any curvilinearity. At loading rates of up to 100 mg P/m^2, data remain close to the linear regression line, but a single datum at 235 mg P/m^2 falls well below the expected removal value, indicating oPO_4 removal is less efficient at high loading. P may prove to be the load-limiting factor in wastewater application to these and other wetlands.[15]

With the exception of TSS, behavior of all parameters was moderately predictable for the warm season, but available winter data did not fit the same regression lines. Separate cold season regressions will require more data.

Corrected Mass Balances and Removal Efficiencies

In previous calculations, "native" materials exchanges and exchanges of wastewater materials were not distinguished. Such distinction is functionally arbitrary, particularly regarding homogeneous parameters, but is of regulatory value. Assuming the same native exchanges seen in controls underlies exchanges observed in treated marshes, tidewater input and control marsh output can be subtracted from total mass inputs and outputs of treated marshes to yield wastewater inputs and outputs that result solely from applied wastewater. Based on such corrected calculations, BOD_5 removal from effluents in the January 22, May 8, and August 23 experiments range from 29% to 100%; TSS, 58% to 108%; total N, 69% to 98%; and total P, 30% to 73%.

These are comparable with performance for other marsh polishing systems,[4-6] and improve activated sludge effluents enough to meet direct discharge standards.

ACKNOWLEDGMENTS

The authors wish to thank the Saltonstall-Kennedy program (U.S. Dept. of Commerce, NOAA, NMFS); Dr. Robert Lippson; the Wetlands Institute of Stone Harbor, New Jersey; and technicians Dominic Dragotta, John Flynn, John Ludlam, and Robert Berardo for their respective efforts on this project, without which this work would not have been possible.

REFERENCES

1. "Development Document for Effluent Limitation Guidelines and New Source Performance Standards for the Fish Meal, Salmon, Bottom Fish, Clam, Oyster, Sardine, Scallop, Herring, and Abalone Segment of the Canned and Preserved Fish and Seafood Processing Industry Point Source Category," U.S. EPA Report-440/1-75/041a (1975).
2. "Utilization of a Saltwater Marsh Ecosystem for the Management of Seafood Processing Wastewater," U.S. EPA (Region IV) Report-904/9-86 142 (1986).
3. Guida, V. G., and I. J. Kugelman. "Feasibility and Modeling of the Use of New Jersey Salt Marshes to Treat Clam Processing Wastewater," Final Report to the National Marine Fisheries Service, prepared by Environmental Studies Center, Lehigh University, Bethlehem, PA (1988).
4. Chan, E., T. A. Bursztynsky, N. Hantzsche, and Y. J. Litwin. "The Use of Wetlands for Water Pollution Control," Draft Report to Municipal Environmental Research Laboratory, Office of Research and Development, U.S. EPA, Cincinnati, OH (1979).
5. Kadlec, R. H., and D. L. Tilton. "The Use of Freshwater Wetlands as a Tertiary Wastewater Treatment Alternative," *CRC Crit. Rev. Environ. Control* 9:185–212 (1979).
6. Kadlec, R. H. "Wetlands for Tertiary Treatment," in *Wetlands Functions and Values: The State of Our Understanding,* P. E. Greeson, J. R. Clark, and J. E. Clark, Eds. (Minneapolis, MN: American Water Resource Association, 1979), pp. 490–504.
7. Valiela, I., J. M. Teal, and W. Sass. "Nutrient Retention in Salt Marsh Plots Experimentally Fertilized with Sewage Sludge," *Estuar. Coastal Mar. Sci.* 1:261–269 (1973).
8. Chalmers, A. G. "The Effects of Fertilization on Nitrogen Distribution in a *Spartina alterniflora* Salt Marsh," *Estuar. Coastal Mar. Sci.* 8:327–337 (1979).
9. DeLaune, R. D., C. J. Smith, and W. H. Patrick, Jr. "Nitrogen Losses from a Louisiana Gulf Coast Salt Marsh," *Estuar. Coastal Shelf Sci.* 17:133–141 (1983).
10. Giblin, A. E., A. Bourg, I. Valiela, and J. M. Teal. "Uptake and Losses of Heavy Metals in Sewage Sludge by a New England Salt Marsh," *Am. J. Bot.* 67:1059–1068 (1980).
11. Woodwell, G. M., D. E. Whitney, C. A. S. Hall, and R. A. Houghton. "The Flax

Pond Ecosystem Study: Exchanges of Carbon in Water Between a Salt Marsh and Long Island Sound," *Limnol. Oceanog.* 22(5):833–838 (1977).

12. Daly, M. A., and A. C. Mathieson. "Nutrient Fluxes Within a Small North Temperate Salt Marsh," *Mar. Biol.* 61:337–344 (1981).
13. *Standard Methods for the Examination of Water and Wastewater,* 16th ed. (Washington, DC: American Public Health Association, 1984).
14. Strickland, J. D. H., and T. R. Parsons. *A Practical Handbook of Seawater Analysis,* 3rd ed. (Ottawa, ON: Fisheries Research Board of Canada, 1980).
15. Richardson, C. J. "Mechanisms Controlling Phosphorus Retention Capacity in Freshwater Wetlands," *Science* 228:1424–1427 (1985).

41c Potential Use of Constructed Wetlands to Treat Landfill Leachate

Ward W. Staubitz, Jan M. Surface, Tammo S. Steenhuis, John H. Peverly, Mitchell J. Lavine, Nathan C. Weeks, William E. Sanford, and Robert J. Kopka

INTRODUCTION

Infiltration of precipitation and migration of water through municipal solid waste landfills produce leachate that contains undesirable or toxic organic and inorganic chemicals. The chemical quality of landfill leachate differs greatly from one landfill to another and fluctuates seasonally within an individual landfill. Leachate composition is waste- and site-specific depending on the waste type, landfill age, and amount of infiltrating water.

Although leachate quality may vary from relatively harmless to extremely hazardous waste, it does have some consistent characteristics. Leachate is generally anoxic and has high biochemical oxygen demand (BOD_5) and high concentrations of organic carbon, nitrogen, chloride, iron, manganese, and phenols, but little or no phosphorus. Following is a list of selected chemical constituents and typical ranges of leachate from municipal solid waste landfills (Shuckrow et al.):[1]

Constituent:	*Typical ranges:*
alkalinity, as $CaCO_3$	21–5400 mg/L
ammonia	0.01–1000 mg/L
arsenic	0.011–10,000 mg/L
barium	0.1–2000 mg/L
benzene	<1.1–7370 μg/L
BOD_5	42–10,900 mg/L
cadmium	5–8200 μg/L
chloride	4–9920 mg/L
chromium	0.001–208 mg/L
DDT	4.28–14.26 μg/L
dieldrin	<2–4.5 μg/L
iron	0.090–678 mg/L
lead	1–19,000 μg/L
manganese	0.010–550 mg/L

pH	3–7.9
phenols	<3–17,000 μg/L
phosphate	<0.01–2.7 mg/L
selenium	3–590 μg/L
specific conductance	1200–16,000 μmhos/cm
toluene	<5–100,000 μg/L
total organic carbon	11–8700 mg/L
vinyl chloride	0.140–32.5 mg/L

Leachate also may have high concentrations of heavy metals, pesticides, chlorinated and aromatic hydrocarbons, and other toxic chemicals, depending on what materials were originally placed in the landfill. In northern, humid climates, quantity and quality of leachate fluctuates seasonally, with larger volumes of dilute concentration generated in early spring and late fall and smaller volumes of higher concentration in the summer and winter.

Landfill leachate has long been a source of groundwater and surface water contamination,[2-4] and in recent years efforts have been directed toward limiting the production and migration of leachate. Most state regulations now require that new landfills be lined, capped, vented, and equipped with a leachate collection system and older operating landfills be retrofitted with ditches or drain systems. In both cases, leachate must be collected and treated.

The most often used method is to haul leachate to a licensed sewage treatment plant. However, many sewage treatment plants are operating at or beyond capacity. They often fail to meet discharge standards and are unwilling to accept leachate. Those that can accept leachate are often distant from landfill sites, entailing expensive hauling costs. Sewage treatment plants also usually require pretreatment to remove excessive concentrations of heavy metals.

Use of on-site, conventional treatment is generally avoided due to high construction and operation costs and requirements for continuous monitoring by a licensed operator. This is a serious problem after the landfill is closed and personnel are not on-site. Because conventional methods of waste treatment are poorly suited for long-term treatment of landfill leachate, alternative methods need to be investigated.

The ideal method would treat a wide range of chemical constituents, be able to accept varying quantities and concentrations of leachate, and be inexpensive to construct and easy to maintain with low energy and manpower requirements. Constructed wetlands used as treatment systems have the potential to meet these criteria.

Constructed wetlands have successfully treated domestic and industrial wastewater and acid mine drainage.[5-8] Although concentrations are different, these wastes have chemical constituents common to landfill leachate. Nutrients (nitrogen, phosphorus, potassium, and micronutrients), heavy metals, pesticides, and other natural and man-made organic chemicals have been significantly reduced in wastewater applied to wetlands.[8-13] Properties of wetland ecosystems that contribute to wastewater renovation include high plant productivity, large adsorptive surfaces on sediments and plants, an aerobic-anaerobic interface, and, most important, an active microbial population.

Wetlands have large, diverse microbial populations that are adapted to a wide range of environmental conditions and perform biochemical transformations of wastewater constituents that take place in conventional wastewater treatment plants.[14,15] In addition, absorption of organics and nutrients by plants, adsorption of metals on sediments and plants, microbially mediated oxidation of metals, and simultaneous aerobic decomposition and anaerobic digestion of organic compounds, transforms and immobilizes a wide range of chemical constituents.[8]

Wetland treatment systems are simple to control and maintain, withstand a wide range of operating conditions, and have relatively low energy and manpower requirements. These systems should therefore prove to be superior alternatives to conventional leachate treatment methods. However, before field-scale operation of wetland treatment systems can be implemented, leachate treatment must be demonstrated and the treatment level of individual constituents quantified. Seasonal, operational, and design variables also need evaluation to optimize operation of future systems, and the dominant treatment processes need to be understood to evaluate the potential for long-term system operation.

The feasibility of constructed wetlands for landfill-leachate treatment is under investigation in New York State by the Water Resources Division of the U.S. Geological Survey, Cornell University's departments of Agricultural Engineering and Agronomy, Tompkins County, and Allegany County. The New York State Energy Research and Development Authority provides the major portion of matching funds for the project. The study is designed to investigate the fate and transport of landfill leachate in a constructed wetland and provide engineering design data for construction and operation of full-size leachate treatment systems. Specific objectives are to examine (1) the efficiency of leachate treatment as a function of substrate material, plant growth, leachate quality, and seasonal change in climate; (2) the effects of leachate application and plant growth on the hydraulic characteristics of the beds; (3) the effect of leachate on plant viability; and (4) the physical and chemical processes by which metals are fixed or transformed.

DESIGN

The wetland systems are modified versions of the root-zone or rock-reed-filter systems.[8,16] They have gently sloping beds lined with an impermeable barrier and planted with emergent hydrophytes such as reeds (*Phragmites*), bulrush (*Scirpus*), or cattails (*Typha*) (Figure 1). In general, the beds have an inlet zone of crushed stone to distribute wastewater evenly over the bed width and an outlet zone of crushed stone to collect and discharge effluent.

Treatment efficiency is partially dependent on the retention time in the bed.

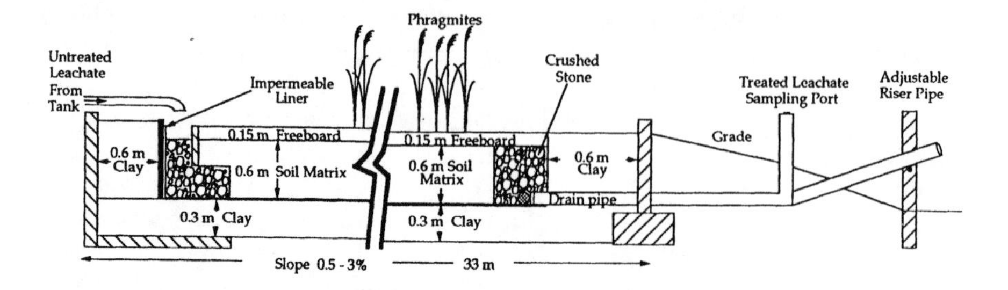

Figure 1. Schematic of a constructed wetland bed.

Table 1 shows the theoretical relationship between the saturated hydraulic conductivity (permeability), bed slope, minimum retention time, and maximum flow through a bed with dimensions of 30 m by 3 m by 0.6 m. The relationship is based on homogeneous flow and Darcy's law. By reducing flow at the top of the bed and blocking flow at the end of the bed, the retention time can be increased and flow reduced from the values in Table 1. However, the reverse is not true without causing surface runoff.

Two wetland systems for treating waste currently in use are the root-zone and the rock-reed filter.[8,16] They differ in hydraulic conductivity of substrate material and the pathway through which wastewater flows. The rock-reed filter uses gravel or crushed stone with a saturated hydraulic conductivity in excess of 30 m/day. Bed slope is small, and water flows evenly through most of the substrate and the root mass. An advantage of this system is the wide range of retention times obtainable by adjusting upstream and downstream control points.

The root-zone system uses a loam or a sandy loam soil with saturated conductivities varying between 1 and 5 m/day. Even at a 6% slope, 0.5 m^3/day of leachate, at most, can be treated by a 30-m-by-3-m-by-0.6-m bed (Table 1). The root-zone system is also dependent on the preferential flow path created by the root mass, which increases the saturated conductivity thereby increasing flow rate (and decreasing retention time). As most root-zone studies are only a

Table 1. Theoretical Relationship Between Saturated Hydraulic Conductivity, Slope of the Bed, Minimum Retention Time, and Maximum Amount of Flow in a Constructed Wetland Bed with Dimensions of 30 m × 3 m × 0.6 m

	1% Slope		2% Slope		3% Slope		6% Slope	
Flow Rate (m^3/day)	Ks[a]	RT[b]	Ks	RT	Ks	RT	Ks	RT
0.5	27	56	14	56	9	56	4.5	56
1.0	55	28	27	28	18	28	9	28
1.5	85	18	42	18	28	18	14	18
2.0	110	14	55	14	37	14	18	14
2.5	140	11	70	11	47	11	23	11
3.0	165	9	82	9	55	9	28	9

[a]Saturated hydraulic conductivity (m/day).
[b]Minimum retention time (days).

Table 2. Experimental Design Factors for Eight Constructed Wetland Beds in Tompkins County and Allegany County, New York

	Bed 1	Bed 2	Bed 3	Bed 4
	Tompkins County			
Vegetation	reeds	bare	reeds	bare
Substrate size	50% 3 cm stone 50% gravel	50% 3 cm stone 50% gravel	40% gravel 60% sand	40% gravel 60% sand
Hydraulic conductivity (m/day)	>250	>250	>250	>250
Slope (%)	0.5	0.5	0.5	0.5
Flow rate[a] (m^3/day)	1.8	1.8	1.8	1.8
Retention time (days)	15	15	15	15
	Allegany County			
Vegetation	reeds	reeds	reeds	bare
Substrate size	50% 3 cm stone 50% gravel	40% gravel 60% sand	sandy loam	sandy loam
Hydraulic conductivity (m/day)	>250	>250	1–5	1–5
Slope (%)	0.5	0.5	3	3
Flow rate (m^3/day)	1.8	1.8	0.05–0.25	0.5–0.25
Retention time (days)	15	15	100–500	100–500

[a]Upstream and downstream controls required.

few years under way and preferential flow paths take a minimum of three years to develop, there is little data on the mechanisms of flow and waste treatment. It has been hypothesized that preferential flow paths have an aerobic microenvironment in which aerobic decomposition of organic matter occurs.

Eight wetland plots are being constructed at municipal solid waste landfills in Allegany County and Tompkins County. The plots are 30 m long, 3 m wide, and 0.6 m deep and have varying combinations of slope, substrate, and planting. The plots are being planted with *Phragmites australis,* a species of reed native to New York State. Two plots are designed as root-zone systems, four plots are designed as rock-reed-filter systems, and two plots have gravel substrates without plants to serve as controls. Engineering design factors, design flow rates, and estimated hydraulic retention time for each plot are listed in Table 2.

Construction of the wetland plots began in the spring of 1988 and will be completed during the summer. The plots will be hand planted with *Phragmites* rhizomes at a density of 2 rhizomes/m^2 in the fall after the plants become dormant. The rhizomes will lie dormant over winter in the plots and leachate will be applied after the plant root mass is established in the spring or summer of 1989.

The quantity of leachate influent and effluent will be continuously measured, and samples of influent and effluent will be collected biweekly and analyzed for organic and inorganic chemical constituents. The water and chemical mass loadings and treatment efficiency of each plot will be calculated.

Accumulation of metals in the wetland plots is of particular importance to long-term use. Metals may accumulate in the wetland plots by (1) bioaccumulation in plants; (2) adsorption to substrate material; (3) microbially mediated oxidation of metals; (4) formation of insoluble sulfide precipitates; and (5) chelation by organic material and incorporation into the biofilm. The first two processes have finite capacities to retain metals and, if dominant in controlling metals retention, may indicate poor long-term treatment performance.

Substrate samples will be taken from each test plot prior to leachate application and at six-month intervals thereafter. The partitioning of trace metals in the substrate material will be studied using selective chemical extractions.[17-22] These procedures determine if metals in the substrate are in clay interlayers or in the crystalline structure of silicate minerals or whether they are organically complexed, held on exchange sites, or bound as carbonate minerals, hydrous manganese oxides, hydrous iron oxides, or sulfides. Speciation of metals in the substrate should indicate the processes by which metals are fixed and the potential for long-term treatment.

Retention time and the contact of water with the soil matrix influence waste treatment. This influence depends on the amount of water applied and the saturated hydraulic conductivity. The hydraulic conductivity may be affected by blockage of pores by bacterial growth and sorting of substrate material during filling of the bed.

In laboratory experiments preceding bed construction, the hydraulic conductivities of the homogeneous mixture were compared to sorted substrate material in test flumes (0.3 m $\times$ 0.3 m $\times$ 2.3 m). Substrates were homogeneously premixed and placed in the test flumes with a shovel, which resulted in interwoven bands of relatively coarse and fine material. The saturated hydraulic conductivity of the homogeneous mixture was 80 m/day, while the banded mixture had a conductivity 20 times higher. Dye tracer experiments showed preferential movement of water through the coarser layers. In a separate experiment with the same substrate material, bacterial growth caused by landfill leachate decreased the saturated hydraulic conductivity as much as 50% after two weeks. Further tests will be carried out in the laboratory and in the field when the beds are constructed.

Greenhouse studies have also been conducted to guide the construction and operation of the field test plots by observing the effects of transplantation and substrate material on rhizome sprouting, the plant-nutrient requirements, and changes in hydraulic conductivity from rhizome growth. *Phragmites* rhizomes were dug from a mature stand by machine after the dead tops were removed in November 1987. The rhizomes were planted at 15 cm depth in rectangular tubs (60-L capacity) filled with coarse, clean gravel or loamy sand, and the substrate was kept moist. The greatest initiation of shoot growth occurred from terminal rhizomes placed in the wet loamy sand, and the least initiation of shoot growth occurred from older nodes placed in saturated coarse stone. Other treatment combinations were intermediate in shoot production. Growth in the loamy sand was excellent, but a complete inorganic fertilizer was required to prevent chlorosis and stunted growth in coarse gravel. Little or no effect of root or rhizome growth on hydraulic conductivities has been noted, though the plants are only three months old.

REFERENCES

1. Shuckrow, A. J., A. P. Pajak, and C. J. Touhill. "Management of Hazardous Waste Leachate," U.S. EPA Report-SW-871 (1980).
2. Cole, J. A. "Groundwater Pollution in Europe," in *Proceedings of a Water Research Association Conference—Reading, England* (New York: Water Information Center, Inc., 1972).
3. Miller, D. W., F. A. DeLuca, and T. L. Tessier. "Groundwater Contamination in the Northeast States," U.S. EPA Report-660/2-74-056 (1974).
4. Clark, D. C. "Remedial Activities for Army Creek Landfill," in *Proceedings of the Fifth Annual EPA Research Symposium,* U.S. EPA Report (1979).
5. Gersberg, R. M., B. V. Elkins, and C. R. Goldman. "Wastewater Treatment by Artificial Wetlands," *Water Sci. Technol.* 17:443-450 (1984).
6. Boon, A. G. "Report of a Visit by Members and Staff of WRc to Germany (GFR) to Investigate the Root Zone Method for Treatment of Wastewaters," Water Research Centre Paper 376-S/1 (1986).
7. Girts, M. A., and R. L. P. Kleinmann. "Constructed Wetlands for Treatment of Mine Water," paper presented at the 1986 Society of Mining Engineers Fall Meeting, St. Louis, MO, September 7-10, 1986.
8. Wolverton, B. C. "Artificial Marshes for Wastewater Treatment," in *Aquatic Plants for Wastewater Treatment and Resource Recovery,* K. R. Reddy and W. H. Smith, Eds. (Orlando, FL: Magnolia Publishing Inc., 1987), pp. 141-152.
9. "The Effects of Wastewater Treatment Facilities on Wetlands in the Midwest," U.S. EPA Report-905/3-83-002 (1983).
10. "The Ecological Impacts of Wastewater on Wetlands—An Annotated Bibliography," U.S. EPA Report-905/3-84-002 (1984).
11. "Freshwater Wetlands for Wastewater Management Handbook," U.S. EPA Report-904/9-85-135 (1985).
12. Godfrey, P. J., E. R. Kaynor, S. Pelczarski, and J. Benforado, Eds. *Ecological Considerations in Wetlands Treatment of Municipal Wastewaters* (New York: Van Nostrand Reinhold Company, 1985).
13. Reddy, K. R., and W. H. Smith, Eds. *Aquatic Plants for Water Treatment and Resource Recovery* (Orlando, FL: Magnolia Publishing Inc., 1987).
14. Kadlec, R. H., and J. A. Kadlec. "Wetlands and Water Quality," in *Wetland Functions and Values: The State of Our Understanding,* Greeson, Clark, and Clark, Eds. (Minneapolis, MN: American Water Works Association, 1978), pp. 436-456.
15. Bastian, R. K., and J. Benforado. "Water Quality Functions of Wetlands, Natural and Managed Systems," paper presented at the International Symposium on Ecology and Management of Wetlands, Charleston, SC, June 16-20, 1986.
16. Cooper, P. F., and A. G. Boon. "The Use of *Phragmites* for Wastewater Treatment by the Root Zone Method: The UK Approach," in *Aquatic Plants for Wastewater Treatment and Resource Recovery,* K. R. Reddy and W. H. Smith, Eds. (Orlando, FL: Magnolia Publishing Inc., 1987), pp. 153-174.
17. Chester, R., and M. J. Hughes. "A Chemical Technique for the Separation of

Ferro-Manganese Minerals, Carbonate Minerals, and Adsorbed Trace Metals from Pelagic Sediments," *Chem. Geol.* 2:249–262 (1967).

18. Gatehouse, S., D. W. Russell, and J. C. VanMoort. "Sequential Soil Analysis in Exploration Geochemistry," *J. Geochem. Explor.* 8:483–494 (1977).
19. Filipek, L. H., and R. M. Owens. "Analysis of Heavy Metal Distribution Among Different Mineralogical States in Sediment," *Can. J. Spectroscopy* 23:31–34 (1978).
20. Tessier, A., P. G. C. Campbell, and M. Bisson. "Sequential Extraction Procedure for the Speciation of Particulate Trace Metals," *Anal. Chem.* 51(7):844–851 (1979).
21. Chao, T. T. "Use of Partial Dissolution Techniques in Geochemical Exploration," *J. Geochem. Explor.* 20:101–135 (1984).
22. Berndt, M. P. "Metal Partitioning in a Sand and Gravel Aquifer Contaminated by Crude Petroleum, Bemidji, Minnesota," Master's Thesis, Syracuse University, Syracuse, NY (1987).

41d Natural Renovation of Leachate-Degraded Groundwater in Excavated Ponds at a Refuse Landfill

James N. Dornbush

INTRODUCTION

Solids waste disposal sites for many years were selected to "reclaim" mined areas and swamps. Into the 1950s, engineering texts suggested low areas such as abandoned borrow pits and swamps as particularly suitable sites. Surface water drainage and pollution was of concern, but groundwater contamination had not yet claimed the environmental status that it does today.

Early emphasis on land reclamation with solid wastes was lasting. Leachate resulting from percolation of precipitation or groundwater movement through the wastes severely contaminated local groundwater resources. Because contaminated groundwater does not mix, moves slowly, is difficult to detect, and is not subjected to aerobic natural treatment cycles as in surface water, it frequently is not discovered until a nearby well has been affected. Then, correction of the problem is difficult and expensive.

Waste disposal sites improperly located, operating or closed, may continue to produce leachate and cause severe damage to the groundwater resource for the next 50 to 100 years. In a report to Congress,[1] EPA listed methods of coping with leachate contamination, including (1) natural attenuation by soil, (2) prevention by restricting entrance of water to the site, (3) collecting leachate for treatment, (4) pretreatment such as incineration to reduce waste volume, or (5) waste detoxification prior to disposal. Additional methods to clean up groundwater at abandoned landfill sites are emerging and needed.

The solid waste landfill at Brookings, South Dakota is typical of many landfills. Its location in an old gravel pit area over a groundwater aquifer with a shallow water table subjects the aquifer to contamination from precipitation-induced leachate from buried solid wastes. Groundwater contamination may also occur when the water table rises to come in contact with deposited wastes. The site would not be considered a suitable location for a new solid wastes landfill, and many similar landfills have been closed.

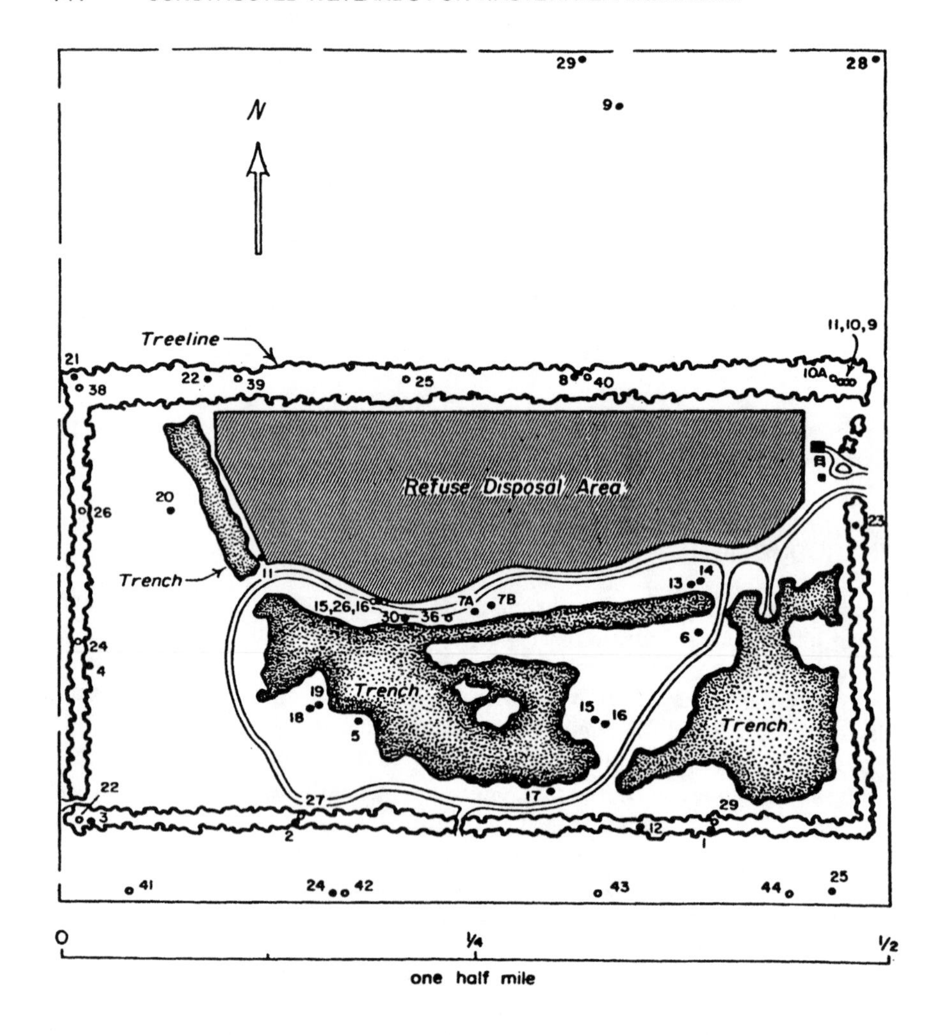

Figure 1. Brookings, South Dakota landfill area in 1983 showing trench and monitoring wells. Black dots are PVC wells constructed in 1981–82 by South Dakota Geological Survey.[2]

In several respects, however, the Brookings landfill is not typical. Up to 75 monitoring wells have been installed to evaluate groundwater quality. A trench and ponds have been constructed in the downstream direction to intercept, treat, and dilute the degraded groundwater (Figure 1). Also, monitoring results have not demonstrated excessive groundwater degradation "downstream" from the trench. The landfill continues to operate despite legal challenges. Considerable water quality data is available through graduate research back to the start of solid waste disposal in 1960. Much of this data deals with the more common parameters such as chlorides and hardness. More recently, data has been collected concerning (1) toxic organic and inorganic chemicals and (2) pesticides that are included on the Hazardous Substance List for waste disposal facilities.

Because the trench and wetland ponds at the Brookings landfill have remediated the effect of excessive contaminant concentrations in the "downstream" groundwater, it is hoped that other landfills (active and closed) might benefit by the use of man-made wetlands in the form of trenches and ponds to protect against, or possibly correct, excessive groundwater degradation. By our presenting this case history, an additional method of coping with leachate may be demonstrated.

EARLY LANDFILL RESEARCH

In 1960, the solid waste disposal operations for Brookings, South Dakota were moved to a 65-ha site containing an abandoned gravel pit located in alluvial soils, primarily glacial outwash. The early operations of the landfill, the soil conditions of the site, and early water quality data have been summarized.[3,4] The landfill is located on the edge of a valley train over a well-graded sand and gravel outwash from 3 m to 10 m thick. Below the sand and gravel is a thick clay till deposit considered a hydraulic boundary for the outwash aquifer. A sandy loam alluvium overlies the outwash to a depth of 0.3 to 1 m. Depth to groundwater varies from near the surface to depths of 3 to 4 m. Average annual precipitation is about 56 cm, with seasonally high water tables occurring in spring and early summer.

Groundwater degradation was occurring at the site, and chloride, conductivity, and sodium were the most useful tests for tracing leachate plumes. Groundwater flowed directly toward the southwest and became grossly polluted in the disposal area. A pond located directly downstream from the disposal area was important in reducing hardness and alkalinity of the leachate-polluted groundwater.[3]

Evaluation of the areal extent and seasonal quality variation of the leachate-degraded groundwater detected a narrow plume extending 300 m from the disposal area. Peaks in chlorides and conductivity demonstrated intermittent movement of the plumes after seasonal rainfall and snowmelt. An existing small pond demonstrated a beneficial effect of reduced hardness of the degraded water.

When gravel mining resumed at the site in 1966, Brookings financed the excavation of an open-water trench about 15 m wide and to a depth of about 5 m below the water table to the till layer below the gravel. The trench was located in a gravel excavation on the downstream side of the landfill to intercept the leachate plume. Higher water tables created a connecting pond in 1967 and subsequent years. Mixing, dilution, and biological treatment were to improve the downstream water qualities. Precise measurements of water table elevations demonstrated that the trench and ponds intercepted leachate and

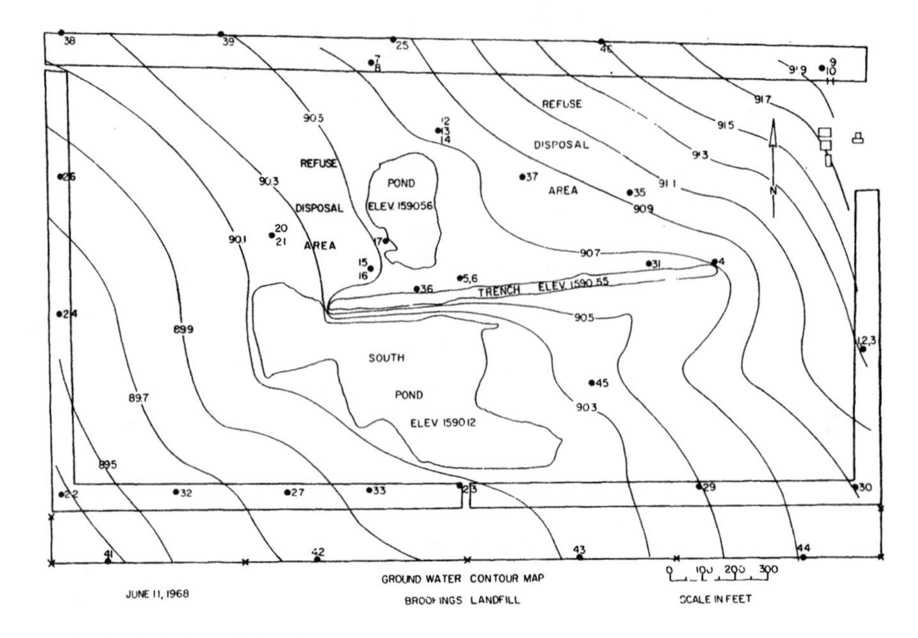

Figure 2. Location of ponds and wells at the landfill and groundwater elevations on June 11, 1968.[4]

attracted groundwater, but downstream groundwater assumed the original direction of flow (Figure 2).

The primary significance of the trench and pond was that the leachate plume was destroyed. Data from 1965 to 1966 (before construction of the pond and trench) had revealed that degraded water flowed in a clearly defined plume moving in the direction of groundwater flow. After construction of the pond and trench, similar data revealed that the well-defined plume was no longer present, and downstream water quality was quite similar in most respects to that of the control wells upstream. Construction of the trench also modified downstream organic concentrations.[4]

MONITORING

From 1975 to 1979, city personnel monitored groundwater quality in the area each fall and spring to evaluate effectiveness of the trench in providing groundwater treatment and to determine the quality of groundwater leaving the landfill site. Selected indicator parameters included chlorides, conductivity, total hardness, and nitrates (Table 1).

After a legal dispute from 1974 ending in a 1979 Supreme Court decision,[5] the city was issued a permit to operate the landfill at the existing site. Attached conditions included (1) a prohibition on disposal of hazardous wastes at the

Table 1. Mean Concentration of Chemical Constituents (1975–1979)

Zone[a]	Nitrates (mg/L as N)	Chlorides (mg/L)	Conductivity (μmhos/cm)	Hardness (mg/L)
Control wells (wells 10, 38, 29)	2.71	17	619	313
Landfill area (wells 5,[b] 15, 20[b])	1.52	79	1010	443
Trench	0.43	44	769	342
Downstream (wells 22, 27, 41, 42)	0.29	41	704	311

[a]The location of the specific wells are shown in Figure 1 as circles.
[b]Wells 5 and 20 were near wells 7b and 11 on south side of disposal area.

Table 2. Maximum Concentrations of Health-Related Inorganics (1979–1983)

Location	Number of Samples	Maximum Concentration Detected (μg/L)			
		Cadmium	Lead	Mercury	Selenium
Control wells (10, 10A, 8)	11	5.5	18	<0.5	2.2
Trench	5	1.6	24	<0.5	4.0
Downstream (27, 2, 24, 4)	16	1.0	7	<0.5	2.8
MCL		10	50	2	10

site, (2) maintenance of 2 m above the area's normal high water table at the lowest portion of the fill, and (3) collection of samples from the trench and three wells (10 as a control and 22 and 27 near the southwest edge of the site) each spring and fall. Selected data for 1979–1983 are shown in Table 2.

When EPA adopted regulations for solid waste disposal facilities, maximum contaminant levels (MCL) were adopted for health-related chemical contaminants in affected groundwater.[6] As a result, previous research, legal involvements, city monitoring data, and nearby land use changes from 1960 through 1983 were reviewed.[7] Available water quality data revealed violations only in excess nitrates at points upstream from the landfill area if compliance of the landfill was measured downstream from the trench and ponds.

In 1981 and 1982, intensive monitoring of the landfill was undertaken by the state as part of an aquifer evaluation. After installation of 30 PVC wells, 127 samples were evaluated for the common inorganic ions and several toxic metals. Well 26, located in the southwest edge of the landfill, was leachate-contaminated with high sodium, bicarbonate, chloride and ammonia levels, and low sulfates. Arsenic violated drinking water standards. Although violations downstream from the trench were not found, the report[2] concluded:

> In summary, the landfill is unquestionably very poorly sited. The role of the intercepting trenches at the Brookings landfill has been a very controversial topic. Even though the sodium and chloride data indicate that dilution is occurring in the trenches, a leachate plume remains down gradient from the trenches

> and is degrading the groundwater. . . . In the final analysis, it must be recognized that the trenches are in fact only partially correcting a bad situation which never should have existed in the first place.

In August 1984 during a period of high water table, wells 26 and 5 were sampled for priority pollutants by state personnel. Arsenic (6700 μg/L at well 26 upstream from the trench) was the only toxic chemical that exceeded the standards. Many inorganic levels indicated leachate at this location (Table 3). Volatile organics, acid compounds, and pesticides were below the detection limit, and only diethyl phthalate at 32 μg/L was detected among the base/neutral compounds. Water quality downstream from the wetlands was nearly comparable to the control.

In February 1987, representatives of EPA conducted a Superfund site inspection of the Brookings landfill. The field investigation team (FIT) collected groundwater samples from four wells (10A, 26, 2, and 4) and two leachate samples from the ice-covered trench. Samples were analyzed for Hazardous Substance List (HSL) organics, pesticides, and dissolved metals.

Although several HSL compounds were detected at very low concentrations, metal contamination was not present at any well. The report[8] to EPA concluded:

> This investigation has shown that no release to groundwater has occurred at the Brookings Landfill based on the samples collected by FIT. The collection trenches constructed by the city of Brookings appear to be intercepting leachate migration leaving the landfill. Based on the analytical results from the Brookings Landfill site investigation, the FIT recommends no further action at this site.

LEACHATE TREATMENT CONCEPT

The trench at the Brookings landfill has protected shallow groundwater in the area by several mechanisms including dilution mixing, gas interchange, metal oxidation, and biological treatment. Analytical data in Table 3 demonstrate these protection mechanisms.

Interception of shallow groundwater moving through a trench/pond would destroy the leachate plume and dilute concentrated contaminants with natural groundwater moving into the trench. Depending upon the porosity of the sands and gravels, the excavated volume below the water table will contain about three to four times the volume of water that would be stored in a similar volume of saturated sands. With intermittent leachate production, continuous movement of groundwater into and through the trench occurs quickly (Figure 3). Chloride, a conservative pollutant, was reduced from 285 to 48 mg/L immediately downstream of the trench/pond. Native groundwater, however, contained 9 mg/L chloride.

Gases produced in the landfill, such as hydrogen sulfide, methane, and carbon dioxide, and carried in leachate in high concentration were exposed to natural aeration in the trench. Escape of volatile gases to the atmosphere and

Table 3. Quality of Groundwater at Brookings Landfill Relative to Trench/Pond Treatment, August 9, 1984

Parameter	Suggested Limit/Units	Upstream (Well 26)	Downstream (Well 5)	Control Well[a]
Total dissolved solids	500 mg/L	1551	444	470
Conductivity @25°C	μmhos/cm	2970	748	693
Alkalinity (as $CaCO_3$)	mg/L	1286	304	255
Bicarbonate	mg/L	1569	371	311
Chloride	250 mg/L	285	47.5	9.1
Sulfate	250 mg/L	<12	29	109
Calcium	mg/L	324	60	87
Magnesium	mg/L	160	45	34
Sodium	mg/L	151	33.2	15.3
Potassium	mg/L	102	6.3	2.5
Iron	0.3 mg/L	1.15	0.09	2.22
Manganese	0.5 mg/L	0.05	0.38	0.36
pH (field)	(6.5–8.5)	6.58	7.08	7.19

[a]Located approximately 1 mile upstream from landfill.

introduction of dissolved oxygen would occur naturally, providing benefits similar to those with aeration of well waters.

Iron and manganese that occur in high concentrations in leachate are oxidized in the trench and precipitate from the water. Similar oxidation and precipitation of trace metals would be expected. Release of high concentrations of carbon dioxide from the water to the atmosphere (see bicarbonate

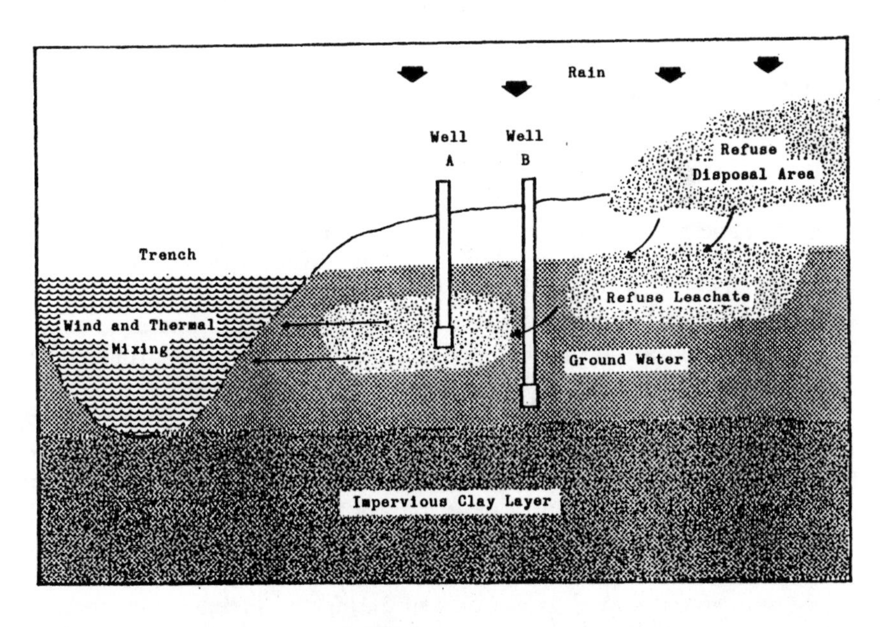

Figure 3. Treatment concept is illustrated as trench intercepts groundwater leachate. Dilution, gas interchange, metal precipitation and biological treatment occurs in trench/ponds.

reduction) might also be expected to result in precipitation of calcium and magnesium from the water. Reductions in sulfates are explained by biochemical reduction and release as hydrogen sulfide from the anoxic leachate plume.

Exposure of the leachate-contaminated groundwater to an aerobic environment will provide a natural biological treatment system. Organic products of anaerobic decomposition moving from the anoxic landfill with leachate into an aerobic pond would be similar to septic tank and aerobic soil adsorption systems for rural wastewater treatment. Carbon, nitrogen, and sulfur of landfill organics could recycle in the complex biological ecosystem treating the leachate-contaminated groundwater. Algae and other photosynthetic plants that develop in the pond would assimilate nutrients and raise the pH of the water to enhance precipitation of calcium, iron, and manganese. Toxic organic chemicals emanating from the landfill would occur in extremely low concentrations, allowing the development of microorganisms capable of degrading the unnatural complex materials.

In total, the trench and pond system complete the natural physical, chemical, and biological cycles that normally exist within the total hydrological cycle but which are delayed when water is moving underground in an anoxic environment. Overall chemical quality parameters (total dissolved solids and conductivity) are comparable at downstream and control locations even though some specific constituents increase (Cl, Na) and some decrease (Ca, SO_4).

SITE AND LAND USE CHANGES

By the late 1970s, the small pond that was so prominent in the early studies had been filled with concrete rubble. Solid waste deposits now extend further to the west and exceed an elevation of 11 m. In the 1980s, the disposal area has been extended through the windbreak tree line toward the northwest. The trench has also been extended to the west and northwest to intercept the groundwater as it moves to the southwest. Additional gravel excavations for refuse cover have created extensive open bodies of water. A softball complex has been constructed in the northeast corner of the site (Figure 4).

Land use changes in the 1970s outside the disposal site may influence groundwater hydrology, quality, and interpretation of data. Approximately 90 new residences, mostly mobile homes, have been located within 1 mile in the upstream, or northeast, direction from the site. Wastewater from these units is discharged to the aquifer via septic tank drainfields. Also, a 65-ha, center-pivot irrigation unit has been installed directly south of the disposal site. Heavy pumping from the irrigation well during the summer may alter the direction of groundwater flow through the landfill area. Gravel has also been removed from a 20-ha area directly east of the landfill site. Recent groundwater elevations indicate that these gravel excavations may be raising the water table at the landfill.

Figure 4. Landfill area in 1987 showing wetlands that will be developed to a nature park in 1990s. Softball complex is located in northeast quadrant.

CONCLUSIONS

A large measure of the successful monitoring history at the landfill site must be afforded the groundwater wetlands that have been constructed at the site. Severely contaminated groundwaters have been reduced to acceptable contaminant levels. The significance of this case study lies in the potential for construction of similar comparatively inexpensive trenches and ponds to intercept and treat leachate-degraded plumes in shallow groundwater at abandoned or poorly located solid waste sites.

With continued landfill operation, a 10-m hill has been formed in the refuse disposal area, and the trench has expanded into sizable wetland ponds. After the landfill closes in the 1990s, the hill, ponds, and wooded area will become features of a future nature park. Activities could include canoeing, fishing, and swimming for day camps and overnight youth groups. Its value is enhanced by location at 0.8 km from the city limits, and the park will be a destination point on a planned recreation trail extending from the city.

REFERENCES

1. *The Report to Congress, Waste Disposal Practices and Their Effects on Groundwater*, U.S. EPA, Office of Water Supply, Office of Solid Waste Management Programs (January 1977), pp. 144–185.

2. Stach et al. "Draft Final Report, Big Sioux Aquifer Study Part I," South Dakota Geological Survey, Vermillion, SD (May 1984).
3. Andersen, J. R., and J. N. Dornbush. "Influence of Sanitary Landfill on Ground Water Quality," *Journal AWWA* 59:457–470 (1967).
4. Andersen, J. R., and J. N. Dornbush. "Quality Changes of Shallow Ground Water Resulting from Refuse Disposal at a Gravel Pit," Water Resources Institute Report-SRI 3553, SDSU, Brookings, SD (November 1968).
5. "City of Brookings vs Department of Environmental Protection South Dakota #12239," Supreme Court of South Dakota, February 1, 1979, Northwestern Reporter 2nd Series, Vol. 274, p. 887.
6. "Criteria for Classification of Solid Waste Disposal Facilities and Practices, Final, Interim Final, and Proposed Regulations," *Federal Register*, 40 CFR, Part 257, September 13, 1979.
7. Dornbush, J. N. "Treatment of Leachate-Contaminated Groundwater with an Interception Trench," Report A-078-SDAK, Water Resources Institute, SDSU, Brookings, SD (April 1984).
8. "Analytical Results Report, Brookings Landfill, Brookings South Dakota TDD F08–8704–05," by ecology and environment inc., FIT Zone II, August 14, 1987.

CHAPTER 42

Control of Acid Mine Drainage Including Coal Pile and Ash Pond Seepage

42a Biology and Chemistry of Generation, Prevention and Abatement of Acid Mine Drainage

Marvin Silver

INTRODUCTION

The importance of microorganisms as catalysts of inorganic chemical reactions has been recognized in commercial metal recovery[1,2] and soil mineral transformations.[3,4] More recently, the importance of microbially mediated reactions with respect to environmental concerns has been documented.[5–7] These reactions are presented with their relevance to the generation, prevention, and abatement of acidic drainage. Reactions involved in the solubilization and reprecipitation of polluting metals such as Fe, Cu, Zn, Mn, and Al will also be discussed.

CHEMICAL AND BIOLOGICAL ACID GENERATION REACTIONS

Bacterial sulfide mineral oxidation is the principal cause of sulfuric acid generation in base metal tailings, coal spoils, and quarry effluents. Requirements for sulfide mineral oxidation are the same as for any other type of oxidation—an oxidizable substrate, an oxidant, and a promoter. In biological mineral oxidations, the mineral is the oxidizable substrate, oxygen is generally the oxidant, and an enzyme system of microorganisms is the promoter. The microorganisms most commonly associated with biological mineral oxidation are iron-oxidizing autotrophs such as *Thiobacillus ferrooxidans, Leptospirillum ferrooxidans, Sulfolobus brierleyii,* and the thermophilic thiobacilli. Other autotrophs such as *Thiobacillus thiooxidans* and certain heterotrophic bacteria may also be implicated.

Bacterial oxidation of sulfide minerals has been attributed to two mechanisms that may not be mutually exclusive: direct and indirect. In direct mechanisms, either or both of the Fe and S moieties are oxidized by bacteria, with the formation of metal sulfates according to the generalized equation:

$$MS + O_2 \rightarrow MSO_4 \quad (1)$$

Because of the extremely low solubilities of the metal sulfide minerals, intimate contact was believed to be a requirement for bacterial oxidation to occur. Although intimate contact has been demonstrated in the presence of pyrite, chalcopyrite, galena, and elemental sulfur, Torma and Sakaguchi[8] showed that the solubilization of metals depended upon the solubility constants of the relevant minerals. Research on oxidation of natural and synthetic metal sulfides devoid of Fe proved that direct oxidation could occur. Further evidence for the direct mechanism was demonstrated by leaching and respirometric studies on chalcopyrite and pyrite using specific metabolic inhibitors: N-ethylmaleimide inhibits S^0 oxidation, and azide inhibits ferrous iron oxidation.

In the direct mechanisms, ferric iron acts as the oxidant, forming metal sulfates, ferrous sulfate, and elemental sulfur, as illustrated by the equation:

$$MS + Fe_2(SO_4)_3 \rightarrow MSO_4 + 2FeSO_4 + S^0 \quad (2)$$

Bacteria can then reoxidize the ferrous iron to ferric iron:

$$4FeSO_4 + O_2 + 2H_2SO_4 \rightarrow 2Fe_2(SO_4)_3 + 2H_2O \quad (3)$$

and the elemental sulfur to sulfuric acid:

$$2S^0 + 3O_2 + 2H_2O \rightarrow 2H_2SO_4 \quad (4)$$

The indirect mechanisms are not limited to metal sulfides; arsenides, carbonates, oxides, and silicates may also be leached. Thus, metals that do not occur as sulfides (Al, Cr, Mn, U, and the alkaline earths) may be solubilized by the indirect mechanism.

Oxidation of Iron Sulfide Minerals

Pyrite, the most common sulfide mineral, has been the most extensively studied with respect to bacterial mineral oxidation. Marcasite has the same chemical composition but a different crystal structure. Both can be oxidized by direct mechanisms, as illustrated by the following series of equations:

$$4FeS_2 + 4H_2O + 14O_2 \rightarrow 4FeSO_4 + 4H_2SO_4 \quad (5)$$

$$4FeSO_4 + O_2 + 2H_2SO_4 \rightarrow 2Fe_2(SO_4)_3 + 2H_2O \quad (6)$$

$$4FeS_2 + 15O_2 + 2H_2O \rightarrow 2Fe_2(SO_4)_3 + 2H_2SO_4 \quad (7)$$

The pyrite is first oxidized directly with formation of ferrous sulfate, which is then oxidized to ferric sulfate. Ferric iron formed can then oxidize the mineral according to the equation:

$$2FeS_2 + 2Fe_2(SO_4)_3 \rightarrow 6FeSO_4 + 4S^0 \quad (8)$$

Ferrous iron can then be oxidized biologically to ferric iron, and the elemental sulfur to sulfuric acid.

Oxidation of Nonsulfide Minerals

Although direct bacterial oxidation of the uranium minerals uraninite (pitchblend) and brannerite have been reported, the principal mechanism seems to be that of indirect oxidation by ferric iron according to the following equations:

$$UO_2 + Fe_2(SO_4)_3 \rightarrow UO_2SO_4 + 2FeSO_4 \quad (9)$$

$$UTiO_2 + Fe_2(SO_4)_3 \rightarrow UO_2SO_4 + TiO_2 + 2FeSO_4 \quad (10)$$

Carbonate minerals such as siderite can also be oxidized by iron-oxidizing bacteria:

$$4FeCO_3 + O_2 + 6H_2O \rightarrow 4Fe(OH)_3 + 4CO_2 \quad (11)$$

and smithsonite by sulfur-oxidizing bacteria:

$$2ZnCO_3 + 3O_2 + 2H_2O + 2S^0 \rightarrow 2ZnSO_4 + 2H_2CO_3 \quad (12)$$

Both autotrophic and heterotrophic bacteria have been implicated in weathering processes in which aluminum is solubilized.[3,9,10] *T. ferrooxidans* attacks silicate minerals such as glauconite, illite, and microcline, with destruction and neoformation of minerals. Fungi transform silicates and other minerals, possibly by providing a sink for K thus stimulating cation loss from those minerals. Chelating compounds such as organic acids, which are the excretion products of microbial metabolism, or humic and fulvic acids, which are the products of plant tissue degradation, are also capable of solubilizing Al from silicate and oxide minerals.

Mn occurs only as 0.1% of the total mass of the earth, but concentrations in rocks range to over 10%.[4,11,12] Mn is found in oxide minerals with two common valences: a highly soluble divalent and the highly insoluble tetravalent species, both of which tend to be very stable. Mn reduction is stimulated by acidity and reducing conditions and occurs in the presence of active heterotrophic bacterial metabolism, plant root exudates, and autotrophic S, thiosulfate, and sulfide mineral oxidation.

Anaerobic Oxidation of Elemental Sulfur and Sulfur Compounds

Sulfate-reducing bacteria, either alone or in association with other bacteria, reduce sulfur oxides to sulfide with intermediate formation of sulfite. Sulfite and thiosulfate, formed by the condensation of sulfite and elemental sulfur, may migrate to regions in which free oxygen is available for thiobacilli to oxidize these compounds to sulfuric acid. The facultative anaerobic *Thiobacillus denitrificans* can also oxidize sulfur compounds and, perhaps, elemental sulfur to sulfuric acid. Thus, ample evidence exists to show that under appropriate conditions, sulfuric acid can be formed anaerobically.

Another anaerobic mechanism for the sulfuric acid generation from elemental sulfur has been proposed in which ferric iron serves as oxidant in a two-step reaction[13] (illustrated below):

$$S^0 + 4FeCl_3 + 3H_2O \rightarrow H_2SO_3 + 4FeCl_2 + 4HCl \quad (13)$$

$$H_2SO_3 + 2FeCl_3 + H_2O \rightarrow H_2SO_4 + 2FeCl_2 + 2HCl \quad (14)$$

$$S^0 + 6FeCl_3 + 4H_2O \rightarrow H_2SO_4 + 2FeCl_2 + 6HCl \quad (15)$$

This mechanism occurs in the thiobacilli, including the iron-oxidizing bacteria, and the lobate thermophile *Sulfolobus acidocaldarius*. Results of unpublished laboratory leaching tests suggest that such a mechanism occurs in oxidized sulfide mineral tailings.

Jarosite Formation by Ferric Iron Hydrolysis

Ochre deposits, "yellow boy," coatings, crusts, and other infillings by reddish or yellowish secondary iron minerals are caused principally by the formation of jarosites by hydrolysis of ferric sulfate. Although formed abiologically, these minerals are associated with biological iron oxidation. The hydrolysis reaction in the presence of cations is illustrated by the following equation:

$$3Fe_2(SO_4)_3 + 2AOH + 10H_2O \rightarrow 2A[Fe(SO_4)_2 \cdot 2Fe(OH)_3] + 5H_2SO_4 \quad (16)$$

where A is, in order of decreasing stability, K^+, Na^+, NH_4^+, and H^+. Formation of jarosites is controlled by concentrations of ferric iron, sulfate, and monovalent ions; pH; and temperature. Below pH 1.8 jarosites tend to dissolve, and above pH 2 other secondary iron precipitates such as goethite form. Jarosite formation removes ferric and monovalent cations from solution and increases free sulfate concentrations that cause pH values to decrease. Jarosites may also armor mineral surfaces, thus protecting them from oxidation or dissolution.

PREVENTION OF ACID GENERATION REACTIONS

The three components of acid generation reactions are the mineral substrate, the oxidant, and the biological catalyst. Eliminate any one of these components and acid generation ceases. The concept is simple; the application is not.

Removal of the mineral mass to another location[14] relocates the problem but does not necessarily solve it, is expensive, and has limited or specific application. Therefore, preventive measures focus on the exclusion of oxygen or the inhibition of bacteria responsible for acid generation, using chemical reagents that either disrupt the metabolism of these bacteria[15] or their environment.[16,17] Limited success has been obtained using phosphates, silicates, chlorides, lime, or detergents. Research using these reagents is continuing. These methods result mostly in temporary amelioration and require repeated application for optimum benefit. Formulations comprising combinations of quick-acting liquid and slow-release pellets appear to increase efficacy.[17]

Attempts to prevent acid generation by the establishment of vegetation or the placement of inert materials on mineral impoundment surfaces have not been successful[18] because iron-oxidizing bacteria can survive within the mineral mass.[19] Laboratory tests[20] have shown that an oxygen-scavenging layer might be effective in preventing acid generation and have led to the suggestion that disposal sites be designed so that oxygen induction is prevented. Other unpublished laboratory tests indicate this procedure to be effective only if a water table is established and maintained above the original mineral surface. Alternative disposal options under water or as layered or thickened slurries have been discussed.[21] Little attention has been given to using acid-generating rock as a resource either as mine backfill or as a source of acid and ferric iron for hydrometallurgical metal recovery operations.

ABATEMENT OF ACID DRAINAGE

Of chemical, physical, and biological abatement options, only those with a substantial biological component offer freedom from perpetual maintenance. Permanence depends on self-regenerative properties of biological systems that are integral functions of metabolism and interactions of all the components of these systems. Operational objectives of these systems are decreases in acidity through neutralization or sulfate removal and precipitation of metallic pollutants.

Sulfate is removed from acid drainage by three mechanisms: reduction to sulfide, biological uptake, and formation of organic esters on plant decomposition products. Bacterial sulfate reduction is the principal method[22] and requires anaerobic conditions, $pH > 5$, and suitable organic substrates. Given these conditions, sulfate-reducing bacteria reduce sulfate with pyruvic or lactic acids by the reactions:

$$\text{pyruvic acid} + 4H_2SO_4 \rightarrow \text{acetic acid} + CO_2 + 4H_2S \quad (17)$$

$$\text{lactic acid} + 8H_2SO_4 \rightarrow \text{acetic acid} + CO_2 + 8H_2S \quad (18)$$

Some sulfate reducers use molecular hydrogen for this reduction:

$$4H_2 + H_2SO_4 \rightarrow 4H_2O + H_2S \quad (19)$$

Sulfides formed may combine with heavy metals, causing their precipitation as sulfides. Sulfate esterification occurs in the presence of plant residues such as humic and fulvic acids, lignin, and cellulose and appears to be mediated by microbes; this mechanism may be responsible for the observed neutralization when acid drainage passes through decomposing wood.[23] Decreasing acidity causes amphoteric metals such as Al to precipitate.

Reactions that precipitate Mn may be enzymatic or nonenzymatic and remain poorly understood.[4] Enzymatic oxidations are catalyzed by manganese oxidase present in such heterotrophic genera as *Arthrobacter, Pseudomonas,* and *Citrobacter,* which require organic substrates, neutral conditions, and oxygen:

$$2MnSO_4 + O_2 + 2H_2O \rightarrow 2MnO_2 + 2H_2SO_4 \quad (20)$$

Manganese may also be oxidized by *Leptothrix pseudoochraceae, Arthrobacter siderocapsulatus,* and *Metallogenium* by a peroxidase:

$$MnSO_4 + H_2O_2 \rightarrow MnO_2 + H_2SO_4 \quad (21)$$

Nonenzymatic manganese oxidation occurs at pH $\sim$5 in the presence of oxygen and excretions from fungi or anaerobically at pH $\sim$7 in the presence of certain strains of *Bacillus* and *Pseudomonas.*

REFERENCES

1. Hutchins, S. R., M. S. Davidson, J. A. Brierley, and C. L. Brierley. "Microorganisms in Reclamation of Metals," *Ann. Rev. Microbiol.* 40:311–336 (1986).
2. Lundgren, D. G., and M. Silver. "Ore Leaching by Bacteria," *Ann. Rev. Microbiol.* 34:263–283 (1980).
3. Robert, M., and J. Berthelin. "Role of Biological and Biochemical Factors in Soil Mineral Weathering," in *Interactions of Soil Minerals with Natural Organics and Microbes*, P. M. Huang and M. Schnitzer, Eds. (Madison, WI: Soil Science Society of America, Inc., 1986), p. 497.
4. Silver, M., H. L. Ehrlich, and K. C. Ivarson. "Soil Mineral Transformation Mediated by Soil Microbes," in *Interactions of Soil Minerals with Natural Organics and Microbes,* P. M. Huang and M. Schnitzer, Eds. (Madison, WI: Soil Science Society of America, Inc., 1986), p. 497.

5. Kleinmann, R. L. P., D. A. Crerar, and R. R. Pacelli. "Biogeochemistry of Acid Mine Drainage and a Method to Control Acid Formation," *Min. Eng.* (March 1980), pp. 300–305.
6. Knapp, R. A. "The Biogeochemistry of Acid Generation in Sulphide Tailings and Waste Rock," in *Proceedings: Acid Mine Drainage Seminar/Workshop, Halifax, Nova Scotia, 23–26 March 1987,* B. Blakeman, J. Scott, M. Guilcher, K. Ferguson, and J. Worgan, Eds. (Ottawa, ON: Environment Canada, 1987), p. 47.
7. Silver, M. "Aquatic Plants and Bog Covers to Prevent Acid Generation-Base Metal Tailings," in *Proceedings: Acid Mine Drainage Seminar/Workshop, Halifax, Nova Scotia, 23–26 March 1987,* B. Blakeman, J. Scott, M. Guilcher, K. Ferguson, and J. Worgan, Eds. (Ottawa, ON: Environment Canada, 1987), p. 411.
8. Torma, A. E., and H. Sakaguchi. "Relation Between the Solubility Product and the Rate of Metal Sulfide Oxidation by *Thiobacillus ferrooxidans*," *J. Ferment. Technol.* 56:173–178 (1978).
9. Tan, K. H. "Degradation of Soil Minerals by Organic Acids," in *Interactions of Soil Minerals with Natural Organics and Microbes,* P. M. Huang and M. Schnitzer, Eds. (Madison, WI: Soil Science Society of America, Inc., 1986), p. 1.
10. Huang, P. M., and A. Violante. "Influence of Organic Acids on Crystallization and Surface Properties of Precipitation Products of Aluminum," in *Interactions of Soil Minerals with Natural Organics and Microbes,* P. M. Huang and M. Schnitzer, Eds. (Madison, WI: Soil Science Society of America, Inc., 1986), p. 159.
11. Nealson, K. H. "The Microbial Manganese Cycle," in *Microbial Geochemistry,* W. E. Krumbein, Ed. (Oxford: Blackwell Scientific Publications, 1983), p. 191.
12. Ehrlich, H. L. *Geomicrobiology* (New York: Marcel Dekker, Inc., 1981).
13. Sugio, T., C. Domatsu, O. Munakata, T. Tano, and K. Imai. "Role of Ferric Ion-Reducing System in Sulfur Oxidation of *Thiobacillus ferrooxidans*," *J. Bacteriol.* 49:1401–1406 (1985).
14. Sturm, J. W. "Materials Handling for Mine Spoil and Coal Refuse," in *Proceedings: Acid Mine Drainage Seminar/Workshop, Halifax, Nova Scotia, 23–26 March 1987,* B. Blakeman, J. Scott, M. Guilcher, K. Ferguson, and J. Worgan, Eds. (Ottawa, ON: Environment Canada, 1987), p. 321.
15. Watkin, E. M., and J. Watkin. "Inhibiting Pyrite Oxidation Can Lower Reclamation Costs," *Can. Min. J.* (December 1983), pp. 29–31.
16. Carruccio, F. T., and G. Geidel. "The *In-Situ* Mitigation of Acidic Drainages—Management of Hydro-Geochemical Factors," in *Proceedings: Acid Mine Drainage Seminar/Workshop, Halifax, Nova Scotia, 23–26 March 1987,* B. Blakeman, J. Scott, M. Guilcher, K. Ferguson, and J. Worgan, Eds. (Ottawa, ON: Environment Canada, 1987), p. 479.
17. Sobek, A. A. "The Use of Surfactants to Prevent AMD in Coal Refuse and Base Metal Tailings," in *Proceedings: Acid Mine Drainage Seminar/Workshop, Halifax, Nova Scotia, 23–26 March 1987,* B. Blakeman, J. Scott, M. Guilcher, J, Ferguson, and J. Worgan, Eds. (Ottawa, ON: Environment Canada, 1987), p. 357.
18. Dubrovsky, N. M., J. A. Cherry, E. J. Reardon, and A. J. Vivyurka. "Geochemical Evolution of Inactive Pyritic Tailings in the Elliot Lake Uranium District," *Can. Geotech. J.* 22:110–128 (1985).
19. Silver, M. "Distribution of Iron-Oxidizing Bacteria in the Nordic Uranium Tailings Deposit, Elliot Lake, Ontario, Canada," *Appl. Environ. Microbiol.* 53:846–852 (1987).

20. Silver, M., and G. M. Ritcey. "Effects of Iron-Oxidizing Bacteria and Vegetation on Acid Generation in Laboratory Lysimeter Leaching Tests on Pyrite-Containing Uranium Tailings," *Hydrometallurgy* 15:255–264 (1985).
21. Bell, A. V. "Prevention of Acid Generation in Base Metal Tailings and Waste Rock," in *Proceedings: Acid Mine Drainage Seminar/Workshop, Halifax, Nova Scotia, 23–26 March 1987,* B. Blakeman, J. Scott, M. Guilcher, K. Ferguson, and J. Worgan, Eds. (Ottawa, ON: Environment Canada, 1987), p. 391.
22. Herlihy, A. T., and A. L. Mills. "Sulfate Reduction in Freshwater Sediments Receiving Acid Mine Drainage," *Appl. Environ. Microbiol.* 49:179–186 (1985).
23. Tuttle, J. H., P. R. Dugan, C. B. Macmillam, and C. I. Randles. "Microbial Dissimilatory Sulfur Cycle in Acid Mine Water," *J. Bacteriol.* 97:594–602 (1969).

42b Design and Construction of a Research Site for Passive Mine Drainage Treatment in Idaho Springs, Colorado

Edward A. Howard, John C. Emerick, and Thomas R. Wildeman

INTRODUCTION

The Idaho Springs–Central City mining district in the foothills of the Colorado Front Range has massive waste rock dumps, mill tailings piles, and abandoned mine shafts and tunnels from precious metal ore production. Tunnel drainage typically has low pH and high metal concentrations that adversely affect the regional aquatic resources. Several sites, including the Big Five Tunnel, are on the National Priorities List under the Comprehensive Environmental Response, Compensation, and Liability Act of 1980 (CERCLA, or Superfund).

Natural wetlands have raised pH and reduced metals concentrations of acid mine drainages,[1,2] but only a few wetlands have been constructed to treat drainage from noncoal mines at higher elevations in Colorado.[3] At the Big Five Tunnel, a demonstration treatment system was built to determine the fate of metals in the system.[4,5] Other objectives of the research were to determine vegetation survival with exposure to elevated metals in a mountain climate, to study function and distribution of bacteria in the system, and to identify appropriate organic substrates and plant species. System design was based on suggestions of experienced investigators.[1,6-8] Their suggestions were modified for our research objectives and a harsh mountain climate.

DESIGN AND CONSTRUCTION

A reinforced concrete structure 3.05 m by 18.3 m was initially divided into three 6.1-m sections, with provisions to subdivide the structure into six 3.05-m sections at some later time if desirable (Figure 1). Each compartment was fitted with two drains, one active and one reserve. Drains were 15-cm polyvinyl chloride (PVC) pipe, and active drains consisted of standpipes initially set at 1-

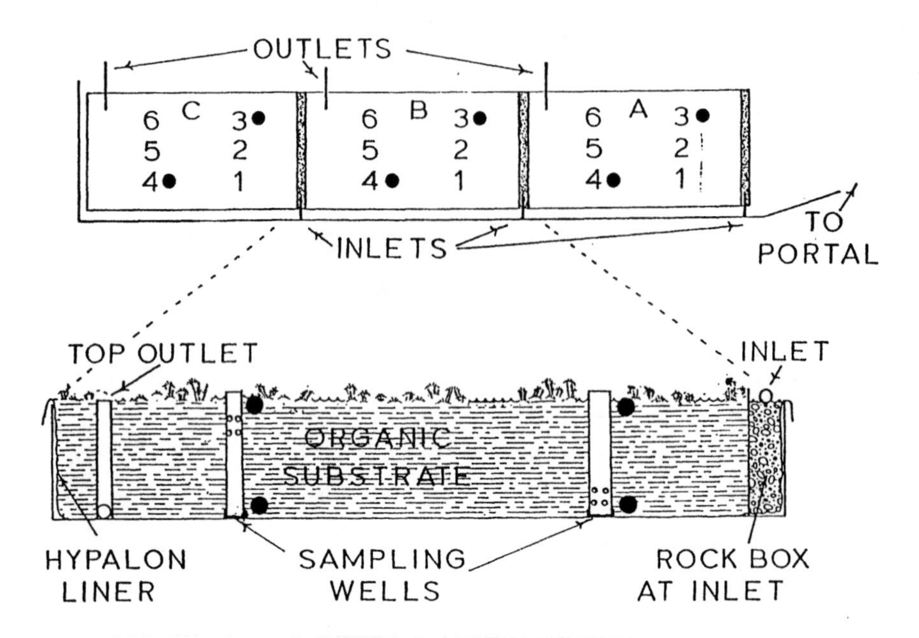

Figure 1. Diagram of the mine drainage treatment structure.

m depth. Drains deliver the overflow water to the existing pond. A 0.76-mm Hypalon liner separated one compartment from another to prevent chemical reactions between construction materials, organic substrates, and mine drainage. Baskets with 10- to 15-cm river rock at the upstream end exposed a maximum cross section of organic substrate to mine drainage. Six sampling wells of 15-cm PVC in each compartment allowed water to enter from the lowest 30 cm, the middle 30 cm, and the upper 30 cm of the organic substrates. A small concrete dam inside the tunnel provided head to distribute water to each compartment through insulated PVC lines fitted with ball valves for flow control. Perforated drain pipe (10-cm PVC) across the width of each compartment distributed water to each compartment, and excess water drained into the pond. Compartment sections were filled with 1 m of organic substrates: the first had mushroom compost; the second had equal parts of peat, aged steer manure, and decomposed wood shavings and sawdust; and the third had 10–15 cm of 5- to 8-cm limestone rock below the same mixture as the second compartment. Substrates were initially saturated with municipal water at 3.78 L/min.

TRANSPLANTING VEGETATION TO THE DEMONSTRATION SITE

Cattails (*Typha latifolia*) and sedges (*Carex utriculata*; *C. aquatilis*) were transplanted from a lake at a similar elevation in Grand County, Colorado into

Table 1. Analyses Performed on Waters, Substrates, and Plants

Waters	DTPA Extracts	Substrates	Plants
RAS[a]	Cd	RAS[a]	RAS[a]
CO_3^{2-}	Co	B	B
HCO_3^-	Cu	NH_3	
Cl^-	Fe	N_{tot}	
F^-	Pb	C_{tot}	
SO_4^{2-}	Mn	P_{tot}	
NO_3^-/NO_2^-	Ni	S_{forms}	
NH_3			
TSS			
TDS			
B			
TOC			
Acidity			

[a]RAS (Routine Analytical Services) includes Al, Sb, As, Ba, Be, Cd, Ca, Cr, Co, Cu, Fe, Pb, Mg, Mn, Hg, Ni, K, Se, Ag, Na, Tl, V, and Zn.

25–30% of each treatment bed. On September 5, sedges (*C. aquatilis*) and rushes (*Juncus arcticus*) transplanted from a fen 6.5 km from the demonstration site covered 25% of the treatment beds. Cattails from a foothills wetland and sedges from an adjacent wetland transplanted on September 18 covered an additional 35% of the treatment beds.

SAMPLING AND PREPARATION OF PLANTS, SUBSTRATES, AND WATERS

At the transplant sites, samples of live leaves, dead leaves, flowers, and roots were collected and analyzed for metals concentrations before plants were subjected to mine drainage. Samples of organic substrates were similarly analyzed. Water sampling of mine drainage, outputs from three cells, and waters from six access wells in each compartment were initiated on October 13, 1987. Temperature, pH, Eh, conductivity, and dissolved oxygen content of the waters were measured in the field. Laboratory analyses are listed in Table 1.

REFERENCES

1. Holm, J. D., Colorado Division of Mined Land Reclamation. Personal communication (1985).
2. Emerick, J. E., Colorado School of Mines, Golden, CO. Personal communication (1985).
3. Holm, J. D. "Passive Mine Drainage Treatment: Selected Case Studies," in *Proceedings, 1983 National Conference on Environmental Engineering,* A. Medine and M. Anderson, Eds. (American Society of Civil Engineers, 1983).
4. Howard, E. A., and T. R. Wildeman. "Passive Treatment Pilot System at the Big Five Tunnel, Idaho Springs, Colorado," unpublished technical proposal to Camp Dresser and McKee, Denver, CO (1987).

5. Guertin, deF., J. C. Emerick, and E. A. Howard. "Passive Mine Drainage Treatment Systems: A Theoretical Assessment and Experimental Evaluation," unpublished report submitted to the Colorado Mined Land Reclamation Division, Cooperative Agreement No. 202–317 (1985).
6. Kleinmann, R. L. P. "A Low-Cost, Low-Maintenance Treatment System for Acid Mine Drainage Using Sphagnum Moss and Limestone," *Symposium on Surface Mining, Hydrology, Sedimentology and Reclamation* (Lexington: University of Kentucky, 1983).
7. Kleinmann, R. L. P. Personal communication (1987).
8. Hiel, M. Personal communication (1987).

42c Manganese and Iron Encrustation on Green Algae Living in Acid Mine Drainage

S. Edward Stevens, Jr., Kim Dionis, and Lloyd R. Stark

INTRODUCTION

Many species of algae are known to tolerate acid mine drainage (AMD) resulting from coal mining.[1] We have observed several filamentous algae in the division Chlorophyta (green algae) in AMD with rust-colored or dark brown encrustations on their cell walls. Because of high iron and, sometimes, manganese concentrations found in AMD, we suspected that the colored encrustations might be composed of iron, manganese, or both. Based on analytical electron microscopy, we found that metallic components of the encrustations on these algae are indeed those metals. Energy dispersive spectroscopy (EDS) with the scanning electron microscope (SEM) discerns between different algal species, between parts of cells, and between living and nonliving matter; the beam can be focused directly on the site of interest. All elements are analyzed simultaneously, so even unexpected substances will be detected. A further advantage of this technique is that only small amounts of tissue are necessary for analysis. EDS, also known as X-ray microanalysis, has been used in studies of metal accumulation in algae in situ[2,3] and in the laboratory[3] for water pollution studies.

Algae were collected from wetlands and sediment ponds receiving coal mine drainage, and if metals are significantly accumulated by these algae, they may play a role in water improvement.

MATERIALS AND METHODS

Samples used in the present study were collected as grab samples from various locations in Pennsylvania and Ohio (Table 1). Filaments of algae were prepared one of two ways, depending on whether they were to be analyzed for metal content by EDS or analyzed and photographed with the scanning electron microscope.

Table 1. Algae with Encrustations from Coal Mine Drainage

Alga	Collection Site	Collection Date	Water Quality	Description of Encrustation	Element(s)
Oedogonium sp.	Clarion Co., PA	15 JUL 1987 30 OCT 1987 23 NOV 1987	pH 6–7 Mn 10–30 Fe 0–5	Blackish-brown Irreg. border	Mn
Oedogonium sp.	Jefferson Co., PA	21 JUL 1987	pH (ND) Mn 19.1 Fe <0.1	Blackish-brown Irreg. border	Mn
Oedogonium sp.	Clarion Co., PA	21 JUL 1987	pH (ND) Mn 21.4–36.4 Fe <0.1	Blackish-brown Irreg. border	Mn, Fe
Microspora floccosa	Armstrong Co., PA	24 JUL 1984	pH 3.3 Mn 130 Fe 47	Golden-brown to rust colored Irreg. border	Fe, S
Microspora quadrata	Coshocton Co., OH	27 APR 1988	pH 6–6.5 Mn 1.1 Fe 39.9	Rust-colored Smooth cylinders	Fe
Mougeotia sp.	Armstrong Co., PA	24 JUL 1987	pH 3.3 Mn 130 Fe 47	Blackish-brown Irreg. border	Fe, S
Microspora abbreviata	Coshocton Co., OH	2 MAR 1988	pH 6.3 Mn 2 Fe 102		
Microspora pachyderma	Clarion Co., PA	5 OCT 1986 15 JUL 1987 30 OCT 1987	pH 6–7 Mn 10–30 Fe 0–5		
Microspora tumidula	York Co., PA	26 FEB 1988	pH 5.3–6.5 Mn 0.1–9 Fe 0–23		
Trachelomonas ehrenberg	Clarion Co., PA	30 OCT 1987	pH 6.1 Mn 2.9 Fe 3.4		
Tribonema bombycinum	York Co., PA	26 FEB 1988	pH 5.3–6.5 Mn 0.1–9 Fe 0–23		
Microspora sp.	Clarion Co., PA	4 AUG 1987	pH 3.2–3.4 Mn 21–27 Fe 0		
Microspora sp.	Clarion Co., PA	MAY 1988	pH 6–7 Mn 12 Fe 0–10		
Mougeotia sp.	Clarion Co., PA	4 AUG 1987	pH 2.5–3.4 Mn 21–40 Fe 0–6		
Mougeotia sp.	Jefferson Co., PA	13 MAY 1988	pH 3 Mn 62.9 Fe 18.2		
Ulothrix sp.	Clarion Co., PA	4 AUG 1987	pH 3.2 Mn 27 Fe 0		

Note: Total Fe and Mn concentrations in water are given in mg/L.

Sample Preparation

Method 1: SEM and EDS

Sample preparation procedures were based on a SEM protocol described for *Oedogonium*.[4] Filaments were washed in buffer (0.025 M sodium phosphate, pH 6.8) to remove associated sediments. Solutions used in subsequent processing of the sample were prepared with Milli-Q water. Filaments were prefixed for 1 hr in 1.5% glutaraldehyde in 0.025 M sodium phosphate buffer (pH 6.8) at room temperature. They were fixed in 5% glutaraldehyde in the same buffer at room temperature for 16 hr, then washed four times in cold buffer (4°C) in the following manner: centrifugation at 4°C for two to three min in a Sorvall RC-5B Refrigerated Superspeed Centrifuge (Du Pont Instruments, Wilmington, DE) at 1075 *g* in order to get a loose pellet at the bottom of the tube. The supernatant was decanted and replaced with cold buffer. Samples were kept at 4°C for 2 hr, and the process was repeated. Filaments were postfixed in 1% osmium tetroxide in the same buffer at 4°C for 16–18 hr. They were washed with cold buffer four times in the same manner as before and then left in cold buffer at 4°C.

The sample was split into two parts, one for SEM, the other for transmission electron microscopy (TEM); the results from TEM will be reported at a later date. Filaments were removed from the buffer solution, placed on 13-mm Nucleopore filters (Nucleopore Corp., Pleasanton, CA), and dehydrated in a gradient of increasing 10-fold ethanol concentrations (10% through 100%). Filters were inserted into specimen-processing capsules (Polaron Instruments, Inc., Cambridge, MA) and placed in a Model H Pelco Critical Point Dryer (Ted Pella, Inc., Tustin, CA). Filters were then placed on double-stick tape, attached to carbon rods, mounted on SEM stubs, and gold-coated with a sputter coater (Anatech Inc., Alexandria, VA).

Method 2: EDS

Filaments were either preserved in 70% ethanol or prepared from freshly collected material, depending on how quickly the work could be completed. We found no differences in the results between preserved and fresh specimens. If fresh specimens were used, they were kept on ice for 3 hr until processing began. Filaments were washed gently to remove sediments, air-dried on Nucleopore filters, and mounted on SEM stubs, as described above. Specimens were carbon-coated in a Carbon-Evaporator (Edwards Hi Vacuum, Inc., Grand Island, NY) and observed in an ISI SX-40 Scanning Electron Microscope (International Scientific Instruments, Milpitas, CA). Carbon rods used for carbon coating were determined to be free of metal contamination by EDS. Elemental analysis was accomplished with a Kevex S100 energy dispersive spectrometer (Kevex Corp., Foster City, CA) interfaced with the ISI SX-40 SEM. SEM observations and EDS analyses were done at 15 or 30 kV. Photographs were made using Kodak Type 55 positive/negative film.

Figure 1. Scanning electron micrograph of critical point dried, whole-cell preparation of *Oedogonium* sp. depicting morphological characteristics of encrustations on algal filaments. Arrows indicate locations of some encrustations. Magnification was 425×. Bar = 40 μm.

RESULTS AND DISCUSSION

Those algae from AMD with observable encrustations are listed in Table 1 along with collection dates, sites, and the general water quality of the site. Of these, several have been studied by EDS, and one has been subjected to both EDS and SEM. We found *Oedogonium, Mougeotia,* and *Microspora* to be the most commonly encrusted genera of algae growing in AMD. Algal encrustations varied in color from rust red to blackish brown and in shape from rough, irregular globules to smooth cylinders. In general, encrustations contain mostly iron if the iron concentration in the water is moderate to high regardless of the manganese concentration. However, if the iron concentration is low, encrustations contain manganese.

We did do a more detailed analysis of an *Oedogonium* sp. collected from a site in Clarion County, Pennsylvania with low iron but relatively high manganese. An SEM micrograph of a typical sample of *Oedogonium* from this site is shown in Figure 1. *Oedogonium* filament encrustations were blackish brown in the natural state. One cell or many cells in tandem were encrusted, and encrustations were randomly distributed among filaments. Elemental composition of *Oedogonium* filament was analyzed by EDS, and a typical result is shown in

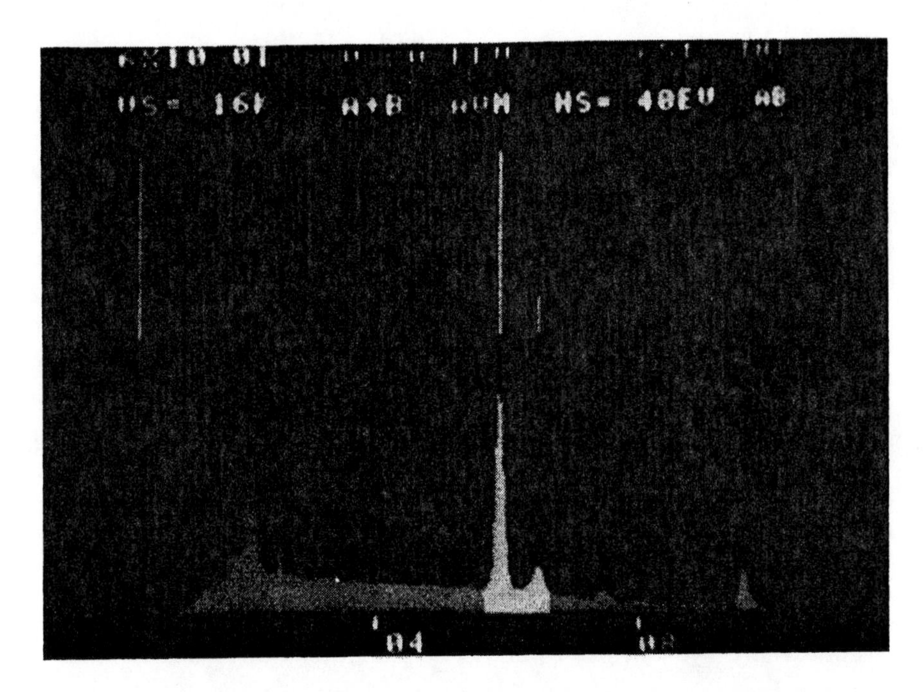

Figure 2. EDS spectrum derived from an encrustation like those shown in Figure 1. The leftmost peak is from gold used in coating algal filaments for SEM. The other large peak in the middle of the photograph is the Mn-Kα peak. Mn is the predominant element in the encrustation. The small peak immediately to the right of the Mn-Kα peak is the Mn-Kβ peak, which overlaps the Fe-Kα position. Thus, if Fe is present in this sample, it is not apparent, because such low amounts would be masked by the Mn-Kβ peak.

Figure 2. The leftmost peak is gold used for gold-coating the sample for SEM. The other peak is that of manganese. In *Oedogonium* samples from sites in Clarion and Jefferson counties, Pennsylvania, encrustations were composed of manganese; in another sample from Clarion County, Pennsylvania, they were composed of both manganese and iron. A typical filament of *Oedogonium* from the same sample with a large manganese encrustation is shown with a filament of the same alga that lacked an encrustation in Figure 3A. An X-ray map of the manganese distribution in these same algal filaments is shown in Figure 3B. Manganese was concentrated in the encrustation with no more than background manganese found on the clean filament.

In *Microspora floccosa*, the encrustations were golden brown to rust-colored and irregular in shape. Encrustations appeared to begin at the cross walls between cells and to cover most of the cells in a filament and were composed of iron and sulfur. In *Mougeotia*, encrustations were rough-bordered, blackish-brown collars. Like *M. floccosa*, collected at the same site, these encrustations were composed of iron and sulfur. The manganese and iron

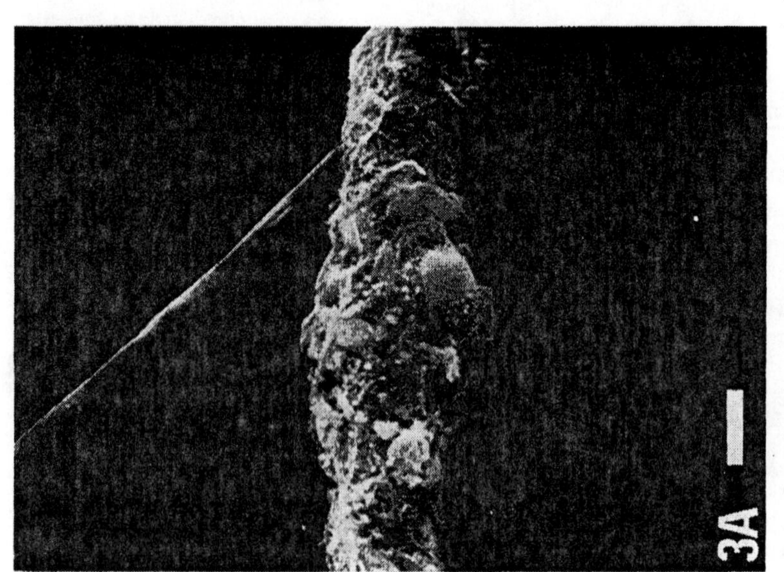

Figure 3. Scanning electron micrograph and Mn X-ray map of critical point dried, whole-cell preparation of *Oedogonium* sp. (A) In the foreground is a manganese encrustation on a filament of *Oedogonium*; in the background is a filament of the same alga free of encrustations. Magnification was 833×. Bar = 20 μm. (B) X-ray map of Mn in the filaments in Figure 3A. Mn is concentrated in the encrustation and is only at background levels in the algal filament lacking encrustations.

concentrations of the water were elevated (130 and 47 mg/L, respectively), and pH was low (3.3).

In *Microspora quadrata*, encrustations were strikingly regular and appeared to begin at the cross walls and form smooth, rust-colored cylinders around single cells. In some cases, many cells in sequence were affected. Distribution appeared regular; generally every fourth cell was girdled with an encrustation. The cylinders were composed of iron; no other metals were present. The iron concentration in the water was 39.9 mg/L; manganese concentration, 1.1 mg/L; and pH near neutrality.

Many algae deposit iron and manganese in their cell walls. The taxonomic key of Ramanathan[5] uses iron encrustation as a characteristic of *Microspora pachyderma*. *Trachelomonas*, a chlorophycean alga that we found growing in acid mine drainage (Table 1), was previously reported to deposit iron and manganese as determined by EDS, presumably as ferric and manganic hydroxides.[6] The chrysophycean algae *Pseudokephyrion pseudospirale* and *Kephyrion rubri-claustri* deposited both iron and manganese in their cell walls; EDS coupled with ultrastructural examination suggested that manganese was deposited as electron-dense, needlelike structures and iron was deposited as dense, granular bands. If both metals were deposited, they seemed to be spatially segregated.[7] Color of euglenoid envelopes has been correlated with metal content: dark golden to brown envelopes are high in manganese[8] and hyaline to light golden envelopes are higher in iron content.[9,10] In envelopes containing both metals, color varied from hyaline to dark brown.[9] Ultrastructure of these deposits was similar to those found in the chrysophycean loricas, and Dunlap et al.[7] suggested that processes involved in iron and manganese biomineralization might be similar in all Fe- or Mn-depositing organisms. Anderson et al.[11] studied iron distribution in the cell wall of the desmid *Staurastrum luetkemuelleri* using EDS coupled with electron microscopy and found iron accumulation only on distinct locations on the cell wall. Accumulation was dependent on Fe content of the culture medium in addition to other environmental or growth factors.

The mechanism by which metals are deposited on cell wall material is unknown. In iron-rich waters, iron salts will coat submerged surfaces. However, this does not appear to play a dominant role in encrustation formation in the algae we have studied. Within established filaments, encrustations appear to be restricted to single cells or groupings of cells, with areas in between which remain clear. In *Microspora*, nodules frequently form in a regular pattern and seem to begin at cross walls. Seemingly, deposition is at least partly controlled by biological processes. Two possibilities can be suggested: deposition could result from a structural or physiological property of the alga, or deposition could result from activities of epiphytic microorganisms. *Siderocystis confervarum* is a bacterium that grows epiphytically on green algal filaments and forms large, round iron oxide aggregates.[12] Other bacteria in the family Siderocapsaceae will also form large masses of manganese oxides. Little is known about their physiology, but the following has been deduced from environmental information: they are aerobic; optimal growth takes place at the beginning of the change from very reduced to oxidized conditions in neutral to lightly alkaline environments; and they are dependent on organic material for growth. *S. confervarum* not only grows on the surface of algae but within their cells as well. There have also been many other Fe- or Mn-oxidizing microbes described that grow in acidic to near-neutral waters and which could potentially colonize the surfaces of algae.[13,14]

The role of metal-encrusting algae in metal removal from AMD is unclear. Ultimately, the significance of these algae in the treatment of AMD will depend on metal removal efficiencies, the relative dissolution of the precipitated metals upon algal filament death, and algae density within the treatment system.

ACKNOWLEDGMENTS

We thank Tom Doman for his assistance with EDS and SEM, E. J. de Veau for his constructive comments on the manuscript, and C. J. Hillson for assistance with the taxonomy of algae. This project is part of a research program examining the use of wetlands in the treatment of acid mine drainage at The Pennsylvania State University funded by The Ben Franklin Challenge Grant Program of Pennsylvania, American Electric Power Service Corporation, Pennsylvania Power and Light Company, The Department of Environmental Resources of Pennsylvania, U.S. Department of the Interior (Bureau of Mines) contract PO373662, and by small donations from 16 coal operating companies or engineering consulting firms.

REFERENCES

1. Bennett, H. D. "Algae in Relation to Mine Water," *Castanea* 34:306–328 (1969).
2. Lindahl, G., K. Wallstrom, G. M. Roomans, and M. Pedersen. "X-ray Microanalysis of Planktic Diatoms in *In Situ* Studies of Metal Pollution," *Botanica Marina* 26:367–373 (1983).
3. Pedersen, M., G. M. Roomans, M. Andren, A. Lignell, G. Lindahl, K. Wallstrom, and A. Forsberg. "X-ray Microanalysis of Metals in Algae—A Contribution to the Study of Environmental Pollution," *Scanning Electron Microscopy* II:499–509 (1981).
4. Pickett-Heaps, J. D., and L. C. Fowke. "Cell Division in *Oedogonium.* I. Mitosis, Cytokinesis, and Cell Elongation," *Aust. J. Biol. Sci.* 22:857–894 (1969).
5. Ramanathan, K. R. *Ulotrichales* (New Delhi: Indian Council of Agricultural Research, 1964).
6. Leedale, G. F. "Envelope Formation and Structure in the Euglenoid Genus *Trachelomonas,*" *New Phytologist* 52:93–113:238–266 (1952).
7. Dunlap, J. R., P. L. Walne, and H. R. Preisig. "Manganese Mineralization in Chrysophycean Loricas," *Phycologia* 26:394–396 (1987).

8. West, L. K., and P. L. Walne. "*Trachelomonas hispida* var. *coronata* (Euglenophyceae). II. Envelope Substructure," *J. Phycol.* 16:498–506 (1980).
9. Dunlap, J. R., P. L. Walne, and J. Bentley. "Microarchitecture and Elemental Spatial Segregation of Envelopes of *Trachelomonas lefevrei* (Euglenophyceae)," *Protoplasma* 117:97–106 (1983).
10. Barnes, L. S. D., P. L. Walne, and J. R. Dunlap. "Cytological and Taxonomic Studies of the Euglenales. I. Ultrastructure and Envelope Elemental Composition in *Trachelomonas*," *Br. Phycol. J.* 21:387–398 (1986).
11. Anderson, S., M. Heldal, and G. Knutsen. "Cell Wall Structure and Iron Distribution in the Green Alga *Staurastrum luetkemuelleri* (Desmidiaceae)," *J. Phycol.* 23:669–672 (1987).
12. Hanert, H. H. "The Genus *Siderocapsa* (and Other Iron- or Manganese-Oxidizing Eubacteria)," in *The Prokaryotes*, M. P. Starr, H. Stolp, H. G. Truper, A. Balows, and H. G. Schlegel, Eds. (New York: Springer-Verlag, 1981), pp. 1049–1059.
13. Ghiorse, W. C. "Biology of Iron- and Manganese-Depositing Bacteria," *Ann. Rev. Microbiol.* 38:515–550 (1984).
14. Harrison, A. P. "The Acidophilic Thiobacilli and Other Acidophilic Bacteria that Share Their Habitat," *Ann. Rev. Microbiol.* 38:265–292 (1984).

42d Determining Feasibility of Using Forest Products or On-Site Materials in the Treatment of Acid Mine Drainage in Colorado

Edward A. Howard, Martin C. Hestmark, and Todd D. Margulies

INTRODUCTION

Acid mine drainage is a common problem associated with coal and metal mining. The low pH and high dissolved metal concentrations characteristic of these effluent streams affect aquatic ecosystems and are hazardous to human health. Effluents from active operations are regulated, but abandoned or inactive mine effluents are not controlled and inactive sites have limited or no funds for construction and maintenance of elaborate treatment systems. In Colorado, many high-elevation mines have adverse climatic conditions and limited accessibility that preclude construction and maintenance of traditional water treatment systems. Inexpensive alternatives with limited maintenance requirements and natural processes for metal ion removal may be the only feasible methods for these situations.

BACKGROUND

Investigations of alternative treatment systems have been and are currently being conducted at the Colorado School of Mines and elsewhere. These systems rely upon natural processes of filtration, cation exchange, sorption, coprecipitation, complexation, and biological extraction to remove heavy metal ions, and aeration and/or addition of limestone to modulate pH. Because investigation of the heavy metal removal processes is a relatively new field, removal mechanisms, relationships among the processes, and their relative importance are not well understood.

Some studies are exploring continual metal extraction by systems that first remove metals, primarily iron, with natural processes in self-perpetuating, artificially created peat bogs. In the second phase, cascades exsolve carbon dioxide, and coarse limestone reduces acidity.[1,2] Treatment with crushed limestone alone suffers an 80% loss of treatment efficiency due to extensive coating of reactive surfaces by iron hydroxides. Coating disruption by enhancing surface erosion through mechanical abrasion or high pressure and reducing particle size has limited applicability for most Colorado drainage sites. Peat and live *Sphagnum* systems rely on ion exchange capabilities of organic materials to remove heavy metals and enhance neutralization efficiency by preventing coating by metal hydroxides.

Metal ion concentrations in acid mine drainage flowing through *Sphagnum*-dominated peat bogs and natural wetlands have been reduced by cation exchange processes. These cation exchange properties are attributed to carboxyl (COOH) functional groups found in humic acids of peats and pectic compounds in plant cellular tissue.[3,4] Pectic compounds are polymers composed primarily of galacturonic acid found in the middle lamella between plant tissue cell walls.[5,6] A nearly 1:1 relationship between cation exchange capacity (CEC) of various types of plant tissues and content of unesterified polyuronic acids in the tissues was shown by Knight et al.[7]

Humic acid is a complex product of plant cell degradation, representing a major component of peat. Due to its complexity, no universally accepted chemical definition of humic acid exists nor has the structure of humic acid molecules been adequately described. However, carboxyl groups are known to be present, and ion exchange properties of humus are attributed to these carboxyl groups.[4,8]

Since forest litter has similar decomposition products and large quantities are produced, we thought these humus materials may also be effective in reducing heavy metal concentrations in acid mine drainage. Our objective was to analyze forest materials for suitable characteristics as media for treatment of acid mine drainage.

METHODS AND PROCEDURES

Materials analyzed included litter and humus layers beneath three forest types and charcoal because it is easily made on-site. Tests performed included cation exchange capacity (CEC) and metal ion removal efficiency for Mn, Zn, and Cd. A stock solution was prepared with a 75:100:20 ratio of Mn to Zn to Cd, similar to concentrations found in Colorado acid mine drainages. Iron was excluded because it would have masked the desired results by coprecipitating other metals.

Sample Preparation

In cooperation with the Forest Service, samples of the litter layer beneath spruce-fir (*Picea engelmannii; Abies bifolia*), lodgepole pine (*Pinus contorta*), and aspen (*Populus tremuloides*) forest types were obtained from locations near Idaho Springs, Colorado. Common charcoal briquettes were purchased.

Each litter and humus sample was prepared by grinding larger materials to pass through a 6.35-mm mesh. Each sample (about 0.06 m^3) was repeatedly passed through a sample splitter for uniform mixing and to reduce the total sample to 1 kg. Standard soil sieve analysis of the 1-kg samples determined size distribution of the materials. Laboratory analysis of CEC was performed on 4- to 6-g samples, and 100-g samples were used for metal removal analyses. The CEC for each of the materials was evaluated using a sodium saturation procedure described by Chapman[9] and Rhoades.[10]

For metal ion removal, a stock solution was prepared with 750 mg/L Mn, 1000 mg/L Zn, and 200 mg/L Cd to approximate relative concentration ratios in water from some Colorado metal mines. Mn was obtained from a 50% manganese nitrate ($Mn(NO_3)_2$) solution, and Zn and Cd were obtained from hydrated nitrate salts of these metals ($Zn(NO_3)_2 \cdot 6H_2O$ and $Cd(NO_3)_2 \cdot 4H_2O$). Concentration ratios were increased to ensure saturation of exchange sites and so the decay rate in metal uptake could be estimated. Sample solutions were analyzed for concentrations of Mn, Zn, and Cd by atomic absorption spectroscopy.

Samples were placed in 3-cm-diameter glass columns with stopcocks to control the flow rate and glass wool to prevent stopcock blockage. Because forest materials have good porosity, flow rate through the columns was controlled by the stopcock to minimize effects of different residence times between tested samples and the stock solution. The amount of stock solution was adjusted to reflect the forest material CEC. Because decrease in the efficiency of metal ion uptake was an objective, each column was washed five times with stock solution equivalent to 30, 60, 90, 120, and 170% of the total CEC. Removal efficiencies were the difference between initial concentrations and concentrations after passage through the column. A local analytical laboratory analyzed for concentrations of Mn, Zn, and Cd by atomic absorption spectroscopy. Quality control analyses were performed on nine of 25 samples.

RESULTS AND DISCUSSION

Sieve Analysis

Litter and humus samples from ponderosa and lodgepole forest types had 80% less than 1.18 mm. Spruce-fir litter and humus had a greater percentage of materials in the 0.85- to 6.3-mm size ranges, and a lower percentage in the size ranges below 0.64 mm. Aspen litter and humus was intermediate between the distributions of spruce-fir and ponderosa or lodgepole samples.

Cation Exchange Capacities

Ponderosa and aspen litter and humus had the highest CEC (121.8 and 116.9 meq/100 g, respectively), and the lodgepole pine had an intermediate

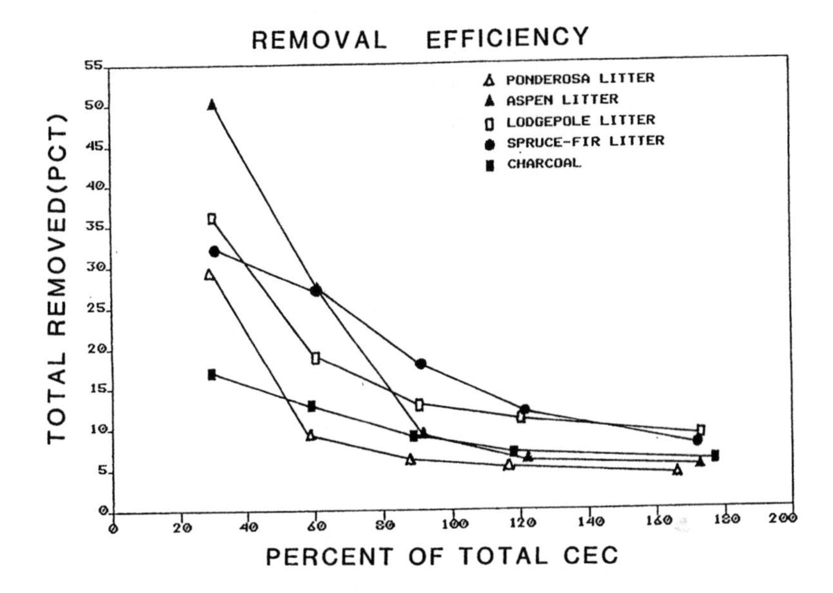

Figure 1. Percent of all three metals removed in 103-g samples of the litter and humus layer from forest types and charcoal. The percent removal efficiency is plotted vs the percent of total cation exchange capacity as determined by sodium saturation. The total concentration of metal ions in the initial solution was 1950 ppm.

CEC (79 meq/100 g). Spruce-fir had a lower CEC (58.6 meq/100 g), and charcoal had less than all the forest materials at 30.2 meq/100 g. The results were consistent with the degree of soil development and litter fermentation within the expected range for humus materials, which may be as high as 200 meq/100 g.[11]

Efficiency of Metal Ion Removal

After the first solution sample, approximately 30% of the total CEC, relative removal efficiency of litter and humus followed a consistent and decreasing sequence: aspen, lodgepole, spruce-fir, ponderosa, and charcoal.

The second solution sample (ca. 60% of CEC) showed a marked drop in the amount of metal ions retained. Aspen had the highest retention for Cd and was nearly the same as spruce-fir in the retention of Zn. Retention of Mn by spruce-fir exceeded aspen, and retention by aspen was approximately the same as the lodgepole sample. Retention of Mn and Cd by ponderosa approximated the charcoal sample, but for Zn, retention by ponderosa was less than in charcoal. All three metals are graphed together in Figure 1.

Curves for forest materials begin to flatten after receiving 90% of the total CEC, indicating saturation of exchange sites. More metal retention by spruce-fir may be due to larger particle sizes in that sample.

The percent of three metal ions removed in 103 g of litter and humus samples and charcoal is shown in Figure 1. Removal efficiency of forest materials and charcoal for three combined metals was consistent with relative removal efficiency for individual metal ions.

Cumulative percent removal was also determined. After 30% of the total CEC had passed through litter and humus materials, results were consistent in that aspen litter and humus had the greatest removal efficiency, followed by lodgepole, spruce-fir, ponderosa, and the charcoal sample. After 60% of the total CEC, the spruce-fir sample overtook the lodgepole sample, with the relative efficiency of other materials remaining in the same order. As the total CEC was approached, aspen, spruce-fir, and lodgepole samples showed cumulative percentages removed of 20–27%. Removal by ponderosa and charcoal had decreased to 13%. At 170% of total CEC, spruce-fir, aspen, and lodgepole were nearly the same (16%), and ponderosa and charcoal decreased to 9%.

CONCLUSIONS

Ponderosa and aspen had better ion exchange than lodgepole and spruce-fir litter and humus. However, the CEC, from sodium saturation, did not adequately estimate the cumulative percent removal capacities of litter and humus samples. For example, the ponderosa litter and humus had the highest CEC, but ponderosa had the lowest cumulative percent metal removal of the forest types, and at 90% of total CEC, ponderosa was not much better than charcoal. The spruce-fir sample had the lowest CEC of the forest types, but its cumulative percent removal was similar to aspen at 90% of total CEC.

Overall trends in the efficiency are consistent and decrease markedly as 60–90% of CEC was reached. All four forest materials tested removed metal ions from mine drainage, but spruce-fir and aspen decomposition materials had higher cumulative removal efficiencies in laboratory studies and should be tested in pilot-scale field situations. Reduction of metal ion concentrations in abandoned mine drainages with on-site forest litter and humus appears promising.

REFERENCES

1. Holm, J. D. "Passive Mine Drainage Treatment: Selected Case Studies," in *Proceedings, 1983 National Conference on Environmental Engineering,* A. Medine and M. Anderson, Eds. (American Society of Civil Engineers, 1983).
2. Kleinmann, R. L. P. "A Low-Cost, Low-Maintenance Treatment System for Acid Mine Drainage Using Sphagnum Moss And Limestone," *Symposium on Surface Mining, Hydrology, Sedimentology and Reclamation* (Lexington: University of Kentucky, 1983).
3. Moore, P. D., and D. J. Bellamy. *Peatlands* (New York: Springer-Verlag, 1974).
4. MacCarthy, P., Colorado School of Mines. Personal communication (1984).

5. Devlin, R. M. *Plant Physiology,* 2nd ed. (New York: Van Nostrand Reinhold Company, 1969).
6. Salisbury, F. B., and C. Ross. *Plant Physiology* (Belmont, CA: Wadsworth Publishing Co., Inc., 1969).
7. Knight, A. H., W. M. Crooke, and R. G. E. Inkson. "Cation Exchange Capacities of Tissues of Higher and Lower Plants and Their Related Uronic Acid Contents," *Nature* 192:142–143 (1961).
8. Guertin, deF., J. C. Emerick, and E. A. Howard. "Passive Mine Drainage Treatment Systems: A Theoretical Assessment and Experimental Evaluation," unpublished report submitted to the Colorado Mined Land Reclamation Division, Cooperative Agreement No. 202–317 (1985).
9. Chapman, H. D. "Cation Exchange Capacity," in *Methods of Soil Analysis,* C. A. Black et al., Eds. *Agronomy* 9:891–901 (1965).
10. Rhoades, J. D. "Cation Exchange Capacity," in *Methods of Soil Analysis, Part 2. Chemical and Microbiological Properties,* Agronomy Monograph No. 9, 2nd ed. (Madison, WI: American Society of Agronomy and Soil Science Society of America, 1982).
11. Buckman, H. O., and N. C. Brady. *The Nature and Properties of Soils,* 7th ed. (New York: Macmillan Publishing Co., Inc., 1969).

42e Use of Wetlands to Remove Nickel and Copper from Mine Drainage

Paul Eger and Kim Lapakko

INTRODUCTION

Drainage from mineralized Duluth Complex stockpiles located at LTV's Dunka Mine in northeastern Minnesota contains elevated concentrations of nickel (Ni), copper (Cu), cobalt (Co), and zinc (Zn) and has increased metal concentrations in nearby receiving waters to levels up to 400 times the natural background concentrations. These levels exceed water quality guidelines established by the Minnesota Pollution Control Agency (MPCA), and in 1985 the company signed a stipulation agreement with the MPCA which established a timeline for achieving water quality goals.

As part of this agreement, the company conducted a feasibility study[1] that analyzed treatment options ranging from a full-scale treatment plant (lime precipitation with reverse osmosis) to various passive treatment alternatives (e.g., stockpile revegetation). The analysis indicated that a treatment plant capable of treating all the drainage (6×10^8 L/yr) could generally achieve guidelines but would have a capital cost of $8.5 million and an annual operating cost of $1.2 million. However, similar reductions in flow and concentration might be achieved through a series of passive (low cost, low maintenance) procedures combining infiltration reduction, alkaline treatment, and wetland treatment. Capital costs were estimated at $4.0 million, with annual operating costs of $40,000. Wetland treatment is a crucial aspect of this passive approach, and although previous work[2-4] has demonstrated peat effectiveness in removing trace metals from mine drainage, an actual treatment system has not been built. In 1986, LTV Steel Mining Company and the Minnesota Department of Natural Resources began a cooperative program to develop data on optimal treatment techniques and system life, to be used to design full-scale treatment systems for the stockpile drainages at the Dunka Mine.

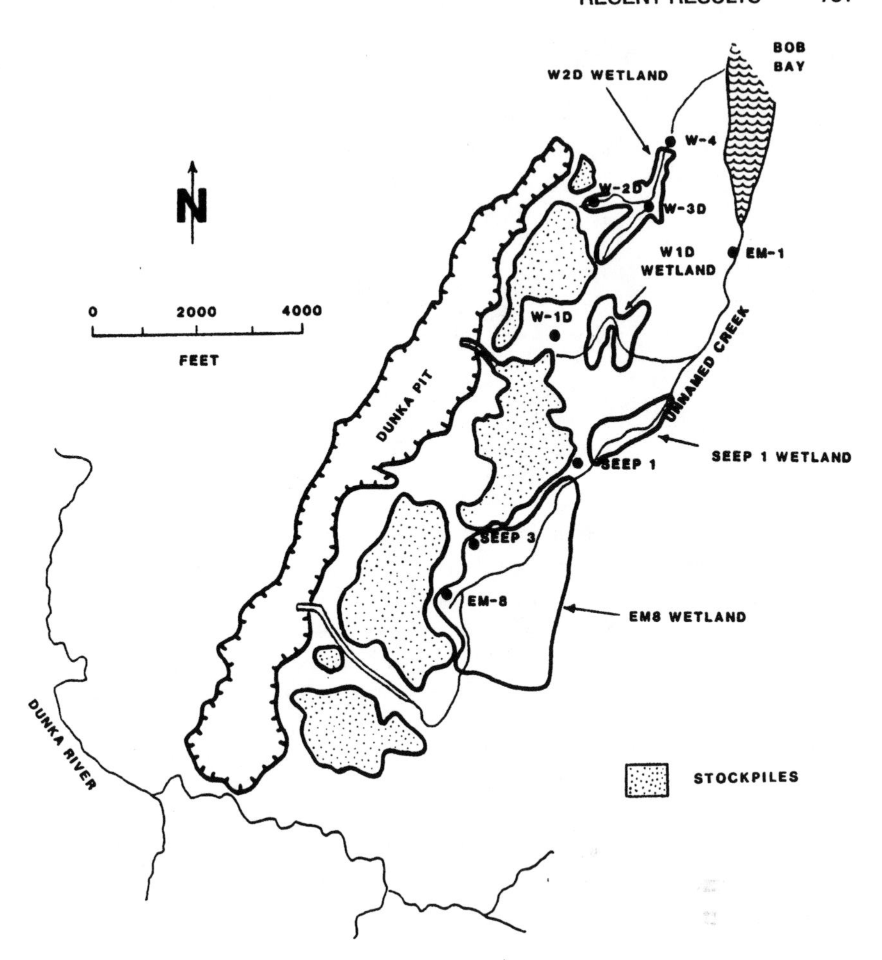

Figure 1. Site and wetland map, Dunka Mine (1988).

SITE DESCRIPTION

The Dunka Mine is a large open-pit taconite operation, covering approximately 160 ha. The pit is 4 km long and 0.4 km wide and has a maximum depth of 110 m. At this mine, the Duluth complex, an igneous intrusion overlies the taconite ore and must be removed and stockpiled. Duluth complex material has been separated on copper content and stockpiled along the east side of the open pit.

Stockpiles are presently built in 13-m lifts with material piled at the angle of repose (45°). Stockpiles contain a total of 32 million metric tons and cover about 120 ha. Drainage from all stockpiles flows to Unnamed Creek and Flamingo Creek and eventually to Bob Bay. Discrete seepages identified along the stockpile bases are delineated in Figure 1 as Em-8, Seep 3, Seep 1, W-1D,

Table 1. Dunka Mine Water Quality (1986)

	Average Concentrations, mg/L				
	pH	Cu	Ni	Co	Zn
Stockpiles[a]	5.4 – 7.5	0.02 – 0.2	1.6 – 14.5	0.05 – 1.0	0.03 – 1.4
Receiving streams[b]	6.8 – 7.2	0.008 – 0.01	0.06 – 0.3	0.02 – 0.03	0.003 – 0.06
Water quality standards[c]	6.0 – 9.0	0.03	0.13	0.01	0.047
Natural background[d]	6.9	0.001	0.001	0.0004	0.002

[a]Stockpile drainage (EM-8, Seep 3, W-1D, Seep 1).
[b]Stream stations (EM-1, W4).
[c]National Pollutant Discharge Elimination System permit.
[d]Regional Copper Nickel Study, median values.

and W-2D. Limited diffuse seepage occurs along the toe of some stockpiles but is not a major contributor to overall metal load to the watershed. Seeps begin to flow during spring thaw (the beginning of April) and flow continuously until freeze up at the end of November. Average flows vary from 0.5 L/s to 14 L/s, but maximum flows exceeding 100 L/s have been observed.

Nickel is the major trace metal in the drainage, with annual median concentrations of 1.6–14.5 mg/L; Cu, Co, and Zn are present but at less than 5% of nickel values (Table 1). Median pH ranges from 5.4 to 7.5, but most stockpile drainages have pH greater than 6.5.

Wetlands located near stockpiles and appearing to offer treatment area for each seepage (Figure 1) are typical of many small lowland areas in northern Minnesota and would be associated with any mining area.

METHODS

To determine the capability of each of the wetlands to treat stockpile drainage, a survey was conducted on each area. Survey lines were established, peat depths were measured every 50–100 m, and the peat was characterized every 100–200 m. Number and spacing of samples depended on the size and shape of the wetland. Characterization included a botanical description of the peat, measures of degree of decomposition (Von Post scale),[5] pH, percent ash, metal and nutrient concentrations, and cation exchange capacity. A 5-cm Macaulay sampler was used to collect samples for a continuous peat description and for chemical analysis at intervals of 0–20, 20–50, 50–100, and 100–200 cm.

Peat pH was measured in 0.01 M $CaCl_2$ solution, and peat samples were dried at 105°C for 24 hr and processed in a blender. Peat was digested with a concentrated HCl-HNO_3 mixture, and metal concentrations were determined by atomic absorption.

RESULTS

The wetlands ranged in size from 1 to 20 ha, with average depths ranging from 0.2 to 1.2 m (Table 2). Peat was generally well decomposed, with decomposition increasing with depth, and was primarily a sedge peat with wood fragments. Metal concentrations varied depending on the sites' proximity to existing stockpile drainage. In individual samples, copper concentrations ranged from 40 to 724 mg/kg, while Ni ranged from 19 to 740 mg/kg. The pH ranged from 5.25 to 7.45, which is within the expected range for shallow peat deposits of this origin. Ash content ranged from 10% to 60%, with highest ash content in the areas with shallowest peat. The average values for each wetland are presented in Table 2.

Metal retention capacity of each wetland was estimated to determine if wetland treatment was a viable treatment alternative. Laboratory and field data from previous studies were used to develop an Ni retention capacity. The maximum Ni concentration observed in field samples was 6400 mg/kg, while concentrations as high as 20,000 mg/kg were measured in laboratory experiments.[2,3] From these data, an Ni removal of 10,000 mg Ni/kg dry peat was used to determine total removal potential of each area. Because most flow in peatlands occurs across and within 30 cm of the surface,[6] an active removal depth of 20 cm was selected. Data collected from a white cedar peatland area demonstrated that metal concentrations in peat decreased with depth and that more than 80% of the metal removal occurred in the upper 20 cm.[2]

Total removal capability was calculated and divided by annual mean Ni loading to determine treatment system lifetimes ranging from 20 to several hundred years (Table 3). Not included were the effects of other reclamation activities that would reduce the annual Ni loading and increase wetland treatment system life. For example, covering a stockpile with soil and establishing vegetation typically reduces annual flow and mass load by 40%[7] and would extend minimum treatment system life from 20 to 33 years.

Based on the estimated lifetimes of the treatment systems, wetland treatment appeared to be a useful mitigation method, but additional test work was needed. An area for additional research was selected to have:

1. A peat depth of at least 50 cm—although most metal removal occurs within the top 20 cm, peat from the white cedar peatland had elevated metal concentrations in the 20-to-50-cm interval.
2. Peat metal concentrations less than several hundred mg/kg—wetland areas that received stockpile drainage had metal concentrations greater than 1000 mg/kg, and removal potential was significantly reduced.
3. A low degree of decomposition (H3 or H4 on the Von Post scale)—the less-decomposed peat is more fibrous, and flow resistance through peat is reduced. As the hydraulic conductivity increases, more stockpile drainage can contact the peat.
4. Leachate chemistry typical of most stockpile drainages on-site and adequate

Table 2. Wetland and Stockpile Drainage Characteristics

								Drainage Associated with Wetland	
Wetland	Total Area (ha)	Mean Depth (m)	pH	CEC (meq/100 g)	Cu (mg/kg)	Ni (mg/kg)	% Ash	Mean [Ni] (mg/L)	Ave. Daily Flow (L/s)
EM-8	20	1.2	5.8	180	92	38	16	1.7	14
W-1D	4.2	0.25	5.65	90	239	57	60	5.9	2.5
Seep 1	1.2	0.2	5.50	130	180	740	33	15.4	0.38
W-2D	11	0.5	5.45	120	620	90	38	0.9	1.5

Table 3. Treatment System Life

Wetland	Peat Mass in Top 20 cm (metric tons)	Annual Nickel Loading (metric tons)	Treatment System Life (yr)[a]
EM-8	4000	0.47	90
W-1D	840	0.43	20
Seep 1	240	0.12	20
W-2D	2200	0.03	780

[a]Assuming that 100 g of dry peat can sequester 1 g of nickel.

flow for input to several test plots—for the Dunka Mine, drainage should have Ni concentrations of 1–5 mg/L, with flow rates of 0.6–0.9 L/s.

5. An open area for test cell construction—large areas of open wetland exist at this mine and construction costs are minimized if tree clearing is avoided.
6. Good access for construction and monitoring activities.

The W-2D wetland met these criteria, and the test cells were established. Objectives of the tests are to (1) optimize treatment efficiency with different flow distribution methods and vegetation types; (2) measure system life; and (3) develop data for application to full-scale treatment systems.

DESIGN

Each 6-m-by-30.5-m cell was surrounded by compacted peat berms (1 m high and 6 m wide at base). Stockpile drainage will be collected by a small dam near the toe of the stockpile, piped to the plots, and distributed to each cell through a series of valves and pipes.

Four treatments were selected, two low-water plots (5-cm depth) with natural vegetation and two high-water plots (15-cm depth) with cattails (*Typha*). Cell 1 will be a control cell containing native vegetation, sedges (*Carex* sp.), and grasses (*Calamograstis* sp.). In cell 2, the peat will be trenched perpendicular to the flow path to increase peat-drainage contact. After trenching, sedge and grass from the surrounding area will be transplanted into this cell (Figure 2).

Because cattail has been used extensively in constructed wetlands and quickly produces a dense stand and large biomass, it will be used in the two high-water cells. Cell 3 will use a serpentine flow path and will contain a 5-cm layer of straw. Peat is not easily decomposed, and straw will be added to stimulate sulfate reduction. Cell 4 uses peat berms perpendicular to the flow to distribute flow and increase the peat-stockpile drainage contact.

Most metal removal in laboratory kinetic studies occurred within 24 hr and is our estimate for the minimum allowable residence time in each cell.[4] Initial flow rates will establish residence times in all cells of 48 hr. Because input flow rates are valve-controlled, residence times can be changed during the experiment to determine effects on metal removal.

Cell inflows and outflows will be recorded, and weekly composite water

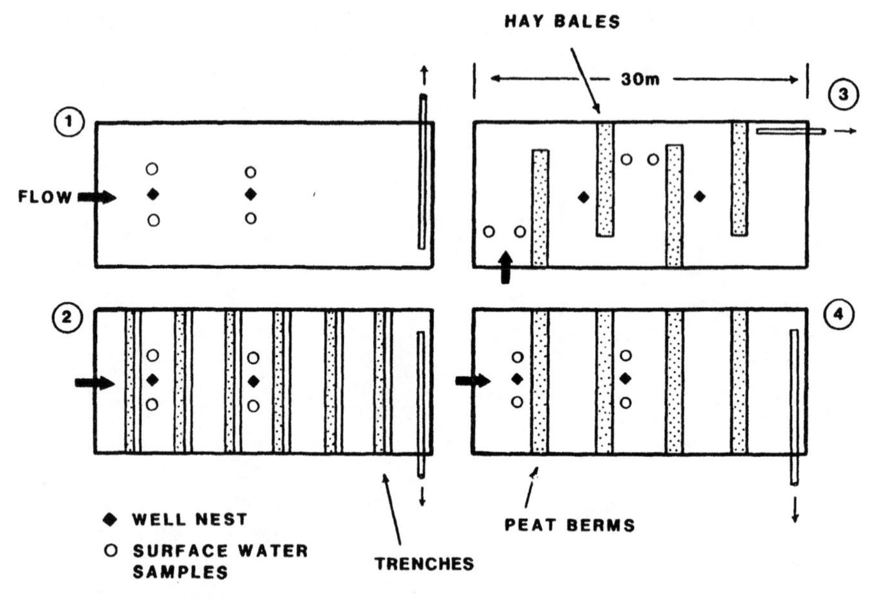

Figure 2. Design of test plots.

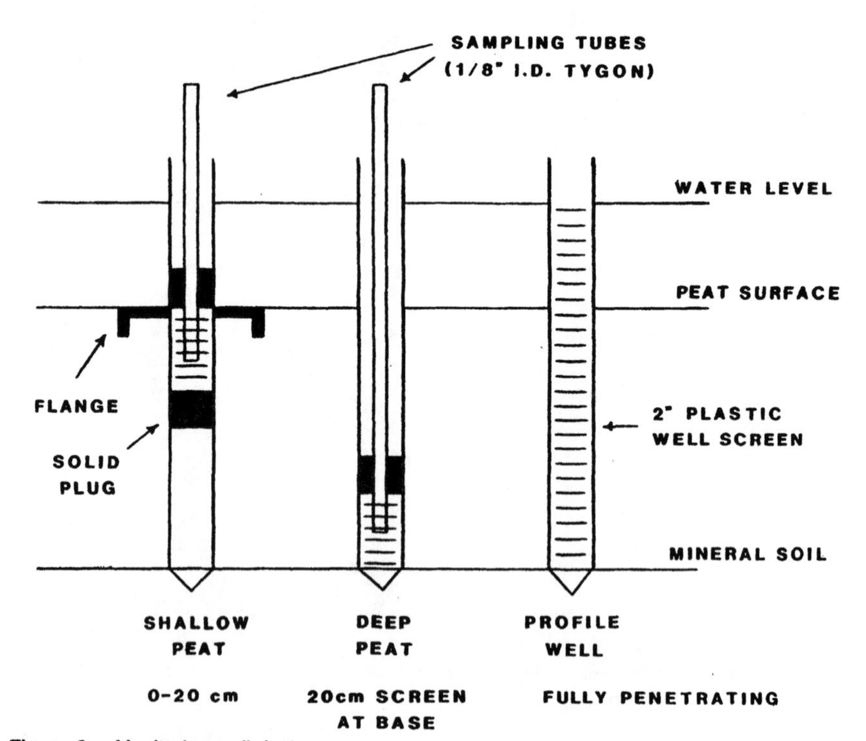

Figure 3. Monitoring well design.

quality samples will be collected and analyzed. Two nests of groundwater monitoring wells will be installed in each plot. Each nest will have three wells: one for the active zone of water movement (0–20 cm); one for deeper water in the peat (20 cm above mineral soil contact); and one fully penetrating well to collect profile measurements (Figure 3). In addition, two sets of surface water samples will be collected within the plot. Wells and surface water will be sampled on a monthly basis. These cells should be operational in 1989.

ACKNOWLEDGMENTS

The authors acknowledge assistance of staff members of the Division of Minerals, who have helped collect survey data and provided chemical analyses of samples. We are also grateful for assistance and support of LTV Steel Mining Company and for design and construction details provided by Ted Frostman of STS Consultants, Green Bay, Wisconsin.

REFERENCES

1. "Feasibility Assessment of Mitigation Measures for Gabbro and Waste Rock Stockpiles, Dunka Pit Area," Barr Engineering Company (1986).
2. Eger, P., and K. Lapakko. "Nickel and Copper Removal from Mine Drainage by a Natural Wetland," in *Mine Drainage and Surface Mine Reclamation, Vol. I, Mine Water and Mine Waste,* Bureau of Mines Information Circular 9183 (1988), pp. 301–309.
3. Lapakko, K., and P. Eger. "Trace Metal Removal from Mine Drainage," in *Mine Drainage and Surface Mine Reclamation, Vol. I; Mine Water and Mine Waste,* Bureau of Mines Information Circular 9183 (1988), pp. 291–300.
4. Lapakko, K., P. Eger, and J. Strudell. "Low Cost Removal of Trace Metals from Copper-Nickel Mine Stockpile Drainage, Vol. 1, Laboratory and Field Investigations," USBM Contract Report J0205047 (1986).
5. Robinson, D. W., and J. G. D. Lambs, Eds. *Peat in Horticulture* (New York: Academic Press, Inc., 1972).
6. Romanov, V. V. "Hydrophysics of Bogs," Israel Program for Scientific Transactions (1968).
7. Eger, P., and K. Lapakko. "The Leaching and Reclamation of Low Grade Mineralized Stockpiles," in *Proceedings of 1981 Symposium on Surface Mining, Hydrology, Sedimentology, and Reclamation* (Lexington: University of Kentucky, 1981), pp. 157–166.

42f Windsor Coal Company Wetland: An Overview

Ronald L. Kolbash and Thomas L. Romanoski

INTRODUCTION

Several hundred wetlands have been constructed in the coal-bearing states of Maryland, West Virginia, Pennsylvania, and Ohio to minimize impacts from acid mine drainage. Few, if any, have a synthetic liner to protect slope stability. Conventional treatment methods, particularly chemicals and mechanical aeration, can cost in excess of $60,000/yr. High operating cost and lack of potential bond release have encouraged the coal industry to consider wetlands a reclamation alternative to conventional treatment. American Electric Power (AEP) Service Corporation's Fuel Supply Department is actively involved in the overall reclamation plan for its abandoned Simco No. 4 mine, in which the wetland is an important component. Based upon the success of the Simco No. 4 Wetland, AEP Fuel Supply decided to build a wetland at Windsor Coal Company. The wetland minimizes a refuse pile seep at the Schoolhouse Hollow Refuse Area as well as enhances the area wildlife and environmental quality.

BACKGROUND

Windsor Coal Company's Beech Bottom Mine is one of the oldest continuously operating mines in West Virginia. It began production in 1899, and AEP purchased an interest in the property in 1918. The Schoolhouse Hollow Refuse Area, on which the wetland was constructed, received refuse from the early 1900s until 1980, when the refuse pile encompassed 30 acres.

Windsor Coal Company, with assistance from AEP Fuel Supply, began reclaiming Schoolhouse Hollow in the fall of 1981 and completed the project 18 months later. Among many project unknowns, particularly regarding subsurface drainage, the number of seeps encountered during reclamation far exceeded our estimates. The sharp V shape of Schoolhouse Hollow at State Route 2 also complicated reclamation efforts.

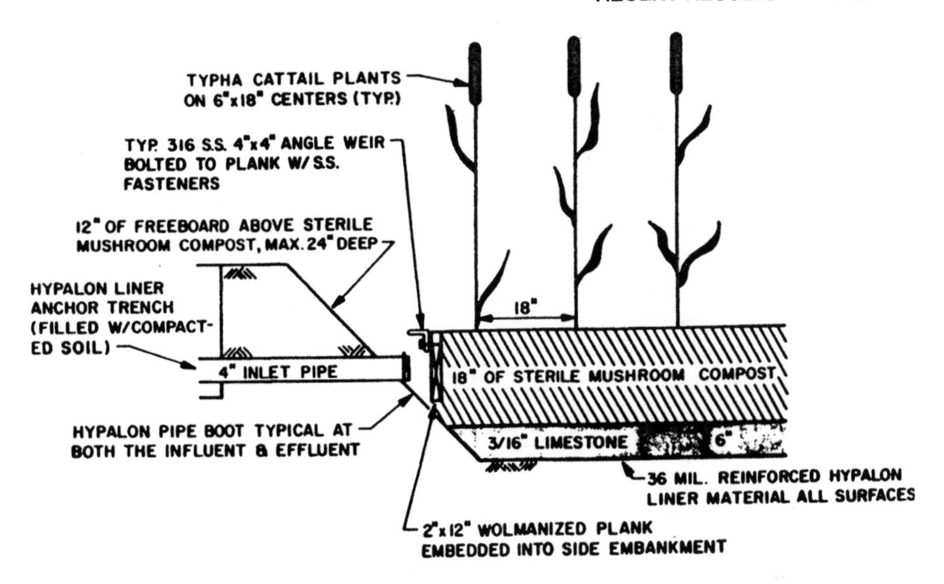

Figure 1. Wetland treatment area detail.

Standard reclamation practices were used at Schoolhouse Hollow with 6-to 9-m benches constructed for each 15-m rise in elevation. The pile was also reshaped so that runoff and seepage water drained to the north side, where it was easier to collect for discharge or treatment. Off-site drainage ditches were constructed around the hollow to prevent runoff water from entering the refuse site. Significantly less water enters the refuse area, and seep water quality has improved slightly over the past few years. Seep quantity has been reduced but not eliminated.

Contaminated seepage is collected and channeled to a small pond at the toe of the refuse pile near State Route 2. Collected seepage is treated with caustic soda, although maintaining a set pH limit has been difficult and fluctuating pH levels have caused occasional elevations of iron. Pond location and size contribute to the difficulty in meeting iron limits on a continuous basis because the pond cannot be enlarged and requires frequent, expensive cleanout.

Several other alternatives considered by Windsor Coal prior to wetland installation were eliminated because of high capital cost, maintenance, and lack of prior success.

WETLAND CONSTRUCTION

Wetland construction began on August 10, 1987, and was completed on August 14, 1987. It consists of typical wetland components—0.5-cm limestone, sterile mushroom compost, and cattail (*Typha*) plants (Figure 1). The wetland was installed on the reclaimed disposal area 60 m below the hill crest on a 4:1 slope. Because of stability problems, the wetland was not placed upon the bench, but installing the wetland on the slope could saturate a section of the slope face and create a stability problem.

Available options for minimizing the stability problem included excavating the wetland area and installing a synthetic Hypalon liner or installing a clay liner. Suitable clay was not available nearby, and delivery to the wetland would have been difficult. Consequently, a 36-mil Hypalon liner was chosen to protect against slope saturation.

Construction had to be conducted during dry weather to minimize the damage caused by equipment. The material was delivered to the wetland by a rubber-tired, four-wheel-drive John Deere 310 hoe. Wetland excavation was accomplished with a John Deere 450-D dozer. Topsoil was stripped and saved to redress the area. The wetland was excavated to 2:1 side slopes, with cut material used as fill for the down-slope side. Total excavation was 3.4 m by 26 m, for a total surface area of 117 m^2. A rock/pipe drain was installed below the liner to carry water from a 4 L/min seep at bottom grade of the excavation. A one-piece Hypalon liner was placed and keyed into the side berms of the wetland.

A small, lined sump was constructed beneath the existing seep collection pipe discharging into the existing riprap ditch (Figure 2), and a 4-in. PVC conveyance pipe was installed from the sump to the wetland.

Cattails were installed on 46-cm centers, and seep water was directed into the wetland. Minor weir adjustments ensured even water distribution and flow across the wetland.

CONSTRUCTION COSTS

Equipment Use:	
40 hr—John Deere 310 hoe @ $60/hr	$2,400.00
40 hr—John Deere 450-D dozer @ $60/hr	2,400.00
(above includes mobilization, operator, and fuel)	
Materials:	
Mushroom compost—30 yd @ $21/yd	630.00
Limestone substrate—17.8 tons @ $7.25/ton	129.05
Underdrain stone—12.23 tons @ $15.30/ton	187.12
Type 316 S.S. weir	912.37
36-mil Hypalon liner	2,896.20
Miscellaneous pipe and fittings	680.86
Cattail (*Typha*) plants	2,275.00
Total:	$12,510.60

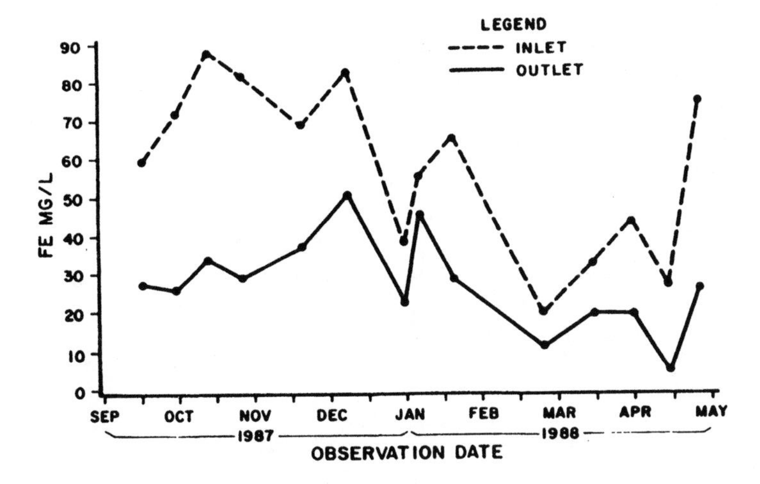

Figure 2. Windsor wetland, total iron.

DISCUSSION

Even with less than one growing season, the Windsor wetland has removed 50% of the incoming iron (Figure 2). Initial, within-wetland water sampling indicated a gradual reduction of dissolved iron without sharp transition zones. Seep manganese ranges from 1.5 to 3.8 mg/L (Figure 3), and inlet manganese has increased since February 1988. Except for three incidents, the pH (Figure 4) does not usually change from inlet to outlet.

The Hypalon liner seems to be enhancing algae growth. The effect of preventing interaction between wetland plants and the substrate is not known.

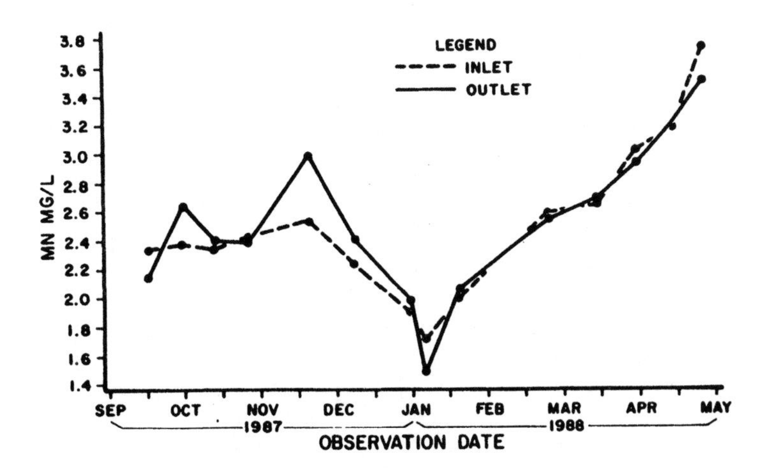

Figure 3. Windsor wetland, total manganese.

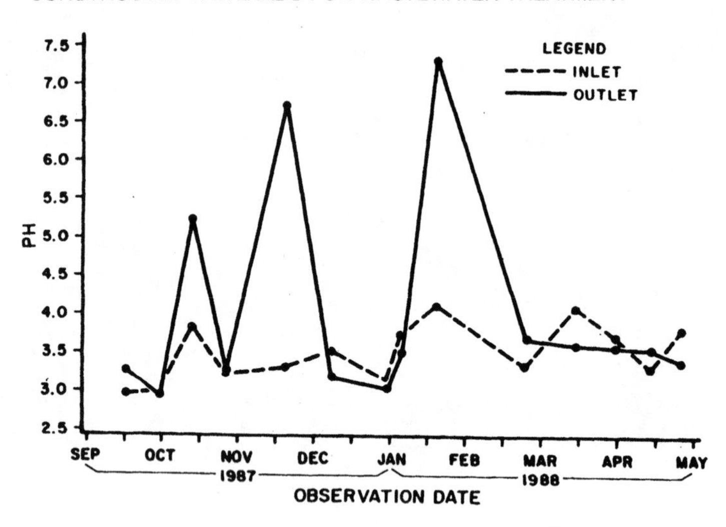

Figure 4. Windsor wetland, pH.

Continued monitoring will measure sulfate, inlet flow, and will evaluate Hypalon liner long-term effects on Windsor's wetland.

AEP Fuel Supply considers wetlands a valuable natural tool in their reclamation programs. Not only do the wetlands improve water quality, but also provide wildlife valuable habitat and provide for possible bond release in a more timely manner.

42g Wetland Treatment of Coal Mine Drainage: Controlled Studies of Iron Retention in Model Wetland Systems

Jacqueline Henrot, R. Kelman Wieder, Katherine P. Heston, and Marianne P. Nardi

INTRODUCTION

Following reports of field situations in which chemistry of acid coal mine drainage improved upon passage through naturally occuring wetlands,[1,2] the possibility that man-made wetland systems might provide a low-cost, low-maintenance alternative to chemical treatment of mine drainage led to a considerable amount of interest and research. Since 1984, at least three conferences[3-5] have had sessions devoted to wetland treatment of mine drainage. Although some man-made wetlands seem to be effective in improving water quality of mine drainage, many have been ineffective.[6-8] Causes for observed variability in apparent effectiveness among man-made wetland treatment systems are not well understood. As a result, considerable uncertainty remains regarding design criteria, predictions of effective treatment lifetime, and *a priori* estimates of cost vs potential benefit.

Optimally, studies designed to evaluate effectiveness of wetland systems constructed for mine drainage treatment should monitor hydrologic fluxes, water chemistry, and wetland substrate chemistry.[9] However, cost and personnel commitment involved in this type of monitoring are often in excess of the resources available to the investigator. Even with an ambitious field monitoring scheme, short-term changes in wetland efficiency (e.g., diel patterns or transient responses to major precipitation events) are likely to be overlooked.

For evaluating the processes involved in chemical modification of mine drainage in wetland systems, smaller scale laboratory studies may be more useful than the field monitoring of man-made wetland systems. Laboratory studies provide ability to control hydrologic fluxes and influent water chemistry, minimizing sources of variability often confounding interpretation of results from field studies.

In this chapter, we report the results of a laboratory pilot study in which replicate model wetland systems were subjected to inputs of water at uniform flow rates but differing Fe concentrations. Measurements of hydrologic fluxes, water chemistry, and substrate chemistry were used to evaluate effects of Fe concentration in influent waters on Fe retention within these model wetland systems.

METHODS

Model Wetlands, Water Sampling, and Chemical Analysis

Six model wetland systems (2.40 × 0.15 m) were filled to a depth of 0.15 m with *Sphagnum* plants (Mosser Lee Co., Millston, WI). Water adjusted to pH 4 with H_2SO_4 was added by a pump at one end of each system to establish a water flow through the wetlands. Thereafter, duplicate systems received water (pH 4) containing Fe (as $FeSO_4$) at a concentration of either 0, 50, or 100 mg/L. Water entered the wetlands at a flow of 200 mL/min for 6 hr per day for 8 days. When not receiving iron solution, the model wetlands (located in a greenhouse) were covered with plastic to prevent water losses by evaporation.

Outflow water samples were collected from the systems at hourly intervals and analyzed immediately for pH. After acidification with HNO_3, Fe, Ca, and Mg concentrations were measured by atomic absorption spectroscopy (AAS).

Daily retention of Fe by model wetlands was calculated as the difference between Fe entering the wetland (Fe concentration of influent water × total influent water volume over 6 hr) and Fe leaving the wetland (hourly effluent Fe concentrations × hourly effluent water volumes).

Peat Sampling and Chemical Analysis

Duplicate 10-g samples of *Sphagnum* substrate were collected at inflow and outflow ends of each model wetland at zero, four, and eight days. They were subjected to four successive extractants: deionized distilled water (20 mL, 30 min); 1 M potassium nitrate (50 mL, overnight); 0.1 M sodium pyrophosphate (50 mL, overnight); and citrate-bicarbonate-dithionite (50 mL of 0.3 M sodium citrate, 6.25 mL of 1 M sodium bicarbonate, and 1.25 g sodium hydrosulfide, 30 min at 80°C) to recover water-soluble metals, exchangeable metals, organically bound metals, and iron oxide bound metals, respectively (procedure modified from Miller et al.[10]). Samples were filtered and rinsed with deionized distilled water after each step. The combined filtrate and rinsings were analyzed for Fe by AAS. Separate substrate samples were collected to determine fresh-to-dry-weight ratios.

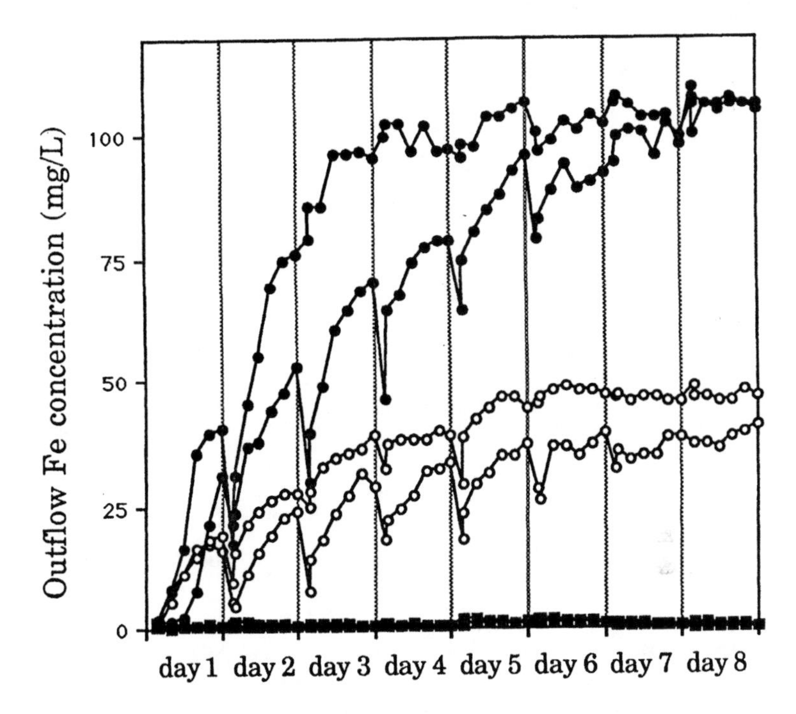

Figure 1. Iron concentrations in the effluent water of model wetlands. Influent Fe concentrations were either 0 (squares), 50 (open circles), or 100 mg/L (solid circles).

RESULTS

Model wetlands efficiently removed Fe from solution for about five days. Afterwards, effluent Fe concentrations were approximately equal to influent Fe concentrations in all treatments (Figure 1).

Iron solution was pumped through the wetlands for six hours each day. Iron removal from solution showed a pronounced diel pattern characterized by a relatively low effluent Fe concentrations at onset of pumping and a subsequent increase in effluent Fe concentrations over the 6-hr period (Figure 1).

Retention of Fe in *Sphagnum* substrates (Figure 2) was calculated from removal of Fe from solution. Total amounts of Fe retained in model wetlands were 12.4–13.2 and 16.0–24.6 mg Fe/g dry *Sphagnum* substrate for input Fe concentrations of 50 and 100 mg/L, respectively. However, Fe retention was not homogeneous throughout the model wetland but more pronounced at the inflow end (Table 1).

Initial Fe speciation in the *Sphagnum* substrate was 0.0 mg water-extractable Fe, 0.64 mg exchangeable Fe, 2.18 mg organically bound Fe, and 1.56 mg Fe oxides per g dry substrate (mean values of four samples). Iron speciation in the control (no Fe input) model wetlands remained unchanged over time (data not shown). In wetlands subjected to Fe solutions, Fe accumulated principally as

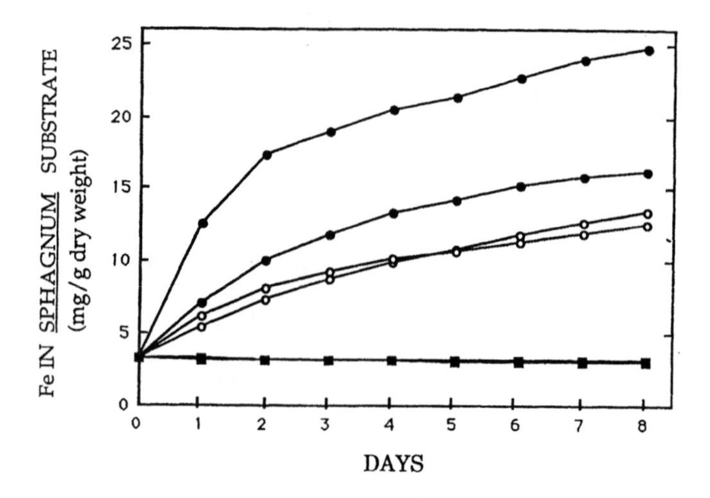

Figure 2. Iron retention in model wetlands receiving influent Fe concentrations of either 0 (squares), 50 (open circles), or 100 mg/L (solid circles).

organically bound Fe and Fe oxides: mean values of 63.8% and 29.1% of total Fe, respectively (Figure 3). Lack of substantial increase in exchangeable Fe indicates that Fe retention by cation exchange does not contribute significantly to overall potential Fe retention (Figure 3). Even though model wetlands that received higher concentrations of Fe (100 mg/L) retained more Fe, there was no remarkable difference in Fe speciation between wetlands that received Fe solutions of 50 or 100 mg/L (Figure 3).

DISCUSSION

Possible mechanisms responsible for Fe retention in *Sphagnum* moss are exchange of Fe with other cations on negatively charged sites; specific binding of Fe to organic substrates; formation of Fe oxides (abiotically or from Fe-oxidizing bacteria); and formation of Fe sulfides (precipitation with H_2S pro-

Table 1. Total Fe Concentrations in *Sphagnum* Moss Samples Collected at Inflow and Outflow Ends of Model Wetlands, at Days 4 and 8

Fe Input to the Model Wetland (mg/L)	Fe Concentration in the *Sphagnum* Moss (mg/g dry moss) Day 4		Day 8	
	Inflow End	Outflow End	Inflow End	Outflow End
0	4.38	4.58	3.34	5.25
0	2.81	1.50	2.18	3.03
50	16.07	9.09	22.46	10.41
50	13.68	6.28	19.76	7.30
100	22.74	10.26	39.86	12.99
100	18.76	11.30	16.02	15.12

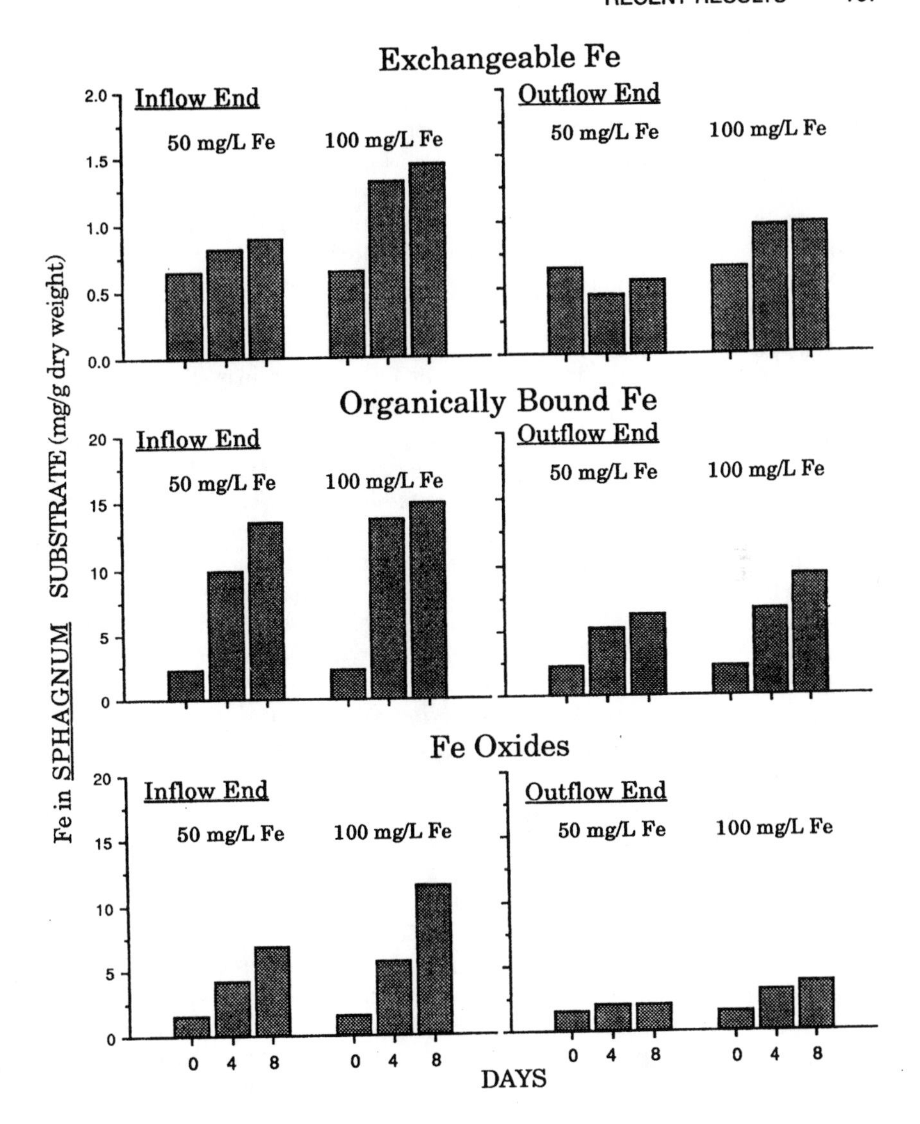

Figure 3. Chemical speciation of Fe in *Sphagnum* substrate samples collected at inflow and outflow ends of model wetlands. (Note scale change for exchangeable Fe concentrations.)

duced by sulfate-reducing bacteria).[9] In this pilot study, retention mechanisms for Fe in model wetlands were probably limited to abiotic interactions between the *Sphagnum* substrate and Fe. Microbial activity in model wetlands was likely inhibited by elevated temperatures (often exceeding 38°C) in the greenhouse containing the model wetlands. Iron-oxidizing bacteria, for example, are particularly sensitive to temperature, with optimum activity near 20°C, and iron-oxidizing efficiency declines sharply with higher temperatures.[11]

Effectiveness of *Sphagnum* model wetlands for Fe removal from solution declined over time (Figure 2), suggesting that Fe retention by abiotic processes approached a maximum capacity. In field situations, "effective life" of wetlands constructed to improve the quality of acid mine drainage might be longer because previously described biologically mediated mechanisms for Fe retention would supplement abiotic processes presumed operational in our greenhouse systems.

Our data and a field study[12] suggest that Fe oxide formation might be more important than Fe sulfide precipitation for Fe retention in wetland systems. Indeed, the highest published Fe sulfide concentration from a wetland subjected to acid mine drainage is 0.597 mg Fe/g dry peat, representing only 1.2% of the total Fe concentration, whereas Fe oxides represented 55.8%.[12]

Iron accumulation in *Sphagnum* substrates over the eight-day treatment reached 39.8 mg Fe/g dry moss (Table 1), a 10-fold increase over initial Fe concentration in the substrate. For comparison, maximum Fe accumulation measured in laboratory microcosms filled with *Sphagnum* peat was 4 mg/g dry peat.[13] In contrast, in a man-made wetland created for removing Fe from seeps on an active coal mine[14] and in a naturally occurring wetland receiving inputs of acid mine drainage from an adjacent abandoned coal mine,[12] mean total Fe concentrations of 36.6 and 47.8 mg Fe/g dry *Sphagnum* peat were observed. The highest reported Fe concentration in a single sample from Tub Run Bog was 150 mg/g peat.[12]

The chemical form of Fe accumulated in the substrate differed between laboratory and field-collected samples. In the present study and in *Sphagnum* peat microcosms,[13] Fe was principally retained as organically bound Fe, whereas *Sphagnum* peat from the man-made wetland[14] and Tub Run Bog[12] contained primarily iron oxides (85% and 56% of total Fe, respectively). As total Fe concentration in the substrate increases, Fe oxide precipitation probably becomes the dominant mechanism for Fe accumulation. The oxide fraction in our samples represented 29% (mean value) of the total Fe in *Sphagnum* moss. If elevated temperatures in the greenhouse inhibited activity of iron-oxidizing bacteria, a nonnegligible amount of Fe oxide precipitated abiotically. Given the apparent importance of Fe oxide formation to Fe retention in man-made wetlands, a study has been initiated to determine to what extent Fe oxides formation in *Sphagnum* peat results from biotic or abiotic processes.

Iron retention in model wetlands was not homogeneous but more pronounced near inflow ends (Table 1), which suggests a progressive removal of Fe as water flows through the wetland. One might therefore expect Fe retention in man-made wetlands created for treating mine drainage to be spatially heterogeneous, with increased Fe concentrations in the substrate located near the source of mine drainage. Comparison between Fe concentration in individual substrate samples (Table 1) and overall Fe retention in the wetland (Figure 2) shows that actual Fe concentrations in substrates can be higher than the maximum retention capacity of the wetland calculated from budgets of Fe entering and leaving the system (e.g., 39.86 > 25.0 mg Fe/g dry moss).

Iron removal from solution exhibited a definite diel pattern, with initial effluents having the lowest Fe concentration (Figure 1). This pattern could have two explanations. First, at night (no water was applied), lack of water flow allowed better contact between Fe and the substrate and promoted Fe removal from solution. The lower Fe concentration of the first daily effluent would reflect the decrease in soluble Fe concentration that occurred at night. During the day, as Fe solution entered the wetland and mixed with the water present in the system, Fe concentration in the effluent increased.

The second explanation calls on a cycle of photoreduction of Fe during the day and Fe oxidation during the night, as demonstrated in streams.[9,15] In our study, because most of the water flowed at the surface of the model wetlands, photoreduction could have maintained Fe in a soluble form during the day.

CONCLUSIONS

A pilot study was initiated to evaluate Fe retention in wetlands under controlled conditions. Model wetlands filled with a *Sphagnum* substrate were subjected to influent water at pH of 4.0 and Fe concentrations of either 0, 50, or 100 mg/L. At the end of the 8-day treatment period, effluent Fe concentrations were approximately equal to influent Fe concentrations. Peat chemical analysis indicated that (1) Fe retention was more pronounced at the inflow end, and (2) accumulated Fe was mainly present as organically bound Fe (63.8% of total Fe) and Fe oxides (29.1%), with little accumulation as exchangeable Fe.

This study and field studies suggest that a key process involved in Fe retention in wetland systems is Fe oxide formation. Microbially mediated Fe oxidation, which was probably inhibited in this pilot study by high greenhouse temperatures, may play an important role in Fe retention and increase the "effective" life of wetlands created for the treatment of acid mine drainage.

ACKNOWLEDGMENT

The work was supported by a U.S. Environmental Protection Agency grant (R812379) to R. K. Wieder.

REFERENCES

1. Huntsman, B.E., J.G. Solch and M.D. Porter. "Utilization of *Sphagnum* species dominated bog for coal acid mine drainage abatement," Abstracts of the 91st Annual Meeting of the Geological Society of America, Ottawa, Ontario, Canada (1978).
2. Wieder, R.K., and G.E. Lang. "Modification of acid mine drainage in a freshwater

wetland," in *Symposium on Wetlands of the Unglaciated Appalachian Region, West Virginia University, Morgantown, WV.*, B.R. McDonald, Ed. (1982), pp. 43–53.

3. Burris, J.E., Ed. *Treatment of Mine Drainage by Wetlands.* Proceedings of a Conference. (Contribution No. 264, Department of Biology, The Pennsylvania State University, University Park, PA., 1984), p. 49.
4. Brooks, R.P., D.E. Samuel and J.B. Hill, Eds. "Wetlands and Water Management on Mined Lands," Proceedings of a Conference, October 23–24, 1985, The Pennsylvania State University, University Park, PA.
5. "Mine Drainage and Surface Mine Reclamation. Volume I. Mine Water and Mine Waste," U.S. Department of the Interior, Bureau of Mines Information Circular 9183 (1988), p. 413.
6. Kleinmann, R.L.P., and M.A. Girts. "Constructed wetlands for treatment of mine water—successes and failures," in *Proceedings of the Eighth Annual Abandoned Mine Lands Conference*, Billings, MT. (1986), pp. 63–73.
7. Girts, M.A., and R.L.P. Kleinmann. "Constructed wetlands for treatment of acid mine drainage : a preliminary review," in National Symposium on Surface Mining, Hydrology, Sedimentology, and Reclamation, University of Kentucky, Lexington, KY (1986), pp. 165–171.
8. Girts, M.A., R.L.P. Kleinmann and P.M. Erickson. "Performance data on *Typha* and *Sphagnum* wetlands constructed to treat coal mine drainage," Eighth Annual Surface Mine Drainage Task Force Symposium, Morgantown, WV (1987).
9. Wieder, R.K. "Determining the capacity for metal retention in man-made wetlands constructed for treatment of coal mine drainage," in: *Mine Drainage and Surface Mine Reclamation*, (U.S. Department of the Interior, Bureau of Mines Information Circular 9183, 1988), vol. I, pp. 375–381.
10. Miller, W.P., W.W. Mc Fee and J.M. Kelly. "Mobility and retention of heavy metals in sandy soils," *J. Environ. Qual.* 12:579–584 (1983).
11. Madsen, E.L., M.D. Morgan and R.E. Good. "Simultaneous photoreduction and microbial oxidation of iron in a stream in the New Jersey Pinelands," *Limnol. Oceanogr.* 31:832–838 (1986).
12. Wieder, R.K., and G.E. Lang. "Fe, Al, Mn, and S chemistry of *Sphagnum* peat in four peatlands with different metal and sulfur input," *Water, Air, and Soil Pollution* 29:309–320 (1986).
13. Wieder, R.K., G.E. Lang and A.E. Whitehouse. "Metal removal in *Sphagnum*-dominated wetlands: experience with a man-made wetland system," in: *Wetlands and Water Management on Mined Lands* (The Pennsylvania State University, University Park, PA, 1985), pp. 353–364.
14. Tarleton, A.L., G.E. Lang and R.K. Wieder. "Removal of iron from acid mine drainage by *Sphagnum* peat: Results from experimental laboratory microcosms," in: *Symposium on Surface Mining, Hydrology, Sedimentology, and Reclamation* (U. of Kentucky, KY, 1984), pp. 413–420.
15. McKnight, D.M., B.A. Kimball and K.E. Bencala. "Iron photoreduction and oxidation in an acidic mountain stream," *Science* 240:637–640 (1986).

42h Tolerance of Three Wetland Plant Species to Acid Mine Drainage: A Greenhouse Study

William R. Wenerick, S. Edward Stevens, Jr., Harold J. Webster, Lloyd R. Stark, and Edward DeVeau

INTRODUCTION

Wetlands have been constructed to ameliorate acid mine drainage (AMD) in the bituminous coal region of the eastern United States.[1-3] Most wetlands have made liberal use of a variety of eucaryotic wetland plant species. However, the tolerance of wetland plants to AMD is not well understood.

Our study of wetland plant tolerance was instigated by a central Pennsylvania coal company with AMD problems at two mine sites. They were interested in constructed wetlands as an alternative to continuing use of chemicals in combination with settling ponds for treatment. Discharge water from one site had 411 mg/L dissolved Fe, 71 mg/L dissolved Mn, and a pH of 3.4–3.5, near the limits of toleration observed in the field for wetland plants (unpublished observations). The purpose of our work was to determine tolerance levels of three wetland plants to AMD under semicontrolled conditions in a greenhouse simulation study.

MATERIALS AND METHODS

Seep water was collected at the mine site and trucked to the Penn State Agricultural Research Center at Rock Springs. Four water treatments were used: a tap water control (similar to an uncontaminated spring at the mine site from water analysis); mine water diluted 1:4 (1 vol mine water with 3 vol tap water); mine water diluted 1:2; and undiluted mine water. Dilution with tap water increased pH and reduced dissolved metal concentrations slightly more than expected.

Typha latifolia (cattail), *Sphagnum recurvum* (peat moss), and *Pohlia nutans* (a subaquatic turf-forming moss) were chosen for study because they are common in wet areas affected by AMD. *Typha* and *Sphagnum* have often been used for constructed wetlands. *P. nutans* grows in association with the green alga *Ulothrix* and has been found on exposed coal seams in water with low pH (2.65) and high metal concentration.[4] Cattails were collected from an uncontaminated wetland near Pine Grove Mills, Pennsylvania. Peat moss was collected from Bear Meadows National Landmark near Boalsburg, Pennsylvania. *Pohlia* was collected from an exposed coal seam seep in Clearfield County, Pennsylvania. All plants were transplanted to the greenhouse and grown hydroponically until the experiment began.

We used multiple-lane, fiberglass-resin-coated wooden "wetland simulators" introduced by Burris et al.[5] for our determination of plant tolerance. Water was pumped through each lane from a reservoir by a submerged electric pump at a flow rate of 250 mL/min. After the water passed through a lane, it returned to the reservoir and was continually recirculated. Plants were collected in early to mid-August and had about 1.5 months to become established before the experiment began. During this time, tap water adjusted to the pH of water at the collection locality was recirculated.

We received weekly water deliveries from the mine site for five weeks. Between deliveries, we initially believed Fe and Mn levels could be maintained by daily addition of metal salts ($FeSO_4$ and $MnSO_4$) and pH kept constant by daily addition of acid (H_2SO_4) or base (NaOH). The pH was not a problem, but maintaining metal levels was difficult, and the plants did not receive the equivalent of undiluted mine water for the whole period.

After two weeks of trying various ways, we found that pH and metal concentrations of mine water did not change when stored less than a week. During the last three weeks, mine water was replaced three times a week, and only one pH adjustment was made.

To assess the impact of mine water on plant health, initial and final samples were taken to measure chlorophyll concentration. Moss samples were collected with a large cork borer. A small cork borer was used to collect samples from *Typha* leaves to allow leaves to remain relatively intact, and the same plants were sampled at the end of the experiment to reduce sample variation. Chlorophyll was extracted from plant tissue with 80% acetone in H_2O (v/v), and absorbance of the resulting extract was determined at 645 and 663 nm. Chlorophyll concentration was calculated by the method of Arnon.[6]

Each week the number of *T. latifolia* living shoots in each lane was counted. In addition, for 10 young plants of similar size selected along the length of each lane, the lengths (heights) of their central leaf were measured each week and any new leaves were counted. Percent leaf growth was calculated as change in leaf length divided by starting length multiplied by 100, so that a leaf that doubled in length was said to have grown 100%.

Table 1. Water Chemistry of the Undiluted Acid Mine Drainage Used in the "Wetland Simulators"

Parameter	$\bar{X}$	S.D.	Range	N
Acidity as $CaCO_3$ (mg/L)	1604	±43.9	1570–1680	5
Alkalinity as $CaCO_3$ (mg/L)	<1	±0	—	5
Dissolved Fe (mg/L)	411.0	±60.9	323–462	4
Total Fe (mg/L)	440.8	±53.9	365–496	5
Dissolved Mn (mg/L)	71.2	±6.3	64.9–78.2	4
Total Mn (mg/L)	76.4	±9.5	67.0–92.1	5
pH	3.45	±0.04	3.40–3.50	5
Total suspended solids (mg/L)	6.8	±5.3	1–14	5
Specific conductance (μmhos)	4088	±358.2	3470–4360	5
SO_4 (mg/L)	4070	±182.6	3830–4330	5

RESULTS AND DISCUSSION

Water chemistry of undiluted AMD water used in the simulation experiments is shown in Table 1. In most cases, five samples were independently analyzed. The water was heavily contaminated with Fe, Mn, and sulfate and had high acidity—among the most severe AMD water observed in the field. Thus, we believed it provided an excellent test water for the study of plant tolerance to AMD.

Initial and final total chlorophyll content was determined for each plant species in a control and three treatment conditions (Figure 1). In the control, total chlorophyll trended slightly upward during the course of the experiment. This was also generally true for the 1:4 dilution of AMD. In the 1:2 dilution, total chlorophyll content of *Sphagnum* and *Pohlia* declined during the experiment but remained unchanged in *Typha*. Total chlorophyll declined substantially over time in all three plant species with undiluted AMD. On the basis of total chlorophyll content, *Typha* was most resistant to AMD, followed by *Pohlia* and then *Sphagnum*. However, a better measure of resistance to stress is the chlorophyll a/b or chlorophyll/accessory pigment ratio.[7-9]

Chlorophyll a/b ratios were determined in the same samples used for total chlorophyll (Figure 2). No differences in chlorophyll a/b ratios in controls or in the 1:4 dilution of AMD were observed. The a/b ratio of *Typha* declined in the 1:2 dilution and in the undiluted AMD, a result of plant stress imposed by these two levels of AMD.

Interestingly, the chlorophyll a/b ratio increased in both *Sphagnum* and *Pohlia* in the 1:2 dilution and in undiluted AMD. This result is not interpretable as the result of plant stress in the classical sense. The only known cause of an increase in the chlorophyll/accessory pigment ratio in a photosynthetic organism is starvation of cells for manganese.[10,11] Perhaps high total and dissolved iron in AMD used in our experiments prevented or significantly interfered with uptake of manganese by *Sphagnum* and *Pohlia*.

We also determined change in shoot number, number of new leaves, and

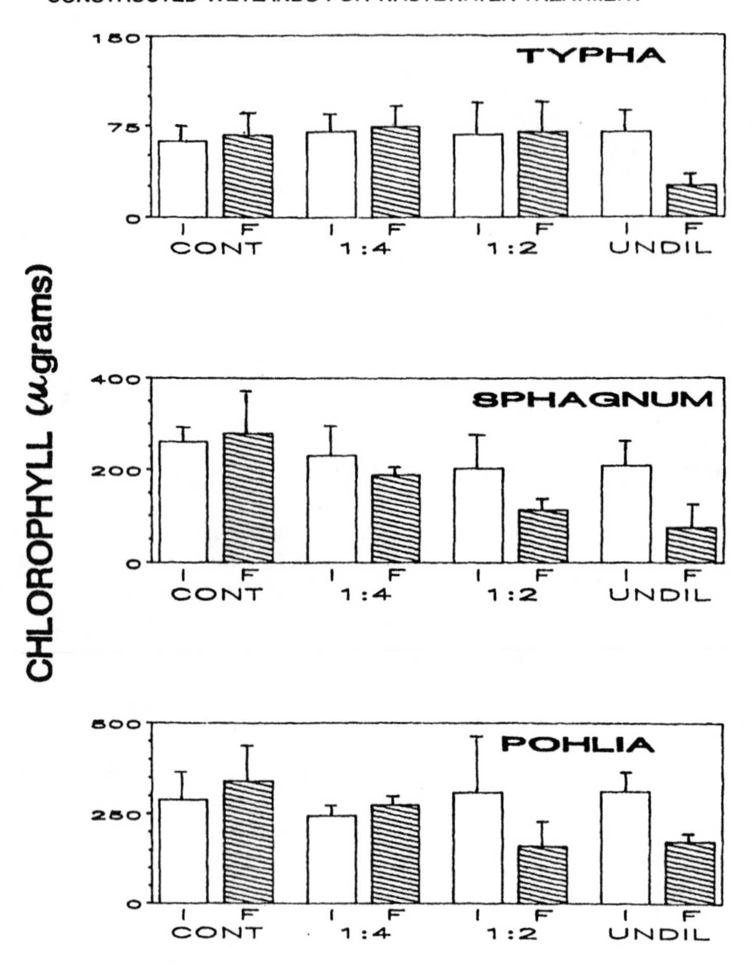

Figure 1. Total chlorophyll content of *Typha, Sphagnum,* and *Pohlia.* I = initial; F = final; cont = control; 1:4 = 1:4 dilution of AMD; 1:2 = 1:2 dilution of AMD; undil = undiluted AMD. Vertical lines = one standard deviation.

change in length of central leaves for *Typha* (Table 2). Similar determinations were not feasible for *Sphagnum* and *Pohlia*. Initially, there were no significant differences in cattail leaf length ($p > 0.05$; Table 2), but there were significant differences ($p < 0.05$; Table 2) in percent growth of cattail leaves after five weeks. Percent growth of cattail leaves in the control lane was 107; AMD water diluted 1:4 supported nearly twice as much growth as the control (193%). AMD water diluted 1:2 supported a percent growth of 108, about the same as the control, while undiluted AMD water supported a percent growth of −3.3. Undiluted mine water had a significant negative effect on cattails. Leaves in the 1:2 dilution grew about the same as in the control and grew better

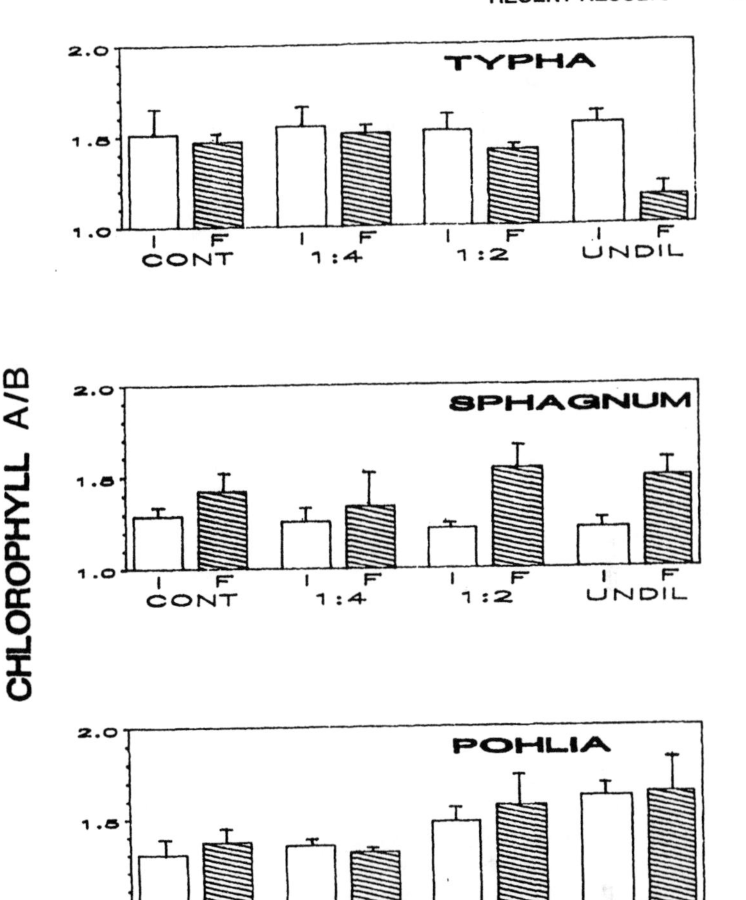

Figure 2. Chlorophyll a/b ratio of *Typha, Sphagnum,* and *Pohlia*. I = initial; F = final; cont = control; 1:4 = 1:4 dilution of AMD; 1:2 = 1:2 dilution of AMD; undil = undiluted AMD. Vertical lines = one standard deviation.

in the 1:4 dilution of mine water than in uncontaminated tap water. This may have been due to a nutrient effect.

The average number of new leaves produced per plant paralleled the percent growth pattern (an average of 0.8 leaves per plant for the control, 1.4 leaves for the 1:4 dilution, 0.8 leaves for the 1:2 dilution, and 0.2 leaves for the undiluted AMD) and is statistically significant ($p < 0.05$; Table 2). Finally, shoot counts of cattails yielded only slightly different results (control, 27% increase in shoot number; 1:4 dilution, 39% increase; 1:2 dilution, 41% increase; undiluted, 60% decrease), with the plants in the 1:2 dilution having slightly higher counts than those in the 1:4 dilution of AMD.

Taken as a whole, our results suggest that *Typha* was the most generally

Table 2. ***Typha* Growth Measures**

	Treatment			
Health Measure	**Undiluted AMD**	**1:2 Dilution**	**1:4 Dilution**	**Control**
Initial shoot number	52	56	49	48
Final shoot number	21	79	68	61
Percent change	–60	41	39	27
Number of new leaves per plant[a]	0.2 ± 0.42	0.8 ± 1.0	1.4 ± 1.0	0.8 ± 0.79
Sample size	10	10	9	10
Initial length of central leaf (cm)[b]	15.4 ± 9.73	15.0 ± 5.38	18.6 ± 7.32	18.1 ± 6.46
Percent growth[c]	–3.3 ± 10.2	108 ± 160	193 ± 135	107 ± 104
Sample size	10	7	8	10

[a]AOV: $p < 0.05$, $F = 5.86$.
[b]AOV: $P > 0.05$, $F = 14.75$.
[c]AOV: $P < 0.05$, $F = 4.75$.

tolerant of three plant species tested. However, if we knew the physiological meaning of the increase in chlorophyll a/b ratio for *Sphagnum* and *Pohlia,* a different conclusion might be necessary.

ACKNOWLEDGMENTS

We thank Dr. R. P. Brooks for his assistance in the initiation of this work and J. Derr for statistical advice. This project is part of a research program examining the use of wetlands in the treatment of acid mine drainage at The Pennsylvania State University funded by The Ben Franklin Challenge Grant Program of Pennsylvania, American Electric Power Service Corporation, Pennsylvania Power and Light Company, The Department of Environmental Resources of Pennsylvania, J. M. Stott Coal Company, U.S. Department of the Interior (Bureau of Mines) contract PO373662, and by small donations from 16 coal operating companies or engineering consulting firms.

REFERENCES

1. Kleinmann, R. L. P. "Treatment of Acid Mine Water by Wetlands," in *Control of Acid Mine Drainage*, Bureau of Mines Information Circular 9027 (1985), pp. 48–51.
2. Kleinmann, R. L. P., and M. A. Girts. "Acid Mine Water Treatment in Wetlands: An Overview of an Emergent Technology," in *Aquatic Plants for Water Treatment and Resource Recovery,* K. R. Reddy and W. H. Smith, Eds. (Orlando, FL: Magnolia Publishing Inc., 1987), pp. 255–261.
3. Stark, L. R., R. L. Kolbash, H. J. Webster, S. E. Stevens, Jr., K. A. Dionis, and E. R. Murphy. "The Simco #4 Wetland: Biological Patterns and Performance of a Wetland Receiving Mine Drainage," in *Mine Drainage and Surface Mine Reclamation, Vol. 1, Mine Water and Mine Waste,* Bureau of Mines Information Circular 9183 (1988), pp. 332–344.

4. Webster, H. J. "Elemental Analyses of *Pohlia nutans* Growing on a Coal Seep in Pennsylvania," *J. Hattori Bot. Lab.* 58:207–224 (1985).
5. Burris, J. E., D. W. Gerber, and L. E. McHerron. "Removal of Iron and Manganese from Water by *Sphagnum* Moss," in *Treatment of Mine Drainage by Wetlands,* J. E. Burris, Ed. (University Park: Pennsylvania State University, 1984), pp. 1–13.
6. Arnon, D. I. "Copper Enzymes in Isolated Chloroplasts. Polyphenol Oxidase in *Beta vulgaris,*" *Plant Physiol.* 24:1–15 (1949).
7. Emerson, R. "The Relation Between Maximum Rate of Photosynthesis and Concentration of Chlorophyll," *J. Gen. Physiol.* 12:609–622 (1929).
8. Fleischer, W. E. "The Relation Between Chlorophyll Content and Rate of Photosynthesis," *J. Gen. Physiol.* 18:573–597 (1935).
9. Hardie, L. P., D. L. Balkwill, and S. E. Stevens, Jr. "The Effects of Iron Starvation on the Physiology of the Cyanobacterium, *Agmenellum quadruplicatum,*" *Appl. Environ. Microbiol.* 45:999–1006 (1983).
10. Somers, I. I., and J. W. Shive. "The Iron-Manganese Relation in Plant Metabolism," *Plant Physiol.* 17:582–602 (1942).
11. Cheniae, G. M., and I. F. Martin. "Photoreactivation of Manganese Catalyst in Photosynthetic Oxygen Evolution," *Plant Physiol.* 44:351–360 (1969).

42i Control of the Armyworm, *Simyra henrici* (Lepidoptera: Noctuidae), on Cattail Plantings in Acid Drainage Treatment Wetlands at Widows Creek Steam-Electric Plant

Edward L. Snoddy, Gregory A. Brodie, Donald A. Hammer, and David A. Tomljanovich

INTRODUCTION

Constructed wetlands for wastewater treatment is a relatively new and inexpensive technology used for treating sewage, certain industrial discharges, and acid mine drainage.[1,2] Some management problems of these biological systems include control of plant pests, mosquitoes, and other biting flies or their exclusion from the system.[3] Even with the best designs, a certain vigilance must be maintained for pests associated with these systems.

Due to the monocultural nature of the macrophytes used in constructed wetlands, some plants, particularly cattail (*Typha latifolia*),[4] are subject to damage by lepidopterous insect pests,[4,5] mainly the armyworm complex. This group can cause serious damage to a wetland planting in a very short time. Control is dependent upon insecticides applied immediately following the discovery of this pest. The only guidelines and data on armyworm control in cattails must be extrapolated from data on related species attacking agronomic crops. To further complicate control procedures, determination of flow, volume, and organic content must be considered when determining the pesticide dosage necessary for effective control. Measures for controlling armyworms in a constructed wetlands treating acid waters are described in this chapter.

RESULTS

A constructed wetlands was established at Widows Creek Fossil Fuel Plant early in May 1986 to aid in the control of acid drainage from an abandoned ash pond.[6] Cattails[7] were planted in two of the three cells, with cattail and rush

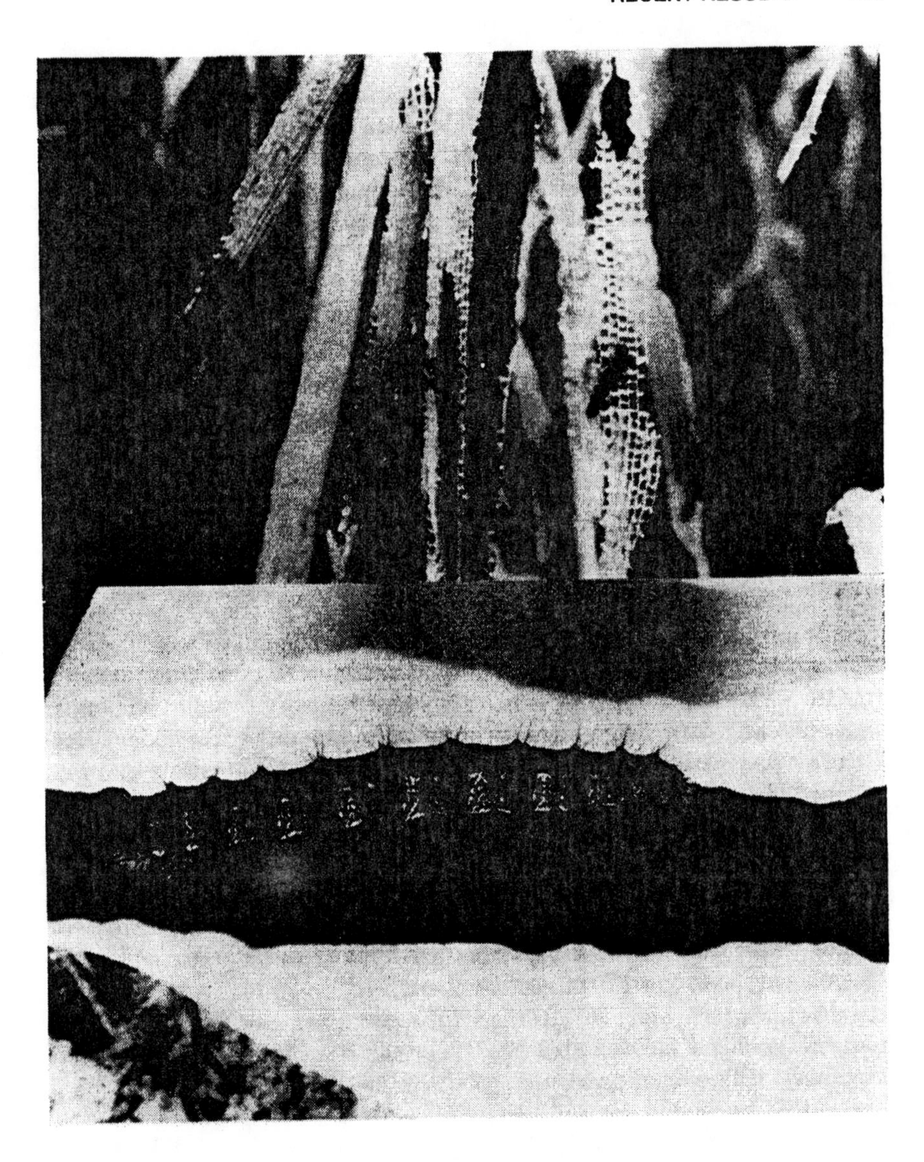

Figure 1. Larvae of the cattail armyworm, *Simyra henrici* (Grt.) feeding on *Typha latifolia* in wetland planting.

(*Juncus effusus*) in the third cell. An infestation of the cattail armyworm, *Simyra henrici* (Grt.),[8,9] was observed in mid-August 1986 (about four months after the wetlands was established). Examination of the plants revealed heavy vegetative destruction (Figure 1). Plant parts and armyworm frass were observed falling on the surface of the water as they wreaked their destructive march. Approximately 60–80% of the plants in the center of the outbreak had significant damage.

Examination revealed all four larval instars were present along with eggs, pupae, and an occasional adult moth flying among the plants. Female moths deposited eggs on plant leaves in long, overlapping shinglelike rows appearing silvery grey but turning dark as they approached maturity. Young armyworm larvae, upon hatching, devoured the eggshell and began rasping the surface of the plant leaf beneath the eggshell.

Due to the rapid and extensive damage on the plantings, the insecticide Lorsban was selected due to its known effectiveness for many species of armyworms. Lorsban was applied with a Bell 206B helicopter using twin Beecomist spray heads on a Simplex belly-tank spray system at a rate of 1.15 kg/ha. This control procedure resulted in >95% control.[10]

DISCUSSION

It is unclear why the population density of caterpillars was much higher at Widows Creek compared to nearby natural wetlands. Widows Creek is unique in that a very dense stand of cattails developed rapidly (within two months) after planting. Armyworms are noted for their boom and crash populations. Because they have many parasites and predators, their strategy is to reproduce and expand rapidly before population regulation factors develop. A complex web within the arthropod community governs population levels.[5] Because of this, it is possible that important components of this ecosystem (parasites and predators)[11] had not fully developed in this stand of cattails, and thus a controlling factor was lacking. Aphids, ladybird beetles, and several species of Diptera were observed in the plantings, and redwings (*Ageliaus phoeniceus*) fed on insects associated with cattail plants.

A routine inspection of plantings for insect pests beginning early in the season should be incorporated into management procedures for constructed wetlands. This should prevent a rapid buildup of these pests. General plans should be made for control action if an outbreak of armyworms occurs. Research on agricultural crop monocultures indicates stressed plants have a far greater attraction and/or are preferred by many insects over strong, viable, rapidly growing plants, as was initially observed at the Widows Creek planting. Trapping borders of stressed cattails or other wetland plants may be an alternative to minimize destruction.

Edge-effect planting may create an attractive barrier, concentrating insect pests, thus buffering feeding on the main vegetative plantings. However, this technique may not be a feasible or practical alternative to basic chemical insecticidal control.

REFERENCES

1. Brodie, G. A., D. A. Hammer, and D. A. Tomljanovich. "Constructed Wetlands for Acid Drainage Control in the Tennessee Valley," in *Mine Drainage and Surface Mine Reclamation, Vol. I, Mine Water and Mine Waste,* Bureau of Mines Information Circular 9183 (1988), pp. 325–331.
2. Brodie, G. A., D. A. Hammer, and D. A. Tomljanovich. "Treatment of Acid Drainage from Coal Facilities with Man-Made Wetlands," in K. R. Reddy and W. H. Smith, Eds. *Aquatic Plants for Water Treatment and Resource Recovery* (Orlando, FL: Magnolia Publishing, Inc., 1987).
3. Hammer, D. E., and R. H. Kadlec. "Design Principles for Wetland Treatment Systems," U.S. EPA Report-PB-83-188-722 (1983).
4. Claassen, P. W. "*Typha* Insects: Their Ecological Relationships," Cornell University Agr. Expt. Sta. Mem. 47:1921, pp. 457–531.
5. Cole, A. C. "*Typha* Insects and Their Parasites," *Ent. News* 42:6–11 (1931).
6. Brodie, G. A., D. A. Hammer, and D. A. Tomljanovich. "Constructed Wetlands for Treatment of Ash Pond Seepage," Chapter 16, this volume.
7. Davis, S. M. "Cattail Leaf Production, Mortality, and Nutrient Flux in Water Conservation Area 2A," South Florida Water Mgmt. Dist. Tech. Pub. 84–8 (1984).
8. Crumb, S. E. "Larvae of the Phalaenidae," USDA Tech. Bull. 1135 (1956), p. 38.
9. Judd, W. W. "The White Veined Dagger, *Simyra henrici* Grt. (Phalaenidae), and Its Parasites Reared from Cattail, *Typha* spp.," *Nova Scotian Inst. Sci. Proc. and Trans.* 23:115–119 (1952).
10. Snoddy, E. L., and J. C. Cooney. "Insecticides for Pest Control in Constructed Wetlands for Wastewater Treatment: A Dilemma," in *Proceedings National Pesticide Conference,* Richmond, VA (1989).
11. Krombein, K. V., P. D. Hurd, Jr., D. R. Smith, and B. D. Burks. "Catalog of *Hymenoptera* in America North of Mexico," (Washington, DC: Smithsonian Institution Press, 1979).

List of Authors

M. Stephen Ailstock, Anne Arundel Community College, 101 College Parkway, Arnold, Maryland 21012

Hollis H. Allen, Environmental Laboratory, U.S. Army Engineer Waterways Experiment Station, Vicksburg, Mississippi 39180-0631

Robert K. Bastian, Office of Municipal Pollution Control, U.S. Environmental Protection Agency, Washington, DC 20460

Wafa Batal, Department of Environmental Sciences and Engineering Ecology, Colorado School of Mines, Golden, Colorado 80401

H. J. Bavor, Water Research Laboratory, Faculty of Food and Environmental Sciences, University of Western Sydney, Hawkesbury, Richmond, N.S.W. 2753, Australia

Hans Brix, Botanical Institute, University of Århus, Nordlandsvej 68, DK-8240 Risskov, Denmark

Gregory A. Brodie, Power Operations, Tennessee Valley Authority, BR 2S 50-B, Chattanooga, Tennessee 37402

P. S. Burgoon, Department of Soil Science, University of Florida, Gainesville, Florida 32611

Thomas E. Conway, formerly of Indiana University of Pennsylvania, R.D. #2, Box 1490, Homer City, Pennsylvania 15748

P. F. Cooper, Water Research Centre, Elder Way, Stevenage, Hertfordshire, United Kingdom SG1 1TH

C. J. Costello, Lough Gara Farms Ltd., Lurgan Lodge, Kingsland, Boyle, County Roscommon, Ireland

Paula Daukas, IEP, Inc., P.O. Box 780, Northborough, Massachusetts 01532

Randal L. Davido, Indiana University of Pennsylvania/Moon Township Municipal Authority, 1114½ Ferree St., Coraopolis, Pennsylvania 15108

T. A. DeBusk, Reedy Creek Energy Services, Inc., P.O. Box 10,000, Lake Buena Vista, Florida 32830

Charles H. Dill, Marin Sonoma Mosquito Abatement District, 556 N. McDowell Blvd., Petaluma, California 94952

James N. Dornbush, South Dakota State University, Brookings, South Dakota 57007

Paul Eger, Minnesota Department of Natural Resources, Minerals Division, 500 Lafayette Rd., St. Paul, Minnesota 55155–4045

Stephen P. Faulkner, School of Forestry & Environmental Studies, Duke University, Durham, North Carolina 27706

J. Scott Feierabend, Fisheries and Wildlife Division, National Wildlife Federation, 1412 Sixteenth Street N.W., Washington, D.C. 20036–2266

R. A. Gearheart, Humboldt State University, Environmental Resources Engineering Department, Arcata, California 95521

Richard M. Gersberg, Graduate School of Public Health, San Diego State University, San Diego, California 92182–0405

Michelle A. Girts, CH2M HILL, 2020 S.W. Fourth Avenue, Portland, Oregon 97201

Wolfgang Grosse, Botanical Institute, University of Cologne, Gyrhofstrasse 15, D-5000, Cologne 41, F.R.G.

Vincent G. Guida, Environmental Studies Center, Chandler-Ullmann Hall #17, Lehigh University, Bethlehem, Pennsylvania 18015

G. R. Guntenspergen, Department of Botany, Louisiana State University, Baton Rouge, Louisiana 70803–1705

Raimund Haberl, Institut für Wasserwirtschaft, Universität für Bodenkultur Wien, Gregor-Mendelstrasse 33, A-1180 Vienna, Austria

Donald A. Hammer, Waste Technology Program, Tennessee Valley Authority, Knoxville, Tennessee 37902

Joe W. Hardy, Mississippi Sandhill Crane National Wildlife Refuge, Gautier, Mississippi 39553

Robert S. Hedin, Bureau of Mines, U.S. Department of the Interior, P.O. Box 18070, Pittsburgh, Pennsylvania 15236

Jacqueline Henrot, Department of Biology, Villanova University, Villanova, Pennsylvania 19085

Delbert B. Hicks, U.S. Environmental Protection Agency, Region IV, Environmental Services Division, College Station Road, Athens, Georgia 30613-7799

John A. Hobson, Water Research Centre, Elder Way, Stevenage, Hertfordshire, United Kingdom SG1 1TH

Robert D. Hoffman, Great Plains Regional Office, Ducks Unlimited, Inc., Bismarck, North Dakota 58501

Edward A. Howard, Department of Environmental Sciences, Colorado School of Mines, Golden, Colorado 80401

Yuch-Ping Hsieh, Wetland Ecology Program, P.O. Box 239, Florida A&M University, Tallahassee, Florida 32307

JoAnn Jackson, Post, Buckley, Schuh & Jernigan, Inc., 800 N. Magnolia Avenue, Suite 600, Orlando, Florida 32803

Beverly B. James, James Engineering Inc., 6329 Fairmount Ave., El Cerrito, California 94530

Robert H. Kadlec, Wetland Ecosystem Research Group, Department of Chemical Engineering, University of Michigan, Ann Arbor, Michigan 48109-2136

J. B. Kingsley, Agricultural Research Department, Tennessee Valley Authority, National Fertilizer Development Center, NFD 2F 237K, Muscle Shoals, Alabama 35660

Ronald L. Kolbash, American Electric Power, P.O. Box 700, Lancaster, Ohio 43130

Lewis C. Linker, U.S. Environmental Protection Agency Chesapeake Bay Liaison Office, 410 Severn Avenue, Annapolis, Maryland 21403

Donald K. Litchfield, Amoco Oil Refinery, P.O. Box 5000, Mandan, North Dakota 58554

Eric H. Livingston, Florida Department of Environmental Regulation, 2600 Blair Stone Road, Tallahassee, Florida 32399-2400

Edgar F. Lowe, St. Johns River Water Management District, P.O. Box 1429, Palatka, Florida 32078-1429

J. J. Maddox, Agricultural Research Department, Tennessee Valley Authority, National Fertilizer Development Center, Muscle Shoals, Alabama 35660

Cecil V. Martin, California Water Resources Control Board, P.O. Box 944212, Sacramento, California 94244

Emy Chan Meiorin, Association of Bay Area Governments, P.O. Box 2050, Oakland, California 94604

Susan Copeland Michaud, P.O. Box 2008, Oak Ridge National Laboratory, Oak Ridge, Tennessee 37831-6230

Gordon Miller, JE Hanna Associates Inc., 1886 Bowler Drive, Pickering, Ontario, Canada L1V 3E4

Thomas J. Mingee, Nolte and Associates, 1730 I Street, Suite 100, Sacramento, California 95814

Ralph J. Portier, Aquatic and Industrial Toxicology Laboratory, Institute for Environmental Studies, Louisiana State University, Baton Rouge, Louisiana 70803

J. Henry Sather, Institute for Environmental Management, Western Illinois University, Macomb, Illinois 61455

Phillip R. Scheuerman, Department of Environmental Health, East Tennessee State University, Johnson City, Tennessee 37614

A. Larry Schwartz, Grand Strand Water and Sewer Authority, P.O. Box 1537, Conway, South Carolina 29526

Marvin Silver, Marvin Silver Scientific Limited, 78 Village Crescent, Bedford, Nova Scotia, Canada B4A 1J2

Gary S. Silverman, Environmental Health Program, Bowling Green State University, Bowling Green, Ohio 43403

Robert L. Slayden, Jr., Tennessee Division of Water Pollution Control, 150 9th Avenue North, Nashville, Tennessee 37219-5404

Richard C. Smardon, Institute for Environmental Policy and Planning, College of Environmental Science and Forestry, State University of New York, Syracuse, New York 13210

Al J. Smith, Water Management Division, Region IV, U.S. Environmental Protection Agency, Chattanooga, Tennessee

Edward L. Snoddy, Vector and Plant Management Program, Tennessee Valley Authority, Muscle Shoals, Alabama 35660

Ward W. Staubitz, U.S. Geological Survey, 1201 Pacific Avenue, Suite 600, Tacoma, Washington 98402

Gerald R. Steiner, Water Quality Department, Tennessee Valley Authority, HB 2S 270C, 311 Broad Street, Chattanooga, Tennessee 37402-2801

Eberhard Stengel, Institute for Biotechnology 3, Kernforschungsanlage Jülich GmbH, P.O.B. 1913, D-5170 Jülich, F.R.G.

S. Edward Stevens, Jr., Department of Biology, 509 Life Sciences Building, Memphis State University, Memphis, Tennessee 38152

Takao Suzuki, Biological Institute, Tohoku University, Sendai 980, Japan

Rudolph N. Thut, Technology Center, Weyerhaeuser Company, Tacoma, Washington 98477

David A. Tomljanovich, Tennessee Valley Authority, Norris, Tennessee 37828

Nancy M. Trautmann, New York State Water Resources Institute, Center for Environmental Research, Cornell University, Ithaca, New York 14853–3501

Jan Vymazal, Water Research Institute, Department 252, Podbabská 30, 160 62 Praha 6, Czechoslovakia

James T. Watson, Water Quality Department, Tennessee Valley Authority, HB 2S 270C, 311 Broad Street, Chattanooga, Tennessee 37402–2801

William R. Wenerick, The Penn State Biotechnology Institute and The Department of Biology, Pennsylvania State University, University Park, Pennsylvania 16082

Kelly J. Whalen, The Lombardo Group of Dames & Moore, 46 Church Street, Boston, Massachusetts 02116

R. Kelman Wieder, U.S. Office of Surface Mining, Reclamation, and Enforcement, Eastern Field Operations, Pittsburgh, Pennsylvania 15220 and Department of Biology, Villanova University, Villanova, Pennsylvania 19085

Thomas R. Wildeman, Department of Chemistry and Geochemistry, Colorado School of Mines, Golden, Colorado 80401

Mel Wilhelm, Apache-Sitgreaves National Forests, Lakeside, Arizona 85929

B. C. Wolverton, National Aeronautics and Space Administration, John C. Stennis Space Center, Building 2423, Stennis Space Center, Mississippi 39529–6000

A. Wood, Steffen, Robertson & Kirsten Consulting Engineers, Inc., P.O. Box 8856, Johannesburg 2000, South Africa

Poster and Other Presentations

WATER QUALITY AND WETLAND MITIGATION PLAN FOR COMMERCIAL DEVELOPMENT—Susan Bitter, Greenspring Environmental Design & Contracting, Incorporated, Brooklandville, Maryland

AVIAN BOTULISM: FACTORS TO CONSIDER DURING WETLAND CONSTRUCTION—Kathryn A. Converse, U.S. Department of the Interior, Madison, Wisconsin

CHEMICAL AND PHYSICAL PROPERTIES OF SOME FRESHWATER WETLANDS IN NORTH FLORIDA—Charles L. Coultas, Havana, Florida

LEMNA SYSTEM—Hendrik Erenstein, Lemna Corporation, Mendota Heights, Minnesota

TENNECO'S USE OF A ROCK-REED FILTER AT A NATURAL GAS PIPELINE COMPRESSOR STATION—Bob Honig, Tenneco Gas Transportation, Houston, Texas

EXPERIENCE WITH WETLAND PLANTS SUITABLE FOR WASTE MANAGEMENT—J. Kvët, Institute of Botany, Czechoslovak Academy of Sciences, Dukelska 145, CS-379 82 Trebon, Czechoslovakia

USING ARTIFICIAL WETLANDS/PONDS TO CONTROL NUTRIENT AND SEDIMENT RUNOFF FROM AGRICULTURAL OPERATIONS—Robert Wengrzynek, U.S. Soil Conservation Service, Orono, Maine

WETLANDS FOR UPGRADING A WASTEWATER DISCHARGE IN HAINES, Alaska—Ricardo Saavedra, City of Haines, Alaska

Index

Abatement 290, 674, 753, 757
Absorption 12, 98, 140, 177, 330, 332, 346, 470, 471, 517, 629, 680, 681, 737, 776, 782, 794
Access 183, 188, 190, 203, 307–309, 312, 316, 399, 401, 411, 452, 547, 550, 591, 644, 674, 699, 763, 785
Accretion 459, 465
Acid drainage 12, 17, 201, 202, 211, 222–224, 301, 312, 320, 337, 364, 368, 381, 383, 427, 757, 758, 808
Acid mine drainage 4, 47, 58, 60, 61, 64, 221–223, 228, 270, 283, 300, 307, 310, 319, 329, 331, 332, 333, 346, 347, 374, 410, 427, 508, 550, 736, 753, 765, 771, 772, 774, 775, 788, 798, 799, 801, 806, 808
Acids 98, 135, 226, 331, 333, 406, 755, 757, 758, 775
Activated 146, 148, 233, 279, 284, 395, 437, 438, 719, 727, 728, 729, 733
Adaptations 73–78, 80, 109
Adsorption 44, 56–58, 61, 62, 81, 100, 150, 224, 226, 233, 247, 248, 255, 312, 320, 330, 331, 337, 346, 363, 375, 376, 424, 432, 437–439, 493, 550, 618, 658, 672, 691, 692, 696, 737, 740, 750
Aeration 58, 79, 141, 143, 175, 255, 329, 368, 369, 373, 421, 436, 469, 477, 478, 480, 483, 484, 489, 491, 582, 623, 657, 727, 728, 748, 749, 774, 788
Aerobic 12, 14, 15, 41, 46, 47, 58, 60, 64, 78, 80, 95, 101, 143, 148, 149, 154, 158, 173–175, 226, 320, 329, 341, 342, 343, 373, 381, 382, 407, 496, 497, 505, 510, 511, 550, 556, 565, 583, 598, 652, 737, 739, 743, 750, 772
Aerobic conditions 173, 174, 341, 381, 382, 407, 497, 510, 511
Aesthetics 297, 334, 354, 355, 593, 696
Agricultural 4, 5, 8, 12, 16, 148, 179, 187, 191, 265, 270, 289, 297, 307, 310, 367, 381, 383, 401, 411, 450, 500, 515, 517, 542, 564, 578, 622, 644, 648, 669, 674, 677, 695, 696, 699, 703, 711, 713, 715, 737, 801, 810
Agricultural runoff 270, 578
Alabama 12, 193, 201, 203, 211, 213, 279, 280, 332, 542, 546, 547
Albedo 25, 26
Algae 12, 14, 80, 81, 90, 100, 130, 191, 227, 229, 272, 281, 288, 451, 460, 553, 555, 561, 598, 600, 603, 623, 625, 715, 750, 765, 768, 771, 772, 791
Algal 126, 129, 224, 227, 290, 320, 331, 333, 369, 421, 440, 556, 558, 559, 585, 598, 600, 648, 765, 768, 769, 771, 772
Alkalinity 96, 218, 310, 330, 333, 346, 374, 420, 508, 510, 582, 583, 586, 711, 735, 745
Aluminum 44, 312, 330, 346, 375, 376, 427, 515, 565, 625, 659, 755
Ammonia 53, 77, 130, 142, 146, 148, 149, 168, 197, 198, 239, 241, 242, 243, 245, 247–250, 261, 266, 282, 283, 285, 329, 330, 342, 346, 368, 369, 375, 424, 426, 443, 477, 480, 481, 493, 569, 576, 583, 585, 600, 603, 628, 680, 681, 707, 709, 729, 735, 747
Ammonia nitrogen 130, 142, 197, 198, 247–250, 330, 342, 346, 369, 375, 424, 426, 480, 481, 603
Ammonification 143, 480, 496, 498, 499
Amoebae 99

Anaerobic bacteria 58, 509
Anaerobic conditions 6, 9, 41, 63, 78, 79, 148, 298, 328, 341, 407, 465, 497, 509, 518, 543, 757
Anaerobic digestion 737
Anaerobic environment 15, 751
Animals 6, 10, 11, 90, 107, 109–111, 114, 115, 143, 288, 289, 353, 354, 403, 412, 447, 449, 450, 454, 542, 598, 727
Anoxia 78, 330
Application methods 574
Application rates 146, 148, 332, 337
Aquatic macrophytes 393, 396, 398, 487, 528, 542, 565
Aquatic plants 74, 76, 77, 81, 82, 148, 149, 175, 239, 259–261, 354, 394, 413, 432, 450, 469–472, 528, 534, 536, 542, 548, 607, 613, 722, 725
Arcata, California 116, 121, 122, 127, 132, 134, 150, 270, 280, 329, 394, 396, 431, 434, 436, 439, 443
Arizona 179–184
Aromatic hydrocarbons 89, 100, 101, 736
Ash ponds 213, 218
Aspect ratio 624
Assimilation capacity 424
Australia 157, 239, 646
Austria 157, 158, 320, 606, 619
Autecology 82
Autoflocculation 131
Azolla 77

Bacteria 14, 52, 58, 61, 90–92, 95, 96, 99–103, 142, 146, 148–150, 168, 173, 174, 222, 226, 229, 239, 261, 266, 298, 329, 332, 346, 431, 432, 434–438, 443, 449, 460, 477, 481, 509, 510, 513, 516–518, 520, 528, 536, 547, 550, 551, 552, 555, 556, 628, 641, 646, 657, 658, 707, 753, 754–757, 761, 771, 796–798
Baffles 125, 143, 201, 203, 257, 260, 283, 421, 439, 593, 594, 596, 648
Basins 4, 125, 183, 258, 297, 298, 319, 397, 411, 515, 686, 688, 689, 691, 692, 697, 700
Beaver 114
Bedrock 203, 209, 312
Benthic 110, 192, 281, 450, 451, 453
Benton, Kentucky 145, 270, 281, 282
Biochemical oxygen demand (BOD) 122, 266, 271, 320, 425, 525, 546, 566, 622, 735
  removal of 122, 131, 132, 134, 145, 146, 148, 149, 167, 320, 341, 368, 526, 528, 536, 538, 642, 650, 732
Biodegradable 89, 102, 343, 379, 523
Biomass harvesting 198
Biomass production 12, 16, 21, 122, 540
Biotransport 16
Bog 5, 8, 12, 27, 50, 58, 62, 109, 111, 113, 114, 226, 287, 331, 464, 502, 508, 702, 703, 798
Brackish marsh 5, 517
Brookhaven National Laboratory 141, 477
Budget 121, 181, 355, 448, 594, 715, 720, 798
Buffer 309, 333, 436, 516, 767
Bureau of Mines 300, 337, 509, 512, 772, 806

Cadmium 228, 249, 600, 681, 692, 735
Calcium 98, 198, 240, 249, 330, 346, 375, 406, 407, 495, 496, 565, 587, 750
California 109, 116, 121, 146, 150, 280, 281, 328, 329, 341, 359, 360, 393–398, 431, 434, 438, 439, 443, 526, 597, 622, 664, 669, 672–674, 677
Canna 177
Carbon 531
Carex 7, 29, 81, 247, 298, 411, 460, 494, 626, 762, 785
Case studies 418, 427, 574
Cation exchange capacity (CEC) 43, 775
Channelization 142, 143, 289, 380, 420, 421, 546, 720
Chemistry 41, 48, 98, 142, 202, 203, 211, 221, 223, 228, 300, 301, 309, 354, 417, 420, 509, 511, 513, 753, 783, 793, 794, 803
Chlorination 132, 180, 369, 431, 434, 438, 442, 478, 597, 624
Chromium 234, 236, 249, 600, 692, 735

Classification 5, 6, 41, 73, 74, 90, 308, 312, 408
Clean Water Act 3, 187, 188, 267, 270–274, 313, 394, 395, 623
Climate 15, 26, 41, 63, 187, 256, 299, 303, 342, 370, 411, 421, 459, 603, 622, 636, 640, 642, 728, 737, 761
Climatic influences 255
Climax 223
Coal mine drainage 223, 297, 509, 765, 793
Coal mining 221, 226, 501, 765
Coastal wetlands 664
Commercial 11, 186, 193, 260, 410, 453, 517, 542, 543, 545, 597, 622, 664, 677, 703, 753
Composition 9, 77, 79, 91, 108, 300, 312, 353, 373, 374, 376, 413, 451, 559–561, 564, 647, 735, 754, 768
Composting 122, 628, 629
Conductivity 42, 159, 374, 629, 680, 738
Configuration 101, 254, 299, 334, 363, 364, 367, 369–371, 375, 388, 403, 432, 636, 689
Connecticut 12
Construction cost 140, 143, 146, 301, 595, 596, 622, 626
Consumption 21, 37, 114, 246, 396, 453, 463, 491, 603, 623, 692
Conventional 4, 16, 17, 21, 90, 139, 146, 148, 209, 234, 266, 284, 299, 300, 301, 307, 319, 346, 380, 395, 396, 421, 424, 431, 432, 437, 443, 447, 477, 548, 550, 565, 574, 588, 612, 623, 628, 636, 643, 646, 722, 736, 737, 788
Conversion 57, 101, 395, 407, 453, 496, 704
Copper 12, 98, 99, 227, 228, 236, 249, 331, 332, 600, 692, 780, 781, 783
Cost estimates 299, 301, 401
Creation 4, 179, 265, 268, 269, 290–292, 355, 356, 360, 577, 669, 672, 675, 688, 695–697, 710, 711, 715, 716, 720, 722, 723
Crustaceans 10, 110, 114
Culex 666, 667
Culture 96, 191, 193, 195, 287, 437, 542, 545, 556, 659, 771
Cycles 14, 50, 58, 418, 649, 679, 743, 750
Cycling 52, 55, 57, 58, 61, 63, 64, 103, 121, 255, 267
Cyperinus 203, 207, 212, 381, 502–505
Cypress dome 255, 425

Darcy's law 22, 158, 159, 340–343, 371, 374, 381, 383, 629, 631, 632, 738
Decay 12, 224, 226, 261, 332, 421, 459, 463, 465, 466, 587, 660, 662, 776
Decomposition 12, 55, 78, 101, 114, 128, 130, 248, 288, 330, 345, 373, 459, 463–465, 705, 707, 737, 739, 750, 757, 775, 778, 782, 783
Deficiency 96, 103
Definition 5, 13, 96, 270, 271, 303, 775
Degradation 100–102, 175, 265, 270, 273, 293, 309, 310, 337, 360, 375, 437, 542, 593, 611, 669, 672, 696, 744, 745, 755, 775
Denitrification 15, 52, 53, 55, 64, 148, 166, 242, 248, 329, 330, 368, 375, 376, 424, 467, 477, 478, 480, 481, 483–485, 491, 493, 496–499, 565, 578, 583–587, 590, 594, 648, 652, 654
Denmark 154, 159, 320, 376, 565, 571
Deposition 8, 50, 58, 223, 255, 320, 381, 384, 462, 467, 610, 611, 695, 771
Destruction 11, 18, 108, 114, 183, 270, 354, 509–511, 645, 664, 755, 809, 810
Detection 454, 638, 657, 681, 748
Detrital 110, 122, 126, 128, 451, 518, 642
Dewatering 227, 725
Diffusion 52, 76, 328, 408, 469, 470, 472, 474, 489, 491, 501, 502, 503, 722
Dike construction 213
Discharge sampling 213, 332
Discharge structure 255, 256, 260, 578
Disease vectors 268, 289
Diseases 431, 439, 450, 666
Disinfection 142, 281, 337, 434, 437, 438, 443, 590, 593, 622, 625
Dispersion 689
Disposal 4, 16, 80, 139, 140, 177, 180,

183, 184, 202, 203, 245, 246, 265, 267, 268, 280, 303, 309, 310, 312, 364, 394, 447, 493, 574, 577, 581, 597, 598, 623, 664, 697, 725, 743, 744–747, 750, 751, 757, 789
Dissimilatory sulfate reduction 58, 508, 509, 513, 551
Dissolved oxygen 12, 50, 96, 111, 128, 142, 202, 241, 290, 330, 346, 367, 369, 397, 406, 439, 480, 502, 505, 546, 643, 749, 763
  effects of 122, 266, 271, 320, 425, 525, 546, 566, 622, 735
Dissolved solids 197, 680, 750
Distribution piping 400
District of Columbia 695
Disturbance 8, 9, 216, 218, 270, 303, 420, 422, 427, 510, 642
Diversity 11, 102, 108, 110, 111, 115, 255, 266, 346, 353, 360, 405, 407, 410, 448, 451, 453, 515, 517, 520, 523, 645, 665, 689, 713
Drawdown 406, 411, 421
Dyes 380
Dynamics 8, 17, 21, 22, 38, 109, 354, 355, 459, 460

Ecological 74, 75, 108, 115, 227, 258, 270, 275, 276, 290, 291, 293, 353, 354, 355, 413, 447, 448, 472, 491, 702, 720
Economic 11, 12, 108, 192, 198, 233, 290, 308, 397, 700, 716
Ecosystem 25, 46, 49, 63, 79, 102, 179, 226, 255, 266, 268, 272, 353, 355, 360, 410, 424, 451, 552, 588, 750, 810
Effluent 127, 129, 161, 167, 180, 193, 228, 234, 241, 243, 249, 299, 375, 381, 420, 432, 442, 478, 537, 538, 543, 559, 566, 569, 585–587, 591, 626, 638, 639, 649, 739
Effluent quality 122, 125, 126, 154, 157, 165–167, 170, 181, 182, 183, 271, 280, 370, 526, 535, 565, 569, 576, 579, 585, 608, 623, 642
Eichhornia 75, 79, 281, 393, 525, 588, 713
Eleocharis 191, 192, 212, 413, 502, 542
Emergent macrophytes 25, 525, 646, 654
Emergent plants 7, 75–77, 80, 81, 183, 266, 297, 298, 345, 354, 381, 406, 536, 709
Emergent wetlands 80, 81
Endangered species 11, 115, 116, 187, 188, 202, 203, 274, 313, 397, 410, 411
Energy consumption 246, 396
Energy costs 140, 291
England 8, 382, 686
Enrichment 180, 435, 438, 466
Environmental issues 313
Environmental Protection Agency 5, 150, 180, 237, 397, 431, 574, 644, 662, 670
Ergun's equation 22, 342, 371, 383
Erosion 8, 108, 142, 158, 159, 183, 190, 191, 203, 253–255, 259, 260, 303, 309, 312, 316, 399, 403, 405, 412, 696, 775
Escherichia coli 100, 431, 586, 658
Ethanol 78, 122, 767
Evapotranspiration 21, 25–27, 29, 31, 35, 38, 328, 340, 344, 381, 383, 421, 538, 594, 603, 604, 608, 610
Everglades 5, 111
Evergreen 603
Evolution 74, 90
Export 261, 673, 713, 728

Facultative 142, 188, 477, 756
Fecal coliform removal 125, 135, 651
Fecal streptococci 438, 638, 639, 658, 660, 662
Federal regulations 80
Fermentation 101, 122, 146, 438, 705, 777
Ferric 45, 56, 61, 222, 333, 510, 551, 552, 754–757, 771
Ferrous 45, 46, 56, 76, 333, 501, 510, 551, 552, 754, 755
Fertilizer 192, 193, 198, 212, 213, 401, 407, 420, 543, 545, 548, 722, 740
Fiber content 192
Field crops 622
Filters 173, 177, 271, 299, 312, 373, 435, 581, 588, 590, 591, 593, 594, 623, 658, 767

Fire 8, 186, 422
Fish 5, 6, 10, 11, 18, 110–113, 142, 180, 188, 191–193, 195, 197, 198, 237, 265, 271–273, 288, 290, 292, 298, 299, 313, 353, 396, 397, 447, 448, 450, 452, 453, 584, 599, 645, 665, 667, 673, 674, 681, 682, 684, 715
Fixation 16, 50, 76, 551, 600
Fixed 26, 78, 284, 285, 437, 493, 558, 646, 655, 737, 740, 767
Floating macrophytes 126, 525, 528
Floating plants 7, 75
Flood control 11, 266, 672–674, 696
Flooding 8, 9, 17, 41, 45–48, 56, 74, 75, 78, 159, 165, 213, 222, 253, 258, 309, 354, 383, 405, 406, 413, 450, 517, 623, 626, 674, 688, 727–729
Floodplain 289, 313, 425, 711
Floodplain Management Executive Order 313
Florida 6, 26, 31, 60, 111, 254, 258, 259, 270, 273, 281, 410, 424, 425, 493, 494, 500, 525, 537, 574, 578, 657, 658, 710, 711, 715
Flow rate 22–24, 79, 125, 157, 158, 196, 215, 255, 342, 344, 439, 449, 537, 546–548, 630, 631, 738, 776, 802
Flushing 192, 193, 198, 386, 420, 421, 453, 518, 520, 576, 681
Food chain 191, 356, 452, 684
Food webs 109
Forested wetlands 81
Forests 108, 179, 180, 269
France 157
Freshwater marsh 5, 7, 394
Freshwater swamp 5, 518, 520
Frogs 110, 111, 113, 114
Fungi 14, 99, 101, 102, 516–518, 520, 755, 758

Georgia 12
Glyceria 77, 411, 413
Grass 81, 142, 189, 236, 248, 369, 406, 411, 478, 583, 664, 689, 713, 785
Gravel marsh 145, 363, 369
Groundwater monitoring 426, 785
Gustine, California 270, 280, 341, 396, 622, 624–626
Gypsum 587

Habitat creation 696
Habitat enhancement 186, 191, 292, 399
Hardin, Kentucky 281, 282
Hardwood swamp 575, 577, 578, 657
Harvesting plants 299
Heavy metals 16, 77, 80, 82, 98–100, 211, 255, 259, 261, 267, 268, 281, 298, 310, 330, 355, 395, 600, 617–619, 657, 679, 681, 682, 684, 691–693, 736, 758, 775
Histosol 42
Houghton Lake, Michigan 24, 25, 29, 31, 38, 340, 422, 424, 425, 460, 462, 465, 466
Hydraulic conductivity 14, 22, 42, 158, 159, 166, 247, 249, 250, 341, 342, 343, 370, 371, 373–376, 382–384, 573, 629–633, 654, 738, 740, 783
Hydraulic considerations 341, 628, 675
Hydraulic gradient 340–343, 382, 383, 629–633, 648, 716
Hydraulic loading 122, 134, 209, 329, 333, 337, 339, 343, 346, 347, 370, 418, 420, 421, 424–427, 432, 439, 449, 450, 525, 537, 566, 567, 569, 571, 591, 624, 692, 710, 716
Hydric 6, 259
Hydrocotyle 439, 525, 526
Hydrodynamics 21, 29, 33
Hydrogen sulfide 298, 502, 508, 509, 748, 750
Hydrology 9, 79, 202, 223, 253–256, 300, 308–310, 316, 405, 427, 450, 578, 638, 698, 727, 728, 731, 750
Hydroperiod 187, 255, 259, 309, 420, 495

Ice cover 284, 320, 330, 370, 608, 638, 642
Illinois 289, 290, 528
Immobilization 224, 346, 465
Indiana 139, 140, 142, 360, 477
Industrial wastes 332, 343
Industrial wastewater 12, 270, 297, 725, 736
Infiltration 22, 31, 36, 140, 141, 196, 254, 256, 310, 340, 341, 383, 425, 478, 608, 610, 672, 735, 780
Influent distribution 299, 654

Inlet distribution 371, 403
Inorganic 10, 16, 41, 50, 55–58, 63, 80, 89, 92, 94, 97, 168, 226, 267, 407, 543, 545, 548, 562, 659, 684, 710, 735, 739, 740, 745, 747, 748, 753
Input/Output 61, 63
Insects 10, 15, 109–112, 114, 173, 183, 268, 303, 396, 397, 422, 808, 810
Interspersion 111, 353, 354, 356
Invasion 183, 188
Ireland 165, 171
Iron removal 329, 381, 795, 798

Japan 530
Juncus 6, 81, 207, 211, 381, 411, 502, 578, 689, 763, 809

Kentucky 145, 281, 282, 369
Kinetics 34, 46, 97, 121, 125, 131, 135, 299, 319, 337, 343, 345, 347, 379, 380, 439, 497, 648, 649, 655

Lagoons 17, 60, 145, 166, 177, 188, 189, 192, 233, 234, 236, 245, 284, 334, 341, 345, 426, 629, 679, 681
Lake Apopka, Florida 710, 711, 715, 716
Lake Michigan 29
Lakeland, Florida 574, 576, 580
Land requirements 198, 627
Land treatment 186–189, 247, 266, 275, 424, 643, 644
Land use 187, 202, 203, 253, 254, 303, 308, 309, 316, 427, 698, 747, 750
Landfill leachate 167, 245, 247–250, 310, 319, 735–737, 740
Larix 7
Larvae 110, 298, 299, 396, 664, 810
Laws 45, 274, 313, 394, 397, 697
Lead 27, 79, 180, 228, 236, 249, 332, 360, 404, 411, 448, 600, 674, 680, 681, 692, 735
Leersia 257
Lemna 76, 142, 439, 460, 478
Levees 289, 309, 625, 627
Light 75, 76, 79, 134, 341, 360, 381, 405, 406, 412, 420, 470, 471, 494, 502, 503, 516, 597, 681, 771, 772, 806
Lignin 101, 758
Limitation 77, 330, 406, 510
Liners 141, 268, 310, 312, 376, 401, 402, 478
Listowel, Ontario 29, 31, 270, 329, 330, 340, 341, 370, 380, 432, 636, 639, 642
Litter 22, 26, 27, 38, 55, 76, 170, 192, 247, 298, 330, 332, 363, 370, 373, 374, 375, 382, 451, 459, 460, 463–467, 518, 578, 579, 611, 618, 619, 695, 775–778
Loading rates 15–17, 53, 57, 58, 122, 135, 209, 256, 284, 299, 319, 329, 333, 334, 337, 341, 346, 347, 367, 369, 420, 427, 449, 450, 537, 566, 567, 569, 571, 582, 586, 645, 647, 730, 732
Louisiana 99, 174, 515, 523

Macrophytes 25, 80, 81, 125, 126, 393, 396, 398, 405, 450, 451, 484, 487, 491, 506, 525, 528, 542, 565, 571, 583, 585, 646, 654, 808
Magnesium 98, 100, 249, 750
Maintenance costs 300, 301, 312, 394, 395
Mammals 114
Management criteria 196
Manganese 12, 45, 60, 63, 79, 80, 92, 98, 211, 222, 226, 228, 245, 248–250, 331, 337, 427, 501, 550–553, 555, 556, 680, 681, 682, 735, 740, 749, 750, 758, 765, 768, 769, 771, 776, 791, 803
Mangroves 6, 74
Manitoba 26, 359
Mapping 6, 312
Marsh construction 180, 407
Marsh creation 710
Marsh system 180, 183, 241, 242, 280, 307, 604, 627, 638, 681, 684
Marsh/pond/meadow 477
Massachusetts 12, 50, 58, 590, 686
Meadow system 140
Mercury 98, 99, 600, 692
Metabolism 75, 78, 80, 89, 92, 99–102, 142, 363, 559, 755, 757
Metal concentration 681, 802

Metal removal 62, 227–229, 247, 258, 331, 332, 550, 551, 772, 774, 776, 778, 783, 785
Metallic ions 15, 80, 211, 375
Michigan 29, 270, 422, 424, 657
Microbial 9, 14–17, 21, 45, 57, 61, 62, 80, 81, 89, 90, 92, 95–97, 100, 101, 134, 142, 145, 173, 175, 228, 267, 312, 320, 331, 346, 373, 374, 408, 421, 437, 450, 463, 467, 499, 508, 509, 515–518, 520, 523, 550, 552, 565, 618, 652, 657, 660, 662, 737, 755, 796
Micronutrients 99, 736
Microspora 768, 769, 771
Mineral 42–44, 55, 56, 58, 61, 75, 77, 80, 92, 121, 221, 224, 226, 375, 465, 466, 494, 498, 499, 502, 753–757, 787
Minerotrophic 62
Minnesota 60, 227, 331, 359, 780, 782
Mississippi 8, 174, 186–188, 239, 241, 282, 411
Missouri 233, 234, 282, 283
Models 46, 292, 418, 438, 646, 652, 655, 692
Monitoring 102, 125, 140, 142, 157, 181, 189, 191, 201–204, 213, 258, 259, 261, 266, 267, 270, 274, 276, 280, 310, 397, 398, 417, 424, 426, 427, 447–454, 518, 523, 580, 600, 607, 644, 648, 658, 676, 678, 692, 693, 705, 736, 744, 746, 747, 751, 785, 792, 793
Montana 227, 228, 359
Morphology 77, 90, 91, 98, 528, 613
Mosquito control 298, 299, 367, 369, 371, 394, 396, 397, 593, 664, 666, 667
Mosquito production 121, 125, 386, 393, 394, 396, 398, 525, 666
Moths 810
Mounding 22, 24, 38, 342, 383, 648
Movement 21, 42, 75, 79, 81, 109, 143, 187, 243, 510, 513, 515, 648, 713, 740, 743, 745, 748, 785
Mt. View Sanitary District, California 269, 597
Municipal wastewater 3, 82, 121, 179, 186, 191, 245, 250, 265, 268, 271, 273, 274, 276, 285, 297, 346, 383, 400, 431, 437, 438, 459, 466, 669
Muskrat 110, 114, 143, 209, 354, 355, 402
Myriophyllum 80, 81, 646

National Pollutant Discharge Elimination System 125, 203, 233, 270, 283, 310, 448, 674, 693
National Science Foundation 82
National Wildlife Federation 237
Natural wetlands 8–10, 12, 13, 16, 18, 53, 55, 63, 64, 73, 116, 145, 180, 253, 258, 265–268, 270, 271, 273–276, 281, 333, 355, 356, 360, 405, 427, 451, 459, 465, 501, 509, 510, 556, 578, 579, 581, 587, 684, 715, 761, 775, 810
Nematodes 109, 150
Nevada 26
New England 8, 686
New Jersey 727, 728, 733
New York 12, 245, 247, 250, 288, 406, 407, 410, 411, 737, 739
Nickel 227, 236, 249, 331, 332, 600, 780, 782
Nitrate 45, 46, 52, 55, 148, 189, 239, 241, 245, 248–250, 261, 329, 333, 426, 477, 480, 484, 485, 489, 491, 495, 496, 497, 499, 561, 600, 681, 709, 729, 776, 794
  removal of 330, 560
Nitrification 15, 52, 55, 64, 128, 148, 149, 329, 330, 337, 346, 367, 368, 369, 374, 375, 407, 424, 477, 478, 480, 481, 483, 493, 496, 498, 499, 536, 565, 569, 571, 583–588, 648, 652
Nitrite 148, 189, 249, 261, 477, 480, 600, 729
Nitrogen 49, 50, 53, 77, 89, 92, 100, 121, 130, 142, 146, 150, 166, 175, 192, 198, 239, 241–243, 247, 248, 249, 261, 266, 298, 312, 320, 330, 342, 343, 346, 355, 367–369, 407, 424–426, 460, 466, 467, 477, 481, 485, 495, 496, 502, 530, 532, 540, 558, 559, 564–566, 569, 576, 590,

591, 594, 596, 600, 603, 628, 646, 649, 652, 655, 680, 691, 692, 695, 696, 710, 711, 727, 729, 731, 735, 736, 750
removal of 148, 189, 197, 250, 329, 364, 367, 375, 424, 480, 493, 499, 595, 650
Nonpoint source 3, 4, 289, 290, 581, 670, 673, 674, 695
North Carolina 27, 60, 112, 283
North Dakota 8, 233, 237, 359
Nuphar 77, 78, 142, 257, 469, 470, 703
Nutrients 8, 13–16, 42, 49, 50, 57, 73, 96, 98, 103, 121, 180, 249, 254–256, 260, 265, 288–290, 333, 355, 373, 406, 407, 421, 460, 462, 463, 465, 466, 477, 546, 550, 551, 558–561, 580–582, 587, 598, 695, 696, 711, 715, 740, 782, 805
removal of 43, 80, 125, 254, 270, 273, 369, 374, 424, 534, 558, 560, 564, 571, 578, 580, 582, 699, 710, 713, 716
uptake of 76–78, 102, 142, 168, 258, 275, 481
Nymphaea 469, 473
Nyssa 6, 7, 78, 494

Odor control 591, 594, 596
Odors 173, 268, 288, 309, 384, 525
Oedogonium 767–769
Oil 233, 234, 259 260, 288, 310, 314, 502, 517, 520, 588, 670, 686, 688, 722
Oklahoma 112
Ontario 27, 31, 150, 227, 270, 292, 329, 636
Open water 25–27, 122, 125, 130, 135, 353, 369, 420, 526, 598, 646, 647, 648, 654, 674
Operating cost 780, 788
Operation and maintenance 4, 140, 145, 260, 364, 367, 369, 380, 394, 395, 399, 422, 447, 454
Optimal design 316
Organic content 518, 562, 808
Organic loading 128, 135, 299, 341, 345, 346, 381, 543, 569
Organics 80, 82, 89, 100, 103, 143, 175, 247, 249, 250, 298, 320, 342, 343, 346, 375, 453, 562, 585, 657, 691, 737, 748, 750
Organisms 12, 14, 16, 41, 90, 94, 101–103, 111, 114, 192, 267, 330, 360, 363, 373, 397, 435, 452, 453, 559, 641, 713, 771
Orlando, Florida 153, 261, 425, 574, 577, 578, 580, 711
Orthophosphate 310, 561, 586, 681, 709, 729
Otter 110, 111
Outlet distribution 400
Outlet structures 257, 258, 380, 384, 386, 403, 689
Overland flow 21, 24, 38, 157–159, 165, 170, 247, 250, 339, 340, 566, 569, 571, 674, 677, 680, 684
Oxidation 15, 45, 52, 55, 56, 58, 60–64, 80, 81, 92, 94, 95, 101, 102, 121, 122, 127–132, 134, 135, 148, 150, 175, 226, 227, 242, 298, 329–331, 333, 393, 394, 396, 406, 408, 421, 432, 437, 472, 501, 503, 506, 510–513, 581, 582, 588, 622–625, 628, 629, 652, 737, 740, 748, 749, 753–756, 758, 799
Oxidizing bacteria 222, 229, 261, 550–552, 556, 755–757, 796, 797, 798
Oxygen diffusion 469, 474, 489
Oxygen requirements 78, 345
Oxygen transfer 330, 345, 506, 585, 652

Paper mill waste 310
Particulate removal 467
Pathogenic 14, 89, 150, 173, 298, 397, 437
Pathogens 247, 265, 266, 268, 320, 332, 346, 363, 397, 431, 432, 435, 587, 657
Peat filters 593
Peatlands 27, 61, 62, 783
Peltandra 257
Pembroke, Kentucky 281, 282, 369
Pennsylvania 12, 139–143, 213, 283, 300, 369, 414, 477, 512, 695, 765, 768, 769, 772, 788, 801, 802, 806
Percolation 236, 247, 312, 603, 743

Performance expectations 319, 320
Periphyton 128, 558–561, 564
Permeability 92, 145, 150, 255, 310, 312, 343, 374, 376, 400, 402, 565, 571, 578, 582, 586–588, 608, 610, 654, 738
Permits 203, 273, 283, 343, 411, 435, 493, 579, 645, 672, 673
Pesticides 89, 192, 289, 397, 452, 600, 665, 688, 736, 745, 748
Pests 15, 17, 422, 808, 810
Petrochemicals 722
pH 42–49, 52, 56–58, 60, 61, 80, 82, 95–98, 130, 135, 189, 197, 198, 201, 202, 204, 211, 213, 215, 216, 222, 224, 226, 228, 234, 241, 249, 281, 300, 301, 310, 330–333, 337, 346, 347, 369, 396, 403, 406, 427, 437, 448, 478, 495–497, 499, 501, 502, 509, 510, 517, 546–548, 551, 583, 600, 618, 658, 659, 682, 711, 719, 722, 736, 750, 756, 757, 758, 761, 763, 767, 771, 774, 782, 783, 789, 791, 794, 799, 801, 802
Phalaris 81, 142, 248, 369, 478, 689
Phosphate 56, 57, 94, 100, 159, 213, 247, 310, 516, 574, 575, 582, 583, 584, 585, 587, 588, 658, 736, 767
Phosphorus 55, 56, 64, 77, 121, 142, 165, 175, 189, 198, 241–243, 247–250, 298, 320, 355, 369, 374–376, 407, 424–426, 460, 466, 467, 495, 532, 540, 558, 559, 564–567, 576, 578, 600, 628, 638, 639, 642, 649, 655, 681, 692, 695, 696, 710, 711, 727, 729, 735, 736
  removal of 166, 197, 239, 243, 261, 266, 312, 330, 345, 346, 376, 424, 425, 530, 539, 579, 580, 646
Photosynthesis 75–77, 79, 191, 370, 405, 406
Phragmites 7, 25, 143, 145, 153, 159, 166, 239, 240, 247, 257, 328, 375, 381, 460, 472, 484, 487, 489, 491, 530, 531, 532, 534, 535, 537–539, 542, 565, 583–587, 607, 613, 617, 628, 629, 737, 739, 740
Physical 4, 15, 41–43, 61, 63, 100, 101, 142, 145, 149, 150, 166, 170, 255, 266, 267, 273, 288, 290, 292, 293, 297, 299, 363, 373, 399, 425, 437, 447, 469, 470, 481, 550, 633, 672, 691, 696, 702, 737, 750, 757
Pinus 179, 186, 187, 494, 598, 775
Planning 3, 115, 116, 140, 143, 254, 260, 284, 285, 289, 337, 395, 398, 401, 418, 590, 667, 670, 674, 699
Plant detritus 528
Plant harvest 528
Plant production 548, 720
Plant selection 193, 502
Plant tissue 198, 242, 755, 775, 802
Plant uptake 242, 247, 329, 331, 550, 578, 629, 681
Planting methods 411
Playas 6
Point source 3, 177, 520, 670, 696
Policies 265, 266, 275, 279, 285, 448, 673–675
Pollutants 3, 12, 14–16, 18, 145, 197, 224, 226, 239, 241, 243, 249, 253, 254–256, 266, 271, 310, 319, 347, 369, 373, 374, 449, 501, 520, 542, 548, 649, 670, 672, 677, 684, 691, 695, 696, 698, 748, 757
Pollution 3, 4, 12, 15, 18, 101, 149, 180, 221, 223, 254, 256, 261, 265, 272, 282, 284, 288, 291, 310, 395, 431, 443, 448, 451, 453, 550, 577, 657, 662, 664, 669, 670, 674, 695, 702, 705, 711, 719, 743, 765, 780
  abatement of 290
Polygonum 412, 638
Ponds 121, 132, 180, 182–184, 203, 204, 213, 218, 227, 233, 234, 236, 245, 260, 271, 299, 303, 333, 345, 393, 394, 396, 400–402, 437, 469, 472, 474, 477, 481, 502, 511, 525, 566, 574, 575, 581, 596, 598, 600, 622–625, 657, 667, 674, 686, 688, 691, 743–745, 747, 751, 765, 801
Pontederia 407
Populus 6, 775
Porosity 22, 23, 339, 342, 344, 379–381, 536, 537, 540, 748, 776
Potamogeton 80, 134, 406, 411
Potassium 98, 100, 175, 249, 407, 707, 736, 794

Prairie potholes 7
Predators 113, 354, 396, 397, 432, 451, 452, 665, 810
Pretreatment 189, 198, 247–249, 255, 259, 260, 268, 270, 281, 285, 298, 299, 334, 360, 364, 420, 421, 424, 425, 427, 564, 566, 619, 623, 625, 677, 736, 743
Primary production 80, 255, 450, 512
Primary sewage effluent 537
Primary treatment 233, 234, 334, 345, 363, 364, 367, 585
Productivity 9, 10, 16, 18, 76, 80, 182, 183, 191, 255, 420, 523, 720, 736
Pseudomonas 102, 551, 758

Quality standards 115, 245, 266, 270–275, 332, 448, 692
Quercus 6, 578

Raccoons 110, 111, 237
Recharge 12, 18, 64, 115, 265–267, 310, 426, 657
Reclamation 146, 150, 191, 201, 203, 204, 223, 268, 270, 300, 313, 395, 438, 512, 590, 595, 604, 703, 743, 783, 788, 789, 792
Recovery 16, 78, 186, 191, 192, 198, 256, 310, 395, 435, 542, 596, 757
Recreation 116, 132, 179, 254, 260, 288, 431, 604, 673, 691, 696, 751
Recycling 16, 50, 78, 299, 367, 477, 720
Redox potential 45, 46, 52, 56, 437, 472
Reed bed 153, 158, 165–167, 170, 171, 320, 343, 363, 566, 569, 571, 586, 628–630, 632
Refractory organics 298, 343
Regulations 6, 80, 184, 245, 256, 270, 271, 273, 281, 313, 400, 404, 410, 448, 670, 674, 675, 686, 697, 727, 736, 747
Regulators 282, 286
Regulatory considerations 203, 308, 313
Reliability 135, 284, 427
Removal
  by adsorption 346
  of BOD5 646
  of metals 556
  of nitrogen 239, 250, 493, 650
  of phosphorus 239, 250, 628
  of solids 649
Removal efficiency 57, 58, 146, 150, 228, 240, 242, 259, 261, 330, 374, 432, 439, 483, 567, 569, 608, 669, 692, 710, 716, 775, 777, 778
Removal performance 579, 649, 650, 655
Renewal rate 255
Replacement 195, 396, 412, 518, 648
Residence time 36, 150, 233, 259, 260, 339, 340, 343, 344, 347, 379, 380, 420, 421, 426, 439, 679, 691, 730, 731, 785
Resource recovery 191, 192, 198, 542
Respiration 45, 58, 75, 77–79, 92, 95, 98, 101, 148, 149
Restoration 258, 268, 269, 290, 293, 355, 359, 675, 710, 715, 716, 720, 722, 725
Retention pond 696
Retention time 141, 174, 193, 195, 239–243, 330, 346, 367, 374, 439, 484, 547, 558, 571, 587, 610, 611, 636, 649, 652, 710, 716, 719, 737–740
Reuse 80, 434, 443, 623
Revegetation 780
Rhizosphere 15, 52, 77, 80, 81, 96, 145, 148, 150, 154, 328, 373, 408, 474, 501, 528, 565, 566, 571, 582, 585, 628, 629
Rock filters 623
Rocky Mountains 8
Root zone 145, 153, 157, 247–250, 328–330, 342, 344, 407, 408, 491, 502, 543, 628, 654
Root-zone method 298, 320, 370, 371, 374
Roots 12, 15, 52, 75–81, 148, 154, 158, 159, 175, 177, 224, 240, 241, 243, 247, 250, 320, 329, 330, 343, 373–375, 382, 403, 407, 408, 412, 413, 460, 464, 465, 469, 471, 472, 474, 481, 501–503, 534, 540, 586, 593, 598, 613, 628, 652, 681, 703, 763
Runoff 12, 16, 21, 82, 167, 189, 190,

201, 202, 211, 227, 234, 236, 253, 254, 258–260, 265, 267–270, 307–309, 319, 388, 403, 447, 493, 515, 523, 571, 578, 603, 608, 667, 669, 670, 672–675, 677, 679, 684, 686, 688, 691–693, 696, 698, 729–731, 738, 789

Sagittaria 142, 257, 411, 526, 537–540, 578, 689
Salix 7, 207, 411, 574, 689
Salt marsh 76, 396, 511, 513, 674, 727, 728, 731
Sampling wells 762
San Diego, California 150, 159, 281, 394, 396
San Francisco Bay, California 665, 666, 669, 670, 672, 673, 675, 677
Sand filtration 193, 195, 198, 530
Santee, California 146, 148, 150, 270, 280, 328, 329, 342, 431, 432, 438, 443
Saturation 41, 42, 62, 76, 149, 189, 346, 375, 424, 453, 776, 777, 778, 790
Scirpus 7, 25, 79, 122, 129, 145, 146, 177, 203, 207, 212, 247, 250, 257, 260, 298, 329, 354, 375, 380, 396, 407, 411, 432, 502, 525, 526, 537–540, 578, 583–587, 591, 593, 598, 625, 638, 646, 681, 682, 689, 724, 737
Sealing 402, 403, 587, 722
Seasonality 27, 405, 650
Seasonally flooded 74
Secondary treatment 146–148, 180, 233, 234, 266, 271–273, 281, 346, 431, 437, 525, 528, 600, 622, 623, 642, 643, 677
Sediment 50, 52, 55–57, 62, 64, 77, 79, 108, 201, 202, 226, 255, 256, 259, 260, 267, 289, 421, 451, 459, 460, 462, 465, 466, 474, 502, 515–518, 523, 618, 619, 642, 649, 658, 660, 680, 682, 684, 691, 695, 696, 703, 765
Septic tanks 177, 266, 299, 334, 345, 364, 566, 581
Sewage 53, 60, 139, 140, 143, 145, 153, 154, 157–160, 165–167, 174, 175, 177, 180, 181, 279, 288, 307–310, 312, 320, 329, 346, 360, 431, 537, 542, 543, 546, 547, 565–567, 569, 571, 582–587, 597, 606–608, 610–612, 617, 629, 636–639, 641, 642, 646, 650, 657, 659, 660, 662, 691, 710, 727, 736, 808
Sheet flow 259
Sinks 58, 355
Site evaluation 307, 314
Site selection 5, 13, 14, 16, 41, 57, 64, 80–82, 89, 107, 115, 116, 117, 121, 180, 187, 233, 247, 254, 265, 267, 268, 270, 272–274, 276, 279, 281–283, 289, 290, 293, 297, 299, 303, 307, 316, 353–356, 375, 379, 393, 395, 399, 405, 406, 408, 424, 447, 515, 574, 581, 595, 622, 636, 643, 808
Sludge 146, 148, 160, 165, 166, 173, 203, 233, 279, 284, 303, 329, 364, 384, 395, 426, 437, 438, 480, 483, 543, 545–548, 618, 619, 654, 727–729, 733
Soil chemistry 41, 48
Soil substrate 370, 374, 459
Solids deposition 384
Solar 25, 27, 29, 38, 370, 487, 598
Solids removal 329, 345, 364, 581, 588
Solubility 80, 407, 682, 754
South Carolina 407, 425, 643–645
South Dakota 284, 285, 359, 360, 743, 745
Sparganium 76, 77, 411, 502
Spartina 6, 78, 81, 240, 406, 517, 723
Species composition 9, 79, 108, 353, 413, 451, 559
Spills 422, 686
Stabilization 18, 192, 271, 282, 364, 566, 578, 579, 582, 583, 684, 689, 724
  by vegetation 8, 82, 115, 201, 254, 256, 260, 282, 307, 310, 384, 396, 397, 402, 422, 590, 591, 593, 596, 608, 636, 669, 675, 678–680, 691, 692, 695, 698, 722
Stabilization ponds 271
Storage area 213, 259
Storage time 410, 437
Stormwater 82, 236, 253–256, 258–261,

265, 270, 307, 310, 578, 608, 669, 670, 672, 673, 674–677, 679, 680, 684, 686, 688, 695, 723
Streambed 289
Stress 62, 75, 76, 78, 79, 102, 108, 114, 310, 360, 369, 412, 420, 421, 435, 451, 454, 613, 636, 642, 803
Submerged plants 75–77, 79, 256
Submergent 13, 126, 183, 413
Submersed 77, 80, 109
Subsidence 312, 313, 711
Substrate design 363
Substrate effects 79
Substrate selection 48, 373
Substrate types 331, 582
Subsurface flow 24, 280, 282, 284, 297, 298, 300, 319, 320, 328, 329, 330,332–334, 337, 341–343, 345–347, 363, 380, 381, 382–384, 386, 402, 403, 448, 525, 565, 566, 571, 587, 632
Succession 8, 425
Sulfate 45, 58, 60, 64, 100, 218, 226, 229, 298, 312, 330, 333, 489, 495, 508–513, 550–552, 555, 556, 754–758, 785, 792, 796, 803
Sulfide 45, 58, 60, 61, 80, 99, 222, 298, 502, 508–511, 513, 551, 740, 748, 750, 753–757, 798
Sulfur 58, 60, 100, 132, 222, 288, 298, 489, 508–510, 551, 750, 754, 755, 756, 769
Surface flow 334
Surface runoff 21, 189, 234, 309, 571, 738
Surface water 6, 121, 309, 426, 449, 574, 632, 736, 743, 787
Survival 11, 75, 102, 113, 166, 179, 186, 403, 413, 414, 436, 439, 472, 543, 659, 660, 662, 761
Suspended solids 122, 197, 266, 271, 320, 525, 546, 600, 622
Swamp 5, 6, 50, 111, 112, 114, 148, 186, 287, 494–497, 499, 517, 518, 520, 575, 577, 578, 643–645, 657, 688
System configuration 334
System design 216, 226, 234, 255, 259, 319, 417, 418, 420, 421, 517, 575, 577, 761
Taxodium 6, 79, 186, 494, 578, 713
Technology 21, 64, 73, 115–117, 140, 188, 209, 239, 270–272, 274, 275, 283, 285, 304, 314, 319, 334, 337, 359, 371, 383, 393, 422, 565, 571, 581, 582, 608, 636, 640, 642, 670, 696, 808
Temperature effects 122, 266, 271, 320, 425, 525, 546, 566, 622, 735
Tennessee 5, 12, 142, 157, 191, 201, 211, 213, 279, 284, 301, 313
Tennessee Valley Authority 5, 157, 201, 211, 279, 301
Tertiary treatment 56, 121, 149, 154, 165, 180, 564, 565, 657
Test pits 312
Testing 146, 193, 221, 239, 274, 281, 291, 293, 397, 403, 439, 453, 674, 677
Tests 125, 146, 167, 195, 298, 312, 438, 485, 489, 504, 515, 550, 552, 740, 745, 756, 757, 775, 785
Texas 112, 285
Thermodynamics 45
Thiobacillus 61, 550, 753, 756
Tides 11
Tilapia 192, 193, 195
Topographic maps 309, 312
Total suspended solids 425, 576
Toxicity 98–101, 298, 449, 452, 600, 727
Toxins 432
Trace metal 331, 451, 782
Trace organics 80, 320, 346, 453, 657
Transfer 45, 330, 345, 438, 506, 585, 628, 652
Transformation 41, 56, 64, 89, 99, 255, 273, 312, 343, 355, 374, 496, 499, 652
Transition 5, 11, 115, 613, 791
Translocation 149, 408
Trapping 143, 404, 667, 810
Treatment efficiency 21, 218, 239, 243, 255, 384, 417, 418, 420-422, 427, 449, 526, 528, 582, 713, 716, 722, 737, 739, 775, 785
Trickling filter 330, 339, 437, 536, 646, 647
Turbidity 241, 290, 310, 711, 722
Turnover 78, 95, 121, 523

Typha 7, 25, 29, 61, 76, 77, 141, 143, 145, 157, 177, 203, 207, 211, 239, 240, 247, 248, 250, 257, 298, 328, 329, 331, 354, 369, 375, 380, 396, 411, 432, 460, 478, 502, 509, 512, 537–539, 542, 566, 574, 578, 585–587, 598, 625, 636, 646, 647, 674, 681, 682, 689, 703, 713, 737, 762, 785, 789, 790, 801–805, 808

U.S. Army Corps of Engineers 414
U.S. Department of Agriculture 179
U.S. Environmental Protection Agency 5, 150, 180, 397, 431, 662, 670
U.S. Fish and Wildlife Service 5, 6, 188, 191, 313, 353, 645
United Kingdom 153, 154, 157, 159, 375, 629, 632, 633
Uptake by plants 15, 226
Urban runoff 12, 308, 319, 447, 669, 670, 672, 674, 675, 677, 691, 692, 698
Utah 26

Vascular 73, 175
Vectors 268, 289
Vegetation effects 38
Vegetation harvesting 122, 125, 126, 256, 593, 684
Vegetation management 299, 371, 625
Vegetation planting 17, 371, 374
Vegetation selection 664, 665
Virginia 12, 285, 300, 478, 511, 695, 788
Virus removal 437, 439, 440, 442
Volatile organics 250, 748
Volatilization 53, 60, 101, 248, 493

Washington (State) 177
Waste treatment 4, 175, 191, 192, 197, 199, 402, 403, 413, 542, 548, 705, 736, 739, 740
Wastes 89, 191, 227, 245, 307, 310, 319, 332, 334, 343, 346, 394, 395, 448, 528, 542, 702, 727, 736, 743, 746
Wastewater discharge 274, 424, 448
Water level 81, 146, 167, 259, 288, 329, 354, 381, 383, 403, 406, 408, 413, 414, 449, 515, 593, 594, 596, 630, 638, 642
  control of 183, 184, 189, 320, 386, 632
Water management 189, 664, 666, 667, 711
Water movement 21, 42, 79, 81, 187, 243, 785
Water pollution control 3, 284, 291, 577, 670
Water quality improvement 12, 18, 34, 197, 290, 291, 300, 459, 540, 700
Water quality standards 115, 245, 266, 270–275, 332, 448, 692
Water source 79, 197, 267, 269, 597
Water storage 21
Waterfowl 8, 9, 11, 107, 110, 113, 116, 134, 180–183, 236, 265, 268, 271, 354, 359, 360, 598, 623, 715
Watershed 691
West Germany 157, 320, 342, 343
West Virginia 300, 788
Wetlands creation 672, 675, 722, 723
White-tailed deer 114, 115
Wildlife management 233, 234, 359
Wildlife habitat 5, 16, 18, 107, 108, 115, 184, 236, 266, 270, 290, 292, 303, 354, 399, 578, 597, 604, 664, 677, 689, 696, 699
Wildlife use 267, 420
Wisconsin 177, 270, 285, 410, 787

Zinc 98, 227, 228, 240, 249, 332, 600, 682, 684, 692, 780